Microorganisms in Home and Indoor Work Environments

Diversity, Health Impacts, Investigation and Control

SECOND EDITION

T0225565

Microorganisms in Home and Indoor Work Environments

Diversity, Health Impacts, Investigation and Control

SECOND EDITION

Edited by

Brian Flannigan
Robert A. Samson
J. David Miller

CRC Press
Taylor & Francis Group
Boca Raton London New York

CRC Press is an imprint of the
Taylor & Francis Group, an **informa** business

CRC Press
Taylor & Francis Group
6000 Broken Sound Parkway NW, Suite 300
Boca Raton, FL 33487-2742

First issued in paperback 2017

© 2011 by Taylor and Francis Group, LLC
CRC Press is an imprint of Taylor & Francis Group, an Informa business

No claim to original U.S. Government works

ISBN-13: 978-1-4200-9334-6 (hbk)
ISBN-13: 978-1-138-07241-1 (pbk)

Visit the Taylor & Francis Web site at
http://www.taylorandfrancis.com

and the CRC Press Web site at
http://www.crcpress.com

Preface

Since the preface to the first edition was drafted ten years ago there have been major changes in the recognition of the economic and health importance of fungal damage in the built environment. A decade ago, the transition from a focus on outdoor air quality to an understanding that most allergic disease is associated with contaminants found in homes was just under way. We noted then that the National Institute for Occupational Health studied the impact of poor indoor air quality on productivity in the USA. The authors of the study estimated that for around 20% of the US working population a benefit approaching what in 2007 would have been 88 billion US dollars might be realized (*American Journal of Public Health*, 2002, 92,1430-1440). In 2007, estimates of the costs of disease associated with mould and dampness were developed which took into consideration the entire population and residential environments. Again in the US, the direct costs were estimated to be in the 4 billion dollar range (*Environmental Health Perspectives*, 2007, 115, 971-975; *Indoor Air*, 2007, 17, 226-235). The indirect costs of lost work and school days are much larger. The Institute of Medicine of the US National Academy of Sciences produced two expert reports that deal with the effects of mould and dampness on asthma in 2000 and on mould and dampness and health in 2004, as did Health Canada (2004), and in 2004, and again in 2007, the World Health Organization also indicated the importance of mould and dampness to public health. Most recently, an expert panel commissioned by the US Centers for Disease Control found that the elimination of moisture and mouldy materials in homes resulted in improved health [D.E. Jacobs *et al.* (2010) *Journal of Public Health Management and Practice* **16**: S5-S10].

These are only some of the landscape changes that have occurred in this field in less than a decade. The first edition contained information that served as a benchmark for defining future progress. Although much was known in the scientific community about the growth of mould and dampness on building materials and contents, and of epidemiological evidence that such growth represented a public health hazard, there were major scientific challenges. We noted that "Fully elucidating their effects on human health has however been bedevilled by problems of accurate assessment of exposure to microorganisms and precise identification of those present in the environment." This remains true.

As in the first edition, the first three sections of the book review to date the types of microorganism in outdoor and indoor air, their growth and control in home and work environments, and their role in respiratory disease. An entirely new chapter on pollen in indoor environments and its allergenic effects has been added. The remaining sections of the book are given over to addressing the twin problems of exposure assessment and identification, discussing the methodology for and conduct of investigations of indoor environments. As before, the book includes information on key fungi and actinobacteria that reflects advances in knowledge of their occurrence in buildings in different parts of the world, as well as changes in taxonomic status.

What is entirely new is treatment of issues that were emergent at the time of writing the first edition. Epidemiological studies had demonstrated an association of mould with respiratory disease not associated with allergic mechanisms. In this second edition, there is a chapter on the emerging picture of the mechanistic basis for this phenomenon, i.e. of the effect of toxins and inflammatory agents on lung biology and other systems. Similarly, there is a new chapter on the use of molecular methods for determining microbial contaminants. Much new material is also found on the problems of remediation, control and quality assurance; occupational exposures in a wider range of work environments and among remediation workers; infectious fungi in the built environment; and endotoxin. The nomenclature of some common indoor fungi has been recently changed and these changes have been applied lea-ving the "old " name between brackets.

We think that the availability of information on the microorganisms that grow in the built environment, together with information on the limitations of the methods currently available to measure them, will be useful to researchers, public health officials and industrial hygienists. Together with reviewers, the authors from Canada, Sweden, the Netherlands, the United Kingdom and USA have worked hard to produce material that is accurate and timely. In addition to thanking reviewers and authors for their efforts, we should also like to express our gratitude to Margaret Flannigan for invaluable editorial assistance.

Brian Flannigan (Edinburgh), Robert A. Samson (Utrecht), J. David Miller (Ottawa)

PREFACE TO THE FIRST EDITION

While much of the concern about air pollution in the past has been focused on the outdoor environment, in recent years indoor air quality (IAQ) has moved up the agenda. Over the period between 1987 and 1999, more than $1 billion of federal government money was spent on research into indoor air pollution in USA. In March 2000 the Environmental Protection Agency released a report on "Healthy Buildings, Healthy People: A Vision for the 21st Century", which set the objective of achieving major health gains by improving indoor environments. The National Institute for Occupational Health has studied the impact of poor indoor air quality on productivity. The median estimate of these losses is $100 billion per year. Other countries, including Canada, Denmark, Finland, Netherlands and Sweden, also have substantial programmes on residential housing and health. However, there is wide variation in the research effort and expenditure on measures to improve IAQ, and IAQ in some countries is very much lower on the order of priorities.

A document produced in July 2000 by a WHO European working group has further emphasized the global importance of IAQ as a determinant of population health and well being. This document, *The Right to Healthy Indoor Air*, sets out nine principles (derived from the general principles in the International Bill of Human Rights). These are intended to inform all those who have an influence on public health of their obligations to honour the right of every individual to breathe healthy indoor air, and influence those national governments that do not have plans for future action on healthy indoor air to put it on their agenda.

Despite the large amount of money spent on research into pollution of the indoor environment, the US General Accounting Office has confirmed that what has been done has pointed to the complexity of the problem and to major gaps in knowledge. Among these gaps are accurate knowledge of the identities and sources of pollutants and of the effects of prolonged exposure to indoor pollutants on health. This book considers one such group of pollutants, namely microorganisms, and more particularly heterotrophic bacteria and fungi. Advances have certainly been made in our knowledge of microorganisms in the home and indoor work environment as research has accelerated in the last decade. Fully elucidating their effects on human health has however been bedevilled by problems of accurate assessment of exposure to microorganisms and precise identification of those present in the environment. The first three sections of the book review the types of microorganism in outdoor and indoor air, their growth and control in home and work environments, and their role in respiratory disease. The remaining sections of the book are given over to addressing the twin problems of exposure assessment and identification, discussing the methodology for and conduct of investigations of indoor environments and providing keys and colour illustrations to assist in the identification of approaching 100 mould, yeast and actinomycete contaminants.

We think that the availability of information on the microorganisms that grow in the built environment, together with information on the limitations of the methods currently available to measure them, will be useful to researchers, public health officials and industrial hygienists. Together with reviewers, the authors from Canada, Sweden, the Netherlands, the United Kingdom and USA have worked hard to produce material that is accurate and timely. In addition to thanking reviewers and authors for their efforts, we should also like to express our gratitude to Margaret Flannigan for editorial assistance, Karin van den Tweel for helping with the drawings and Ans Spaapen-de Veer for preparing the index. Thanks are given to Dr Brian Crook (HSL, Sheffield, UK) for substantive assistance in the preparation of Chapter 3.2.

We have dedicated this book to the memory of a fellow microbiologist, John Lacey, who was internationally recognized for his unique expertise at the interface of stored product microbiology, aerobiology, occupational hygiene and medicine, and was the author of more than 300 publications. The aerobiological, taxonomic and ecological studies of John and his co-workers not only clarified the role of microorganisms in a number of occupational lung diseases, but have a relevance to home and non-industrial work environments as well as to the storage and processing of the particular agricultural materials with which they were first concerned. In his friendly collaboration and links with many workers in overseas institutes, he was generous in passing on his knowledge and expertise to others, and his work and the influence that he has had are reflected in this book.

List of Contributors

Dr Denis Corr
Corr Associates
7 Spruceside Ave, Hamilton L8P 3Y2, Ontario
Canada

Dr Brian Crook
Health Improvement Group, Health and Safety Laboratory
Buxton, SK17 9JN
UK

Prof. James H. Day, MD
Division of Allergy and Immunology
Department of Medicine, Queen's University
Kingston K7L 3N6
and
Kingston General Hospital
76 Stuart St,
Kingston K7L 2V7, Ontario
Canada

Prof. Anne K. Ellis, MD
Division of Allergy and Immunology
Department of Medicine, Queen's University
Kingston, K7L 3N6
and
Kingston General Hospital
76 Stuart St,
Kingston K7L 2V7, Ontario
Canada

Dr Brian Flannigan
Scottish Centre for Pollen Studies
Napier University
10 Colinton Road, Edinburgh EH10 5DT
UK

Dr Jonathan M. Gawn
Health Improvement Group,
Health and Safety Laboratory
Buxton, SK17 9JN
UK
Present address:
Biological Agents Unit
Health and Safety Executive
Redgrave Court
Bootle
Merseyside L20 7HS
UK

Dr Brett J. Green
Sporometrics, Inc.
219 Dufferin Street, Suite 20C
Toronto, Ontario M6K 1Y9
Canada
Present address: Allergy and Clinical Immunology Branch,
Health Effects Laboratory Division
National Institute for Occupational Safety and Health
Centers for Disease Control and Prevention
1095 Willowdale Road MS L-4020
Morgantown, WV 26505
USA

Dr Michael J. Hodgson, MD, MPH
Office of Public Health and Environmental Hazards
Veterans Health Administration
810 Vermont Avenue, NW
Washington, DC 20420
USA

Ing. Jos Houbraken
CBS-KNAW Fungal Biodiversity Centre
P.O. Box 85167, 3508 AD Utrecht
The Netherlands

Dr Timo Hugg
South Karelia Allergy and Environment Institute
Joutseno
and
Institute of Health Sciences
University of Oulu, Oulu
Finland

Prof. J. David Miller
Department of Chemistry
Carleton University,
Ottawa K1S 5B6, Ontario
Canada

Dr Philip R. Morey
ENVIRON International Corporation
2235 Baltimore Pike
Gettysburg, PA 17325
USA

Dr John Mullins
The Centre
33 The Parade, Rooth, Cardiff CF24 3AD
UK

Prof. Thomas G. Rand
Department of Biology
Saint Mary's University
923 Robie Street
Halifax B3H 3C3, Nova Scotia
Canada

Dr Auli Rantio-Lehtimäki
Aerobiology Unit,
Department of Biology
University of Turku, FIN-20014 Turku
Finland

Prof. Timothy J. Ryan
School of Health Sciences
E 344 Grover Center
Ohio University
Athens, OH 45701-2979
USA

Prof. Dr Dr h.c. Robert A. Samson
CBS-KNAW Fungal Biodiversity Centre
P.O. Box 85167, 3508 AD Utrecht
The Netherlands

Dr James A. Scott
Dalla Lana School of Public Health
University of Toronto
223 College Street, Toronto M5T 1R4
and
Sporometrics, Inc.
219 Dufferin Street, Suite 20C
Toronto, Ontario M6K 1Y9
Canada

Dr Richard C. Summerbell
Dalla Lana School of Public Health
University of Toronto
223 College Street
Toronto M5T 1R4
and
Sporometrics, Inc.
219 Dufferin Street, Suite 20C
Toronto, Ontario M6K 1Y9
Canada

Dr Jillian R.M. Swan
Health Improvement Group,
Health and Safety Laboratory
Buxton, SK17 9JN
UK

Dr Eija Yli-Panula
Aerobiology Unit
Department of Biology
University of Turku, FIN-20014 Turku
Finland

Contents

Chapter 1. Microorganisms in air

Chapter 1.1

MICROORGANISMS IN OUTDOOR AIR

John Mullins[1] and Brian Flannigan[2]

[1] The Centre, Rooth, Cardiff, UK;
[2] Scottish Centre for Pollen Studies, Napier University, Edinburgh, UK.

EARLY STUDIES OF OUTDOOR AIR

It had long been believed that the air could bring disease to humans and crops, but it was not until the invention of the microscope in the 17th century that it was possible to observe the array of particles that are carried in the air. With his lens, Antonie van Leeuwenhoek (Dobell 1932) was just able to observe bacteria. It gradually became recognized that the air carried bacteria, yeasts, fungal spores, spores of mosses and ferns, algae, pollen grains and even protozoa. Initial studies were concerned with the controversy surrounding spontaneous generation of organisms and it was Pasteur (1861) who, by drawing air through gun cotton and then dissolving the gun cotton and examining the deposit under a microscope, discovered that the air contained a variety of different particles. However, he did not pursue these studies and the realisation that the air contained a variety of microbes resulted in a concerted effort by medical men to discover the microbes that caused disease (Bulloch 1938). The original work of Miquel (1899) in Paris into airborne bacteria stands as one of the most sustained series of volumetric measurements of the microbial population of the air ever attempted. Samples were collected over a 16-year period in plugs of gun cotton, and after this was dissolved the filtrate was cultured in flasks of filtered saline beef extract. From his studies he discovered that in a park 5 km from the centre of Paris bacteria were nearly three times as numerous in summer as in winter; in the centre of Paris counts were twice those in the park, but with a similar seasonal fluctuation. He also sampled a narrow unhygienic street and the main sewer of Paris, in which the air proved to be no more contaminated than in the streets outside. On average, in the park there were

290 bacteria m^{-3} air, in the centre of Paris 7480 m^{-3}, in the unhygienic street 5550 m^{-3} and in the sewer 3835 m^{-3}. He also noted a steady annual decline, which he attributed to improved street cleaning and washing to lay dust. Miquel came to the conclusion that the source of most outdoor airborne bacteria is the surface of the ground. He also studied the variations during the day, and attributed increases during the course of the day to mechanical causes such as road sweeping and traffic. Miquel lost interest in fungal spores however, and developed media that selectively discouraged mould growth. Fungal spores in the air were then largely ignored until investigated by Cadham (1924), who confirmed spores of cereal rust fungi as a cause of asthma and rekindled an interest in airborne fungal spores driven by allergists.

FUNGAL SPORES IN OUTDOOR AIR

Types and sources of fungal spores

Fungal spores are present in outside air throughout the year (the air spora) with scarcely any exceptions, and in the senior author's experience concentrations in the centre of the British city Cardiff have reached as high a 24-h mean as nearly 85000 m^{-3} air. It is virtually impossible to take a breath without inhaling a quantity of fungal spores. Estimates of the volume of air inspired at rest suggest a value of 10 l min^{-1}, the rate of sampling adopted for Hirst spore traps (Hirst 1952). At this rate, 1 m^3 air would be inspired in 100 min, but any increase in activity would dramatically increase the volume of air inspired, resulting in a greater intake of fungal spores.

The presence of so many fungal spores in the air is a consequence of the mechanism possessed by many fungi as a means of dispersal, viz. the production and release into the air of enormous numbers of spores. The spores of some fungi are dispersed in water and may only become airborne in spray thrown up by wave action. Insects also are responsible for the dispersal of some fungal spores, but the majority of fungi release their spores directly into the air.

Fungi produce both sexual and asexual spores: some produce only asexual spores, some produce only sexual spores, and some produce both. Asexual spore types include sporangiospores, conidiospores (conidia), pycnidiospores, teliospores of the cereal pathogens in the phylum Basidiomycota (basidiomycetes) that are known as smut fungi, and the teliospores, uredospores and aeciospores produced at different stages of the life cycle of the other basidiomycete plant pathogens referred to as rust fungi. The sexual spores include those of the Zygomycota (zygospores, which are mostly sessile), the Ascomycota (ascospores) and the Basidiomycota (basidiospores).

Release and aerosolization of spores

As fungi are relatively small, certain barriers to spore dispersal exist which have to be overcome. There is a static layer of air known as the laminar boundary layer varying in thickness from 1 m to 10 cm from the ground; it is thicker in still conditions and becomes thinner with increased wind speed. If spores are released into this layer they will not be dispersed. Accordingly, mechanisms have evolved in the fungi that ensure that their spores are released into the turbulent layer, which extends above the laminar layer.

Extensive work was carried out by Ingold and his co-workers on spore release mechanisms, to which reference should be made for a more exhaustive treatment of the subject (Ingold 1971). Most of the asexual spores are released relatively passively and rely on wind currents and turbulence to carry them away. Among these, sporangiospores are produced within a sporangium, which is raised up on a sporangiophore, exposing spores to the scouring effects of air currents as the sporangial wall bursts on maturity, exposing the spores for dispersal. In some species of *Mucor* the sporangial wall appears to dissolve leaving a mass of spores in a sporangial drop exposed to drying air currents. In the majority of sporangial fungi the sporangial wall ruptures as the sporangiospores, initially packed into the sporangium in polyhedral

shapes, round off as they mature and increase the pressure on the sporangial wall. Although this method of spore release is found in *Mucor* and other fungi in the order Mucorales within the Zygomycota (see Chapter 1.2), despite the abundance of this group in nature relatively few of their spores are isolated from outside air.

The asexual aeciospores of rusts are released by a similar method of rounding off from a polyhedral shape as they mature, exposing them to erosion by air currents. However, by their situation as biotrophic parasites on the leaves of plants, rusts and their spores are effectively raised above the laminar layer, facilitating spore release.

Perhaps the most common asexual spore (or mitospore) produced by fungi is the conidiospore or conidium. These conidia are produced externally, rather than within a sporangium, and the spores are individually budded off or may be formed in chains but raised up on conidiophores, allowing for erosion of the spores by wind currents. This latter mechanism is found in *Cladosporium*, the genus that contributes most to the air spora in temperate countries, including UK (Harvey 1967, 1970). *Cladosporium* and a number of other common airborne fungi, such as *Alternaria*, *Botrytis* and *Epicoccum*, colonize the surface of leaves (the phylloplane), stems and other plant organs, particularly as they senesce, so that these fungi are often referred to as phylloplane fungi. In contrast to these aerially dispersed dry-spored fungi, in those such as *Fusarium*, *Gliocladium* and *Trichoderma* the conidia occur in minute droplets of aqueous slime and are not directly detached by wind currents. Dispersal in these wet-spored fungi is by water — rain-splash and surface water — and in some cases by insects and other arthropods.

Sexual spores are produced by fruiting bodies, or sporocarps, of the Ascomycota and Basidiomycota. The members of these two phyla are frequently referred to as ascomycetes and basidiomycetes. In the Ascomycota, members of the large subphylum Pezizomycotina (the cup fungi) produce ascospores in groups of eight within a sac known as an ascus. Asci may be produced within, or over the surface of the sporocarp, the ascocarp (or ascoma). In response to the appropriate stimuli the ascus releases its spores explosively into the air. The stimulus may be a change in relative humidity (RH) or even a response to light (Walkey and Harvey 1966 1968a). In the Basidiomycota, members of the subphylum Agaricomycotina (which includes mushrooms, toadstools and bracket fungi), the whole of the sporocarp, the basidiocarp, may be raised above the ground on a thick stalk or it may grow out from decaying wood, as in the bracket fungi. The basidiospores develop externally, usually as a group of four, each on a separate sterigma, on a basidium. Depending on the type of basidiomycete, the basidia cover the surface of sheet-like gills or surround pores in the basidiocarp. They are forcibly ejected by a mechanism that is not yet fully understood into the space between the gills or into pores, and thereafter fall by gravity into the turbulent airflow below the cap. In puffballs and the earth star fungi a different mechanism has evolved in which the spores are produced inside a thin papery capsule with an apical opening, the ostiole. The dry spores are either forced out of the capsule by the impact of raindrops on the surface of the capsule or are drawn out as the air passes across the ostiole.

Rain affects the numbers of not only puffball and earth star spores in the atmosphere. For instance, over a period of five days in central London Battarbee *et al.* (1997) recorded the size of airborne particles impacting on the sticky tape of a Burkard automatic volumetric spore trap (see below under Sampling the Air Spora). As well as pollen grains and occasionally diatoms, conidia of *Cladosporium*, *Alternaria* and *Epicoccum*, ascospores of *Leptosphaeria* and lichens, and basidiospores were among the more common recognizable spores on the tape. On two consecutive dry summer days 1-4% of particulate matter of an aerodynamic diameter up to 10 µm (PM10) consisted of such spores. Light rain on the day after and heavy rain the day after that raised the level to 3-6%. On the day following the heavy rain, 23-27% of the PM10 comprised fungal spores, this increase being attributed to initiation of spore release by the rain. Collectively then, the various types of fungal spore may form a sizeable proportion of particulate matter in outdoor air. Based on a Canadian study of glycerophospholipids in airborne particles <2.5 µm in size (PM2.5), it has been suggested that at three sites in the Toronto area fungal spores and pollen grains between them accounted for 12-22% of the organic carbon fraction in the outdoor air, or 4-11% of the total mass on a fresh weight basis (Womiloju *et al.* 2003).

The length of time spores remain in the air will depend in part on the size of the spore, as large spores will naturally tend to be deposited from the air more rapidly than small spores. The rate of fall of a spherical spore in still air is given by Stoke's Law:

$$V = \frac{\sigma - \rho}{\mu} gr^2$$

where

V is terminal velocity in cm sec^{-1}
σ is the density of the spore
ρ is the density of air
g is acceleration due to gravity (981 cm sec^{-1})
μ is the viscosity of air (1.8 x 10^{-4} g cm^{-1} sec^{-1} at 18°C)
r is the radius of the spore.

Since the density of most spores is 1.0 and the density of air is so small that it can be ignored, this can be simplified to:

$$V = \frac{2gr^2}{9\mu}$$

The rate of fall of a spore is then proportional to the square of the radius, giving rates of fall of 2.0-2.8 cm sec^{-1} for the very large spores of the microfungus *Cochliobolus sativus* (*Helminthosporium sativum*), which are 80 x 15 µm, compared with 0.05 cm sec^{-1} for the spores of the puffball *Lycoperdon pyriforme* (4 µm diameter).

Long distance transport of spores

Fungal spore distribution from a point source has always been of interest to plant pathologists who wish to predict the distance which spores will travel, and to workers attempting to predict concentrations of smoke screens, gas clouds, radioactive particles or pollen released from genetically modified crops (Emberlin *et al.* 1999).

Gregory (1973) described the extensive work done in this field, both in his own studies at Rothamsted and by others. It is apparent that producing a mathematical model to predict the distance to which a spore will travel is extremely difficult. Unlike gas clouds, spores are eroded from the cloud by deposition, but Chamberlain (1966) considered that for grass pollen (with grains around 20 µm in diameter) and other pollen released at about knee height the evidence suggests ranges in the order of 1 km, and for pollen released at tree top level distances of travel in the order of tens of kilometres. Raynor *et al.* (1970) found that 1% of ragweed pollen released from a point source remained airborne at a distance of 1 km from the point of release. As agents by which plant disease may spread, spores and the distances they may travel have also long been of economic interest. Stakman and Hamilton (1939) reported uredospores of stem rust of wheat (*Puccinia graminis tritici*) at distances of some 970 km from the source, which was a vast area of winter wheat in the southern USA. Hirst *et al.* (1967) had the opportunity to sample airborne particles from an aircraft travelling over the North Sea, and some 400-500 km from the English coast recorded unexpected clouds of fungal spores. These corresponded to spores, which had been released the previous day over the land and had been carried eastward by the prevailing winds. There was a further increase between 500 and 600 km from the coast of spores characteristic of those released at night, again which had been released over land, but the previous night. As Gregory (1973) noted, the extent of spore dispersal is sufficient to spread plant diseases between countries and across continents.

Sampling the air spora

Estimates of spore concentrations in the air are obtained by air sampling. Two methods are widely used to carry out a census of the atmosphere. The first of these involves collecting spores onto culture plates, counting the colonies which develop from these and identifying the fungi from the characteristics of the colonies. Although spores may comprise the bulk of airborne fungal material, there can also be pieces of the mycelium, hyphal fragments, which may also give rise to colonies (see below). Aggregated clumps of spores, not just individual spores (or in the case of yeast, not only individual cells but clumps), may also be trapped on culture plates. To allow the spores simply to settle on the plate is to invite sampling errors because of the different rates of fall of different spores. Consequently, an efficient sampling device such as the Andersen sampler (Andersen 1958), based on the cascade impactor devised by May (1945), is required to ensure volumetric sampling. Because of the danger of overloading the culture plates, the time for which the plates can be exposed for sampling purposes is limited. Therefore, sampling has to be restricted to "spot" or "grab" samples or a series of samples taken at intervals throughout the day and night, with the attendant problems of incubating and subsequently examining a large number of culture plates. There is also the problem that different culture media tend to be somewhat selective in the fungi which will grow on them, and many fungal spores will not germinate on culture plates, including most of the sexual spores which may form a significant part of the air spore.

As one cannot be certain as to what exactly colonies have arisen from – individual spores, spore clumps or hyphal fragments – counts on culture plates are expressed volumetrically as propagules or colony forming units (CFU) m^{-3} air.

The alternative is to collect the spores on a glass microscope slide using a volumetric trap such as the Hirst trap (Hirst 1952) or the Burkard automatic volumetric spore trap (AVST) into which it evolved. Irrespective of whether the spores collected for visual identification under the microscope are viable (culturable) or non-viable, identification is carried out and is dependent on the morphological characteristics of the spores. Counts are expressed as spores m^{-3} air, and not as CFU m^{-3} as in culture-based methods. Some spores are easy to identify but others are very difficult to distinguish, either because of their small size or their lack of distinguishing features. In consequence, many spores tend to be counted according to categories such as colour and shape. This is particularly true of the ascospores and basidiospores, which frequently occur in the air in high numbers and are likely to have originated from many different species. It may therefore be necessary to use a culture method when it is the only way of confidently estimating the concentration of airborne spores of a particular species of fungus, particularly when visual identification is not possible (Mullins *et al.* 1976).

With the advent of PCR (polymerase chain reaction) and other modern molecular methods, advances have been made in identifying specific organisms, such as those that only occur in the air in small numbers, e.g. the human pathogen *Pneumocystis carinii* (Wakefield *et al.* 1998), so that such techniques are proving to have more widespread applications. For example, PCR-based techniques have also been used for prediction of crop disease epidemics by monitoring the pathogenic fungi in the ambient air (see West *et al.* 2008). A very recent demonstration of the use of modern methods is a study carried out in Mainz, Germany, by Fröhlich-Nowoisky *et al.* (2009), who used a high-volume dichotomous sampler to separate airborne particles of aerodynamic diameter >3 µm from those <3 µm in samples of approx. 3000 m^3 of the ambient air, which is representative of a mixture of urban and rural continental boundary layer air encountered in central Europe. On analysis, DNA extracted and amplified from the two fractions suggested to the authors that >1000 fungal species were present in the sampled air, although around 70% of these were found only once. More plant pathogens were found in the coarser material (>3 µm) and more

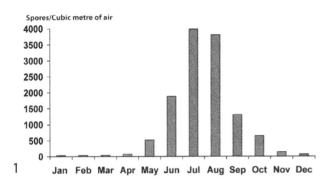
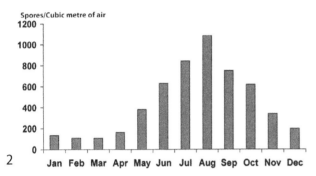
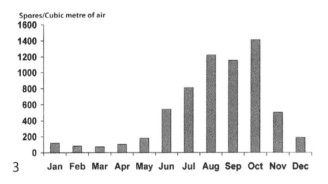
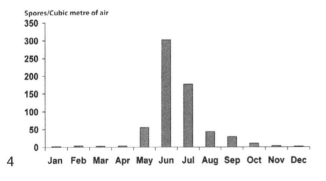
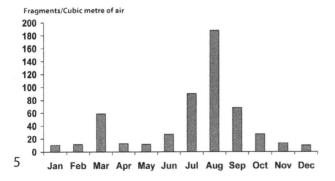

fungi allergenic and pathogenic for humans in the finer material. Basidiomycetes accounted for 64% of all fungi identified and ascomycetes (including mitosporic species) only 34%, leading the authors to conclude that in earlier studies, particularly those using culture-based methods, the presence of basidiomycetes may have been underestimated.

Factors affecting the composition of the air spora

The variety and concentration of spores in outdoor air are subject to continuous diurnal and seasonal variation. Contributory factors include availability of substrate, activities such as mowing grass and harvesting grain, and climatic factors, particularly temperature and rainfall (see, for example, Rodriguez-Rajo *et al.* 2005, Stepalska and Wolek 2005), which have a direct effect on the release of spores into the air. Rain and warmth also promote the development of vegetation on which parasitic and saprophytic fungi subsequently develop. Broad correlations between the composition of the air spora and climatic factors enable predictions to be made that certain spore types will be more abundant in warmer, drier summers, whereas others will be more abundant during damp weather.

The outdoor air spora is largely derived from spores produced by moulds and other fungi growing on natural and cultivated vegetation and on surface vegetable debris. Not all fungi on leaves have spores that are easily aerosolized, e.g. yeasts and *Phoma*, which are dispersed by rain-splash. However, the colonization of leaf surfaces by those with dry spores that are readily dispersed into the air, such as *Cladosporium* and *Alternaria*, can make a major contribution to the fungal burden in the air (Levetin and Dorsey 2006). In a review on fungal endophytes, i.e. fungi that grow intercellularly or invade single cells in leaves and other plant organs, Schulz and Boyle (2005) noted that the majority in temperate habitats belonged to more or less ubiquitous genera. These include species in genera that are known to colonize the phylloplane and are commonly isolated from air, not only

Fig. 1. Incidence of airborne *Cladosporium* spores trapped by an AVST at roof top level in Cardiff.

Fig. 2. Incidence of airborne ascospores trapped by an AVST at roof top level in Cardiff.

Fig. 3. Incidence of airborne basidiospores trapped by an AVST at roof top level in Cardiff.

Fig. 4. Incidence of airborne rust and smut spores trapped by an AVST at roof top level in Cardiff.

Fig. 5. Incidence of airborne hyphal fragments trapped by an AVST at roof top level in Cardiff.

Cladosporium and *Alternaria* but also *Acremonium,* *Coniothyrium, Epicoccum, Fusarium, Phoma, Pleospora* and others. Unterseher and Schnittler (2009) recently reported that the endophytes that they most frequently isolated from healthy beech leaves were unspecified ascomycete cup-fungi, and species of *Phomopsis,* the pink yeast *Rhodotorula, Paecilomyces,* and *Nodulisporium. Alternaria alternata, Aureobasidium pullulans, Cladosporium cladosporioides, C. herbarum, Chrysosporium* sp. and *Epicoccum nigrum.* While *Alternaria* and *Cladosporium* are well-known as being al-

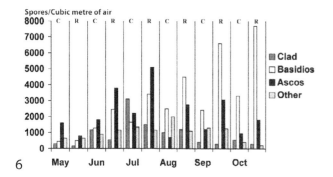

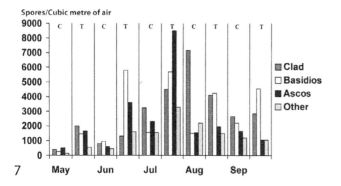

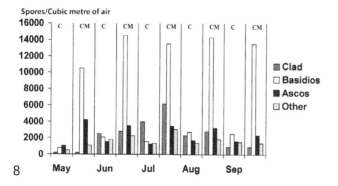

lergenic, it should be mentioned that the anamorphic ascomycete *Nodulisporium* (teleomorph *Xylaria*) has been noted as causing allergic fungal sinusitis (Cox *et al.* 1994).

Surveys of the air spora at any site tend to be dominated by spores of local origin, with others of more distant origin forming only a smaller part of the total census. Some species are practically ubiquitous, whereas others are more or less confined to certain localities. It is therefore not surprising that the air spora of towns and cities will tend to be less abundant than that of the surrounding countryside, where such agricultural activities as mowing and grain harvesting result in the aerosolization of large numbers of spores and hyphal fragments.

It appears that most hyphal fragments have cross walls, or septa, and can be simple or branched. Pady and Kramer (1960) reported that they can vary considerably in size, with most 5-15 μm in length, but occasional fragments can be up to 100 μm in length. In their study in Kansas, these authors found that the majority of fragments were dark-coloured (mostly brown) and thick-walled, but hyaline fragments were also present. The fragments were frequently the terminal portions of conidiophores, sometime comprising only a single cell and often with an immature spore attached. They were noted throughout the year, but were more numerous in summer (175-1800 m^{-3} air) than in winter (35-210 m^{-3}). When collected on water agar in a slit-sampler, 29-82% of such fragments germinated and gave rise to colonies of *Alternaria, Cladosporium* and *Penicillium,* leading Pady and Kramer (1960) to consider that they would be an important means of asexual reproduction.

With the advent of scanning electron microscopy (SEM), it became possible to detect extremely small aerosolized particles of hyphae and spores. These particles are <1 μm in size and are referred to as submicron particles or fragments, and will be discussed in relation to the indoor environment in Chapter 1.2.

AVST surveys of fungi in outdoor air

Local variation in the temperate air spora

To illustrate the main characteristics of the outdoor air spora and some factors, which have a bearing on it, data gathered by the senior author from air sampling in and around Cardiff in UK will be examined. Cardiff, the capital city of Wales, lies in the southern part and is a large seaport on the north side of the Bristol Channel, approximately 220 km west of Lon-

Fig. 6. Comparison between airborne fungal spores at a city centre site in Cardiff and at a coniferous woodland site at Resolven.
Fig. 7. Comparison between airborne fungal spores at a city centre site in Cardiff and at a mixed deciduous/coniferous woodland site at Tintern.
Fig. 8. Comparison between airborne fungal spores at a city centre site in Cardiff and at a mixed deciduous woodland site at Cefn Mably.

don. Air in Cardiff has been sampled continuously since 1954 using an AVST at roof top level in the centre of the city.

The data obtained can be regarded as being typical for a north temperate climate and the incidence of different spores throughout the year in the UK and comparable climatic areas of mainland Europe are broadly similar.

City centre site

In Figs 1-5, the incidence in the air of *Cladosporium*, ascospores, basidiospores, rust and smut spores, and particles of hyphae throughout the year is shown. It will be noticed that *Cladosporium* reaches its peak in July and August (Fig. 1), hyphal fragments reach their highest levels in August (Fig. 5), and rusts and smuts are most numerous in June when smut fungi infect the flowers of grasses (Fig. 4). Ascospores also reach their highest levels in August (Fig. 2), but basidiospores continue at high levels through the months of September and October (Fig. 3). Relative to *Cladosporium* conidia, ascospores and basidiospores (Figs 1-3), hyphal fragments (Fig. 5) form a minor component of the airborne fungal burden. Other investigations have also shown that hyphal fragments account for only a few per cent of fungal particles in outdoor air, e.g. those of Li and Kendrick (1995) and Delfino *et al.* (1997). Despite this, they can be a significant factor in asthma symptom severity (Delfino *et al.* 1997).

Comparison of the data obtained in urban Cardiff with counts obtained during the summer months at various outstations at woodland, coastal, upland valley and coastal plain sites shows the influence of local factors on spore concentrations. Since only one site distant from the city centre site was sampled in any particular year, it is not however possible to make direct comparisons between the outstations except as they contrast with Cardiff.

At Resolven (R), where the trap was positioned on the edge of a field of rough grass surrounded by plantations of Sitka spruce, Douglas fir and Japanese larch, spore concentrations were consistently higher than at Cardiff (C) after May (Fig. 6), because of the release of large numbers of ascospores and basidiospores in the plantations. Overall, the basidiospore totals were 217% and ascospore totals 206% of those at Cardiff. *Cladosporium* totals at Resolven were 61% of those at Cardiff and the other categories of fungi, including other imperfect fungi and rusts and smuts, were 83% of those recorded at Cardiff. This was due to relative paucity of local substrates for growth of these fungi within the plantations.

At Tintern, alongside a stream in a field surrounded by mixed deciduous and coniferous woodland with intervening fields and small orchards, the total spore concentrations (Fig. 7) were greater than at the city centre site each month with the exception of August. During that month there was a very high urban concentration of *Cladosporium*, which was 175% of that at Tintern, with an average daily concentration of 7000 spores m^{-3} air. Over the survey period basidiospore totals were 337% and ascospore totals 254% of those in Cardiff.

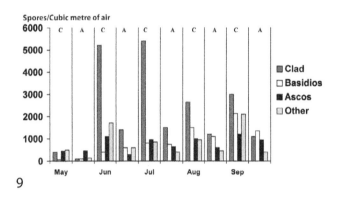

9

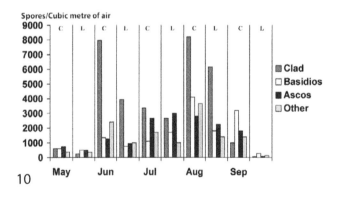

10

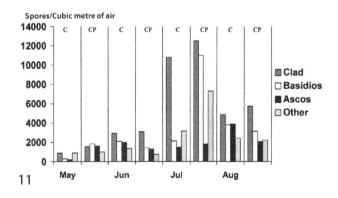

11

Fig. 9. Comparison between airborne fungal spores at a city centre site in Cardiff and at a seafront site at Aberystwyth.
Fig. 10. Comparison between airborne fungal spores at a city centre site in Cardiff and at an upland valley site at Llwynypia.
Fig. 11. Comparison between airborne fungal spores at a city centre site in Cardiff and at a farmland site at Cleppa Park.

Table 1. Mean concentrations of airborne fungal spores collected outdoors by Kramer-Collins drum sampler in and around Manhattan, Kansas, over a 10-day sampling period in 1979 (Kramer and Eversmeyer 1984).

	Mean concentration (spores m⁻³ air)			
Location	*Cladosporium*	*Alternaria*	Ascospores	Basidiospores
Prairie	3852a	307ab	509ab	426a
Experimental farm	995b	183b	94b	121b
City park	2750a	452ab	894a	468a
New suburbs	4751a	561a	689ab	503a
Old suburbs	2872ab	403ab	688ab	465a
City centre	4063a	608a	685ab	555a

Mean values in the same column followed by the same letter are not significantly different from each other.

A trap at Cefn Mably (CM) was surrounded for a distance of 500-900 m by mixed deciduous woodland, parts of which had been neglected for some time and included large quantities of decaying timber. Total spore concentrations at this site were >300% greater than at Cardiff (Fig. 8). Basidiospore totals were 755% greater than at Cardiff and ascospores 375% greater. *Cladosporium* totals were also greater at this woodland site, averaging 130% of those at Cardiff. Owing to the proximity of many basidiocarps associated with decaying timber, the usual annual periodicity seen in Cardiff over a period of years was not replicated at Cefn Mably, where basidiospore levels were consistently high throughout the summer months.

These woodland surveys indicate that woodlands are, as might be expected, a prolific source of ascospores and basidiospores, the unmanaged mixed woodland (Fig. 8) being a greater source of spores than the managed coniferous or managed mixed coniferous and deciduous woodland. Herxheimer *et al.* (1969) demonstrated that basidiospores and ascospores can cause allergy and induce asthma attacks. Many cases of late summer asthma can be attributed to the inhalation of spores originating in woodlands.

Coastal site

Spore concentrations over the sea are generally much lower than those over the land, particularly at some distance from the land, owing to deposition of the spores on the sea without any compensating addition of spores (Hirst *et al.* 1967), so that traditionally a sea cruise was considered an effective way of avoiding seasonal allergens.

As the prevailing wind direction in UK is from the west, a west-facing coast should benefit from on-shore air currents comparatively free of spores, unlike the east coast, which will receive the spore load picked up as the air currents cross the land. Fig. 9 shows a comparison between Cardiff city centre and a west coast site on Cardigan Bay.

This roof top site was at Aberystwyth (A), 120 km NW of Cardiff, and faced west near the seafront, from which town streets containing relatively few trees and gardens extended 600-900 m to the east, with agricultural land lying beyond that. Spore concentrations were consistently lower over the period of the survey, with total concentrations being only 44% of those in Cardiff. While both ascospore and basidiospore concentrations in Aberystwyth were around 80% of those in Cardiff, *Cladosporium* totals were only 32% of the corresponding Cardiff totals. The Aberystwyth counts confirm the trend to lower spore counts near a west-facing coast, particularly so in the case of *Cladosporium*.

Another example of this phenomenon has been recorded by Rodriguez-Rajo *et al.* (2005), who observed that a rooftop AVST in the west-coast city of Vigo, in the province of Galicia, NW Spain, gave lower spore counts than at two inland Galician locations. The total number of spores trapped in Vigo was less than 60% of the counts at the other two sites. At all three locations, *C. herbarum* predominated over the year, accounting for 57-68% of the annual total, with *C. cladosporioides* comprising 30-44% and *Alternaria* spp. no more than 1%. *Cladosporium* spore counts were again particularly low relative to the inland sites. As will be discussed later, another instance illustrating the difference between the air at coastal and at inland sites was noted by Prospero *et al.* (2005) on the Caribbean island of Barbados.

Upland valley site

At Llwynypia Hospital, in the Rhondda Valley 25 km northwest of Cardiff, a sampler was sited on the roof top of a building on a west facing slope covered for the most part by rough pasture and scrub vegetation with some hedgerows. On the opposite side of the valley was an extensive conifer plantation, as well as more pasture and scrub vegetation. The valley, like all South Wales valleys, runs roughly in a north-south direction, and Davies (1969a,b) suggested that such a valley might mimic the situation of upland valleys in Switzerland. The relatively poor vegetation plus wind flows across the Swiss valleys produce environments with low spore concentrations, which have favoured the construction of sanitoria in these areas for the relief of asthma.

During the period from May to September (Fig. 10), overall spore counts (L) at Llwynypia were found to be 57% of those recorded in Cardiff, with basidiospores being 49%, ascospores 73% and *Cladosporium* 65% of the corresponding Cardiff totals. When Williams and Higgins (1959) carried out a survey of the incidence of asthma in the Rhondda Valley they did in fact observe a lower incidence of asthma compared with Cardiff, and lower concentrations of grass pollen, another cause of asthma, were also noted in the survey at Llwynypia mentioned above.

Coastal plain site

South of the valleys in South Wales is a coastal plain leading to the Bristol Channel. A trap was situated at Cleppa Park (CP), which is a rural site 14 km east of Cardiff. The trap was surrounded by fields used for cereal and vegetable production and for pasture, with small woodlands of mixed conifer and broad-leaved trees some 550-1190 m away.

The incidence of spores (Fig. 11) was higher than at Cardiff overall, with counts of basidiospores being 209%, ascospores 90% and *Cladosporium* 118% of those in Cardiff. However, in June and August, ascospore and basidiospore concentrations at Cardiff exceeded those at Cleppa Park. Surprisingly for a rural site, rust and smut spores in total were marginally lower than in the city, but together rusts and smuts accounted for only some 2-3% of the total air spora recorded at both sites. During July, the spore concentration at Cleppa Park was particularly high in relation to Cardiff, being some 186% of the city concentration.

The comparisons between the air spora at Cardiff and at the different outstations around the city illustrate, within one geographical area, the importance of local sources of spores in the concentration and make-up of the air spora. When Kramer and Eversmeyer (1984) sampled the air at four sites within the city of Manhattan, Kansas, and two sites outside the city, all within a radius of 10 km of the city, they found no significant difference between mean spore concentrations in major spore types at the different

sites when tested with Duncan's multiple range test (Table 1). However, the authors noted that variations in concentration of the air spora between different sites were primarily due to differences in environmental conditions which influence the numbers and kinds of fungi that develop, sporulate and release spores, although the site descriptions do not indicate large vegetational differences between the environs of the different sites. However, the differences noted by Kramer and Eversmeyer (1984) were most pronounced in periods during which rain was sufficient to allow for abundant growth and sporulation of fungi locally. Conversely, during dry periods development of fungi in the area is slowed severely, if it occurs at all, so that overall the greatest percentage of the air spora at all these sites is composed primarily of spores transported through the atmosphere from remote sources. Unfortunately, the time of year during which the concentrations shown in Table 1 were recorded is not given, so comparisons cannot by made with data from other sites. The high numbers of *Alternaria* spores in the air suggest that the prairie is the likely source of spores found in the city.

Another instance in which high numbers of *Alternaria* in an urban environment were attributed to the growth of cereals in surrounding areas occurred when Corden *et al.* (2003) compared daily *Alternaria* records in the coastal city of Cardiff with those in urban Derby for the period 1970-1996. They noted that there had been a marked upward trend in the seasonal total for Derby, while the trend in Cardiff had been downwards. It was suggested that the difference could have been due to increased cereal production, together with higher midsummer temperatures, in the area around Derby and the smaller amount of arable production round Cardiff. Latterly, *Alternaria* counts had exceeded 10^3 spores m^{-3} air on some days in Derby, possibly triggering asthma attacks in *Alternaria*-sensitive patients, as they are known to provoke asthma attacks. Despite such differences, however, an investigation in Madrid, which has a continental Mediterranean climate, illustrates the broad similarity between the air spora in Cardiff and that in other European urban environments. In Madrid, Herrero *et al.* (2006) used a rooftop AVST and collected 70 different spore types, but three-quarters of the annual spore total was accounted for by four categories of spore. These were conidia of *Cladosporium* (41.1%); teliospores of the rust *Tilletia* (0.1%) and the smut fungus *Ustilago* (17.6%); basidiospores of *Coprinus* (8.7%) and other members of the Basidiomycota (5.9%); and ascospores of *Pleospora* (1.5%), *Leptosphaeria* (1.5%)

Table 2. Dominant fungal spores during 1992 from outdoor air of Waterloo, Ontario, expressed as mean concentration and percentage of total airspora (after Li and Kendrick, 1995).

Type	Whole year		May-October		November-April	
	Spores m^{-3}	% of total	Spores m^{-3}	% of total	Spores m^{-3}	% of total
Alternaria	62	1.8	139	1.9	13	1.3
Aspergillus/Penicillium	130	3.8	139	1.9	125	12.4
Cladosporium	1425	41.0	3018	40.8	421	41.9
Epicoccum	46	1.3	64	0.8	35	3.5
Leptosphaeria	183	5.3	454	6.1	12	1.1
Unidentified ascospores	185	5.3	403	5.5	48	4.7
Coprinaceae	174	5.0	444	6.0	5	0.5
Ganoderma	249	7.2	641	8.7	0.8	0.1
Unidentified basidiospores	422	12.1	992	13.4	62	6.2
Other unidentified spores	294	8.5	563	7.6	125	12.4
Hyphal fragments	171	4.9	258	3.5	116	11.6

and others (1.5%). An eight-year AVST study of airborne ascospores in another city with a true Mediterranean climate, Heraklion on the island of Crete, showed *Leptosphaeria* to be much more prominent in the air spora (Gonianakis *et al.* 2005). It formed 6.5% of the total fungal spora and 47.1% of the ascospore load, which also included spores of *Chaetomium* and, more sporadically, *Didymella*, *Leptosphaerulina* and *Pleospora*. Overall, the mean ascospore concentration was 30 spores m^{-3} air day^{-1}, which approximates to 13.9% of the total airborne mycobiota (Gonianakis *et al.* 2005).

The predominance of *Cladosporium* can again be seen in the southern hemisphere, where in a comparable temperate climate Mitakakis and her colleagues (Mitakakis *et al.* 1997, Mitakakis and Guest 2001) surveyed the air spora in the Australian city of Melbourne. Among the spores of 29 genera and five other groups collected by AVST, *Cladosporium* conidia (41.7% of the total spores), *Leptosphaeria* ascospores (14.9%) and *Coprinus* basidiospores (14.6%) comprised nearly three-quarters. Rust and smut teliospores amounted to 12.0% of the total, ascospores other than *Leptosphaeria* 5.5%, and *Ganoderma* basidiospores 2.1%. None of the other fungi, including *Alternaria*, *Periconia*, *Drechslera* and other phylloplane fungi such as *Epicoccum*, *Fusarium* and *Stemphylium*, amounted to more than 1% of the total. Small hyaline spores, including those of *Aspergillus* and *Penicillium* were not recorded, although it was known from Andersen sampler culture plates that they were present in the air. The total counts showed marked annual variation, with the count one year being nearly three times those of the succeeding two years (Mitakakis

et al. 1997). In the third year, *Cladosporium* numbers were approximately 1.5 times those in the two earlier years.

Seasonal and diurnal variation in the air spora

Seasonal variations similar to those in Cardiff (Figs 1-5) have been recorded in Waterloo, Ontario (Table 2), where Li and Kendrick (1995) sampled outside air on a balcony over a period of a year using a particle sampler, which took 10-min samples of the air every two hours. Compared with the spore concentrations recorded in Cardiff in 1992, the concentrations of *Cladosporium*, ascospores and basidiospores recorded in Ontario were all higher, with respective average concentrations of 1425, 368 and 845 spores m^{-3} air in Ontario and 871, 161 and 248 m^{-3} in Cardiff, and the percentages of the total air spora respectively 41, 10.6 and 24.1% in Ontario compared with 55, 10.2 and 15.7% in Cardiff. However, concentrations for Cardiff in 1992 differ from the 30-year averages for the city, in which *Cladosporium* accounts for 41% of the total air spora, ascospores for 17.2%, and basidiospores 20.6%.

Trends broadly similar to those shown in Fig. 1 have been noted for *Cladosporium* elsewhere in Europe. Gravesen and Schou (1997) noted that in a Danish study *Cladosporium* spores were most abundant in late July/August (Fig. 12), whilst *Alternaria* peaked in late August, both much later than allergenic tree pollens, which variously peaked between March and May, and grass pollen, which was most abundant in June (Fig. 12). In the Spanish province of Galicia, Rodriguez-Rajo *et al.* (2005) observed that the highest counts of *C. herbarum*, the more abundant of two

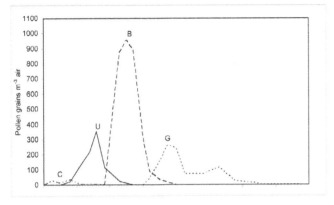

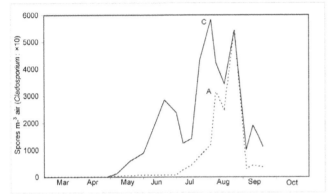

Fig. 12. The seasonal occurrence in outdoor air of pollen and mould spores relevant to allergy (adapted from Gravesen and Schou 1997). Upper panel: B, *Betula* (birch); C, *Corylus* (hazel); G, *Gramineae* (grasses); U, *Ulmus* (elm). Lower panel: A, *Alternaria*; C, *Cladosporium*.

species of *Cladosporium*, occurred in the summer months, particularly June-September, with peak levels in August. On the other hand, *C. cladosporioides* levels did not begin to rise until August and the highest counts were encountered in September-October. For *Alternaria*, the greatest counts were recorded between July and September (Rodriguez-Rajo *et al.* 2005). In contrast, in the city of Leon in the neighbouring province *Cladosporium* counts peaked in October one year, but during the following year maximum levels were reached in July (Fernández *et al.* 1998). *Alternaria* counts were at their greatest in August of the first year and in July of the second year.

That such variation in seasonality between different studies is largely a result of climatic differences can be seen from three European studies previously mentioned. In Madrid (Herrero *et al.* 2006), with its continental Mediterranean climate, the maximum occurrence of the four main spore types differed from those presented in Figs 1-4 for Cardiff. Teliospores peaked earlier (May-June), and ascospores (September/October), *Cladosporium* (October) and basidiospores (October/November) all peaked later than in Cardiff. In the true Mediterranean climate of Heraklion (Gonianakis *et al.* 2005), total ascospore numbers

were relatively high in mid-spring and remained so throughout the summer, with *Leptosphaeria* reaching maxima of up to 70 spores m⁻³ air in June. In an AVST survey in the continental climate of the Polish city of Cracow (Stepalska and Wolek 2005) counts of the spores of most of 13 selected fungi peaked in August. These included *Alternaria* and *Epicoccum* conidia and *Ganoderma* basidiospores, but numbers of *Didymella* ascospores were highest in July. In contrast to both Cardiff (Fig. 1) and Madrid, *Cladosporium* numbers peaked during June.

In addition to recording seasonal variation, Li and Kendrick (1995) noted diurnal periodicity in ascospore and basidiospore concentrations, with an increased prevalence during the early hours of the morning, similar to that for the prevalence of *Didymella* ascospores (Allitt 1986) observed in UK (Richardson 1996). Richardson derived his data using an AVST for continuous sampling on a rooftop and found an association between increased spore concentrations and relative humidity during the early hours of the morning. In addition to the diurnal periodicity of ascospores and basidiospores, Li and Kendrick (1995) observed periodicity in the prevalence of conidia of *Cladosporium* and other imperfect fungi, which are released passively. These conidia tended to increase in number during the afternoon, with the highest concentration of *Cladosporium* at 16.00 h and of *Alternaria* at 18.00 h. Three AVST studies in Spain indicate, however, that there can be marked temporal variation in peak concentrations between different locations. In the first of these, *Alternaria* was described by Giner *et al.* (2001) as being a "late afternoon taxon" in SE Spain; its "maximum" numbers in the air were reported to occur between 13.00 and 21.00 h. However, the data appear to show an absolute maximum at around 19.00 h on each of six years. In the north-western province of Galicia, Rodriguez-Rajo *et al.* (2005) observed the greatest numbers of *C. cladosporioides*, *C. herbarum* and *Alternaria* spp. between 19.00 and 22.00 h, whereas in the neighbouring province Fernández *et al.* (1998) noted maximum numbers of *Cladosporium* and *Alternaria* conidia between 12.00 and 14.00 h.

Culture plate surveys of fungi in outdoor air

Temperate climates

The limitations of trapping airborne inoculum on agar plates has already been mentioned, but it is worth considering the mistaken impressions of the air spora which may result from relying on sampling methods based on culture. When comparing AVST and volumetric culture plate sampling by a wind-orientated Andersen sampler, Burge *et al.* (1977) noted that *Cladosporium* spore concentrations assessed by both methods varied directly. However, as AVST spore levels rose the culture plate data progressively underestimated prevailing concentrations, giving low estimates of prevalence (20-40%) at levels below 100 spores m^{-3} and falling to below 5% at levels above 500 spores m^{-3}. Several additional taxa were also substantially understated in abundance and regularity. However, in a recent study in the Greek city of Athens, Pyrri and Kapsanaki-Gotsi (2007) found that for *Cladosporium* a single-plate Burkard sampler gave counts only 5% less than corresponding total counts on a Burkard personal volumetric air sampler, and *Alternaria* 35% less.

Sampling for culturable airborne fungi in Copenhagen using a slit sampler (Table 3) indicated a much higher percentage incidence of *Cladosporium* at 68.9% (Larsen 1981) and 77.8% (Larsen and Gravesen 1991) than was found using the AVST in Cardiff or the particle sampler in Ontario (Li and Kendrick 1995). Whereas it is possible to differentiate a much greater number of imperfect fungi in the air and to record a range of yeasts by culture-based methods, it is important to realise that the large numbers of ascospores and basidiospores in the air are not recorded on culture plates, and so sampling onto agar medium is likely to have biased the values for the percentage incidence of *Cladosporium* recorded by the Danish workers.

Despite their limitations, culture plate methods are frequently used in assessing the indoor air spora and have also been used to effect in surveys of outdoor air. For example, using the Andersen N6 single-plate sampler (Shelton *et al.* 2002) carried out a countrywide survey over three years, sampling inside and outside some 1700 buildings in USA, and taking 12,000 samples in all, 2,400 of which were outdoors. There were considerable differences between regions: the highest levels of culturable fungi outdoors (in summer and autumn) were found in the Southeast, Southwest and Far West regions, and the lowest in the Northwest region. The commonest of the culturable fungi in the outdoor air during all seasons and in all regions were *Cladosporium* (when detected, median concentration approx. 200 CFU m^{-3} air), *Penicillium* (approx. 50 CFU m^{-3}), *Aspergillus* (including *Eurotium* and *Emericella*, approx. 20 CFU m^{-3}) and non-sporulating fungi (approx. 100 CFU m^{-3}). A wide range

Table 3. Principal viable fungi collected by slit sampler from outdoor air in Copenhagen, as percentage of total colonies on slit-sampler collection plates.

Type	1977-79[a]	1978-87[b]
Cladosporium	68.9	77.8
Alternaria	9.4	3.0
Penicillium	6.0	2.5
Aspergillus	2.5	0.8
Non-sporing filamentous fungi	6.3	n.s.
Yeasts	4.1	n.s.

a Larsen (1981); b Larsen and Gravesen (1991).

of other taxa were detected, including the toxigenic genus *Stachybotrys*, which was present in the outdoor air at 1% of the buildings under study. Just how wide the range of culturable fungi in outdoor air can be is evident in a qualitative study in Turin, Italy (Airaudi and Marchisio 1996). In a yearlong survey employing a Surface Air Systems single-plate sampler, these workers isolated 170 different species. However, when Lugauskas *et al.* (2003) used a combination of a liquid impinger, an open-faced filter sampler and settle plates to examine the airborne mycobiota near busy streets in five Lithuanian cities, some 430 species were recorded. Among these, the vast majority (83%) were mitosporic fungi, i.e. those for which no sexual stage has been found. Some 45 species were in the Zygomycota (more than one-half in the family Mucoraceae) and 21 species in the Ascomycota, but no members of the Basidiomycota were mentioned in their paper. Surprisingly, *Aspergillus fumigatus* and *A. niger* were among the species most frequently isolated (detection frequencies approx. 57 and 84%, respectively), together with *Alternaria alternata* (63%), *Aureobasidium pullulans* (57%), *Botrytis cinerea* (43%), *Cladosporium herbarum* (64%), *C. cladosporioides* (57%), *C. sphaerospermum* (28%), *Geotrichum candidum* (33%), and *Penicillium funiculosum* (46%). The two active sampling methods used revealed that propagule counts were greatest in the industrial areas of Vilnius, and those close to heavily used highways, reaching approx. 6400 CFU m^{-3} air in summer. Corresponding counts for the four other (smaller) cities ranged from approx. 500 to 4500 CFU m^{-3}.

As mentioned above, culture-based sampling methods enable a wider range of yeasts to be identified than can be differentiated by cell morphology after AVST. For example, when Rantio-Lehtimäki (1988) sampled the air specifically for yeasts in southern Finland over a one-year period using an Andersen sampler with yeast-peptone-D-glucose agar culture plates, she found that concentrations of yeasts in the air never exceeded 50 CFU m^{-3} air. The predominant yeast genus isolated on the agar plates was *Sporobolomyces,* followed by *Rhodotorula* and *Cryptococcus,* all of which are basidiomycetous yeasts. Ascomycetous yeasts including *Candida* were caught infrequently.

Employing a slit sampler, Larsen (1981) found that yeasts comprised only 4.1% of the air spora in Copenhagen. The mirror yeasts *Sporobolomyces* and *Tilletiopsis* can be identified on AVST slides or tapes, and Gregory and Sreeramulu (1958) reported *Sporobolomyces* concentrations of up to 10^6 spores m^{-3} air over an estuary in the south of England; and routine monitoring of the air at Cardiff over 30 years showed an average *Sporobolomyces* spore concentration of 167 m^{-3} air, with a maximum of 27297 m^{-3}, and an average for *Tilletiopsis* of 48 m^{-3} air, with a maximum of 7920 m^{-3}. This reinforces the view of Burge *et al.* (1977) that culture plate methods seriously underestimate the spore content of the atmosphere.

Dust events

In the same era as Miquel observed a boost in the air spora with dust-raising activities such as street cleaning, Carnelley *et al.* (1887) recorded that in dry weather the raising of dust by wind increased numbers of microorganisms overall, and greatly increased the ratio of bacteria to fungi in the streets of two large Scottish towns. Over the years since then, aerobiological studies have continued to demonstrate the importance of risen dust as a contributor to both the outdoor and the indoor air spora. Two types of dust event that can result in augmentation of the outdoor air spora are discussed below.

The first type occurs during building construction and destruction, which are significant in altering the nature and magnitude of the air spora locally. Airborne dust and associated spores generated and dispersed externally during constructional work can penetrate into nearby buildings. Such penetration is of particular relevance in hospitals, since it presents a potential health risk to patients. Among the spores associated with dust particles that are generated and dispersed during construction those of thermotolerant species are of particular concern. These fungi, able to grow at body temperature (37°C), include some aspergilli that can behave as opportunistic pathogens and are a particular hazard to immunocompromised patients. *Aspergillus fumigatus* is the most important of these pathogens, but *A. flavus, A. niger* and *A. terreus* may also present a risk (Fitzpatrick *et al.* 1999).

Demolition of buildings may have an even greater effect than construction activities in elevating microbial numbers in outside air. In reporting the effect of explosive demolition of a hospital building in Minnesota, Streifel (1983) noted that relative to levels before the explosion samples taken 15 m from the building 2 min after the explosion revealed a 1.8-log increase in thermotolerant fungi to 1.6 × 10^5 CFU m^3 air. Among these thermotolerant fungi, *A. fumigatus* and *A. niger* showed 3.3- and 1.5-log increases to 8.4 ×10^2 and 1.4 × 10^4 CFU m^3, respectively. At a sampling site about 60 m away from the building, the collective counts, and those for *A. fumigatus* and *A. flavus*, remained high for 45 min before declining. In another study of hospital demolition, in Madrid (Bouza *et al.* 2002), concentrations of airborne fungi were found to be much lower than in the Minnesota investigation, however. During the five days before demolition by controlled explosion the thermotolerant average count was 17.6 CFU m^3 air; after the explosion it was 70.2 CFU m^3; and after falling to little over half that level it rose again to 74.5 CFU m^3 four days later during a second, mechanical phase of demolition; not until 11 days after the explosion had counts fallen back to levels roughly the same as before demolition. Although the identity of two-thirds of the isolates was not established by Bouza *et al.* (2002), the principal identified taxon was *A. fumigatus,* and *A. niger, A. flavus, Mucor, Alternaria, Fusarium* and unspecified penicillia were also isolated.

While building and demolition activities have a local effect on the air spora, it has been estimated that dust events of the second type, desert dust storms, result in 2.2 ×10^9 metric tons of dust becoming airborne and being transported annually across continents and oceans (Goudie and Middleton 2001). Satellite imagery has shown long distance migration of dust clouds from desert areas in Asia, Africa, North America and also Australia and South America. It is those originating in the 9 × 10^6 km^2 of the Sahara Desert which have the greatest global impact, with the dust reaching Southern, and much less frequently, Northern Europe and the British Isles; the Eastern Mediterranean; Caribbean islands; and North and South America (Griffin et al. 2001b). With the knowledge that spore dispersal can spread plant diseases between countries and across continents (Gregory 1973), it has been a concern that microorganisms in the surface layers of the soil that is aerosolized might have implications for human health and ecosystems downwind of desert storms. It is this concern that has driven research into these dust events, especially so in relation to the deposition of Saharan dust in the

Caribbean islands and south-eastern USA (Griffin *et al.* 2001a, 2003, Prospero *et al.* 2005).

In an investigation on the west coast of St John in the US Virgin Islands airborne microorganisms were collected on membrane filters during dust events, cultured on agar plates and subsequently identified by comparison of 16S and 18S rDNA sequences with those in GenBank (Griffin *et al.* 2001a). Very few airborne fungi were found. *Cladosporium cladosporioides*, *Coccodinium bartschii*, *Gibberella pulicaris* and *Pleospora rudis* were isolated on a clear day, but in a dust event one week later only the first two species were detected. Three days after that, during a second dust event, *C. bartschii*, *Cochliobolus sativus* and *P. rudis* were the only fungi isolated. However, using traditional culture-based identification methods Prospero *et al.* (2005) isolated a much wider range of species on the east coast of Barbados. Although nearly 60% of the isolates were not identified, *Arthrinium* and *Periconium* together comprised three-quarters of those that were, with (in descending order) *Penicillium*, *Curvularia*, *Cladosporium*, aspergilli (principally *Aspergillus niger*, but also *A. fumigatus*, *A. clavatus*, *A. terreus* and *A. flavus*), *Neurospora* (*Chrysonilia*) and *Alternaria* being minor components. Inland, numbers were much greater and the composition was rather different, with *Cladosporium* amounting to one-third, and in descending order *Aspergillus* (*A. niger*, *A. flavus*), *Bipolaris*, *Curvularia* and *Penicillium* collectively comprising one-half of the total. In a follow-up to Griffin *et al.* (2001a), airborne culturable fungi at a site on the south coast of the neighbouring island of St Thomas were collected during an African dust event (Griffin *et al.* 2003). The fungi isolated belonged mainly to the genera *Cladosporium*, *Aspergillus* and *Penicillium*. Smaller numbers of *Microsporium*, *Bipolaris*, *Paecilomyces*, *Acremonium* and *Nigrospora* were encountered. In none of these investigations is there clearcut evidence that the source of the fungi detected was Africa; the evidence from Prospero *et al.* (2005) suggests that the composition of the air spora at the sampling sites on St John and St Thomas (Griffin *et al.* 2001a, 2003) could have been influenced by *Cladosporium* and other phylloplane fungi from local sources being entrained by the trade winds crossing the islands from the east.

Nevertheless, Griffin *et al.* (2006) found that the airborne mycobiota at a mid-Atlantic research site in the transatlantic pathway from West Africa included common penicillia and phylloplane fungi such as *Alternaria* spp., *Aureobasidium pullulans* and *Phoma herbarum* (Table 6). The most frequently isolated spe-

Table 4. Percentage frequency of viable airborne fungi collected during 1988 on settle plates exposed to outdoor-air in the city of Natal in tropical Brazil (after de Oliviera *et al.* 1993).

Type	Dry season	Rainy season
Aspergillus	77.7	55.3
Penicillium	62.9	40.0
Fusarium	16.6	44.6
Cladosporium	25.9	10.7
Curvularia	11.1	20.0
Rhizopus	11.i	16.9
Rhodotorula	12.9	9.2
Neurospora	12.9	3.0
Drechslera	1.8	10.7
Aureobasidium	5.5	4.0
Trichoderma	-	6.1
Cunninghamella	1.8	3.0

cies were the *Cladosporium*-like ascomycete, *Lojkania enalia*, and another ascomycete, *Neotestudina rosatii*, which is common in tropical soils and is a cause of mycetoma in humans. There was a statistically significant relationship between the number of culturable microorganisms recovered and atmospheric dust levels. However, the concentration of culturable fungi in the air was only around 1% of that recorded during desert storms in Mali, West Africa (Kellogg *et al.* 2004), where *Alternaria* sp., *Cladosporium cladosporioides*, *Aspergillus niger* and *A. versicolor* were identified by their micro-morphology. This could be expected because of the extended period of exposure to UV light, desiccation and other stresses during long-distance transport, and would suggest that an even smaller proportion of microorganisms reaching the Americas would be culturable.

On the eastern Mediterranean coast it was also found that dust storms in North Africa lead to increased concentrations of microorganisms in the atmosphere in Haifa, Israel (Schlesinger *et al.* 2006). Counts for culturable fungi during separate dust events were approximately two and seven times those on clear days, but bacterial counts were around nine times greater. The commonest species during the dust events was *Penicillium chrysogenum*, which was 20-30 times more abundant than on clear days. *Aspergillus fumigatus* and *P. griseoroseum* were also prominent but not detected on clear days, while *Alternaria alternata*, *A. niger*, *A. thomii* and *P. glabrum* counts were greater for one or other event than on the clear days. On the Mediterranean coast of Turkey, Griffin *et al.* (2007) isolated *Acremonium*, *Alternaria*,

Cladosporium, Fusarium, Microsporum, Penicillium and *Trichophyton* from samples of airborne dust of regional and North African origin. The maximum number of culturable fungi in the air recorded on dust event days was more than twice the non-dust day maximum and, unusually, around 12 times greater than the corresponding bacterial count. During a cruise across the Mediterranean from Tel Aviv to Istanbul, however, Waisel *et al.* (2008) detected very little particulate matter from a dust cloud originating in North Africa and moving eastwards at an altitude of 3000 m. The mean daily AVST spore count in mid-sea was 300-750 spores m^{-3} air, but was much higher nearer the Turkish and Israeli coasts, repectively 1200-2400 and 340-1695 m^{-3} air. The major component of these counts was *Cladosporium*. The authors inferred that north-northwesterly winds at lower altitudes were probably responsible for bringing these spores from Turkey, Greece and the Balkans across the Mediterranean to Israel.

Tropical and subtropical climates

The air spora in tropical and subtropical regions is likely to be substantially different from that in temperate regions of the world. This is exemplified by a study of Oliveira *et al.* (1993), who used settle plates to study the air spora at five locations around the city of Natal in tropical Brazil over a one-year period. A similar incidence was observed at all of the sites, each of which was sampled for 15 min every two weeks. The lack of volumetric sampling did not allow estimations of spore concentrations, and the incidence of fungi is expressed as percentage frequency of fungal isolation (Table 4), but from this it is apparent that the relative incidence of fungi in Natal is quite different from that reported in Copenhagen (Larsen 1981, Larsen and Gravesen 1991). *Aspergillus* and *Penicillium* were isolated with a very high percentage frequency especially during the dry season and there was a dramatic increase in *Fusarium* during the rainy season.

In contrast, *Cladosporium*, which dominates the air spora in temperate regions had a comparatively low frequency, although this was not true for all cities in Brazil and the authors quote percentage frequencies from other cities where the incidence of *Cladosporium* was much greater, but overall the incidence of *Aspergillus* and *Penicillium* throughout the surveys quoted tends to be higher than that reported in temperate regions.

In subtropical Taiwan, Li and Hsu (1995) used an Andersen sampler with malt extract agar culture plates to compare spore concentrations within and

Table 5. Concentration of viable fungi (colony forming units) collected by Andersen N6 impaction sampler from outdoor air of homes of asthmatic, atopic and control children in Taiwan (after Li and Hsu 1995). Geometric mean concentration (CFU m^{-3} air)

Type	Asthmatic (n = 46)	Atopic (n = 20)	Control (n = 26)
Alternaria	1.7	1.5	1.8
Aspergillus	17.2	22.5	23.6
A. clavatus	1.1	1.1	1.2
A. flavus	4.7	2.7	5.9
A. fumigatus	1.1	1.1	1.2
A. niger	4.7	6.7	5.9
Cladosporium	19.5	7.3	3.8
Curvularia	1.7	1.3	1.9
Fusarium	4.3	1.8	5.5
Paecilomyces	2.2	1.5	2.0
Penicillium	47.6	55.5	101.0
Trichoderma	1.1	1.1	1.2
Yeasts	70.3	73.8	72.5
Total	547	449	668

outside the Taipei homes of atopic and non-atopic children between July and September (Table 5). The numbers of airborne CFU m^{-3} air recorded were comparatively low, with *Cladosporium* having a lower incidence than *Aspergillus* and *Penicillium*. Hurtado *et al.* (1989) sampled the air in the mildly tropical climate of Bogota in Colombia for pollen and spores for a year and found that *Cladosporium* was the most frequently recorded genus, with the combination *Penicillium/ Aspergillus* being next. The findings in the tropical and subtropical surveys show a different pattern of fungal spore incidence, and spores of *Aspergillus* and *Penicillium* appear to be more prevalent in the air than in temperate regions of Europe and North America.

Although aspergilli and penicillia are generally less abundant in outdoor air in temperate regions, a range of species in both genera may be isolated by culture plate sampling for spores in the atmosphere (Hudson 1969, Fradkin *et al.* 1987). The opportunistic respiratory pathogen *Aspergillus fumigatus*, for example, may be detected. In a comparative study of *A. fumigatus* in the air of St Louis in USA and Cardiff in UK (Fig. 13) over a 12-month period using Andersen samplers, sampling onto Blakeslee's malt extract agar culture plates which were incubated at 37°C to eliminate other fungi, average concentrations recorded were 13.5 CFU m^{-3} in St Louis and 11.3 CFU m^{-3} in Cardiff and the seasonal incidence of the fungus was similar

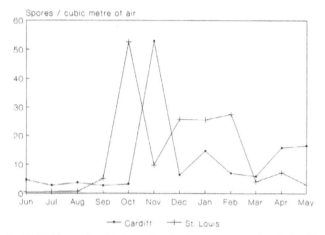

Fig. 13. Outdoor culturable *Aspergillus fumigatus* concentrations in Cardiff and St. Louis (after Mullins, Hutcheson and Slavin 1984)

at both sites (Mullins *et al.* 1984).

In order to sample for *A. fumigatus* in the air both culture media and incubation conditions were modified to favour growth of this fungus at the expense of any competing fungi. In particular, richer media, such as Blakeslee's malt extract agar, favour *Aspergillus* and *Penicillium*. Blakeslee's 1915 formula for the agar contains 2% malt extract, 2% glucose and 0.1% peptone, as opposed to the unamended 2% malt extract usually recommended for isolation of the general run of fungi from air. In all surveys of the air spora the sampling method chosen will favour sampling of certain fungi and it should be remembered, as mentioned previously, that the sampling techniques used can have a major influence on the results of the survey.

Owing to the important effect of weather conditions on spore release, daily spore counts can vary considerably; in some fungi spore discharge is associated with drying whereas in others damp conditions are required. Raindrops can discharge basidiospores from puffballs, and Walkey and Harvey (1968a,b) observed that ascospore discharge through the ostiole of the ascocarps in pyrenomycetes (flask fungi) was stimulated by rainfall. Many spores are thrown up by rain splash, e.g. in *Chaetomium* (Harvey *et al.* 1969), so that certain spore types are characteristically associated with damp conditions and others with dry conditions. The variability of conditions during the day can affect spore concentrations and it was found in UK that fungi such as the yeast *Sporobolomyces* tend to occur in high concentrations during the night when humidity is highest, whereas *Cladosporium* has its highest concentration in the day when humidity is at its lowest (Hirst 1953, Gregory and Sreeramulu 1958). This was also noted in Ontario by Li and Kendrick (1995). Even light can act as a contributory fac-

tor in spore release, and Ingold (1971) has described light and dark induced rhythms of spore discharge in a number of fungi.

The regular seasonal cycles of the year determine which fungal spores are likely to be available for release into the air. Thereafter the development of the fungi and the subsequent release of their spores will depend on weather conditions. The concentration will further be dependent on location and prevailing winds.

Table 6. Culturable fungi isolated from tropical air in mid-Atlantic Ocean (~15° N, 45°W) and identified using nucleic acid sequencing methodology (Griffin *et al.*, 2006).

GenBank closest relative
Alternaria brassicae, Embellisia sp., *Pezizaceae*
Alternaria dauci
Alternaria sp.
Aureobasidium pullulans or *Discosphaerina fagi*
Cladosporium sp.
Cladosporium sp. or *Trimmatostroma macowanii*
Dendryphion sp.
Lojkania enalia
Lithothelium septemseptatum
Massaria platani
Myriangium duriaei
Myrothecium sp., *Letendraea helminthicola, Cucurbidothis pityophila, Paraphaeosphaeria* sp. *(P. pilleata* or *P. michotii)*
Neotestudina rosatii
Penicillium chrysogenum, P. glabrum, Penicillium sp. or *Talaromyces leycettanus, Penicillium* sp., *Uscovopsis* sp., *Thysanophora* sp., *Chromocleista malachitea, T. leycettanus* or *Eupenicillium* sp.
Phoma herbarum
Pleosporaceae
Preussia terricola
Setosphaeria monoceras, Pleospora herbarum or *Embellisia* sp.
S. monoceras, S. rostrata or *Cochliobolus sativus*
S. rostrata
Stachybotrys kampalensis
Trichophyton mentagrophytes or *T. rubrum*
Ulocladium botrytis
U. botrytis, Clathrospora diplospora or *Alternaria* sp.

BACTERIA IN OUTDOOR AIR

Surveys of outdoor air for bacteria are much less common than for fungi. The interest in airborne bacteria tends to be confined to the possible presence of pathogens, but even when sewage sludge is being applied to land (Pillai *et al*. 1996) bacterial pathogens appear to be absent from the outdoor air, although they may possibly be present at levels below the limits of detection of the methods employed.

Counts of airborne culturable bacteria vary widely, and great differences can be noted over very short time intervals. For example, Lighthart and Shaffer (1994) observed as much as a 14-fold difference over 2 min. Numbers are affected by variables such as location, the time of day and the season, meteorological conditions, and the type of sampler and isolation medium employed. Differences between locations were demonstrated in Oregon by Shaffer and Lighthart (1997). The average count for a major thoroughfare in Corvallis (725 CFU m^{-3} air) was higher than for a ryegrass field and a fir forest, and more than six times greater than at a headland on the Pacific coast that was exposed to onshore sea breezes during sampling days. In the rye-grass field, when the grass was mature the average count was 127 CFU m^{-3} air, but after harvesting the seed by combine it had risen to 704 CFU m^{-3}. Chaff, seed and cut straw scattered on the ground by combining were likely to have been sources contributing to the increased numbers. Contrasting with these findings, when Köck *et al*. (1998) sampled at seven sites around Graz in Austria they found that, at 327 CFU m^{-3} air, bacterial counts in an agricultural area were four times higher than in a suburban residential area. In the vicinity of a composting facility the counts were 29% higher than in the residential area, but at an industrial and business site affected by heavy traffic numbers were twice those at the composting facility.

The well-known shortcomings of methods involving culture in assessing the fungal burden in air that have already been mentioned (Burge *et al*. 1977) also apply to airborne bacteria. Lighthart (2000) concluded from studies by his group using a wet cyclone bioaerosol sampler in Oregon that the culturable numbers were on average only around 1% of the total bacterial burden in the atmosphere, but ranged from as low as 0.01% to as much as 75% of the total. In a dust plume downwind of a combine harvesting grain, 73% of the 2.9 × 10^6 cells m^{-3} air were culturable (Lighthart and Tong 1998). Other counts in midsummer were 8.8 × 10^2 to 5.9 × 10^5 m^{-3}, but only 0.02-10.6% of the bacte-

Table 7. Culturable bacteria isolated from air on a clear day and during dust events in St John, US Virgin Islands, and identified to genus or species by 16S/18S rDNA gene sequencing (Griffin et al., 2001).

GenBank closest relative	Gram stain	No. of isolates	
		Clear day	Dust events
Arthrobacter globiformis	G+ve	2	
Arthrobacter sp.	G+ve		1
Bacillus megaterium	G+ve	1	1
B. pumilis	G+ve		1
Curtobacterium albidum	G+ve		1
C. citreum	G+ve	2	5
C. luteum	G+ve		4
Curtobacterium sp.*	G+ve		1
Kocuria erythromyxa	G+ve		1
Microbacterium arborescens	G+ve		3
M. testaceum	G+ve		6
*M. testaceum**	G+ve		1
Microbacterium sp	G+ve		2
Paracoccus aminovorans	G-ve		1
Paracoccus sp.	G-ve		1
Pseudomonas alcalophila	G-ve		1
Ps. oleovorans	G-ve		1
Ps. riboflavina	G-ve		1
Sinorhizobium sp.	G-ve		1
Sphingomonas pruni	G-ve		4
S. trueperi	G-ve		1
Sphingomonas sp.	G-ve	1	4
Sphingomonas sp.*	G-ve		1

*Unidentified isolates with homology to named taxon

ria were culturable. Diurnal periodicity was observed, with both total counts and the culturable percentage (39%) being highest during the day and lowest (0.5-2% culturable) at dawn and dusk (Lighthart and Tong 1998). However, in other studies (Lighthart 2000) peak concentrations were noted at dawn in inland forested, rural-agricultural and urban areas, although not at coastal sites.

Most investigations have indicated that culturable bacteria in outdoor air are predominantly Gram-positive. For example, Fang *et al*. (2007) recently reported that in Beijing, China, counts at three locations within the city ranged from 71 CFU m^3 to 2.2 × 10^4 CFU m^3, and 165 species in 47 genera were identified. Gram-positive bacteria comprised on average 84% of the total, 53% being cocci and 31% rods. At nearly 27% of

Table 8. Species of culturable bacteria isolated from air in Bamako, Mali, during four dust events and identified to genus or species by 16S rNA/18S rDNA gene sequencing (Kellog *et al.*, 2004)

Acinetobacter calcoaceticus[e]	*Gordonia terrae*[a]
Acinetobacter sp. phenon 2[e]	*Kocuria erythromyxa*[b]
Agrococcus jenensis	*K. polaris*[c]
Arthrobacter nicotianae	*K. rosea*
A. protophormiae	*Kocuria* sp.
Aureobacterium liquifacens[d]	*Microbacterium barkeri*
Bacillus aminovorans	*Micrococcus luteus*
B.endophyticus	*Micrococcus* sp. (2 different)
B. flexus	*Paenibacillus illinoiensis*
B. firmus	*Paenibacillus* sp.
B. kangii	*Paracoccus* sp.
B. megaterium	*Planococcus* sp.
B. mycoides	*Planomicrobium koreense*
B. niacini	*P. mcmeekinii*
B. pumilus[b]	*Rhodococcus ruber*
B. subtilis[b]	*Saccharococcus* sp.
Bacillus sp. (5 different)	*Staphylococcus gallinarum*
Corynebacterium cf. *aquaticum**[c]	*S. xylosus*
Corynebacterium sp.	*Streptomyces* sp.[a]
Deinococcus erythromyxa[d]	*Zoogloea ramigera*[e]
Dietzia sp.[a]	

[a] Actinobacteria, [b] Also isolated from non-dust air sample, [c] Only isolated from non-dust air sample, [d] Identified by fatty acid profiling, [e] Gram-negative.

the total, the dominant genus was *Micrococcus*, with *Staphylococcus* (12%), *Bacillus* (7%), *Corynebacterium* (4%) and the Gram-negative genus *Pseudomonas* (4%) the most prominent among the other genera.

The previously mentioned investigations of the air spora associated with Saharan dust clouds has also provided us with much additional information on the nature of bacteria in the atmosphere. In the first of these (Griffin *et al.* 2001a), only seven strains of bacteria were cultured before dust events on the island of St John, all pigmented and therefore likely more resistant to solar radiation (Table 7). More than 40 isolations were made during the dust events, 60% of which were Gram-positive. Most were pigmented, including six of the 10 Gram-negative types, i.e. isolates of *Sphingomonas* sp., *Pseudomonas alcalophila* and *Ps. oleovorans*. On St Thomas, Griffin *et al.* (2003) isolated a wider range of bacteria, 60% of these being Gram-positive, including four species in the Ac-

tinobacteria. The members this phylum have a higher GC content in their DNA than other Gram-positive bacteria. Within the phylum a range of species form branching filaments that resemble the hyphae of fungi and are classified in the order Actinomycetales. This name derives from them formerly being classified among the fungi as the Actinomycetes, and they are still frequently referred to as actinomycetes. The relatively large proportion of Gram-negative types in this and the earlier study appears to suggest local rather than Saharan origin. This would be supported by the rather different balance in mid-Atlantic (Griffin *et al.* 2006), where only two isolates (possibly of marine origin) were Gram-negative. The average count of airborne culturable bacteria here was only 0.1 - 0.4 CFU m^{-3} air, representing a roughly 10^4 reduction from levels measured during dust storms in Mali, West Africa (Kellogg *et al.* 2004). With a single exception, the levels in Mali were also one to two orders of magnitude greater than those during dust events in the Virgin Islands (Griffin *et al.* 2001a, 2003). Many of the bacterial colonies were highly pigmented, and 96% of isolates were Gram-positive (Table 8). Many of the Gram-positive species were spore formers commonly found associated with soil and dust. The predominance of Gram-positive bacteria in investigations where culture methods are involved can be attributed to the greater susceptibility of Gram-negative bacteria to solar radiation and desiccation.

As a review by Peccia and Hernandez (2006) shows, our perception of the composition of the bacterial burden in the atmosphere changes markedly when PCR-based methods are applied directly in identification, i.e. without any cultivation step and therefore being based on dead as well as living cells. For example, when Brodie *et al.* (2007) investigated urban air in two cities in Texas, San Antonio and Austin, they found at least 1,800 types of bacteria, a phylogenetic diversity approaching that in some soilborne communities. Statistical analysis revealed that location was less important as a factor in explaining variability in the composition of the bacterial burden than were temporal and meteorological influences. An investigation in Colorado, during which 4-h samples were collected into liquid in an SKC Biosampler and DNA was extracted directly from the samples, further illustrates this bacterial diversity (Fierer *et al.* 2008). A survey of the small-subunit ribosomal RNA gene sequences generated indicated a preponderance of Gram-negative bacteria in the samples. The putative identities of the most abundant of these were *Flavobacterium/Chryseobacterium, Flexibacter,* and *Hymenobacter* in the *Cytophaga-Flavobacterium-Bacteroides* (CFB) group, *Acidovorax, Acinetobacter, Bradyrhizobium, Burkholderia, Caulobacter,* Comamonadaceae, Enterobacteriaceae, *Methylobacterium, Nitrosovibrio, Pseudomonas, Rhizobium, Rhodobacter* and *Sphingomonas*. The Gram-positive genera *Arthrobacter* (an actinobacterium) and *Planococcus* (*Planomicrobium*) were among the most abundant bacteria on one of the five sample days. In relation to Gram-negative bacteria, the Gram-positive genera *Micrococcus* and *Bacillus* were relatively rare (<2% of bacterial sequences). It should be mentioned that, applying the same methodology, Fierer *et al.* (2008) also noted that the airborne mycobiota was dominated by fungi in the Hypocreales.

Clearly, these findings show that the differences in results that are associated with employment of different sampling and analytical procedures not only for fungi apply equally to airborne bacteria. They also confirm the conclusions of Miquel (1899) from his sampling in Paris, that the main source of outdoor bacteria is the surface of the ground.

CONCLUSION

Hyde (1969) characterized the air spora as being an "expression of climate, a reflex of the vegetation as a whole, and an essential factor or complex of factors in the general environment". As this chapter has shown, such essential factors include weather, prevailing winds, the local vegetation, topography and human activity.

What we learn about the outdoor air spora depends very much on sampling techniques, which have a major influence on the results obtained. The use of different techniques in different surveys can make comparisons difficult. There is, as yet, and there may never be, a sampling technique to satisfy all the requirements of aerobiologists, and the sampling technique chosen will depend on resources, the nature and number of sites to be sampled, the facilities available and the information which is sought from the survey. It continues to be necessary to have regard to the sampling technique when considering the results from any survey, and to be cautious when making comparisons between surveys.

REFERENCES

Airaudi, D. and Marchisio, V.F. (1996) Fungal biodiversity in the air of Turin. *Mycopathologia,* **136**, 95-102.

Allitt, U. (1986) Identity of airborne hyaline, one-septate ascospores and their relation to inhalant allergy. *Trans. Br. Mycol. Soc.,* **87**, 147-154.

Andersen, A.A. (1958) New sampler for the collection, sizing and enumeration of viable airborne particles. *J. Bact.,* **76**, 471-484.

Battarbee, J.I., Rose, N.L., and Long, X. (1997) A continuous high resolution record of urban airborne particulates suitable for retrospective microscopical analysis. *Atmos. Environ.,* **31**, 171-181.

Bouza, E., Pelaez, T., Perez-Molina, J., *et al.* (2002) Demolition of a hospital building by controlled explosion: the impact on filamentous fungal load in internal and external air. *J. Hosp. Infect.,* **52**, 234-242.

Brodie, E.L., DeSantis, T.Z., Moberg Parker, J.P., *et al.* (2007) Urban aerosols harbor diverse and dynamic bacterial populations. *PNAS,* **104**, 299-304.

Bulloch, W. (1938) *The History of Bacteriology*. Clarendon Press, Oxford.

Burge, H.P., Boise, J.R., Rutherford, J.A., and Solomon, W.R. (1977) Comparative recoveries of airborne fungus spores by viable and non-viable modes of volumetric collection. *Mycopathologia,* **61**, 27-33.

Cadham, F.T. (1924) Asthma due to grain rusts. *JAMA,* **83**, 27.

Carnelley, T., Haldane, J.S., and Anderson, A.M. (1887) The carbonic acid, organic matter and micro-organisms in air, more especially in dwellings and schools. *Phil. Trans. Royal Soc., Series B,* **178**, 61-111.

Chamberlain, A.C. (1966) Transport of *Lycopodium* spores and other small particles to rough surfaces. *Proc. Roy. Soc., A,* **296**, 45-70.

Corden, J.M., Millington, W.M., and Mullins, J. (2003) Long-term trends and regional variation in the aeroallergen *Alternaria* in Cardiff and Derby UK – are differences in climate and cereal production having an effect. *Aerobiologia,* **19**, 191-199.

Cox, G.M., Schell, W.A., Scher, R.L., and Perfect, J.R. (1994) First report of involvement of *Nodulisporium* species in human disease. *J. Clin. Microbiol.,* **32**, 2301-2304.

Davies, R.R. (1969a) Aerobiology and the relief of asthma in an Alpine valley. *Acta Allergol.,* **24**, 377-395.

Davies, R.R. (1969b) Climate and topography in relation to aeroallergens at Davos and London. *Acta Allergol.,* **24**, 296-409.

Delfino, R.J., Zeiger, R.S., Seltzer, J.M., *et al.* (1997) The effect of outdoor fungal spore concentrations on daily asthma severity. *Environ. Health Perspect.,* **105**, 622-635.

Dobell, C. (1932) *Antonie van Leeuwenhoek and his "Little Animals"*. Bale and Danielsson, London.

Emberlin, J., Adams-Groom, B., and Tidmarsh, J. (1999) *The Dispersal of Pollen from Maize, Zea mays. A Report Based on Evidence Available from Publications and Internet Sites*. The Soil Association, Bristol, UK.

Fang, Z., Ouyang, Z., Zheng, H., *et al.* (2007) Culturable airborne bacteria in outdoor environments in Beijing, China. *Micro. Ecol.,* **54**, 487-496.

Fernández, D., Valencia, R.M., Molnár, T., *et al.* (1998) Daily and seasonal variations of *Alternaria* and *Cladosporium* airborne spores in León (North-West Spain). *Aerobiologia,* **14**, 215-220.

Fierer, N., Liu, Z., Rodriguez-Hernandez, M., *et al.* (2008) Short-term temporal variability in airborne bacterial and fungal populations. *Appl. Environ. Microbiol.,* **74**, 200-207.

Fitzpatrick, F., Prout, S., Gilleece, A., *et al.* (1999) Nosocomial aspergillosis during building work – a multidisciplinary approach. *J. Hosp. Infect.*, **42**, 170-171.

Fradkin, A., Tobin, R.S., Tarlo, S.M., *et al.* (1987) Species identification of airborne molds and its significance for the detection of indoor pollution. *JAPCA*, **37**, 51-53.

Fröhlich-Nowoisky, J., Pickersgill, D.A., Després, V.R., and Pöschl, U. (2009) High diversity of fungi in air particulate matter. *PNAS*, **106**, 12814-12819.

Giner, M.M., Garcia, J.S.C., and Camacho, C.N. (2001) Airborne *Alternaria* spores in SE Spain (1993-98): occurrence patterns, relationship with weather variables and prediction models. *Grana*, **40**, 111-118.

Gonianakis, M., Neonakis, I., Darivianaki, E., *et al.* (2005) Airborne Ascomycotina on the island of Crete: seasonal patterns based on an 8-year volumetric survey. *Aerobiologia*, **21**, 69-74.

Goudie, A.S., and Middleton, N.J. (2001) Saharan dust storms: nature and consequences. *Earth. Sci. Rev.*, **56**, 179-204.

Gravesen, S., and Schou, C. (1997) *Allergi og Anden Overfølsomhed*. Munksgaard, Copenhagen, Denmark

Gregory, P.H. (1973) *Microbiology of the Atmosphere*, 2nd Ed. Leonard Hill, Aylesbury, Bucks, UK.

Gregory, P.H., and Sreeramulu, T. (1958) Air spora of an estuary. *Trans. Br. Mycol. Soc.*, **41**, 145-156.

Griffin, D.W., Garrison, V.H., Herman, J.R., and Shinn, E.A. (2001a) African desert dust in the Caribbean atmosphere: microbiology and health. *Aerobiologia*, **17**, 203-213.

Griffin, D.W., Kellog, C.A., and Shinn, E A. (2001b) Dust in the wind: long range transport of dust in the atmosphere and its implications for global public and ecosystem health. *Global Change & Human Health*, **2**, 20-33.

Griffin, D.W., Kellog, C.A., Garrison, V.H., *et al.* (2003) Atmospheric microbiology in the northern Caribbean during African dust events. *Aerobiologia*, **19**, 143-157.

Griffin, D.W., Westphal, D.L., and Gray, M.A. (2006) Airborne microorganisms in the African desert dust corridor over the mid-Atlantic ridge, Ocean Drilling Program, Leg 209. *Aerobiologia*, **22**, 211-226.

Griffin, D.W., Kubilay, N., Kocak, M., *et al.* (2007) Airborne desert dust and aeromicrobiology over the Turkish Mediterranean coastline. *Atmos. Environ.*, **41**, 4050-4062.

Harvey, R.R. (1967) Air spora studies at Cardiff. I. *Cladosporium. Trans. Br. Mycol. Soc.*, **50**, 479-495.

Harvey, R.R. (1970) Spore productivity in *Cladosporium. Mycopath. Mycol. Appl.*, **41**, 251-256.

Harvey, R.R. Hodgkiss, J.J., and Lewis, P.N. (1969) Air-spora studies at Cardiff. II. *Chaetomium. Trans. Br. Mycol. Soc.*, **53**, 269-278.

Herrero, A.D., Ruiz, S.S., Bustillo, M.G., and Morales, P.C. (2006) Study of airborne fungal spores in Madrid. *Aerobiologia*, **22**, 135-142.

Herxheimer, H., Hyde, H.A., and Williams, D.A. (1969) Allergic asthma caused by basidiospores. *Lancet 2*, 131-133.

Hirst, J.M. (1952) An automatic volumetric spore trap. *Ann. Appl. Biol.*, **39**, 257-265.

Hirst, J.M. (1953) Changes in atmospheric spore content: diurnal periodicity and the effects of weather. *Trans. Br. Mycol. Soc.*, **36**, 375-393.

Hirst, J.M., Stedman, O.J., and Hurst, G.W. (1967) Long distance spore transport: vertical sections of spore clouds over the sea. *J. Gen. Microbiol.*, **48**, 357-377.

Hudson, H. J. (1969) Aspergilli in the air-spora at Cambridge. *Trans. Brit. Mycol. Soci.*, 52, 153-159.

Hurtado, I., Leal Quevedo, F.J., Rodriguez Ciodaro, A., *et al.* (1989) A one year survey of airborne pollen and spores in the neotropical city of Bogota (Colombia). *Allergol. Immunopathol. (Madrid)*, **17**, 95-104.

Hyde, H.A. (1969) Aeropalynology in Britain – an outline. *New Phytol.*, **68**, 579-590.

Ingold, C.T. (1971) *Fungal Spores: Their Liberation and Dispersal*. Clarendon Press, Oxford.

Kellogg, C.A., Griffin, D.W., Garrison, V.H., *et al.* (2004) Characterization of aerosolized bacteria and fungi from desert dust events in Mali, West Africa. *Aerobiologia*, **20**, 99-110.

Köck, M., Schlachter, R., Pichler-Semmelrock, F.P., *et al.* (1998) Airborne microorganisms in the metropolitan area of Graz, Austria. *Cent. Eur. J. Public Health*, **6**, 25-28.

Kramer, C.L., and Eversmeyer, M.G. (1984) Comparisons of airspora concentrations at various sites within a ten kilometer radius of Manhattan, Kansas, USA. *Grana*, **23**, 117-122.

Larsen, L.S. (1981) A three-year-survey of microfungi in the air of Copenhagen 1977-79. *Allergy*, **36**, 15-22.

Larsen, L.S., and Gravesen, S. (1991) Seasonal variation of outdoor viable microfungi in Copenhagen, Denmark. *Grana*, **30**, 467-471.

Levetin, E., and Dorsey, K. (2006) Contribution of leaf surface fungi to the air spora. *Aerobiologia*, **22**, 3-12.

Li, C.S., and Hsu, L.Y. (1995) Fungus allergens inside and outside the residences of atopic and control children. *Arch. Environ. Health*, **50**, 38-43.

Li, D.W., and Kendrick, B. (1995) A year-round outdoor aeromycological study in Waterloo, Ontario, Canada. *Grana*, **34**, 199-207.

Lighthart, B. (2000) Mini-review of the concentration variations found in the alfresco atmospheric bacterial populations. *Aerobiologia*, **16**, 7-16.

Lighthart, B., and Shaffer, B. T. (1994) Bacterial flux from chaparral into the atmosphere in mid-summer at a high desert location. *Atmos. Environ.*, **28**, 1267-1274.

Lighthart, B., and Tong, Y. (1998) Measurements of total and culturable bacteria in the alfresco atmosphere using a wet-cyclone sampler. *Aerobiologia*, **14**, 325-332.

Lugauskas, A., Šveistytė, L., and Ulevičius, V. (2003) Concentration and species diversity of airborne fungi near busy streets in Lithuanian urban areas. *Ann. Agric. Environ. Med.*, **10**, 233–239.

May, K.R. (1945) The cascade impactor: an instrument for sampling coarse aerosols. *J. Sci. Instrum.*, **22**, 187-195.

Miquel, P. (1878-99) Annual reports, in *Annu. Obs. Montsouris*. Baulthier-Villars, Paris.

Mitakakis, T., Ong, E.K., Stevens, A., *et al.* (1997) Incidence of *Cladosporium*, *Alternaria* and total fungal spores in the atmosphere of Melbourne (Australia) over three years. *Aerobiologia*, **13**, 83-90.

Mitakakis, T.Z., and Guest, D.I. (2001) A fungal spore calender for the atmosphere of Melbourne, Australia, for the year 1993. *Aerobiologia*, **17**, 171-176.

Mullins, J., Harvey, R., and Seaton, A. (1976) Sources and incidence of airborne *Aspergillus fumigatus* (Fres.). *Clin. Allergy*, **6**, 209-217.

Mullins, J., Hutcheson, P., and Slavin, R.G. (1984) *Aspergillus fumigatus* spore concentrations in outside air: Cardiff and St. Louis compared. *Clin. Allergy*, **14**, 351-354.

Oliveira, M.T.B., Santos Braz, R.F., and Ribeiro, M.A.G. (1993) Airborne fungi isolated from Natal, State of Rio Grande do Norte – Brazil. *Rev. Microbiol., São Paulo*, **24**, 198-202.

Pady, S.M., and Kramer, C.L. (1960) Kansas aeromycology, VI: hyphal fragments. *Mycologia*, **52**, 681-687.

Pasteur, L. (1861) Mémoire les corpuscles organisés qui existent dans l'atmosphère. Examen de le doctrine des générations spontanées. *Ann. Sci. Nat. (Zool.), 4e Sér.*, **16**, 5-98.

Peccia, J., and Hernandez, M. (2006) Incorporating polymerase chain reaction-based identification, population characterization, and quantification into aerosol science: a review. *Atmos. Environ.*, **40**, 3941-3961.

Pillai, S.D., Widmer, K.W., Dowd, S.E., and Ricke, S.C. (1996) Occurrence of airborne bacteria and pathogen indicators during land application of sewage sludge. *Appl. Environ. Microbiol.*, **62**, 296-299.

Prospero, J.M., Blades, E., Mathison, G., and Naidu, R. (2005) Inter-hemispheric transport of viable fungi and bacteria from Africa to the Carribean with soil dust. *Aerobiologia*, **21**, 1-19.

Pyrri, I., and Kapsanaki-Gotsi, E. (2007) A comparative study on the airborne fungi in Athens, Greece, by viable and non-viable sampling methods. *Aerobiologia*, **23**, 3-15.

Rantio-Lehtimäki, A. (1988) Yeasts in rural and urban air in southern Finland. *Grana*, **27**, 313-319.

Raynor, G.S., Ogden, E.C., and Hayes, J.V. (1970) Dispersal and deposition of ragweed pollen from experimental sources. *J. Appl. Meteorol.*, **9**, 885-895.

Richardson, M.J. (1996) The occurrence of airborne *Didymella* spores in Edinburgh. *Mycol. Res.*, **100**, 213-216.

Rodriguez-Rajo, F.J., Iglesias, I., and Jato, V. (2005) Variation assessment of airborne *Alternaria* and *Cladosporium* spores at different bioclimatical conditions. *Mycol. Res.*, **109**, 497-507.

Schlesinger, P., Mamane, Y., and Grishkan, I. (2006) Transport of microorganisms to Israel during Saharan dust events. *Aerobiologia*, **22**, 259-273.

Schulz, B., and Boyle, C. (2005) The endophytic continuum. *Mycol. Res.*, **109**, 661-686.

Shaffer, B.T., and Lighthart, B. (1997) Survey of culturable airborne bacteria at four diverse locations in Oregon: urban, rural, forest, and coastal. *FEMS Microbiol. Ecol.*, **34**, 167-177.

Shelton, B.G., Kirkland, K.H., Flanders, W.D., and Morris, G.K. (2002) Profiles of airborne fungi in buildings and outdoor environments in the United States. *Appl. Environ. Microbiol.*, **68**, 1743-1753.

Stakman, E.C., and Hamilton, L.M. (1939) Stem rust in 1938. *Pl. Dis. Reptr.*, **117** (Suppl.), 69-83.

Stepalska, D., and Wolek, J. (2005) Variation in fungal spore concentrations of selected taxa associated to weather conditions in Cracow, Poland, in 1997. *Aerobiologia*, **21**, 43-52.

Streifel, A. J., Lauer, J. L., Vesley, D., *et al.* (1983) *Aspergillus fumigatus* and other thermotolerant fungi generated by hospital building demolition. *Appl. Environmental Microbiol.*, **46**, 375-378.

Unterseher, M., and Schnittler, M. (2009) Dilution-to-extinction cultivation of leaf-inhabiting endophytic fungi in beech (*Fagus sylvatica* L.) – Different cultivation techniques influence biodiversity assessment. *Mycol. Research*, **113**, 645-654.

Waisel, Y., Ganor, E., Epshtein, V., *et al.* (2008) Airborne pollen, spores, and dust across the Mediterranean Sea. *Aerobiologia*, **24**, 125-131.

Wakefield, A.E., Stringer, J.R., Tamburrini, E., and Dei-Cas, E. (1998) Genetics, metabolism and host specificity of *Pneumocystis carinii*. *Med. Mycol.*, **36** (Suppl. 1), 183-193.

Walkey, D.G.A., and Harvey, R. (1966) Spore discharge rhythms in Pyrenomycetes. *Trans. Brit. Mycol. Soc.*, **49**, 583-592.

Walkey, D.G.A., and Harvey, R. (1968a) Spore discharge rhythms in Pyrenomycetes. IV. The influence of climatic factors. *Trans. Brit. Mycol. Soc.*, **51**, 779-786.

Walkey, D.G.A., and Harvey, R. (1968b) Spore discharge rhythms in Pyrenomycetes. V. The effect of temperature on spore discharge in *Sordaria macrospora*. *Trans. Brit. Mycol. Soc.*, **51**, 787-789.

West, J.S., Atkins, S.D., and Fitt, B.D.L. (2008) Detection of airborne plant pathogens: halting epidemics before they start. In J.S.West and P.J. Burt, (eds.), *Aspects 89: Applied Aspects of Aerobiology*, Association of Applied Biologists, Wellesbourne, Warwick, UK, pp. 11-14.

Williams, D.A., and Higgins, I.T.T. (1959) The prevalence of asthma in two communities in South Wales. *Acta Allerg.*, **13**, 126-127.

Womiloju, T.O., Miller, J.D., Mayer, P.M., and Brook, J.R. (2003) Methods to determine the biological composition of particulate matter collected from outdoor air. *Atmos. Environ.*, **37**, 4335-4344.

Chapter 1.2

MICROORGANISMS IN INDOOR AIR

Brian Flannigan

Scottish Centre for Pollen Studies, Napier University, Edinburgh, UK.

INTRODUCTION

Bioaerosols in indoor environments comprise a range of microscopic biological particles of animal, plant and microbial origin that can be inhaled and impinge on human health. Allergenic animal particulates include dander and dried saliva of pets, such as cats and dogs, and other furred animals; fragments of insects such as cockroaches, fleas and clothes moths; fragmented bodies and faeces of storage and house dust mites; and amoebae and other unicells belonging to the Protozoa. Plant particulates include unicellular algae and fragments of multicellular algal filaments; spores of lower plants such as mosses and ferns; and minute fragments and whole pollen grains of higher plants, which are discussed in Chapter 1.3. Although protozoans and unicellular algae are microorganisms, the most medically important microbial particulates in indoor air belong to neither of these groups, but are the cells and spores of fungi and bacteria, and viruses. In this chapter, the different categories of microorganism (including algae and protozoa) present in indoor air are described in general terms. Because of the diversity of forms, fungi are dealt with in more detail than the other categories. Where the health implications of individual groups are not dealt with in detail in later chapters, they are discussed briefly here.

COMPONENTS OF BIOAEROSOLS IN INDOOR ENVIRONMENTS

Algae

Nature of algae in indoor air and house dust
Although it has been shown that they can behave as inhalant allergens and cause rhinitis and asthma, microalgae comprise a largely ignored category of microorganism in indoor environments. These organisms contain chlorophyll and are therefore photosynthetic, unlike the heterotrophic fungi and bacteria to be considered later. The types most often isolated are green (the plant division Chlorophyta) and blue-green algae (now recognized as being bacteria and therefore classified as Cyanobacteria), and diatoms (Chrysophyta). They may be unicells, simple chains of cells, cylindrical filaments or colonies (composed of up to 50,000 unicells organized into coherent structures). Some are macroscopic membranous or tubular structures composed of very large numbers of cells. Unlike the other types, the unicellular diatoms are enclosed by a two-piece wall of crystalline silica, which has been likened to a Petri dish. The shape of these diatoms varies, with centric diatoms being radially symmetrical and pinnate diatoms being bilaterally symmetrical.

Although algae may be generated indoors, the most usual sources of the organisms in indoor environments are unicells or fragments of filaments in air infiltrating from outside and dust blown in or carried in on clothing or animal fur. These algae normally grow in water or in soil, and on rocks and the bark of trees. They are aerosolized by wind, mechanical disturbance of soil or dust, bubble phenomena or forced aeration of aqueous environments ranging from aquaria to sewage disposal plants (Sharma *et al.* 2007). The role of the wind in dispersal has been observed by Torma *et al.* (2001) when they continuously sampled the outdoor air of the city of Badajoz (SW Spain). Most of the airborne algae were members of the order Chlorococcales and small centric diatoms (particularly *Cyclotella*). Some filamentous green algae and even coenobia (a coenobium is a colony of undifferentiated cells) of the colonial type *Pediastrum*. The concentrations of airborne algae correlated positively with temperature, and negatively with relative humidity, but wind speed also appeared to have a positive effect on concentrations of diatoms and the coenobia of members of the Chlorococcales. When Chrisostomou *et al.* (2009) investigated diversity of air-dispersed algae from a Greek river-reservoir system, they also found evidence that, at least over short distances, the wind was an important agent of dispersal of the algae, of which *Chlorella* was among the most frequent and abundant, with the larger types *Mougeotia* and *Ulothrix* also being frequent.

Counts of algae in the air are not likely to exceed those of pollen grains or fungal spores (Tiberg 1987). In USA, the most prominent types detected by cultural methods in house dust collected during the indoor heating season in homes were the unicellular green algae *Chlorella*, *Chlorococcum*, *Chlamydomonas* and *Planktosphaeria* and the filamentous cyanobacteria *Anabaena* and *Schizothrix* (Bernstein and Safferman 1970). In Sweden, 88% of dust samples collected by Tiberg *et al.* (1984) from carpets and beds in the homes of atopics contained viable algae. The genera were mostly the same as those found outdoors, but with cyanobacteria being found in slightly higher frequency. The most frequent types were the unicellular green algae *Chlorococcum*, *Chlorella* and *Stichococcus* and the cyanobacteria *Nostoc*, *Phormidium* and *Anabaena*. The same types dominated in dust from a health care centre, a recreation centre and several nursery schools. In the nursery schools, the diatoms *Hantzschia* and *Navicula* were occasionally isolated. Although counts of green algae in these schools peaked in June, the cyanobacteria showed no marked seasonal differences.

Health effects of algae

Since the allergenic properties of algae are discussed in Chapter 3.1, it is sufficient to note here that it is more than 40 years since it was first reported that they were a cause of respiratory allergy in children and that in skin tests they reacted positively in patients with histories of inhalant allergies (McElhenny *et al.* 1962, McGovern *et al.* 1966, Bernstein and Safferman 1967). It is now recognized that, in addition to members of the Chlorophyta such as *Chlorella* and *Chlorococcum* (Tiberg *et al.* 1984,1995), cyanobacteria such as *Anabaena*, *Nostoc* and *Phormidium* are allergenic (Tiberg 1987, Sharma and Rai 2008).

Although there appear to be no reports of algae in air or house dust causing dermatitis, a case of contact dermatitis has been reported in a swimmer exposed to *Anabaena*, one of a number of species forming "algal blooms" in freshwater lakes (Cohen and Reif 1953). There are also rare human cases of cutaneous or disseminated infections by unicellular *Prototheca*. For example, a disseminated infection by achlorophyllous *P. wickerhami* in a male with a specific deficiency in cell-mediated immunity was reported from New Zealand (Cox *et al.* 1974). However, *Prototheca* is generally found in soil and it is thought that soil contact is the major cause of infection.

Protozoa

Nature of Protozoa in indoor environments

Motile free-living members of this group of unicellular animals are found scavenging on organic debris in soil and water outdoors, and may also be found indoors in bodies of water such as aquaria, humidifier reservoirs and HVAC drainage pans (Schlichtung 1969), and also in plumbing systems (Steinert *et al.* 1997). Although some have cilia and others flagella that propel them through water, most of the free-living protozoans that impinge on human health are amoeboid, moving and engulfing nutrient particles by means of protoplasmic extensions, pseudopodia. They become aerosolized by mechanisms ranging from forced aeration of aquatic environments such as sewage ponds or aquarium tanks and dispersion of water droplets by humidifiers to scouring of drying soil by wind. An example of soils being the most likely source of airborne protozoa is provided by an investigation of outdoor air in Rapid City, South Dakota, in which Rogerson and Detwiler (1999) collected the cysts of 25 morphologically different protozoans from the air. The commonest of these were flagellates and naked amoebae, while ciliates were rare. This reflected their natural occurrence in the soil. In general, the concentration of cysts increased as a function of total airborne particulates. The concentration of both particulates and cysts exhibited a wide range of variability, but the highest concentrations of cysts and total particulates were recorded on days with higher winds and lower relative humidity.

The types of protozoan that have been isolated from air in a variety of investigations have been collated by Schlichtung (1969), and the types that may be present in indoor environments have been indicated by Rohr *et al.* (1998). Amoebae in hospital hot water systems were identified as *Hartmannella vermiformis* and species of *Echinamoeba, Saccamoeba*

Table 1. Abbreviated classification of some fungi found as spores in indoor air.

Phylum	Sub-phylum	Genera
Ascomycota	Pezizamycotina	*Emericella, Eurotium, Peziza, Alternaria*, Aspergillus*, Botrytis*, Epicoccum*, Penicillium*, Phoma*, Stachybotrys**
	Saccharomycotina	*Saccharomyces*
Basidiomycota	Agaricomycotina	*Agaricus, Coprinus, Ganoderma, Sistotrema, Serpula, Filobasidiella* (*Cryptococcus*)
	Wallemiomycetes[†]	*Wallemia*
Zygomycota	Mucoromycotina	*Absidia, Cunninghamella, Mucor, Rhizopus*

*Anamorphic genus, [†]Class

and *Valkampfia*, and species of *Acanthamoeba, Hartmanella, Naegleria, Valkampfia* and *Vanella* were isolated from moist areas in bathrooms and showers, sink drains and water taps (Rohr *et al.* 1998). *Hartmannella castellani* has been isolated from the air of a respiratory care unit (Kingston and Warhurst 1969).

Health effects of Protozoa in indoor environments

As far as health is concerned, the protozoans best known as affecting human health are those parasites that are spread by insect bite, causing diseases such as malaria, sleeping sickness, Chaga's disease and leishmaniasis. However, there are free-living protozoans that may occasionally cause disease. The widespread species *Naegleria fowleri* is occasionally pathogenic, causing a fatal meningoencephalitis. It is an amoeboflagellate, existing in flagellate form in water and as an amoeba in human tissues. In nature, it also forms cysts that are protected from adverse environmental conditions by a thick wall. Species of *Acanthamoeba* are also widespread and may cause keratitis and encephalitis, most particularly in immunocompromised individuals (Curry *et al.* 1991). They also cause ulceration of the skin and cornea (Baron *et al.* 1994). The cysts of these amoebae are readily dispersed by air currents, contaminating aqueous environments ranging from brackish water to hot tubs, and also soil and dust. It seems probable that infection with *N. fowleri* is most often the result of contaminated water entering the nose, for example during swimming, but it is considered that in the case of acanthamoebae inhalation may be responsible as well as direct contact with contaminated waters or soil. Rohr *et al.* (1998) found six potentially pathogenic strains of *Acanthamoeba*, but none of *Naegleria*, colonizing shower heads, drains, walls and tiles in bathrooms. They postulated that infection might occur by inhalation of aerosolized *Acanthamoeba* in sprayed water.

Members of the Protozoa growing within buildings have also been implicated in humidifier fever. Edwards *et al.* (1976) isolated both amoebae and ciliates from a contaminated humidifier serving an office in which a number of workers suffered this febrile disease. Extracts of the amoebae, but not the ciliates, reacted with the sera of affected workers, and antigenic identity between extracts of the amoebae, dust on a suspended ceiling and a culture-collection strain of *Naegleria gruberi* was demonstrated. Protozoa such as *Acanthamoeba, Hartmannella* and *Naegleria* have been isolated from plumbing systems contaminated with bacteria in the genus *Legionella* (Steinert *et al.* 1997) and the presence of the bacterium in amoebae such as *A. polyphaga* has been demonstrated (Newsome *et al.* 1998). Some 13 amoebae and two ciliates were shown to provide an environment in which legionellae can replicate prolifically (Kwaik *et al.* 1998), and it was also shown that two species of *Acanthamoeba* can expel vesicles 2.1-6.4 µm in diameter containing living *L. pneumophila* (Berk *et al.* 1998). Aerosolization of such vesicles, which are of respirable size, may be a means of transmission of this agent of legionellosis. Another bacterium for which *Acanthamoeba* can act as environmental host is *Mycobacterium avium*, a pathogen which like *Legionella pneumophila* is widespread in aquatic environments, including municipal drinking water. However, *M. avium*, a primary health threat to AIDS patients, is found in the cyst wall of the amoeba rather than in vesicles (Steinert *et al.* 1998).

The potential for free-living *A. culbertsoni*, and *Paramecium caudatum* and two other ciliated protozoans, to provide a predatory mechanism for removing *Cryptosporidium parvum* from aquatic ecosystems, including wastewater treatment plants, was investigated by Stott *et al.* (2003). While the *Acanthamoeba* ingested the cysts of this waterborne protozoan parasite of the human intestine, the ciliates showed greater predatory activity.

Filamentous Fungi

Structure and classification of filamentous fungi

Like those in outdoor air, the fungi found in indoor air as spores and cells fall into two basic categories, filamentous fungi and yeasts. Yeasts are regarded conventionally as unicells and are dealt with in the next section of this chapter. The typical filamentous fungus develops from a germinating spore, which extends into a germ tube. As a result of continued apical growth this develops into a roughly tubular hypha, which branches as it grows, forming an extending network of hyphae known as a mycelium.

The hyphae are incompletely divided by septa into communicating compartments (the equivalent of cells) along their entire length, giving septate mycelium. In members of the Zygomycota (Table 1), however, the mycelium is described as non-septate, although occasional septa occur. These are in older areas of mycelium and where reproductive structures are formed. The mycelium in members of the Basidiomycota is characterized by clamp connections at the septa which give the hyphae a nodular appearance absent in other septate fungi.

An extremely important fundamental characteristic of fungi is that they produce dispersible spores. In microfungi, including those frequently referred to as moulds, a microscope is needed to examine the spores and the specialized structures on or in which they are produced and thus enable the species to be identified. In other fungi, aggregations of hyphae organized into complex macroscopic sporing structures may arise from a vast mycelium. Because fungi were originally regarded as plants and were described and classified by botanists, these macroscopic structures tend still to be known as fruiting bodies. They include the cups, mushrooms and brackets (or conks) of species that are referred to as macrofungi.

The greatest number of filamentous fungi found indoors were formerly allocated to an artificial grouping of species for which no sexual reproductive phase had been found, but there are in indoor air also spores of species which can be assigned to one or other of three phyla with a sexual phase (Table 1). The fungi lacking sexual (or perfect) stages have at various times in the past been categorized as belonging to the Fungi Imperfecti, Deuteromycetes, Deuteromycotina or Deuteromycota, but these taxonomic categories are now considered not acceptable, since they are artificial groupings of diverse fungi that have not evolved from a common ancestral group (Kirk *et al.* 2001). In modern classification, most of them have

been allocated to the subphylum Pezizomycotina of the Ascomycota. Nevertheless, very often they are still informally referred to as deuteromycetes, and indeed sometimes by the earlier term Fungi Imperfecti. The term "mitosporic fungi" has also been applied to these fungi because the spore is a mitospore, its nuclei being the product of the type of nuclear division associated with vegetative growth and asexual reproduction, i.e. mitosis, and not of the reduction division, meiosis, which is part of sexual reproduction (Kirk *et al.* 2001). However, as was mentioned in Chapter 1.1, the spore is most often called a conidium. The structure bearing conidia is a conidiophore. The wide range of morphological variation in both conidia and conidiophores which enables different mitosporic fungi isolated from the air or from surfaces in buildings to be identified is fully evident from the illustrated descriptions in Chapter 4.2 and Chapter 5. In earlier classifications, fungi in the Deuteromycetes were classed according to whether the conidiophores were naked (Hyphomycetes) or were enclosed in sporing bodies (Coelomycetes), e.g. *Phoma*. There are also some well-characterized fungi which have never been found to produce spores. These sterile fungi, or Mycelia Sterilia, were formerly allocated to the Agonomycetes.

As mentioned in Chapter 1.1, filamentous fungi in the three phyla informally referred to as ascomycetes, basidiomycetes and zygomycetes (Table 1) are recognizable by the characteristics of their sexual phase, the teleomorph. They may also have an asexual phase, or anamorph, in which they produce mitospores. Because the anamorph may be more frequently encountered, the anamorphic name may be more widely used than that of the teleomorph, e.g. in the two ascomycetes mentioned later in this paragraph. Most of the fungi referred to as deuteromycetes are either fungi which in the process of evolution have lost their teleomorphic phase, or are anamorphs for which no teleomorph has yet been found. Based on morphological and/or physiological similarities and particularly comparison of DNA sequences, most of these anamorphic fungi are now allocated to the ascomycete subphylum Pezizomycotina. Although the reproductive structures in ascomycetes such as *Emericella nidulans* (anamorph: *Aspergillus nidulans*) or *Eurotium herbariorum* (anamorph: *Aspergillus glaucus*) are microscopic, the teleomorphs of many others are characterized by complex macroscopic fruiting bodies, ascocarps, e.g. the cup fungi in the genus *Peziza*. Likewise, although the teleomorphs of basidiomycetes like *Sistotrema brinkmannii* are microscopic, those of the dry rot fungus *Serpula lacrymans* indoors

Table 2. Abundance of principal spore types in indoor and outdoor air of 15 homes in Ontario, presented as mean concentration for 132 sampling days over 22 months and percentage of total air spora (after Li and Kendrick 1995).

	Indoor		Outdoor	
	No. m^{-3} air	%	No. m^{-3} air	%
Alternaria	44	1.9	74	2.1
Aspergillus/Penicillium	457	19.8	131	3.8
Cladosporium	895	38.8	1479	42.5
Coprinus (basidiospores)	41	1.8	78	2.3
Epicoccum	7	0.3	20	0.6
Ganoderma (basidiospores)	59	2.6	111	3.2
Leptosphaeria (ascospores)	182	7.9	547	15.7
Unidentified ascospores	65	2.8	138	4
Unidentified basidiospores	152	6.5	310	8.9
Other unidentified spores	206	8.9	301	8.7
Hyphal fragments	146	6.3	112	3.2

and of the beech bracket *Ganoderma applanatum* outdoors are conspicuously macroscopic basidiocarps. The ascospores and basidiospores produced by ascomycetes and basidiomycetes are dispersed by air currents, but the zygospores in the zygomycetes are larger and their function is regarded as being survival rather than dispersal. Under suitable conditions, a zygospore will germinate to produce an anamorphic structure, the sporangium, which contains sporangiospores that are dispersed.

Outdoor air as a source of filamentous fungi in the indoor air spora

The wide diversity of fungal spores in outdoor air between early summer and autumn, when vegetative growth and sporulation of fungi are at their greatest, has been mentioned in Chapter 1.1. The outdoor air is for indoor air an important source of living spores and hyphal fragments, which act as propagules from which mycelium can develop. These propagules infiltrate naturally ventilated buildings and strongly influence the indoor fungal burden both qualitatively and quantitatively (Sneller and Roby 1979). Factors such as the proximity of a building to vegetation and organic debris that support fungal growth and spore production, and shade that enhances survival, lead to increased numbers of airborne propagules penetrating buildings (Kozak *et al.* 1979). Among other authors, Ackermann *et al.* (1969) and Sneller and Roby (1979) have reported that indoor concentrations of mould propagules broadly parallel those outdoors during the summer, but at a lower level and with a time lag before peak outdoor concentrations are reflected.

In the extensive regional investigation in USA, mentioned in Chapter 1.1, Shelton *et al.* (2002) noted that, overall, the mean and median concentrations of culturable fungi indoors correlated with the corresponding outdoor concentrations. Like those outdoors, the median indoor concentrations were highest in the summer and autumn, and lowest in winter and spring. The indoor/outdoor ratio only ranged from roughly 0.1 to 0.5, and there was no significant variation in the ratio either seasonally or from year to year. Corresponding to the situation outdoors, the highest indoor concentrations occurred in the Southeast, Southwest and Far West regions and the lowest in the Northwest region, with those in the Midwest and Northeast being intermediate. In arctic and temperate climates, of course, outdoor fungal growth and sporulation are minimal in winter because of the lower temperature. Consequently, winter counts of spores indoors are normally higher than those outdoors, particularly when there is snow cover (Reponen *et al.* 1992), which precludes re-entrainment of settled spores in dust and soil and on outdoor surfaces.

Naturally, the degree and type of ventilation employed greatly affects the air spora in buildings. An investigation in UK (Adams and Hyde 1965) found that, as with pollen, simply closing the windows and doors of a room in a naturally ventilated home during summer could exclude 98% or more of the spores present in outdoor air. Provided that it is not circumvented by opening doors and windows, central air-conditioning reduces spore counts in houses by 50% or more (Spiegelman *et al.* 1963). Window air-conditioning units can reduce numbers of *Alternaria* and *Ganoderma* spores, and also pollen grains, to around 5% of the outdoor levels (Solomon *et al.* 1980). It has also been reported that central electrostatic filtration can reduce indoor air concentrations of total propagules to 3% of those experienced without such filtration, and *Cladosporium* to <1% (Kozak *et al.* 1979).

The lower numbers of airborne spores indoors relative to outdoor air in summer is, with some exceptions, well illustrated by a Canadian study (Table 2) in which spores were collected on coated glass slides and counted under a microscope (Li and Kendrick 1995). It should be said here that Table 2 also shows that air contains spores of types which do not appear in counts obtained by so-called viable sampling methods that are more frequently employed in studies of indoor air. Although there are large discrepancies, which will be discussed later (see Chapter 4.1), counts of the phylloplane fungi *Alternaria*, *Cladosporium* and *Epicoccum* can be made by either method. However, numbers of basidiospores of the macrofungi *Coprinus* and *Ganoderma* and the ascospores of *Leptosphaeria* (a plant pathogen and an important cause of late summer asthma) in indoor air are not enumerated by viable count methods. More detailed analysis of the Canadian data (Li and Kendrick 1996) showed that from May to October the relationships between numbers of airborne spores and hyphal fragments (and the diversity of fungi) indoors and outdoors were very strong (especially so for *Alternaria* and *Leptosphaeria*), probably because of windows being opened during the summer months. Path analysis indicated that, as might have been expected, *Alternaria*, *Leptosphaeria*, unidentified ascospores and *Coprinus* and *Ganoderma* basidiospores, came mainly from outdoor sources.

However, because of their large size, the spores of some fungi may not penetrate buildings in any quantity. This is instanced in a study of basidiospore distribution of a white-spored variety of the toadstool known as the fly agaric (*Amanita muscaria* var. *alba*). Li (2005) used an Allergenco MK-3 sampler at ground level 30 cm from basidiocarps, which were 6 m from a house. For the first three days in which the sporing surfaces were fully exposed the daily average concentration of airborne spores adjacent to the basidiocarps was >10^4 m^{-3} air. Small numbers of basidiospores infiltrated the house occasionally each day, the daily averages in the indoor air were <0.1% of the spores dispersed from the basidiocarps. There was a positive correlation between the indoor concentration of airborne spores and windows and doors being left open, but there was no correlation with the activ-

Table 3. Rank order of 30 most abundant airborne fungi inside and outside 50 non-problem houses in Atlanta, GA (after Horner *et al.* 2004).

Indoor (600 air samples)	Outdoor (200 air samples)
Cladosporium cladosporioides	*Cladosporium cladosporioides*
Cladosporium spp.	*Cladosporium* spp.
C. sphaerospermum	*C. sphaerospermum*
Penicillium spp.	*Penicillium chrysogenum*
P. sclerotiorum	*Penicillium* spp.
Epicoccum nigrum	*P. corylophilum*
P. brevicompactum	*P. brevicompactum*
P. decumbens	*Aspergillus niger*
Yeasts	*P. citrinum*
Aspergillus niger	*P. variabile*
Non-sporulating hyaline fungi	Non-sporulating hyaline fungi
Alternaria alternata	*Epicoccum nigrum*
P. pinophilum	*P. commune*
Curvularia spp.	*P. decumbens*
P. corylophilum	*P. glabrum*
P. glabrum	*Curvularia* spp.
Non-sporulating fungi	*P. citreonigrum*
Arthrospore-forming fungus	*P. pinophilum*
Non-sporulating pigmented fungi	Yeasts
P. citrinum	Non-sporulating fungi
P. variabile	*P. sclerotiorum*
Aureobasidium pullulans	*P. aurantiogriseum*
P. crustosum	*Alternaria alternata*
Alternaria sp.	Arthrospore-forming fungus
Bipolaris sp.	*Aspergillus versicolor*
Aspergillus fumigatus	*P. crustosum*
P. purpurogenum	*P. purpurogenum*
P. solitum	*P. paxilli*
P. chrysogenum	Non-sporulating pigmented fungi
Aspergillus versicolor	*P. rugulosum*

ity of the occupants. Li (2005) has suggested that the relatively large size of the basidiospores (in the early phase of release 12.5 ± 2.6 × 7 ± 1.4 μm, and later 10.5 ± 2.2 × 7 ± 1.4 μm) and their shape (broadly ellipsoid to elongate) may be major factors in their rapid deposition oudoors and, consequently, limited potential for infiltrating buildings.

Effect of growth of filamentous fungi indoors on the air spora

Since indoor temperatures are usually favourable for fungi, growth and sporulation of fungal contaminants

Table 4. Species of *Aspergillus* and *Penicillium* reported present as airborne viable particles in homes of asthmatic children and adults in Mexico City (Garcia *et al.* 1995; Rosas *et al.* 1997).

Category	Frequency (Percentage of samples)	
	Adults (n =30)†	Children (n = 8)*
Aspergillus spp.	32	48
A. candidus	3	-
A. flavus	9	10[a]
A. fumigatus	3	-
A. glaucus (Eurotium herbariorum)	1	63
A. melleus	-	27[b]
A. niger	10	16
A. ochraceus	6	-
A. parasiticus	2	-
A. versicolor	4	56[c]
A. wentii	1	-
Penicillium spp.	97	100
P. aurantiogriseum	58	88
P. brevicompactum	13	-
P. chrysogenum	15	79
P. citrinum	9	-
P. crustosum	1	1
P. griseofulvum	4	-
P. janthinellum	2	-
P. mineoluteum	-	2
P. oxalicum	1	-
P. purpurogenum	3	7
P. spinulosum	-	41
P. verrucosum	2	-
P. viridicatum	20	-

Burkard personal sampler, malt extract agar; † Andersen 2-stage sampler, dichloran-18% glycerol agar; - Not reported; [a]Plus *A. sydowi*; [b]Plus *A. flavofurcatis*; [c]Plus *A. petrakii*

are likely to occur in any damp areas in buildings and consequently modify the relationship between the indoor and outdoor aeromycota (see Chapter 2.1). The major exceptions to the tendency for indoor counts in summer to be lower than the corresponding outdoor counts are *Aspergillus* and *Penicillium*. The failure of Li and Kendrick (1996) to detect any functional or causal relationship, winter or summer, between *Aspergillus/Penicillium* conidia in indoor and outdoor air is in line with the generally held belief that the conidia of fungi in these two genera are primarily of endogenous origin. Path analysis also suggests indoor sources of *Cladosporium*, *Epicoccum*, unidentified basidiospores

and "other unidentified spores" (Table 2), although they also have outdoor origins (Li and Kendrick 1996). Along with *Penicillium*, *Cladosporium* is usually one of the commonest genera isolated from mould patches in damp homes (Grant *et al.* 1989).

As in other studies involving counting of spores under the microscope, the category *Aspergillus/Penicillium* was used by Li and Kendrick (1996) because the spores of only a very few species in these two large genera are sufficiently distinctive for them to be recognizable in a field sample under the microscope, and even then only by an extremely skilled mycologist. Although the quantitative data indicate the relative abundance of aspergilli and penicillia (taken together), it should be realized that different species of *Aspergillus* and *Penicillium* are quite different in their physiology, ecology and significance for health. It is therefore important that in investigating indoor environments species should be accurately identified, and that involves culture of the fungi isolated. An example of the diversity in *Penicillium* spp. that may be encountered in indoor air can be seen in Table 3. The results presented are for the fungi in the air inside and outside a set of 50 houses in Atlanta in which there was either minimal or no water damage or associated mould growth (Horner *et al.* 2004). Although both indoors and outdoors *Cladosporium* spp. ranked highest among the 30 most abundant taxa, some 12 species of *Penicillium* were abundant in indoor air and 16 in outdoor air. Table 4 presents the results of two studies carried out in Mexico City using different samplers and different isolation media, illustrating a greater diversity of aspergilli than in Table 3, and also a diverse range of penicillia. The species in these two genera given in Tables 3 and 4 are by no means the only ones that have been reported in studies. For example, in their nationwide study Shelton et al. (2002) noted *A. caespitosus*, *A. carneus*, *A. restrictus*, *A. sydowii*, *A. terreus*, *A. unguis* and *A. calidoustus* (= *A. ustus*) among the aspergilli in indoor air, and yet others are mentioned elsewhere in this volume.

It has to be realized, however, that not all of the species contributing to total counts of *Aspergillus/Penicillium* obtained in investigations such as that of Li and Kendrick (1995), or the collective viable counts for *Aspergillus* spp. or *Penicillium* spp. in many other published investigations may actually be more abundant indoors. Fradkin *et al.* (1987) found that, whilst the collective viable counts of penicillia indoors were twice those outdoors, the very common species *Penicillium chrysogenum* was more abundant outdoors. This study also found that some *Cladosporium* spp.

were less abundant indoors during the summer, but the concentrations of others were double those outdoors. There are notable differences in the ranking of the penicillia in the Atlanta investigation (Horner *et al.* 2004, Table 3) that also point to the fundamental importance of species identification in aerobiological studies. For example, whilst *P. chrysogenum* was the most abundant of the penicillia in outdoor air, it ranked lowest among those in the 30 most abundant taxa in indoor air. Conversely, *P. sclerotiorum* was the most abundant *Penicillium* species in indoor air, but it was of much lower rank in outdoor air. Further, five species among the most abundant taxa in outdoor air – *P. commune*, *P. citreonigrum*, *P. aurantiogriseum*, *P. paxilli* and *P. rugulosum* – were not among those most abundant in indoor air, while *P. solitum* appeared in the top 30 taxa in indoor air, but not in outdoor air.

The sporocarps of macrofungi growing within buildings can also contribute to the air spora of the indoor environment. For example, it has been shown that the concentration of basidiospores near to the basidiocarps of the dry rot fungus, *Serpula lacrymans*, may be as high as $3.6 \times 10^5 \, \mathrm{m^{-3}}$ air (Hirst and Last 1953). This is of the same order as the peak spore concentration of approx. $2.8 \times 10^5 \, \mathrm{m^{-3}}$ recorded by Li (2005) outdoors close to the *Amanita muscaria* var. *alba* basidiocarps mentioned earlier. It has been shown that such concentrations of *S. lacrymans* can be responsible for development of asthma and hypersensitivity pneumonitis (extrinsic allergic alveolitis) in susceptible occupants of buildings where there is extensive dry rot of wood (O'Brien *et al.* 1978).

A further source of microbial contamination has been demonstrated in mechanically ventilated buildings in Finland. If the floor of first-floor apartments is not airtight, air exhausted from the apartments is replaced by air from the crawl space. Airaksinen *et al.* (2004a,b) found that the air in that space usually had a greater concentration of airborne spores than the outside air, and was at its greatest in summer. Spores of an unspecified *Acremonium*, not regarded by these workers as typically having a source indoors, were found to be present in larger numbers during summer in the crawl space than in the air outdoors. The numbers in the apartments correlated with those in the crawl space, and it was concluded that in summer the warm damp unfiltered air entering the crawl space produced conditions favourable for growth and sporulation of *Acremonium*, leading to more spores infiltrating the apartments through the floor. No such correlation was found between the numbers

Table 5. Rank order of most abundant fungi in 2-5 mg samples of indoor dust from 50 non-problem houses in Atlanta, GA, directly plated over entire surface of malt extract agar (MEA) and dichloran-18% glycerol agar (DG18) (after Horner *et al.* 2004).

MEA	DG18
Cladosporium cladosporioides	*Cladosporium* spp.
Yeasts	*C cladosporioides*
C. sphaerospermum	*Penicillium* spp.
Cladosporium spp.	*C. sphaerospermum*
Penicillium spp.	*Aspergillus niger*
Aureobasidium pullulans	*P. chrysogenum*
Aspergillus niger	*P. brevicompactum*
Epicoccum nigrum	*Aspergillus versicolor*
P. chrysogenum	*Aspergillus* spp.
P. glabrum	*P. citrinum*
P. aurantiogriseum	*P. glabrum*
P. sclerotiorum	*P. aurantiogriseum*
P. citrinum	*Aspergillus ochraceus*
P. purpurogenum	Non-sporulating fungi
Alternaria alternata	*Rhodotorula* spp.
Rhodotorula spp.	*P. expansum*
Aspergillus spp.	*P. variabile*
Curvularia spp.	*P. spinulosum*
Aspergillus versicolor	*Aspergillus sydowii*
P. spinulosum	*Aspergillus unguis*
P. decumbens	*P. crustosum*
P. brevicompactum	Yeasts
P. variabile	*Eurotium amstelodami*
P. citreonigrum	*Aureobasidium pullulans*
P. corylophilum	*Alternaria* spp.
Non-sporulating fungi	*Syncephalastrum racemosum*
Non-sporulating pigmented fungi	*Alternaria alternata*
Trichoderma harzianum	Unidentified
Unidentified	*P. solitum*
Pithomyces charatarum	*Eurotium herbariorum*
Bipolaris spp.	Non-sporulating hyaline fungi

of the most abundant category, the large genus *Penicillium*, possibly for reasons mentioned in the second last paragraph to this.

Overall, the range of species that may be encountered in indoor air is wide. Zyska (2001) showed this in compiling from available literature a list of fungi reported as either being present in the air of European indoor environments or growing on structural or other materials within these environments and therefore

Table 6. Dermatophytes and other fungi isolated from floor dust of student houses (Maghraby *et al.* 2008).

Dermatophytes and related fungi:	
Aphanoascus fulvescens	*Chrysosporium lucknowense*
Aphanoascus sp.	*Gymnoascus uncinatus* (anamorph *Chrysosporium merdarium*)
Arthroderma cuniculi	*Trichophyton rubrum*

Other fungi:	
Alternaria alternata	*Mucor circinelloides*
A. citri	*M. hiemalis*
Aspergillus alutaceus	*M. racemosus*
A. flavus	*Nectria haematococca*
A. flavus var. *columnaris*	*Paecilomyces lilacinus*
A.fumigatus	*Penicillium brevicompactum*
A. niger	*P. camemberti*
A. ochraceus	*P. chrysogenum*
A. parasiticus	*P. citrinum*
A. sulphureus	*P. duclauxi*
A. terreus	*P. funiculosum*
A. ustus	*P. griseofulvum*
Candida albicans	*P. oxalicum*
Cladosporium cladosporioides	*P. purpurogenum*
C. sphaerospermum	*P. rubrum*
Cunninghamella echinulata	*Rhizopus stolonifer*
C. elegans	*Rhodotorula rubra*
Emericella nidulans	*Syncephalastrum racemosum*
Geotrichum candidum	Non-sporing isolates
Gibberella pulicaris	

likely to contribute to the airborne fungal burden. Of the 227 species isolated from residential and public buildings (libraries) and work environments (sawmills and deep coal mines) nearly 80 were known mycotoxin producers and 17 had been isolated from human tissues. In a later compilation, Zyska (2004) listed 434 species, of which 73% are anamorphic.

The presence of hyphal fragments in outdoor air has been demonstrated in Table 5 of Chapter 1.1, and in indoor air in Table 2 of the present chapter, and the possibility of these fragments acting as propagules has also been mentioned. However, there are fragments which are so badly damaged that they are nonviable or are too small to contain intact nucleate compartments that will grow into mycelium. Despite this, the emphasis in most investigations of the airborne mycobiota has been on spores, and hyphal fragments and subcellular particles have been ignored. Górny *et al.* (2002), however, examined the release of spores and fragments of three species, *Aspergillus versicolor*, *Penicillium melinii* and *Cladosporium cladosporioides*, from agar plate cultures and ceiling tiles. They found that up to 320 times more fragments were released than spores, and that they contained the same antigens. Fungal particulates smaller than conidia released from ceiling tiles on which the toxigenic species *Stachybotrys chartarum* had grown have been shown to contain trichothecene mycotoxins (Brasel *et al.* 2005). These investigations therefore have implications for assessment of human exposure to allergens and mycotoxins.

Effect of human behaviour and activity on the indoor air spora

Spores and hyphal fragments that have infiltrated buildings by the airborne route add to those brought in on contaminated clothes and fur by humans and their animal pets. Li and Kendrick (1996) suggested that firewood and fuel wood chips, vegetables, and clothes and tools brought indoors after garden work, were possible sources of *Epicoccum*. Unwashed root vegetables and birch firewood were found to be abundant sources of *Cladosporium* and *Penicillium* spores contaminating indoor air Lehtohnen *et al.*

(1993), and fuel wood chips have been reported as a potential source of spores of the toxigenic fungus *Trichoderma* (Miller *et al.* 1982). Pasanen *et al.* (1989) presented circumstantial evidence for carriage on clothes of *Acremonium*, *Alternaria*, *Botrytis* and *Chrysosporium* spores from cow sheds into farm houses during the Finnish winter, and (Lehtonen *et al.* 1993) noted transport of *Aspergillus* on riding clothes. By whatever route they enter, or whatever their indoor source, airborne propagules are rapidly disseminated throughout naturally ventilated buildings by air currents (Christensen 1950), although the closer to an endogenous source the higher are airborne counts likely to be (Hunter *et al.* 1988).

House dust usually contains large numbers of microorganisms that have sedimented out from the air. For example, in a health-related investigation of settled dust in schools during winter, Meyer *et al.* (2004, 2005) recorded levels that ranged from 8.3 × 10^2 to 3.1 × 10^6 CFU g^{-1} dust. One can therefore expect that indoor activities that raise dust will also affect the indoor air spora. This was elegantly demonstrated clearly 120 years ago in a study of houses and schools in a Scottish mill town, Dundee (Carnelley *et al.* 1887). This paper, which incidentally made observations on antibiosis that pre-date Fleming's serendipitous discovery by half a century, should be prescribed reading for anyone carrying out surveys of indoor air. Comparison of corresponding male and female classes in school showed that under normal conditions "boys tend to make the air of a room more impure than girls do" because they "are more restless, and so raise more dust, which necessarily contains micro-organisms". When Carnelley *et al.* (1887) subsequently got a class of boys to stamp on their classroom floor, the total number of microorganisms in the air increased 15-fold. In more recent times, using a continuous recording volumetric spore trap Millington and Corden (2005) reported a five-fold increase in the total count of airborne fungi after re-entry to a closed room in a modern house, and also an elevated total on return from work in the evening to the house after it had been unoccupied during the day. In an older cottage, in which the proportion of *Aspergillus/ Penicillium* spores was much higher than in the modern house, the passage of people to and fro boosted the airborne count by more than 20 times (Millington and Corden 2005). Various other investigators have shown that constructional/demolition work (Maunsell 1952, Hunter *et al.* 1988, Goebes *et al.* 2008), floor sweeping (Lehtonen *et al.* 1993), cleaning carpets using a vacuum cleaner lacking an exhaust filter (Hunter

et al. 1988) and changing bedclothes (Lehtonen *et al.* 1993) all temporarily elevate airborne mould counts. The source strength for house dust raising activity is a function of the type and vigour of the activity and the number of people performing the activity, and also the type of flooring (Ferro *et al.* 2004).

Dybendal *et al.* (1991) noted that carpeted floors accumulated more dust and pollen and mould allergens (and by inference pollen grains and fungal spores/hyphal fragments) than smooth floors, indicating why the same activity raises more particulate matter from a carpeted floor than from a wooden floor (Ferro *et al.* 2004). Cho *et al.* (2006) also found that antigen level was greater in carpeted rooms than in those without carpeting. In a birth cohort study, these workers surveyed the child's primary activity room and found *Alternaria* antigen in the collected floor dust of nearly 90% of 777 homes. The level of *Alternaria* antigen was not associated with visible mould/water damage. Taken together with this, their observing the antigen levels were highest (a) in the autumn and (b) in homes with dogs led Cho *et al.* (2006) to conclude that transport of *Alternaria* from the outdoor environment was the source of the antigen in floor dust. Culturable *Alternaria* has certainly been noted in house dust. Horner *et al.* (2004) found that in 85% of samples of dust from non-problem homes in Atlanta 20% or more of the colonies on isolation plates were *Alternaria* and other phylloplane fungi, although collectively penicillia, aspergilli and other fungi shown in Table 5 predominated. At the other end of the scale, Maghraby et al. (2008) found that among the culturable fungi listed in Table 6 *Alternaria* spp. were only present in 22% of dust samples from university student houses in Egypt and accounted for only 1.6% of the total numbers. Although the species *Aspergillus* and *Penicillium* listed made up about 40% of the mean total of >38500 CFU g^{-1} dry dust, and the yeasts *Candida albicans* and *Rhodotorula rubra* approx. 11%, dermatophytic fungi comprised approx. 33%. The most abundant of these dermatophytes were *Chrysosporium lucknowense* and the two *Aphanoascus* spp. (altogether, around 26%).

Further consideration of the mycobiota in dust is given in Chapter 2.1, in which fungal growth in indoor environments is discussed.

It is clear, then, that human behaviour and activity have a pronounced impact on the indoor air spora. Variations in the degree and type of activity of adults and children, and of pets, in occupied buildings are responsible for large temporal and spatial differences in the numbers of fungi (and bacteria) in indoor air.

Marked differences can be seen even when counts are taken within minutes of each other (Hunter *et al.* 1988, Verhoeff *et al.* 1990, Mouilleseaux and Squinazi 1991).

Health effects of filamentous fungi in indoor environments

The health impact of filamentous fungi, as sources of allergens, mycotoxins and β-glucans, and as pathogens is dealt with in Chapter 3.1-3.3 and Chapter 4.5 of this book.

Yeasts

Nature of yeasts in indoor environments

Yeasts are usually defined as unicellular fungi reproducing vegetatively by budding, i.e. producing buds that develop into daughter cells, which either separate from the mother cell or remain adherent so that clusters or branched chains of cells are formed. In some genera (*Sterigmatomyces* and *Fellomyces*), the buds are produced on short stalks. Not all yeasts bud, however. In *Schizosaccharomyces*, reproduction is by fission (one or more cross-walls divide the cell into segments which split off and develop independently). In some budding yeasts, such as *Candida*, under a range of growth conditions the budded cells are elongated and remain attached in chains, forming pseudohyphae. In *Geotrichum* and *Trichosporon*, there are true hyphae which split up into loosely joined chains of cylindrical cells.

As well as reproducing vegetatively, yeasts such as *Saccharomyces* reproduce sexually, forming ascospores, and are therefore allocated to the ascomycete subphylum Saccharomycotina. Other sexually reproducing yeasts are basidiomycetes. One such is *Rhodosporidium sphaerocarpon*, which is best known because of its anamorphic state *Rhodotorula glutinis*. *R. sphaerocarpon* produces teliospores from which basidiospores are derived. The genera *Sporobolomyces* and *Bullera*, which multiply vegetatively by budding, also produce spores which are forcibly discharged (ballistospores). These two genera are also basidiomycetes; teleomorphs have been found for some species of *Sporobolomyces*.

Although yeasts are frequently mentioned in reports of investigations of the indoor air spora, they are seldom identified even to generic level. The reason for this is that identification of individual species, and even genera, requires both experience and skill and is extremely time-consuming, involving morpho-

logical examination and a battery of physiological or biochemical tests (see Chapter 4.5). Consequently, many reports present only total viable counts, sometimes subdividing them according to colony colour into "pink" and "white" yeasts. Mention of pink or reddish yeasts as *Sporobolomyces* or *Rhodotorula* in reports should always be treated with caution, unless full identification procedures have been followed.

A detailed examination of three urban apartments and a rural farm-house in Finland did, however, reveal that members of the basidiomycetous genera *Cryptococcus*, *Rhodotorula* and *Sporobolomyces* were the most numerous airborne yeasts indoors (Rantio-Lehtimäki 1988). Both indoors and outdoors, other yeasts (mostly *Debaryomyces hansenii*, *Williopsis californica* and *Wingea robertsii*) only accounted for around 2% of the total trapped on Andersen sampler plates. As with outdoor air, numbers of yeasts were greater in the autumn than at other times of the year. *S. roseus*, *S. holsaticus* and *S. salmonicolor* were found indoors during autumn, and small numbers of *S. hispanicus* were noted. Cryptococci appeared to be more abundant in indoor air than outdoors (or in similar numbers in the case of the farmhouse) and showed little seasonal variation, with *C. albidus* and *C. laurentii* as the most frequent species. *Rhodotorula* spp. were also more frequently isolated from indoor air, with *R. graminis* and *R. glutinis* being isolated in summer and *R. pilimanae* in late autumn. Elsewhere, Mouilleseaux and Squinazi (1994) found that airborne yeasts were primarily unspecified *Rhodotorula* and *Torulopsis*, although *Candida* also appeared occasionally, and Solomon (1974) reported that in USA *Sporobolomyces roseus* and *Rhodotorula* spp., and also *Geotrichum candidum*, were abundant in air humidified by a cold-mist vaporizer.

Various authors have reported that yeasts are abundant in house dust, e.g. Flannigan *et al.* (1993) and Verhoeff *et al.* (1994), and the predominant species found in dust from both mattresses and floors have been *R. glutinis*, *R. minuta*, *R. mucilaginosa*, *C. albidus* and *C. laurentii* (Hoekstra *et al.* 1994).

Health effects of yeasts in indoor environments

In general, it appears that very few of the yeasts isolated in homes or non-industrial work places are either fermentative or ascomycetous (Rantio-Lehtimäki 1988), but in work environments such as bakeries, breweries and distilleries, the fermentative ascomycete *Saccharomyces cerevisiae* (baker's or brewer's yeast) may be detected in the air spora in addition to cereal-borne fungi. The enolase of this species is

Table 7. Viable bacteria isolated from air in air-conditioned and naturally ventilated sites in two office buildings (after Austwick *et al.* 1989).

Gram-positive	Gram-negative
Micrococcus spp.	*Acinetobacter calco-aceticus* var. *lwoffii*
Staphylococcus aureus	*Aeromonas hydrophila*
Staph. epidermidis	*Flavobacterium* sp.
Streptococcus spp.	*Moraxella* sp.
	Pasteurella haemolytica
	P. pneumotropica
	Pseudomonas aeruginosa
	Ps. cepacia
	Ps. fluorescens
	Ps. paucimobilis
	Ps. vesicularis

a major allergenic component, and a wider range of patients other than bakery workers with inhalant allergies to fungi show sensitivity to this enzyme (Baldo and Baker 1988), which cross reacts with that from *Candida albicans*. Cross reactivity between allergens from different yeast genera has also been reported by Koivikko *et al.* (1988). Although the range of yeasts which have been implicated in allergic disease is much smaller than for filamentous fungi, *Rhodotorula*, *Sporobolomyces* and *Tilletiopsis* are other yeasts which have been reported to cause allergic reactions (Jackson 1984, Rantio-Lehtimaki and Koivikko 1984, Burge 1989). Hodges *et al.* (1974) noted precipitating antibodies to both *Rhodotorula* sp. and *Cryptococcus* sp. (and a range of other fungi isolated from a home cold-mist vaporizer) in the serum of a patient with hypersensitivity pneumonitis (HP), and a *Rhodotorula* sp. has been recorded as a cause of HP in the occupant of a house with a heavily contaminated basement (Gravesen 1994). Sensitivity to *Candida* has been recorded in children with asthma (Koivikko *et al.* 1991), and *Trichosporon cutaneum* is the cause of a summer HP in Japan (Yoshida *et al.* 1989).

Bacteria

Numbers of bacteria in indoor air are usually, but not always, greater than the numbers of fungi in the air of homes (Nevalainen *et al.* 1988), schools (Mouilleseaux *et al.* 1993) and non-industrial workplaces such as air-conditioned offices (Mouilleseaux *et al.* 1993). Total numbers of bacteria may, for example, be as high as 15×10^6 m^{-3} air in homes (Kujundzic et al. 2006). Ventilation and overcrowding affect the numbers of air-

borne bacteria, as was demonstrated more than 120 years ago in a study of homes and schools by Carnelley *et al.* (1887). Carnelley and his colleagues noted that numbers of viable airborne bacteria were higher in naturally ventilated than in mechanically ventilated schools. The ratio of bacteria to fungi was 132:1 under natural ventilation as opposed to only 29:1 with mechanical ventilation. Counts of culturable bacteria in domestic air were shown to rise with occupation density and the ratio of bacteria to fungi also increased, from 21:1 in spacious houses to 49:1 in overcrowded houses. In more recent times, it has been shown that even in much larger buildings human numbers and activity have an important bearing on counts of airborne bacteria. Despite filtration and air-conditioning in the Vatican, counts of airborne bacteria (and fungi) correlated positively with the numbers of visitors to, and their presence in, the Sistine Chapel (Montacutelli *et al.* 2000). The increased bacterial numbers observed during visiting hours were comprised largely of human-shed *Staphylococcus* spp.

As Tables 7 and 8 show, any investigation of airborne bacteria in buildings generally leads to the isolation of a diverse range of bacteria, but a limited number of types shed by the occupants normally predominate among the culturable bacteria. These dominant types are principally members of the Micrococcaceae; among these Gram-positive bacteria *Staphylococcus epidermidis* associated with skin scales is frequently most prominent. The shedding of staphyloccoci by humans accounts for their airborne numbers increasing proportionally more than other bacteria during visiting hours at the Sistine Chapel (Montacutelli *et al.* 2000). Other bacteria in the Micrococcaceae which are most likely to be present are *Staph. aureus* shed from the nasal membranes and skin and *Micrococcus* spp., such as *M. luteus,* originating on the skin (Table 8). Although Gram-positive endospore-forming rods in the genus *Bacillus* are commonly isolated from a wide range of indoor (and outdoor) environments, they are seldom isolated from indoor air in substantial numbers. Two other Gram-positive elements that may found in indoor air (Austwick *et al.* 1986) are the irregularly shaped, non-sporing rods of coryneform actinobacteria and the spores and mycelial fragments of filamentous actinobacteria in the order Actinomycetales.

Only rather small numbers of culturable mesophilic *Streptomyces* and other filamentous actinobacteria are normally found in indoor air (Nevalainen *et al.* 1988, Nevalainen 1989). They are commoner in homes and other buildings with dampness prob-

Table 8. Culturable airborne bacteria found in 11 homes in Central Scotland (after Flannigan *et al.* 1999).

Gram-positive cocci	Gram-negative cocci
Aerococcus viridans*	Various, unidentified
Micrococcus spp.[1]*	
Staphylococcus aureus*	**Gram-negative rods**
Staph.epidermidis*	*Achromobacter* spp.
Streptococcus spp.	*Acinetobacter* sp.*
	Aeromonas hydrophila*
Gram-positive rods	*Agrobacterium* sp.*
Actinomyces spp.	*Alcaligenes denitrificans*
Arthrobacter sp.	*Enterobacter agglomerans*
Bacillus spp.[2]*	*Flavobacterium* sp.
Corynebacterium spp.*	*Klebsiella* spp.*
Erysipelothrix sp.*	*Moraxella lacunata**
Kurthia sp.	*Proteus* sp.
Lactobacillus sp.	*Pseudomonas* spp.[3]*
Mycobacterium sp.	

[1] *M. luteus, M. roseus, M. varians* and *M. viridans,* [2] including *B. lichenlformis, B. megaterium* and *B. subtilis,* [3] Including *Ps. fluorescens, Ps. mallei, Ps. inendocina, Ps. oryzihabitans, Ps. paucimobilis, Ps. pickettii, Ps. pseudomallei, Ps. putida, Ps. stutzeri* and *Ps. vesicularis,* * Also detected on surface of walls

lems, and have been particularly associated with complaints of odour in a range of types of building (Nevalainen *et al.* 1990). Ström *et al.* (1990) found streptomycetes in 25% of building material samples taken from sick buildings, and Hyvärinen et al. (2002) recorded unspecified mesophilic actinobacteria in association with fungi on a range of visibly damaged building materials. In a house with water damage in a basement bathroom, roof and outdoor walls Peltola *et al.* (2001) detected in the basement approx. 10^3 culturable bacteria m^{-3} air, of which 64 m^{-3} were spore-forming actinobacteria, including toxigenic strains of *Streptomyces* and *Nocardiopsis*. In an investigation of schools with visible moisture/mould problems, Meklin *et al.* (2002) found 0-7 mesophilic actinobacterial CFU m^{-3} air (GM = 0.1) in concrete/brick reference schools and 0-43 CFU m^{-3} (GM = 1.3) in corresponding index schools. In wooden schools the respective ranges were 0-47 (GM = 5.7) and 0-2700 CFU m^{-3} (GM = 6.3). It is not rare to isolate thermophilic actinobacteria such as *Saccharopolyspora* (*Faenia*) *rectivirgula* and *Thermoactinomyces candidus* or *Th. vulgaris* from HVAC equipment (Fink *et al.* 1971, Kreiss and Hodgson 1984).

Culturable Gram-negative bacteria are usually found to be much less abundant than Gram-positive species in investigations of indoor air. Elevated counts of Gram-negatives are usually indicators of conditions wet enough to allow proliferation, e.g. in HVAC humidifier reservoirs and drainage pans or on very damp surfaces. In such cases, *Pseudomonas, Acinetobacter, Alcaligenes* and *Flavobacterium* are likely

to be among the most common culturable bacteria, but other Gram-negative rods such as *Achromobacter, Aeromonas, Agrobacterium, Enterobacter, Klebsiella, Moraxella* and *Proteus*, and also Gram-negative cocci, may also be present (Table 8).

It is well known that the proportion of the bacterial total in indoor air that can be cultured is small; for example, Flannigan *et al.* (1996) found that <1% of the total bacterial burden in a set of houses was culturable. A strategy that can be employed as a marker of the total level of Gram-negative bacteria is to assay air samples collected on membrane filters for lipopolysaccharide (LPS), or endotoxin. This is of medical significance because it is an immunomodulator, and is present in the outer membrane of the cell walls of both culturable and non-culturable Gram-negative bacteria, but not in Gram-positive bacteria. Two examples of the use of endotoxin measurement in investigations of indoor air are those of Kujundzic *et al.* (2006), who used it to assess seasonal differences in the air in non-problem homes, and Rao et al. (2007) to compare the air in residences with different degrees of flood damage.

Endotoxin has also been used as a marker of Gram-negative bacteria in investigations of house dust. In a study of some 400 European houses, Bischof *et al.* (2002) found that endotoxin levels in house dust were higher in old houses; the lower storey of houses; in houses with longer occupancy; high utilisation; infrequent vacuum cleaning of carpets; and an indifferent attitude to ventilation. Among other workers, Wickens *et al.* (2003) and Giovannangelo *et al.* (2007) have noted higher levels in houses with more occupants or pet cats and dogs, and Giovannangelo *et al.* (2007) also reported up to 3.4 times more endotoxin in dust from carpeted floors than from floors without carpets. Instanes *et al.* (2005) and Hyvärinen *et al.* (2006) recorded more endotoxin in floor dust than in dust from mattresses. The presence of pets and contact with animals outside have been found to be factors contributing to greater levels of endotoxin in mattresses (Gehring *et al.* 2004).

The value of PCR-based methods in determining the identity of airborne bacteria, irrespective of whether the bacteria are culturable or not, has been mentioned in connection with outdoor air (Chapter 1.1). These methods have also been employed in the investigation of bacteria in house dust, for instance by Pakarinen *et al.* (2008), who carried out their work in Russian and Finnish Karelia, where living conditions on opposite sides of the national border are fundamentally different. By DNA cloning these workers were able to identify 94 different genera of dustborne bacteria, more than observed in any earlier study in houses. Compared with Finnish Karelia, dust from Russian Karelia contained up to 20 times more muramic acid, a marker for Gram-positive bacteria, and two-thirds of the bacterial DNA clones represented Gram-positives, approximately double the number in Finnish Karelia. They were predominantly members of the Staphylococcaceae and Corynebacteriaceae. Among members of the former family in Russian Karelia were species that are typically associated with animals. In Finnish Karelia, where the number of households with cats was 50% of that on the other side of the border, no such staphylococci were present and the only animal-associated species detected was the actinobacterium *Dietzia maris*. In the Russian houses 14 species (including staphylococci) typically associated with animals were found. In contrast to the Russian houses, Gram-negatives (mainly Proteobacteria) predominated in the house dust from the Finnish houses, and the endotoxin levels were higher. The majority of the protobacterial DNA sequences represented species associated with plants, and the number of clones of plant-associated species was three times that in Russian Karelia.

The effect of the numbers of human occupants and the presence of pets in the premises, the types of floor covering and the nature and intensity of associated activity have been mentioned as being among the factors quantitatively affecting the airborne microbial burden indoors. The investigation carried out by Pakarinen *et al.* (2008) illustrates that variation in the types of bacteria in house dust will lead to qualitative differences between exposures.

Health effects of bacteria in indoor environments
Inhalation of the LPS or endotoxin present in the wall of Gram-negative bacteria is known to cause ill health in work and domestic situations (see Chapters 2.2, 4.1 and 4.6). The Gram-negative bacterium of greatest concern is the respiratory pathogen *Legionella pneumophila*, which has been mentioned above in relation to free-living protozoa. *Legionella* has been the subject of reviews in many recent publications, e.g. Cianciotto *et al.* (2007) and Hoffman *et al.* (2009), to which the reader is referred. Although some coryneforms are harmless saprotrophs in soil and water, many are pathogens. Among these, *Mycobacterium tuberculosis* and related species are responsible for tuberculosis, a disease now reported to claim more victims than malaria and AIDS. Like *Mycobacterium tuberculosis, Corynebacterium diphtheriae* is considered to invade the

host mainly as a result of inhalation of droplet nuclei from a carrier of the disease and probably does not survive long in indoor air. The spores of thermophilic actinobacteria dispersed from massively contaminated HVAC systems may cause HP among building occupants (Banaszak *et al.* 1970, Fink *et al.* 1971).

The importance for health of exposure to various bacteria in work environments is discussed in Chapter 2.2.

Viruses

It is generally assumed that, under normal circumstances, viruses pass from person to person by direct contact or by forcible expulsion of droplets during coughing and sneezing. The largest droplets expelled are of an aerodynamic diameter (d_a) >100 μm and settle in the environment within seconds. They may have a role in infection by settling on fomites such as towels, clothing and upholstery from which viruses may infect victims by contact. Large droplets, d_a =10-100 μm, are expelled by coughing and sneezing are inspirable by individuals in the proximity, but are largely restricted to the upper airways. Airborne transmission of droplet nuclei of d_a <10 μm is still a matter of some contention, but if it occurs it could lead to the lower respiratory tract. Weber and Stilianakis (2008) have suggested that all three modes of transmission are involved in the spread of influenza. Fabian *et al.* (2008) have recently reported that coughing and sneezing are not the only source of infective viruses. Individuals infected with influenza A or B exhaled influenza virus RNA during normal tidal breathing at rates ranging from <3.2 to 20 particles min^{-1}. More than 87% of these particles were <1 μm in diameter and would therefore reach the lower respiratory tract.

Studies in which coughing has been simulated have shown that forcibly expired droplets can be dispersed well beyond a metre, the currently viewed limit of risk of droplet infection. When in a hospital ward equipped with ceiling intake diffusers and exhaust vents Wan *et al.* (2007) simulated human coughs using artificial saliva droplets similar in size distribution (peak size 12 μm) and airflow rate (0.4 L sec^{-1}) to a natural cough, they demonstrated that the exhaust vents had significant impact on the dispersion pattern of expiratory droplets, not unexpectedly enhancing lateral dispersion in the direction of the vents. Wan and Chao (2007) further noted that the time taken for droplets and droplet nuclei to be transported to exhaust vents or deposition surfaces for removal differed according to the ventilation flow pattern. In the same air-conditioned hospital ward as Wan *et al.* (2007), Sze To *et al.* (2008) examined the distribution of a simulated respiratory fluid containing a known concentration of a benign bacteriophage aerosolized in artificial coughs. The air was sampled using a six-stage Andersen sampler containing culture plates seeded with *Escherichia coli* and plaque counts made on incubation of the exposed plates. The direction of coughing had a significant effect on the airborne transport of the phage; plaque counts decreased with lateral distance from the "infector" when the cough was directed vertically up, but were constant or even increased with distance when the cough was directed sideways.

It has been widely held that viruses do not survive in infective form for any length of time outside the host organism, but in reviewing the available literature on environmental inactivation of the influenza A virus Weber and Stilianakis (2008) noted that aerosolized influenza viruses appear to be stable at low RH and low to moderately high temperatures. They noted that whilst the daily inactivation rate constants are of the order of 10^3 on hands they are in the range $1\text{-}10^2$ in aerosols and on inanimate surfaces, i.e. the virus can survive on hands for only minutes, but when airborne it can remain infective for some hours – the half-life of influenza A viruses in aerosols is 1-16 h.

The increased risk of airborne infection that is to be expected with overcrowding and poor ventilation is not confined to buildings. Air travel presents examples. In one instance, 72% of passengers on a commercial flight became infected with influenza during a 3-h delay on the ground, during which the ventilation system was not operational (Moser *et al.* 1979). Another, more recent, example is of a 3-h commercial flight from Hong Kong which carried one passenger symptomatic for severe acute respiratory syndrome (SARS). The result was that 22 of the other 119 passengers became infected with the SARS Coronavirus (SARS-CoV). Illness with this febrile respiratory disease was related to the physical proximity of the victims to the index patient (Olsen *et al.* 2003). In contrast, after a 90-min flight carrying four symptomatic passengers, two of whom had been coughing, only one other person reported fever and respiratory symptoms, but this was not reported as having been a probable case of SARS. On another 90-min flight, this time carrying a presymptomatic SARS patient, no illness was documented among other passengers (Olsen *et al.* 2003).

Robust evidence of airborne transport of viruses as a result of air movement within a building was presented by Wehrle *et al.* (1970), who reported on a nosocomial smallpox outbreak in a 3-storey hospital in Meschede, Germany, in which 17 patients became infected from the index patient in a ground-floor room. Smoke tests indicated that the virus particles were disseminated from the ground floor room to other floors as air currents created by the radiators rose up the stairwell, so infecting patients on the upper floors. There is also evidence that viruses can be distributed by mechanical ventilation systems, and a paper frequently cited as indicating such a spread is that by Brundage *et al.* (1988) concerning an outbreak of acute febrile respiratory disease among trainee soldiers. Significantly greater numbers of recruits in recently constructed mechanically ventilated barracks were affected than in similar, but older, naturally ventilated barracks. It was later found that the risk in mechanically ventilated barracks was reduced significantly by prophylactic treatment with adenovirus vaccine (Brundage *et al.* 1988). In a report of an outbreak of measles in a school, the spread of infection from infected children in one classroom to others in different classrooms was attributed to recirculation of air via the central ventilation system (Riley *et al.* 1978). Another measles outbreak was associated with a paediatric practice, where the disease was transmitted from a vigorously coughing child in an examining room to children elsewhere in the practice (Bloch *et al.* 1985). The authors suggested that that the evidence indicated that the measles virus survived for at least 1 h when airborne, and speculated that tightly insulated modern offices with a substantial proportion of recirculated air might predispose to airborne transmission. In addition to this paper, a paper by Gustafson *et al.* (1982) revealed that the employment of isolation rooms at positive pressure relative to other areas is ill-advised. They reported on an outbreak of chickenpox in a hospital where the airborne spread was from an immunocompromised child kept in strict isolation. The isolation room was at a higher pressure than in a corridor through which the varicella zoster virus was transmitted to other rooms, where susceptible children developed chickenpox. Another, more recent, example of a nosocomial disease outbreak has been reported by Li *et al.* (2004). The index patient in this case had SARS and was in one semi-enclosed cubicle of a ward of a Hong Kong hospital in which 70% of the air supply was recirculated. The non-functional return air outlet for this cubicle enhanced the spread of aerosolized SARS-CoV to other cubicles in the ward.

Whilst there is conflicting evidence, the overall conclusion of an international interdisciplinary panel (Li *et al.* 2007) that reviewed 40 relevant original studies carried out between 1960 and 2005, including some of those mentioned above, is that there is sufficient strong evidence to confirm that there is an association between ventilation, air movement in buildings and the transmission/spread of infectious diseases such as measles, chickenpox, smallpox and SARS.

REFERENCES

Ackermann, H., Schmidt, B., and Lenk, V. (1969) Mycological studies of the outdoor and indoor air in Berlin. *Mykosen*, **12**, 309-320.

Adams, K.F., and Hyde, H.A. (1965) Pollen grains and fungal spores indoors and out at Cardiff. *J. Palynol.*, **1**, 67-69.

Airaksinen, M., Kurnitski, J., Pasanen, P., and Seppanen, O. (2004a) Fungal transport through a building structure. *Indoor Air*, **14**, 92-104.

Airaksinen, M., Pasanen, P., Kurnitski, J., and Seppanen, O. (2004b) Microbial contamination of indoor air due to leakages from crawl space: a field study. *Indoor Air*, **14**, 55-64.

Austwick, P.K.C., Davies, P.S., Cook, C.P., and Pickering, C.A.C. (1986) Comparative microbiological studies in humidifier fever. In C. Molina, (ed.), *Maladies des Climatiseurs et des Humidificateurs*, Colloque INSERM, Paris, pp. 155-164.

Austwick, P.K.C., Little, S.A., Lawton, L., *et al.* (1989) Microbiology of sick buildings. In B. Flannigan, (ed.), *Airborne Deteriogens and Pathogens*, The Biodeterioration Society, Kew, Surrey, UK, pp. 122-128.

Baldo, B.A., and Baker, R.S. (1988) Inhalant allergies to fungi: reaction to bakers' yeast (*Saccharomyces cerevisiae*) and identification of bakers' yeast enolase as an important allergen. *Int. Arch. Allergy Appl. Immunol.*, **86**, 201-208.

Banaszak, E.F., Thiede, W.H., and Fink, J.N. (1970) Hypersensitivity pneumonitis due to contamination of an air conditioner. *N. Engl. J. Med.*, **283**, 271-276.

Baron, E.J., Chang, R.S., Howard, D.H., *et al.* (1994) *Medical Microbiology*. Wiley-Liss, New York.

Berk, S.G., Ting, R.S., Turner, G.W., and Ashburn, R.J. (1998) Production of respirable vesicles containing live *Legionella pneumophila* cells by two *Acanthamoeba* spp. *J. Allergy Clin. Immunol.*, **64**, 279-286.

Bernstein, I.L., and Safferman, R.S. (1967) Sensitivity of skin and bronchial mucosa to green algae. *J. Allergy*, **38**, 166-173.

Bernstein, I.L., and Safferman, R.S. (1970) Viable algae in house dust. *Nature*, **227**, 851-852.

Bischof, W., Koch, A., Gehring, U., *et al.* (2002) Predictors of high endotoxin concentrations in the settled dust of German homes. *Indoor Air*, **12**, 2-9.

Bloch, A.B., Orenstein, W.A., Ewing, W.M., *et al.* (1985) Measles outbreak in a paediatric practice: airborne transmission in an office setting. *Paediatrics*, **75**, 676-683.

Brasel, T.L., Douglas, D.R., Wilson, S.C., and Strauss, D.C. (2005). Detection of airborne *Stachybotrys chartarum* macrocyclic mycotoxins on particulates smaller than conidia. *Appl. Environ. Microbiol.*, **71**, 114-122.

Brundage, J.F., Scott, R.M., Lednar, W.M., *et al.* (1988) Building-associated risk of febrile acute respiratory diseases in army trainees. *J. Amer. Med. Assoc.*, **259**, 2108-2112.

Burge, H.A. (1989) Airborne allergic fungi: classification, nomenclature, and distribution. *Immunol. Allergy Clin. N. Am.*, **9**, 307-319.

Carnelley, T., Haldane, J.S., and Anderson, A.M. (1887) The carbonic acid, organic matter and micro-organisms in air, more especially in dwellings and schools. *Phil. Trans. Roy. Soc., Series B*, **178**, 61-111.

Cho, S.-H., Reponen, T., Bernatein, D.I., *et al.* (2006) The effect of home characteristics on dust antigen concentrations and loads in homes. *Sci. Total. Environ.*, **371**, 31-43.

Chrisostomou, C., Moustaka-Gouni, M., Sgardelis, S., and Lanaras, T. (2009) Air-dispersed phytoplankton in a Mediterranean River-Reservoir System (Aliakmon-Polyphytos, Greece). *J. Phycol. Research*, **31**, 877-884.

Christensen, C.M. (1950) Intramural dissemination of spores of *Hormodendrum resinae*. *J. Allergy*, **21**, 409-413.

Cianciotto, N.P., Kwaik, Y.A., Edelstein, P.H., *et al.* (2007) *Legionella: State of the Art 30 Years after Its Recognition*. ASM Press, Washington, DC.

Cohen, S.G., and Reif, C.B. (1953) Cutaneous sensitization to blue-green algae. *J. Allergy*, **24**, 452-457.

Cox, G.E., Wilson, J.D., and Brown, P. (1974) Prototothecosis: a case of disseminated algal infection. *Lancet* 2, 379-382.

Curry, A., Turner, A.J., and Lucas, S. (1991) Opportunistic protozoan infections in human immunodeficiency virus disease: review highlighting diagnostic and therapeutic aspects. *J. Clin. Pathol.*, 44, 182-193.

Dybendal, T., Wedberg, W.C., and Elsayed, S. (1991) Dust from carpeted and smooth floors: IV. Solid material, proteins and allergens collected in the different filter stages of vacuum cleaners after ten days of use in schools. *Allergy*, **46**, 427-435.

Edwards, J.H., Griffiths, A.J., and Mullins, J. (1976) Protozoa as sources of antigen in humidifier fever. *Nature*, **264**, 438-439.

Fabian, P., McDevitt, J.J., DeHaan, W.H., *et al.* (2008) Influenza virus in human exhaled breath: An observational study. *PLoS ONE*, **3**, e2691.

Ferro, A.R., Kopperud, R.J., and Hildemann, L.M. (2004) Source strengths for indoor human activities that resuspend particulate matter. *Environ. Sci. Technol.*, **38**, 1759-1764.

Fink, J.N., Resnick, A.J., and Salvaggio, J. (1971) Presence of thermophilic actinomycetes in residential heating systems. *Appl. Microbiol.*, **22**, 730-731.

Flannigan, B., McCabe, E.M., Jupe, S.V., and Jeffrey, I.G. (1993) Mycological and acaralogical investigation of complaint and non-complaint houses in Scotland. In *Indoor Air '93, Proceedings of the Sixth International Conference on Indoor Air Quality and Climate*, Vol. 4, Indoor Air '93, Helsinki, pp. 143-148.

Flannigan, B., McCabe, E.M., and Jupe, S.V. (1996). Quantification of air- and dust-borne deteriogenic microorganisms in homes. In *Biodeterioration and Biodegradation, Proceedings of the Tenth International Biodeterioration and Biodegradation Symposium*, DECHEMA, Frankfurt am Main, pp. 377-384.

Flannigan, B., McEvoy, E.M., and McGarry, F. (1999) Investigation of surface and airborne bacteria in homes. In G. Raw, C. Aizlewood, and P. Warren (eds.), *Indoor Air '99, Proceedings*, Vol. 1. Construction Research Communications Ltd., London, pp. 844-849.

Fradkin, A., Tobin, R.S., Tarlo, S.M., *et al.* (1987) Species identification of airborne molds and its significance for the detection of indoor pollution. *J. Air Pollut. Control Assoc.*, **37**, 51-53.

Garcia, B.E., Comtois, P., and Borrego, P.C. (1995) Fungal content of air samples from some asthmatic children's homes in Mexico City. *Aerobiologia*, **11**, 95-100.

Gehring, U., Bischof, W., Borte, M., *et al.* (2004) Levels and predictors of endotoxin in mattress dust samples from East and West German homes. *Indoor Air*, **14**, 284-292.

Giovannangelo, M., Gehring, U., Nordling, E., *et al.* (2007) Determinants of house dust endotoxin in three European countries – the AIRALLERG study. *Indoor Air*, **17**, 70-79.

Goebes, M.D., Baron, E.J., Mathews, K.L., and Hildemann, L.M. (2008) Effect of building construction on *Aspergillus* concentrations in a hospital. *Infect. Control Hosp. Epidemiol.*, **29**, 462-464.

Górny, R.L., Reponen, T., Willeke, K., *et al.* (2002). Fungal fragments as indoor air biocontaminants. *Appl. Environ. Microbiol.*, **68**, 3522-3531

Grant, C., Hunter, C.A., Flannigan, B., and Bravery, A.F. (1989) Water activity requirements of moulds isolated from domestic dwellings. *Int. Biodet.*, **25**, 259-284.

Gravesen, S. (1994) Allergic and non-allergic manifestations related to indoor fungal exposure – management of cases. In R.A. Samson, B. Flannigan, M.E. Flannigan, *et al.*, (eds.), *Health Implications of Fungi in Indoor Environments*, Elsevier, Amsterdam, pp. 241-248.

Gustafson, T.L., Lavely, G.B., Brawner, E.R., *et al.* (1982) An outbreak of airborne nosocomial varicella. *Pediatrics*, **70**, 550–556.

Hirst, J.M., and Last, F.T. (1953) Concentrations of basidiospores of the dry rot fungus (*Merulius lacrymans*) in the air of buildings. *Acta Allergy*, **6**, 168-174.

Hirvonen, M.-R., Huttunen, K., and Roponen, M. (2005) Bacterial strains from moldy buildings are highly potent inducers of inflammatory and cytotoxic effects. *Indoor Air*, **15**, 65-60.

Hodges, G.R., Fink, J.N., and Schleuter, D.P. (1974) Hypersensitivity pneumonitis caused by a contaminated cool-mist vaporizer. *Ann. Intern. Med.*, **80**, 501-504.

Hoekstra, E.S., Samson, R.A., and Verhoeff, A.P. (1994) Fungal propagules in house dust: a qualitative analysis. In R.A. Samson, B. Flannigan, M.E. Flannigan, *et al.* (eds.), *Health Implications of Fungi in Indoor Environments*, Elsevier, Amsterdam, pp. 169-177.

Hoffman, P., Friedman, H., and Bandinelli, M., (eds.) (2009) *Legionella pneumophila: Pathogenesis and Immunity*. Springer, New York.

Horner, W.E., Worthan, A.G., and Morey, P.R. (2004) Air- and dust-borne mycoflora in houses free of water damage and fungal growth. *Appl. Environ. Microbiol.*, **70**, 6394-6400.

Hunter, C.A., Grant, C., Flannigan, B., and Bravery, A.F. (1988) Mould in buildings: the air spora of domestic dwellings. *Int. Biodet.*, **24**, 81-101.

Hyvärinen, A., Meklin, T., Vepsäläinen, A., and Nevalainen, A. (2002). Fungi and actinobacteria in moisture-damaged building materials – concentrations and diversity. *Int. Biodet. Biodeg.*, **49**, 27-37.

Hyvärinen, A., Roponen, M., Tiittanen, P., *et al.* (2006) Dust sampling methods for endotoxin - an essential, but underestimated issue. *Indoor Air*, **16**, 20-27.

Instanes, C., Hetland, G., Bernsten, S., *et al.* (2005) Allergens and endotoxin in settled dust from day-care centers and schools in Oslo, Norway. *Indoor Air*, **15**, 356-362.

Jackson, F. (1984) *Sporobolomyces; Tilletiopsis*. In K. Wilken-Jensen, and S. Gravesen, (eds.), *Atlas of Moulds in Europe Causing Respiratory Allergy*. ASK Publishing, Copenhagen, pp. 46-47.

Kingston, D., and Warhurst, D.C. (1969) Isolation of amoebae from the air. *J. Med. Microbiol.*, **2**, 27-36.

Kirk, P.M., Cannon, P.F., Minter, D.W., and Stalpers, J.A. (2001) *Ainsworth and Bisby's Dictionary of the Fungi*, (10th Ed.), CAB International, Wallingford, Oxford, UK.

Koivikko, A., Kalima, K., Nieminen, E., *et al.* (1988) Allergenic cross-reactivity of yeasts. *Allergy*, **43**, 192-200.

Koivikko, A., Viander, M., and Lanner, A. (1991) Use of the extended Phadebas RAST panel in the diagnosis of mould allergy in asthmatic children. *Allergy*, **46**, 85-91.

Kozak, P.P., Gallup, J., Cummins, L.H., and Gillman, S.A. (1979) Currently available methods for home mold surveys. II. Examples of problem homes surveyed. *Ann. Allergy*, **45**, 167-176.

Kreiss, K., and Hodgson, M.J. (1984) Building-associated epidemics. In P. J. Walsh, C. S. Dudney, and E. D. Copenhaver, (eds.), *Indoor Air Quality*, CRC Press, Boca Raton, FL, pp. 88-106.

Kujundzic, E., Hernandez, M., and Miller, S.L. (2006) Particle size distributions and concentrations of airborne endotoxin using novel collection methods in homes during the winter and summer seasons. *Indoor Air*, **16**, 216-226.

Kwaik, Y.A., Gao, L.-Y., Stone, B.J., *et al.* (1998) Invasion of Protozoa by *Legionella pneumophila* and its role in bacterial ecology and pathogenesis. *Appl. Environ. Microbiol.*, **64**, 3127-3133.

Lee, T., Grinshpun, S. A., Martuzevicius, D., *et al.* (2006) Relationship between indoor and outdoor bioaerosols collected with a button inhalable aerosol sampler in urban homes. *Indoor Air*, **16**, 37-47.

Lehtonen, M., Reponen, T., and Nevalainen, A. (1993) Everyday activities and variation of fungal spore concentrations in indoor air. *Int. Biodet. Biodeg.*, **31**, 25-39.

Li, D.-W. (2005) Release and dispersal of basidiospores from *Amanita muscaria* var. *alba* and their infiltation into a residence. *Mycol. Res.*, **109**, 1235-1242.

Li, D.-W., and Kendrick, B. (1995) A year-round comparison of fungal spores in indoor and outdoor air. *Mycologia*, **87**, 190-195.

Li, D.-W., and Kendrick, B. (1996) Functional and causal relationships between indoor and outdoor airborne fungi. *Can. J. Bot.*, **74**, 194-209.

Li, Y., Duan, S., Huang, X., *et al.* (2004) Role of air distribution in SARS transmission during the largest nosocomial outbreak in Hong Kong. *Indoor Air*, **15**, 83-95.

Li, Y., Leung, G.M., Tang, J.W., *et al.* (2007) Role of ventilation in airborne transmission of infectious agents in the built environment – a multidisciplinary systematic review. *Indoor Air*, **17**, 2-18.

McElhenny, T.R., Bold, H.C., Brown, R.M., and McGovern, J.P. (1962) Algae: a cause of inhalant allergy in children. *Ann. Allergy*, **20**, 739-743.

McGovern, J.P., Haywood, T.J., and McElhenny, T.R. (1966) Airborne algae and their allergenicity. II. Clinical and multiple correlation studies with four genera. *Ann. Allergy*, **24**, 145-149.

Maghraby, T.A., Gherbawy, Y.A.M.H., and Hussein, M.H. (2008) Keratinophilic fungi inhabiting floor dusts of student houses at the South Valley University in Egypt. *Aerobiologia*, **24**, 99-106.

Maunsell, K. (1952) Airborne fungal spores before and after raising dust. *Int. Arch. Allergy Appl. Immunol.*, **3**, 93-102.

Meklin, T., Husman, T., Vepsäläinen, A., *et al.* (2002). Indoor air microbes and respiratory symptoms of children in moisture damaged and reference schools. *Indoor Air*, **12**, 175-183.

Meyer, H. W., Wurtz, H., Suadicani, P., *et al.* (2004) Molds in floor dust and building-related symptoms in adolescent school children. *Indoor Air*, **14**, 65-72.

Meyer, H.W., Wurtz, H., Suadicani, P., *et al.* (2005) Molds in floor dust and building-related symptoms in adolescent school children: a problem for boys only? *Indoor Air*, **15**, 17-24.

Miller, J.D., Schneider, M.H., and Whitney, W.J. (1982) Fungi on fuel wood chips in a home. *Wood and Fiber*, **14**, 54-59.

Millington, W.M., and Corden, J.M. (2005) Long term trends in outdoor *Aspergillus/Penicillium* spore concentrations in Derby, UK from 1970 to 2003 and a comparative study in 1994 and 1996 with the indoor air of two local houses. *Aerobiologia*, **21**, 105-113.

Montacutelli, R., Maggi, O., Tarsitani, G., and Gabrielli, N. (2000) Aerobiological monitoring of "Sistine Chapel": airborne bacteria and microfungi trends. *Aerobiologia*, **16**, 441-448.

Moser, M.R., Bender, T.R., Margolis, H.S., et al. (1979) An outbreak of influenza aboard a commercial airliner. *Amer. J. Epidemiol.*, **110**, 1-6.

Mouilleseaux, A., and Squinazi, F. (1991) Contamination microbienne de l'air: strategie d'etude et exemples de differents environnements. In *Proc. Societe Francais d'Aerobiologie, 3eme Congres National*, Institute Pasteur, Paris.

Mouilleseaux, A., and Squinazi, F. (1994) Airborne fungi in several indoor environments. In R.A. Samson, B. Flannigan, M.E. Flannigan, A.P. Verhoeff, O.C.G. Adan, and E.S. Hoekstra, (eds.), *Health Implications of Fungi in Indoor Environments*, Elsevier, Amsterdam, pp. 155-162.

Mouilleseaux, A., Squinazi, F., and Festy, B. (1993) Air quality in air conditioned office buildings. In *Indoor Air '93, Proceedings of the Sixth International Conference on Indoor Air Quality and Climate*, Vol. 6, Indoor Air '93, Helsinki, pp. 615-620.

Nevalainen, A. (1989) Bacterial Aerosols in Indoor Air. PhD Dissertation (University of Kuopio), National Public Health Institute, Kuopio, Finland.

Nevalainen, A., Jantunen, M.J., Rytkönen, A.-L., et al. (1988) The indoor air quality of Finnish homes with mold problems. In B. Berglund and T. Lindvall, (eds.), *Healthy Buildings '88*, Vol. 2, Swedish Council for Building Research, Stockholm, pp. 319-323.

Nevalainen, A., Heinonen-Tanski, H., and Savolainen, R. (1990) Indoor and outdoor occurrence of *Pseudomonas* bacteria. In *Indoor Air '90, Fifth International Conference on Indoor Air Quality and Climate*, Vol. 2, Canada Mortgage and Housing Corporation, Ottawa, pp. 51-53.

Newsome, A.L., Scott, T.M., Benson, R.F., and Fields, B.S. (1998) Isolation of an amoeba naturally harboring a distinctive *Legionella* species. *Appl. Environ. Microbiol.*, **64**, 1688-1693.

Nilsson, A., Kihlström, E., Lagesson, V., et al. (2004) Microorganisms and volatile organic compounds in airborne dust from damp residences. *Indoor Air*, **14**, 74-82.

O'Brien, I.M., Bull, J., Creamer, B., et al. (1978) Asthma and extrinsic allergic alveolitis due to *Merulius lacrymans*. *Clin. Allergy*, **8**, 535-542.

Olsen, S.J., Chang, H.L., Cheung, T.Y.-Y., et al. (2003) Transmission of the severe acute respiratory syndrome on aircraft. *New Engl. J. Med.*, **349**, 2416-2422.

Pakarinen, J., Hyvärinen, A., Salkinoja-Salonen, M., et al. (2008). Predominance of Gram-positive bacteria in house dust in the low-allergy risk Russian Karelia. *Environ. Microbiol.*, **10**, 3317-3325.

Pasanen, A.-L., Kalliokoski, P., Pasanen, P., et al. (1989) Fungi carried from farmers' work into farm homes. *Am. Industr. Hygiene Assoc. J.*, **50**, 631-636.

Peltola, J.S.P., Andersson, M.A., Haahtela, T., et al. (2001) Toxic-metabolite-producing bacteria and fungus in an indoor environment. *Appl. Environ. Microbiol.*, **67**, 3269-3274.

Rantio-Lehtimäki, A. (1988) Yeasts in rural and urban air in southern Finland. *Grana*, **27**, 313-319.

Rantio-Lehtimäki, A., and Koivikko, A. (1984) *Rhodotorula*. In K. Wilken-Jensen and S. Gravesen, (eds.), *Atlas of Moulds in Europe Causing Respiratory Allergy*, ASK Publishing, Copenhagen, p. 42.

Rao, C.Y., Riggs, M.A., Chew, G.L., et al. (2007). Characterization of airborne molds, endotoxins, and glucans in homes in New Orleans after Hurricanes Katrina and Rita. *Appl. Environ. Microbiol.*, **73**, 1630-1634.

Reponen, T., Nevalainen, A., Jantunen, M., et al. (1992) Normal range criteria for indoor air bacteria and fungal spores in a subarctic climate. *Indoor Air*, **2**, 26-31.

Riley, E.C., Murphy, G., and Riley, R.L. (1978) Airborne spread of measles in a suburban elementary school. *Am. J. Epidemiol.*, **107**, 421-432.

Rogerson, A., and Detwiler, A. (1999) Abundance of airborne heterotrophic protists in ground level air of South Dakota. *Atmos. Research*, **51**, 35-44.

Rohr, U., Weber, S., Michel, R., et al. (1998) Comparison of free-living amoebae in hot-water systems of hospitals with isolates from moist sanitary areas by identifying genera and determining temperature tolerance. *J. Allergy Clin. Immunol.*, **64**, 1822-1824.

Rosas, I., Calderón, C., Martinez, L., et al. (1997) Indoor and outdoor airborne fungal propagule concentrations in Mexico City. *Aerobiologia*, **13**, 23-30.

Schlichtung, H.E. (1969) The importance of airborne algae and protozoa. *J. Air Pollut. Control Assoc.*, **19**, 946-951.

Sharma, N.K., and Rai, A.K. (2008) Allergenicity of airborne cyanobacteria *Phormidium fragile* and *Nostoc muscorum*. *Ecotoxicol. Environ. Safety*, **69**, 158-162.

Sharma, N.K., Rai, A.K., Singh, S., and Brown, R.M. (2007) Airborne algae: their present status and relevance. *J. Phycol.*, **43**, 615-627.

Shelton, B.G., Kirkland, K.H., Flanders, W.D., and Morris, G.K. (2002) Profiles of airborne fungi in buildings and outdoor environments in the United States. *Appl. Environ. Microbiol.*, **68**, 1743-1753.

Sneller, M.R., and Roby, R.R. (1979) Incidence of fungal spores at the homes of allergic patients in an agricultural community. I. A 12-month study in and out of doors. *Ann. Allergy*, **43**, 225-228.

Solomon, W.R. (1974) Fungal aerosols arising from cool-mist vaporizers. *J. Allergy Clin. Immunol.*, **54**, 222-228.

Solomon, W.R., Burge, H.A., and Boise, J.R. (1980) Exclusion of particulate allergens by window air cleaners. *J. Allergy Clin. Immunol.*, **65**, 305-308.

Spiegelman, J., Blumstein, J., and Friedman, H. (1963) Effects of central air conditioning on pollen, mold and bacterial concentrations. *J. Allergy*, **34**, 426-431.

Steinert, M., Emödy, L., Amann, R., and Hacker, J. (1997) Resuscitation of viable but nonculturable *Legionella pneumophila* Philadelphia JR32 by *Acanthamoeba castellani*. *Appl. Environ. Microbiol.*, **63**, 2047-2053.

Steinert, M., Birkness, K., White, E., et al. (1998) *Mycobacterium avium* bacilli grow sporozoically in coculture with *Acanthamoeba polyphaga* and survive within cyst walls. *J. Allergy Clin. Immunol.*, **64**, 2256-2261.

Stott, R., May, E., Ramirez, E., and Warren, A. (2003) Predation of *Cryptosporidium* oocysts by protozoa and rotifers: implications for water quality and public health. *Water Sci. Technol.*, **47**, 77-83.

Ström, G., Palmgren, U., Wessen, B., et al. (1990) The sick building syndrome – an effect of microbial growth in building constructions? In *Indoor Air '90, Fifth International Conference on Indoor Air Quality and Climate*, Vol. 1, Canada Mortgage and Housing Corporation, Ottawa, pp. 173-178.

Sze To, G.N., Wan, M.P., Chao, C.Y.H., et al. (2008) A methodology for estimating airborne virus exposure in indoor environments using the spatial distribution of expiratory aerosols and virus variability characteristics. *Indoor Air*, **18**, 425-438.

Tiberg, E. (1987) Microalgae as aeroplankton and allergens. In G. Boehm and R. Leuschner, (eds.), *Advances in Aerobiology, Proceedings of the Third International Conference on Aerobiology, 1986*, Birkhäuser Verlag, Basel, pp. 171-173.

Tiberg, E., Bergmann, B., Wictorin, B., and Willén, T. (1984) Occurrence of microalgae in indoor and outdoor environments in Sweden. In S. Nilsson and B. Raj, (eds.), *Nordic Aerobiology, Proceedings of the Fifth Nordic Symposium on Aerobiology, Abisko, 1983*, Almqvist and Wiksell International, Stockholm, pp. 24-29.

Tiberg, E., Dreborg, S., and Björkstén, B. (1995) Allergy to green algae (*Chlorella*) among children. *J. Allergy Clin. Immunol.*, **96**, 257-259.

Tormo, R., Recio, D., Silva, I., and Muñoz, A.F. (2001) A quantitative investigation of airborne algae and lichen soredia obtained from pollen traps in south-west Spain. *Eur. J. Phycol.*, 36, 385-390.

Verhoeff, A.P., van Wijnen, J.H., Boleij, J.S.M., *et al.* (1990) Enumeration and identification of airborne viable mould propagules in houses. A field comparison of selected techniques. *Allergy*, **45**, 275-284.

Verhoeff, A.P., van Wijnen, J.H., van Reenen-Hoekstra, E.S., *et al.* (1994) Fungal propagules in house dust. II. Relation with residential characteristics and respiratory symptoms. *Allergy*, **49**, 540-547.

Wan, M.P., and Chao, C.Y.H. (2007) Transport characteristics of expiratory droplets and droplet nuclei in indoor environments with different ventilation airflow patterns. *J. Biomech. Eng.*, 129, 341-353.

Wan, M.P., Chao, C.Y.H., Ng, Y.D., *et al.* (2007) Dispersion of expiratory droplets in a general hospital ward with ceiling mixing type mechanical ventilation system. *Aerosol Sci. Technol.*, **41**, 244-258.

Weber, T.P., and Stilianakis, N.I. (2008) Inactivation of influenza A viruses in the environment and modes of transmission. *J. Infect.*, **57**, 361-373.

Wehrle, P.F., Posch, J., Richter, K.H., and Henderson, D.A. (1970) An airborne outbreak of smallpox in a German hospital and its significance with respect to other recent outbreaks in Europe. *Bull. World Health Organ.*, **43**, 669-679.

Wickens, K., Douwes, J., Siebers, R., *et al.* (2003) Determinants of endotoxin levels in carpets in New Zealand homes. *Indoor Air*, **13**, 128-135.

Yoshida, K., Ando, M., Sakata, T., and Araki, S. (1989) Prevention of summer-type hypersensitivity pneumonitis: effect of elimination of *Trichosporon cutaneum* from the patients' homes. *Arch. Environ. Health*, **44**, 317-322.

Zyska, B. (2001) Fungi in indoor air in European countries. *Mikol. Lek.*, **8**, 127-140. [In Polish.]

Zyska, B. (2004) The mycology of residential buildings, public buildings and indoor work environments, paying special attention to the taxonomy of fungi. In *Proceedings, Indoor Air in Poland 2003*, Politechnika Warszawska. [In Polish, with English summary.]

Chapter 1.3

POLLEN AND POLLEN ALLERGENS IN INDOOR ENVIRONMENTS

Timo Hugg[1], Eija Yli-Panula[2], Auli Rantio-Lehtimäki[2]

[1] South Karelia Allergy and Environment Institute, Joutseno, and Institute of Health Sciences, University of Oulu, Oulu, Finland;
[2] Aerobiology Unit, Department of Biology, University of Turku, Turku, Finland

INTRODUCTION

Allergy is among the major diseases of the world, affecting about 500 million people (Bousquet et al. 2006). Pollen is a well-known cause of allergies and allergic symptoms like rhinorrhea (runny nose), nasal congestion, sneezing, red and itchy eyes, skin itching, flushing and hives (urticaria), and asthma symptoms such as coughing, wheezing and shortness of breath. With people in developed countries nowadays generally spending 90–95% of their time indoors, factors affecting the presence of aeroallergens indoors are important in understanding health impact and potential intervention methods. In general, most indoor allergens are seen to be strongly associated with asthma sensitisation, pollen primarily with rhinitis.

Pollen grains of different plant genera vary in shape, size, surface pattern and number of apertures (Fig. 1). An aperture, or pore, is any thinning, thickening or other modification of the wall of pollen that serves as an exit for its contents or allows shrinking and swelling of the grain in response to changes in moisture content. Two types of allergologically important pollen grain are those of the birch tree, typically with a diameter of 20-22 μm, and grass, in which pollen is 25-60 μm in diameter. Allergens are mostly proteins; the major birch allergen, Bet v 1, is a 17 kDa protein, and the major grass allergen Phl p 5 with its isoforms is a 32/38 kDa protein (Jarolim et al. 1989, Ong et al. 1996). Allergens have been found in several parts of the pollen grain: the cytoplasm; amyloplasts and starch granules; orbicules (small granules or droplets of sporopollenin within the anther where pollen grains mature); and in the mature pollen walls (Grote et al. 1994, Swoboda et al. 1995, El Ghazaly et al. 1996, Emilson et al. 1996, Behrendt et al. 1999, Gro-

te 1999). Antigens similar to pollen allergens are also found in seeds and other parts of birch trees (Fountain et al. 1992).

The most important allergenic pollen types are anemophilous (wind-pollinated). They include some weeds in the family Asteraceae, especially mugwort (Artemisia) and ragweed (Ambrosia), and in the Urticaceae, e.g. pellitory-on-the-wall (Parietaria). Allergenic tree pollens include birch (Betula), alder (Alnus) and hazel (Corylus) in the Betulaceae; ash (Fraxinus) and olive tree (Olea) in the Oleaceae, oak (Quercus) and plane (Platanus) in the Fagaceae, and the coniferous genera Cupressus (tree cypresses), Juniperus (junipers/ mountain cedar) and Thuja (thuyas) in the Cupressaceae. The Japanese cedar or sugi (Cryptomeria japonica) in the Taxodiaceae is another conifer that is an important allergenic type (Takahashi et al. 1993, Enomoto et al. 2004).

SAMPLING AND IDENTIFICATION OF POLLEN AND POLLEN ALLERGENS

For sampling air for pollen, there are many different methods. These include:

- non-volumetric gravitational methods, e.g. sticky slides (Durham 1944);
- impactors such as spore and pollen traps in which air samples are drawn through a slit and particles impact on an adhesive surface, the standard method for outdoor pollen monitoring being a Hirst-type trap (Hirst 1952), such as the Lanzoni VPPS 1000 volumetric pollen and particle sampler or the Burkard automatic volumetric spore trap, or AVST (Fig. 2);

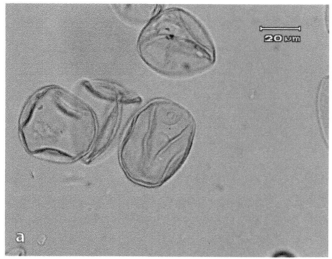

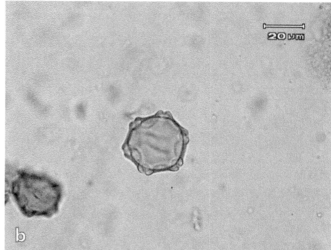

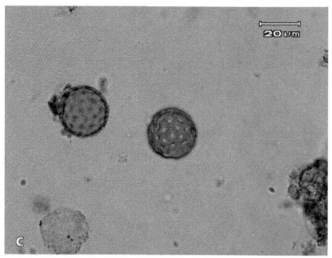

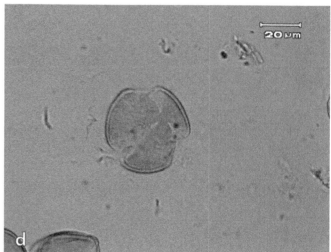

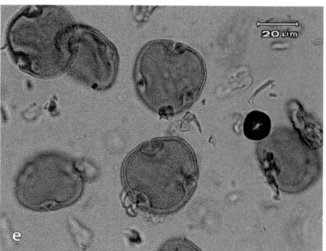

Figs. 1a-e. a. Grass pollen with a single aperture (pore) and a smooth surface. b. Alder pollen with five pores. c. *Chenopodium* pollen with its many apertures giving a golf ball-like appearance. d. Maple pollen with three furrows and a striate surface. e. Lime tree pollen grain with pores, furrows and a finely reticulate surface.

- virtual samplers (Fig. 2), in which the impact surface is replaced with a large stagnant volume (see, for example, Kesavan *et al.* 2008);
- whirling arm impactors (rotorod, rotobar, rotoslide; Perkins 1957) where vertical arms rotate and particles impact on rods or slides (Fig. 3);
- sieve samplers where air is drawn through one or several filter stages, such as the Andersen sampler (Fig. 4) or the SKC button inhalable aerosol sampler (Lee *et al.* (2006a, b);
- impingers where samples are taken into liquid; and
- centrifugal samplers, in which particles are washed into liquid by a vortex system (Fig. 5).

The intake air volume may simulate human respiration, but in samplers with a high intake rate (known as high volume samplers) particles are collected from a large volume of air, so that, for example, there is sufficient pollen for an immunochemical assay. Size-selective samplers separate particles into different size classes mimicking human airways, e.g. in the An-

dersen sampler. Dust samples are often collected on a filter using a vacuum cleaner.

POLLEN AND POLLEN ALLERGENS IN INDOOR AIR

Owing to poor pollen production among commonly used decorative plants, pollen release indoors is generally not significant. However, people with strong and long-lasting occupational indoor exposures, such as florists and others handling flowers, may sometimes be sensitised to pollen of ornamental plants (de Jong *et al.* 1998). In one such case, a florist suffered rhino-conjunctivitis, contact urticaria and IgE-mediated asthma provoked by Easter lily and tulip pollen (Piirilä *et al.* 1999). Another florist first developed urticaria, and later rhinoconjunctivitis and asthma, as a result of exposure to the pollen of members of the family Compositae (Asteraceae), leading to a change of occupation (Uter *et al.* 2001). Twigs of deciduous trees with developing catkins just about to release pollen and used for decorative purposes can also sensitise and exacerbate allergic symptoms.

Allergy to flower pollen is also recognised as a problem among greenhouse workers. For instance, Groenewoud *et al.* (2002a) reported that in Dutch greenhouses 20% of *Chrysanthemum* workers were sensitized to the pollen, the main symptom being rhinitis. Occupational allergy to the pollen of tomato and *Capsicum annum* (sweet or bell pepper, paprika) is also well known (van Toorenbergen *et al.* 2000). Groenewoud *et al.* (2002b) reported that in the Netherlands roughly one-third of greenhouse workers involved in growing bell peppers were sensitized to the pollen. Allergy to the pollen of other plants, such as strawberry (Watanabe *et al.* 2000b) and eggplant or aubergine, *Solanum melongena* (Gil *et al.* 2002), has been reported. The former caused atopic cough, and the latter rhinoconjunctivitis as well as asthma, which is a recognised occupational hazard among greenhouse workers (Jurewicz *et al.* 2007). Monsó *et al.* (2002) concluded that asthma occurred in 8% of growers of greenhouse flowers and/or ornamental plants. It also occurs among those growing greenhouse plants for their fruits, e.g. tomato (Watanabe,

Fig. 2-6. (2). Rooftop Burkard spore trap (left) and two virtual impactors (centre), the latter being used to separate out larger from smaller airborne particles for analysis (see, for example, Kim and Lee 2000, Strawa *et al.* 2003). (3). Rotorod-type sampler. (4). Six-stage Andersen particle sizing sampler (Andersen 1958). (5). A Coriolis Delta centrifugal liquid sampler (Bertin Technologies; analysis by immunochemical methods). (6). ALK device attached to vacuum cleaner for dust sampling.

Table 1. Summary of published studies on concentrations of indoor-outdoor pollen grains in inhalation air.

Author(s), year (country)	Type of indoor space	Indoor/outdoor concentration (pg m⁻³)	Type of sampler
Solomon *et al.* 1980 (USA)	Homes and outpatient clinics examining rooms	16–253/688 (median values[1])	Rotobar and Rotorod samplers
O'Rourke and Lebowitz 1984 (USA)	Private homes	0–4/0–186	Rotorod sampler
Stock and Morandi 1988 (USA)	Private homes	1.0–6.2/3.9–96.1 (geometric means)	Rotorod sampler
O'Rourke *et al.* 1989 (USA)	Private homes	0–600/0–1171 (mean values)	Burkard personal sampler
Morrow-Brown 1991 (UK)	Air filtered car	Almost nothing/1000 –10000	Mobile slit sampler
Sterling and Lewis 1998 (USA)	Mobile homes	2.3–26.3/4.9–59.7 (geometric means)	Rotorod sampler
Tormo Molina *et al.* 2002 (Spain)	Hospital building	10.0/299.3 (mean values)	Portable volumetric sampler
Cariñanos *et al.* 2004 (Spain)	University buildings	< 150/< 2200	Lanzoni VPPS 1000 sampler
Lee *et al.* 2006a (USA)	Single-family homes	1–5/1–1234	Button personal inhalable aerosol sampler
Lee *et al.* 2006b (USA)	Single-family homes	0–2/1–44	Button personal inhalable aerosol sampler
Hugg and Rantio-Lehtimäki 2007 (Finland)	Single-family home, block of flats apartment and central hospital building	0–17/0–855	Rotorod-type sampler

[1] Pollen and spore data combined.

2000a, Vandenplas *et al.* 2008) and eggplant (Gil *et al.* 2002).

Although the outdoor environment is the main source of indoor pollen and pollen allergens, little is known about the efficiency of penetration of pollen and pollen allergens from the outdoor air into indoor environments. Airborne pollen grains and allergens are thought to pass indoors through ventilation ducts and open windows and doors. Instructions for avoidance of exposure are often based on closing the indoor environment off from the outdoors as tightly as possible by special filters in the openings and controlling carefully the air exchange using different penetration barriers, etc. However, people and pets who have been outdoors probably bring considerable quantities of pollen grains inside on, respectively their clothes and fur (see, for example, O´Rourke and Lebowitz 1984, Yli-Panula and Rantio-Lehtimäki 1995). An instance of pollen being brought indoors on textiles other than clothing has been noted by Takahashi *et al.* (2008). These workers measured pollen of the Japanese cedar adhering to clothes, laundry and futon bedding out of doors by quantification of

the antigen Cry j 1. The amount of adherent pollen was particularly high on futons, and >50% of this pollen remained on the surface of this bedding (and also laundry) after shaking or brushing by hand. Takahashi *et al.* (2008) also noted that most pollen entering unoccupied apartments by ventilation remained near the windows. This is in line with the expectation that the highest pollen grain and pollen allergen concentrations in indoor dust are to be found close to the front door and ventilating windows, diminishing with distance from the openings (Hugg and Lehtimäki 2007, Yli-Panula *et al.* 2008).

As has been amply demonstrated (Table 1), pollen grains are found in lower concentrations in most indoor environments than outside. Generally, numbers of pollen grains detected in indoor air have been low, probably causing allergic reactions in only the most sensitive subjects. According to Lee *et al.* (2006a), the indoor pollen concentration in urban homes ranged from 1 to 5 pollen grains m⁻³ air, while outdoor levels were 1-1234 pollen grains m⁻³. Low indoor/outdoor ratios can be partially explained by the size of pollen grains. Owing to the size of the particles (20-60 µm),

Table 2. Summary of the previously published studies on concentrations of indoor pollen antigens or/and allergens in settled dust and inhalation air.

Author(s), year (country)	Type of study place and space	Indoor concentration - units m^{-3} air or g^{-1} dust (range, mean or GM[A]) (type of sampler)
Dybendal et al. 1989a (Norway)	Homes (carpeted/smooth floors) Settled dust	14/9 Bet v RI[B] 20/14 Aln i RI[B] 11/8 Phl p RI[B] (Vacuum cleaner)
Dybendal et al. 1989b (Norway)	Schools (carpeted/smooth floors) Settled dust	18/33 Bet v RI[B] 11/9 Aln i RI[B] 7/4 Phl p RI[B] (Vacuum cleaner)
Dybendal et al. 1991 (Norway)	Schools (carpeted/smooth floors) Settled dust	16/3 Bet v RI[B] 20/5 Aln i RI[B] 13/7 Phl p RI[B] (Vacuum cleaner)
Yli-Panula et al. 1995 (Finland)	Rural and urban single family homes Settled dust	90-2800 Bet v SQ g^{-1} dust (Vacuum cleaner)
D´Amato et al. 1996 (Italy)	Home Indoor air	0.1-5 Par j ng m^{-3} air (High volume air sampler)
Yli-Panula 1997 (Finland)	Rural and urban single family homes Indoor settled dust	0-206 grass antigen SQ g^{-1} dust (vacuum cleaner)
Dotterud et al. 1997 (Norway)	Homes (LR[C]) Settled dust Schools Settled dust	12 Bet v RI[B] 2 Phl p RI[B] 6 Bet v RI[B] 1 Phl p RI[B]
Holmquist & Vesterberg 1999 (Sweden)	Schools, offices Indoor air	242-403 Bet v SQ m^{-3} air (DOSAFE*)
Fahlbush et al. 2000 (Germany)	Homes Settled dust	0.03-81 Phl p 5 μg g^{-1} dust (Vacuum cleaner)
Fahlbush et al. 2001 (Germany)	Homes Settled dust (LR[C], CR[C])	235-4300 ng Phl p 5 m^{-2} (GM[A,D]) 97-1300 ng Phl p 5 m^{-2} (GM[A,E]) (Vacuum cleaner)
Holmquist et al. 2001 (Sweden)	Street level shops Indoor air	20-30 Bet v SQ m^{-3} air (DOSIS[†])
Holmquist & Vesterberg 2001 (Sweden)	Driving compartments of coaches Indoor air	19 Bet v SQ m^{-3} air (GM[A]) (DOSIS[†])
Ohashi et al. 2005 (Japan)	Private house Indoor air	1.6-8.0 Cry j 1 pg m^{-3} air (8-stage Andersen sampler)
Yli-Panula et al. 2008 (Finland)	Private homes Indoor air Settled dust	Bet v pollen and allergen IgE HALO Qualitative, (Nasal air sampler) 4.8-15.3 SQ g^{-1} dust (Vacuum cleaner)

[A] GM, geometric mean; [B] RI, RAST inhibition test, expressed as percent specific IgE inhibition, mean values are given; [C] LR, living room; CR, children's room; GMA[D], geometric mean, samples taken during pollination; GMA[E], geometric mean, samples taken outside pollen period; * Direct on sampling filter estimation; † Direct on sampling filter in solution.

only a small fraction of pollen grains penetrate from outdoors to indoors, especially when doors and windows are closed and the ventilation system is equipped with filters (Solomon et al. 1980).

Typically, the pollen taxa in indoor air are the same as in the air outdoors, but concentrations vary.

Tormo Molina et al. (2002) analysed the air in a hospital outpatient ward using two portable volumetric aerobiological traps, one at the floor level and the other at a height of 1 m. The particle counts were compared with those outside the building. They found 20 types of pollen grain and the concentrations ranged

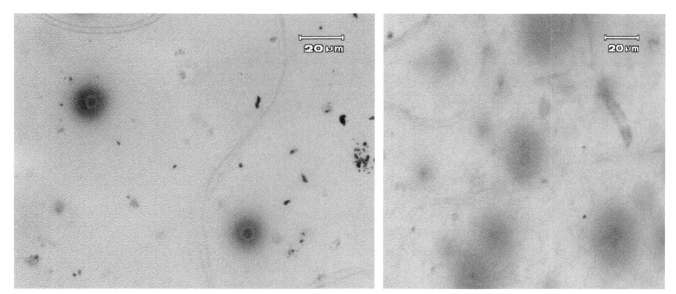

Fig. 7. Birch pollen and birch pollen allergens collected from indoor air using a nasal sampler, and analysed by HALOgen assay using human serum IgE serum (left) and polyclonal rabbit antibodies to Bet v (right). The large stained spots are birch pollen with released allergens around, and several small stained spots are Bet v allergens.

from 2.7 to 25.1 pollen grains m^{-3} air. No significant differences were found between the floor level and the meter-high measurements. Comparison with outdoor levels showed that the indoor concentrations of three most abundant pollen types correlated with the outdoor concentrations.

Hugg and Rantio-Lehtimäki (2007) studied the penetration of pollen into the indoor air by measuring the pollen concentrations in different spaces, throughout the *Betula* pollen season. The pollen concen-trations were measured inside and outside a ground floor apartment in a block of flats, a detached house and the regional central hospital, using rotorod-type samplers. The outdoor concentrations of *Betula* pollen grains ranged between low and abundant (0-855 pollen grains m^{-3}). The corresponding indoor concentrations near to the main front door varied from low to moderate in the central hospital (0-17 pollen grains m^{-3}), and were low in residential buildings (< 10 pollen grains m^{-3}). Concentrations of *Betula* pollen decreased significantly from outdoors to indoors and further toward the centre of the building, indicating relatively poor penetration of pollen grains and/or the short-lived presence of pollen grains in the indoor air (Hugg and Rantio-Lehtimäki 2007).

O'Rourke and Lebowitz (1984) compared indoor *vs* outdoor pollen load in four locations in Tucson, Arizona. Atmospheric pollen was collected in 55 home locations with a rotorod sampler for three consecutive days each season. Pollen was rare inside homes and occurred only when production was high. They found no differences in concentrations between rooms. Outdoor pollen concentrations near homes (local pollen rain) were usually lower than the regional concentrations (regional pollen rain), and their correlations were poor even if pollen taxa were very similar.

Because people spend an increasingly larger part of their time in a variety of different vehicles, it has been considered important to assess pollen concentrations inside such vehicles (Muilenberg *et al.* 1991). Hugg *et al.* (2007) analysed pollen in dust of private cars with rotorods, and in dust using the sticky tape method (Holopainen *et al.* 2002). Pollen deposits on the dashboard and rear shelf were studied in cars both during driving and when parked. Grass (Poaceae) and *Artemisia* pollen was recorded on one day only, and both in low concentrations (<10 pollen grains m^{-3} air), whereas concentrations of *Betula* and *Pinus* were low to moderate, being respectively 0-15 and 0-41 pollen grains m^{-3}. Surprisingly, there were no statistical differences between the mean concentrations of *Betula* and *Pinus* pollen inside a car when the windows were closed compared with when one window was partly open. The number of pollen grains on the inside surfaces of private cars varied with species and ranged from zero to 72 pollen grains cm^{-2}. On average, both *Betula* and *Pinus* pollen was found in greater quantities on the dashboards than on the rear shelves of the cars.

The concentration of pollen allergen in outdoor air is low, and in indoor air even lower. Owing to the lack of suitable sampling and analytical methods, few studies on indoor pollen allergens have been published (Table 2). Just as Rantio-Lehtimäki *et al.* (1994) had reported for the birch antigen Bet v 1 in outdoor air, when validating quantification of pollen allergens in indoor air using two school and two office rooms as experimental models Holmquist and Vesterberg (1999a) found that the substantial amounts of allergens were predominantly in particles of smaller diameter than the pollen grains. In one school room, birch pollen allergen concentrations increased from 242 to 403 SQ units (standard quality units, ALK-Abello) m^{-3} air over the sampling period, although the corresponding outdoor air concentrations decreased from 350 to 90 SQ units m^{-3} air. This might have been due to re-entrainment of accumulated allergenic dust carried in from outdoors on footwear and clothes.

To compare the airborne birch pollen allergen load with the allergen concentration of settled dust, Yli-Panula *et al.* (2008) employed human IgE- and rabbit IgG-HALOgen immunoassays and IgG-ELISA using rabbit antiserum to detect Bet v allergens. Air samples were collected using nasal air samplers (Inhalix Pty Ltd., Australia), and dust samples by adapted vacuum cleaner. The antigenic activity was 4.8-15.3 SQ units g^{-1} dust. Airborne birch pollen allergens were found indoors and outdoors, and antigenic activity was detected in the settled dust over the whole study period of two months. Both small particle and pollen-sized fractions were present in the air samples. In addition to the detection of birch pollen antigens using the rabbit antiserum in the HALOgen assay, the use of the patient's own serum IgE in the HALOgen analysis enabled detection of allergens to which the patient had been sensitized.

D'Amato *et al.* (1996) reported that with the balcony open, there was no great difference between outdoor and indoor allergenic activity, but with the balcony closed, there was a reduction in indoor allergenic activity of about one-third in comparison with outdoor allergenic activity. Thus a significant correlation between outdoor pollen count and indoor allergen levels was found when the balcony was open, but not with the balcony closed. They also found a highly significant correlation between outdoor pollen count and outdoor allergen levels.

POLLEN AND POLLEN ALLERGENS IN HOUSE DUST

Rapid sedimentation in indoor environments may result in pollen numbers in settled dust being as high as 5.5 × 10^6 pollen grains g^{-1} house dust (O'Rourke and Lebowitz 1984). One consequence of this rapid sedimentation is that in sampling indoor air for pol-

len using, for example, rotorod-type samplers, extended sampling time is recommended. Sedimentation varies according to the height and position of the indoor spaces. Using sedimentation plates coated with adhesive, Kiyosawa and Yoshizawa (2001) observed that values for Japanese cedar pollen at floor level were 1.5-2 times greater than in the breathing zone, and values just inside the windows were 5-6 times higher than in the centre of rooms.

Taxonomic variation between the pollens in consecutive dust sampling can be considerable. Yankova (1991) observed 6-22 different tree and shrub pollen taxa and the pollen of 8-27 different herbaceous plants. Daily indoor dust sampling was based on the free sedimentation of pollen onto glass slides at a height of 1.5 m above floor level. The number of indoor pollen grains in dust samples ranged from 48 to 405 pollen grains cm^{-2} area sampled (Yankova 1991). However, a variable proportion of the grains may be devoid of antigenic/allergenic material. Settled pollen on the surfaces is only aerosolised into the human inhalation zone when the airflow created by different types of housework activity exceeds the adhesion force of deposited pollen, e.g. when books are moved from bookshelves or floors are vacuum-cleaned.

Earlier, Chafee (1985) used sedimentation slides during the ragweed season to measure the pollen load in an air-conditioned hospital room, an adjacent room without air-conditioning, and outdoors at roof level. The difference between airconditioned and non-air-conditioned rooms was clear, with the concentrations being much lower in the air-conditioned room, and led Chafee (1985) to consider that the simple way to avoid pollen was to stay indoors and employ air-conditioning, bearing in mind the possible presence of pollen beyond the usual outdoor season. Results indicated that some pollens were almost always present in samples (Poaceae, *Salix*, *Castanea* and *Betula*), reflecting either the occurrence of pollen in the indoor atmosphere outwith the pollination period or re-suspension of previously settled pollen (Chafee 1985).

In indoor environments the main interest in antigens has been in animal antigens – cat, dog, dust mite and cockroach – whereas pollen antigen studies (Table 2) have been relatively marginal (Dybendahl *et al.* 1989, Dotterud *et al.* 1997). These studies documented antigenic activity of birch and grass pollen in settled dust in Norwegian schools. It is typical that the antigenic activity in outdoor dust increased during the pollen period, but indoors the antigen peak appeared later than outdoors, even weeks after the pollen period (Yli-Panula and Rantio-Lehtimäki 1994,1995, Yli-Panula 1997). The explanation put forward for this was that pollen has to settle first and then be carried indoors by human and pets, so that the amount of pollen antigens rises more slowly indoors. Subsequently, Yli-Panula and Ahlholm (1998) found that birch pollen antigens are preserved in dry conditions from one pollen season to the next. Therefore, pollen allergy symptoms may not be restricted to a particular season alone, but can occur perennially. The pollen grains may be ruptured by moisture outdoors and the antigenic materials released. Since indoor environments are drier and more stable than outdoors, antigenicity can be expected to be more persistent indoors.

REFERENCES

Behrendt, H., Tomczok, J., Sliwa-Tomczok, W., *et al.* (1999) Timothy grass (*Phleum pratense* L.) pollen as allergen carriers and initiators of an allergic response. *Int. Arch. Allergy Immunol.*, **118**, 414-418.

Bousquet, J., Bieber, T., and Fokkens, W. (2006) Themes in allergy (Editorial). *Allergy*, **61**, 1-2.

Cariñanos, P., Alcázar, P., Galán, C., *et al.* (2004) Aerobiology as a tool to help in episodes of occupational allergy in work places. *J. Investig. Allergol. Clin. Immunol.*, **14**, 300-308.

Chafee, F.H. (1985) Pollen studies in a hospital air-conditioned room. *N. Engl. Reg. Allergy Proc.*, **6**, 150-152.

D'Amato, G., Russo, M., Liccardi, G., *et al.* (1996) Comparison between outdoor and indoor airborne allergenic activity. *Ann. Allergy Asthma Immunol.*, **77**, 147-152.

Dotterud, L.K., Van, T.D., Kvammen, B., *et al.* (1997) Allergen content in dust from homes and schools in northern Norway in relation to sensitization and allergy symptoms in schoolchildren. *Clin. Exp. Allergy*, **27**, 252-261.

Durham, O.C. (1944) The volumetric incidence of atmospheric allergens. IV. A proposed standard method of gravity sampling, counting and volumetric interpolation of results. *J. Allergy*, **17**, 79-86.

Dybendal, T., Hetland, T., Vik, H., Apold, J., *et al.* (1989a) Dust from carpeted and smooth floors. Comparative measurements of antigenic and allergenic proteins in dust vacuumed from carpeted and non-carpeted classrooms in Norwegian schools. *Clin. Exp. Allergy*, **19**, 217-224.

Dybendal, T., Vik, H., and Elsayed, S. (1989b) Dust from carpeted and smooth floors. II. Antigenic and allergenic content of dust vacuumed from carpeted and smooth floors in schools under routine cleaning schedules. *Allergy*, **44**, 401-411.

Dybendal, T., Wedberg, W.C., and Elsayed, S. (1991) Dust from carpeted and smooth floors. IV. Solid material, proteins and allergens collected in the different filter stages of vacuum cleaners after ten days of use in schools. *Allergy*, **46**, 427-435.

El Ghazaly, G., Nakamura, S., Takahashi, Y., *et al.* (1996) Localization of the major allergen Bet v I in *Betula* pollen using monoclonal labeling. *Grana*, **35**, 369-374.

Emilson, A., Berggren, B., Svensson, A., *et al.* (1996) Localization of the major allergen Bet v I in birch pollen by confocal laser scanning microscopy. *Grana*, **35**,199-204.

Enomoto, T., Onishi, S., Sogo, H., *et al.* (2004) Japanese cedar pollen in floating indoor house dust after a pollinating season. *Allergol. Intern.*, **53**, 279-285.

Fahlbusch, B., Hornung, D., Heinrich, J., *et al.* (2000) Quantification of group 5 grass pollen allergens in house dust. *Clin. Exp. Allergy*, **30**, 1645-1652.

Fahlbusch, B., Hornung, D., Heinrich J., and Jäger L. (2001) Predictors of group 5 grass-pollen allergens in settled house dust: comparison between pollination and nonpollination seasons. *Allergy*, **56**, 1081-1086.

Fountain, D.W., Berggren, B., Nilsson, S., and Einarsson, R. (1992) Expression of birch pollen-specific IgE-binding activity in seeds and other parts of birch trees (*Betula verrucosa* Ehrh.). *Int. Arch. Allergy Immunol.*, **98**, 370-376.

Gil, M., Hogendijk, S., and Hauser, C. (2002) Allergy to eggplant pollen. *Allergy*, **57**, 652.

Groenewoud, G.C., de Jong, N.W., Burdorf, A., *et al.* (2002a) Prevalence of occupational allergy to *Chrysanthemum* pollen in greenhouses in the Netherlands. *Allergy*, **57**, 835-840.

Groenewoud, G.C., de Jong, N.W., van Oorschot-van Nes, A.J., *et al.* (2002b) Prevalence of occupational allergy to bell pepper pollen in greenhouses in the Netherlands. *Clin. Exp. Allergy*, **32**, 434-440.

Grote, M., Dolecek, C., Van Ree, R., and Valenta, R. (1994) Immunogold electron microscopic localization of timothy grass (*Phleum pratense*) pollen major allergens Phl p I and Phl p V after an hydrous fixation in acrolein vapor. *J. Histochem. Cytochem.*, **42**, 427-431.

Grote, M. (1999) In situ localization of pollen allergens by immunogold electron microscopy: allergens at unexpected sites. *Int. Arch. Allergy Immunol.*, **118**, 1-6.

Hirst, J.M. (1952) An automatic volumetric spore trap. *Ann. Appl. Biol.*, **39**, 257-265.

Holmquist, L., and Vesterberg, O. (1999) Quantification of birch and grass pollen allergens in indoor air. *Indoor Air*, **9**, 85–91.

Holmquist, L., and Vesterberg O. (2001) Airborne birch and grass pollen allergens in driving compartments of coaches. *Arbete och Hälsa*, **2**, 1–13.

Holmquist, L., Weiner, J., and Vesterberg, O. (2001) Airborne birch and grass pollen allergens in street-level shops. *Indoor Air*, **11**, 241-245.

Holopainen, R., Asikainen, V., Pasanen, P., and Seppänen, O. (2002) The field comparison of three measuring techniques for evaluation of the surface dust level in ventilation ducts. *Indoor Air*, **12**, 47–54.

Hugg, T., and Rantio-Lehtimäki, A. (2007) Indoor and outdoor pollen concentrations in private and public spaces during the *Betula* pollen season. *Aerobiologia*, **23**, 119–129.

Hugg, T., Valtonen, A., and Rantio-Lehtimäki, A. (2007) Pollen concentrations inside private cars during the Poaceae and *Artemisia* pollen season – a case study. *Grana*, **46**, 110–117.

Jarolim, E., Rumpold, H., Endler, A., *et al.* (1989) IgE and IgG antibodies of patients with allergy to birch pollen as tools to define the allergen profile of *Betula verrucosa*. *Allergy*, **44**, 385-395.

de Jong, N.W., Vermeulen, A.M., Gerth van Wijk, R., and de Groot, H. (1998) Occupational allergy caused by flowers. *Allergy*, **53**, 204-209.

Jurewicz, J., Kouimintzis, D., Burdorf, A., *et al.* (2007) Occupational risk factors for work-related disorders in greenhouse workers. *J. Pub. Health*, **15**, 265-277.

Kesavan, J., Bottiger, J.R., and McFarland, A.R. (2008) Bioaerosol concentrator performance: comparative tests with viable and with solid and liquid nonviable particles. *J. Appl. Microbiol.*, **104**, 285-295.

Kiyosawa, H., and Yoshizawa, S. (2001) Intrusion of airborne pollen into indoor environment and exposure dose. Study on the control of indoor pollen exposure. Part 1. *J. Arch. Plan. Environ. Eng.*, **548**, 63-68. [In Japanese]

Kiyosawa, H., and Yoshizawa, S. (2002) Brought-in pollen into indoor environment by residents' activities. Study on the control of indoor pollen exposure. Part 2. *J. Arch. Plan. Environ.*

Eng., **558**, 37-42. [In Japanese]

Lee, T., Grinshpun, S.A., Martuzevicius, D., *et al.* (2006a) Relationship between indoor and outdoor bio-aerosols collected with a button inhalable aerosol sampler in urban homes. *Indoor Air,* **16**, 37–47.

Lee, T., Grinshpun, S.A., Kim, K.Y., *et al.* (2006b) Relationship between indoor and outdoor airborne fungal spores, pollen, and (1→3)-β-D-glucan in homes without visible mold growth. *Aerobiologia,* **22**, 227–236.

Monsó, E., Magarolas, R., Badorrey, I., *et al.* (2002) Occupational asthma in greenhouse flower and ornamental plant growers. *Am. J. Respir. Crit. Care Med.,* **165**, 954-960.

Morrow Brown, H. (1991) Mobile slit sampler for pollen and spore sampling on motorways. *Aerobiologia,* **7**, 69–72.

Muilenberg, M.L., Skellenger, W.S., Burge, H.A., and Solomon, W.R. (1991) Particle penetration into the automotive interior. I. Influence of vehicle speed and ventilatory mode. *J. Allergy Clin. Immunol.,* **87**, 581–585.

Ohashi, E., Yoshida, S., Ooka, R., and Miyazawa, H. (2005) Indoor concentration of airborne cedar pollen and allergen (Cry j 1); indoor air-pollution by Japanese cedar pollen. *J. Environ. Eng.,* **594**, 39-43. [In Japanese; abstract in English.]

Ong, E.K., Knox, R.B., and Singh, M.B. (1996) Molecular characterization and environmental monitoring of grass pollen allergens. In S.S. Mohapatra and R.B Knox, (eds.), *Pollen Biotechnology: Gene expression and allergen characterization,* Chapman & Hall, New York, pp. 176-210.

O'Rourke, M.K., and Lebowitz, M.D. (1984) A comparison of regional atmospheric pollen with pollen collected at and near homes. *Grana,* **23**, 55-64.

O'Rourke, M.K., Quackenboss, J.J., and Lebowitz, M.D. (1989) An epidemiological approach investigating respiratory disease response in sensitive individuals to indoor and outdoor pollen exposure in Tucson, Arizona. *Aerobiologia,* **5**, 104–110.

Perkins, W.A. (1957) The rotorod sampler. Second Semiannual Report of the Aerosol Laboratory, Department of Chemistry and Chemical Engineering, Stanford University, CML 186, 66 pp.

Piirilä, P., Kanerva, L., Estlander, T., *et al.* (1999) Occupational IgE-mediated asthma, rhinoconjunctivitis, and contact urticaria caused by Easter lily (*Lilium longiflorum*) and tulip. *Allergy,* **54**, 273-277.

Rantio-Lehtimäki, A., Viander, M., and Koivikko, A. (1994) Airborne birch pollen antigens in different particle sizes. *Clin. Exp. Allergy,* **24**, 23-28.

Solomon, W.R., Burge, H.A., and Boise, J.R. (1980) Exclusion of particulate allergens by window air conditioners. *J. Allergy Clin. Immunol.,* 65, 305–308.

Stock, T.H., and Morandi, M.T. (1988) A characterization of indoor and outdoor microenvironmental concentrations of pollen and spores in two Houston neighbourhoods. *Environ. Int.,* **14**, 1–9.

Sterling, D.A., and Lewis, R.D. (1998) Pollen and fungal spores indoor and outdoor of mobile homes. *Ann. Allergy Asthma Immunol.,* **80**, 279–285.

Swoboda, I., Dang, T.C.H., Heberle-Bors, E., and Vicente, O. (1995) Expression of Bet v 1, the major birch pollen allergen, during anther development. An *in situ* hybridization study. *Protoplasma,* **187**, 103-110.

Takahashi, Y., Nagoya, T., Watanabe, M., *et al.* (1993) A new method of counting airborne Japanese cedar (*Cryptomeria japonica*) pollen allergens by immunoblotting. *Allergy,* **48**, 94-98.

Takahashi, Y., Takano, K., Suzuki, M., *et al.* (2008) Two routes for pollen entering indoors: ventilation and clothes. *J. Investig. Allergol. Clin. Immunol.,* **18**, 382-388.

van Toorenbergen, A.W., Waanders, J., Gerth van Wijk, R., and Vermeulen, A.M. (2000) Immunoblot analysis of IgE-binding antigens in paprika and tomato pollen. *Intern. Arch. Allergy Immunol.,* **122**, 246-250.

Tormo Molina, R., Gonzalo Garijo, M.A., Muñoz Rodriguez, A.F., and Silva Palacios, I. (2002) Pollen and spores in the air of a hospital out-patient ward. *Allergol. Immunopathol. (Madrid),* 30, 232-238.

Uter, W., Nöhle, M., Randerath, B., and Schwanitz, H.J. (2001) Occupational contact urticaria and late-phase asthma caused by Compositae pollen in a florist. *Am. J. Contact Dermat.,* **12**, 182-184.

Vandenplas, O., Sohy, C., D'Alpaos, V., *et al.* (2008) Tomato-induced occupational asthma in a greenhouse worker. *J. Allergy Clin. Immunol.,* **122**, 1229-1231.

Watanabe, N., Masuda, H., Sagara, H., and Fukuda, K. (2000a) A case of occupational asthma induced by tomato pollen. *Allergy. Pract.,* **264**, 660-665.

Watanabe, N., Ota, M., Asakura, T., *et al.* (2000b) A case of atopic cough induced by strawberry pollen. *Allergy. Pract.,* **266**, 819-825.

Yankova, R. (1991) Outdoor and indoor pollen grains in Sofia. *Grana,* **30**, 171-176.

Yli-Panula, E., and Rantio-Lehtimäki, A. (1994) Antigenic activity in settled dust outdoors: a study on birch pollen antigen. *Grana,* **33**, 177-180.

Yli-Panula, E., and Rantio-Lehtimäki, A. (1995) Birch pollen antigenic activity of settled dust in rural and urban homes. *Allergy,* **50**, 303-307.

Yli-Panula, E. (1997) Grass-pollen allergenic activity of settled dust in rural and urban homes in Finland. *Grana,* **36**, 306-310.

Yli-Panula, E., and Ahlholm, J. (1998) Prolonged antigenic activity of birch and grass pollen in experimental conditions. *Grana,* **37**, 180-184.

Yli-Panula, E., Järvinen, K., Tovey, E.R., and Rantio-Lehtimäki, A. (2008) Birch pollen allergenic activity and personal allergen exposure in Finnish home environment. *Fourth European Symposium on Aerobiology, Turku, Finland,* Abstracts, p. 52.

Chapter 2. Microorganisms in homes and work environments

Chapter 2.1

Microbial growth in indoor environments

Brian Flannigan[1] and J. David Miller[2]

[1]Scottish Centre for Pollen Studies, Napier University, Edinburgh, UK;
[2]Institute of BioChemistry, Carleton University, Ottawa, Canada.

MICROBIAL NUTRITION

In Chapter 1.2, it was noted that algae differed in their nutrition from the bacteria, fungi and protozoans in indoor environments. Algae are autotrophic; they obtain the carbon that is required for the biosynthesis of the organic molecules that make up their structure solely from carbon dioxide. This they are able to do because they are photosynthetic, utilizing light energy for the reduction and incorporation of the carbon dioxide. Although most of the algae in indoor air are of outdoor origin, they are able to grow indoors on perpetually damp surfaces exposed to light, e.g. on wet window reveals or on the soil of potted plants.

In contrast, bacteria, fungi and protozoans are heterotrophic; they cannot use carbon dioxide as their sole carbon source, but utilize preformed organic molecules as both sources of carbon and energy. The source of these organic molecules is normally other organisms. The Protozoa have two types of heterotrophic nutrition. As indicated in Chapter 1.2, they can engulf and digest organic particles, including bacteria – holozoic nutrition. However, they can also take up nutrients such as amino acids and sugars through their surface – saprozoic nutrition. The two types of nutrition are not mutually exclusive and operate side by side in aquatic environments rich in both particulate and dissolved organic matter, e.g. in heavily contaminated drainage pans and humidifier reservoirs of heating, ventilation and air-conditioning (HVAC) systems.

The bacteria and fungi in buildings are usually saprotrophs; their nutrients are obtained from dead organic matter. Laboratory studies have shown that bacteria such as *Pseudomonas cepacia* can utilize any of more than a hundred different organic compounds as sole carbon source. Saprotrophs can take up any free organic molecules, such as simple sugars or amino acids, which may be present in the environment. Most of the organic molecules available to microorganisms are, however, bound up in polymers that have been synthesized by other organisms. For example, plant polysaccharides such as cellulose, hemicellulose, pectin and starch are constructed from simple sugars, but the sugars only become available when cleaved from the polymers by the hydrolytic action of microbial extracellular enzymes. Amino acids are similarly released from proteins in dead plants, animals and microbes. Because of the particular importance of cellulose in constructional and finishing materials, the breakdown of this polymer by cellulolytic microorganisms will be examined in more detail later in the chapter.

In addition to being able to lead a saprotrophic existence, some bacteria and fungi that grow in indoor environments can also behave as parasites, i.e. they can obtain their nutrients from living organisms. One such facultative parasite is the mould *Aspergillus fumigatus*. It can grow as a saprotroph under a wide range of circumstances, but is particularly notable in indoor environments where there are organic materials that are or have been at temperatures well above ambient, e.g. mushroom compost, or self-heated grain and hay causing farmer's lung. In its role as a parasite it behaves as an opportunistic pathogen (see Chapter 3.3); it does not normally invade the healthy human body, but can be invasive in debilitated or immunosuppressed individuals.

WATER RELATIONS OF MICROORGANISMS

Moisture in materials

Irrespective of the nutrient status of any material, whether a microorganism grows on and in that material or not depends on the availability of water in the material. Each and every microorganism has its own particular moisture requirements. Its growth depends on enough "free" water being available. This water does not include the bound water or water of constitution held tightly in chemical union with the material, but consists of the adsorbed water that is less strongly held by molecular attraction and the absorbed water loosely held in spaces by the weak forces of capillary action.

In older texts, e.g. Block (1953), water availability was expressed in terms of relative humidity (RH), but most microbiologists consider that it is wrong to describe water availability in a material using a term that in the strict sense applies to the moisture status of air. Instead, most biologists now use "water activity" (a_w) for the amount of free or available water in a substrate. This is defined as the ratio of the vapour pressure exerted by water in the (hygroscopic or porous) material to the vapour pressure of pure water at the same temperature and pressure. In effect, it is equivalent to 1/100th of the RH of the air in equilibrium with the material, i.e. a material in equilibrium with an atmosphere of 95% RH has an a_w of 0.95. The a_w can be determined indirectly by using a dew point meter to measure the RH of the atmosphere within a sealed container in which a sample of the material has been allowed to equilibrate. The RH of the atmosphere in this equilibrated system is of course the "equilibrium relative humidity" (ERH), and this term is occasionally used in microbiological publications rather than a_w, e.g. by Rowan *et al.* (1999).

ERH is a term better appreciated by engineers, most of whom use not a_w but "equilibrium moisture" or "equilibrium moisture content" (EMC) to describe the moisture status of a material. Alternatively, the engineer may use "fibre saturation point" or maximum sorption moisture content (Richards *et al.* 1992), EMC or merely MC to describe the moisture-holding properties of materials. As Richards *et al.* (1992) have noted, fibre saturation point (or maximum sorption point) is a useful indicator of the relative amount of water in material. However, because it is difficult to measure directly, this point is usually determined either by extrapolation to 100% RH or at an arbitrarily defined lower RH, say 98%.

The problem with using MC to describe moisture status is the confusion that arises because different materials take up different amounts of water from the ambient atmosphere. Materials with the same MC may have quite different a_w or ERH values (Flannigan 1993). It follows that materials with the same a_w or ERH may have entirely different water contents. An example illustrating this was quoted by an International Society for Indoor Air Quality task force on moisture problems in buildings (Flannigan *et al.* 1996): at ERH 80% (a_w 0.80) the MC of softwood is around 17%, wallpaper 11.3%, cement render 1%, brick 0.1-0.9%, and gypsum plaster 0.7%. Provided there has been sufficient soiling of the last three with organic nutrients, at a_w 0.80 all of these materials are susceptible to growth of some moulds, despite the large differences in MC between them. In a building context, this disparity in MC is very well illustrated by Oxley and Gobert (1983), who described a hypothetical wall with wooden skirting board affixed to gypsum plaster-finished brick, with the three components being in moisture equilibrium. Under normal dry conditions, the brick might have a MC of about 1%, the plaster 0.5% and the skirting 12%. If, however, the wall became wet due to rain penetration and the components equilibrated the corresponding MC's would be 2-5 (brick), 1-3 (plaster) and 22% (skirting).

Clearly there would be no confusion if there were clear documentation available to show the relationship between MC and ERH or a_w for every material present in buildings, but information here lags well behind that available to the food industry concerning microbiologically safe storage of food commodities. There have, however, been useful studies carried out by various workers, e.g. Richards *et al.* (1992) and Burch and Desjarlais (1994), who have constructed sorption curves that show the relationship between RH and EMC for a range of materials, including two types of softwood, several types of board, sheathing, insulation, kraft paper and vinyl covered gypsum board. More recently, Kuishan *et al.* (2009) have produced sorption curves for wall construction materials (concrete, mortar and clay brick) and insulation products used in exterior walls (expanded polystyrene, extruded polystyrene and polyurethane boards) in Chinese buildings. Bearing in mind that a_w is expressed as 1/100th of the ERH and knowing the minimum a_w for particular types of fungi, it is theoretically possible to predict from any such curve which fungi might colonize the material that had absorbed water and reached a particular MC. However, for many materials curves showing the relationship between MC and

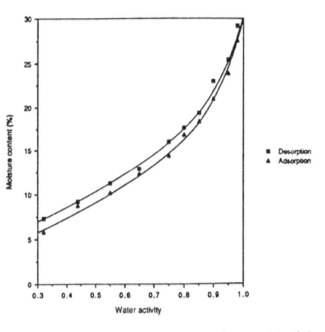

Fig. 1 Relationship between moisture content and water activity of plywood faced with Douglas fir under adsorption and desorption conditions, based on data provided by L. Geraldine Lea, Building Research Establishment, Garston, Watford, UK (after Flannigan 1993).

sponding a_w values (abscissa). The a_w values at any point on the sorption curves correspond to the RH in the sealed vessels in which the plywood was incubated. As applies to all material, there is a difference (hysteresis) between this adsorption curve, where the dry material has taken up water from the ambient atmosphere, and the corresponding desorption curve, where the same material fully imbibed with water has come to equilibrium by losing water to the ambient atmosphere. Consequently, a porous or hygroscopic material, say paper or wood, taking up water from a humid atmosphere and reaching a particular MC will have a slightly higher a_w than the same material drying down to that MC after flooding. This has implications as far as the susceptibility to microbial invasion is concerned. Depending on the precise MC, some organisms that are just able to grow on a material which has arrived at that MC by absorbing water from the atmosphere may not be able to grow on that material when it has reached that MC by drying down from a wetter state.

MOISTURE REQUIREMENTS OF MICROORGANISMS

Each type of microorganism has a minimum and an optimum a_w for growth. Although the a_w minima for different species may differ by 0.30 or more, the differences between the optima are much smaller, with most optima being in the range 0.90-0.99 (Ayerst 1969). For example, the optimum a_w for a great many species of bacteria is c. 0.99, although for some staphylococci it may be around 0.93. Scott (1957) reported that the minimum a_w for *Staphylococcus aureus* was 0.86, but many other bacteria, such as members of the genera *Escherichia* and *Pseudomonas*, fail to grow at a_w <0.95. Although the optimum for growth of filamentous fungi is usually appreciably greater than a_w 0.90, the minimum is normally much lower than for the majority of bacteria. There is a range of fungi that are found in indoor environments that are able to grow at a_w 0.70-0.80 (Ayerst 1969), and some even at a_w <0.65. Although the minimum for most yeasts is a_w 0.88-0.92, the minimum for *Zygosaccharomyces rouxii* is c. 0.65. Fig. 2 shows the overall effect of a_w on the growth of a number of microorganisms, ranging from bacteria in the genus *Salmonella* to the food-borne mould, *Xeromyces bisporus*, which is able to grow at an a_w as low as 0.61.

RH are not readily available. Where they are available, such as those determined in the studies mentioned earlier in this paragraph, they require to be treated with caution since they are usually only based on values for MC arrived at by uptake of water from the ruling atmosphere. This has been emphasised by Miller (2004).

Naturally, the relationship between a_w (or ERH) and MC follows the same pattern as for RH and EMC (see Richards *et al.* 1992). As an example, Fig. 1 displays the relationship between a_w and MC of a plywood sample over a microbiologically relevant range. The measured EMC data (ordinate) are plotted against corre-

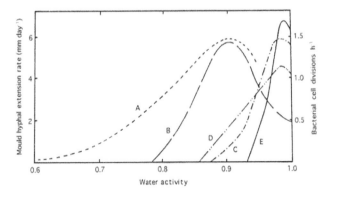

Fig. 2. Relationship between water activity and growth rate of moulds and bacteria: A, *Xeromyces bisporus*; B, *Eurotium herbariorum* (*A. glaucus*); C, *Aspergillus niger*; D, *Staphylococcus aureus*; E, *Salmonella* sp. (after Scott 1957).

Since filamentous fungi are generally more of a problem in buildings than other microorganisms, species for which minimum a_w values for growth have

Table 1. Minimum aw for growth at 25°C of filamentous fungi isolated from building interiors, after Northolt *et al.* (1995) with additional data from authors cited below.

Extremely xerophilic (minimum a_w <0.75)	*Aspergillus penicillioides*	0.73-0.77
	A. restrictus	0.71-0.75
	A. wentii	0.73-0.75
	Eurotium amstelodami	0.71-0.76
	E. chevalieri	0.71-0.73
	E. echinulatum	0.64
	E. repens	0.72-0.74
	E. rubrum	0.70-0.71
	Wallemia sebi	0.69-0.75
Moderately xerophilic (minimum a_w 0.75-0.79)	*Aspergillus candidus*	0.75-0.78
	A. flavus	0.78-0.80
	A. ochraceus	0.76-0.83
	A. sydowii	0.78; 0.81
	A. tamarii	0.78
	A. terreus	0.78
	A. versicolor	0.78; 0.74[e]; 0.75[d]; 0.79[b]
	Exophiala werneckii	0.77-0.78
	Paecilomyces variotii	0.79-0.84
	Penicillium aurantiogriseum	0.79-0.85
	P. brevicompactum	0.78-0.82
	P. chrysogenum	0.78-0.81; 0.85[d]
Slightly xerophilic (minimum a_w 0.80-0.89)	*Absidia corymbifera*	0.88-0.89[d]
	Alternaria alternata	0.85-0.88
	A. clavatus	0.85; 0.88 [d]
	A. fumigatus	0.85-0.94
	Aureobasidium pullulans	0.87-0.89[b]
	Chrysonilia sitophila	0.88-0.90[a]
	Cladosporium cladosporioides	0.86-0.88; 0.83-0.84[b]
	C. herbarum	0.85-0.88
	C. sphaerospermum	0.83-0.84[b]
	Epicoccum nigrum	0.86-0.90
	Fusarium culmorum	0.87-0.91
	F. graminearum	0.89
	F. moniliforme	0.87 *[a]; 0.89-0.91[b]
	F. solani	0.87-0.90
	Mucor plumbeus	0.87-0.93[b]
	Penicillium citrinum	0.80-0.82
	P. commune	0.83
	P. expansum	0.82-0.85
	P. fellutanum	0.80[e]
	P. oxalicum	0.88
	P. rugulosum	0.85; 0.80[d]
	P. viridicatum	0.81[c]

	Ulocladium chartarum	0.89[b]
Hydrophilic (minimum a_w ≥0.90)	Botrytis cinerea	0.93-0.95
	Geomyces pannorum	0.92; 0.89[b]
	Mucor racemosus	0.94; 0.92[d]
	Neosartorya fischeri	0.925
	Rhizopus stolonifer	0.93
	Sistotrema brinkmannii	0.96-0.97[b]
	Stachybotrys chartarum	0.94; 0.91-0.93[b]
	Verticillium lecanii	0.9

*Spore germination, but growth not observed; [a] Armolik and Dickson (1956); [b] Grant *et al.* (1989); [c] Mislivec and Tuite (1970); [d] Panasenko (1967); [e] Snow (1949).

been published and which are encountered in buildings are listed in Table 1. They have been grouped according to a widely accepted scheme of categorization (Lacey *et al.* 1980). Those which have minima greater than a_w 0.90 are classified as **hydrophilic** (Gr., "water-loving") as they are only suited to growing under conditions of high water availability. Those classified as mildly, moderately and extremely **xerophilic** (Gr., "dry-loving") are in that order increasingly tolerant of, and able to grow in, conditions of reduced a_w. At opposite ends of the range of fungi in Table 1 are the extreme xerophile *Eurotium echinulatum* with a minimum a_w of 0.64 and the hydrophilic basidiomycete microfungus *Sistotrema brinkmannii* with its minimum of a_w 0.96-0.97.

In a building, of course, the distribution of moisture within any material is seldom, if ever, uniform. Consequently, different areas of a structural element such as a wall may well select for the growth of different microorganisms. Grant *et al.* (1989) cited the case of a dwelling where owing to a structural fault there was direct ingress of water to the interior surface of a boundary wall. Around the fault, where the wall was extremely wet, the hydrophilic mould *Stachybotrys chartarum* grew profusely, but it was replaced nearer the drier margins of the affected area by more xerophilic species, viz. *Aspergillus versicolor* and various penicillia. Grant *et al.* (1989) also recorded that the mycobiota of the papered surface of a kitchen wall changed as it became increasingly wet with condensation as winter progressed. The first colonists of the paper, when it was still comparatively dry, were *Penicillium* spp. and *A. versicolor*, but with continued condensation they were gradually overtaken by other species, culminating in hydrophiles such as *Ulocladium* spp. and *S. chartarum* when the wallpaper was very wet.

Based on field observations and laboratory studies of a_w minima, Grant *et al.* (1989) classified moulds suc-

cessively colonizing materials as primary, secondary and tertiary colonizers (Table 2). Naturally, if a material is rapidly and massively wetted, so that its a_w is well above 0.90, it is open to immediate colonization by all types of mould, from xerophiles to hydrophiles. Under such immediately favourable conditions, so-called tertiary colonizers can act as primary invaders of the material.

Carlisle and Watkinson (1994) quote a lower limit for growth of most major basidiomycetous wood-rotting fungi as being about a_w 0.97, i.e. they are hydrophilic, but data on decay are usually quoted in terms of MC of the wood in question. A common cause of 'wet rot' of timber is the basidiomycete *Coniophora puteana*; it is believed to account for around 90% of wet rot in buildings in UK. The optimum MC for decay of wood by this species is 50-60% and the minimum in spruce wood is 24% (Eaton and Hale 1993). Although rare in North America, another widespread wood-rotting basidiomycete is *Serpula lacrymans*, known as the dry rot fungus. The optimum MC of wood for decay by this species is variously quoted as somewhere between 30% and 80% and the minimum at 20% (Schmidt and Moreth-Kerbernik 1990). Although *S. lacrymans* does not grow in extremely wet wood, i.e. above about 90% MC, the 'dry' epithet is derived from the fact that the decayed wood breaks up into dry cubes which can easily be reduced to a powder. Although fibre saturation may be needed for initiation of *S. lacrymans* growth, once it is established in wood the organism itself can generate enough moisture metabolically that it does not require an external supply of water. It also has a capacity for translocating water from wetter areas into previously dry wood, causing localized wetting and so extending its area of growth.

It is generally held that only where wood can be guaranteed to remain at 18% (Eaton and Hale 1993) or 20% MC (Oxley and Gobert 1983) is it safe from de-

Table 2. Moisture levels for colonization of construction, finishing or furnishing materials by selected microorganisms (Grant *et al.* 1989; Flannigan 1992; Nevalainen 1993).

Moisture level in material	Category of microorganism
High (a$_w$ >0.90, ERH >90%)	TERTIARY COLONIZERS *Mucor plumbeus* *Alternaria alternata* *Stachybotrys atra* *Ulocladium consortiale* Yeasts, e.g. *Rhodotorula, Sporobolomyces* Actinobacteria
Intermediate (a$_w$ 0.80-0.90, ERH 80-90%)	SECONDARY COLONIZERS *Cladosporium cladosporioides* *C. sphaerospermum* *Aspergillus flavus* *A. versicolor**
Low (a$_w$ <0.80, ERH <80%)	PRIMARY COLONIZERS *A. versicolor†* *Eurotium* spp. *Penicillium brevicompactum* *P. chrysogenum* *Wallemia sebi*

*At 12°C; † at 25°C.

cay by wood-rotting basidiomycetes. Connolly (1993) has recommended adoption of a rather higher maximum safe level, 30% MC (fibre saturation), for prevention of fungal attack and decay, however.

Just as a$_w$ affects the growth rate of moulds (usually measured as the rate of linear extension of hyphae), it also influences germination of mould spores. At any given temperature, a reduction in a$_w$ depresses the rate of germination, firstly by increasing the latent period, i.e. the time before germ tubes initially appear (see Chapter 1.2), and secondly by slowing the rate of elongation of the germ tubes once they have appeared (Scott 1957). The limiting a$_w$ for germination of spores is usually lower than for linear growth of hyphae at optimum temperatures. For example, Magan and Lacey (1984) noted that spore germination of *Alternaria alternata* and other field (phylloplane) fungi occurred at a$_w$ >0.85, but for linear growth of hyphae in colonies of these fungi a$_w$ >0.88 was required.

TEMPERATURE RELATIONS OF MICROORGANISMS

Like a$_w$, environmental temperature also has a great bearing on microbial growth. Microorganisms can be allocated to different classes depending on the temperature range over which they grow.

Psychrophiles are defined as growing well at 0°C, with an optimum temperature for growth at 15°C or

less, and a maximum in the region of 20°C. The bacterium *Micrococcus cryophilus* is a good example of a psychrophilic microorganism. Its cardinal temperatures are -4°C (minimum), 10°C (optimum) and 24°C (maximum).

Thermophiles are defined as being able to grow at temperatures of 50-55°C or higher. Bacteria classed as being thermophilic usually have minima around 45°C, but the thermophilic label is applied to fungi when they cannot grow below 20°C. Hence, the bacterium *Bacillus stearothermophilus* (cardinal temperatures 30°, 60-65° and 75°C) and the mucoraceous mould *Rhizomucor pusillus* (21-23°, 45-50° and 50-58°C) are both in the thermophilic category. In buildings, thermophilic *Bacillus* spp. and actinobacteria such as *Saccharopolyspora* (*Faenia*) *rectivirgula*, *Thermoactinomyces candidus* and *T. vulgaris* may be isolated (and dispersed) from some types of heating system, and from the air during handling of self-heated agricultural materials (Chapters 1.2 and 2.1).

Mesophiles are microorganisms that are neither psychrophilic nor thermophilic. By far the majority of bacteria and fungi in buildings fall into this category. Some of these mesophiles can grow at 0°C, but their optima are higher than for psychrophiles (in the range 20-30°C) and their maxima are around 35°C. The heterotrophic microorganisms that fall into this category are therefore described as **psychrotolerant mesophiles** (although the noun is often dropped from the term). Both psychrotolerant and psychrophilic microorganisms are able to grow in refrigerated environments. Some such bacteria and fungi, the latter including some species of *Cladosporium* and *Penicillium*, are major causes of spoilage in refrigerated foods. Some mesophiles are able to grow at temperatures of 50-55°C, but are excluded from the thermophilic category because their minima are <20°C. They are classified as **thermotolerant mesophiles**. The opportunistic fungal pathogen *Aspergillus fumigatus* (10-12°, 37-40° and 53-57°C) is one such thermotolerant species.

INTERACTIVE EFFECTS OF WATER ACTIVITY AND TEMPERATURE ON MICROORGANISMS

The minimum a_w for growth of the species listed in Table 1 were mostly derived from experiments in which the fungi were grown on agar media at 25°C, the optimum temperature for a large number of mesophilic moulds. In buildings, however, the materials on which growth occurs are almost invariably less nutritionally rich than such agar media. Also, at least in cold temperate climates, the temperature of substrates on which moulds grow, such as hardwall gypsum of outer walls, may be well below the optima for the moulds in question. It has been shown (Grant *et al.* 1989) that the a_w minima for growth of moulds on wallpaper are higher than on agar media. However, experiments in which additional nutrients were provided by coating the paper with a cellulose ether, carboxymethyl cellulose (CMC), to simulate soiling with wallpaper paste, reduced the differential. Painting with mould-susceptible, water-based emulsion paint in which cellulose ethers are the thickening agents also promoted growth at a lower a_w than on unamended wallpaper (Grant *et al.* 1989). Earlier experiments showed that temperature has an important bearing on the minimum a_w for growth. In common with other investigators, Ayerst (1969) observed that the greatest tolerance to low a_w in all the fungi that he studied occurred at temperatures close to the optimum temperature at optimum a_w. For most of these fungi, the optimum a_w for growth was lowest at temperatures close to the maximum for the species concerned. Magan and Lacey (1984) noted that, for a selection of fungi encompassing both hydrophilic and xerophilic mould species, the more divergent the temperature from the optimum the less tolerant of lower a_w were these moulds. Table 3 illustrates this reduced tolerance in a range of fungi with tempera-

Table 3. Effect of temperature on minimum water activity at which selected moulds isolated from air and walls in houses grow on emulsion-painted woodchip paper (after Grant *et al.* 1989).

Species	12°C	18°C
Aspergillus versicolor	0.87	0.79
Penicillium brevicompactum	0.87	0.83
Penicillium chrysogenum	0.87	0.85
Cladosporium sphaerospermum	0.93	0.92
Ulocladium consortiale	0.94	0.92
Stachybotrys chartarum	0.98	0.97

ture optima around 25°C but growing on wallpaper at temperatures that may be encountered at the internal surfaces of perimeter walls during winter in cold temperate regions. Nutrient availability and temperature are seldom likely to be optimal in buildings, so that if heterotrophic moulds and bacteria are to grow there they will almost invariably need damper conditions than indicated by their minimum a_w as determined in the laboratory on agar medium.

Bearing in mind these cautions, detailed observations of spore germination and hyphal growth on malt extract agar (Ayerst 1969) can be used to indicate the limiting conditions for a range of mesophilic mould species which can be found growing in buildings. Figs. 3-7 have been reprinted from the original paper of Ayerst (1969) with permission from Elsevier Science. Each individual diagram depicts (a) a series of isopleths, or lines connecting points of equal value, for rates of linear extension of hyphae up to 10 mm day⁻¹, and (b) a series of points indicating the time taken for spore germination to occur.

The effect of reduced a_w in increasing the time required for germination has already been mentioned, but is illustrated very clearly in the diagrams. For example, in *Eurotium chevalieri* strains tested at 25°C (Fig. 3) germination at a_w 0.95 took at most two days, but at a_w 0.85 the time required was 2-4 days and at a_w 0.75 8-16 days. The rates of linear extension of hyphae in Figs. 3-7 show that growth was generally most rapid under the same conditions as the delay in germination was shortest (Ayerst 1969). Also, linear growth was slowest where the preceding delay in germination was greatest.

The relatively wide range of a_w and temperature under which *Eurotium* spp. can grow (Figs. 3 and 4) is clearly seen. For *E. chevalieri*, *E. intermedium* (Fig. 3) and *E. amstelodami* (Fig. 4) the minimum a_w for growth was 0.71 and the lower and upper temperature limits were 10°C and 42°C. For *E. herbariorum*

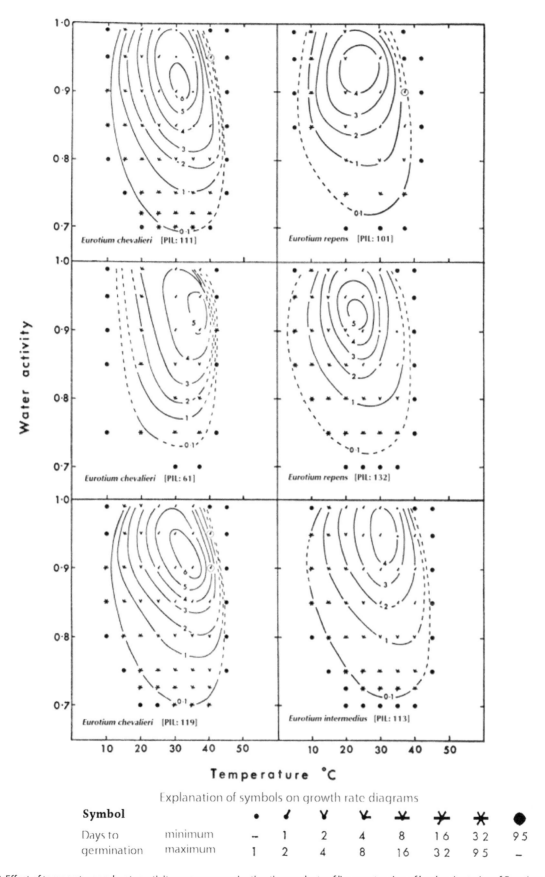

Fig. 3. Effect of temperature and water activity on spore germination time and rate of linear extension of hyphae in strains of *Eurotium chevalieri, E. herbariorum* and *E. intermedium*. Reprinted from Ayerst (1969), with permission from Elsevier Science. Key to symbols used in Figs. 3-7 to indicate germination times. If circled, germination occurred but any further growth did not continue at a steady rate.

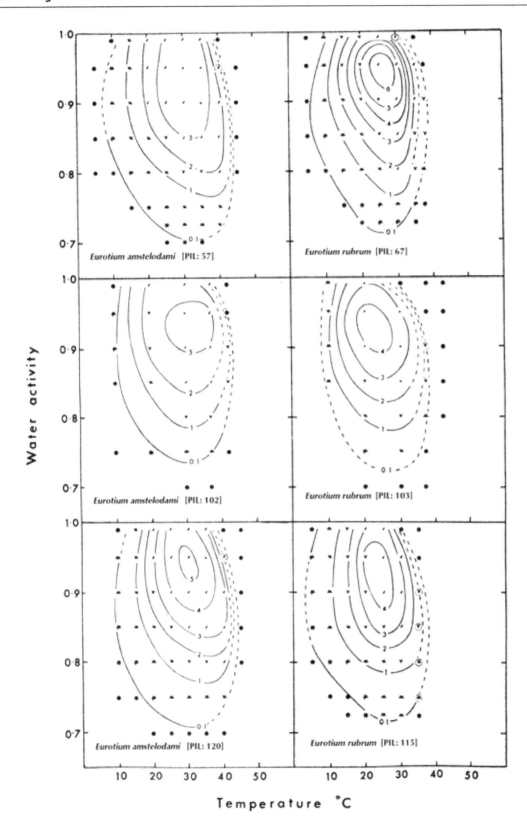

Fig. 4. Effect of temperature and water activity on spore germination time and rate of linear extension of hyphae in strains of *Eurotium amstelodami* and *E. rubrum*. Reprinted from Ayerst (1969), with permission from Elsevier Science.

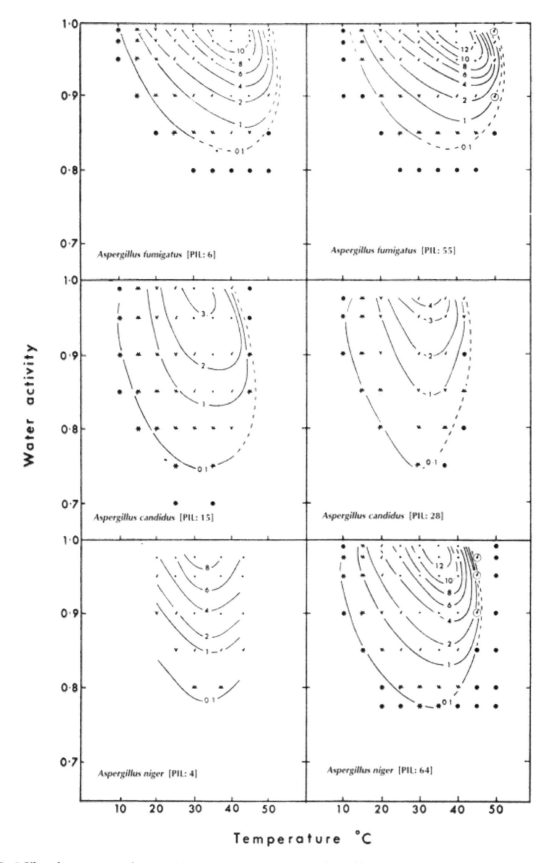

Fig. 5. Effect of temperature and water activity on spore germination time and rate of linear extension of hyphae in strains of *Aspergillus fumigatus*, A. *candidus* and A. *niger*. Reprinted from Ayerst (1969), with permission from Elsevier Science.

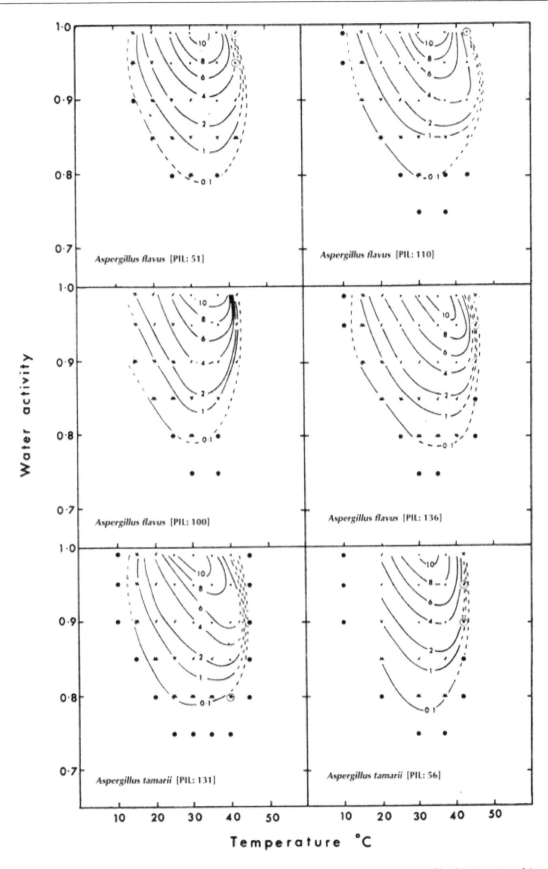

Fig. 6. Effect of temperature and water activity on spore germination time and rate of linear extension of hyphae in strains of *Aspergillus flavus* and *A. tamarii*. Reprinted from Ayerst (1969), with permission from Elsevier Science.

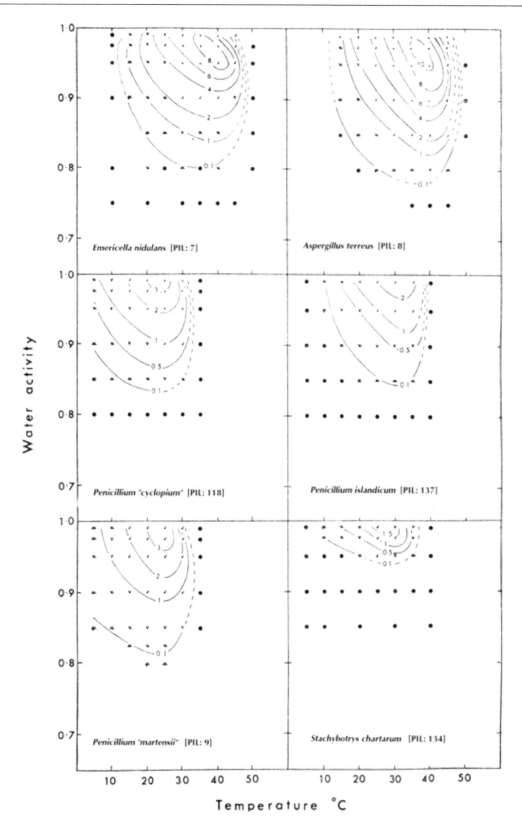

Fig. 7. Effect of temperature and water activity on spore germination time and rate of linear extension of hyphae in strains of *Emericella nidulans, Penicillium "cyclopium"* and *"martensii"* (= *P*. cf. *aurantiogriseum*), *Aspergillus terreus, P. islandicum* and *Stachybotrys chartarum*. Reprinted from Ayerst (1969), with permission from *Elsevier* Science.

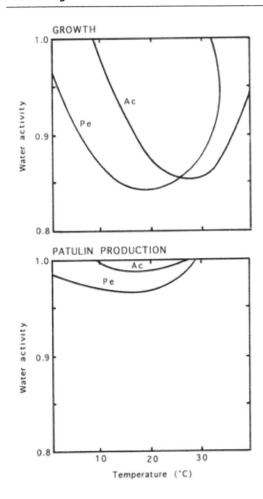

Fig. 8. Temperature and water activity (aw) limits for hyphal growth and patulin production in *Aspergillus clavatus* (Ac) and *Penicillium expansum* (Pe) (after Northolt 1979).

(Fig. 3) and *E. rubrum* (Fig. 4), the limiting a_w was also 0.71, but the temperature limits were 7°C and 38°C. Standing in sharp contrast to the diagrams for these xerophiles are those showing the somewhat more restricted conditions under which certain species grew, e.g. *Penicillium islandicum* and *Stachybotrys chartarum* (Fig. 7). The different temperature relations of psychrotolerant and thermotolerant mesophiles are well illustrated by comparing Fig. 7 for *Penicillium aurantiogriseum* (cardinal temperatures <5°C, 23°C and 32°C) with Fig. 5 for *Aspergillus fumigatus* (12°C, 40°C and 53°C).

The net effect of moisture availability and temperature on growth is, of course, the outcome of their effects on the individual energy release and synthetic processes of primary metabolism in microorganisms. However, it is not only growth-related processes that are subject to their influence. Also affected is the production of compounds that are not essential for growth but may serve some other role, i.e. secondary metabolites such as mycotoxins. When Northolt

(1979) examined the effect of a_w and temperature on the growth of a range of toxigenic moulds and the toxins that they produced, the limiting conditions in some cases appeared to be near to those for mycelial growth. For example, the limits for production of aflatoxin B_1 by *Aspergillus flavus* (and the related species *A. parasiticus*) were relatively close to those found by Ayerst (1969) for growth of *A. flavus* (Fig. 6). In contrast to this, however, the a_w and temperature range within which penicillic acid and patulin accumulated was much more restricted than for growth of the producers (Fig. 8). The restriction of penicillic acid production by *Penicillium aurantiogriseum*, and of patulin by *A. clavatus* (Chapter 3.2) and *P. expansum* (Fig. 8), to conditions of extremely high a_w was particularly marked.

SOURCES OF WATER FOR MICROBIAL GROWTH IN BUILDINGS

Condensation and microbial growth

Condensation on walls

Before discussing the conditions under which growth occurs in buildings, the reader should be cautioned that what we deduce from the presence of different types of microorganism is based very largely on steady-state experiments under different moisture regimes, rather than from well-designed experiments in which there are temperature and RH fluctuations like those that actually occur in buildings. One of the few investigations that has revealed the effects of high:low temperature and RH cycling on fungal growth is presented in the thesis of Adan (1994), who examined the growth of *Penicillium chrysogenum* on gypsum and drew attention to the "time of wetness" for this growth.

The first sign of a condensation problem in a naturally ventilated building is often the appearance of mould growth on the inside surface towards the top of external walls, particularly in winter. This condensation occurs because the warm moist room air coming into contact with the surface is cooled to (or below) the dew point temperature, i.e. the temperature at which it is saturated with water vapour and dew forms. Psychometric charts will show that the dew point of air at 20°C and 70% RH is 14.5°C. For air at the same temperature but with a RH of 60% the dew point is about 12°C. Under these circumstances, condensation will make possible growth of a wide range of moulds, ranging from xerophiles to hydrophiles, and also yeasts and Gram-negative bacteria. However, mould growth can also occur where there is no condensation. Even if the temperature of a surface is only 5° lower than room air at 20°C and 60% RH, the RH at that surface will be 80%. Any porous finish to a wall, such as wallpaper, will equilibrate to a$_w$ 0.80 (ERH 80%), a level at which xerophilic moulds could grow (Table 1), although not hydrophilic species.

In practice, in cold-temperate climates there are many poorly insulated buildings where the temperature of at least some of the inner surface of exterior walls is close to dew point for much of the winter. For example, during winter in Britain mean weekly temperatures of 9-10°C and RH's of 85-91% were recorded at a kitchen wall surface, with those for the air adjacent to the wall being 11.0-11.2°C and 77-83% RH. The surface RH was >70% for 85-95% of the time, and 100% for 15-25% of it; penicillia, *Aspergillus versicolor*, *Cladosporium*, *Ulocladium*, *Stachybotrys* and other moulds were isolated from the wallpaper (Hunter 1989). Although mould growth may be widespread, it can also be limited to scattered patches. This is usually because there is thermal bridging between inner and outer layers of the building envelope at these sites, and as a consequence there are cold spots and condensation or high a$_w$. Structural members such as metal studs or ties, which have a high thermal conductivity, can act as thermal bridges. Irrespective of the cause of condensation, particular areas that favour mould growth because ventilation is locally limited and the air is stagnant include corners of external walls; areas behind large items of furniture placed up to external walls; and cupboards or closets built onto external walls. Another site is of course at the junction of the glass and wooden frame of single-glazed windows, where water accumulates and *Aureobasidium pullulans* and *Sistotrema brinkmannii* are notable. The former has also been found to predominate at the junction in triple-glazed windows, owing to the metal beading forming a thermal bridge to the cold exterior (Käpylä 1985).

Condensation in basements, cellars and under-floor crawl spaces

Condensation may be a problem in both occupied basements and unoccupied cellars. In basements, throughout the year the temperature of the surrounding ground is significantly lower than that indoors. In humid summers, the RH may frequently be high enough for mould to appear on paper and wooden surfaces, and condensation may be found in the envelope and collecting as puddles on the floor (Rowley and Walkinshaw 1995). The temperature

difference between cellars and the occupied space above is one that accounts for much of the frequency of mould growth in cellars. As Kelly (1988) observed, even where the occupied space in a house is at 21°C and the RH of the air is only 50%, the cellar temperature is commonly as low as the dew point, i.e. about 11°C. Another problem area is the crawl space under suspended floors, where evaporation from walls below the damp-proof course and from the soil can produce a very high RH. As a result of taking up water from highly humid sub-floor air, joists and floorboards may have a MC that is anything from 18 to 30%. The higher levels occur where the floor is covered by a waterproof material like linoleum, vinyl or tiles, and at these levels decay by basidiomycetes is common (Oxley and Gobert 1983). Without a waterproof covering, water evaporates from the floor into the rooms above, where it can contribute to condensation problems. Christian (1993) indicated that 40-72 litres of water day^{-1} can enter the occupied space from the underfloor space by this route, but where there is a waterproof solum or ground cover it is likely to be only about 10% of this.

Condensation in the roof space

Another area in which condensation may be a problem is the roof space. As Oxley and Gobert (1983) have pointed out, in naturally ventilated houses which have been draught proofed, or sealed, and decorative vinyl wall coverings applied, the roof space may be virtually the only area to which water vapour can escape. Water vapour will pass up though the ceilings and thermal insulation and, especially where ventilation is impeded, condense on the roof lining whenever its temperature is at or below dew point. Where the lining is impermeable, the condensate may drip onto the rafters, thermal insulation, uninsulated gypsum board ceilings or stored materials, which may consequently be colonized by moulds. Where the lining absorbs the condensation it may also become mouldy. For example, the first named author has examined a house where the lower surface of the gypsum board lining the roof bore profuse growth of *Acremonium* and *Stachybotrys*.

Interstitial condensation

What is known as interstitial condensation can occur within the thickness of a porous wall through which water vapour can pass when there is a temperature gradient between the inner and outer sides. Water vapour in warm air passing out from the occupied space through the fabric of the wall condenses when the dew point is reached. If this is near the outer surface it is not usually a problem, because it evaporates at times when there is no condensation. In cavity walls, the condensation occurs in the space between the inner and outer leaves of the wall, and again is not normally a problem if the cavity is ventilated. However, if the cavity is filled by insulation, dew point may be reached somewhere within the insulation and mould growth may ensue (Oxley and Gobert 1983). Although not necessary for most masonry or brick walls, a vapour retarder applied to the inner surface of the wall reduces the risk of interstitial condensation. However, for wood frame and other building envelope structures correctly sited vapour retarders are essential. For a more detailed discussion of this see Oxley and Gobert (1983), TenWolde (1993) and Lstiburek (1994).

Origin of water vapour for condensation

Major sources of water vapour which condenses in homes are bathing, clothes washing/drying and cooking (Oxley and Gobert 1983, Christian 1993). In combination, they can more than double the amount that enters the air as a result of the respiration and perspiration of the occupants, but the total will vary widely. Obviously, high occupancy load can be a factor contributing to condensation problems (Merrill and Ten-Wolde 1989). Among other factors which will boost the total are gas cookers or heaters which discharge their combustion gases inside, rather than through a flue; kerosene heaters, which generate virtually as much water as the amount of fuel that they consume; storing green wood (which itself may support mould and be a source of airborne spores) indoors for use as fuel; and drying of clothes on a rack, rather than in a drier venting to the outside. As Flannigan *et al.* (1996) have noted, estimates of the total amount of water vapour generated daily in the home of a family of four range from 7.5 to 23.1 litres.

A further point to be borne in mind is the time for which a building is heated. A problem that has emerged in recent years in UK is that many homes are now unoccupied and unheated for much of the day. The house may be heated for only a short period in the morning, during which moisture is generated by showering and cooking, before the occupants leave for work and school. The house then cools, with resultant condensation, before the occupants return in the evening and it is heated again and more moisture is generated prior to overnight cooling. This mode of occupancy has led to an increasing number of condensation and mould problems in the British housing stock. Similar problems may arise in public and other larger buildings where heating is only periodic. In this context, an interesting observation was made by Wälchli and Raschle (1983). They noted that, over a 25-year period, in Swiss churches 75% of cases of dry rot (*Serpula lacrymans*) occurred in Protestant churches and only 25% in Roman Catholic churches. This was attributed to the fact that public services of worship were held throughout the week in the latter, but only on Sunday in the former. That is, Protestant churches were heated for less time, and condensation was greater.

It should also be borne in mind that in newly built or renovated buildings considerable amounts of water evaporate from green timber, concrete, gypsum plaster and types of insulation that are applied wet. It has been estimated that in a North American house the average volume of water released into the indoor air during the first year is 10 litres day^{-1} (Christian 1993), an amount which may equal or exceed the total amount from all other internal sources. However, this input of construction water is halved in the second year and is thereafter negligible. Another source to be mindful of is water that either accumulates in condensate pans of air conditioning systems or drains into the underfloor crawl space, instead of to the outside or into the waste water system. In hot humid summers, re-evaporation of this water can add considerably to problems, with the resultant water vapour migrating into the occupied space (Christian 1993).

Condensation is also a problem in tropical and subtropical climates, but there it occurs on internal surfaces of external walls because of infiltration of the building envelope by humid air from outdoors. In air-conditioned buildings, the cooled conditioning air can lower the temperature at the surface of gypsum board to near or below the dew point of the infiltrating air. Microbial growth may therefore occur. For example, in Singaporean buildings a wide range of fungi contribute to the major problem of biodeterioration and discoloration of painted walls, wallpaper, ceiling tiles and even glass panels (Lim *et al.* 1989). The abundance of the hydrophiles *Fusarium decemcellulare* on ceiling tiles and *F. solani* on wallpaper are witness to conditions of extreme moisture. Where vinyl or other low-permeability decorative covering on the gypsum board impedes inward transfer of moisture, there may be extensive growth of xerophilic and hydrophilic fungi, and Gram-negative bacteria, below the covering. Morey (1993b) has reported that in southern USA growth may be extreme enough to render some buildings unfit for occupation. Further discussion of building envelope problems in tropical and subtropical climates, and also in cold climates, can be found in Flannigan *et al.* (1996).

Standing water

Standing water in evaporative condensers, shower stalls, whirlpool baths and evaporative humidifiers are among aquatic environments listed as having been linked epidemiologically to Legionnaire's disease episodes, although quite clearly any body of static water, including water in inadequately maintained drain pans, drains, cooling towers and cold water humidifiers, is a potential environment for microorganisms. Dust particles and organic debris (including dead microorganisms) in the water will provide a supply of nutrients. Unless the water is severely oxygen depleted, it will support the growth of nearly all types of saprotrophic bacteria and microfungi that are likely to be found indoors. Gram-negative bacteria, yeasts and moulds can grow suspended in the water, in accumulated deposits at the bottom of the vessel and as biofilms on the surfaces. For example, Airoldt and Litsky (1972) reported that cold water humidifiers in use as an adjunct to respiratory therapy in a hospital were heavily colonized by *Flavobacterium* and other Gram-negative bacteria. Staphylococci, *Bacillus*, *Alternaria*, *Aspergillus*, *Mucor* and *Penicillium* were also isolated from the reservoir water, and Gram-negative bacteria were present in the mist emitted by the humidifiers. As indicated in Chapter 1.1, amoebae can ingest bacteria that are present in standing waters. In doing this they may become hosts in which *Legionella pneumophila* amplifies, particularly where the water temperature is greater than 20°C and favours its growth. Also, in Japan members of the medically important fungal genus *Exophiala* are found in bath water and in the sludge in bathroom drains, which are considered important habitats and perhaps significant sources of human infection by these organisms (Nishimura *et al.* 1987).

Other Sources of Water

Indoor sources

Apart from providing an environment for microbial growth, standing water adds by evaporation to existing water vapour, aggravating condensation and dampening the fabric of the building so that it is affected by mould. However, it is only if large amounts of standing water are discharged, e.g. as a result of breakage, that it directly causes readily visible water damage to the fabric of a building. Slow persistent leakages in plumbing which go undetected may have a similar effect, but it is sudden failures leading to widespread flooding of ceilings, floors and walls that are likely to have a catastrophic effect. Not only do they cause clearly visible mould growth on surfaces but they also are likely to lead to colonization of concealed elements such as insulation (Chapters 2.3 and 4.6). The report of Fergusson *et al.* (1984) illustrates the effect that a persistent leak in pipework may have. A fine jet of water spraying onto the underside of a wooden floor from a domestic central heating system led to profuse local growth of two penicillia on the floor boards and their linoleum cover and was responsible for hypersensitivity pneumonitis in the occupants. At the other end of the scale, Morey (1993a) recorded that three months after 10^6 litres of water had been used to extinguish a serious fire on the ninth floor of a 10-storey office building carpets, gypsum board, wood panelling, ceiling tiles, and plaster walls and ceilings in the lower floor of both the fire-damaged building and another building connected to it were still moist. Species of *Cladosporium*, *Penicillium* and *Aspergillus* were widespread on surfaces and airborne propagule concentrations ranged up to 1.3×10^6 m^{-3} air (above ceilings) seven months after the fire.

Outdoor sources

Water can enter the fabric of a building from the outside both above and below ground. Infiltration of the building envelope by water vapour in humid outdoor air has been mentioned as a source of moisture that may condense within. To this source of moisture can be added liquid water in the form of rain, melting snow and ice, and soil water.

The most obvious form of water input above ground is rain penetration. Rain striking and wetting the surface of porous construction materials – concrete, bricks and mortar – moves through the material by capillary action (wicking) and can reach the inside surface of that element of the building envelope. In the normal run of events, during dry weather the wall will lose water to the atmosphere by evaporation from the outer surface and, as a consequence, the water in the pores will move towards the surface because of the difference in vapour pressure. However, this drying process takes much more time than wetting. When a porous wall is exposed to extended periods of rain, the pores may entirely fill with water

Table 4. Common fungi in 1529 samples of urea formaldehyde foam insulation, expressed as percentage of samples contaminated (Miller *et al.* 1984).

Species	%
Penicillium fellutanum	49
Cladosporium cladosporioides	36
Alternaria alternata	35
Penicillium viridicatum	19
Ulocladium chartarum	10
Penicillium brevicompactum	9
Penicillium decumbens	7
Trichoderma viride	7
Paecilomyces variotii	6
Cladosporium herbarum	5
Penicillium glabrum	4
Penicillium janthinellum	4

(capillary saturation) and then the inner surface may be visibly wet. In a solid wall this will lead to mould growth, but a cavity wall is designed to prevent this happening. Any water reaching the inside of the outer layer evaporates into the ventilated space and thence to the outside. However, water will reach the inner layer if the cavity is bridged by mortar or other porous material that acts as a wick. Where the cavity is filled by insulating material, that material should neither allow movement of water across the cavity (Oxley and Gobert 1983), nor should it support microbial growth. As Table 4 indicates, and as will be discussed later, significant problems have occurred in the past with urea-formaldehyde foam insulation (UFFI).

Other factors which may be the cause of, or exacerbate, ingress of rainwater (or melt water from snow and ice) include defects in brick- or stonework, mortar and rendering; badly installed windows; defective sealant joints or caulks; inadequate flashing of walls and roofs; faulty design and maintenance of roofs; and faults in rainwater disposal - undersized, choked or leaking rain gutters and down pipes (Oxley and Gobert, 1983, Christian, 1993, Connolly 1993, Morey *et al.* 2003).

Below ground, improper damp-proofing of foundations and flooding are important causes of moisture problems (Oxley and Gobert 1983, Christian, 1993). Rising damp, the capillary rise of water up through walls from the foundations, occurs where a damp-proof course is faulty, bridged or lacking. In extreme cases water may rise to a height of 50 cm above a suspended floor, resulting in a band of mould growth extending from the skirting board upwards,

and occasionally the development at skirting level of fruiting bodies of "plaster fungi" such as *Peziza* (Ascomycetes).

Christian (1993) has listed sources of water that may penetrate the underground building fabric, including those caused by building and drainage defects. As well as entering by capillary action, water can be forced into foundations and basements under the hydrostatic pressure exerted by a high water table or flood and melt water. Whatever the source of water, flooded basements are subject to bacterial and fungal growth, particularly if the water is contaminated by backed-up sewage. In addition, just like crawl spaces, the basement can behave as a large humidifier (Christian 1993), supplying water vapour which can migrate upwards into the remainder of the building.

GROWTH OF MICROORGANISMS ON CONSTRUCTIONAL, FINISHING AND OTHER MATERIALS IN BUILDINGS

As discussed above, anything that becomes mouldy does so because it contains naturally, or has gained, enough moisture to permit growth and sporulation. Which fungus or fungi dominate on sufficiently moist substrates depends on both the a_w and composition of the building material. Just as, on a worldwide basis, there is surprisingly little variation in the fungi recovered from outdoor air, there appears to be little variation in the mycobiota of building materials in different parts of the world. This reflects (1) the widespread adoption of particular building materials (e.g. gypsum board), (2) the general reduction of air change rates in buildings and (3) the increase in conditioning the indoor environment to 20-25°C regardless of outdoor conditions. As more careful studies of the mycobiota are made there will be a greater understanding of the common fungi associated with mould-damaged building materials. The intention in this section is to illustrate the ecology of the built environment, first considering growth on cellulosic materials in general before turning to gypsum board and then other materials and situations in which growth may occur. Before dealing with growth on individual materials, however, an investigation by Morey *et al.* (2003) of a Californian student residence hall will serve to illustrate the effect of water penetration into a multi-component structure, an external wall. The wall comprised (from the outside inwards) stucco, metal laths, asphaltic building paper, fibrous glass batt insulation with a Kraft paper facing on the inner surface (next to the wall cavity), and a layer of paper-faced gypsum wallboard painted on the surface next to the occupied space. Prior to the investigation, the 300-room residence suffered considerable rainwater penetration during the 1997-1998 El Niño storms owing to cracks in stucco and failure of detailing round windows. In addition to this, water entering the wall failed to drain away. Destructive inspection of the wall (see Chapter 4.1) carried out in 1999 revealed that there was extensive mould growth on the inner surface of the asphaltic building paper; some of the insulation Kraft paper was also visibly mouldy; and, on average, 26% of the the wallboard surface facing into the cavity was almost entirely covered by mould, and where it abutted rusty metal studs it was also rust-stained. The moderately xerophilic species *Penicillium chrysogenum* was the predominant mould cultured from visibly colonized wallboard and building paper, and adhesive tape samples showed that *Stachybotrys chartarum* conidiophores were also abundant on these surfaces. The abundance of *P. chrysogenum* in the air of "leaky" rooms, and its minor presence in unaffected rooms and the outdoor air, indicated that its spores passed into the air of the occupied space from mould growth concealed within walls of the type described above.

Cellulosic Materials

Types of cellulosic material

The constructional and finishing materials in buildings that are most susceptible to microbial growth and biodeterioration are those materials containing natural organic polymers encountered and degraded by microorganisms in their role of recycling nutrients in their natural environments. These polymers include the starch, cellulose, hemicellulose, pectin and lignin in plants. A great many microbes are able to degrade starch, hemicelluloses and pectins; a narrower range can degrade cellulose; and lignin degradation is confined to an even more limited range, mainly wood-rotting basidiomycetes such as those mentioned earlier. Under damp conditions water-based adhesives such as starch and animal glues containing the proteins casein and gelatin are highly susceptible to attack by microbial deteriogens. The ability of wallpaper and cellulose ether adhesives to sustain growth of moulds has already been mentioned (in the previous section), and it is microbial growth on these and the wide range of other materials containing cellulose that gives the greatest cause for concern. The microbial biodeterioration of structural, finishing and furnishing materials

Table 5. Order of abundance of commonest taxa among 4500 isolates of cellulolytic and non-cellulolytic fungi from exposed cotton fabrics (after Siu 1951).

Rank	Name	% of total	Cellulolytic activity
1	*Trichoderma viride*	7.2	+
2	*Penicillium citrinum* (series)	5.1	+
3	*Botryodiplodia theobromae*	3.2	+/-
4	*Aureobasidium pullulans*	3	-
5	*Aspergillus niger*	2.7	-
6	*Aspergillus sydowii*	2	-
7	*Aspergillus flavus*	1.9	-
8	*Penicillium luteum* (*Talaromyces luteus*)	1.7	-
9	*Aspergillus versicolor*	1.7	-
10	*Aspergillus oryzae*	1.7	-
11	*Chaetomiun globosum*	1.4	+
12	*Fusarium moniliforme*	1.2	+
13	*Penicillium funiculosum* (series)	1.1	+/-
14	*Memnoniella echinata*	1.1	+
15	*Penicillium waksmanii*	1	-
16	*Pestalotia virgatula*	0.9	?
17	*Paecilomyces variotii*	0.9	-
18	*Aspergillus terreus*	0.8	+
19	*Fusarium semitectum* var. *majus*	0.6	+/-
20	*Curvularia lunata*	0.6	+

such as timber, wood composites, jute, cotton, cardboard and wallpaper in buildings may have not only serious economic consequences, but may also result in the amplification of species that are of significance for health. For example, in investigating the etiology of asthma in a number of patients Kozak *et al.* (1980) noted the importance of damp wallpaper, jute carpet backing and wicker and straw baskets as environments for the growth of the causative moulds in their homes. Ando *et al.* (1991) reported that damp wood and tatami mats were associated with the development of a summer-type hypersensitivity pneumonitis caused by *Trichosporon cutaneum*.

Cellulolytic microorganisms in air and on exposed cotton in outdoor environments

To get a good idea of the types of airborne microorganism, excepting major wood-rotting species, that may contaminate and cause damage to cellulosic materials in buildings one really has to go back some fifty years to the investigations of the US Army

Table 6. Relative degradative activity of cellulolytic fungi isolated from exposed cotton fabrics (after Siu 1951).

Very Strong	*Chaetomium funicola*	Definite	*Alternaria alternata*
	C. globosum		*Aspergillus oryzae* var. *effusus*
	C. indicum		*A. (Fennellia) flavipes*
	Myrothecium verrucaria		*A. rugulosus*
	Sordaria fimicola		*A. sulphureus*
Strong	*Aspergillus fumigatus*		*A. ustus*
	Chaetomium atrobrunneum		*Chrysonilia sitophila*
	C. cupreum		*Cladosporium herbarum*
	C. succineum		*Fusarium scirpi*
	C. brasiliense		*F. semitectum*
	Corynascus sepedonium		*Penicillium funiculosum*
	Curvularia lunata		*P. glabrum*
	Fusarium lateritium		*P. implicatum*
	Memnoniella echinata		*P. janczewskii*
	Myrothecium roridum	Weak	*Aspergillus japonicus*
	Stachybotrys chartarum		*Aspergillus awamori*
	Trichoderma koningii		*Aspergillus (Emericella) nidulans*
	T. viride		*Aspergillus ochraceus*
Moderate	*Aspergillus (Neosartorya) fischeri*		*Botryodiplodia theobromae*
	Aspergillus niveus		*Penicillium atramentosum*
	A. terreus		*Penicillium purpurogenum*
	Chaetomium aureum		*Penicillium miczynskii*
	C. cochliodes		*Penicillium waksmanii*
	C. murorum		*Scopulariopsis brevicaulis*
	C. spirale		*Trichothecium roseum*
	Cladosporium cladosporioides		
	Eupenicillium javanicum		
	Fusarium avenaceum		
	F. equiseti		
	F. moniliforme		
	F. oxysporum		
	F. solani		
	P. luteum (Talaromyces luteus)		
	P. sclerotiorum		
	P. simplicissimum		
	Ulocladium consortiale		

Quartermaster Corps during and after World War II (Siu 1951). The need for field and laboratory investigations rapidly became apparent because of the conflict in Southeast Asia and the Pacific, where it was found that the useful life of untreated tarpaulins, tents, ropes and sandbags in areas with high temperature and rainfall was very short. For example, in New Guinea (annual rainfall, 250 cm) army tents be-

gan to leak after 3-4 months and had to be replaced after 6-8 months, and in the former Dutch East Indies the average life of a sandbag was only 6 weeks. The Quartermaster Corps amassed a culture collection of some 5000 isolates from microorganisms present on deteriorated materials gathered from military bases in the South and Southwest Pacific, others collected on expeditions in Panama and Florida and cultures from earlier programmes. Among 4500 isolates of fungi from exposed cotton textiles (Siu 1951), the most abundant categories were the genera *Penicillium* (18.2% of total) and *Aspergillus* (14.6%). In rank order, the next most abundant were non-sporing isolates (11.2%), *Fusarium* (8.6%), *Trichoderma* (8.5%), *Pestalotia* (4.4%), *Botryodiplodia* (3.2%), *Aureobasidium* (3.2%), *Chaetomium* (2.9%), *Phoma* (2.2%), *Cladosporium* (1.9%), *Gliocladium* (1.6%), *Curvularia* (1.6%), *Memnoniella* (1.2%) and *Paecilomyces* (1.1%).

The most common species shown in tests to be cellulolytic was *Trichoderma viride* (Table 5), which accounted for 7.2% of all isolates. The only other species that individually comprised 2% or more of the total were cellulolytic *Penicillium citrinum* series isolates; *Botryodiplodia theobromae*, some isolates of which were cellulolytic and others not; and three non-cellulolytic species, *Aureobasidium pullulans*, *Aspergillus niger* and *A. sydowii*. As indicated by the results for *B. theobromae*, whatever the type of function or enzyme under study there is always variation in the degree of activity exhibited by different isolates of an individual species. Based on the collective results for the cellulolytic isolates in Table 5 and many other species not shown, Siu (1951) divided the cellulose-degraders on exposed cotton textiles into five groups. These are presented in decreasing order of activity in Table 6. Since isolations were made from cotton textiles exposed to air and dust or in contact with soil, not unsurprisingly, the mycobiota of exposed cotton was found to be similar to that of soil (Siu 1951).

In a later survey at the Tropical Research Unit in North Queensland, Australia, Upsher (1972) found strong similarities between the fungi on exposed cotton and those in soil. Most fungi in the 46 genera recorded were regarded as being soil-borne. Among the 46 genera, 30 were recovered from soil and vegetation and 31 from exposed cotton fabric, with 24 genera being common to vegetation and cotton fabric. However, only 17 genera were isolated from exposed timber and 13 from paints and plastics.

Breakdown of cellulose

Truly cellulolytic microorganisms have the ability to degrade native cellulose, in which the antiparallel long-chain molecules of the glucose polymer are aligned in micelles. These in turn are aligned in microfibrils that form the skeletal strengthening of the walls of plant cells ranging from thin-walled parenchyma in herbaceous plants to thick-walled fibres in woody plants. Briefly, cellulose molecules are composed of up to 10^4 D-anhydroglucose units and can

be described as a $(1{\rightarrow}4)$-β-D-glucan. Laterally contiguous cellulose molecules are linked by hydrogen bonding, giving highly ordered crystalline regions within microfibrils, but there are also paracrystalline and non-crystalline (amorphous) regions in which the degree of order is less (Eaton and Hale 1993). Complete breakdown of native cellulose by fungi is currently held to depend on their ability to produce three types of hydrolytic enzyme, viz. exoglucanase, endoglucanase and β-D-glucosidase, which act synergistically. Exoglucanase, or $(1{\rightarrow}4)$-β-D-glucan cellobiohydrolase (CBH), cleaves off disaccharide residues (cellobiose) stepwise from the non-reducing end of cellulose molecules in crystalline cellulose; endoglucanase (EG) or endo-$(1{\rightarrow}4)$-β-D-glucan 4-glucanohydrolase, randomly cleaves cellulo-oligosaccharides from amorphous cellulose; and β-D-glucosidase, or β-D-glucoside glucohydrolase, hydrolyses cellobiose, cellotriose and lower oligosaccharides to glucose. A $(1{\rightarrow}4)$-β-D-glucan 4-glucohydrolase may also be involved; this breaks down to glucose any soluble oligomers released during cellulose breakdown (Eaton and Hale 1993).

Non-cellulolytic microorganisms

As can be seen in Table 5, far from all of the isolates that can be made from deteriorating cotton are cellulolytic. *Aureobasidium pullulans* is a very good example of a fungus that is found on plant surfaces, in leaf litter and soil, and on canvas and timber, and yet appears unable to break down cellulose to any extent (Upsher 1972). Species of *Eurotium* and many aspergilli and penicillia are also widespread fungi that are not cellulolytic. Siu (1951) commented that, although *Stachybotrys* (*Memnoniella*) *echinata*, *S. chartarum* and *Chaetomium* spp. were the predominant and destructive fungi on tents and tarpaulins after long exposure in New Guinea, the earliest colonists on the inside of tents (and to some extent on the outside) were species of *Aspergillus*, *Penicillium* and *Rhizopus*. He considered that these latter fungi utilized detritus or finishing substances (waterproofing) as nutriment.

Another explanation for the prevalence of types that lack the full array of synergistic enzymes required to degrade native cellulose to sugars that can be metabolized is that they may, however, be able to degrade soluble or modified forms of cellulose. For example, CMC and other cellulose ethers made from wood cellulose and used as adhesives, as thickeners in paints, or for sizing, are susceptible to degradation by a wider range of fungi than those classified as cellulolytic (Flannigan 1970, Flannigan and Sellars

1972). Any of the species possessing endoglucanases (and glucosidases) which degrade CMC should be able to hydrolyse oligosaccharides released from fibres by fully cellulolytic fungi such as those in Table 6. Members of the genus *Mucor*, which ranked eighteenth in order of abundance of the genera isolated from exposed cotton textiles in the US Army studies (Siu 1951), have no activity against even the most suceptible forms of cellulose and have been described as "sugar fungi". After having utilized free sugars in fresh substrates and rapidly colonized them, for further growth they are dependent on sugars released from cellulose and hemicelluloses by the extracellular hydrolytic enzymes of other fungi present on the substrate. Their relative abundance and that of *Rhizopus* and other sugar fungi on exposed canvas may reflect the availability of sugars released by cellulolytic fungi that assimilate the products of degradation less rapidly than they degrade the cellulose substrate.

Major cellulolytic fungi

Because of their "very strong" or "strong" cellulolytic properties (Table 6), *Chaetomium globosum*, *Stachybotrys* (*Memnoniella*) *echinata*, *Myrothecium verrucaria* and *S. chartarum* have all been used in testing the decay resistance of mould-proofed canvas and other cellulosic materials. However, Siu (1951) noted that, despite its being the most powerful cellulose-destroying species in laboratory tests, *Myrothecium verrucaria* was practically never found growing or sporulating on cotton fabrics in the field. In contrast, *Stachybotrys* (*Memnoniella*) *echinata* was the dominant species on deteriorated canvas tents throughout the Southwest Pacific area, despite pure-culture tests showing that it was much more fastidious in its growth requirements than *S. chartarum*. Further laboratory tests indicated the reason for the dominance of *S. echinata* in the field; it competed well against other species. When cotton duck was inoculated with a mixed spore suspension of *S. chartarum* and *Chaetomium* sp., the two species grew in well-defined areas and did not invade areas already colonized by the other. However, when the suspension contained spores of *S. echinata* as well as *S. chartarum* and *Chaetomium* sp., *M. echinata* became dominant, excluding the other species. Other tests showed that sporulation in *S. echinata* was in fact stimulated by *Stachybotrys*, *Chaetomium* and another 11 fungi (both cellulolytic and non-cellulolytic). These may have provided an exogenous source of biotin, a vitamin found to be a requirement for growth of *S. echinata* (Siu 1951).

Growth on gypsum board

One of the structural components of modern buildings which frequently becomes damp and also provides an ample source of cellulose on which microorganisms can grow is gypsum board, also known as wallboard in North America and plasterboard in UK. It is manufactured by sandwiching a paste of gypsum (hydrated calcium sulphate) between cardboard layers that are held to the paste by starch glue and by gypsum crystals which penetrate the surface of the cardboard as the sheets dry. By the mid-1960's, this cheap material had almost entirely replaced traditional lath and plaster in wall construction. Although gypsum plaster is still applied directly to the inner "leaf" of cavity walls constructed from building blocks or brick, gypsum board is used as a wall component in much of the present built environment.

The cardboard and starch glue of gypsum board are of course highly biodegradable sources of carbon for moulds, and relatively little water is required to raise the a_w of the wrapper to a level which will sustain mould growth. This water can come from condensation or, as cardboard is hygroscopic, can be taken up from the atmosphere. Where there is gradual moistening of the wrapper, as condensation worsens over a winter, for example, a succession of moulds may appear. Given sufficient input of moisture, the succession may culminate in *Stachybotrys chartarum* or other hydrophilic fungi, as described by Grant *et al.* (1989) for wallpaper and discussed earlier in connection with moisture relations.

A representative range of fungi that grow on gypsum board can be gauged from Table 7, which presents records of fungal isolations from more than 2000 samples of mouldy wallboard in six buildings in North America. Four of the buildings in Table 7 were in mid-continental Pacific coastal areas and two in the Atlantic Northeast. Although the samples were taken in areas the width of the continent apart, no major geographical differences were evident in the species found, which are recognized as being cosmopolitan (Ellis 1976, Klich and Pitt 1988, Pitt 1988). However, Table 7 shows that there were differences in the prevalence of some fungi found in the buildings in the Maritime Pacific climatic zone. For example, *Penicillium citrinum* was relatively more prevalent in Building 1 than in the others, as were *P. glabrum* and *S. echinata* in Building 5. Pitt (1988) has suggested that *P. citrinum* may well be one of the most common eukaryotic life forms on Earth and noted that it is a powerful biodeteriogen and is found worldwide in

Table 7. Filamentous fungi isolated from 2134 samples of mouldy gypsum wallboard taken in six buildings in USA and Canada, with taxa in descending order of prevalence overall.

Building:	\#1[a]*	\#2[b]**	\#3[a]†	\#4[a]†	\#5[a]†	\#6[b]†
	Samples contaminated by listed taxa (%)					
Chaetomium globosum	55.2	27.5	8.9	66.7	12.9	7
Penicillium viridicatum	32.1	60.5	49.1	22.2	58.1	-
Non-sporulating isolates	41.7	9.5	14.3	94.4	35.5	51
Eurotium herbariorum	36.3	27.5	5.4	27.8	22.6	8.5
Penicillium aurantiogriseum	22.8	47.3	19.6	22.2	-	45.1
Penicillium citrinum	32.1	3.8	4.5	5.6	3.2	2.8
Stachybotrys chartarum	25.8	1.7	11.6	11.1	77.4	60.1
Aspergillus sydowii	12.7	2	0.9	44.4	3.2	-
Penicillium chrysogenum	6.6	12.9	4.5	2.8	12.9	10
Penicillium commune	0.7	37	-	2.8	35.5	4.2
Basidiomycetes (clamp connections)	3.1	15.2	25.9	19.4	-	-
Eurotium repens	7.4	0.9	-	8.3	6.5	-
Stachybotrys (Memnoniella) echinata	4.3	1.7	4.5	2.8	22.6	-
Aspergillus versicolor	4.2	-	0.9	30.6	16.1	25.4
Paecilomyces variotii	1.9	4.6	5.4	8.3	-	7
Cladosporium sphaerospermum	2.4	1.1	3.6	-	9.7	5.6
Penicillium oxalicum	2.5	0.3	-	19.4	-	-
Aspergillus niger	1.1	5.7	-	8.3	9.7	15.5
Penicillium corylophilum	2.2	1.4	2.7	2.8	3.2	11.3
Penicillium decumbens	0.7	4.3	5.4	-	-	2.8
Trichoderma harzianum	0.7	-	15.2	-	3.2	1.4
Penicillium fellutanum	0.7	0.6	10.7	-	-	4.2
Penicillium glabrum	0.4	2.6	-	-	19.4	1.4
Aspergillus ustus	0.8	-	-	-	22.6	12.7
Penicillium variable	0.7	-	8	-	-	10
Talaromyces flavus	0.3	1.4	-	25	-	-

[a] Located in mid-continental Pacific coastal area; [b] Located in Atlantic Northeast. * Single water event taking approx. 3 months to dry, exposed to rainwater during this period. ** Single water event taking approx. 3 months to dry . † Subjected to more or less chronic water infiltration for months to years.

soil, decaying vegetation and air. He also states that although most monographers consider that *P. glabrum* is also cosmopolitan, and it is widely distributed in soils, he believes that this biodeteriogen is much less common than *P. spinulosum*. As mentioned earlier, *S. echinata* is a strongly cellulolytic species which can occupy the same niche as *Stachybotrys chartarum*, and indeed supplant it on cellulosic substrates in tropical climates.

Other fungi that commonly grow on gypsum wall board in North America (e.g. *Acremonium strictum* and *Emericella nidulans*) and those frequently growing on this substratum (e.g. *Aureobasidium pullulans* and *Scopulariopsis brevicaulis*) are listed in Table 8a. This has been compiled from survey data presented in Prezant *et al.* (2008) and compares the range of

such fungi with those encountered on other cellulose containing materials, viz. solid wood, unspecified manufactured/composite wood products, ceiling tiles, and unspecified fabrics and textiles. Species growing infrequently (e.g. *Aspergillus fumigatus* and, among a number of penicillia, *Penicillium solitum*) or rarely (e.g. *Alternaria alternata* and *Hyalodendron lignicola*) on wallboard are listed in Table 8b. The occurrence of fungi on other substrates is presented in Tables 9a and 9b.

It is worth noting that among the fungi that grow commonly or frequently on gypsum wallboard (Table 8a), the ubiquitous species *Aspergillus versicolor* and *Penicillium chrysogenum* also do so on all the other substrata given in Tables 8a and 9a, including those that are definitely non-cellulosic (UFFI, ceramic tiles

Table 8a. Filamentous fungi commonly (C) or frequently (F) growing on samples of urea-formaldehyde foam insulation and cellulosic materials from North American indoor locations (compiled from data in Prezant et al. 2004).

Species	Insulation	Gypsum wallboard	Solid wood	Composite wood	Ceiling tiles	Fabrics and textiles
Acremonium strictum	C	C	F	C	C	C
Alternaria alternata				F	F	
A. restrictus						F
A. sydowii	F	C		C	F	C
A. ustus	F	C		F		
A. versicolor	C	C	C	C	C	C
Aureobasidium pullulans	C	F	F	C	F	C
Ceratocystis/Ophiostoma spp.			F			
Chaetomium globosum	F	C	C	F	F	C
C. indicum						F
Cladosporium herbarum	F			F		F
C. sphaerospermum	C	C	C	C	C	C
Emericella nidulans	F	C		C		F
Eurotium amstelodami	F			F		F
E. herbariorum	C	C		C	F	F
E. repens	F			F		
E. rubrum		F				
Geotrichum candidum				F		
Stachybotrys (Memnoniella) echinata	F	C				C
Oidiodendron griseum			F	F		
Paecilomyces variotii	C	C	C	C		F
Penicillium aurantiogriseum	C	C	F	F	F	
P. brevicompactum	F	F	C	C		C
P. chrysogenum	C	C	C	C	C	C
P. citrinum		F				F
P. commune	C	C	F	F	F	F
P. corylophilum		F				
P. decumbens	C	F	F	C	F	C
P. fellutanum	C	F	F			
P. glabrum		F			F	
P. janthinellum	F	F				
P. purpurogenum		F		F		
P. simplicissimum	F					F
P. solitum	F					
P. spinulosum	C	F	F		F	F
P. thomii			F			
P. variabile		F				
P. viridicatum	F					
Phoma glomerata			F			
Scopulariopsis brevicaulis		F		F		C
S. candida						F

Species						
Sistotrema brinkmannii			F			
Stachybotrys chartarum	F	C	C	C	C	C
Talaromyces flavus			F	F		
Trichoderma harzianum	C	C	C	C		F
T. koningii			F			
Ulocladium chartarum	F		C	C	C	C

Table 8b. Filamentous fungi infrequently (I) or rarely (R) growing on samples of urea-formaldehyde foam insulation and cellulosic materials from North American indoor locations (compiled from data in Prezant et al. 2004).

Species	Insulation material	Gypsum wallboard	Solid wood	Composite wood	Ceiling tiles	Fabrics and textiles
Acremoniella atra						R
Acremonium charticola		R			R	
A. murorum		I			R	
Alternaria alternata		R				
A. tenuissima		R				
Aphanocladium aranearum		R				
Arthrinium phaerospermum				R		
Arthrographis lignicola			I			
A. cuboidea			R			
Ascotricha chartarum		R	R			
A. erinacea		R	R			
Aspergillus caespitosus		R		R		
A. candidus				I	I	
A. flavipes		I				R
A. flavus	I	I	R	I		I
A. fumigatus	I	I	R	I	R	I
A. niger	I	I		I	R	I
A. niveus	I	R	R		R	
A. ochraceus	I	I		I		
A. penicillioides		R		R		
A. restrictus				R		
A. sydowii			I			
A. terreus		R		I		R
A. unguis	R	I	R			
A. ustus					I	
A. wentii		R				
Botryotrichum piluliferum		I				R
Botrytis cinerea			R			R
Cephaloascus fragrans			R			
Chaetomium cochlioides		R				
C. elatum		I	R	R	R	R
C. funicola		R		R		
C. indicum		I		I		
C. murorum		R				
C. thermophila				R		

Chrysonilia sitophila		R				R
Chrysosporium pannorum		I	R		I	R
Clonostachys rosea				R		
Cunninghamella elegans	R					
Curvularia inaequalis		R				
C. lunata						R
Doratomyces microsporus						R
Emericella nidulans					R	
Epicoccum nigrum			R			
Eurotium amstelodami		I	R			
E. chevalieri	I					I
E. herbariorum			R			
E. repens		I			I	
E. rubrum			R			
Exophiala jeanselmei		R	R[a]	R		
Geomyces pannorum		I		R		R
Geotrichum candidum	I		R		R	I
Gliocladium catenulatum			R			R
G. roseum						R
G. viride		I	I			
Gliomastix murorum		I				
Hyalodendron lignicola		R	R			
Stachybotrys (Memnoniella) echinata				I	R	
Monodictys castaneae			R			
M. levis			R			
Myrothecium olivaceum		R				
M. verrucaria						R
Oedocephalum glomerulosum		R				
O. cerealis		R			R	
O. griseum		I	R		R	
O. tenuissimum			R			
Paecilomyces liliacinus			R	R		
P. variotii					R	
Penicillium aurantiogriseum						I
P. canescens	R			R		
P. citreonigrum			R			
P. citrinum			R	I	R	
P. corylophilum	I		I	I	I	I
P. crustosum		I	I	I		I
P. expansum		I				
P. funiculosum		R	I			
P. glabrum	I		R	I		
P. griseofulvum	I	I	I		R	
P. implicatum			R		R	R
P. janthinellum				I	R	
P. mineoluteum		I		R		R

Species						
P. olsonii	I			I		
P. oxalicum		I	R	I		I
P. purpurogenum					I	R
P. roquefortii	I		I		I	
P. rugulosum		I		I		
P. simplicissimum		I	I			
P. solitum		I	I	I	I	I
P. spinulosum				I		
P. thomii	R	I				R
P. variabile			R	I	R	R
P. verrucosum						R
P. waksmanii				R		
Peziza domiciliana				R		R
Phialophora americana			R			
P. bubakii	R					
P. fastigiata		R	I	I		
P. melinii	I		I			
P. parasitica			R			
P. richardsiae		R	I			
P. verrucosa			I	I		
Phoma fimeti		R	R			
P. glomerata		R		I		
P. herbarum		I		I	R	
P. leveillei		R	R	R		
Pithomyces chartarum		R	R		I	
Rhinocladiella atrovirens			R			
Scolecobasidium constrictum		R				
Scopulariopsis brevicaulis			I		I	
S. brumptii		R	R			R
S. candida		I	R	I	R	
S. chartarum		R				
S. fusca		R				
Scytalidium lignicola			I	I		
Serpula lacrymans			R			
S. incrassata			R			
Sistotrema brinkmannii		R				
Sordaria fimicola						R
Sphaeropsis sapinea			R			
Sporothrix schenckii			R			
Syncephalastrum racemosum	R					I
Taeniolella rudis			R			
Talaromyces flavus					I	R
Thamnidium elegans				R		
Thysanophora canadensis			R			
T. penicillioides			R			
Torulomyces lagena			R			

Trichoderma hamatum	R	R				
T. harzianum					R	
T. koningii		I	I	I		I
T. viride		R	R			I
Trichothecium roseum		R				
Tritirachium oryzae	I			I	I	
Ulocladium botrytis			I			
Verticillium lecanii		R	I	I	I	I
Wallemia sebi						I

[a]*Exophiala jeanselmei* var. *heteromorpha*

and concrete efflorescences). *Acremonium strictum, Aureobasidium pullulans, Cladosporium sphaerospermum, P. commune* and *P. decumbens* are common or frequent on other substrates that contain or may contain cellulose, including unspecified air handling unit (AHU) filters, which can range from pleated paper and cellulosic-polyester to electrostatic types. *Eurotium herbariorum* has been noted on all of the substrata except solid wood, and *Stachybotrys chartarum* in all but ceramic tile and concrete efflorescence samples. *Chaetomium globosum* is common or frequent on wallboard and the other cellulosic materials (and UFFI) in Table 8a, but infrequent or rare on AHU filters, carpets and painted surfaces (Table 9b).

When Pearson correlations (Ludwig and Reynolds, 1988) were run comparing the six previously mentioned North American buildings and their species lists (Table 7), an association was noted between the mycobiota on the wallboard of two buildings in Maritime Pacific climatic areas, viz. Buildings 1 and 4 (r^2 = 0.648). As noted, Building 1 underwent a major flood, which was followed by rainwater penetration for several months afterwards, whereas Building 4 had a chronic water leak. The mycobiota of Buildings 2 and 3 were also associated (r^2 = 0.740). In this case, Building 2 had suffered a single water incursion, which was followed by slow drying without any further ingress. The history of Building 3 was quite different; over several years there had been chronic cycles of water incursion followed by periods of dryness.

Analysis of the more xerophilic species isolated (Table 7) showed that *Eurotium repens* was associated with *Aspergillus sydowii* (r^2 = 0.614) and *E. herbariorum* (r^2 = 0.749), apparently reflecting similarities in a_w and substrate preferences. Among less xerophilic species (Table 1), *Penicillium commune* was associated with *P. chrysogenum* (r^2 = 0.821) and *Paecilomyces variotii* (r^2 = 0.649). The less xerophilic species *P. chrysogenum, Paecilomyces variotii* and *P. aurantiogriseum* were negatively correlated with the more xerophilic *A. sy-*

dowii, A. versicolor and *E. repens* (r^2 = -0.666, -0.688 and -0.600 respectively). *A. sydowii* was not associated with *A. versicolor*, indicating that these closely related species, which are common on gypsum wallboard, have different a_w requirements for growth on this substrate. The published minimal a_w for growth of the former under laboratory conditions is 0.78-0.81, and the latter 0.74-0.79 (Table 1).

Among four species which were more hydrophilic and were frequently isolated from the wallboard in the six buildings, viz. *Chaetomium globosum, Cladosporium sphaerospermum, Stachybotrys (Memnoniella) echinata* and *Stachybotrys chartarum*, the first-mentioned was negatively correlated with *C. sphaerospermum* (r^2 = -0.603) and not associated with either *M. echinata* or *S. chartarum*. Although *C. sphaerospermum* was isolated less frequently than the other three species, it was strongly associated with both *S. chartarum* and *M. echinata* (r^2 = 0.919 and 0.778, respectively). *S. chartarum* was associated with the closely related species *M. echinata* (r^2 = 0.649). This suggests a clear ecological difference between the most common species on gypsum wallboard, *C. globosum*, and the other species. For both hydrophilic species and moderate xerophiles the statistical analyses divided up the clusters of organisms according to their known biology. The clear separation of *C. globosum* from the other hydrophilic species is perhaps surprising. It has, however, been suggested by Prezant *et al.* (2008) that the difference in prevalance of this species and *S. chartarum* might be due to its being more strongly cellulolytic than *S. chartarum* (Siu 1951).

While cluster analysis of standardized species prevalence data has presented evidence that firstly a_w and secondly nutritional differences are factors determining which fungi growth on wet gypsum board (Prezant *et al.* 2008), the concentration of calcium salts and some other salts may also enter into the equation. Prezant *et al.* (2008) noted that, although *Penicillium aurantiogriseum* is similar in its a_w requirements

Table 9a. Filamentous fungi commonly (C) or frequently (F) growing on AHU filter, carpet, painted surface, ceramic tile and concrete efflorescence samples from North American indoor locations (compiled from unpublished data of Miller et al.).

Species	AHU filters	Carpets	Painted surfaces	Ceramic tiles	Concrete florescences
Acremonium strictum	C	C	F		F
Aspergillus candidus		F			F
A. restrictus			F		
A. sydowii	F		F		
A. ustus		F			
A. versicolor	C	C	C	F	C
Aureobasidium pullulans	F	C	C		
Cladosporium herbarum			F		
C. sphaerospermum	C	C	C	C	
Emericella nidulans	F	F	F		
Eurotium amstelodami		F	F	F	
E. herbariorum	F	C	C	F	F
E. rubrum		C			
Paecilomyces variotii			F		
Penicillium aurantiogriseum	F	F			C
P. brevicompactum	F	C			
P. chrysogenum	C	C	C	C	F
P. commune	C	C	F		F
P. corylophilum	F	F	F		
P. decumbens	F	C	F	C	
P. fellutanum		F			
P. griseofulvum	F				
P. olsoni		F			
P. restrictum					F
P. rugulosum		F			
P. simplicissimum		F			
P. solitum	F				
P. spinulosum	F	F	F		
Scopulariopsis brevicaulis		F	F	F	
S. candida				F	
Stachybotrys chartarum	F	F	F		
Trichoderma harzianum	F	F			
Ulocladium chartarum	F	F	C		

Table 9b. Filamentous fungi infrequently (I) or rarely (R) growing on AHU filter, carpet, painted surface, ceramic tile and concrete efflorescence samples from North American indoor locations.

Species	AHU filters	Carpets	Painted surfaces	Ceramic tiles	Concrete efflorescence
Absidia corymbifera	R	R			
Acremoniella atra	R				
Acremonium charticola		R			
A. murorum			R		
Alternaria alternata			I		
Aspergillus candidus	I		I		
A. flavipes		R			
A. flavus	I	I			
A. fumigatus	I	I			
A. niger	I	I			
A. ochraceus	I	R		R	R
A. penicilliodes		I			I
A. restrictus		I			
A. sydowii		I		I	
A. terreus		R			
Botrytis cinerea		I or R			
Chaetomium elatum		R			
C. globosum	I	R	I		
Chrysosporium pannorum		R			
Cladosporium herbarum		I		I	I
C. sphaerospermum					I
Curvularia lunata		R	I		
Eurotium amstelodami	I				
E. repens		I	I		I
Gliocladium catenulatum		R			
G. roseum		R			
Hormoconis resinae			R		
Stachybotrys echinata		R			
Paecilomyces variotii		R			
Penicillium brevicompactum			I		
P. citrinum			R		
P. commune				I	
P. corylophilum				I	I
P. crustosum	I	I			
P. decumbens					I
P. fellutanum				I	R
P. glabrum		I			
P. implicatum		R	R		
P. janthinellum	I	I		R	
P. mineoluteum		R			
P. olsonii	I				

	Col1	Col2	Col3	Col4
P. oxalicum		I		
P. purpurogenum	I	R		
P. simplicissimum	I			
P. solitum		I		R
P. thomii		R	R	
P. variabile		R		
P. verrucosum		R		
P. viridicatum	I	R		R
Peziza domiciliana		R		
Phoma glomerata			I	
P. herbarum			I	
Pithomyces chartarum		R		
Scopulariopsis brumptii			R	
S. candida			I	
Syncephalastrum racemosum	R			
Talaromyces flavus	I	I	I	R
Trichoderma koningii		R		
T. viride	R	R		
Tritirachium oryzae		I		
Ulocladium chartarum				I
Verticillium lecanii		I		
Wallemia sebi	R	I	I	

to three other moulds found in the study of six buildings (Table 7), viz. *P. chrysogeonum, P. commune* and *Paecilomyces variotii,* it did not cluster with them. Like *Aspergillus sydowii, Chaetomium globosum, Eurotium herbarum* and *Stachybotrys chartarum, P. aurantiogriseum* was reported by Steiman *et al.* (1997) to tolerate high concentrations of calcium, but in addition it showed similar tolerance of magnesium salts. Based on this, Prezant *et al.* (2008) have suggested that *P. aurantiogriseum* stands separate from the cluster mentioned because of difference in tolerance of the salts present in wallboard.

Although there have been many studies on damp/wet cellulosics and other building materials, in many cases identification of the organisms present has not been taken to species level. This can be seen, for example, in the review of insulation material by Van Loo *et al.* (2004) and elsewhere in the text of this chapter. In Chapter 1.2 it was observed that while the quantitative data obtained using AVST's indicated the relative abundance of aspergilli and penicillia (taken together), it did not differentiate the species of *Aspergillus* and *Penicillium* that could be quite different in their physiology and ecology. In this light, the importance of precise identification is brought home by

the study of Fradkin *et al.* (1987), who recorded that some *Cladosporium* spp. were less abundant indoors during the summer, but the concentrations of others were double those outdoors. *C. cladosporioides* and *C. herbarum* are most common members of this genus in outdoor air, but *C. macrocarpum, C. oxysporum* and *C. sphaerospermum* are also present. However, the commonest species growing indoors on wet building materials is generally *C. sphaerospermum* (Tables 7, 8a and 9a).

When Gravesen *et al.* (1997) investigated water-damaged buildings in Denmark they found that gypsum board and other materials containing cellulose were the constructional components most vulnerable to attack by moulds. The genera of microfungi that they isolated most frequently from gypsum board and other materials were *Aspergillus* and *Penicillium* (Table 10). They were present in 56 and 68% of samples, respectively. About one-fifth of samples were contaminated by *Ulocladium* and the strongly cellulolytic genera *Chaetomium* and *Stachybotrys.* The species encountered most frequently were *S. chartarum, P. chrysogenum* and *A. versicolor.* In subsequent studies in Copenhagen, Nielsen *et al.* (1998a,b) obtained evidence that gypsum board could support

Table 10. Principal mould genera isolated from 72 samples of mould-affected building materials (Gravesen *et al.* 1997).

Genus	Samples (%)
Penicillium	68
Aspergillus	56
Chaetomium	22
Ulocladium	21
Stachybotrys	19
Cladosporium	15
Acremonium	14
Mucor	14
Paecilomyces	10
Alternaria	8
Verticillium	8
Trichoderma	7

mycotoxin production as well as growth of *S. chartarum*. Hydrolysed extracts of heavily *S. chartarum*-contaminated gypsum board from either side of a wall frame in a school and from a basement wall in a house contained verrucarol, a strong indicator of toxic macrocyclic trichothecenes, probably satratoxins (Nielsen *et al.* 1998a). When old and new gypsum board was artificially inoculated with this species, there was again evidence of toxin production by *S.*

chartarum (Nielsen *et al.* 1998b), but none of a range of trichothecenes was produced by eight strains of another cellulolytic genus, *Trichoderma*, which has also been found on gypsum board in the field (Tables 7, 8a and 10). Although not notable as a cellulolytic species, *A. versicolor* grew and produced large quantities of sterigmatocystin and 5-methoxysterigmatocystin on gypsum board, and also on wallpaper, pine wood and chipboard pieces (Nielsen *et al.* 1998b).

Also in Europe, the species composition of mould growths on gypsum board and gypsum-plastered solid walls finished with wallpaper or cellulose ether-based paints in damp houses was recorded by Hunter and Bravery (1989). They observed that *Cladosporium* was present in about 70% of mould patches (and a predominant component in more than four-fifths of these), in contrast to occurring in only 15% of samples of mould-affected materials examined by Gravesen *et al.* (1997). Hunter and Bravery (1989) recorded that *Ulocladium* spp. (74%) and *Penicillium* spp. (58%) were also frequently isolated (Table 11), but *Aspergillus* (mainly *A. versicolor*) and *Stachybotrys* were much less frequent. *Chaetomium* was apparently absent, but it is possible that its isolation would not have been favoured by the method employed, i.e. surface swabbing followed by plating of serial dilutions prepared from the swabs (Hunter *et al.* 1988).

Table 11. Principal fungi in mould growths on walls in dwellings, according to percentage frequency of isolation from growths (after Hunter and Bravery 1989).

Species	Predominant (%)	Present (%)
Acremonium strictum	8	11
Alternaria alternata	3	3
Aspergillus spp. (mainly *A. versicolor*)	8	18
Aureobasidium pullulans	8	16
Cladosporium spp. (*C. cladosporioides* and/or *C. sphaerospermum*)	61	71
Fusarium moniliforme	-	13
Geomyces pannorum	5	18
Geotrichum candidum	-	5
Mucor spp.	-	3
Non-sporing isolates	3	24
Paecilomyces variotii	-	5
Penicillium spp. (incl. *P. brevicompactum, P. chrysogenum, P. glabrum, P. janczweskii* and *P. viridicatum*)	39	58
Phoma herbarum	5	18
Scopulariopsis brevicaulis	-	3
Stachybotrys chartarum	5	13
Ulocladium spp. (*U. chartarum* and/or *U. consortiale*)	39	74
Yeasts	26	76

Although there is no definitive proof, it would appear likely that some of the bacteria noted in Table 8 of Chapter 1.2 as having been isolated from mould affected areas of damp walls in houses will be able to proliferate on damp gypsum board. As Flannigan *et al.* (1999) found that Gram-negative bacteria were associated with the wettest (*Acinetobacter* and *Agrobacterium*) or the intermediate areas (*Pseudomonas)* and Gram-positive micrococci with the least wet areas of active mould patches, it might reasonably be expected that they would show the same association on gypsum board. *Arthrobacter, Micrococcus* and *Mycobacterium, Bacillus pumilis, Paenibacillus macerans* and *P. polymyxa* were isolated from water-damaged areas of gypsum wallboard in Swedish day care centres for children by Andersson *et al.* (1997), but it is not known whether there was active proliferation of these mesophilic Gram-positive species. Andersson *et al.* (1997) also isolated the thermophilic Gram-positive species *B. amyloliquefaciens*, the Gram-negative mesophiles *Chryseomonas luteola* and species of *Agrobacterium, Caulobacter* and *Stenotrophomonas,* and the actinobacteria *Gordona* and *Streptomyces*. Hyvärinen *et al.* (2002) observed a higher-than-expected co-occurrence of actinobacteria with *Acremonium* (p<0.001) and unspecified aspergilli (*p* < 0.01).

Roponen *et al.* (2001) found that a strain of *Streptomyces anulatus* isolated from indoor air sporulated strongly on inoculated gypsum board, the spores inducing the most intense production of NO, TNFα and IL-6 in mouse RAW264.7 macrophages among the spores on a range of building materials. Using another species from indoor air, *S. californicus*, Murtoniemi *et al.* (2003b) found that growth and spore activity against RAW264.7 depended on the brand of gypsum board. Where the brands supported profuse growth the spore activity was low, and conversely where there was rather limited growth on other brands the spore activity against the macrophages was high. Murtoniemi *et al.* (2003a) also found that gypsum board composition had a bearing on growth and spore activity of the toxic mould *Stachybotrys chartarum*. In another experimental study by Murtoniemi *et al.* (2003c) it was found that the presence of starch, used as binder between the cardboard wrapper and the gypsum within, was the major factor enabling another species of *Streptomyces, S. californicus*, to grow and produce biologically active metabolites on gypsum board. Other *Streptomyces* strains that have been isolated from water-damaged buildings by Suutari *et al.* (2002) have shown >99.7% 16S rDNA sequence similarity to *S. griseus, S. albidoflavus* and *S. coelicolor*.

It is probable that these will also grow on gypsum wallboard.

Wood and other cellulosic materials

As was mentioned earlier, a sizeable number of species commonly or frequently found growing on gypsum wallboard in North America are also to be found on a range of other materials containing cellulose (Tables 8a and 9a). Taking these two categories of occurrence together with the corresponding infrequent and rare (Table 8b), the cellulosics on which the diversity of species can be found are gypsum board and solid wood, with almost one-half of the fungi being common to the two materials. The fungi on manufactured or composite wood products, which are not specified but might be taken to include products such as veneered blockboard, oriented strand board, chipboard, particle board or plywood (Lstiburek 2007), are in total about 80% of those on solid wood. Of the species on the manufactured composite wood, nearly one-half were also among those on solid wood.

Whilst *Sistotrema brinkmannii* has been found growing on solid wood it has not been recorded on composite wood (Table 8a). This resupinate homobasidiomycete is widespread on wood and plant debris outdoors and is associated particularly with decaying window joinery, although it has not been regarded as causing decay itself (Eaton and Hale 1993), and was first reported as part of the indoor airborne mycobiota by Hunter *et al.* (1988), who found it in the air of one-half of damp houses that they examined. Another basidiomycete recorded in Table 8a as not being found on composite wood, but present on solid wood on rare occasions, is the dry rot fungus *Serpula lacrymans*, which was mentioned earlier as an important agent of decay in wood. Other fungi recorded on solid, but not composite, wood are the ascomycete sapstain fungi *Ceratocytis/Ophiostoma* spp. These are found growing on unseasoned sawn timber, staining the sapwood black/dark blue. What has been found to be the most often encountered fungus disfiguring wood with blue stain, *Aureobasidium pullulans* (Eaton and Hale 1993), has been found both frequently on solid wood and commonly on wood composites. However, blue stain fungi in the genus *Phialophora* are among the fungi less often encountered (Table 8b), more likely on solid than on composite wood. As well as disfigurement, *Phialophora* spp. cause soft rot (Eaton and Hale 1993). Another important soft rot organism growing on both solid and composite wood (Table 8a) is the ascomycete *Chaetomium globosum*.

The fungi found growing on ceiling tiles in North America are apparently less diverse than those on gypsum board and solid or composite wood. Although ceiling tiles may commonly or frequently have growing on them, the same nine species on gypsum board and all of the other materials in Table 8a, viz. *Acremonium strictum, Aspergillus versicolor, Aureobasidium pullulans, Cladosporium sphaerospermum, Penicillium chrysogenum, P. commune, P. decumbens* and *Stachybotrys chartarum*, it appears that there are few other fungi in these categories on the tiles (Table 8a). There are also fewer species in the infrequent and rare categories (Table 8b), and the total number recorded (Tables 8a and 8b) is little more than half of those found on gypsum board.

Although the range of fungi is less restricted than on ceiling tiles, the overall range of fungi for the unspecified fabrics and textiles is less wide than on gypsum board and solid or composite wood. While the number of species commonly or frequently found growing on fabrics and textiles is of the same order as on gypsum board and both types of wood, the total number of species (Tables 8a and 8b) is only about two-thirds of those on gypsum board. This is also the case with the carpets (Tables 9A and 9B). The nature of the fabrics and textiles, and also the carpets, is unspecified, but it is reasonable to assume that while cellulose might have been present in some, possibly in combination with synthetic fibres, some may have contained wool and others may have been entirely synthetic. This also applies to the carpet category in in Tables 9a and 8b. Reference to Tables 5 and 6 reveals that among the species growing on fabrics and textiles some, such as *Chaetomium globosum, Stachybotrys chartarum* and *Trichoderma viride*, are potent cellulase producers but others, such as *Emericella nidulans* and *Scopulariopsis brevicaulis*, are only weakly cellulolytic. In a recent review, the first three of these species are also included in a catalogue of fungi isolated from or degrading keratinous substrates (Blyskal 2009). *C. globosum* has been isolated from wool and contemporary woollen textiles, causing structural deterioration of both and discoloration of dyed and undyed textiles, although it is listed as having only low keratinolytic activity; *S. chartarum* has also been isolated from wool, which it decomposes, and causes discoloration of undyed woollen textiles as it grows; and *T. viride* causes extensive destruction of wool. Of the other fungi listed in Table 8a as being common or frequent on fabrics and textiles (Table 8a), *Acremonium strictum, Chaetomium indicum, Cladosporium herbarum, Geotrichum candidum, Penicillium chrys-*

ogenum and *Scopulariopsis brevicaulis* have been recorded by Blyskal (2009) as being keratinolytic. Among the species less often found (Table 8b), *Aspergillus fumigatus* and *A. terreus* are strongly keratinolytic, *Geotrichum candidum* is keratinolytic and *Eurotium chevalieri* causes slight damage and marked discoloration in wool and woollen textiles (Blyskal 2009).

Although there are differences in the prevalence of different taxa between these various sets of data, it is not surprising that there are strong qualitative resemblances in the mycobiota on gypsum board and other cellulosic constructional and finishing materials. As in the studies of fungi on exposed cotton fabrics in tropical environments (Siu 1951, Upsher 1972), there may be some that are regarded as "hot climate" types (Upsher 1972), but most are ubiquitous. What develops on the substrate depends not only on a_w and nutrient availability, but also on the presence of toxic substances in the substrate, on temperature and on moisture. These factors also modify the inter-specific interactions that are important in the development of the mycobiota.

Other materials

Plastics

It was stated earlier that the materials which are most susceptible to microbial growth and biodeterioration contain organic polymers found in their natural environments. This does not mean, however, that synthetic polymers will not support or be degraded by microbes. For example, various types of synthetic rubber and plastic suffer microbial biodeterioration. Caneva et al. (2003) note that fungi are the main microorganisms degrading such synthetics, and cite *Aspergillus niger, A. flavus, Aureobasidium pullulans, Chaetomium* spp., *Penicillium funiculosum, P. luteum* and *Trichodema* spp. as examples. Although natural rubber is one of the most resistant of natural polymers, some microorganisms are able to metabolize its long chains of repeating isoprene units. Based on experiments with a nocardioform member of the Actinobacteria and two other coryneform bacteria (Linos and Steinbüchel 1996), synthetic styrene-butadiene rubber appears to be more susceptible to attack than other types of synthetic rubber tested. Nitrile, silicone and neoprene rubber appear to be extremely resistant to microbial attack. Among plastics, polyesters such as polycaprolactone and polybutylene adipate are susceptible to microbial attack, but others such as those derived from phthalates or aromatic hydrocarbons are not. Polyurethanes comprise a wide range of materials - elastomers, surface coatings, adhesives and both flexible and rigid foams - in which urethane is the predominant component and other groups such as esters and ethers are incorporated. Polyether-containing polyurethanes are more susceptible than polyester polyurethanes, with fungi being known to cause cracking in some of the latter type (e.g. poly-caprolactone polyurethane) in as little as 10 days. Localized cracking round fungal colonies may be the result of loss of plasticity caused by attack on susceptible plasticizers, based on adipates of sebacates, which are included in the finished product (Caneva *et al.* 2003, Allsopp *et al.* 2004).

Thermal insulation

Urea-formaldehyde resin, which is used in adhesives, laminates, and finishes for textiles, is another example of a substance "new" to microorganisms. Another form, urea-formaldehyde foam insulation (UFFI), has been noted earlier in this chapter as being colonized by fungi. Because of a possible health hazard due to off-gassing, in Canada use of UFFI was banned at the end of 1982, but not before it had been installed in an estimated 80,000 homes, either in the cavities in masonry walls of older houses or the spaces between studs in frame houses. The reaction between the urea-formaldehyde resin and the acidic hardening agent in the injected mixture is reversible, and the foam slowly deteriorates after installation. Although this may contribute to fungal growth, freshly injected UFFI may contain 75% water by weight and the release of water during curing, condensation on the surface of the foam and/or leaks are considered to be the main reasons for growth. Urea is of course a source of nitrogen, which is usually the limiting nutrient for the growth of fungi in nature, and can be used as such for the growth of the wide range of microorganisms associated with it.

As the report of Bissett (1987) and Table 4 illustrate, UFFI may be extensively contaminated by phylloplane and soil-borne moulds. Table 4 shows that, as for a variety of insulation materials (Dudney *et al.* 1982), in UFFI which had been in service for 5-10 years the common phylloplane species *Cladosporium cladosporioides*, *Alternaria alternata* and *C. herbarum* were frequently present (Miller *et al.* 1984). However, one-half of the samples were contaminated by *Penicillium fellutanum*, an organism associated with decaying organic material and, by inference, tolerant of high concentrations of formaldehyde. The presence of *P. fellutanum* in the samples was negatively correlated with wall cavity formaldehyde concentration. This species

also occurs on gypsum board (Table 7), but it is not a major component of the mycobiota of that substrate, although it is frequently found on it (Table 8a). In contrast, *P. viridicatum*, the most common species associated with wallboard (Table 7), was only present on about 20% of the UFFI samples (Table 4), and there was a lesser level of contamination by other species. This illustrates that fungal growth on wetted building materials is modulated not just by nutrients and water, but also by the presence of toxic compounds, in this case formaldehyde. As can be seen from Tables 8a and 8b, which in the case of UFFI is derived from the data of Miller *et al.* (1984) and Bissett (1987), the total number of species in UFFI was only slightly more than half that in gypsum board samples. However, three-quarters of the species in the common/frequent categories for UFFI are also found in gypsum board samples (Table 8a). UFFI is, of course, not the only type of insulation used in walls, and Pessi et al. (2002), for instance, found that *Aspergillus versicolor* and unidentified species of *Acremonium*, *Aureobasidium*, *Cladosporium* and *Penicillium* were prominent in the mineral (rock) wool and fibreglass wool insulation between outer and inner concrete panels forming the external walls of Finnish apartment buildings. They did not observe any statistically significant infiltration of the indoor air by these fungi, but infiltration by the smaller spores of actinobacteria growing in the insulation was significant. In experimental studies, Klamer *et al.* (2004) inoculated a range of insulation materials commercially available in Denmark with a mixed suspension of *Alternaria alternata*, *Aspergillus versicolor*, *Cladosporium sphaerospermum*, *Penicillium chrysogenum* and *Trichoderma viride*. As was to be expected, they observed that after four weeks at 26°C fungal growth (biomass) was greater on the two cellulosic materials, in particular flax, than the limited growth on the glass and rock wools, which have been labelled as being poor in nutrients and non-supportive of microbial growth (Tye *et al.* 1980). There were substantial dry mass losses in paper (c. 18-39%) and flax insulation (15-17%), but the maximum loss in glass and rock wools was <3%. Since paper is flammable, fire retardents are usually included in loose-fill insulation manufactured from recycled paper. Most formulations are borate based (Herrera 2005, 2008) and inhibitory of fungal growth in the presence of moisture, but if the insulation becomes wet in service to that extent that borate leaches out it will become susceptible to fungal growth. Klamer *et al.* (2004) suggested that such leaching may have occurred in their experiments, and hence the loss in dry mass. Koivula *et al.* (2005) also noted rapid proliferation of moulds in flax, but in naturally contaminated commercial flax matting and loose-fill insulation treated with fire retardants. In laboratory experiments, before a three-week period at 90% RH, 10^2-10^3 mould CFU g^{-1} (dry weight) and total counts of 10^4-10^6 bacteria g^{-1} were recorded, but by the end of this period the mould count had risen to 10^7-10^8 CFU g^{-1}, and the number of bacteria in the only case counted was 10^8 g^{-1}. In contrast, neither initially nor after the high humidity exposure were any moulds detected in paper insulation, presumably owing to inclusion of inhibitory substances, or in rock wool and glass wool, the nutritionally poor mineral products (Koivula *et al.* 2005). The moulds associated with the flax insulation (Koivula *et al.* 2005, Nykter 2006) include unidentified species of *Acremonium, Aspergillus, Chaetomium, Penicillium* and *Ulocladium*, which as has been discussed above are genera which contain species that are widespread in the natural and building environments, and on other cellulosic and building materials (Tables 8a, 8b, 9a and 9b).

Paint

Mould growth on in-service films of water-based paints in which cellulose ethers are the thickeners has already been mentioned, but it can also occur on oil- or solvent-based paints. The fungi found growing on indoor painted surfaces in North America are listed in Tables 9a and 9b, but are in total little more than one-third of those on gypsum board (Tables 8a and 8b). Other species which have been isolated from paint films include *Cladosporium cladosporioides* and *C. oxysporum; Cochliobolus geniculatus; Epicoccum nigrum; Fusarium oxysporum; Geomyces pannorum; Penicillium expansum, P. janczewskii* and *P. viridicatum; Pestalotia macrocarpa; Phoma violacea; Pithomyces chartarum; Trichoderma viride;* and *Ulocladium atrum* and *U. consortiale* (Ross *et al.* 1968, Hunter 1989, Allsopp *et al.* 2004).

As Allsopp *et al.* (2004) have pointed out, it may not be clear whether moulds on the surface are actually attacking the paint film or not. While some may be utilizing paint components, others may only be taking nutrients from dirt on the surface. Eveleigh (1961) concluded from a study of fungal disfigurement of oil paint containing white lead applied to wood- and metal-framed windows inside 30 greenhouses that, whether contamination originated from above or below the paint film, primary colonization depended on the presence of external nutrient sources. The commonest colonists of the paint in this study were *Cladosporium herbarum* (63% of samples), *Aureobasidium pullulans* (37%) and *Phoma violacea* (33%). Other *Phoma* spp. (12%) and *Alternaria alternata* (4%) were much less frequent. Fungi may disfigure paint films because of their inherent pigmentation, e.g. *Alternaria, Cladosporium* or *Ulocladium* or by producing a soluble pigment that is taken up by the paint, e.g. *Aureobasidium* (Allsopp *et al.* 2004). *P. violacea* causes a characteristic red-violet staining of paint films, and it was noted by Eveleigh (1961) in both UK and Canadian homes on the paintwork of wooden window frames, where it may be deep-seated in the wood. In cases where mycelium is growing on the substratum under the paint the damage that the fungi cause may be largely physical. For example, it is well known that as a result of mechanical pressure as the organism grows below and within the paint films *Aureobasidium pullulans* can cause cracking, thereby opening underlying wood to invasion by other species. *P. violacea* is another organism that causes mechanical damage to paint films. Eveleigh (1961) found that it could penetrate paint films from the exposed (top) surface, but its penetration and proliferation was most rapid where the paint has been applied to wood that had already been colonized. This could cause the film to fracture and pycnidia erupt through the cracks.

Bacteria which have been isolated from oil-based paint include *Alcaligenes, Bacillus* spp. (*B. cereus, B. mycoides* and *B. sphericus*), *Flavobacterium* and *Micrococcus* (Ross *et al.* 1968). The most common form of paint failure caused by bacteria appears to be peeling, i.e. loss of adhesion to the substratum due to biodeterioration of resinous binders in the paint. Where conditions are extremely wet and light is available, paint films of all types may be disfigured by green algae such as *Chlorella, Chlorococcum, Stichococcus* and *Trentepohlia*, and by cyanobacteria such as *Lyngbya, Nostoc, Oscillatoria* and *Synechococcus* (Allsopp *et al.* 2004).

Ceramic materials

It was mentioned earlier that Prezant *et al.* (2008) recorded that *Aspergillus versicolor* and *Penicillium chrysogenum* grew on all of the cellulosic materials examined and on ceramic tiles and concrete efflorescences, and *Eurotium herbariorum*, which grew on all but one of cellulosics (solid wood), also grew on these two ceramic materials. However, it can be seen from Tables 8a, 8b, 9a and 9b that, overall, substantially fewer species grew on the inorganic substrates than on the cellulosic materials. Wallboard supported the

Table 12. Filamentous fungi (in descending order of occurrence) isolated from settled dust in 369 homes in Wallaceburg, Ontario, Canada, using intermediate and high a_w culture media.

Common	Frequent	Rare
Aureobasidium pullulans	*Stachybotrys chartarum*	*Penicillium oxalicum*
Alternaria alternata	*Ulocladium chartarum*	*Penicillium implicatum*
Penicillium chrysogenum	*Paecilomyces variotii*	*Aspergillus ustus*
Aspergillus versicolor	*Penicillium griseofulvum*	*Penicillium crustosum*
Eurotium herbariorum	*Aspergillus sydowii*	*Pithomyces chartarum*
Epicoccum nigrum	*Penicillium citreonigrum*	*Scopulariopsis candida*
Cladosporium cladosporioides	*Penicillium raistrickii*	*Penicillium echinulatum*
Penicillium spinulosum	*Penicillium vulpinum*	*Eurotium repens*
Penicillium corylophilum	*Scopulariopsis brevicaulis*	*Penicillium miczynskii*
Penicillium commune	*Aspergillus fumigatus*	*Penicillium italicum*
Cladosporium sphaerospermum	*Penicillium simplicissimum*	*Penicillium digitatum*
Trichoderma viride	*Penicillium decumbens*	*Fusarium oxysporum*
Aspergillus niger		*Aspergillus candidus*
Cladosporium herbarum		*Chaetomium globosum*
Penicillium expansum		*Penicillium glandicola*
Penicillium viridicatum		*Penicillium purpurogenum*
Penicillium brevicompactum		*Penicillium islandicum*
Wallemia sebi		*Geomyces pannorum*
Phoma herbarum		*Penicillium verrucosum*
Aspergillus ochraceus		*Penicillium variabile*
Penicillium aurantiogriseum		*Emericella nidulans*
Penicillium citrinum		*Alternaria tenuissima*

growth of 94 species, but only 15 species were found on ceramic tiles and 20 on concrete efflorescences. Of course, growth is likely only to have been possible on these inorganic materials because of soiling of the surface, with organic materials in dirt or settled dust being the source of carbon for heterotrophic growth.

Dust

While a selection of materials that are susceptible to microbial biodeterioration has been discussed above, it has to be remembered that virtually no material in a building is immune from microbial growth if there is sufficient moisture available.

In the previous section and earlier in the chapter, the role of dirt (dust and organic debris) in supplying nutrients for microbial growth has been referred to. It enables microorganisms that cannot degrade particular organic materials, e.g. cellulose, to grow on the surface of these materials. Even where the material is entirely inorganic (e.g. steel, glass, brick, concrete or stone), soiling with dust, organic debris and volatile substances which condense on its surface can allow microbial growth to take place. As well as being a source of nutrients, settled dust is an important reservoir of dormant microorganisms which may be dispersed into the air. For example, Cole *et al.* (1994) isolated 1.6×10^4 - 1.2×10^5 colony forming units (CFU) of fungi g^{-1} of carpet dust in a four-storey multi-use building, as well as 1.4×10^6 - 8.8×10^6 mesophilic bacteria and 5×10^2 - 7×10^3 thermophilic bacteria (including actinobacteria). The mixture of fungi present in settled dust includes species exhibiting varying degrees of xerophily and also others regarded as being more hydrophilic.

Table 12 shows that the species prevalence in settled dust in family dwellings is normally dominated by fungal propagules originating outdoors. However, three kinds of change appear. Firstly, since these data are gathered from studies of culturable fungi, species with long-lived propagules gain in apparent importance, including various types of yeast. Secondly, soil-borne fungi are tracked into buildings. Thirdly, food-associated fungi become important components of the mycobiota. In the list of common fungi in settled dust in Table 12, *Penicillium expansum*, *P. digitatum* and

P. islandicum are most likely to have come from fruit brought into the house. Another category in Table 12 comprises species such as *P. chrysogenum*, which may originate from food, but may sometimes also grow on building materials. Among the houses surveyed, the ratio of the combined total of *Aspergillus*, *Eurotium* and *Penicillium* to the phylloplane species or of all others to the phylloplane species approached unity in the settled dust of those in which fungal growth was absent (Miller 1995). Concentrations of fungi in settled dust on media for xerophilic fungi are typically 10^4-10^5 CFU g^{-1} dust m^{-2} floor (Verhoeff 1994, Dales *et al.* 1997, Lawton *et al.* 1998), with concentrations on carpeted floors (CFU m^{-2}) being 3-4 times higher (Verhoeff 1994). Comparison of the list in Table 12 with, for example, species isolated in Japan (Takatori *et al.* 1994) and the Netherlands (Hoekstra *et al.* 1994, Verhoeff 1994) confirms that there is a strong similarity in the floor dust mycobiota of buildings in different geographical regions.

Notably absent from Table 12 is *Chaetomium globosum*, although it is common on mouldy gypsum board (Tables 7 and 8a). This may reflect its relative absence from the particular set of houses surveyed, but what is more likely is that it reflects the difficulty of isolating this organism and also the short survival time of its propagules. In contrast, one of what are known as black yeasts, *Aureobasidium pullulans*, and various other yeasts are exceptionally common in house dust, as it appears that their propagules survive for very long periods. Black yeasts are actually yeast-like states of various filamentous fungi such as *Aureobasidium*, *Cladosporium*, *Moniliella* and especially anamorphs of ascomycete genera in the Herpotrichiellaceae, including the pathogen *Exophiala*. Among phylloplane fungi, *Alternaria alternata* and *Epicoccum nigrum* propagules must also survive longer than *Cladosporium cladosporioides* and *C. herbarum* (and *C. sphaerospermum*), because their prevalence is much greater than that indicated in outdoor air. Data such as these strongly reinforce the importance of evaluating the relative persistence of fungal propagules when considering the order of species in laboratory reports.

Aspergillus versicolor, *Eurotium herbariorum*, *Penicillium commune*, *P. viridicatum* and *P. aurantiogriseum* are all common in dust (Table 12) and also on mouldy gypsum board (Tables 7 and 8a), indicating the likelihood that this wall material is an important source of these species in the settled dust. Although it is not certain, moulds such as *P. spinulosum* and *P. brevicompactum* probably grow on some other damp material,

such as carpets. The last of the common moulds in house dust (Table 12), *Wallemia sebi*, is an extreme xerophile. This species has been reported as common in house dust from Japan (Sakamoto *et al.* 1989) and the Netherlands (Verhoeff 1994). In Japan, it was associated with floor mats (Torii *et al.* 1990), but the source of this fungus in the Canadian and Dutch investigations is not known.

Qualitatively, the culturable mycobiota of dust in upholstered furniture (Schober 1991) and mattresses (Hoekstra *et al.* 1994) is similar to that of floor dust. Schober (1991) did, however, note that hydrophilic fungi such as *Alternaria* and *Rhizopus* were more abundant in carpet dust. Counts of slightly xerophilic types (dominated by *Penicillium brevicompactum* and *P. chrysogenum*) were significantly greater in the case of upholstered furniture, where conditions were drier and more closely related to the ambient climate than in carpets, hence fewer hydrophiles in upholstery. Notwithstanding, Schober (1991) did make the point that regular use of upholstered furniture modifies conditions as a result of uptake of moisture generated by the users. This would clearly apply to mattresses, although Hoekstra *et al.* (1994) found that the higher frequency of isolation of *Scopulariopsis brevicaulis* from mattress dust was the only noteworthy difference from the moulds and yeasts isolated from floor dust.

It should be remembered that not all viable fungi occurring in dust will be culturable, e.g. obligate plant pathogens such as rust and smut fungi. Also, as was mentioned in Chapter 1.2, Maghraby *et al.* (2008) found that dermatophytic fungi, particularly species of *Chrysosporium* and *Aphanoascus*, comprised approx. one-third of the fungi in floor dust in Egyptian student residences. It can reasonably be assumed that dermatophytes will also be present in carpet dust and on upholstery, but to isolate these normally requires use of rich clinical media such as Kimmig or Sabouraud dextrose agar.

Although many of the microorganisms in dust on floors, shelves, upholstered furniture, etc., are there because their propagules have sedimented out of the air, they may also have proliferated in the dust. Not only are skin scales, animal dander and other organic debris a source of nutrients for growth, but to varying degrees the debris is hygroscopic and can absorb moisture from a humid atmosphere. In water-saturated atmospheres, total numbers of viable bacteria in floor dust have been found to increase 100-fold in eight weeks (Flannigan *et al.* 1992). In addition to such quantitative change, the qualitative nature of the microbiota changes, with pseudomonads replacing as the predominating bacteria the Gram-positive types (*Staphylococcus, Micrococcus* and *Bacillus*) characteristic of dry dust. Although the species of *Penicillium* and *Eurotium* in dry dust may still dominate the mycobiota, the range of moulds widens as the dust absorbs increasing amounts of water from the atmosphere, with members of the Mucorales being prominent where the atmosphere is saturated.

Just as counts of culturable fungi increase when dust is damp enough to allow proliferation of moulds already resident in it, so they also increase when propagules of fungi growing on damp building materials sediment out into dust. For example, in houses where an area of mould growth visually assessed as 1.2 m^2 was present there was a 100-fold increase in counts of culturable propagules enumerated on media for xerophilic fungi relative to where there was no such mould growth (Lawton *et al.* 1998). In parallel with observations that fungi growing on building materials alter the ratio of phylloplane to other fungi in the air (Nathanson and Miller 1989, Miller 1993), Lawton *et al.* (1998) found that such growth also had an effect on the mycobiota of settled dust. The ratio of the combined total of *Aspergillus, Eurotium* and *Penicillium* to the total for phylloplane fungi rose to around 10.

Another study in which elevated levels of fungi were found in dust in water-damaged houses, although whether they had grown in the dust or had sedimented out from growths on damp building materials is not known, has been reported by Lignell (2008). This study evaluated microbial concentrations in dust from Finnish schools and residences using both a quantitative polymerase chain reaction (qPCR) and a culture method. Total microbial concentrations were found by the qPCR method to be at least two orders of magnitude greater than by culture. The highest concentrations in both schools and homes were for a group assay for *Aspergillus* spp., *Penicillium* spp. and *Paecilomyces variotii*, and for *A. penicillioides*, *Aureobasidium pullulans, Cladosporium cladosporioides, P. brevicompactum/P. stoloniferum* and actinobacteria in the genus *Streptomyces*. In homes, there were significant increases in concentrations of *P. brevicompactum/stoloniferum, Wallemia sebi, Trichoderma viride/atroviride/koningii, C. sphaerospermum, Eurotium amstelodami/chevalieri/herbariorum/repens/rubrum* and the *Aspergillus/Penicillium/ Paecilomyces* assay group parallel with increasing moisture damage. *Streptomyces* concentrations showed a similar trend, but the difference in concentrations was not significant.

Heating, ventilation and air-conditioning (HVAC) systems

This section on microbial growth material in buildings turns to HVAC systems because of the range of niches that they present, ranging from standing water to absorptive porous material.

Central and window air-conditioning has the general effect of reducing the number of airborne microorganisms relative to outdoor air (Flannigan *et al.* 1991). However, if incorrectly operated or maintained the complex HVAC systems such as are installed in many large public buildings and non-industrial workplaces may become a source of microorganisms which contaminate the conditioned space. Surfaces such as those of the filters, coils, mix box areas, drainage pans, areas where flow is not continuous, and exposed porous insulation, are sites where dust and other particles may accumulate and be a nutrient base for microorganisms. Microbial growth becomes possible where there is standing water, as in drain pans and humidifiers, or moisture is absorbed by porous insulation or dust, particularly in the vicinity of cooling coils and humidifiers and on airstream surfaces downstream of cooling coils.

Table 11 illustrates the viable bacteria isolated from a range of sites and in an air-handling system lacking humidifiers and duct insulation. The range of bacterial types may be regarded as typical of air-handling systems, although it is the first record of the budding bacterium *Blastobacter*, an organism associated with outdoor aquatic environments. Among bacteria that may be found in humidifiers and other HVAC equipment are *Saccharopolyspora* (*Faenia*) *rectivirgula* and *Thermoactinomyces vulgaris*, thermophilic actinobacteria that have also been isolated from domestic heating systems (Fink *et al.* 1971). Although Heinemann *et al.* (1994) did not find thermophilic actinobacteria among the bacteria on filters, they noted that *T. candidus*, a species related to *T. vulgaris*, was more abundant than the latter in HVAC spray humidifiers. As with the bacteria, fungi isolated from filters were not numerous, but included the thermotolerant opportunistic pathogen of man, *Aspergillus fumigatus*. Another such pathogen, *Exophiala jenselmei*, was isolated from nearly 50% of humidifier reservoirs, but along with *Phoma, Acremonium* and other species in the humidifier water was never found in the air of the offices served by the HVAC systems.

Table 9a illustrates a range of moulds that may grow commonly or frequently on AHU filters in North America. These include *Acremonium strictum, Asper-*

Table 13. Bacteria in a well-maintained air-handling system (AHS) in Brisbane, Australia (after Hugenholz and Fuerst 1992).

Category of bacterium	Location in system
Blastobacter sp. (pink)	Surface of supply coils
Blastobacter sp. (yellow)	Drain pan water Post-coil air Supply air
Flavobacterium spp. (esp. Fl. odoratum)	Evaporative condenser sump water Surface of air intake grill Return-fresh air mix Surface of return coils Drain pan water Supply air
Bacillus spp., Pseudomonas spp., Species of Acinetobacter, Arthrobacter, Cedecea, Corynebacterium, Staphylococcus	Various sites -- air, water and/or surfaces Sporadically throughout AHS -- air, water and/or surfaces

gillus versicolor, Aureobasidium pullulans, Cladosporium sphaerospermum, Eurotium herbariorum, Penicillium chrysogenum, P. commune and P. decumbens, and Stachybotrys chartarum, which were specifically mentioned earlier as growing on gypsum wallboard and other cellulosics. Various other species are listed in Table 9b. By microscopy, Ljaljević-Grbić et al. (2008) found in Serbian schools that dust on filters which had not been cleaned or replaced for two years most often bore Cladosporium herbarum and some of the fungi listed as being less frequently encountered in North American AHU filters, viz. Alternaria alternata, Aspergillus flavus, A. fumigatus, A. niger, A. ochraceus, Botrytis cinerea and Epicoccum nigrum (Table 9b), and only one listed as common in Table 9a, i.e. P. aurantiogriseum. On the filters of a surgical ward of a hospital, where the filters were cleaned monthly, the only fungi in the dust were A. versicolor and an unidentified ascomycete in the Sphaeriales (Ljaljević-Grbić et al. 2008). In an earlier report, Ahearn et al. (1997) noted Cladosporium and Penicillium spp. on both sides of filters in AHU's, not just the load side. On examining preservative-treated and untreated high efficiency particulate arrestance (HEPA) filters and cellulosic-polyester filters in HVAC systems in hospital and commercial buildings over a period of eight years, Price et al. (2005) found that cellulosic filters in systems where there were water entrainment problems were either already colonized or rapidly became so on incubation in moisture chambers. Acremonium, Aspergillus and Cladosporium were the commonest moulds observed under the microscope. In HEPA and 90-95% arrestance filters, colonization was usually restricted to the load surface, but Aspergillus flavus grew throughout one untreated example and an unidentified species of Cladosporium on a second.

Porous acoustic or thermal insulation installed on the airstream surface of HVAC components and ducts may be colonized by moulds. Where the insulation is covered by a metal foil or plastic-type facing material, and this remains intact, mould growth appears to be limited to the surface. Relatively few fungi develop on metal foil, but plastic facings may be densely covered by heat-resistant and xerophilic fungi such as Byssochlamys and Eurotium (Ahearn et al. 1993). However, where the facings are breached, the exposed fibreglass matrix becomes contaminated with dust and organic debris and will absorb moisture from the system. It has been reported that in atmospheres with a RH >90%, such insulation can be colonized by the moulds Acremonium obclavatum, Aspergillus niger, A. versicolor, Cladosporium herbarum and Penicillium miczynskii (Ezeonu et al. 1994). Laboratory experiments also showed that aqueous extracts of fibreglass insulation, containing urea, formaldehyde and unidentified organic compounds, supported spore germination and growth in a representative species, A. versicolor, with urea serving as a nitrogen source (Ezeonu et al. 1995). With the potential of HVAC systems for dispersing microorganisms throughout buildings, the importance of preventing and controlling microbial contamination of HVAC systems cannot be stressed enough. Another reason for these measures being necessary is that microbial growth in such systems may generate VOC's that are then distibuted thoughout the building. For example, in a building where there had been complaints about "mouldy air" and VOC's had been detected in the air of the building by Ahearn et al. (1996), the filters and the fibreglass duct liner were extensively colonized by C. herbarum and Penicillium spp. and was found to emit some of these VOC's. A combination of removal of the colonized duct liner and continuous operation of the air distribution system reduced the number of complaints (Ahearn et al. 1997). For further reading on remediation, the reader can consult Shaughnessy *et al.* (1999). More extended treatments of moisture problems in general are to be found in Rose and Tenwolde (1993) and Flannigan *et al.* (1996).

Cultural heritage collections

Manuscripts, documents and books

Large collections of cellulosic materials that are susceptible to microbial growth are to be found in libraries, archives and museums. These materials include paper, papyrus, cotton fabrics and photographic/cine film. In addition to cellulosics, other susceptible materials include vegetable glues, which are mostly starch- or dextrin-based mixtures with gums and resins, and animal products, such as parchment, leather and protein colloid glues (fish, hide, hoof and rabbit-skin).

There have been numerous investigations of the air spora in libraries, aimed either at determining whether it contains allergenic or toxigenic fungi that could present a risk of rhinitis, asthma or other respiratory symptoms in library workers (e.g. Zielińska-Jankiewicz *et al.* 2008, Apetrei *et al.* 2009), or determining whether it includes fungi that could put books, manuscripts and documents at risk of biodeterioration (e.g. Ruga *et al.* 2008). As could be expected, although these investigations have shown that allergenic, toxigenic and deteriogenic fungi are present in the air spora in non-problem libraries, the air spora is somewhat similar to that outdoors or in buildings other than libraries. For example, when Burge *et al.* (1978) investigated the airborne mycobiota in 11 non-problem libraries in Michigan, they reported that the fungi isolated were similar to those in domestic buildings in the area and to the outdoor air, with the indoor CFU counts being lower than outdoors, particularly where the libraries were air-conditioned. In view of this, Burge *et al.* (1978) suggested that factors other than fungi might account for respiratory symptoms among library workers. Again, in Brazil Gambale *et al.* (1993) reported that in 28 university libraries in São Paulo, although some of the librarians reporting rhinitis and a few of them registered positive in skin-prick tests with a pool of 20 fungi (and the individual components) most often isolated from the libraries, the fungi were likely to be encountered in the air anywhere in the city, but not in the higher concentrations encountered in the libraries. Dust mites may be a factor in such allergies in librarians (Sánchez-Medina *et al.* 1996, Solarz 2001).

As in other indoor environments, the airborne mycobiota in the library is boosted by human activity that raises settled dust and spores. Burge *et al.* (1978) noted that taking books from shelves, paging through and returning them to the shelves raised counts of airborne CFU by 240%, and Singh *et al.* (1995) in India and Apetrei *et al.* (2009) in Rumania have also reported that manipulation of books increases airborne concentrations. The spores associated with dust raised by disturbance are redistributed in the air and as a result may settle on fresh surfaces, and so spread viable inoculum onto previously uncontaminated books and other surfaces. For example, Apetrei *et al.* (2009) noted deposition rates of 419-1,677 CFU m^{-2} of surface. The predominant fungi in dust on shelves, books and archived materials are most often species of *Cladosporium*, *Aspergillus* and *Penicillium*, but other types such as strongly cellulolytic *Chaetomium* spp. may also be present (Maggi *et al.* 2000). Using contact plates, Zielińska-Jankiewicz *et al.* (2008) isolated *Acremonium murorum*, *Aspergillus fumigatus*, *A. sydowii*, *Cladosporium cladosporioides*, *C. herbarum*, *Paecilomyces variotii*, *Penicillium chrysogenum*, *P. corylophilum* and *Rhizopus nigricans* from the surface of archived 19th and 20th century documents in storage rooms.

Such superficial contamination means that in order to prevent microbial growth it is important to maintain a low humidity level in libraries, and therefore a low a_w in the documents, manuscripts and books that they contain, especially so in hot, humid tropical climates. This is amply illustrated by Temby (2001), who reported a major mould problem in a marine science library at Townsville in tropical Australia. During the wet season the average temperature in Townsville is 31°C and the RH is usually 75% or more. The problem arose in the wet season of 1999 when the library air-conditioning unit failed and the RH rose from the usual 50% to about 80%. Subsequently, isolated mould growths appeared and spread across the book collection.

Ingress of water as a result of leakage or flooding is another hazard for books and documents in libraries and archives. Flooding can cause extremely serious damage in libraries, as was illustrated in 1997 at Colorado State University in Fort Collins, which "suffered the greatest water-related disaster in US history" (Silverman 2004). Water forced its way into the below-grade floor of the library rising to roughly 8 ft (c. 2.4 m) and immersing the book stacks. Visible signs of mould growth appeared on surfaces about three days after the floodwater was pumped out. Mould growth on the book collection increased each day

Fig. 9. Foxed familial photograph, 1901. Spots are scattered on the cardboard mount, and especially on the top left of the image. (Reproduced with permission of Margaret E. Flannigan.)

that the books remained wet in the basement; while books retrieved from the basement during the first few days showed little or no staining of the text, damage in those collected during the last few days of the recovery operation had frequently extended to 20 or more pages in from both covers. The fungi isolated from the damaged books were species of *Absidia, Alternaria, Aspergillus, Botrytis, Chrysonilia, Cladosporium, Curvularia, Fusarium, Paecilomyces, Penicillium, Stachybotrys* and *Trichothecium*, and unidentified yeasts. The occurrence of *Botrytis* and *Stachybotrys* among these fungi is a mark of how wet the materials were, since they are both hydrophilic fungi, i.e. are only able to grow on materials which are at $a_w > 0.90$ (Table 1). Exposure to the toxigenic species *S. chartarum* may have implications for the health of library workers. There is a well-documented case of acute health problems among workers who spent two or three days without respiratory protection cleaning and removing water-damaged materials from the sub-basement of an art museum in New York, in which about 30% of the employees began to complain of headaches, chronic fatigue and respiratory

ailments (Johanning 1995). The sub-basement had a history of recurrent flooding (Johanning *et al.* 1993) and was heavily contaminated by *S. chartarum*, with the surfaces of the boxes, books and wallboard being profusely colonized (Johanning 1995). The exposed workers were diagnosed as having chronic laryngitis, sinusitis, bronchitis, asthma and toxic encephalopathy (Johanning *et al.* 1996, Johanning 1998). This case led to the formulation of the New York City Guidelines on Assessment and Remediation of *Stachybotrys atra* in Indoor Environments (NYC 1993), which are discussed in detail in Chapter 2.3.

Data from investigations by a number of authors into fungal and bacterial degradation of paper and a selection of other materials, some cellulosic and others not, in libraries and archives were collated by Gallo (1993). The principal genera of fungi that appear to be associated with damage to these materials are shown in Table 14. Among the authors was Belyakova (1964), who reported that in a survey of 441 water/fungus-damaged books the principal fungi isolated were penicillia and aspergillli, the former comprising 32.6% and the latter 26.8% of the total number

Table 14. Fungal genera containing species associated with deterioration of materials in libraries (after Gallo 1993).

Genus	Paper/Cardboard	Parchment	Leather	Inks	Glues (animal/plant)	Synthetics	Fabrics	Wax Seals	Photographs	Magnetic Tapes
Alternaria	X	X	X		X	X	X	X		X
Aspergillus	X	X	X	X	X	X	X	X	X	X
Aureobasidium	X		X			X	X	X		
Cephalosporium	X	X				X	X			
Chaetomium	X	X	X			X	X	X	X	X
Cladosporium	X	X	X			X	X	X	X	X
Doratomyces	X					X	X			
Fusarium	X	X	X		X	X	X	X	X	
Gymnoascus	X									
Helminthosporium	X				X					
Monilia	X	X				X	X			
Mucor	X	X	X				X			
Paecilomyces	X		X				X		X	
Penicillium	X	X	X	X	X	X	X	X	X	X
Phoma	X						X			
Rhizopus	X	X	X				X		X	
Rhodotorula	X		X							
Scopulariopsis	X	X	X				X	X	X	
Stachybotrys	X				X	X	X		X	
Stemphylium	X		X		X	X	X	X		X
Trichoderma	X	X	X				X	X	X	
Trichothecium	X	X			X				X	
Verticillium	X						X			

of isolates. Belyakova also recorded that the species most often isolated from books, manuscripts and documents were *Aspergillus versicolor*, *Penicillium notatum*, *Eurotium rubrum* and *P. chrysogenum*. In a later review in which data on unspecified book material, consolidated books and microfilm were added to the materials listed in Table 13, Zyska (1997) tabulated 84 genera, some of which had only been isolated from one of two of the materials. In addition, 40 species from within these genera were named, including *A. flavus*, *A. fumigatus*, *A. niger*, *A. ochraceus*, *A. terreus*, *A. versicolor*, *Chaetomium bostrychodes*, *C. elatum*, *S. globosum*, *C. indicum*, *C. murorum*, *P. aurantiogriseum*, *P. brevicompactum*, *P. chrysogenum*, *P. decumbens*, *P. expansum*, *P. funiculosum*, *P. glabrum*, *P. oxalicum*

and *P. roquefortii*. As can be seen from Table 14, Gallo (1993) has listed 14 genera able to grow on and damage leather bindings. Zyska (1997) has listed species within these genera, but there are fewer species than those that degrade paper. Orlita (2004) has listed a number of aspergilli, penicillia and some other species that attack finished leather, which presumably would be capable of attacking damp bookleather.

More recently, Shamsian et al. (2006) found that in a major cultural collection at an Iranian library approaching one-quarter of nearly 500 randomly selected items including unidentified and consolidated books, papers and papyrus, some which were more than 1000 years old, had visible signs of fungal growth or damage. Microscopical examination confirmed the presence of fungi in >40% of the damaged items, while 70% were found to bear culturable filamentous fungi. The commonest of these fungi were unspecified aspergilli and penicillia, which together comprised 70% of all fungi isolated. Collectively, *Mucor*, *Cladosporium* and *Trichoderma* accounted for >20%, the other fungi including *Rhizopus*, *Alternaria* and, in four cases, yeasts. Most of the manuscripts were contaminated by only one taxon, but no evidence was presented in the report that these were any more than casual superficial contaminants. In an investigation of archived items, Mesquita *et al.* (2009) examined ancient parchment and laid-paper documents and more recent documents manufactured from wood-pulp paper in a Portuguese university. The fungi present in areas showing discoloration, altered texture or patent colonization were identified from micro- and macro-morphological characteristics and DNA sequencing of cultures. As can be seen in Table 15, fungi associated with the damage were common environmental species, but also present were the basidiomycetous types *Coprinus* sp., *Thanetophorus cucumeris* (the telemorph of the common plant pathogen *Rhizoctonia solani*), *Phlebiopsis gigantea* and *Skeletocutis* sp.; the ascomycete *Chromelosporium carneum*; and the mitosporic producer of irritant volatile compounds, *Toxicocladosporium irritans*.

So-called "foxing" of stored paper and photographic prints in archives and museums, i.e. development of brown, reddish-brown or yellowish spots (Fig. 9), is a long-standing problem associated with storage conditions (Caneva *et al.* 2003). The cause of foxing has been disputed, with on the one hand chemical phenomena such as oxidation and heavy metal deposition being invoked, and on the other microbial activity. In reviewing his previous work, during which the extreme xerophilic fungi *Aspergillus penicillioides*,

Table 15. Filamentous fungi isolated from documents in a university archive in Portugal (Mesquita *et al.* 2009).

Parchment (13th-18th C)	Laid paper (16th-18th C)	Wood-pulp paper (19th-20th C)
Cladosporium cladosporioides	*Alternaria alternata*	*Aspergillus fumigatus*
Epicoccum nigrum	*Aspergillus nidulans*	*Aspergillus versicolor*
Penicillium chrysogenum	*Botrytis cinerea*	*Chaetomium globosum*
Phlebiopsis gigantea	*Chaetomium globosum*	*Chromelosporium carneum*
*Thanetophorus cucumeris**	*Chromelosporium carneum*	*Cladosporium cladosporioides*
	Cladosporium cladosporioides	*Penicillium canescens*
	Coprinus sp.*	*Penicillium chrysogenum*
	Penicillium chrysogenum	*Penicillium* sp.
	Penicillium helicum	
	Penicillium sp.*	
	Phlebia subserialis	
	Skeletocutis sp.*	
	*Toxicocladosporium irritans**	

*Identified only on ribosomal DNA loci amplification and sequencing.

A. restrictus and *Eurotium* spp. were found in association with foxing, Arai (2000) recorded not only isolating these fungi from foxed areas of paper but also extracting from the spots various organic acids, oligosaccharides and amino acids produced as a result of fungal growth and activity. On the basis of experiments in which foxing was induced by certain combinations of the extracted substances, Arai proposed that cello-oligosaccharides, released from celluloses by the action of fungal cellulases, and amino acids such as γ-aminobutyric acid produced during growth underwent a Maillard reaction, producing melanoidins responsible for the browning. In examining old prints from a Genoese museum, Montemartine Corte (2003) isolated from foxed maps and test papers a number of fungi, which included extremely xerophilic *Eurotium pseudoglaucum*, moderately xerophilic *Penicillium chrysogenum*, slightly xerophilic *Cladosporium sphaerospermum* and *Epicoccum nigrum* (Table 1), and hydrophilic *Chaetomium globosum* (Yang and Li 2008) and yeasts. *C. globosum* was the most frequently isolated species. From the same museum, Zotti *et al.* (2008) isolated further species from brown spots, including the moderately xerophilic species *A. flavus* and *Paecilomyces variotii,* slightly xerophilic *Aureobasidium pullulans* and hydrophilic *Geomyces pannorum*. Staining similar to foxing on paper also occurs on paintings and textiles.

Very recently, Abe (2010) made use of xerophilic fungi associated with foxing to determine whether environmental conditions were favourable for fungal growth in storage rooms of an art museum in which a Japanese painted screen stored for two years in one room showed fungal growth and foxing. Detectors in which dried spore suspensions of sensor fungi were sandwiched between permeable membranes were exposed in all of the rooms to germinate and grow. The sensor organisms were the extreme xerophiles *A. penicillioides, A. restrictus, Erotium herbariorum, E. amstelodami* and *Wallemia sebi,* and the moderately xerophilic species *Cladosporium cladosporioides* and *A. niger* (see Fig. 5). As *A. penicillioides* showed the greatest growth response in the room in which the painting was stored and *E. herbariorum* the greatest response in the other rooms, these two species have been selected by Abe (2010) as the two sensor fungi for assessing museum environments.

Photographic plates and film

The light sensitive emulsion on photographic plates is, like that of its successor the photographic film, an emulsion of silver salt particles in gelatin, which can be degraded by a considerable range of microorganisms. While the support for the emulsion in photographic plates is glass or metal, in film it is cellulose acetate. Thus, photographic film presents two substrates that are potentially susceptible to damage by microorganisms (Abrusci *et al.* 2004). Microbial degradation of cellulose acetate involves firstly deacetylation and then breakdown of the long-chain cellulose backbone as described earlier. The susceptibility to microbial attack depends on the degree of substitution (DS) of hydroxyl groups in the long-chain cellulose molecules by acetyl groups. For instance, Buchanan (1993) found that in enrichment culture degradation was more rapid at DS 1.7 than at DS 2.5;

at the lower DS 80% degradation took 4-5 days, but it took 10-12 days at the higher DS. Using an *in vitro* aerobic mixed culture system, Komarek *et al.* (1993) demonstrated that the microorganisms involved degraded labelled cellulose acetate of DS ranging from 1.85 to 2.57 in 14-31 days. More than 80% of the original labelled polymeric carbon was degraded to CO_2 at a DS of 1.85, and there was 60% conversion at DS 2.07 and 2.57. It was reported that the bacterium *Sphingomonas* (*Pseudomonas*) *paucimobilis* was able to utilize insoluble cellulose acetate as sole carbon source (Nelson *et al.* 1993), and Abrusci *et al.* (2009) demonstrated that a strain of *S. paucimobilis* isolated from cinematographic film actively degraded the cellulose triacetate (DS 2.7) in film that had been stripped of the gelatin emulsion. Sakai *et al.* (1996) isolated from soil two strains of *Neisseria sicca* that degraded cellulose acetate membrane filters (DS 2.8 + 2.0) and textiles (DS 2.34), and among 35 strains of environmental bacteria (in 15 genera) that were able to degrade cellulose acetate the most efficient was *Bacillus* sp. S2055, which produced significant weight loss in cellulose-acetate plastic film (DS 1.7) within 35 days (Ishigaki *et al.* 2000).

Degradation of the gelatin emulsion is a particular problem, both for manufacturers of film and for conservators (Stickley 1986). The air spora in archives tends to be dominated by *Penicillium*, *Cladosporium* and *Aspergillus* (Opela 1992, Borrego *et al.* 2010), and most of these fungi can be included in the very wide range of microorganisms that are able to degrade gelatin. It is therefore not surprising that all filamentous fungi isolated by Abrusci *et al.* (2005, 2006) from samples of black-and-white cinematographic film in Spanish archives in Madrid, Barcelona and Gran Canaria were gelatinase positive. They were *Alternaria alternata*; *Aspergillus ustus* (2 strains), *A. nidulans* var. *nidulans* and *A. versicolor*; *Cladosporium cladosporioides*; *Mucor racemosus*; *Penicillium chrysogenum* (7 strains); *Phoma glomerata*; and *Trichoderma longibrachiatum*. The only yeast isolated, *Cryptococcus albidus*, did not degrade gelatin. *T. longibrachiatum*, *A. nidulans* var. *nidulans* and *A. ustus* showing the greatest activity against gelatin at 25°C, but at 4°C *P. glomerata*, *C. cladosporioides* and *Alternaria alternata* were the most active species. The bacteria isolated were *Bacillus amyloliquefaciens*, *B. megaterium* (2 strains), *B. pichinotyi*, *B. pumilus*, *B. subtilis*, *Kocuria kristinae*, *Pasteurella haemolytica*, *Sphingomonas paucimobilis*, and *Staphylococcus epidermidis*, *Staph. hominis*, *Staph. lentus*, *Staph. haemolyticus* and *Staph. lugdunensis*,

but only the six *Bacillus* isolates and *Staphylococcus hominis* were able to degrade gelatin. A recent study in which film was artificially inoculated with an unidentified *Penicillium* sp. and an *Aspergillus* sp. in the *A. versicolor* group has provided some evidence that gelatin emulsion is more susceptible in colour film than in black-and-white film (Lourenço and Sampaio 2009). As well as blemishes such as small colonies appearing on the image, fungal growth may affect the dyes in the emulsion, producing stains in colour images.

As with the other archived materials discussed, the importance of maintaining a suitable storage environment is paramount. The effect that increased temperature and RH in storage has on the lifetime of a photographic slide has been tabulated by Dr Saulo Guths (see p. 206, Allsopp *et al.* 2004). At 25°C, the lifetime at 50% RH could be expected to be about 24 years, at 70% RH about 14 years and at 80% RH about 11 years, while at 33°C the corresponding times would be around 9, 6 and 4 years, respectively.

Mural and easel paintings

In paint for mural paintings, the pigments are typically suspended in either water or oil. Although organic substances such as casein or milk may be used as binders, the paint is largely inorganic and is applied to damp lime plaster. Consequently, the microbiota that colonizes murals differs, at least initially, from that colonizing easel paintings, which are susceptible to microbial damage because the paint contains biodegradable elements, most obviously the linseed oil, or less often poppyseed, walnut or safflower oil, that binds the pigments, but also additives such as thickeners, glues and emulsifiers. As an example of the susceptibility of such materials, tests have shown that oxidation of diterpenes occurs in coatings of colophony (rosin) and Venetian turpentine varnishes inoculated with *Chrysonilia sitophila*, *Streptomyces celluloflavus*, *Bacillus amyyloliquefaciens* and *Arthrobacter oxydans*, all known to have been isolated from paintings (Romero-Noguera, *et al.* 2008). Plant and animal glues used to "size" the support to which the paint is applied are also biodegradable, as are paper, canvas, wood, parchment and other organic supports to which the paint is applied. With time, both types of painting acquire other sources of nutrient, e.g. dust, soot and other environmental contaminants, and also materials added during re-touching or restoration, which can further promote the aesthetic and structural changes wrought by contaminating organisms

Table 16. Genera of fungi isolated from mural paintings (Nugari *et al.* 1993).

Acremonium	*Enygiodontium*	*Phoma*
Alternaria	*Epicoccum*	*Scopulariopsis*
Anixiopsis	*Exophiala*	*Stachybotrys*
Aspergillus	*Geomyces*	*Stemphylium*
Beauveria	*Geotrichum*	*Torula*
Botryotrichum	*Gliocladium*	*Trichocladium*
Botrytis	*Gliomastix*	*Trichoderma*
Chaetomium	*Humicola*	*Ulocladium*
Circinella	*Mucor*	*Verticillium*
Cladosporium	*Paecilomyces*	
Cunninghamella	*Penicillium*	

(Strzelczyk 2004).

In reviewing the literature on mural paintings, Nugari *et al.* (1993) collated data from investigations in 22 European buildings, mostly churches or monasteries, noting that nearly 90 different species of fungi had been isolated. These are not listed by species here, but they were members of the 31 genera listed in Table 16. A typical example of the types of fungi that can be found is provided by Guglielminetti *et al.* (1994) who found that the fungal colonization of the two 17th century frescoes in an Italian monastery was mainly by *Cladosporium, Aspergillus, Alternaria, Penicillium* and *Fusarium*. In decreasing order, *Cladosporium, Penicillium, Chaetomium* and *Acremonium*, were the most frequently isolated taxa. *Cladosporium*, represented by *C. cladosporioides* and less frequently *C. herbarum* and *C. sphaerospermum*, was present in almost all sampled areas of the frescoes. *Penicillium* was very frequent, with *P. expansum* predominant among the 11 penicillia isolated. *Alternaria* was detected in certain areas, and *Chaetomium* spp. were often found in association with *Acremonium, Arthrinium, Aureobasidium, Epicoccum* and *Drechslera*. Again, when Gorbushina *et al.* (2004) investigated 16th century murals in a German parish church 20 years after they had been exposed after having been covered with whitewash for 260 years, they reported profuse development of microbial biofilms and discoloration. Gorbushina and her colleagues examined the surface of the frescoes by scanning electron microscopy (SEM) and found ample evidence of fungal growth and sporulation, both on and below the paint surface. In addition to unspecified representatives of the genera *Cladosporium, Penicillium, Acremonium, Fusarium* and *Chrysosporium*, the fungi isolated included *Aspergillus sydowii, C. sphaerosporium, Eupenicillium javanicum, Neosartorya fischeri, Scopulariopsis char-*

tarum, Ulocladium oudesmanii and also unidentified melanized non-sporing fungi. In addition to identifying the fungal taxa present on the deteriorated murals, Gorbushina *et al.* (2004) reported that 16S rDNA sequencing of bacteria isolated from the murals into culture revealed that the majority were in the genera *Bacillus* and *Arthrobacter*.

One of the fungi most frequently isolated from paintings is the cosmopolitan species, *Penicillium chrysogenum,* and an example of the type of damage that it may cause has been described by Milanesi *et al.* (2006b). This species is known to colonize the organic glue and binders mixed with the mineral pigments used commonly in the 14th century and was grown from dormant spores when millimetre-sized fragments of a medieval fresco in the Chapel of the Holy Nail in Siena were incubated in a mineral solution. As this *P. chrysogenum* strain grew on the fragments of fresco, the hyphae caused the breakup of the surface, while endogenous carbon in the fragments halved and the copper in the fragments was leached from the inorganic pigment, azurite. The copper concentrated fivefold in the hyphae, showing that the strain was particularly tolerant of copper.

The prime source of deteriogenic microorganisms is, of course, the airborne microbiota in the buildings housing the paintings, but the range of fungi and bacteria that can be isolated is likely to be smaller than in the ambient air. For example, Pangallo *et al.* (2009) have shown this for bacteria. Using PCR-based technology on isolates from a deteriorated medieval fresco in Slovakia, and from the air surrounding it, they identified from their 16S rDNA sequences 51 different strains in the air, but only 15 of these were found on the surface of the fresco. The prevailing bacteria on the fresco were *Bacillus* (8 strains) and in the air *Staphylococcus* (4). Among those strains present in the air but absent from the fresco were two strains of *Pseudomonas*. However, when Gurtner *et al.* (2000) investigated the directly extracted bacterial DNA from the murals in a German church mentioned earlier (Gorbushina *et al.* 2004) and in an Austrian castle by amplifying it and identifying 16S rDNA sequences, a total of 23 sequences corresponded to 14 actinobacterial genera, including *Arthrobacter, Frankia, Geodermatophilus* and *Nocardioides*; 27 Protobacteria, including *Pseudomonas, Halomonas, Chromohalobacter* and unidentified γ-Protobacteria; and 19 unidentified Cytophagales. The only sequences common to the two different murals were *Arthrobacter, Frankia, Geodermatophilus* and unidentified members of the Cytophagales. It seems likely that differences in the materials used in the painting and in the environmental microclimate could have accounted for these differences in the bacterial communities. Sequences for bacteria isolated into culture from the paintings corresponded to *Bacillus, Paenibacillus, Micrococcus, Staphylococcus, Methylobacterium* and *Halomonas.* The sequence for the cultured *Halomonas* was different from those *Halomonas* sequences derived directly from the murals.

Bacteria on paintings may grow as extensive biofilms, but not all do so. For instance, the biodeterioration of the lower part of an 18th century fresco in a church in Siena, was at first attributed by Milanesi *et al.* (2006a) to a combination of the presence of two bacteria that were isolated from a damaged area of the fresco and dampness caused by capillary rise of water from below. The bacteria, identified by sequencing the 16S rDNA gene, were *Kocuria erythromyxa* (formerly *Micrococcus roseus*) and *Sphingomonas echinoides. S. echinoides* extensively colonized the surface of small fragments taken from the fresco, with its cells being surrounded by water-retentive extracellular polymeric substance (EPS) giving attachment to the surface. However, *K. erythromyxa* did not appear to be involved in biofilm formation, although the possibility that it was a minor component was not excluded. The investigation of Gurtner *et al.* (2000) bears witness to the increasing recognition that actinobacteria may form an important element in the microbiota. Giacobini *et al.* (1988) signposted their importance in an examination of historically important frescoes in crypts, tombs and grottos at various locations in Italy. She and her colleagues isolated 19 different species of *Streptomyces.* As well as *Nocardia* spp., the most frequently found were *S. rectus flexibilis, S. albus, S. cinereoruber, S. griseolus* and *S. vinaceus.* Under the environmental conditions at the different sites, where the RH remained at >90% and the temperature ranged from 6 to 7°C in winter to 16 to18°C in summer, these actinobacteria produced on the frescoes a powdery white mycelium or a colourless patina with a scattering of small, light grey granular areas. White growths in which *Pseudonocardia* was the largest component detected by molecular methods were noted by Stomeo *et al.* (2008) on deteriorating palaeolithic paintings on the walls of two humid caves in different climatic regions of Spain. In addition to other actinobacteria, bacteria in the Planctomycetes and Chloroflexi were present in both cases. Protobacteria and Firmicutes (mainly *Bacillus* and *Streptococcus*) were prominent in the microbiota of the cooler cave.

Where humidity and light are favourable for their growth, microalgal and cyanobacterial photoautotrophs may assume considerable importance, sometimes leading to closure of caves containing prehistoric paintings because of *maladie verte*, as in those at Lascaux and Altamira. In an investigation of the 9th century wall paintings in The Crypt of the Original Sin, a cave near Matera in Italy that was used as a church in the Middle Ages, Nugari *et al.* (2009) demonstrated the importance of these photoautotrophs. The murals were discoloured in areas, from brilliant green, dark green, brown, black with a powdery appearance to rose-pink patinas. The dark green patinas were predominantly comprised of the cyanobacteria *Chlorogloea microcystoides, Gloeocapsa bisporus, G. kuuetzingiana* and *G. rupestris,* and the green algae *Chlorococcum* sp. and *Muriella terrestris.* The microbiota of black patinas was dominated by the four cyanobacteria; the brown by the three species of *Gloecapsa;* and the brilliant green by the green algae *Chlorococcum* sp., *M. terrestris* and *Apatococcus lobatus* (with *Chlorella vulgaris* being scarce). The authors considered that the pink patina might have owed its colour to carotenoids produced by photoheterotrophic actinobacteria related to *Rubrobacter radiotolerans,* but none were isolated.

REFERENCES

Abe, K. (2010) Assessment of the environmental conditions in a museum storehouse by use of a fungal index. *Int. Biodet. Biodegr.,* **64**, 32-40.

Abrusci, C., Allen, N.S., Del Amo, A., *et al.* (2004) Biodegradation of motion picture film stocks. *J. Film. Preserv.,* **67**, 37-54.

Abrusci, C., Marquina, D., Del Amo, A., *et al.* (2006) A viscometric study of the biodegradation of photographic gelatin by fungi isolated from cinematographic films. *Int. Biodet. Biodegr.,* **58**, 142-149.

Abrusci, C., Martin-González, A., Del Amo, A., *et al.* (2005) Isolation and identification of bacteria and fungi from cinematographic films. *Int. Biodet. Biodegr.,* **56**, 58-68.

Abrusci, C., Marquina, D., Santos, A., *et al.* (2009) Biodeterioration of cinematographic cellulose triacetate by *Sphingomonas paucimobilis* using indirect impedance and chemiluminescence techniques. *Int. Biodet. Biodegr.,* **63**, 759-764.

Adan, O.C.G. (1994) On the Fungal Defacement of Interior Finishes, PhD Thesis, University of Eindhoven, The Netherlands.

Ahearn, D.G., Price, D.L., Simmons, R.B., and Crow, S.A. (1993) Colonization studies of various HVAC insulation materials. In *IAQ '92: Environments for People,* American Society of Heating, Refrigerating and Air-Conditioning Engineers, Atlanta, GA, pp. 179-184.

Ahearn, D.G., Crow, S.A., Simmons, R.B., Price, D.L., Mishra, S.K., and Pierson, D.L. (1996) Fungal colonization of fiberglass insulation in the air distribution system of a multi-story office building: VOC production and possible relationship to a sick building syndrome. *J. Industr. Microbiol. Biotechnol.,* **16**, 280-285.

Ahearn, D.G., Crow, S.A., Simmons, R.B., Price, D.L., Mishra, S.K., and Pierson, D.L. (1997) Fungal colonization of air filters and insulation in a multi-story office building: production of volatile organics. *Current Microbiol.*, **35**, 305-308.

Airoldt, T. and Litsky, T. (1972) Factors contributing to the microbial contamination of cold-water humidifiers. *Am. J. Med. Technol.*, 38, 491-495.

Allsopp, D., Seal, K., and Gaylarde, C. (2004) *Introduction to Biodeterioration*, 2nd ed., Cambridge University Press, Cambridge, UK.

Andersson, M.A., Nikulin, M., Koljalg, U., Andersson, M.C., Rainey, F., Reijula, K., Hintikka, E.L., and Salkinoja-Salonen, M. (1997) Bacteria, molds, and toxins in water-damaged building materials. *Appl. Environ. Microbiol.*, 63, 387-393.

Ando, M., Arima, K., Yoneda, R., and Tamura, M., (1991) Japanese summer-type hypersensitivity pneumonitis. Geographic distribution, home environment, and clinical characteristics of 621 cases. *Am. Rev. Respir. Dis.*, **144**, 765-769.

Apetrei, I.C., Drăgănescu, G.E., Popescu, I.T., *et al.* (2009) Possible cause of allergy for the librarians: books manipulation and ventilation as sources of fungus spores spreading. *Aerobiologia*, **25**, 159-166.

Arai, H. (2000) Foxing caused by fungi; twenty-five years of study. *Int. Biodet. Biodegr.*, **46**, 181-188.

Armolik, N., and Dickson, J.G. (1956) Minimum humidity requirements for germination of conidia of fungi associated with storage of grain. *Phytopathology*, **46**, 462-465.

Ayerst, G. (1969) The effects of moisture and temperature on growth and spore germination in some fungi. *J. Stored Prod. Res.*, **5**, 127-141.

Belyakova, L.A. (1964) The mold species and their injurious effect on various book materials. In *Restoration and Preservation of Library Resources, Documents and Books*, Oldbourne Press, London, UK, pp. 183-194.

Bissett, J. (1987) Fungi associated with urea-formaldehyde foam insulation in Canada. *Mycopathologia*, **99**, 47-56.

Block, S.S. (1953) Humidity requirements for mold growth. *Appl. Microbiol.*, **1**, 287-293.

Blyskal, B. (2009) Fungi utilizing keratinous substrates. *Int. Biodet. Biodegr.*, **63**, 631-653.

Borrego, S., Guiamet, P., Gómez de Saravia, S., *et al.* (2010) The quality of air at archives and the biodeterioration of photographs. *Int. Biodet. Biodegr.*, **64**, 139-145.

Buchanan, C.M., Gardner, R.M., and Komarek, R.J. (1993) Aerobic biodegradation of cellulose acetates. *J. Appl. Polymer Sci.*, **74**, 1709-1719.

Burch, D.M., and Desjarlais., A.O. (1995) *Water-Vapor Measurements of Low-Slope Roofing Materials*, NISTIR 5681. National Institute of Standards and Technology, Gaithersburg, MD.

Burge, H.P., Boise, J.R., Solomon, W.R., and Bandera, E. (1978) Fungi in libraries: an aerometric survey. *Mycopathologia*, **64**, 67-72.

Caneva, G., Maggi, O., Nugari, M.P., *et al.* (2003) The biological aerosol as a factor of biodeterioration. In P. Mandrioli, G. Caneva, and C. Sabbioni, *Cultural Heritage and Aerobiology*. Kluwer, Dordrecht, The Netherlands, pp. 1-29.

Carlisle, M.J., and Watkinson, S.C. (1994) *The Fungi*. Academic Press, London.

Christian, J.E. (1993) A search for moisture sources. In W.B. Rose and A. TenWolde, (eds.), *Bugs, Mold & Rot II*, National Institute of Building Sciences, Washington, DC, pp. 71-81.

Cole, E.C., Foarde, K.K., Leese, K.E., Green, D.E., Franke, D.L., and Berry, M.A. (1994) Assessment of fungi in carpeted environments. In R.A. Samson, B. Flannigan, M.E. Flannigan, A.P. Ver-

hoeff, O.C.G. Adan, and E.S. Hoekstra, (eds.), *Health Implications of Fungi in Indoor Environments*, Elsevier, Amsterdam, pp. 103-128.

Connolly, J.D. (1993) Humidity and building materials. In W.B. Rose and A. TenWolde, (eds.), *Bugs, Mold & Rot II*, National Institute of Building Sciences, Washington, DC, pp. 29-36.

Dales, R.E., Miller, J.D., and McMullen, E. (1997) Indoor air quality and health: validity and determinants of reported home dampness and moulds, *Int. J. Epidemiol.*, **26**, 1-6.

Dudney, C.S., Hinke, N.F., and Becker, J.M. (1982) *On the Occurrence of Fungi in Loose-fill Attic Insulation in Typical Single Family Dwellings*, ORNL/CON-93. Oak Ridge National Laboratory, TN.

Eaton, R.A., and Hale, M.D.C. (1993) *Wood – Decay, Pests and Protection*, Chapman & Hall, London.

Ellis, M.B. (1976) *Dematiaceous Hyphomycetes*. CAB International, London.

Eveleigh, D. (1961) The disfigurement of painted surfaces by fungi, with special reference to *Phoma violacea. Ann. Appl. Biol.*, **49**, 403-411

Ezeonu, I.M., Noble, J.A., Simmons, R.B., Price, D.L., Crow, S.A., and Ahearn, D.G. (1994) Effect of relative humidity on fungal colonization of fiberglass insulation. *Appl. Environ. Microbiol.*, **60**, 2149-2151.

Ezeonu, I.M., Price, D.L., Crow, S.A., and Ahearn, D.G. (1995) Effects of extracts of fiberglass insulations on the growth of *Aspergillus fumigatus* and *A. versicolor. Mycopathologia*, **132**, 65-69.

Fergusson, R.J., Milne, L.J.R., and Crompton, G.K. (1984) *Penicillium* allergic alveolitis: faulty installation of central heating. *Thorax*, **39**, 294-298.

Fink, J.N., Resnick, A.J. and Salvaggio, J. (1971) Presence of thermophilic actinomycetes in residential heating systems. *Appl. Microbiol.*, **22**, 730-731.

Flannigan, B. (1970) Degradation of arabinoxylan and carboxymethyl cellulose by fungi isolated from barley kernels. *Trans. Br. Mycol. Soc.*, **55**, 277-281.

Flannigan, B. (1993) Approaches to assessment of the microbial flora of buildings. In *IAQ '92: Environments for People*, American Society of Heating Refrigerating and Air-Conditioning Engineers, Atlanta, GA, pp. 136-146.

Flannigan, B., and Sellars, P.N. (1972) Activities of thermophilous fungi from barley kernels against arabinoxylan and carboxymethyl cellulose. *Trans. Br. Mycol. Soc.*, **58**, 338-341.

Flannigan, B., McCabe, E.M., and McGarry, F. (1991) Allergenic and toxigenic micro-organisms in houses. *J. Appl. Bact.*, **67**, 61S-73S.

Flannigan, B., McEvoy, E.M. and McGarry, F. (1999) Investigation of airborne and surface bacteria in homes. In G. Raw, C. Aizlewood and P. Warren (eds.) *Indoor Air 99, Proceedings*, Vol. 1. Construction Research Communications Ltd, London, pp. 884-889.

Flannigan, B., Morey, P.R., Broadbent, C., Brown, S.K. Follin, T., Kelly, K.M., Miller, J.D., Nathanson, T., Walkinshaw, D.S., and White, W.C. (1996) *ISIAQ Guideline, Task Force 1: Control of Moisture Problems Affecting Biological Indoor Air Quality*. International Society for Indoor Air Quality and Climate, Ottawa, Canada.

Flannigan, B., Pasanen, A.-L., Pasanen, P., and Vicars, S. (1992) Assessment of bioparticles in airborne dust. In T.K. Pierson and D.F. Naugle, (eds.), *Sampling and Analysis of Biocontaminants and Organics in Non-Industrial Environments*. Report of NATO/CCMS Pilot Study on Indoor Air Quality, Fifth Plenary Meeting, Research Triangle Institute, Research Triangle Park, NC, pp. 35-43.

Fradkin, A., Tobin, R.S., Tarlo, S.M., Tucic-Porretta, M., and Malloch,

M. (1987) Species identification of airborne molds and its significance for the detection of indoor pollution. *J. Air Pollut. Control Assoc.*, **37**, 51-53.

Giacobini, C., De Ciccio, M.A., Tiglie, I., and Accardo, G. (1988) Actinomycetes and biodeterioration in the field of fine art. In D.R. Houghton and H.O.W. Eggins, (eds.), Biodeterioration 7. Elsevier, Applied Science, London, UK, pp. 418-423.

Gallo, F. (1993) Aerobiological research and problems in libraries. *Aerobiologia*, **9**, 117-130.

Gambale, W., Croce, J., Costa-Manso, E., *et al.* (1993) Library fungi at the University of São Paulo and their relationship with respiratory allergy. *J. Investig. Allergol. Clin. Immunol.*, **3**, 45-50.

Gorbushina, A.A., Heyrman, J., Dornieden, T., *et al.* (2004) Bacterial and fungal diversity and biodeterioration problems in mural painting environments of St. Martins church (Greene-Kreisen, Germany). *Int. Biodet. Biodegr.*, **53**, 13-24.

Grant, C., Hunter, C.A., Flannigan, B., and Bravery, A.F. (1989) Water activity requirements of moulds isolated from domestic dwellings. *Int. Biodeterioration*, **25**, 259-284.

Gravesen, S., Nielsen, P.A., and Nielsen, K.F. (1997) *Microfungi in Water Damaged Buildings*, SBI Report No. 282. Danish Building Research Institute, Copenhagen, Denmark.

Guglielminetti, M., De Giuli Morghen, C., Radaelli, A., *et al.* (1994) Mycological and ultrastructural studies to evaluate biodeterioration of mural paintings. Detection of fungi and mites in frescos of the monastery of St. Damian in Assisi. *Int. Biodet. Biodegr.*, **33**, 269-283.

Gurtner, C., Heyrman, J., Piñar, G., *et al.* (2000) Comparative analyses of the bacterial diversity on two biodeteriorated wall paintings by DGGE and 16S rDNA sequence analysis. *Int. Biodet. Biodegr.*, **46**, 229-239.

Heinemann, S., Beguin, H., and Nolard, N. (1994) Biocontamination in air-conditioning. In R.A. Samson, B. Flannigan, M.E. Flannigan, A.P. Verhoeff, O.C.G. Adan, and E.S. Hoekstra, (eds.), *Health Implications of Fungi in Indoor Environments*, Elsevier, Amsterdam, pp. 179-186.

Herrera, J. (2005) Assessment of fungal growth on sodium polyborate treated cellulose insulation. *J. Occup. Environ. Hygiene*, **2**, 626-632.

Herrera, J. (2008) Sodium polyborate-based additives on recycled cellulose insulation kill or prevent germination of common indoor fungi. *Building Enclosure Science and Technology (BEST) 2008 Conference, Minneapolis*, Proceedings, 10 pp. [http://www.thebestconference.org/best1/program.htm].

Hoekstra, E.S., Samson, R.A., and Verhoeff, A.P. (1994) Fungal propagules in house dust: a qualitative analysis. In R.A. Samson, B. Flannigan, M.E. Flannigan, A.P. Verhoeff, O.C.G. Adan, and E.S. Hoekstra, (eds.), *Health Implications of Fungi in Indoor Environments*, Elsevier, Amsterdam, pp. 169-177.

Hugenholz, P., and Fuerst, J.A. (1992). Heterotrophic bacteria in an air-handling system. *Appl. Environ. Microbiol.*, **58**, 3914-3920.

Hunter, C.A. (1989) Factors Affecting Mould Growth and the Air Spora in Houses. PhD Thesis, Heriot-Watt University, Edinburgh, UK.

Hunter, C.A., Grant, C., Flannigan, B., and Bravery, A.F. (1988) Mould in buildings: the air spora of domestic dwellings. *Int. Biodeterioration*, **24**, 84-101.

Hunter, C.A., and Bravery, A.F. (1989) Requirements for growth and control of surface moulds in dwellings. In B. Flannigan, (ed.), *Airborne Deteriogens and Pathogens*, The Biodeterioration Society, Kew, Surrey, UK, pp. 174-182.

Hyvärinen, A., Meklin, T., Vepsäläinen, A., and Nevalainen, A. (2002) Fungi and actinobacteria in moisture-damaged building materials - concentrations and diversity. *Int. Biodet. Biodegr.*, **49**, 27-37.

Ishigaki, T., Sugano, W., Ike, M., and Fujita M. (2000) Enzymatic degradation of cellulose acetate plastic by novel degrading bacterium *Bacillus* sp. S2055. *J. Biosci. Bioeng.*, **90**, 400-405.

Johanning, E.(1995) Health problems related to fungal exposure – the example of toxigenic *Stachybotrys chartarum* (*atra*). In E. Johanning and C.S. Yang, (eds.), Fungi and bacteria in indoor air environments. East New York Occupational Health Program, 12110, Latham, NY, pp. 207-218.

Johanning, E. (1998) *Stachybotrys* revisited. *Clin. Toxicol.*, **36**, 629-631.

Johanning, E., Morey, P.R., and Jarvis, B.B. (1993) Clinical-epidemiological investigation of health effects caused by *Stachybotrys atra* building contamination. *Indoor Air '93*, Proceedings of the Sixth International Conference on Indoor Air Quality and Climate, Vol. 1, Helsinki, Indoor Air '93, pp. 225-230.

Johanning, E., Biagini, R., Hull, D., *et al.* (1996) Health and immunology study following exposure to toxigenic fungi (*Stachybotrys chartarum*) in a water-damaged office environment. *Int. Arch. Occup. Environ. Health.*, **68**, 207-218.

Käpylä, M. (1985) Frame fungi on insulated windows. *Allergy*, **40**, 558-564.

Kelly, K.M. (1988) *Wet House Checklist*. Jay-K Independent Lumber Co. Inc., New Hartford, NY.

Klamer, M., Morsing, E., and Husemoen, T. (2004) Fungal growth on different insulation materials exposed to different moisture regimes. *Int. Biodet. Biodegr.*, **54**, 277–282.

Klich, M. and Pitt, J.I. (1988) *A Laboratory Guide to Common Aspergillus Species and Their Teleomorphs*. CSIRO, North Ryde, Sydney, Australia.

Koivula, M., Kymäläinen, H.-R., Virta, J., *et al.* (2005) Emissions from thermal insulations – part 2: evaluation of emissions from organic and inorganic insulations. *Build. Environ.*, **40**, 803-814.

Komarek R.J., Gardner, R.M., Buchanan, C.M., and Gedon, S. (1993) Biodegradation of radiolabeled cellulose acetate and cellulose propionate. *J. Appl. Polymer Sci.*, **50**, 1739-1746.

Kozak, P.P., Gallup, J., Cummins, L.H., and Gillman, S.A. (1980) Currently available methods for home mold surveys. II. Examples of problem homes surveyed. *Ann. Allergy*, **45**, 167-176.

Lacey, J., Hill, S.T., and Edwards, M.A. (1980) Micro-organisms in stored grains: their enumeration and significance. *Trop. Stored Prod. Inform.*, **39**, 19-33.

Lawton, M.D., Dales, R.E., and White, J. (1998) The influence of house characteristics in a Canadian community on microbiological contamination. *Indoor Air*, **8**, 2-11.

Lignell, U. (2008) *Characterization of Microorganisms in Indoor Environments*, Publication A3 2008. National Public Heath Institute, Kuopio, Finland.

Lim, G., Tan, T., and Toh, A. (1989) The fungal problems in buildings in the humid tropics. *Int. Biodeterioration*, **25**, 27-37.

Linos, A., and Steinbüchel, A. (1996) Investigations on the microbial breakdown of natural and synthetic rubbers. In *Biodeterioration and Biodegradation, Proceedings of Tenth International Biodeterioration and Biodegradation Symposium*, DECHEMA, Frankfurt am Main, pp. 279-286.

Ljaljević-Grbić, M., Vukojević, J., and Stupar, M. (2008) Fungal colonization of air-conditioning systems. *Arch. Biol. Sci. Belgrade*, **60**, 201-206.

Lourenço, M.J.L., and Sampaio, J.P. (2009) Microbial deterioration of gelatin emulsion photographs: differences of susceptibility between black and white and colour materials. *Int. Biodet. Biodegr.*, **63**, 496-502.

Lstiburek, J. (1994) *Mold, Moisture and Indoor Air Quality. A Guide for Designers, Builders, and Building Owners.* Building Science Corporation, Chestnut Hill, MA.

Lstiburek, J.W. (2007) The material view of mold. *ASHRAE J.,* **49**, 81-83.

Ludwig, J.A., and Reynolds, J.F. (1988) *Statistical Ecology.* John Wiley, New York.

Magan, N., and Lacey, J. (1984) Effect of temperature and pH on the water relations of field and storage fungi. *Trans. Br. Mycol. Soc.,* **82**, 71-81.

Maggi, O., Persiani, A.M., Gallo, F., *et al.* (2000) Airborne fungal spores in dust present in archives: proposal for a detection method, new for archival materials. *Aerobiologia,* **16**, 429-434.

Maghraby, T.A., Gherbawy, Y.A.M.H., and Hussein, M.H. (2008) Keratinophilic fungi inhabiting floor dusts of student houses at the South Valley University in Egypt. *Aerobiologia,* **24**, 99-106.

Merrill, J.L., and TenWolde, A. (1989) Overview of moisture-related damage in one group of Wisconsin manufactured homes. *ASHRAE Trans.,* **95**, 405-414.

Mesquita, N., Portugal, A., Videira, S., *et al.* (2009) Fungal diversity in ancient documents. A case study on the archive of the University of Coimbra. *Int. Biodet. Biodegr.,* **63**, 626-629.

Milanesi, C., Baldi, F., Borin, S., *et al.* (2006a) Biodeterioration of a fresco by biofilm forming bacteria. *Int. Biodet. Biodegr.,* **57**, 168-179.

Milanesi, C., Baldi, F., Vignani, R., *et al.* (2006b) Fungal deterioration of medieval wall fresco determined by analyzing small fragments containing copper. *Int. Biodet. Biodegr.,* **57**, 7-13.

Miller, J.D. (1993) Fungi and the building engineer. In *IAQ '92: Environments for People,* American Society of Heating Refrigerating and Air-Conditioning Engineers, Atlanta, GA, pp. 147-158.

Miller, J.D. (1995) Quantification of health effects of combined exposures: a new beginning. In L.L. Morawska, N.D. Bofinger and M. Maroni, (eds.), *Indoor Air Quality: An Integrated Approach,* Elsevier, Amsterdam, The Netherlands, pp. 159-168.

Miller, J.D. (2004) Mold growth on insulation materials: from UFFI to modern products. *Book of Papers, International Nonwovens Technical Conference, Toronto, 2004.* INDA, Cary, NC, 11 pp.

Miller, J.D., Dussault, R., Shirtliffe, C., Laflamme, A.M., and Richardson, M. (1984) *Studies Concerning the Association of Filamentous Fungi and Urea Formaldehyde Foam Insulation in Some Canadian Houses.* Consumer and Corporate Affairs, Ottawa, Ontario, Canada.

Mislivec, P.B., and Tuite, J. (1970) Temperature and relative humidity requirements of species of *Penicillium* isolated from yellow dent corn kernels. *Mycologia,* **62**, 75-88.

Montemartine Corte, A., Ferroni, A., and Salvo, V.S. (2003) Isolation of fungal species from test samples and maps damaged by foxing, and correlation between these species and the environment. *Int. Biodet. Biodegr.,* **51**, 167-173.

Morey, P.R. (1993a) Microbiological events after a fire in a high-rise building. *Indoor Air,* **3**, 354-360.

Morey, P.R. (1993b) Use of hazard communication standard and general duty clause during remediation of fungal contamination. In *Indoor Air '93, Proceedings of the Sixth International Conference on Indoor Air Quality and Climate,* Vol. 4, *Particles, Microbes, Radon,* Indoor Air '93, Helsinki, pp. 391-395.

Morey, P.R., Hull, M.C., and Andrew, M. (2003) El Niño water leaks identify rooms with concealed mould growth and degraded indoor air quality. *Int. Biodet, Biodegr.,* **52**, 197-202.

Murtoniemi, T., Nevalainen, A., and Hirvonen, M.-R. (2003a) Effect of plasterboard composition on *Stachybotrys chartarum* growth and biological activity of spores. *Appl. Environ. Micro-*

biol., **69**, 3751-3757.

Murtoniemi, T., Hirvonen, M.-R., Nevalainen, A., and Suutari, M. (2003b) The relation between growth of four microbes on six different plasterboards and biological activity of spores. *Indoor Air,* **13**, 65-73.

Murtoniemi, T., Keinänen, M.M., Nevalainen, A., and Hirvonen, M.-R. (2003c) Starch in plasterboard sustains *Streptomyces californicus* growth and bioactivity of spores. *J. Appl. Microbiol.,* **94**, 1059-1065.

Nathanson, T., and Miller, J.D. (1989) Studies of fungi in indoor air in large buildings. In B. Flannigan, (ed.), *Airborne Deteriogens and* Pathogens, Biodeterioration Society, Kew, Surrey, UK, pp. 129-138.

Nelson, M., McCarthy, S.P., and Gross, R.A. (1993) Isolation of a *Peseudomonas paucimobilis* capable of using insoluble cellulose acetateas a sole carbon source. *Proc. ACS Div., Polym. Mat. Sci. Eng.,* **67**, 139-140.

Nevalainen, A. (1993) Microbial contamination of buildings. In *Indoor Air '93, Proceedings of the Sixth International Conference on Indoor Air Quality and Climate,* Vol. 4, *Particles, Microbes, Radon,* Indoor Air '93, Helsinki, Finland, pp. 3-13.

Nielsen, K.F., Hansen, M.Ø., Larsen, T.O., and Thrane, U. (1998a) Production of trichothecene mycotoxins on water damaged gypsum boards in Danish buildings. *Int. Biodet. Biodegr.,* **42**, 1-7.

Nielsen, K.F., Thrane, U., Larsen, T.O., Nielsen, P.A., and Gravesen, S. (1998b). Production of mycotoxins on artificially inoculated building materials. *Int. Biodet. Biodegr.,* **42**, 9-16.

Nishimura, K.M., Miyaji, M., Taguchi, H., and Tanaka, R. (1987) Fungi in bathwater and sludge of bathroom drainpipes. 1. Frequent isolation of *Exophiala* species. *Mycopathologia,* **97**, 17-23.

Northolt, M. D. (1979) *The Effect of Water Activity and Temperature on the Production of Some Mycotoxins.* PhD Thesis, University of Wageningen, The Netherlands.

Nugari, M.P., Realini, M., and Roccardi, A. (1993) Contamination of mural paintings by indoor airborne fungal spores. *Aerobiologia,* **9**, 131-139.

Nugari, M.P., Pietrini, A.M., Caneva, G., *et al.* (2009) Biodeterioration of mural paintings in a rocky habitat: the Crypt of the Original Sin (Matera, Italy). *Int. Biodet. Biodegr.,* **63**, 705-711.

NYC (1993) *Guidelines on Assessment and Remediation of* Stachybotrys atra *in Indoor Environments.* New York City Department of Health, New York City Human Resources Administration, and Mount Sinai-Irving J. Selikoff Occupational Health Clinical Center, New York.

Nykter, M., (2006) *Microbial Quality of Hemp (Cannabis sativa* L.) and Flax *(Linum usitatissimum* L.) from Plants to Thermal Insulation. Academic Dissertation, University of Helsinki, Helsinki, Finland.

Opela, V. (1992) Fungal and bacterial attack on motion picture film. Joing Technical Symposium, Ottawa, Canada, 1992, pp. 139-144.

Orlita, A. (2004) Microbial biodeterioration of leather and its control: a review. *Int. Biodet. Biodegr.,* **53**, 157-163.

Oxley, T.A., and Gobert, E.G. (1983) *Dampness in Buildings.* Butterworths, London.

Panasenko, V.T. (1967) Ecology of microfungi. *Bot. Rev.,* **33**, 189-215.

Pangallo, D., Chovanová, K., Drahovska, H., *et al.* (2009) Application of fluorescence internal transcribed spacer-PCR (f-ITS) for the cluster analysis of bacteria isolated from air and deteriorated fresco surfaces. *Int. Biodet. Biodegr.,* **63**, 868-872.

Pessi, A.-M., Suonketo, J., Pentti, M., Peltola, K., and Rantio-Lehtimäki, A. (2002) Microbial growth inside insulated walls as an indoor air biocontamination source. *Appl. Environ. Microbiol.*, **69**, 963-967.

Pitt, J.I. (1988) *Laboratory Guide to Common* Penicillium *Species*. CSIRO, North Ryde, Sydney, Australia.

Prezant, B., Weekes, D., and Miller, J.D. (eds.) (2008) *Recognition, Evaluation and Control of Indoor Mold*. American Industrial Hygiene Association, Fairfax, VA.

Price, D.L., Simmons, R.B., Crow, S.A., and Ahearn, D.G. (2005) Mold colonization during use of preservative-treated and untreated air filters, including HEPA filters from hospitals and commercial locations over an 8-year period (1996–2003). *J. Ind. Microbiol. Biotechnol.*, **32**, 319-321.

Richards, R.F., Burch, D.M. and Thomas, W.C. (1992) Water vapor sorption measurements of common building materials. *ASHRAE Technical Data Bulletin*, **8**, 58-68.

Romero-Noguera, J., Bolivar-Galiano, F.C., Ramos-López, J.M., *et al.* (2008) Study of biodeterioration of diterpenic varnishes used in art painting: colophony and Venetian turpentine. *Int. Biodet. Biodegr.*, **62**, 427-433.

Roponen, M., Toivola, M., Meklin, T., Ruotsalainen, M., Komulainen, H., and Nevalainen, A. (2001) Differences in inflammatory responses and cytotoxicity in RAW264.7 macrophages induced by *Streptomyces anulatus* grown on different building materials. *Indoor Air*, **11**, 179-184.

Rose, W.B., and TenWolde, A. (eds.) (1993) *Bugs, Mold & Rot II*. National Institute of Building Sciences, Washington, DC.

Ross, R.T., Sladen, J.B., and Wienert, L.A. (1968) Biodeterioration of paint and paint films. In A.H. Walters and J.J. Elphick, (eds.), *Biodeterioration of Materials: Microbiological and Allied Aspects*, Barking, U.K., pp. 317-325.

Rowan, N., Johnstone, C.M., McLean, R.C., Anderson, J.G., and Clarke, J.A. (1999) Prediction of toxigenic fungal growth in buildings by using a novel modelling system. *Appl. Environ. Microbiol.*, **65**, 4814-4821.

Rowley, B., and Walkinshaw, D.S. (1995) *Basement Moisture Problem Survey*, Ottawa Carleton Home Builders Association, Ottawa.

Ruga, L., Bonofiglio, T., Orlandi, F., *et al.* (2008) Analysis of the potential fungal biodeterioration effects in the "Doctorate Library" of the University of Perugia, Italy. *Grana*, **47**, 60-69.

Sakai, K., Yamauchi, T., Nakasa, F., and Obe, T. (1996) Biodegradation of cellulose acetate by *Neisseria sicca*. *Biosci. Biotech. Biochem.*, **60**, 1617-1622.

Sakamoto, T., Urisu, A., and Yamanda, M. (1989) Studies on the osmophilic fungus *Wallemia sebi* as an allergen evaluated by skin prick text and radioallergosorbent test. *Int. Arch. Allergy Appl. Immunol.*, **90**, 368-372.

Sánchez-Medina, M., Zarante, I., Rodríguez, M., Ramírez, M. (1996) Respiratory allergy in library workers. Are mites implicated? *J. Allergy Clin. Immunol.*, **97**, 226.

Schmidt, O., and Moreth-Kebernik, U. (1990) *Old and New Facts on the Dry Rot Fungus* Serpula lacrymans, Document No. IRG/WP/1470. International Research Group on Wood Preservation. Stockholm, Sweden.

Schober, G. (1991) Fungi in carpeting and dust. *Allergy*, **46**, 639-643.

Scott, W.J. (1957) Water relations of food spoilage microorganisms. *Adv. Food Res.*, **7**, 83-127.

Shamsian, A., Fata, A., Mohajeri, M., and Ghazvini, K. (2006) Fungal contaminations in historical manuscripts at Astan Quds Museum Library, Mashhad, Iran. *Int. J. Agric. Biol.*, **8**, 420-422.

Shaughnessy, R., Morey, P.R., and Cole, E.C. (1999) Remediation of microbial contamination. In J. Macher, H.A. Ammann, H.A.

Burge, *et al.*, (eds.), *Bioaerosols: Assessment and Control*. American Conference of Government Industrial Hygienists, Cincinnati, OH, pp. 15.1-15.7.

Silverman, R. (2004) The day the university changed. *Idaho Librarian*, **55**, 3. (Online at www.idaholibraries.org/newidaholibrarian/200402/ColStateFloodPF.htm).

Singh, A., Ganguli, M., and Singh, A.B. (1995) Fungi are an important component of library air. *Aerobiologia*, **11**, 231-237.

Siu, R.G.H. (1951) *Microbial Decomposition of Cellulose*. Reinhold, New York.

Snow, D. (1949) The germination of mould spores at controlled humidities. *Ann. Appl. Biol.*, **36**, 1-13.

Solarz, K. (2009) Indoor mites and forensic acarology. *Exp. Appl. Acarol.*, **49**, 135-142.

Steiman, R., Guiraud, P., Sage, L., and Seigle-Murandi, F. (1997) Soil mycoflora from the Dead Sea oases of Ein Gedi and Einot Zuqim (Israel). *Antonie van Leeuwenhoek*, **72**, 261–270.

Stickley, F.L. (1986) The biodegradation of gelatin and its problems in the photographic industry. *J. Photogr. Sci.*, **34**, 111-112.

Stomeo, F., Portillo, M.C., Gonzalez, J.M., *et al.* (2008) *Pseudonocardia* in white colonizations in two caves with Paleolithic paintings. *Int. Biodet. Biodegr.*, **62**, 483-486.

Strzelczyk, A.B. (2004) Observations on aesthetic and structural changes induced in Polish historic objects by microorganisms. *Int. Biodet. Biodegr.*, **53**, 151-156.

Suutari, M., Rönkä, E., Lignell. U., Rintala, H., and Nevalainen, A. (2002) Characterisation of *Streptomyces* spp. isolated from water-damaged buildings. *FEMS Microbiol. Ecol.*, **39**, 77-84.

Takatori, K., Lee, H.-J., Ohta, T., and Shida, T. (1994) Composition of the house dust mycoflora in Japanese houses. In R.A. Samson, B. Flannigan, M.E. Flannigan, A.P. Verhoeff, O.C.G. Adan, and E.S. Hoekstra, (eds.), *Health Implications of Fungi in Indoor Environments*, Elsevier, Amsterdam, pp. 93-101.

Temby, M.A. (2001) Mould: the invasive intruder. *Aust. Library J.*, **50**, 175-180.

TenWolde, A. (1993) Indoor humidity and the building envelope. In W.B. Rose and A. TenWolde, (eds.), *Bugs, Mold & Rot II*, National Institute of Building Sciences, Washington, DC, pp. 37-41.

Torii, S., Sakamoto, T., and Matsuda, Y. (1990) Significance of xerophilic fungi in indoor environment – allergenic and antigenic activities of the xerophilic fungi in asthmatic patients. In *IUMS Congress of Bacteriology and Mycology, Osaka, Japan, Abstracts*, p. 30.

Tye, R.P., Ashare, E., Guyer, E.C., and Sharon, A.C. (1980) An assessment of thermal insulation materials for building applications. In D.L. McElroy and R.P. Tye (eds.), *Thermal Insulation Performance*, ASTM STP 718. American Society for Testing and Materials, West Conshohocken, PA, pp. 9–26.

Upsher, F.J. (1972) Microfungi at the Joint Tropical Research Unit, Innisfail, Queensland. In A.H. Walters and E.H. Hueck-van der Plas, (eds.), *Biodeterioration of Materials*, Vol. 2, Applied Science, London, pp. 27-34.

Van Loo, J.M., Robbins, C.A., Swenson, L., and Kelman, J. (2004) Growth of mold on fiberglass insulation building materials - a review of the literature. *J. Occup. Environ Hyg.* **6**, 349-354.

Verhoeff, A.P. (1994) *Home Dampness, Fungi and House Dust Mites and Respiratory Symptoms in Children*. PhD Thesis, Erasmus University, Rotterdam, The Netherlands.

Wälchli, O., and Raschle, P. (1983) The dry rot fungus – experience on causes and effects of its occurrence in Switzerland. In T.A. Oxley and S. Barry, (eds.), *Biodeterioration 5*, John Wiley, Chichester, UK, pp. 84-96.

Yang, C.S., and Li, D. (2008) Fungal spores in the air: how to use the results of spore count? Prestige EnviroMicrobiology Inc., Voorhees, NJ. (Available online at www.prestige-em.com/tech/FungalSporesInTheAir).

Zielińska-Jankiewicz, K., Kozajda, A., Piotrowska, M., Szadkowska-Stańczyk, I. (2008) Microbiological contamination with moulds in work environment in libraries and archive storage facilities. *Ann. Agric. Environ. Med.*, **15**, 71-78.

Zotti, M., Ferroni, A., and Calvini, P. (2008) Microfungal biodeterioration of historic paper: preliminary FTIR and microbiological analyses. *Int. Biodet. Biodegr.*, **62**, 186-194.

Zyska, B. (1997) Fungi isolated from library materials: a review of the literature. *Int. Biodet. Biodegr.*, **40**, 43-51.

Chapter 2.2

BACTERIA, VIRUSES AND OTHER BIOAEROSOLS IN INDUSTRIAL WORKPLACES

Brian Crook, Jonathan M. Gawn and Jillian R.M. Swan

Health Improvement Group, Health and Safety Laboratory, Buxton, U.K.

INTRODUCTION

At the outset, it is probably of value to define what shall and shall not be covered in this chapter in the context of other chapters. Firstly, the definition of "industrial workplaces" may overlap with other environments. For example, there may be no clear distinction between indoor work environments handling produce and livestock, i.e. the point at which they cease to be agricultural and become food production and/or industrial work is unclear. Similarly, hospitals and laboratories as work environments may have much in common with non-industrial (domestic and office) environments with regard to air quality, while at the same time overlapping with industrial workplaces if these are taken to include pilot- and small-scale biotechnology operations.

The potential health consequences of exposure to bioaerosols in industrial workplaces remain the same across the broad spectrum of the definition, respiratory infection, allergic respiratory disease and toxicological response to microbial cell components.

This present chapter aims to describe:
• the circumstances in which workers may be at risk of exposure to bioaerosols responsible for these respiratory syndromes;
• the agent(s) involved; and
• typical concentrations, as measured in various research investigations of respiratory ill-health.

The methods used to measure exposure to bioaerosols in industrial workplaces have been reviewed elsewhere (Crook and Olenchock 1995, Eduard and Heederik 1998, Muilenberg 2003, Levetin 2004) and therefore will not be repeated here.

RESPIRATORY INFECTIONS CAUSED BY EXPOSURE TO BIOAEROSOLS IN INDUSTRIAL WORKPLACES

There are many work sectors in which exposure to airborne viral pathogens is likely. This is principally in healthcare and laboratories, but could be any setting where viruses or potentially contaminated body fluids are handled. In the healthcare sector, frontline staff will be exposed to aerosols emanating from patients shedding respiratory pathogens such as influenza virus or SARS-Coronavirus (SARS-Cov). In addition, some enteric viruses such as Norovirus may be aerosolised during vomiting attacks (Barker *et al.* 2001). In some situations, this may require the use of respiratory protective equipment. In diagnostic and research laboratories, high titres of infectious viruses may be handled in significant volumes and many routine operations, e.g. pipetting, are likely to result in infectious aerosol formation that must be contained. In practice, handling infectious virus pathogens when aerosols can be produced will necessitate the use of a microbiological safety cabinet.

Aside from the presence of ubiquitous respiratory viruses, such as influenza (see Table 1), in industrial workplaces, infections caused by exposure to aerosolised viruses are expected to be rare. There are some industrial settings where infection due to exposure to aerosolised viruses may be possible, at least in principle, and these are discussed below.

Some bacterial infectious agents transmitted via bioaerosols can be encountered in the workplace and in some circumstances may be life-threatening. These are also discussed below.

Table 1. Common human viruses that may be naturally transmitted via the air.

Adenovirus	Measles virus	SARS coronavirus
Cocksackie viruses	Mumps virus	(SARS-CoV)
Influenza virus	Rubella virus	Variola virus
Lymphocytic choriomeningitis virus	Rhinovirus	Vaccinia virus
		Varicella zoster virus (chickenpox)

LEGIONNAIRES' DISEASE

The infection probably causing the greatest impact in industrial workplaces is Legionnaires' disease, a life-threatening pneumonia caused by inhalation of the bacterium *Legionella pneumophila*. More detailed descriptions of the biology of *L. pneumophila* lie outside of the scope of this chapter but can be found in detail in Heinsohn (2001).

This ubiquitous water-borne bacterium is readily capable of multiplying in ambient to hot water systems. It therefore presents a significant hazard in industries where hot processes are controlled using cooling waters, e.g. in cooling towers and evaporative condensers, if the workplace (or its surroundings) is exposed to aerosols or airborne droplets of contaminated cooling water. Extensive research has increased understanding of the ecology and pathology of this organism, and therefore should help to control exposure. It is still a major cause of respiratory infection and death each year, however, as exemplified by occasional outbreaks in industrial premises (Castellani-Pastoris *et al.* 1997, Heng *et al.* 1997, Anon. 1998, Breiman and Butler 1998, Boshuizen *et al.* 2001, Fry *et al.* 2003). The bacterium can proliferate in biofilms which give protection from physical and chemical control agents. In industrial water sources, particular problems are associated with stagnation points such as dead-legs in pipework and unused taps or faucets where *L. pneumophila* can multiply, or in poorly maintained packing elements in cooling towers.

ZOONOTIC AGENTS CAUSING PNEUMONIAS

Other bioaerosols capable of causing respiratory infection in industrial workplaces are limited and are found especially at the interface between industry and agriculture. Here, workers may be exposed to zoonotic agents, i.e. those capable of spreading from animals to humans. Examples are described below.

Q fever is a life-threatening respiratory disease caused by the rickettsial bacterium *Coxiella burnettii*, and workers may be at risk where they are exposed to contaminated dusts. This disease presents acutely with fever, headache and other influenza-like symptoms, or as atypical pneumonia. Less frequently, it may present as chronic infection including endocarditis. *C. burnettii* can survive for a long period in dust, which can assist its spread and infectivity. It is highly infective, as is evident from an outbreak attributed to wind-borne spread of dust from agricultural land into a conurbation (Hawker *et al.* 1998). However, industrial exposure is limited, although a cluster of cases in postal workers was attributed to their exposure to contaminated dust in mail sacks from rural postal delivery (Winner *et al.* 1987) and in office workers exposed to contaminated straw boards being handled in the vicinity (van Woerden *et al.* 2004).

Another zoonotic bacterium, *Chlamydophila* (previously *Chlamydia) psittaci*, can cause respiratory infection in humans exposed to dusts contaminated by infected birds, especially psittacines and pigeons. As with Q fever, psittacosis presents as flu-like symptoms or an atypical pneumonia. Outbreaks, including fatalities, have been reported in poultry process workers whose work involved evisceration (Newman *et al.* 1992, Petrovay and Balla 2008); in pet shop workers exposed to infected cage birds (Moroney *et al.* 1998); and in workers at bird parks (Matsui *et al.* 2008; see also review by Smith *et al.* 2005). One study has shown that a significant number of wild birds may carry *C. psittaci* (Olsen *et al.* 1998), so that there is a theoretical risk of infection for workers clearing up debris, for example in building roof spaces, but to date no cases of infection under such circumstances have been published.

Inhalation anthrax

Industrial anthrax can be contracted via the inhalation route. The spore-forming Gram-positive bacterium *Bacillus anthracis*, the causal agent of anthrax, causes endemic infection in domestic animals, especially cattle, sheep and goats, in a broad belt of the tropical zones spreading from Africa and Eastern Europe through Asia to China (Hugh-Jones 1996). Infected animals shed spores of *B. anthracis* into soil and vegetation, where they can survive for many years (Turnbull 1996), and their presence in dust can result in contamination of animal hair, especially goat hair, and hides. The spores can become airborne in dust during manufacturing processes, and goat hair (such as cashmere) has been particularly associated with anthrax spore contamination in the past (Crook *et al*. 1996).

In the early 1900s, anthrax was an important industrial disease in UK, especially in the Yorkshire woollen mills, where it was referred to as "wool sorter's disease" (Metcalfe 2004). Inhalation anthrax affects the mediastinal lymph system and the lungs (Brachman 1986). Inhaled spores are phagocytised by alveolar macrophages and transported into the lymph system where they germinate, multiply and produce toxins, which cause toxaemia. Clinical effects include non-specific upper respiratory tract infection, a general malaise, fatigue and mild fever with cough, followed by sudden onset of severe respiratory distress with a rise in temperature and respiratory and pulse rates, and a fall in blood pressure. Death usually occurs within 24 h of the onset of this acute phase. The workers who were at greatest risk were those sorting fleeces by hand to remove debris. However, improved hygiene conditions and diligent control measures, including exhaust ventilation, reduced to a minimum the incidents of reported human infections in industrialised countries. Anthrax spores have been shown still to be present in animal hair at the early stages of processing (Wattiau *et al*. 2008), but in England and Wales only 18 confirmed cases of the infection, these being cutaneous rather than inhalation infection, were reported between 1981 and 2006 (Health Protection Agency 2008). However, earlier case reports and classic animal inhalation exposure studies verified its position as an infection potentially caused by bioaerosol exposure in the industrial workplace.

An example of incidental exposure was the death from inhalation anthrax of a man working in a factory next door to a goat-hair processing plant (LaForce 1969). Spores were recovered not only from machinery, hair and the air in the plant, but also from the factory where the infected worker was employed. In Switzerland, during the late 1970s and early 1980s, 25 workers in one textile factory handling goat hair from Pakistan contracted anthrax (Pfisterer 1991). Of these, 24 had cutaneous anthrax and one inhalation anthrax. Studies at a mill in USA where outbreaks of anthrax had occurred revealed airborne *B. anthracis* at concentrations up to 300 spores m^{-3} air, this being between 0.1 and 0.9% of the total number of airborne

bacteria detected (Dahlgren *et al*. 1960). Infectivity was demonstrated by drawing contaminated air from the process into exposure chambers housing monkeys. The disease therefore remains a hazard in the textile industry, although because of improved control measures it is one with low risk.

Other potential sources of industrial anthrax are contaminated animal skins and hides in tanneries, and, because of the long survival potential of the spores, also in dusts created during remediation of old tannery sites. The presence of anthrax spores has been detected in soil along drainage courses from old tanneries (Böhm and Otto 1996). In building work, it was the practice until less than 100 years ago to reinforce plaster work with animal hair. Therefore, any demolition of old buildings may expose workers to anthrax spores surviving in dusts if the animal hair was contaminated. Although the likelihood of exposure is small, it should nevertheless be recognized and suitable control measures taken, since spores have been identified in plasterwork (Turnbull *et al*. 1996).

The status of inhalation anthrax as an industrial workplace infection was further emphasised by its use as a bioterrorist weapon in USA in 2001. Letters containing gram quantities of anthrax spore powder were posted to a US senator in Washington and to news agencies in New York and Florida, resulting in 11 cases of inhalation anthrax and 11 cases of cutaneous anthrax. As a consequence of anthrax spores leaking from the letters being sorted at mail handling facilities, the majority of the five fatalities through inhalation anthrax occurred in postal workers (Jernigan *et al*. 2002, Sanderson *et al*. 2004). Investigations into the perpetrator in the USA were only concluded in 2008 following the suicide of the accused person (CNN 2008), but the implications of this act were widespread. In the intervening years, considerable resources were spent firstly in decontaminating the affected industrial premises, and then in improving control and detection measures in vulnerable workplace facilities such as mail handling centres in USA and in many other countries around the world. Much further effort has also been devoted to tighter controls on access to pathogenic cultures in laboratories, as this was almost certainly the way in which the bioterrorist gained access to the anthrax spores.

Two recent cases of inhalation anthrax, one in USA (CDC 2006) and one in Scotland (Health Protection Scotland 2006), have revealed a further source of the infection. In both cases, fatalities occurred as a result of preparing contaminated animal skins for use in ethnic drums. It was assumed in both cases that aero-

Table 2. Human enteric viruses that may be present in wastewater (derived from Bosch 1998).

Virus name	Disease caused
Poliovirus	Paralysis, meningitis, fever
Coxsackie viruses A and B	Herpangina, meningitis, fever, respiratory disease, hand-foot-and-mouth disease, myocarditis, heart anomalies, rush, pleurodynia, diabetes?
Echovirus	Meningitis, fever, respiratory disease, rush, gastroenteritis
Hepatitis A virus	Hepatitis
Human reovirus	Unknown
Human rotavirus	Gastroenteritis
Human norovirus	Gastroenteritis (winter vomiting disease)
Human adenovirus	Gastroenteritis, respiratory disease, conjunctivitis
Human astrovirus	Gastroenteritis
Human parvovirus	Gastroenteritis
Human coronavirus	Gastroenteritis, respiratory disease
Human torovirus	Gastroenteritis

sols generated by the preparation process, with the infected persons in close proximity, were sufficient to be an infectious dose. There is still much conjecture as to what represents an infectious inhalation dose. In some estimates it is considered to be thousands of spores (Fenelly *et al.* 2004), but clearly in some circumstances it may be very few.

Zoonotic viruses

Viral bioaerosols capable of causing respiratory infections may be found at the interface between agriculture and industry, where workers may be exposed to zoonotic agents. For instance, some strains of avian influenza can cause infections in humans. Handling of sick or dying poultry is a known risk factor and transmission via the air is a recognised possibility (reviewed in Abdel-Ghafar *et al.* 2008).

Infectious bioaerosols from wastewater treatment

Enteric viruses are important human pathogens that are shed in human faeces and are probably responsible for the majority of water-borne gastrointestinal disease outbreaks. The term "enteric virus" is a general one that encompasses various viral species, including those listed in Table 2 (for comprehensive reviews see Bosch 1998, Fong and Lipp 2005). These viruses may be present in large quantities in sewage. For instance, rotaviruses can reach levels of 10^{12} particles g^{-1} stool (Barker *et al.* 2001, Morawska 2006). Furthermore, these viruses are highly stable in the aquatic environment and can be resistant to standard treatment processes such as chlorination.

Some waterborne pathogens may be naturally transmissible by the aerosol route and can cause respiratory infection, most notably adenoviruses and cocksackie viruses. Most enteric viruses are transmitted via the faecal-oral route. However, when present in large quantities it may be possible for them to be transmissible by aerosol. Inhalation of other species is unlikely to result in respiratory disease, although subsequent ingestion of virus may result in gastrointestinal disease, as some have low infectious doses, e.g. rotavirus, and exposure to very few virus particles may be sufficient to result in an infection (Haas *et al.* 1993).

Wastewater treatment processes such as aeration may generate significant aerosols. Indeed, infectious reoviruses and enteroviruses have been detected in the air up to 50 m from wastewater aeration tanks (Carducci *et al.* 1995, Sigari *et al.* 2006). Despite this, there is no evidence that wastewater workers have contracted enteric viral infections via exposure to aerosols. Instances of infections in waste water workers have been reported, such as Hepatitis A infection (Brugha *et al.* 1998), but the precise mechanism of transmission, e.g. direct contact rather than aerosol, is not clear.

Exposure to coliform bacteria in waste water treatment bioaerosols may cause infection, but again the role of inhalation (as distinct from ingestion) as a route of infection is not clear. Waste water treatment may also represent a toxigenic bioaerosol hazard which will be discussed later.

Viral infection risk in biotechnology

Manufacture of live-attenuated, killed and split viral vaccines usually involves large-scale culture of an appropriate virus strain. For example, product information literature confirms that measles, mumps and rubella vaccine strains are grown to high titres in cell culture. Similarly, current influenza vaccine strains are cultured in embryonated hen eggs and harvested from the amniotic fluid. In both cases, large volumes of cultured virus are harvested before being purified for use as a therapeutic vaccine. These processes may result in aerosol formation and workers may be exposed to aerosols containing vaccine viruses, virus proteins or other contaminants, such as egg proteins from cultures. In most cases, the virus itself can be considered a low risk as the strains used have been attenuated, e.g. measles, mumps and rubella virus vaccines. However, some vaccine strains may represent a greater risk of harmful infection, particularly if they are destined to be inactivated as part of the vaccine manufacturing process, e.g. influenza virus vaccines. Clearly, the risks associated with exposure to these agents must be carefully assessed to ensure that proper controls are included to protect the safety of workers.

In recent years, there has also been an increase in the number of genetically modified viruses that have been constructed for use as vaccines or "gene therapy" agents. Many of these have entered clinical trials and this has necessitated manufacture in large amounts before processing and purification for therapeutic use in humans. The risks associated with these agents will vary, depending on the nature of both the modified organism and the modification itself. To date, the large majority of these agents have been considered to pose little risk and have been classified as being in the least hazardous categories of biological agents (Gene Therapy Advisory Committee 2008). In the UK, 17% of these agents have been modified adenoviruses and 24% have been modified poxviruses. Both of these virus types are believed to be transmissible via the aerosol route (Couch 1981). Even so, given the large volumes and high titres of viruses grown for these applications, there may be a residual aerosol risk from harvesting and purification processes, whether or not the virus species that has been modified can be transmitted naturally by aerosols.

Infection risk from fungal bioaerosols

Although we are concerned in this chapter prima-rily with bacteria and viruses, it should be noted at this point that fungal bioaerosols represent an unusual respiratory infection hazard, mainly associated with immunocompromised hosts. Probably the most prominent is *Aspergillus fumigatus*, a species that is ubiquitous in the environment, and is both a potential allergen and a pathogen (Latge 1999).

Occasional cases are reported of infection following exposure to contaminated dusts generated during building demolition or renovation (Anderson *et al.* 1996b). This can create an infection risk for hospitalized patients exposed to bioaerosols generated from such building work. However, construction work can also lead to exposure to bioaerosols which have been implicated in influenza-like illness, but this was considered not to be infection but allergic response (Epling *et al.* 1995). *Penicillium* spores were isolated from the air at various points during reconstruction work at an airport in concentrations up to 4,000 colony forming units (CFU) m^{-3} air sampled. It was speculated that exposure to these spores, or mycotoxins associated with them, or to some other bacterial bioaerosol present, contributed to workers' respiratory ill health.

Rarely, there have been cases of lung infection caused by *A. fumigatus* associated with handling composted material (Conrad *et al.* 1992, Allmers *et al.* 2000), but the major risk is of respiratory sensitization as described below.

BIOAEROSOLS IN INDUSTRIAL WORKPLACES AS ALLERGENS AND TOXICANTS

Any industrial work process that can expose the workforce to large concentrations of airborne microorganisms or their products can potentially elicit respiratory syndromes such as occupational asthma, bronchitis or rhinitis, mucous membrane irritation, allergic alveolitis (hypersensitivity pneumonitis) or inhalation fever (ODTS). The potential sources therefore are many and varied. Broadly, industrial workplace biological hazards can be divided into three categories:

- workplace activities which involve purposeful handling of products containing biological agents, as in biotechnology or composting, for example;
- where workplace materials are contaminated by biological agents; or
- where sources of bioaerosols are incidental to the main work activity, e.g. contamination of humidifiers and air conditioners.

Biotechnology

In industrial biotechnology, workers may be exposed to the process microorganisms or their components at various stages, particularly at the downstream processing stage, including centrifugation, product concentration and waste handling (Crook 1996). Occasional cases of occupational asthma and mucous membrane irritation have been reported (Topping *et al*. 1985, Carvalheiro *et al*. 1994) in "low technology" biotechnology, i.e. more conventional brewing and enzyme production. Exposure to bacterially derived proteases in detergent production is a recognized cause of industrial respiratory allergy (Flindt 1995, Johnsen *et al*. 1997). Reports of ill-health among workers in biotechnology are rare, even although the potential exists because of the process organisms used (Bennett *et al*. 1991). For example, a common process organism is *E. coli*. This could give rise to exposure to endotoxin (see Chapter 4.5), a component of the cell walls of Gram-negative bacteria (Palchak *et al*. 1988, Crook 1996). However, because containment is a requirement, not only to ensure product purity but also for environmental safety, especially if the process organism is genetically modified, this provides the worker with protection.

Composting

Exposure to bioaerosols is less controlled in compost production, where the presence of microorganisms is fundamental to the composting process. Composting involves enhanced multiplication of a broad range of thermotolerant fungi and thermophilic actinomycetes (actinobacteria in the order Actinomycetales) to degrade the organic matter present. A succession of microbial groups predominates as the microorganisms present break down lignocelluloses in the substrate and their metabolic activity increases the heat within the compost. Temperatures can reach 80°C in active composting and enhance the growth of thermotolerant and thermophilic species, many of which are respiratory allergens. Workers handling the compost therefore are potentially exposed to these microorganisms (Lacey *et al*. 1992, Swan *et al*. 2003). Exposure can occur at various critical points in commercial compost production or associated activities.

Turning of windrows (heaps of composting material) to ensure efficient composting is likely to be carried out using tractors or mechanical diggers, and the dust made airborne by this activity will be heavily laden with actinomycete and fungal spores (Swan

et al. 2003). Even where these vehicles have integral cabs with air filters, they may provide limited protection from bioaerosols. In agricultural vehicles, where workers may be exposed to similar bioaerosol concentrations, poorly maintained cab filters lead to limited protection, and opening the cab door even briefly negates any protective effect (Thorpe *et al*. 1997).

In mushroom production, composts typically are prepared from horse manure and straw and undergo a two-phase composting process before being inoculated with spawn (mycelium of the common mushroom *Agaricus bisporus*). This may involve mechanical agitation of the compost, to mix in the spawn, and this can generate bioaerosols of the compost microbiota (actinomycetes and fungi). Later in the mushroom production process, disturbance of the compost during mushroom picking can generate bioaerosols of the compost microbiota and contaminant moulds on the surface of the compost, as well as spores from the mushrooms themselves as the caps mature and open (Crook and Lacey 1991). Other mushroom species, such as shiitake (*Lentinus edodes*) and *Pleurotus*, may themselves contribute greatly to the bioaerosol as they liberate their spores early in the growth phase (Michils *et al*. 1991, Tarvainen *et al*. 1991, Lenhart and Cole 1993).

Wood bark is subject to a composting process to stabilize its metabolic activity prior to being prepared for horticultural use. Sorting and bagging of prepared composts may involve mechanical screening into different sizes - the largest chips for soil cover and the finer material for soil conditioner - then being put into bags for sale. The screening, bagging and any other movement by tractor or conveyor belt can generate bioaerosols of the compost microbiota (Crook *et al*. 1994).

In northern European countries and North America, wood chips and bark are used for fuel in homes and as energy sources for power stations. The number of facilities using wood chips is likely to increase as part of renewable energy initiatives. However, workers handling the fuel chips may be exposed to bioaerosols generated from microorganisms degrading the wood chips (Hedenstierna *et al*. 1986, Eduard *et al*. 1994, Millner *et al*. 1994).

Other initiatives to use alternatives to fossil fuel include coal-fired power stations that combust a proportion of biological material such as palm kernel or olive waste following oil extraction. Other facilities combust refuse-derived fuel pellets of recycled domestic waste. However, in both instances if the material is stored damp microbial activity will increase and

Table 3. Total numbers of microorganisms m^{-3} air in bioaerosols associated with composting.

Activity	Bacteria, incl. actinomycetes	Fungi	Predominant taxa	Reference
Mushroom compost - preparation and mixing	10^7-10^8	10^5 10^5	*Thermomonospora, Actinomadura*	Crook and Lacey 1991, Van de Bogart *et al.* 1993
Mushroom picking - *Agaricus*	10^5	10^5	*Penicillium, Trichoderma, Plicaria*	Crook and Lacey 1991
Mushroom picking - *Pleurotus*	ND	ND	*Pleurotus* spores	Mori *et al.* 1998
Mushroom picking - shiitake (*Lentinus* sp.)	ND	>10^6	*Lentinus* spores	Sastre *et al.* 1990
Bark composting - screening	10^5	10^7	*Rhizomucor, Paecilomyces*	Crook *et al.* 1994
Bark composting - bagging	10^4	10^6	*Rhizomucor, Paecilomyces*	Crook *et al.* 1994
Fuel chip preparation	ND	10^4-10^7	*Penicillium*	Blomquist *et al.* 1984, Hedenstierna *et al.* 1986
Domestic waste	10^4-10^6	10^4-10^6	*Aspergillus fumigatus, Thermoactinomyces, Thermomonospora*	Lacey *et al.* 1992, Marchand 1995, Nielsen *et al.* 1997b

ND, not determined

workers handling the material could be exposed.

The microbiota, in which thermophilic actinomycetes may be prominent, and typical concentrations in bioaerosols measured in composting-associated activities are summarized in Table 3.

Waste disposal

Microbiological activity naturally occurs in any organic waste material, therefore workers disposing of waste may be exposed to bioaerosols generated at various stages of the process (see review by Wouters *et al.* 2006). Bioaerosols may be created by domestic waste disposal during collection from domestic and industrial premises and when the material is emptied from waste collection vehicles at waste transfer stations. Subsequently, material may be transferred into bulk containers and taken for disposal at landfill sites, or may be disposed of by incineration. Each process will involve further mechanical handling and the potential for generation of dust aerosols, and any delays during this process will increase the opportunity for multiplication of microorganisms present.

Initiatives to increase resource recovery have a potential impact on worker exposure. In the European Union, targets have been set for recycling of disposed waste (Defra 2008). This means that more waste streams are being diverted into activities such as materials recycling and reclamation that could increase the potential for workers being exposed to contaminant microorganisms. Some schemes involve separa-

tion at source of recyclable materials such as plastics, glass and metal, but further sorting is required at materials recovery facilities (MRF's). In other schemes recyclable materials are separated from organic materials.

Much of this involves manual sorting, which can expose workers to bioaerosols. The organic fraction of domestic waste may be composted, as described earlier. Table 4 summarizes typical reported exposures to bioaerosols during all aspects of domestic waste disposal.

A likely future development is landfill mining, i.e. opening up old landfill sites to recover non-degraded plastics, metals etc. for recycling (van der Zee *et al.* 2004). Although the initial mining is likely to be done with mechanical diggers where workers' exposure to bioaerosols can be controlled, downstream sorting could lead to greater exposure.

Wastewater treatment

Treatment of wastewater and human sewage involves many processes that are capable of generating bioaerosols, e.g. pumping, screening and aeration. Much of this is outdoors, but some can be indoors or in confined areas, so that there is the potential for workers to be exposed to toxigenic bacteria. In a study of airborne bacteria and endotoxin at wastewater treatment plants, total bacterial levels as measured by direct fluorescence microscopy were in the range of

Table 4. Components of domestic waste bioaerosols as airborne concentrations in CFU m^{-3}.

Activity	Concentration in air (CFU m^{-3})				
	Total bacteria	Gram-nega-tive bacte-ria	Total fungi	Airborne endo-toxin (EU m^{-3})	Reference
Door-to-door collection	10^3-10^4	ND	10^4-10^5	50	Heldal *et al.* 1997, Nielsen *et al.* 1997a
Waste transfer station	up to 10^6	10^5	10^6	ND	Crook *et al.* 1987
Incineration	10^7	10^4	10^7	ND	Crook *et al.* 1987
Materials recycling (unseparated waste)	10^5	10^4	10^5	9.9×10^3	Malmros *et al.* 1992, Poulsen *et al.* 1995
Materials recycling (paper source separated)	10^4	10^4	10^4	1.1×10^2	Poulsen *et al.* 1995, Würtz and Breum 1997
Landfill	10^6	10^4	10^5	ND	Rahkonen *et al.* 1987
Background	10^3	10^1	10^3	ND	Crook *et al.* 1987

ND, not determined.

10^7-10^9 cells m^{-3} air, viable bacteria 10^2-10^5 CFU m^{-3} and up to 3000 endotoxin units (EU) m^{-3} air (Laitinen *et al.* 1992, Melbostad *et al.* 1994). They found good correlation between airborne viable bacteria and endotoxin concentrations.

De-watered sewage sludge (biosolids) may be disposed of by incineration, landfill, application to land, or composting in combination with green waste. However, this may increase the potential for worker exposure to bioaerosols similar to that described earlier. Measurements taken at a sewage sludge composting facility in USA showed that it was a very dusty process, exceeding occupational standards for general dust exposure, and that endotoxin levels were high, ranging from 28.9 ng m^{-3} (289 EU m^{-3}) to 5930 ng m^{-3} (59300 EU m^{-3}) when screening and sweeping operations took place (Darragh *et al.* 1997).

Biosolid application to land has been shown to generate significant aerosols of bacteria and endotoxin in the close vicinity of operations, therefore potentially exposing workers, although downwind spread of bioaerosol is limited (Brooks *et al.* 2004, 2005)

Engineering and manufacturing industries

In manufacturing industry, exposure to bioaerosols may result either from working with contaminated materials or from indirect sources, such as humidifiers or air conditioning where contaminated water sources can lead to gross exposure to waterborne bacteria other than legionellae. Complex fever symptoms and immunological responses have been associated with contaminated air conditioning and humidification systems used in some printing and textile works, the syndrome being described generally as "humidifier

fever" (Burge *et al.* 1985, Finnegan *et al.* 1985). However, although in some instances a relationship has been demonstrated between symptoms and antibody response to microbial contaminants (McSharry *et al.* 1993), in other instances exposure to microbial contaminants and endotoxin is low. This suggests a multifactorial etiology, in which microbial exposure may be a single component (Kateman *et al.* 1990).

Studies by one of the present authors examined the relationship between water sources and bioaerosols. Bioaerosol samples taken next to humidifiers in textile mills yielded 10^2-10^4 CFU of bacteria m^{-3} air, the lowest concentrations being where water was delivered directly from a piped supply and the greatest where a reservoir of standing water was involved. Airborne fungi were in the range of 10^2-10^3 CFU m^{-3} (Crook 1995). Bioaerosol samples taken in a print works immediately prior to a reservoir-fed humidifier being switched on yielded 10^2 CFU m^{-3}, but shortly after the humidifier was switched on airborne bacterial concentrations in the vicinity rose to 10^5 CFU m^{-3}. This demonstrated the potential influence that a source of contamination can have upon bioaerosol concentrations (Crook 1995, Crook and Sherwood-Higham 1997). Gram-negative bacteria predominated in humidifier water and bioaerosol in both instances, typical isolates being pseudomonads and *Flavobacterium* spp.

Exposure to bacterial contaminants occurs in engineering where metals are being cut, drilled and shaped using metalworking fluid (MWF; also referred to as metal removal fluids, cutting fluids or metalworking emulsions) to cool and lubricate machine tools (Simpson *et al.* 2003). These mineral-oil-based, synthetic or semi-synthetic oil-in-water emulsions are highly susceptible to attack by bacteria, typically waterborne Gram-negative bacteria such as *Pseudomonas* and related species (Kriebel *et al.* 1997). Because the MWF are recirculated via a sump which either supplies a single machine, or a reservoir which services a number of machines, the opportunity arises for bacterial contamination to multiply in the MWF. This is exacerbated by continued use of the same recirculated MWF, with periodic topping up, often for several weeks. Most MWF has incorporated biocide or has biocide added separately to the sump, but this may fail to fully control bacterial growth. Aerosols of the MWF are generated as it is delivered to the machine parts. Contamination levels have been measured in sumps in the order of 10^7-10^9 microorganisms ml^{-1} MWF, of which Gram-negative bacteria accounted on average for 65% of the total (Thorne *et al.* 1996). Similar prevalence was found in the air (Woskie *et al.* 1996). Bioaerosols in engineering facilities measured in the same study showed workers to be exposed to these bacteria at up to 550 CFU m^{-3} of air sampled. The predominant bacteria are pseudomonads, including *Ps. pseudoalcaligenes* and *Ps. alcaligenes*, and Gram-positive *Micrococcus luteus*. Exposure to the pseudomonads present has been associated with immunological response in engineering workers (Matsby-Balzer *et al.* 1989, Travers-Glass and Crook 1994, Fishwick *et al.* 2005).

Because of the predominance of Gram-negative bacteria, endotoxin is present in large quantities in such systems. Endotoxin levels in sumps may be as high as 166×10^3 EU ml^{-1} MWF and airborne endotoxin in engineering works has been reported at concentrations of 787 EU m^{-3} (Thorne *et al.* 1996). This has been associated with workers' respiratory ill health such as bronchitis and fever and flu-like symptoms (Kriebel *et al.* 1997, Rosenman *et al.* 1997).

Fungal contamination is not usually a problem in MWF, because fungi are out-competed by bacteria. In the study by Thorne *et al.* (1996), most bulk samples yielded ≤ 50 CFU ml^{-1}, with airborne levels up to 3.8×10^3 CFU m^{-3}, but mostly less than 60 CFU m^{-3} air. However, biocidal additives or changes to some formulations, including the use of synthetic oil emulsions, to suppress Gram-negative bacteria could encourage the growth of fungi or even Gram-positive bacteria and acid-fast bacilli. More recent studies in USA identified other potential etiological agents associated with clusters of hypersensitivity pneumonitis (also known as extrinsic allergic alveolitis) among workers exposed to MWF. Strong candidates were the acid-fast Gram-positive bacilli *Mycobacterium chelonae*,

M. fortuitum and *M. abscessus* (Kreiss and Cox-Ganser 1997, Fox *et al.* 1999) and the more recently described *M. immunogenum* (Wallace *et al.* 2002). Despite evidence from animal models that *Mycobacterium* species in MWF can induce hypersensitivity pneumonitis (Thorne *et al.* 2006), doubts continue regarding its role in respiratory disease aetiology (Trout *et al.* 2003).

A major outbreak of allergic respiratory disease was reported at a car component factory in the UK (Dawkins *et al.* 2006: Robertson *et al.* 2007). Altogether, 101 cases of occupational asthma or hypersensitivity pneumonitis were diagnosed in an 800-strong workforce. Although some large sumps supplying several metalworking machines had little bacterial contamination, smaller individual sumps and component washing machines were heavily contaminated with up to 10^8 bacteria and 10^7 EU ml^{-1}. Extracts prepared from some of the predominant bacterial contaminants and from used MWF demonstrated immunological response in some of the affected workers. However, contrary to US investigations of hypersensitivity pneumonitis, no evidence of *Mycobacterium* contamination was found, and there was no immunological response to extracts of *Mycobacterium* in any of the workers. Consequently, no single aetiological agent has emerged in over 20 years of reported cases of respiratory ill health associated with MWF mists, but the balance of evidence suggests a bioaerosol involvement, possibly a combination of response to bacterial protein and endotoxin.

Other industrial process waters

Other process waters in industry have led to workers being exposed to bioaerosols consisting of bacteria and endotoxin. A detailed study was made at a fibreglass manufacturing plant in USA (Milton *et al.* 1995, 1996a,b). Here cooling water and wash water used in the manufacturing process were found to be highly contaminated with bacteria and endotoxin. While Gram-negative bacteria predominated, no legionellae were detected. Gram-positive *Bacillus* spp. were few, no thermophilic actinomycetes were detected and fungi were few. The process waters were found to be contaminated by $>10^7$ Gram-negative bacteria ml^{-1} and 340×10^3 EU ml^{-1}. Workers were found to be exposed to concentrations of more than 10^3 bacteria and 278×10^3 EU m^{-3} air. This exposure was associated with the symptoms of severe flu-like illnesses and asthma.

A similar problem was found with cooling water held in a large sump in an electronics facility (Anderson *et al*. 1996a). The water was found to be heavily contaminated by Gram-negative bacteria, including *Pseudomonas putrefaciens*, *Ps. testosteroni* and *Alcaligenes* spp., yielding endotoxin concentrations in the sump of up to 57×10^3 EU ml^{-1}. Exposure of workers to bacterial bioaerosol concentrations in excess of 10^4 CFU m^{-3} air was associated with work taking place in the process area, and exposed workers suffered fever symptoms, a syndrome described by the authors as "sump bay fever". Few fungi were isolated - in the region of 50-100 CFU m^{-3} air – and these were mainly *Penicillium* spp.

Soluble materials from paper and wood pulp provide a rich nutrient source for multiplication of bacteria, creating slimes in process waters. These comprise a complex mixture of bacteria (Vaisanen *et al*. 1994, 1998). Recycled process water in paper and pulp mills, and in saw mills, have led to workers being exposed to considerable endotoxin levels (Liesivuori *et al*. 1994). As well as 10^4 bacteria m^{-3} air, endotoxin at levels up to 360 EU m^{-3} air was measured in the air during a study of workers complaining of respiratory and mucosal symptoms at a paper mill where they were exposed to recycled water (Sigsgaard *et al*. 1994). Other studies have revealed in nasal swabs taken from mill workers the excessive presence of *Klebsiella pneumoniae*, a bacterium predominant in paper mill slime (Niemala *et al*. 1985).

Contamination of process waters in food production has led to workers being exposed to bacteria and endotoxin in bioaerosols. Forster *et al*. (1989) reported that water used to transport sugar beet from storage hoppers to chopping machines and on into sucrose extraction equipment was recycled via settling lagoons. However, as the beet-harvesting season progressed the bacterial contamination levels built up in the water, and were added to by the deterioration in microbial quality of the stored beet. Workers were exposed to airborne Gram-negative bacteria and endotoxin concentrations up to 2×10^5 total bacteria, 4×10^3 Gram-negative bacteria and 320 EU m^{-3} air before installation of local exhaust ventilation, which controlled exposure.

In potato processing, workers have suffered flu-like symptoms and acute lung-function changes. The processing of potatoes for industrial starch, protein and fibre production is similar to that described above for sugar beet, and creates several points where workers can be exposed to bacterial bioaerosols. Again, recycling of process water increases the likelihood of bacterial contamination. As a consequence, workers may be exposed to as many as 9×10^4 bacterial CFU m^{-3} air,

of which up to 3×10^4 m^{-3} can be Gram-negative bacteria, and 62×10^3 EU m^{-3} inhalable and 1.4×10^3 EU m^{-3} respirable endotoxin (Hollander *et al*. 1994, Zock *et al*. 1995, 1998).

Textile industries

Raw materials used in textile manufacture are contaminated by soil bacteria from harvesting, and the numbers may increase during storage if conditions allow. The most widely researched aspect of this is the role of inhaled bioaerosols in the etiology of a chronic bronchitis affecting textile workers in the cotton industry, byssinosis. Related to this is a syndrome, referred to as Monday morning fever, in which there is an increased bronchial reactivity following initial exposure after a period of absence from work, and progressive reduction in symptoms with regular repeated exposure. This response is similar to the symptoms of chronic exposure to endotoxin and studies have proved the link between the exposure of workers to endotoxin in airborne cotton dust and their respiratory symptoms (Rylander *et al*. 1985, Rylander and Bergstrom 1993, Niven *et al*. 1997). Especially when handling raw cotton at the early stages of the process (combing, carding and weaving), workers may be exposed to up to 10^5 total bacteria, 10^5 Gram-negative bacteria (*Enterobacteriaceae*) and 10^5 fungi m^{-3} air, and endotoxin at a level of 7×10^3 EU m^{-3} air has been recorded (Haglind and Rylander 1984, Castellan *et al*. 1987, Kennedy *et al*. 1987, Sigsgaard *et al*. 1992). Studies of lung function in workers have shown a decline in response as the working week progresses, suggesting habituation with repeated exposure (Raza *et al*. 2000, Fishwick *et al*. 2002). Wool processing exposes workers to bioaerosols containing predominantly fungal spores. Up to 800 CFU m^{-3} have been reported (Sigsgaard *et al*. 1992). Byssinosis was not diagnosed in wool workers, although other studies have reported respiratory ill health among wool textile workers (Love *et al*. 1991, Zuskin *et al*. 1995), possibly stimulated more by the dust than the microbial component in this instance (Brown and Donaldson 1996).

Processing man-made textiles can also result in exposure to bioaerosols. An investigation of respiratory ill health, including humidifier fever, at a nylon production plant revealed significant exposure to *Cytophaga* and associated endotoxin in the process water (Nordness *et al*. 2003).

Processing of hemp leads to intense exposure to dust and bioaerosols for workers, causing immunological response and reduction in lung function

Table 5. Bioaerosols associated with handling of stored products.

Stored product	Total bacteria, incl. actinomycetes	Total fungi	Taxa present	Reference
	Microorganisms m⁻³ air			

Stored product	Total bacteria, incl. actinomycetes	Total fungi	Taxa present	Reference
Tea	10^2	10^3	Predominantly *Aspergillus*	Cartier and Malo 1990, Muller *et al.* 1991, Zuskin *et al.* 1993, Jayawardana and Udupihille 1997
Tobacco	10^3	10^4	Includes *A. fumigatus* and *Rhizomucor pusillus*	Huskonen *et al.* 1984, Ogundero and Aina 1989, Uitti *et al.* 1998, Nordman *et al.* 1984
Citrus fruits	ND	10^5-10^7	Predominantly *Penicillium*	Ström and Blomquist 1986
Spices	10^2	10^3	Predominantly *Aspergillus*	Zuskin *et al.* 1993, Delcourt *et al.* 1994, Sastre *et al.* 1996
Cork	ND	106	Predominantly *Penicillium*	Lacey 1973, Avila and Lacey 1974, Alegre *et al.* 1990

ND, not determined

(Zuskin *et al.* 1992, 1994). A recent study of a British hemp processing factory revealed inhalable bacteria ranging in number from 4.7×10^6 to 190×10^6 CFU m⁻³ air and endotoxin at levels between 4.7×10^3 and 59.8 $\times 10^3$ EU m⁻³ air (Fishwick *et al.* 2001).

Flax, used for weaving linen, relies on growth of bacteria during the early stages of preparation (retting) to soften the fibres. Subsequent handling of the flax can expose workers to bioaerosols, predominantly bacteria and endotoxin (Cinkotai *et al.* 1988). Although airborne concentrations were not measured, settled dust was examined microbiologically by Buick *et al.* (1994), who found Gram-negative bacteria at 10^9 CFU g⁻¹ and endotoxin at 920×10^3 EU g⁻¹ dust.

Stored food and other products

Bagasse, sugar cane fibre used to produce constructional board, is much like flax, in that microbial activity at an early stage of the process softens the fibres prior to use. This means, however, that workers can be exposed to bioaerosols, in this case predominantly thermophilic actinomycetes. Khan *et al.* (1995) reported that spores of the *Thermoactinomyces sacchari* account for >50% of the total population of thermophilic actinomycetes, followed by *T. vulgaris*, *T. thalpophilus*, *Saccharomonospora viridis* and *Saccharopolyspora rectivirgula*. Both this investigation and other studies demonstrated correlation between exposure to these spores and immunological response in exposed workers (Phoolchund 1991).

The storage of products in bulk quantity, and subsequent handling and movement, mean that any microbial contaminants present may be aerosolized. Also, if storage conditions allow, multiplication of the contaminants can occur, resulting in workers being exposed to very high bioaerosol concentrations. This is particularly the case in agriculture, e.g. stored hay and grains, but in other more industrially related work areas there is the potential for bioaerosol exposure. Organic dust from coffee beans, stored and subsequently handled for roasting and preparing coffee on an industrial scale, has resulted in workers reporting allergic respiratory symptoms. The cause is likely to be a combination of a number of factors, including antigens from the coffee beans and from castor beans, as old castor bean sacks are sometimes recycled for use to store coffee (Romano *et al.* 1995). However, bioaerosol measurements in a large-scale coffee factory revealed that workers were exposed to airborne fungi at a concentration of 2×10^4 CFU m⁻³ air, predominantly species of *Penicillium* and *Aspergillus*, and bacteria at 2×10^3 CFU m⁻³ (Thomas *et al.* 1991).

Handling of stored products in animal feed mills has resulted in workers being exposed to endotoxin at concentrations up to 4700 EU m⁻³ air, and this exposure is strongly related to reduction in lung function (Smid *et al.* 1992, 1994, Heederik *et al.* 1994). Bakery workers may be exposed to large concentrations of flour dust and enzymes such as amylases which can trigger asthmatic response (Burdorf *et al.* 1994, Baur *et al.* 1998, Houba *et al.* 1998, Sande *et al.* 2007), but when the raw products are handled there may also be exposure to microbial aerosols in which fungal taxa such as *Cladosporium* and *Penicillium* predominate (Crook *et al.* 1988, Musk *et al.* 1989).

Some other examples of bioaerosols to which workers can be exposed during handling of particular stored products are given in Table 5.

SUMMARY

A diversity of industrial work environments can expose workers to a wide range of bioaerosols. The extent of exposure is dependent on the contamination source and the work activity. This is a function of the potential for creation of aerosols and the amount of protection the worker has from the aerosols, e.g. engineering control or personal protective equipment. Exposure to bioaerosols can cause infection, allergy or toxicosis, dependent on the microorganisms in the bioaerosol and the exposure dose. Only limited data are available on dose-response relationships for bioaerosols. No occupational standards exist at present, and although occupational limits for endotoxin were previously proposed in Europe (Heederik and Douwes 1997), there is currently no plan for their implementation (Heederik and Wouters 2007). In the absence of such standards, it is important to raise awareness of the work practices associated with significant exposure to bioaerosols and the composition of those bioaerosols. These data can then be used both to correlate with development of respiratory disease and to use in risk assessment tools to control disease.

REFERENCES

Abdel-Ghafar, A.N., Chotpitayasunondh, T., Gao, Z., *et al.* (2008) Update on avian influenza A (H5N1) virus infection in humans. *N. Engl. J. Med.,* **358,** 261-273.

Alegre, J., Morell, F., and Cobo, E. (1990) Respiratory symptoms and pulmonary function of workers exposed to cork dust, toluene diisocyanate and conidia. *Scand. J. Work Environ. Health,* **16,** 175-181.

Allmers, H., Huber., H., and Baur, X. (2000) Two year follow-up of a garbage collector with allergic bronchopulmonary aspergillosis (ABPA). *Am. J. Ind.. Med.,* **37,** 438-442.

Anderson, K., McSharry, C.P., Clark, C., *et al.* (1996a) Sump bay fever: inhalational fever associated with a biologically contaminated water aerosol. *Occup. Environ. Med.,* **53,** 106-111.

Anderson, K., Morris, G., Kennedy, H., *et al.* (1996b) Aspergillosis in immunocompromised paediatric patients; associations with building hygiene, design and indoor air. *Thorax,* **51,** 256-261.

Anon. (1998) Recent cases of Legionnaires' disease associated with London. *Commun. Dis. Rep. CDR Wkly,* **8,** 361.

Avila, R., and Lacey, J. (1974) The role of *Penicillium frequentans* in suberosis (respiratory disease in workers in the cork industry). *Clin. Allergy,* **4,** 109-117.

Barker, J., Stevens, D., and Bloomfield, S.F. (2001) Spread and prevention of some common viral infections in community facilities and domestic homes. *J. Appl. Microbiol.,* **91,** 7-21

Baur, X., Sander, I., Posch, A., and Raulf-Heimsoth, M. (1998) Baker's asthma due to the enzyme xylanase – a new occupational allergen. *Clin. Exp. Allergy,* **28,** 1591-1583.

Bennett, A.M., Hill, S.E., Benbough, J.E., and Hambleton, P. (1991) Monitoring safety in process biotechnology. In *Genetic Manipulation,* Society for Applied Bacteriology, Technical Series, Vol. 28, Academic Press, London, pp. 361-376.

Blomquist, G., Ström, G., and Stromquist, L.-H. (1984) Sampling of high concentrations of airborne fungi. *Scand. J. Work Environ. Health,* **10,** 109-113.

Bohm, R., and Otto, J. (1996) Risk assessment of former tannery sites. *Salisbury Med. Bull.* (Special Supplement, Proceedings of an International Anthrax Workshop), **87,** 65S.

Bosch, A. (1998) Human enteric viruses in the water environment: a minireview. *Int. Microbiol.,* **1,** 191-6.

Boshuizen, H.C., Neppelenbroek, S.E., van Vliet, H., *et al.* (2001) Subclinical *Legionella* infection in workers near the source of a large outbreak of Legionnaires disease. *J. Infect. Dis.,* **184,** 515-518.

Brachman, P.S. (1986) Inhalation Anthrax. In J.A. Merchant (ed.), *Occupational Respiratory Diseases,* DHHS NIOSH Publications, Washington, pp. 693-698.

Breiman, R.F., and Butler, J.C. (1998) Legionnaires' disease: clinical, epidemiological, and public health perspectives. *Semin. Respir. Infect.,* **13,** 84-89.

Brooks, J.P., Tanner, B.D., Josephson, K.L., *et al.* (2004) Bioaerosols from the land application of biosolids in the desert southwest USA. *Water Sci. Technol.,* **50,** 7-12.

Brooks, J.P., Tanner, B.D., Josephson, K.L., *et al.* (2005) A national study on the residential impact of biological aerosols from the land application of biosolids. *J. Appl. Microbiol.,* **99,** 310-322.

Brown, D.M., and Donaldson, K. (1996) Wool and grain dusts stimulate TNF secretion by alveolar macrophages *in vitro. Occup. Environ. Med.,* **53,** 387-393.

Brugha, R., Heptonstall, J., Farrington, P., *et al.* (1998). Risk of hepatitis A infection in sewage workers. *Occup. Environ. Med.,* **55,** 567-569.

Buick, J.B., Lowry, R.C., and Magee, T.R.A. (1994) Isolation, enumeration and identification of Gram-negative bacteria from flax dust with reference to endotoxin concentration. *Am. Ind. Hyg. Assoc. J.,* **55,** 59-61.

Burdorf, A., Lillienberg, L., and Brisman, J. (1994) Characterization of exposure to inhalable flour dust in Swedish bakeries. *Ann. Occup. Hyg.,* **38,** 67-78.

Burge, P.S., Finnegan, M., Horsfield, N., *et al.* (1985) Occupational asthma in a factory with a contaminated humidifier. *Thorax,* **40,** 248-254.

Carducci, A., Arrighi, S., and Ruschi, A. (1995) Detection of coliphages and enteroviruses in sewage and aerosol from an activated sludge wastewater treatment plant. *Lett. Appl. Microbiol.,* **21,** 207-209.

Cartier, A., and Malo, J.L. (1990) Occupational asthma due to tea dust. *Thorax,* **45,** 203-206.

Carvalheiro, M.F., Marques Gomes, M.J., Santos, O., *et al.* (1994) Symptoms and exposure to endotoxin among brewery employees. *Am. J. Ind. Med.,* **25,** 113-115.

Castellan, R.M., Olenchock, S.A., Kinsley, K., and Hankinson, J.L. (1987) Inhaled endotoxin and decreased spirometric values, an exposure response relation for cotton dust. *N. Engl. J. Med.,* **317,** 605-610.

Castellani-Pastoris, M., Ciceroni, L., Lo-Monaco, R., *et al.* (1997) Molecular epidemiology of an outbreak of Legionnaires' disease associated with a cooling tower in Genova-Sestri Ponente, Italy. *Eur. J. Clin. Microbiol. Infect. Dis.,* **16,** 883-892.

Centers for Disease Control and Prevention (CDC) (2006) Inhalation anthrax associated with dried animal hide. Pennsylvania and New York City, 2006. *MMWR Morb. Mortal. Wkly Rep.,* **55,** 280-282.

Cinkotai, F.F., Emo, P., Gibbs, A.C., *et al.* (1988) Low prevalence of byssinotic symptoms in 12 flax scutching mills in Normandy, France. *Brit. J. Ind. Med.*, **45**, 325-328.

CNN News (2008) Report: anthrax suspect kills self before filing of criminal charges. (Online at http://www.cnn.com/2008/CRIME/08/01/anthrax.suicide.ap/index.html?eref=rss_top-stories #cnnSTCText).

Conrad, D.J., Warnock, M., Blanc, P., *et al.* (1992) Microgranulomatous aspergillosis after shoveling wood chips: report of a fatal outcome in a patient with chronic granulomatous disease. *Am. J. Ind. Med.*, **22**, 411-418.

Couch, R.B. (1981) Viruses and indoor air pollution. *Bull. NY Acad. Med.*, **57**, 907-921.

Crook, B. (1995) Airborne microorganisms in humidified textile mills and print works. In A. Bousher, M. Chandra, and R. Edyvean, (eds.), *Biodeterioration and Biodegradation 9*, Institute of Chemical Engineers, Rugby, UK, pp. 328-333.

Crook, B. (1996) Methods of monitoring for process microorganisms in biotechnology. *Ann. Occup. Hyg.*, **40**, 245-260.

Crook, B., and Lacey, J. (1991) Airborne allergenic microorganisms associated with mushroom cultivation. *Grana*, **30**, 446-449.

Crook, B., and Olenchock, S.A. (1995) Industrial workplaces. In C.S. Cox and C.M. Wathes, (eds.) *Bioaerosols Handbook*. CRC/Lewis Publ., Boca Raton, FL, pp. 531-545.

Crook, B., and Sherwood-Higham, J.L. (1997) Sampling and assay of bioaerosols in the work environment. *J. Aerosol. Sci.*, **28**, 417-426.

Crook, B., Venables, K.M., Lacey, J., Musk, A.W., and Newman Taylor, A.J. (1988) Dust exposure and respiratory symptoms in a UK bakery. In W.D. Griffiths and N.P. Vaughan, (eds.), *Aerosols, Their Generation, Behaviour and Applications, Proceedings of the Second Aerosol Society Conference*, The Aerosol Society, London, pp. 341-345.

Crook, B., Higgins, S., and Lacey, J. (1987) *Airborne Microorganisms Associated with Domestic Waste Disposal.* HSE Contract 1/MS/126/643/82. Final Report, AFRC Rothamsted Experimental Station.

Crook, B., Botheroyd, E.M., Travers-Glass, S.A., and Gould, J.R.M. (1994) The exposure of Scottish wood bark chip handlers to microbially contaminated dust. *Ann. Occup. Hyg.*, **38**, Suppl. 1, 903-906.

Crook, B., Hoult, B., and Redmayne, A.C. (1996) Workplace health hazards from anthrax-contaminated textiles. *Salisbury Med. Bull.*, 87 (Special Supplement, Proceedings of an International Anthrax Workshop), 66S-67S.

Dahlgren, C.M., Buchanan, L.M., Decker, H.M., *et al.* (1960) *Bacillus anthracis* aerosols in goat hair processing mills. *Am. J. Hyg.*, **72**, 24-31.

Darragh, A.H., Buchan, R.M., Sandfort, D.R., and Coleman, R.O. (1997) Quantification of air contaminants at a sewage sludge composting facility. *Appl. Occup. Environ. Hyg.*, **12**, 190-194.

Dawkins, P., Robertson, A., Robertson, W. *et al.* (2006) An outbreak of extrinsic allergic alveolitis at a car engine plant. *Occup. Med. (Lond.)*, **56**, 559-565.

Defra (2008) Recycling and Waste; Strategy and Legislation – Waste Strategy for England and Waste Strategy Annual Progress Reports. (At http:// www.defra.gov.uk/ environment/waste/strategy/index.htm).

Delcourt, A., Rousset, A., and Lemaitre, J.P. (1994) Microbial and mycotoxic contamination of peppers and food safety. *Boll. Chim. Farm.*, **133**, 235-238.

Eduard, W., and Heederik, D. (1998) Methods for quantitative assessment of airborne levels of non-infectious microorganisms in highly contaminated work environments. *Am. Ind. Hyg. Assoc. J.*, **59**, 113-127.

Eduard, W., Sandven, P., and Levy, F. (1994) Exposure and IgG antibodies to mould spores in wood trimmers; exposure response relationships with respiratory symptoms. *Appl. Occup. Environ. Hyg.*, **9**, 44-48.

Epling, C.A., Rose, C.S., Martyny, J.W., *et al.* (1995) Endemic work-related febrile respiratory illness among construction workers. *Am. J. Ind. Med.*, **28**, 193-205.

Fennelly, K.P., Davidow, A.L., Miller, S.L., *et al.* (2004) Airborne infection with *Bacillus anthracis* - from mills to mail. *Emerg. Infect. Dis.*, **10**, 996-1002.

Finnegan, M.J., Pickering, C.A.C., Davies, P.S., and Austwick, P.K.C. (1985) Factors affecting the development of precipitating antibodies in workers exposed to contaminated humidifiers. *Clin. Allergy*, **15**, 281-292.

Fishwick, D., Allan, L.J., Wright, A., and Curran, A.D. (2001) Assessment of exposure to organic dust in a hemp processing plant. *Ann. Occup. Hygiene*, **45**, 577-583.

Fishwick, D., Raza, S.N., Beckett, P., *et al.* (2002) Monocyte CD14 response following endotoxin exposure in cotton spinners and office workers. *Am. J. Ind. Med.*, **42**, 437-442.

Fishwick, D., Tate, P., Elms, J., *et al.* (2005) Respiratory symptoms, immunology and organism identification in contaminated metalworking fluid workers. What you see is not what you get. *Occup. Med. (Lond.)*, **55**, 238-241.

Flindt, M.L.H. (1995) Biological washing powders as allergens. *Brit. Med. J.*, **310**, 195.

Fong, T.T., and Lipp, E.K. (2005) Enteric viruses of humans and animals in aquatic environments: health risks, detection, and potential water quality assessment tools. *Microbiol. Mol. Biol. Rev.*, **69**, 357-371.

Forster, H.W., Crook, B., Platts, B.W., *et al.* (1989) Investigations of organic aerosols generated during sugar beet slicing. *Am. Ind. Hyg. Assoc. J.*, **50**, 44-50.

Fox, J., Anderson, H., Moen, T., *et al.* (1999) Metal working fluid-associated hypersensitivity pneumonitis; an outbreak investigation and case control study. *Amer. J. Ind. Med.*, **35**, 58-67.

Fry, A.M., Rutman, M., Allan, T., *et al.* (2003) Legionnaires' disease outbreak in an automobile engine manufacturing plant. *J. Infect. Dis.*, **187**, 1015-1018.

Gene Therapy Advisory Committee (2008) *Fourteenth Annual Report Covering the Period from January 2007 to December 2007.* Health Departments of the United Kingdom, London.

Haas, C.N., Rose, J.B., Gerba, C., and Regli, S. (1993). Risk assessment of virus in drinking water. *Risk Anal.*, **13**, 545-552.

Haglind, P., and Rylander, R. (1984) Exposure to cotton dust in an experimental cardroom. *Brit. J. Ind. Med.*, **10**, 340-345.

Hawker, J.I., Ayres, J.G., Blair, I., *et al.* (1998) A large outbreak of Q fever in the West Midlands; windborne spread into a metropolitan area? *Commun. Dis. Publ. Health*, **1**, 180-187.

Health Protection Agency (2008) Human anthrax in England and Wales; epidemiological data. (At http://www.hpa.org.uk/web/HPAweb&HPAwebStandard/HPAweb_C/1218180268521).

Health Protection Scotland (2006) Probable human anthrax death in Scotland. *HPS Weekly Report*, **40**, 177. (At http://www.documents.hps.scot.nhs.uk/ewr/pdf 2006/ 0633.pdf).

Hedenstierna, G., Alexandersson, R., Belin, L., *et al.* (1986) Lung function and *Rhizopus* antibodies in wood trimmers. *Int. Arch. Occup. Environ. Health*, **58**, 167-177.

Heederik, D., and Douwes, J. (1997) Towards an occupational exposure limit for endotoxins? *Ann. Agric. Environ. Med.*, **4**, 17-19.

Heederik, D., and Wouters, I.M. (2007) Endotoxin exposure and health effects: re-evaluating the evidence and new research initiatives. *Gefahrstoffe Reinhaltung der Luft*, **67**, 357-360.

Heederik, D., Smid, T., Houba, R., and Quanjer, P.H. (1994) Dust-related decline in lung function among animal feed workers. *Amer. J. Ind. Med.*, **25**, 117-119.

Heinsohn, P. (2001) Respiratory tract infections caused by bacteria. In B. Flannigan, R.A. Samson and J.D. Miller, (eds.), *Microorganisms in Home and Work Environments*. Taylor & Francis, London, U.K., pp. 155-194.

Heldal, K., Eduard, W., and Bergum, M. (1997) Bioaerosol exposure during handling of source separated household waste. *Ann. Agric. Environ. Med.*, **4**, 45-52.

Heng, B.H., Goh, K.T., Ng, D.L., and Ling, A.E. (1997) Surveillance of legionellosis and *Legionella* bacteria in the built environment in Singapore. *Ann. Acad. Med. Singapore*, **26**, 557-565.

Hollander, A., Heederik, D., and Kauffman, H. (1994) Acute respiratory effects in the potato processing industry due to a bioaerosol exposure. *Occ. Environ. Med.*, **51**, 73-78.

Houba, R., Doekes, G., and Heederik, D. (1998) Occupational respiratory allergy in bakery workers: a review of the literature. *Am. J. Ind. Med.*, **34**, 529-546.

Hugh-Jones, M.E. (1996) World situation 1993/94. *Salisbury Med. Bull.* (Special Supplement, Proceedings of an International Anthrax Workshop), **87**, 1S-2S.

Huuskonen, M.S., Husman, K., Jarvisalo, J., *et al.* (1984) Extrinsic allergic alveolitis in the tobacco industry. *Brit. J. Ind. Med.*, **41**, 77-83.

Jayawardana, P.L., and Udupihille, M. (1997) Ventilatory function of factory workers exposed to tea dust. *Occup. Med. (Lond.)*, **47**, 105-109.

Jernigan, D.B., Raghunathan, P.L., Bell, B.P., *et al.* [National Anthrax Epidemiologic Investigation Team] (2002) Investigation of bioterrorism-related anthrax, United States, 2001: epidemiologic findings. *Emerg. Infect. Dis.*, **8**, 1019-1028.

Johnsen, C.R., Sorensen, T.B., and Larsen, A.I. (1997) Allergy risk in an enzyme producing plant. *Occup. Environ. Med.*, **54**, 671-675.

Kateman, E., Heederik, D., Pal, T.M., *et al.* (1990) Relationship of airborne microorganisms with the lung function and leucocyte levels of workers with a history of humidifier fever. *Scand. J. Work Environ. Health*, **16**, 428-433.

Kennedy, S.M., Christiani, D.C., Eisen, E.A., *et al.* (1987) Cotton dust and endotoxin exposure response relationships in cotton textile workers. *Am. Rev. Respir. Dis.*, **125**, 194-200.

Khan, Z.U., Gangwar, M., Gaur, S.N., and Randhawa, H.S. (1995) Thermophilic actinomycetes in cane sugar mills: an aeromicrobiologic and seroepidemiologic study. *Antonie van Leeuwenhoek*, **67**, 339-344.

Kreiss, K., and Cox-Ganser, J. (1997) Metalworking fluid-associated hypersensitivity pneumonitis: A workshop summary. *Am. J. Ind. Med.*, **32**, 423-432.

Kriebel, D., Samam, S.R., Wosk, S., *et al.* (1997) A field investigation of the acute respiratory effects of metal working fluids. 1. Effects of aerosol exposures. *Am. J. Ind. Med.*, **31**, 756-766.

Lacey, J. (1973) The air spora of a Portuguese cork factory. *Ann. Occup. Hyg.*, **16**, 223-230.

Lacey, J., Williamson, P.A.M., and Crook, B. (1992) Microbial emissions from composts made for mushroom production and from domestic waste. In D.V. Jackson, J.M. Merillot, and P. L'Hermite, (eds.), *Composting and Compost Quality Assurance Criteria*. Office for Official Publications of the European Community, Luxembourg, EUR14254, pp. 117-130.

LaForce, F.M. (1969) Epidemiological study of a fatal case of inhalation anthrax. *Arch. Environ. Health*, **18**, 798-805.

Laitinen, S., Nevalainen, A., Kotimaa, M., *et al.* (1992) Relationship between bacterial counts and endotoxin concentrations in the air of wastewater treatment plants. *Appl. Environ. Microbiol.*, **58**, 3774-3776.

Latgé, J.P. (1999) *Aspergillus fumigatus* and aspergillosis. *Clin. Microbiol. Rev.*, **12**, 310-350.

Lenhart, S.W., and Cole, E.C. (1993) Respiratory illness in workers of an indoor shiitake mushroom farm. *Appl. Occup. Environ. Hyg.*, **8**, 112-119.

Levetin, E. (2004) Methods for aeroallergen sampling. *Curr. Allergy Asthma Rep.*, **4**, 376-383.

Liesivuori, J., Kotimaa, M., Laitinen, S., *et al.* (1994) Airborne endotoxin concentrations in different work conditions. *Am. J. Ind. Med.*, **25**, 123-124.

Love, R.G., Muirhead, M., Collins, H.P., and Soutar, C.A. (1991) The characteristics of respiratory ill health of wool textile workers. *Brit. J. Ind. Med.*, **48**, 221-228.

Malmros, P., Sigsgaard, T., and Bach, B. (1992) Occupational health problems due to garbage sorting. *Waste Manage. Res.*, **10**, 227-234.

Marchand, G., Lavoie, J., and Lazure, L. (1995) Evaluation of bioaerosols in a municipal solid waste recycling and composting plant. *J. Air Waste Manage. Assoc.*, **45**, 778-781.

Matsby-Balzer, I., Edebo, L., Jarvholm, B., and Lavenius, B. (1989) Serum antibodies to *Pseudomonas pseudoalcaligenes* in metal workers exposed to infected metal working fluids. *Int. Arch. Allergy Appl. Immunol.*, **88**, 304-311.

Matsui, T., Nakashima, K., Ohyama, T., *et al.* (2008) An outbreak of psittacosis in a bird park in Japan. *Epidemiol. Infect.*, **136**, 492-495.

McSharry, C., Anderson, K., Speekenbrink, A., *et al.* (1993) Discriminant analysis of symptom pattern and serum antibody titres in humidifier related disease. *Thorax*, **48**, 496-500.

Melbostad, E., Eduard, W., and Skogstad, A. (1994) Exposure to bacterial aerosols and work related symptoms in sewage workers. *Am. J. Ind. Med.*, **25**, 59-63.

Metcalfe, N. (2004) The history of woolsorters' disease: a Yorkshire beginning with an international future? *Occup. Med. (Lond.)*, **54**, 489-493.

Michils, A., De Vuyst, P., Nolard, N., *et al.* (1991) Occupational asthma to spores of *Pleurotus cornucopiae*. *Eur. Respir. J.*, **4**, 1143-1147.

Millner, P.D., Olenchock, S.A., Epstein, E., *et al.* (1994) Bioaerosols associated with composting facilities. *Compost Sci. Util.*, **2**, 1-57.

Milton, D.K., Amsel, J., Reed, C.E., *et al.* (1995) Cross-sectional follow-up of a flu-like respiratory illness among fiberglass manufacturing employees: endotoxin exposure associated with two distinct sequelae. *Am. J. Ind. Med.*, **28**, 469-488.

Milton, D.K., Walters, M.D., Hammond, K., and Evans, J.S. (1996a) Worker exposure to endotoxin phenolic compounds, and formaldehyde in a fiberglass insulation manufacturing plant. *Am. Ind. Hyg. Assoc. J.*, **57**, 889-896.

Milton, D.K., Wypij, D., Kriebel, D., *et al.* (1996b) Endotoxin exposure-response in a fiberglass manufacturing facility. *Am. J. Ind. Med.*, **29**, 3-13.

Morawska, L. (2006) Droplet fate in indoor environments, or can we prevent the spread of infection? *Indoor Air*, **16**, 335-347.

Mori, S., Nakagawa-Yoshida, K., Tsuchihashi, H., *et al.* (1998) Mushroom worker's lung resulting from indoor cultivation of *Pleurotus ostreatus*. *Occup. Med. (Lond.)*, **48**, 465-468.

Moroney, J.F., Guevara, R., Iverson, C., *et al.* (1998) Detection of chlamydiosis in a shipment of pet birds, leading to recognition of an outbreak of clinically mild psittacosis in humans. *Clin. Infect. Dis.*, **26**, 1425-1429.

Muilenberg, M.L. (2003) Sampling devices. *Immunol. Allergy Clin. North Am.*, **23**, 337-55.

Muller, J., Halweg, H., Podsiadlo, B., and Radwan, L. (1991) Symptoms and functional disorders of the respiratory system caused by exposure to tea dust. *Pneumonol. Allergol. Pol.*, **59**, 210-217.

Musk, A.W., Venables, K.M., Crook, B., *et al.* (1989) Respiratory symptoms, lung function and sensitisation to flour in a British bakery. *Brit. J. Ind. Med.*, **46**, 636-642.

Newman, C.P., Palmer, S.R., Kirby, F.D., and Caul, E.O. (1992) A prolonged outbreak of ornithosis in duck processors. *Epidemiol. Infect.*, **108**, 203-210.

Nielsen, E.M., Breum, N.O., Nielsen, B.H., *et al.* (1997a) Bioaerosol exposure in waste collection: A comparative study on the significance of collection equipment, type of waste and seasonal variation. *Ann. Occup. Hyg.*, **41**, 325-344.

Nielsen, B.H., Würtz, H., and Poulsen, O.M. (1997b) Microorganisms and endotoxin in experimentally generated bioaerosols from composting household waste. *Ann. Agric. Environ. Med.*, **4**, 159-168.

Niemela, S.I., Vaatanen, P., Mentu, J., *et al.* (1985) Microbial incidence in upper respiratory tracts of workers in the paper industry. *Appl. Environ. Microbiol.*, **50**, 163-168.

Niven, R.M., Fletcher, A.M., Pickering, C.A.C., *et al.* (1997) Chronic bronchitis in textile workers. *Thorax*, **52**, 22-27.

Nordman, H., Zitting, A., and Mantyjarvi, R. (1984) Extrinsic allergic alveolitis in the tobacco industry. *Brit. J. Ind. Med.*, **41**, 77-83.

Nordness, M.E., Zacharisen, M.C., Schlueter, D.P., and Fink, J.N. (2003) Occupational lung disease related to cytophaga endotoxin exposure in a nylon plant. *J. Occup. Environ. Med.*, **45**, 385-392.

Ogundero, V.W., and Aina, J.O. (1989) Storage temperature and viability of sporangiospores of potentially human pathogenic species of *Rhizomucor* from Nigerian tobacco. *J. Basic Microbiol.*, **29**, 171-175.

Olsen, B., Persson, K., and Broholm, K.A. (1998) PCR detection of *Chlamydia psittaci* in faecal samples from passerine birds in Sweden. *Epidemiol. Infect.*, **121**, 481-484.

Palchak, R.B., Cohen, R., Ainslie, M., and LaxHoerner, C. (1988) Airborne endotoxin associated with industrial-scale production of protein products in Gram-negative bacteria. *Am. Ind. Hyg. Assoc. J.*, **8**, 420-421.

Petrovay, F., and Balla, E. (2008) Two fatal cases of psittacosis caused by *Chlamydophila psittaci*. *J. Med. Microbiol.*, **57**, 1296-8.

Pfisterer, R.M. (1991) Eine Milzbrandepidemie in der Schweiz; Klinische, diagnostische und epidemiologische Aspekte einer weitgehend vergessenen Krankheit. *Schweiz. Med. Wschr.*, **121**, 813-825.

Phoolchund, H.N. (1991) Aspects of occupational health in the sugar cane industry. *J. Soc. Occup. Med.*, **41**, 133-136.

Poulsen, O.M., Breum, N.O., Ebbehoj, N., *et al.* (1995) Collection of domestic waste. Review of occupational problems and their possible causes. *Sci. Tot. Environ.*, **170**, 1-19.

Rahkonen, P., Ettala, M., and Loikkanen, I. (1987) Working conditions and hygiene at sanitary landfills in Finland. *Ann. Occup. Hyg.*, **31**, 4A, 505-513.

Raza, S.N., Fletcher, A.M., Francis, H.C., *et al.* (2000) Relationship between CD14 expression on monocytes, spirometry and symptoms across the working week in endotoxin exposed cotton workers. *Proceedings, Beltwide Cotton Dust Research Conference*. National Cotton Council, Memphis, TN, pp. 228-230.

Robertson, W., Robertson, A.S., Burge, C.B., *et al.* (2007) Clinical investigation of an outbreak of alveolitis and asthma in a car engine manufacturing plant. *Thorax*, **62**, 981-990.

Romano, C., Sulotto, F., Piolatto, G., *et al.* (1995) Factors related to the development of sensitization to green coffee and castor bean allergens among coffee workers. *Clin. Exp. Allergy*, **25**, 643-650.

Rosenman, K.D., Reilly M.J., and Kalinowski, D. (1997) Work related asthma and respiratory symptoms among workers exposed to metal working fluids. *Am. J. Ind. Med.*, **32**, 325-331.

Rylander, R., and Bergstrom, R. (1993) Bronchial reactivity among cotton workers in relation to dust and endotoxin exposure. *Ann. Occup. Hyg.*, **37**, 57-63.

Rylander, R., Haglind, P., and Lundholm, M. (1985) Endotoxin in cotton dust and respiratory function decrement among cotton workers in an experimental cardroom. *Am. Rev. Resp. Dis.*, **131**, 209-213.

Sande, I., Zahradnik, E., Bogdanovic, J., *et al.* (2007) Optimized methods for fungal alpha-amylase airborne exposure assessment in bakeries and mills. *Clin. Exp. Allergy*, **37**, 1229-1238.

Sanderson, W.T., Stoddard, R.R., Echt, A.S., *et al.* (2004) *Bacillus anthracis* contamination and inhalational anthrax in a mail processing and distribution center. *J. Appl. Microbiol.*, **96**, 1048-1056.

Sastre, J., Ibanez, M.D., Lopez, M., and Lehrer, S.B. (1990) Respiratory and immunological reactions among Shiitake (*Lentinus edodes*) mushroom workers. *Clin. Exper. Allergy*, **20**, 13-19.

Sastre, J., Olmo, M., Novalvos, A., *et al.* (1996) Occupational asthma due to different spices. *Allergy*, **51**, 117-120.

Sigari, G., Panatto, D., Lai, P., *et al.* (2006). Virological investigation on aerosol from waste depuration plants. *J. Prev. Med. Hyg.*, **47**, 4-7.

Sigsgaard, T., Pedersen, O.F., Juul, S., and Gravesen, S. (1992) Respiratory disorders and atopy in cotton, wool and other textile mill workers in Denmark. *Am. J. Ind. Med.*, **22**, 163-184.

Sigsgaard, T., Malmros, P., Nersting, L., and Pedersen, C. (1994) Work related symptoms and lung function among Danish refuse workers. *Am. Rev. Respir. Dis.*, **149**, 1407-1412.

Simpson, A.T., Stear, M., Groves, J.A., *et al.* (2003) Occupational exposure to metalworking fluid mist and sump fluid contaminants. *Ann. Occup. Hyg.*, **47**, 17-30.

Smid, T., Heederik, D., Houba, R., and Quanjer, P.H. (1992) Dust and endotoxin-related respiratory effects in the animal feed industry. *Am. Rev. Respir. Dis.*, **146**, 1474-1479.

Smid, T., Heederik, D., Houba, R., and Quanjer, P.H. (1994) Dust and endotoxin-related acute lung function changes and work-related symptoms in workers in the animal feed industry. *Am. J. Ind. Med.*, **25**, 877-888.

Smith, K.A., Bradley, K.K., Stobierski, M.G., and Tengelsen, L.A. (2005) Compendium of measures to control *Chlamydophila psittaci* (formerly *Chlamydia psittaci*) infection among humans (psittacosis) and pet birds. *J. Am. Vet. Med. Assoc.*, **226**, 532-5399.

Ström, G., and Blomquist, G. (1986) Airborne spores from mouldy citrus fruit – a potential occupational health hazard. *Ann. Occup. Hyg.*, **30**, 455-460.

Swan, J.R.M., Kelsey, A., and Crook, B. (2003) Occupational and environmental exposure to bioaerosols from composts and potential health effects – A critical review of published data. Health and Safety Executive Research Report 130.)Online at http://www.hse.gov.uk/research/rrhtm/rr130.htm).

Tarvainen, K., Salonen, J.-P., Kanerva, L., *et al.* (1991) Allergy and toxicodermia from shiitake mushrooms. *J. Am. Acad. Dermatol.*, **24**, 64-66.

Thomas, K.E., Trigg, C.J., Bennett, J.B., *et al.* (1991) Factors relating to the development of respiratory symptoms in coffee process workers. *Brit. J. Ind. Med.*, **48**, 314-322.

Thorne, P.S., DeKoster, J.A., and Subramanian, P. (1996) Environmental assessment of aerosols, bioaerosols, and airborne endotoxins in a machining plant. *Am. Ind. Hyg. J.*, **57**, 1163-1168.

Thorne, P.S., Adamcakova-Dodd, A., Kelly, K.M., *et al.* (2006) Metal working fluid with mycobacteria and endotoxin induces hypersensitivity pneumonitis in mice. *Am. J. Respir. Crit. Care Med.*, **173**, 759-768.

Thorpe, A., Gould, J.R.M., Brown, R.C., and Crook, B. (1997) Investigation of the performance of vehicle cab filtration systems. *J. Agric. Engineering Res.*, **66**, 135-149.

Topping, M.D., Scarisbrick, D.A., Luczynska, C.M., *et al.* (1985) Clinical and immunological reactions to *Aspergillus niger* among workers in a biotechnology plant. *Brit. J. Ind. Med.*, **42,** 312-318.

Travers-Glass, S.A., and Crook, B. (1994) Respiratory sensitisation of workers' exposure to microbially contaminated oil mists. *Ann. Occup. Hyg.*, **38** (Suppl. 1), 907-910.

Trout, D., Weissman, D.N., Lewis, D., *et al.* (2003) Evaluation of hypersensitivity pneumonitis among workers exposed to metal removal fluids. *Appl. Occup. Environ. Hyg.*, **18**, 953-960.

Turnbull, P.C.B. (1996) Guidance on environments known to be or suspected of being contaminated with anthrax spores. *Land Cont. Reclam.*, **4**, 37-45.

Turnbull, P.C.B., Bowen, J.E., Gillgan, J.S., and Barrett, N.J. (1996) Incidence of anthrax, and environmental detection of *Bacillus anthracis* in the UK. *Salisbury Med. Bull.* (Special Supplement, Proceedings of an International Anthrax Workshop), **87**, 5S-6S.

Uitti, J., Nordman, H., Huuskonen, M.S., *et al.* (1998) Respiratory health of cigar factory workers. *Occup. Environ. Med.*, **55**, 834-839.

Vaisanen, O.M., Nurmiaho-Lassila, E.L., Marmo, S.A., and Salkinoja-Salonen, M.S. (1994) Structure and composition of biological slimes on paper and board machines. *Appl. Environ. Microbiol.*, **60**, 641-653.

Vaisanen, O.M., Weber, A., Bennasar, A., *et al.* (1998) Microbial communities of printing paper machines. *J. Appl. Microbiol.*, **84**, 1069-1084.

van den Bogart, H.G.G., van den Ende, G., van Loon, P.C.C., and van Griensven, L.J.L.D. (1993) Mushroom workers' lung; serologic reactions to thermophilic actinomycetes in the air of compost tunnels. *Mycopathologia*, **122**, 21-28.

van der Zee, D.J., Achterkamp, M.C., and de Visser, B.J. (2004) Assessing the market opportunities of landfill mining. *Waste Manag.*, **24**, 795-804.

van Woerden, H.C., Mason, B.W., Nehaul, L.K., *et al.* (2004) Q fever outbreak in industrial setting. *Emerg. Infect. Dis.*, **10**, 1282-1289.

Wallace, R.J., Zhang, Y., Wilson, R.W., *et al.* (2002) Presence of a single genotype of the newly described species *Mycobacterium immunogenum* in industrial metalworking fluids associated with hypersensitivity pneumonitis. *Appl. Environ. Microbiol.*, **68**, 5580-5584.

Wattiau, P., Klee, S.R., Fretin, D., *et al.* (2008) Occurrence and genetic diversity of *Bacillus anthracis* strains isolated in an active wool-cleaning factory. *Appl. Environ. Microbiol.*, **74**, 4005-11.

Winner, S.J., Eglin, R.P., Moore, V.I.M., and Mayon-White, R.T. (1987) An outbreak of Q fever affecting postal workers in Oxfordshire. *J. Infect.*, **14**, 255-261.

Wouters, I.M., Spaan, S., Douwes, J., *et al.* (2006) Overview of personal occupational exposure levels to inhalable dust, endotoxin, beta(1→3)-glucan and fungal extracellular polysaccharides in the waste management chain. *Ann. Occup. Hyg.*, **50**, 39-53.

Woskie, S.R., Virji, M.A., Kriebel, D., *et al.* (1996). Exposure assessment for a field investigation of the acute respiratory effects of metalworking fluids. 1. Summary of findings. *Am. Ind. Hyg. Assoc. J.*, **57**, 1154-1162.

Würtz, H., and Breum, N.O. (1997) Exposure to microorganisms during manual sorting of recyclable paper of different quality. *Ann. Agric. Environ. Med.*, **4**, 129-135.

Zock, J.P., Heederik, D., and Kromhout, H. (1995) Exposure to dust, endotoxin and micro-organisms in the potato processing industry. *Ann. Occup. Hyg.*, **39**, 841-854.

Zock, J.P., Hollander, A., Heederik, D., and Douwes, J. (1998) Acute lung function changes and low endotoxin exposures in the potato processing industry. *Am. J. Ind. Med.*, **33**, 384-391.

Zuskin, E., Kanceljak, B., Mustajbegovic, J., *et al.* (1995) Immunological reactions and respiratory function in wool textile workers. *Am. J. Ind. Med.*, **28**, 445-156.

Zuskin, E., Kanceljak, B., Schachter, E.N., *et al.* (1992) Immunological findings in hemp workers. *Environ. Res.*, **59**, 350-361.

Zuskin, E., Schachter, E.N., Kanceljak, B. *et al.* (1993) Organic dust disease of airways. *Int. Arch. Occup. Environ. Health*, **65**, 135-140.

Zuskin, E., Mustajbegovic, J., and Schachter, E.N. (1994) Follow-up study of respiratory function in hemp workers. *Am. J. Ind. Med.*, **26**, 103-115.

Chapter 2.3

REMEDIATION AND CONTROL OF MICROBIAL GROWTH IN PROBLEM BUILDINGS

Philip R. Morey

ENVIRON International Corp. Gettysburg, Pennsylvania, USA.

INTRODUCTION

In the context of microbial growth, a problem building may mean that chronic leaks or damp conditions exist, or that building HVAC and water systems are poorly designed, operated or maintained. In such buildings, filamentous fungi may grow on biodegradable water-damaged or damp finishing and construction materials. In addition to fungi, mites may reproduce on damp interior surfaces with abundant human skin scales. Fungi such as *Fusarium* and yeasts, and also Gram-negative bacteria, may grow in stagnant water or on wet surfaces in components of heating, ventilation and air-conditioning (HVAC) systems, and *Legionella* may grow in potable water systems and cooling towers.

The necessity for microbial remediation in a building implies that extensive biodeterioration or growth has already occurred. Almost all microbial problems in buildings are caused by failure to keep the infrastructure clean and dry, and/or by failure in the design, operation and maintenance of building systems. This chapter first reviews microbial remediation procedures, with the emphasis on fungal contamination. Actions that can be taken to prevent microbial colonization, primarily through moisture control, are then described. The chapter concludes with checklists that should be considered when inspecting a building for moisture and microbial growth problems. Sampling as a component of the building evaluation is reviewed in Chapter 4.6 (see also AIHA 2005, 2008, Morey 2007a).

REMEDIATION OF FUNGAL GROWTH IN BUILDINGS

It has long been recognized that extensive growth of fungi on interior surfaces is unacceptable. Over 3000 years ago, the Book of Leviticus (Blomquist 1994, Heller *et al.* 2003) described the removal of fungus-affected materials from buildings. Leviticus Chapter 14 tells us that materials affected by the "plague" should be removed to an unclean place outside the city. The plaster, mortar and timbers should be removed if the plague continues to grow. Additionally, persons living in the unclean house should wash their clothing. The unacceptability of fungal growth in interior environments and the general principles set down in Leviticus for dealing with fungal growth have been reaffirmed, beginning in 1993 with the first edition of the New York City Guidelines (NYC 1993) and continuing to the most recent documents on mould remediation (AIHA 2008, NYC 2008).

While fungal growth problems in buildings have been known since ancient times, some modern constructional practices have presented new habitats for colonization. In North America, traditional construction practice prior to about 1950 emphasized the use of stone, brick, plaster and wood. These materials are broadly resistant to microbial growth. Surfaces in pre-1950 buildings were mostly non-porous and therefore easily cleaned. In recent decades, the use of amorphous cellulose products, porous coverings for floors, walls, and ceilings, porous insulation and pressed-wood products have greatly increased. These modern construction and finishing products are generally susceptible to biodeterioration when there is chronic dampness or wetness in building components. In addition, porous construction and finishing materials are difficult to clean. The presence of dust and dirt (nutrient for microbes) in some porous materials allows microbial growth when moisture is non-limiting.

The location of modern air-conditioned buildings in warm, moist geographical locations such as Florida and Singapore increases the potential for condensation on cold surfaces and therefore, for microbial growth on moisture intolerant construction and finishing materials. Vigilance in keeping modern building systems dry and clean is essential if microbial growth is to be prevented.

REVIEW OF CONSENSUS PUBLICATIONS

A number of publications during the past 15 years (NYC 1993, 2000, 2008, Samson *et al.* 1994, ISIAQ 1996, Health Canada 1998, 2004, ACGIH 1999, EPA 2001, AIHA 2004, 2008, Canadian Construction Association 2004, Institute of Medicine 2004, California Research Bureau 2006) have provided general guidance on clean-up of mould contamination. Some of these publications are now reviewed.

International workshop on health implications of fungi in indoor environments (1992)

This workshop held in Baarn in the Netherlands in 1992 (Samson *et al.* 1994) made the following recommendations regarding fungal growth in buildings:

- The growth of fungi on interior surfaces of non-industrial buildings is unacceptable on both health and hygiene grounds.
- Condensation on building surfaces and penetration of water into building envelopes should be prevented.
- The health risks of biocides are not fully understood, and therefore biocides should be used only as a last resort for controlling fungal growth indoors.
- The inhalation of fungal spores and other products such as microbial volatiles should be avoided when handling contaminated materials.

NYC *Stachybotrys* guidelines (1993)

A panel in 1993 in New York City discussed appropriate assessment and remediation actions to be taken when *Stachybotrys chartarum* growth was visible on interior surfaces (NYC 1993). The panel was convened because of several cases of *Stachybotrys* colonization of paper-fibre gypsum board and library materials in New York City buildings. Panel recommendations included the following:

- Visibly mouldy materials should be removed by persons equipped with appropriate personal protective equipment including respirators and gloves.
- The use of containment barriers (plastic sheeting) and negative pressurization is required for removal of mouldy materials with a surface area greater than about 3 m², the approximate surface area of one side of a paper-fibre gypsum board used in North American buildings. Smaller patches of mouldy materials can be removed by simpler methods.
- Building maintenance personnel can remove small patches (<3 m²) of colonized materials. Persons trained in the removal of hazardous dusts and the appropriate use of personal, protective equipment should be employed when larger areas of colonized materials are removed.
- Defects causing incursion of water into the building infrastructure must be repaired.

ACGIH Bioaerosols Assessment and Control (1999)

This publication (ACGIH 1999) affirmed the general principles of other guides (NYC 1993, Health Canada 1995, ISIAQ 1996), including removal of porous materials with extensive colonization; removal of surface colonization from non-porous materials; and elimination of moisture problems that initially allowed growth to occur. The ACGIH classified the extent of fungal colonization as minimal, moderate and extensive, without assignment of numerical surface-area guidelines. It stated that plastic-sheeting barriers and negative pressurization should be used for containment of dusts when extensively colonized material is removed. Regardless of the extent of fungal colonization, at a minimum N-95 respirators should be used by those performing remediation. Personnel employed in large-scale fungal remediation should be informed of the potential health risks associated with exposure to bioaerosols.

Canadian Construction Association Guidelines (2004)

The Canadian Construction Association Standard 82

(CCA 2004) is unique in recommending that principles of "universal precautions" and "controlled conditions" be used during mould remediation. "Universal precautions" implies that, unless proved otherwise, when visible mould growth occurs it should be assumed that an exposure risk occurs, so that, for example, clean-up workers must use respirators. "Controlled conditions" means that regardless of the extent or amount of visible mould growth, dust suppression and containment techniques should be used to prevent the aerosolization of bioaerosols from the area undergoing remediation.

The guide recommends a series of detailed clean-up actions that should be followed, depending upon the extent of visible mould growth on interior surfaces (small-scale or <1 m^2; medium-scale or <1-10 m^2; large-scale or >10 m^2). For example, containment and depressurization (-5 pascals) are recommended for medium and large-scale mould remediation. In addition, for large-scale remediation, an occupational health professional should provide independent oversight to verify that appropriate work practices are followed during remediation.

Institute of Medicine – Damp Indoor Spaces and Health (2004)

Chapter 6 of this Institute of Medicine (IOM) publication presents a review of the "Prevention and Remediation of Damp Indoor Environments". Destructive inspection of wall, ceiling and floor systems in order to estimate the extent of visible mould growth, while often necessary as a key aspect of the building evaluation process, should be carried out with prudent dust control actions. Because cutting into visibly mouldy materials may release dusts containing microbial particulates, dust suppression methods such as using HEPA filter vacuum cleaners attached to power tools should be considered during the inspection process.

The chapter discusses the logic behind recommendations in other guidelines on increasing personal protective equipment (PPE) requirements and dust suppression (containment) methods when large or extensive amounts of visible mould growth are found during building inspection. The presence of visible mould colonies on a refrigerator gasket is mentioned as unlikely to be a significant source of exposure, but the presence of extensive mould growth (100 m^2 in the example given on p. 296) is of significant concern for occupant or worker exposure and for dust suppression during remediation. Recent data published elsewhere (Morey 2009) shows that, even

when medium-scale visible mould growth (1-3 m^2) is being removed in depressurized containment, worker exposure to *Penicillium*, *Aspergillus* or *Stachybotrys* may exceed 10^4 spores or spore equivalents m^{-3}. It is therefore prudent to consider use of depressurized containment or other verifiable dust suppression action to protect both occupants and clean-up workers during mould remediation.

The review discusses quality-assurance actions to verify that visible mould growth has been successfully remediated. Concentrations of fungal spores, hyphae or metabolites on remediated surfaces, or in the air within containments, are considered problematic as indicators of successful remediation because of the absence of health-based standards for these microbial components (see also ACGIH 1999, UCONN 2004). The review then indicates that mould particulates are invariably present on interior surfaces (see discussion of "normal deposition" in Horner *et al.* 2004). The presence of abundant fungal elements, such as abundant spores, hyphae, etc., as seen on remediated surfaces by using transparent adhesive tape (Cellotape/Sellotape) sampling and analysis (AIHA 2005, Morey 2007b) is, however, an indication of unsuccessful cleaning. It is recommended that fixing the moisture (dampness) problem and verifying that there is no visible contamination remaining be key principles of mould remediation. The review refers to the AIHA Microbial Task Force report (AIHA 2001) as an important source of other key quality-assurance indicators for verification of the effectiveness of mould remediation.

New York City Guidelines (2008)

The third edition of the New York City mould guidelines reaffirms the investigative and remediation principles found in early editions (NYC 1993, NYC 2000), with some important changes. This 2008 edition states more clearly than earlier editions that the physical inspection of the building to define the mould problem and to provide a basis for the remediation strategy is the most important action in the investigative process. Defining the location and extent of water damage and mould growth, including concealed damage, is a key investigative action necessary for the successful outcome of future remediation activities. The 2008 Guidelines state that environmental sampling (analysis of air, surface, dust or bulk materials) is…"not necessary to undertake a remediation…" as long as the inspection process for visible mould growth and water damage is thorough.

The 2008 guidelines note that the surface area of visible mould growth is an important factor in determining the scope of the mould remediation process. These guidelines enumerate abatement actions that should be considered when the area of mould growth on interior surfaces ≤1 m² (Level I), 1-10 m² (Level II) and >10 m² (Level III). NYC Guidelines (2000) had listed abatement actions under four levels, viz. Level I (<1 m²), Level II (1-3 m²), Level III (3-10 m²), and Level IV (>10 m²). The 2008 edition of the NYC Guidelines continues to address abatement actions for HVAC systems separately from remediation procedures used in occupied or interior spaces in buildings. Separate lists of abatement actions are provided when the area of mould growth in the HVAC system is small (<1 m²) or large (>1 m²). The 2008 guidelines also mention that HVAC systems should be shut down whenever mould abatement is carried out, regardless of the extent of growth. This approach to mould remediation is appropriate because HVAC systems are designed to transport ventilation air, so that particulates that may be aerosolized from ductwork and air-handling units are transferred directly to the occupant-breathing zone.

Guidance on appropriate cleaning procedures during mould abatement is provided in the 2008 NYC Guidelines. Non-porous materials affected by mould growth are regarded as being readily cleanable with detergent or soapy water. Porous or rough-surface materials such as insulation, ceiling tiles, and paper-faced wallboard with more than a trivial amount of mould growth are discarded. Guidance on the cleaning of porous, semi-porous, and non-porous materials affected by mould growth is also given in the Institute of Inspection Cleaning and Restoration Certification (IICRC) Standard S520 (2008), especially in Chapter 13 on "Contents Remediation".

The 2008 edition of the NYC Guidelines recommends that disinfectants should not to be used for mould abatement. Additionally, these 2008 guidelines state that use of aerosolized antimicrobials "is not recommended". The rationale for this recommendation includes the potential adverse health effects of antimicrobials on people as well as the ineffectiveness of antimicrobials with regard to physically removing "non-viable mould". In other words, dead mould can still cause allergies.

AIHA – Recognition, Evaluation, and Control of Indoor Mold (2008)

This publication from the American Industrial Hygiene Association (AIHA 2008) presents a comprehensive coverage of diagnostic and remedial strategies that can be employed in dealing with indoor mould problems. Many of the authors are practitioners and as a result this book has a "how to approach" treatment of building diagnostics and remediation not found in other publications.

Chapter 16 presents a pathway for assessing the severity of mould contamination. Thus, when planning a remediation strategy, factors such as (a) the moisture tolerance of constructional material and (b) pathways through which bioaerosols can reach occupants should be considered in addition to the surface area extent (m²) of mould growth. Accordingly, mould growth that occurs on surface dust/dirt on concrete or sheet metal surfaces is less of a concern than mould growth on paper-faced wallboard or fibreboard. The presence of transport pathways in a building that can lead to occupant exposure can greatly influence the complexity of the recommended remediation process. For example, remediation of a crawlspace with a moderately extensive amount of mould growth on wood framing is more demanding (or complex) if the return air ductwork of a HVAC system is present in the crawlspace. Chapter 16 notes that the degree of disturbance anticipated during removal of mouldy materials can add to the precautions required in the remediation process.

Chapter 17 provides advice on the remediation of concealed mould growth. It recommends that concealed mould growth be remediated regardless of its extent. However, it recognizes that scientific studies have not determined with certainty how small an area of concealed growth can be ignored. Encapsulation of small areas of mould growth that would otherwise require extraordinary remediation efforts, e.g. mould growth on a wood panel housing a complex electrical circuit system, is appropriate in some cases.

One of the most controversial aspects of mould remediation, viz. judging its effectiveness, is discussed in Chapter 18. As pointed out in an earlier AIHA publication (AIHA 2001) effective quality assurance does not require air or surface sampling to demonstrate that visible mould has been removed. Following the remediation plan, fixing the moisture problem, using correctly functioning HEPA vacuum cleaners during clean-up and reducing "residual dust" levels on cleaned surfaces are some of the key quality assurance actions. Chapter 18 discusses the measurement of residual dust as a key quality assurance indicator, noting that gravimetric levels of dust on non-porous surfaces can be reduced to 10-20 mg m⁻² after a thorough HEPA vacuum cleaning.

Chapter 18 also contains a discussion on the scientific utility of air and surface sampling in determining the effectiveness of mould remediation. When surface or air sampling is employed as the primary verification method for mould remediation interpretation, difficulties such as the following may occur:

- Health-based guidelines are unavailable for acceptable levels of microbial agents, including spores, hyphal fragments or metabolites that may occur in the air or on interior surfaces.
- The sensitivity of various analytical methods for microbial agents, while useful in diagnostic studies, may be too great to be useful during verification of mould remediation. A key quality assurance indicator for mould remediation is whether or not visible growth has been effectively removed, which can be verified by physical inspection.
- Sampling of clearance air and surfaces is generally unnecessary, provided the diagnostic inspection process for moisture and mould damage, and the subsequent clean-up process, are thorough.
- Water indicator fungi such as *Aspergillus versicolor* may still be dominant among the moulds found in trace amounts of residual dusts present on interior surfaces after clean-up.
- An inadequate number of samples may be collected for making valid comparison between remediated and non-remediated/control areas.

Finally, this chapter notes that cellulose adhesive tape sampling and analysis are useful in identifying locations where mould remediation has been deficient, as evidenced by, for example, the presence of mycelium/conidiophores detected by direct microscopy. A caution in Chapter 18 calls attention to the investigative process detecting the presence of moisture and mould damage. If the initial diagnostic inspection is deficient, microbial sampling can be useful in detecting the presence of hidden mould growth. A more thorough inspection of the building and additional remediation is then appropriate.

BASIC PRINCIPLES FOR MOULD REMEDIATION

Important components of mould remediation are (a) the physical removal of mycelium from interior surfaces, (b) removal of spores in settled dusts that may have previously been dispersed from mouldy surfaces, (c) prevention of dusts associated with clean-up from entering occupied or clean areas, and (d) use of personal protective equipment by clean-up workers. A plan to prevent future leaks or dampness that caused colonization is essential for effective remediation.

Porous materials such as paper-fibre gypsum board, ceiling tiles, insulation, wallpaper, carpet, etc. that are visibly mouldy should be discarded. Because it is difficult to see fungal micro-colonies, it is prudent to remove some of the sound material adjacent to the visibly colonized surface. Cellulose adhesive tape sampling may be used to distinguish colonized surfaces (with hyphae or sporing structures present) from sound surfaces. Visible mould that may be present on non-porous surfaces such as sheet metal, ceramic tiles, glass, etc. is physically removed by cleaning. Tap water with detergents should be effective for most cleanings.

The method used to remove visible mould growth from semi-porous materials such as wood depends on the degree to which hyphae have penetrated the substrate. Wood that is rotted by dry- or wet-rot basidiomycete fungi (Lloyd and Singh 1994) is discarded. Wood that is sound except for colonization of the outer surface (to a depth of around 1 mm) may be sanded, planed, refinished, and reused. The guiding principle for reuse should be absence of hyphae and sporing structures, over and above those normally present in sound timber, in the wood cells of the timber being salvaged. Encapsulation in lieu of physical removal of mould growth is considered unacceptable (CRB 2006).

During large-scale mould remediation (Figs. 1 and 2), the airborne spore concentration can exceed 10^6 m^{-3} air (Morey and Ansari 1996, Rautiala *et al.* 1996). Even during mid-scale mould remediation where only 1-3 m^2 of mouldy wallboard is removed (Fig. 3), airborne concentrations of spores can exceed background levels by 2-4 orders of magnitude (Table 1). An N-95 respirator and gloves are adequate for small-scale clean-ups (ACGIH 1999). For remediation involving mid-scale and large-scale clean-ups, the use of disposable full-body protective clothing and respirators (P-100 or HEPA cartridge) must be considered. Chapter 6 of the IICRC S520 Standard (IICRC 2008) discusses respiratory protection and personal protective equipment that should be considered for use during mould remediation.

Dust suppression methods should be used whenever visible mould growth is removed, regardless of

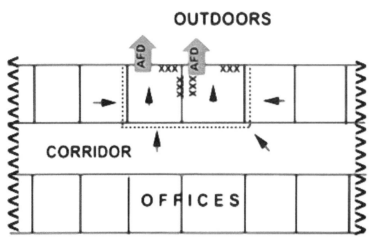

Fig. 1. A floor plan of a large-scale remediation where dusts from colonized surfaces in two rooms are prevented from entering the rest of the building by a contaminant barrier. Key: (-----), barrier; xxx, colonized surfaces; AFD = Air filtration device or negative air machine; → direction of airflow.

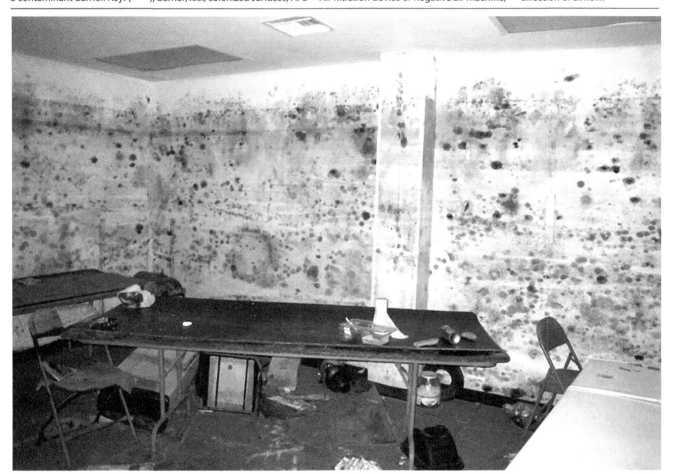

Fig. 2. Extensive mould growth on the walls throughout a room. A depressurized containment is necessary to prevent the transport of bioaerosols into clean areas of the building (photograph courtesy of G. Crawford).

the surface area or the extent of the growth. Thus, the nozzle of a HEPA vacuum cleaner is used to collect dust emissions at their source even for small-scale removal (Fig. 4). Power tools used during mould removal should be equipped with integral HEPA vacuum attachments. Depressurized containments (-5 pascals) should, depending on professional judgment, be used for mid-scale and large-scale removals. The

S520 (IICRC 2008) provides remediators with information on construction and operation of depressurized containments and dust control methodologies. Some additional general principles to be followed during mould removals include:

• At the completion of mould remediation, verify by physical inspection and cellulose adhesive

Table 1. Airborne spores within a depressurized containment during a mid-size mould removal.

spore trap sampling[a]	
Aspergillus/Penicillium	700,000 spores m^{-3}
Stachybotrys	580,000 spores m^{-3}
Filter cassette sampler for culturable moulds[b]	
Stachybotrys chartarum	77,000 CFU m^{-3}
Aspergillus sydowii	16,000 CFU m^{-3}
A. ustus	6,000 CFU m^{-3}

Spore trap sampler (airflow rate 10 l min-1; sampling time <1.0 min; N = 6). [b] Filter cassette (sterile PVC filter; airflow rate 2.5 l min-1; sampling duration 2.4 h; N = 2; filter eluted onto malt extract agar).

tape sampling that mycelum has been removed.
- Using a particulate meter verify that HEPA vacuum cleaners are working properly (absence of particle emissions in and around seals and filters).
- The amount of residual dust on interior surfaces should be sufficiently low to eliminate the need for re-vacuum cleaning. A white/black glove test or collection of residual dust can be used to verify the effectiveness of cleaning (Chapter 18 in AIHA 2008).
- The use of disinfectants and antimicrobial agents is discouraged because the objective of clean-up is to physically remove mould from surfaces. The production of sterile, interior surfaces is not the objective of mould remediation (Morey 2009, IOM 2004).

MICROBIAL REMEDIATION - SPECIAL SITUATIONS

Hospitals

For more than two decades it has been known that, relative to natural ventilation, when air is mechanically ventilated (filtered) the incidence of fungal infection (aspergillosis) among immunosuppressed patients is reduced (Rose and Hirsch 1979). Other studies have shown that epidemics of aspergillosis in immunocompromised patients were associated with the presence of fungal growth on surfaces in HVAC systems and patient rooms and with emission of dusts during soil excavation and/or new construction (Walsh and Dixon 1989).

Because of aspergillosis outbreaks in some medical centres, very stringent guidelines have been recommended for controlling moisture incursion and preventing exposure of immunocompromised pa-

Fig. 3. Mid-scale mould growth (arrowed) hidden in wall cavity. During wall demolition, which was within a depressurized containment, airborne spore levels for various moulds exceeded 10^4-10^5 m^3 air (see Table 1).

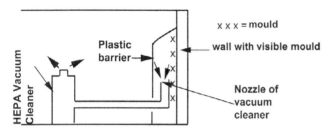

Fig. 4. The nozzle of a HEPA vacuum cleaner can be used to capture dusts at their source during small-scale mould removal. Mouldy particulates flow into nozzle. The air that exits a properly operating HEPA vacuum cleaner is spore free. Key: xxx, colonized surface; → direction of airflow.

tients essentially to all culturable fungal spores (Streifel 1996, Morey 2004). Procedures such as those that follow are recommended during hospital renovation or remedial work:

(a) Isolation and negative-pressurization of the mould remediation/clean-up area; impervious, rigid floor-to-ceiling barriers to isolate patient areas from potential sources of viable fungi;
(b) Employment of administrative procedures to prohibit tracking of dusts by clean-up workers into patient areas;
(c) Provision of high quality air (spores absent) in patient areas by point-of-discharge HEPA filtration in supply air ductwork; and
(d) Positive pressurization of patient rooms relative to areas in which there is fungal colonization or the presence of construction and renovation dusts.

The stringent actions used to reduce incidence of fungal infection among highly immunocompromised patients provide a framework for remediation guidelines that may be necessary in special clean-up situations arising when highly susceptible people may be present in non-medical facilities.

HVAC Airstream Surface

A niche exists for fungal growth in HVAC air-conditioning systems when airstream surfaces are both moist and dirty (Yang 1996). Remediation of fungal growth in HVAC systems is dependent on the extent and location of the colonized surface. Dirt and fungal growth on non-porous surfaces such as sheet metal can be removed by cleaning (Morey 1996). However, fungal colonization of a dirty, porous substrate, such as insulation, cannot be removed by cleaning. Physical removal of the porous insulation itself is required once hyphae have become entwined in and among the insulation fibres (Fig. 5).

When fungal growth in HVAC systems has to be removed, provision must be made during clean-up to deactivate and isolate HVAC system components, and to prevent dusts and spores from following air pathways into occupied spaces (NYC 2008)

Because of the difficulty associated with remediation of fungal colonization on porous HVAC insulation materials, interest has been shown in the incorpora-

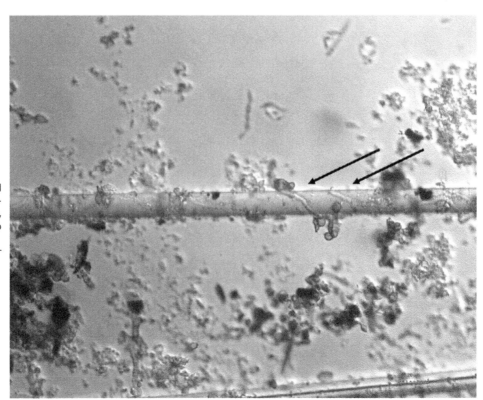

Fig. 5. Airstream surface with entwined hyphae (arrowed) around a fibre of fibreglass. It is difficult, if not impossible, to remove hyphae that have grown into fibrous duct liner.

tion of antifungal compounds into insulation (Yang and Ellringer 1996) and the manufacture of insulation surfaces that are smooth, cleanable, hydrophobic, and resistant to biodeterioration (Morey 1994, ASHRAE 2007).

Additional measures that can be used to minimize the growth of moulds on HVAC airstream surfaces include:

- Using highly efficient filters to reduce the accumulation of fine dusts on airstream surfaces. A minimum efficiency reporting value (MERV) of 11 provides good protection. Filters with a MERV rating of 13 or 14 provide better protection.
- Avoiding the use of fibrous/porous duct liner on airstream surfaces.
- Designing air handling units and ductwork for easy access for periodic inspection and future cleaning as necessary.

Books, paper and libraries

The clean-up of books, paper and archives damaged by floods and dampness involves discarding mouldy items, drying out wet materials and removing settled dusts containing spores. Fungi can grow rapidly on these materials because of the adhesives, gums, starch, etc. often present in jackets and bindings, and because of the presence of delignified cellulosic (paper) substrates.

Because of the susceptibility of books and paper to biodeterioration, the drying of water-damaged or damp materials is of critical importance in restoration. Special techniques such as freeze-drying of water soaked material are important in restoration because low temperatures arrest fungal colonization and evaporation of water molecules (subliming) lowers available moisture, so that growth is minimized (Peterson 1993, Conservation Center 1996, Florian 1997).

The dominating presence of xerophilic moulds such as *Eurotium herbariorum* and *Aspergillus penicillioides* growing on paper materials indicates the previous occurrence of damp, but not wet, conditions (Arai 2000). The occurrence of moulds such as *Stachybotrys* and *Chaetomium* indicates previous wet or flooded conditions. Evidence of arthropod faecal pellets (such as from mites) on the mouldy surface (see Fig. 1, Chapter 4.6) suggests that environmental conditions favourable for fungal growth have been present for a significant period of time, since some mites may feed on mould spores.

Conservators have developed several simple techniques for removing superficial colonization from valuable paper materials. Miniature aspirators are used to provide gentle suction through a pipette nozzle, which is used to carefully remove spores from the surface of the paper (Conservation Center 1996). A larger vacuum device can be used to remove spores if a fine screen is placed firmly over the paper to protect the fragile material being cleaned (Conservation Center 1996).

The cleaning of library materials which are not visibly colonized, but which have been stored in buildings with problems of moisture and mould growth on ceilings, walls and floors, is a challenge because of the enormous amount of paper surface potentially involved. Books, for example, may be covered with a fine layer of dusts, containing abundant mould spores that originated on water-damaged cellulosic ceiling tiles overhead. The following can be effective in cleaning non-colonized library materials that have been stored in dusty, mouldy environments:

- HEPA-vacuum cleaning of the top, bottom and sides of books to remove settled dusts;
- HEPA-vacuum cleaning and damp wiping of the surfaces of shelves, file cabinets, desks and other non-porous surfaces in the library. The presence of visible dust on books and on non-porous surfaces such as library shelves indicates unsuccessful cleaning;
- Fanning the pages of the books, files, and other paper materials in the immediate vicinity of the suction orifice of a HEPA vacuum cleaner. The object is to reduce the amount of settled dust present on surfaces of library materials. However, this cleaning will probably not change the qualitative make-up of fungi in residual dust that remains on surfaces.

Residences

Remediation of fungal colonization in residences differs from that in most commercial and public buildings for reasons of occupancy and construction. In homes, occupants may be present for the entire day, every day of the week. In some residences, occupants may be present who are especially sensitive or susceptible, such as infants, the elderly and those with immunosuppressive conditions.

Most single-family residences are smaller in volume than commercial and public buildings. This means that compared with a large office building a residence has a greater ratio of envelope surface (roof, exterior walls, and basement) to air volume. Consequently, there is a greater envelope surface through which moisture from precipitation or the soil may enter. The association between dampness, mould growth and respiratory symptoms in homes (Health Canada 2004) may be related to the relatively large envelope surface in small residential buildings. Problems of fungal growth in residences are exacerbated by the use of porous biodegradable materials, such as carpet and paper-faced wallboard, in damp locations, where moisture control is difficult, e.g. basements.

The greater use of timber and pressed wood products in residences is an important difference between residences and commercial buildings. In consequence, wood-rotting fungi such as the dry rot fungus *Serpula lacrymans* and soft-rot fungi such as *Trichoderma* spp. are more likely to be present in residences with persistent moisture problems. Remediation of fungal growth in residences is often logistically difficult because of problems of access to biodeteriorated wood structural members in crawl spaces and attics.

Another important difference between residences and commercial buildings is the greater amount of porous textiles and furnishings in the former. Section 14 of the IICRC S520 Standard (2008) contains an extensive discussion on microbial remediation of porous contents including clothing, carpet, upholstery, mattresses and textiles.

Sewage backflow or black water

Backflow of sewage can make a residence or commercial building uninhabitable. Effective restoration requires that sewage waters be removed from building infrastructure and that potentially pathogenic viruses, bacteria (e.g. *Escherichia coli* O157/H7), fungi, protozoa (e.g. *Giardia lamblia*) and invertebrates be inactivated by disinfection (Berry *et al.* 1994). Residual moisture from sewage waters in building finishing materials must be removed to prevent the subsequent growth of filamentous fungi.

During building restoration, attention must be given to the porosity of building materials directly contaminated by sewage waters. When porous materials of minimal value such as insulation, paper-fibre gypsum board, clothing and carpet are contaminated by sewage waters, they should be discarded (Berry *et al.* 1994). Semi porous materials such as timber, wood furniture, concrete, and pressed wood products can usually be cleaned and disinfected. Some

semi-porous materials, such as particleboard flooring, may have to be replaced where sewage waters may have penetrated beneath this material. Non-porous materials such as metal, glass, linoleum can be easily cleaned and disinfected (Berry *et al.* 1994).

A difficult aspect of building restoration following a backflow is to identify areas where sewage waters have penetrated into building infrastructure such as wall cavities, floor systems, ventilation ductwork, and crawl spaces. Some removal of undamaged and nonporous finishes may be required to gain access for disinfection.

The IICRC Standard S500 (2006) defines sewage backflow as "Category 3" water, implying that it is contaminated with harmful biological, chemical or toxic agents. This standard provides guidelines on the restoration and drying of building components affected by Category 3 water, as well as restoration and drying guidelines after damage by Category 2 water, such as washing machine and dishwasher discharge, and Category 1 water, such as discharge from potable water piping. In addition to sewage, Category 3 water includes floods from river and sea water, and "contaminated" wind-driven rain from tropical storms and hurricanes. The Standard S500 (2006), does not, however, give a microbiological definition of "contaminated" for wind-driven rain.

MOULD REMEDIATION – IMPORTANT CONSIDERATIONS

Unlike guidelines on removal of hazardous chemical and physical agents, e.g. asbestos, where rigid and standardized inspection and removal protocols are followed and specific numerical guidelines are used for documenting remediation effectiveness, the removal of mould growth from buildings is a complex process affected by many variables. In order to carry out building restoration successfully, a considerable amount of information must be obtained, including sites of chronic moisture incursion and visible mould growth, and the location of occupants, including those with special susceptibilities. The location of biodegradable and porous finishing and construction materials affected by chronic dampness and water incursion must also be established.

The objective of mould remediation is not to disinfect or sterilize interior surfaces, but rather to restore surfaces to conditions characteristic of non-problem buildings. Surfaces in non-problem buildings are characterized by the occurrence of settled dusts con-

Fig. 6. Biofilm (arrowed) present in the drain pan of a poorly maintained air-handling unit.

taining mould spores predominantly derived from outdoor sources such as vegetation. The fungi in settled dusts in non-problem buildings generally exhibit a diversity of species normally characteristic of outdoor air (see Chapter 4.6, Horner *et al.* 2004, AIHA 2005). The finding of colonization by any mould on interior surfaces is an indicator of an unsuccessful cleaning or restoration.

Important points to be taken into consideration regarding mould remediation include the following:

- The extent and location of moisture damage and visible mould growth is determined by inspection.
- The moisture problem is identified and fixed and a plan is developed for removal of visible mould growth.
- Emphasis is placed on quality assurance that the mould remediation protocol is followed (AIHA 2001) and that visible growth (mycelium) is removed.
- Because mould growth problems in buildings differ in extent, pattern and ecology, protocols for restoration will necessarily be complex and varied.

REMEDIATION OF MICROBIAL GROWTH IN DRAIN PANS AND ON COOLING COILS

Building-related health symptoms in air-conditioned environments have been consistently associated with the occurrence of water, biofilm and dirt in or on HVAC system drain pans and cooling coils (see Fig. 6; Harrison *et al.* 1987, Seppanen and Fisk 2002). Studies in problem buildings in the USA have found defects such as stagnant water and microbial growth

in drain pans and on cooling coils of HVAC systems (Crandall and Sieber 1996, Sieber *et al.* 1996). Control of both water and dirt accumulation in wet portions of HVAC systems is important in reducing building-related symptoms in air-conditioned buildings (see also earlier section on HVAC Airstream Surface). In order that the drain pan be self-draining, the pan should be sloped with the drain hole flush with the bottom of the pan (Chapter 10 in ACGIH 1999, ASHRAE 2007). Additionally, the air-handling unit should be mounted in the mechanical equipment room with adequate clearance so that the drain line can be properly sealed (Trane Applications Manual 1998).

Procedures for removal of biofilm from drain pans and coils are well known (Brundrett 1979) and include (a) turning off the HVAC component, (b) draining stagnant water, (c) physically removing the biofilm, using detergents and disinfectants in the cleaning process, (d) removing cleaning chemicals and (e) re-commissioning the HVAC component. The introduction of aerosolized biocides and disinfectants to ventilation air by an active HVAC system is not appropriate (Brundrett 1979, ISIAQ 1996, NYC 2008).

Although better maintenance and physical cleaning of HVAC airstream surfaces, combined with installation of more highly efficient filters, will reduce microbial growth, ultraviolet germicidal irradiation (UVGI) of drain pan and coil surfaces provides an additional option for control of fungal and bacterial growth (Menzies *et al.* 2003, Martin *et al.* 2008). However, caution must be used with UVGI because of safety issues (e.g., exposure of maintenance workers to lamps) and because UVGI can degrade filter media, pipe jackets and ductliners (ASHRAE 2008).

LEGIONELLA IN WATER SYSTEMS

There are approximately 48 known species of *Legionella*, the bacterium which causes Legionnaires' disease and Pontiac fever (AIHA 2005). In the USA there are approximately 18,000 cases of Legionnaires' disease with a case fatality rate in the range of 20-40% resulting in an annual total mortality of 4,000 - 5,000 (Benin *et al.* 2002, Squire *et al.* 2005). *Legionella* can colonize cooling towers and potable water piping as well as vegetable misters, water fountains, whirlpools and humidifiers (ASHRAE 2000).

Implementation of appropriate design and maintenance principles are effective in controlling *Legionella* amplification in building water systems (HSE 1994, ASHRAE 2000, WHO 2007, OSHA 2008). For cooling towers, the following are important:

(a) Cooling towers should be located as far as possible from building openings or outdoor areas where people may congregate. ASHRAE Standard 62.1-2007 requires a minimum separation distance of 15 feet (4.5 m) between the water basin and HVAC system outdoor air inlets, and 25 feet (7.5 m) between the cooling tower exhaust and outdoor air inlets.

(b) Attention to biocide, scale, and corrosion control is especially important for tower start-up and for start-up of towers containing stagnant waters (ASHRAE 2000).

(c) Enhanced *Legionella* control is achieved by use of non-porous construction materials and provision of easy access to tower components for purposes of cleaning.

Recommendations for achieving *Legionella* control in potable water systems include:

(a) The hot water storage temperature should be at 60°C (140°F) with point of discharge temperature at taps and shower heads of 50°C, or 124°F (Fields and Moore 2006).

(b) Design considerations to be taken into account include avoidance of rubber washers, elimination of stagnant plumbing lines, and reduction of dirt and debris in city and well water entering the building (WHO 2007).

Emergency disinfection of water systems occurs when one or more cases of Legionnaires' disease occurs among building occupants and when the species/serotype of *Legionella* that caused illness are the same as those of *Legionella* in the building water system. Emergency disinfection of potable water systems may involve thermal pasteurization (at 71-77°C, or 160-170°F) of the hot water tank and tap outlets (Fields and Moore 2006). Care must be taken during thermal disinfection to avoid scalding of occupants. Shock chlorination (free residual chlorine at 10-50 mg l^{-1}) may be used for emergency disinfection of cooling tower water systems. Dispersants are circulated in cooling tower water systems during emergency disinfection to dislodge *Legionella* that may be present in biofilm.

During emergency disinfection of cooling towers, it is essential to protect clean-up workers from *Legionella* aerosols (at a minimum with an N-95 respirator, which should be frequently changed), and from disinfectant splash or aerosols (an acid gas respirator cartridge is required for chlorine disinfectants, and skin and eye protection are also required).

HOUSE DUST MITES

In buildings where hypersensitivity or allergic disease is caused by house dust mite allergen (ACGIH, 1999, Chapter 25) remediation involves reduction of the relative humidity (RH) near or on interior surfaces, removal of nutrients (human skin scales) and less reliance on use of fleecy finishing materials, including upholstered furniture (IOM 1993). The optimum RH for mite reproduction is 75-85% (Arlian 1992, IOM 1993, Chapter 22 in ACGIH 1999). Lowering the RH to <50% by air-conditioning, dehumidifier operation or raising the dry bulb temperature of surfaces results in desiccation and death of mites (Arlian 1992, Chapter 22 in ACGIH 1999). Use of non-porous covering for bedding and non-porous floor covering reduces the number of niches for accumulation of skin scales and mite faeces. Exposure to mite allergens is also reduced by removal of settled dust from the building by cleaning with a highly efficient vacuum cleaner. The vacuum cleaner bag should be discarded, because captured mites can reproduce and eventually produce more allergen (Chapter 22 in ACGIH 1999).

PREVENTION OF MOULD GROWTH

General requirements

As has been mentioned in Chapter 2.1, favourable temperatures, nutrients (substrate) and moisture are required for mould growth. Since the range of temperatures in buildings that is acceptable for human comfort is also suitable for microbial growth, and since virtually all buildings contain some biodegradable materials, control of moisture in construction and finishing materials is the most critical variable for preventing mould growth in indoor environments. There are many publications on moisture control in buildings (e.g. Rose and TenWolde 1993, Trechsel 1994, ASHRAE 2005a,b, IICRC 2006).

The critical role of moisture in relation to mould colonization of materials is considered here in relation to remediation. A case of flooding can be taken as an example. If the RH in a room that has been recently flooded remains consistently in the 90-100% range, equilibrated porous finishing materials will have a water activity (a$_w$) >0.90 or equilibrium relative humidity (ERH) >90%. Consequently, actinomycetes and other bacteria, as well as hydrophilic fungi (Chapter 2.1) such as *Fusarium*, *Phoma*, *Stachybotrys*, *Chaetomium*, *Trichoderma* and yeasts, might be expected to grow on the nearly saturated (a_w >0.9) finishing materials. However, xerophilic moulds such as *Eurotium* spp. or *Wallemia sebi* are less moisture-demanding and can grow on surfaces that are consistently damp (a_w 0.65-0.85) rather than wet. Since virtually no fungi can grow on interior surfaces at a_w <0.65, even with optimal nutrient conditions (ISIAQ 1996), the prevention of mould growth is achieved by lowering the available moisture to levels equivalent to an a_w <0.65 as rapidly as is feasible. Failure to dry materials rapidly enough or completely enough will result in only partial control, with growth of hydrophiles being prevented, but not xerophiles. Hunter and Sanders (1991) have noted that when the a_w of biodegradable material exceeds 0.80 for several days, mould growth is likely, although it will be of xerophilic rather than hydrophilic species. Thus, if wallpaper in the flooded room is only dried to the extent that its a_w is 0.80 (moisture content approximately 11.3%; Flannigan 1992) it will support growth of *Eurotium* spp. or *Wallemia sebi*, but not hydrophilic fungi such as those mentioned above.

Assuming normal building temperatures to be 20-30°C, prevention of colonization of biodegradable materials that have become water saturated by flooding is achieved by drying the materials within 24-48 h (Chapter 10 in ACGIH 1999). Simply lowering the RH in room air to 50% does not mean that the a_w at the surface of biodegradable materials will be below 0.65. On several occasions, the author encountered an actively photosynthesizing green plant growing on the flooring of a building subject to previous chronic flooding. Although the RH of the air was <65% and the surface of the carpet felt dry to the touch, the presence of the healthy plant showing no signs of wilting indicated that the pad beneath the carpet and the underlying flooring were still moist. The roots of plants require an a_w of at least 0.97-0.98 for root hairs to carry out their normal function of extracting from the substrate the water necessary for growth and survival (Flannigan 1992). This example, therefore, illustrates the necessity of removing moisture from hidden layers within building components if biodeterioration of susceptible materials is to be avoided. IICRC Standard S500 (2006) provides practical recommendations by which water damage restoration is achieved.

Preventing mould growth in the building envelope

Mould growth can occur in the envelope of buildings in both cold (heating) and warm (cooling) climates. Mould growth may occur in the envelope when water vapour from a warm source, which can be either indoors or outdoors, encounters a relatively cool surface, resulting in localized dampness and condensation. In a cold climate, this kind of localized dampness or condensation occurs on or in the envelope when moisture from warm room air encounters a surface that is cool because of thermal bridging or poor insulation (ASHRAE 2005b). In tropical climates or seasons, moisture-laden warm air can migrate through the envelope of an air-conditioned building causing dampness or condensation and consequent mould colonization of the cool inner surface of the wall.

Regardless of geographical location, buildings must be constructed so that any water entering the envelope drains to the outside. Buildings should be designed so that water drains away from the foundations and so that below-grade portions of the envelope are adequately damp-proofed to prevent entry of capillary moisture (Lstiburek and Carmody 1994). When wet materials such as concrete and plaster are used in construction, they must be allowed to dry before closing the envelope, so that biodegradable materials that may be near do not take up moisture from them.

If a vapour diffusion retarder is used in a cold climate, it should be installed on the insulation facing the occupied space. Dampness and condensation in cold climates can be prevented by reducing indoor moisture-producing activities and by mechanical ventilation with relatively dry outdoor air.

In air-conditioned buildings in warm humid climates, if vapour diffusion and air retarders are used they should be installed in the external portion of the envelope. Buildings should be positively pressurized to minimize the infiltration of humid outdoor air into the envelope. Indoor air must not be cooled to a temperature below that of the mean monthly outdoor dew point. Finishes like vinyl wall coverings that are impermeable to water vapour should not be used on the inner surface of the envelope.

THE BUILDING INSPECTION – FOR MOULD AND WATER DAMAGE

Prevention and remediation of microbial growth begin with an inspection of the building and its infrastructure. The location and cause of moisture damage, as well as the location and extent of visible mould growth, must be determined by physical inspection. It should be realized that water damage and mould growth may be concealed within building infrastructure (Miller *et al.* 2000, Morey *et al.* 2003). Literature on identification of moisture and mould growth problems should be consulted as part of the inspection process (Chapter 4 in ACGIH 1999, CCA 2004, AIHA 2008). Items in the building inspection which are important are listed in the sections that follow.

Inspection checklist for moisture damage

1. Inspect the building for evidence of current water leaks. Work with an architect or construction specialist to identify the cause of the water intrusion (Fig. 7).
2. Inspect the building for evidence of current condensation (Fig. 8), such as on the outer surface of chilled water pipe jackets and in other locations, including the building envelope, where thermal gradients can lead to dampness.
3. Use moisture meters or other appropriate methodologies (Chapter 7 in IICRC 2006) to document the location of damp or wet finishing and construction materials (Fig. 9). Moisture meter readings should be taken in potentially damp or wet areas and compared with readings taken from similar materials in locations thought to have normal moisture content. It is important to understand the limitations of moisture measuring equipment and methodologies. Comprehensive reviews of moisture measurement and control in buildings can be found elsewhere (ASHRAE 2005a,b, IICRC 2006).
4. Determine whether interior finishes that are currently dry have suffered past water damage. Look for signs of water damage beneath carpeting (e.g. stained wood tack strips), on the upper surface of ceiling tiles and, in cases where substantial water damage may have occurred in the past, inspect wall cavities.
5. Determine whether seasonal moisture sources exist. For example, poorly insulated cold or chilled water pipes in walls and ceiling plenums may be affected by condensation (Fig. 8) only during the humid season. Also find out whether or not humidifiers are used during the heating season.
6. Ascertain whether unusual water sources are present such as waterfalls in atrium landscaping. Bioaerosols can be generated from poorly maintained water systems (Rose 1996).
7. Find out whether wet methods are used for cleaning of carpet or other interior finishes. Wet or damp finishes provide possible niches for microbial growth.
8. Determine whether materials that are highly sus-

Fig. 7. Water splash points to a water leak during a rainstorm. Following the pathway of the water leak can determine the cause of this construction defect (photograph courtesy of C. Colladay).

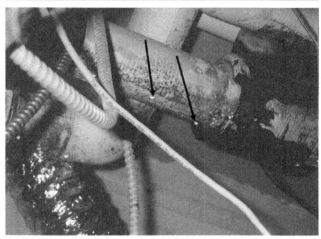

Fig. 8. Condensation (arrowed) on pipe jacket insulation. The surface temperature of the jacket is below the dew point temperature of the surrounding air, resulting in condensation.

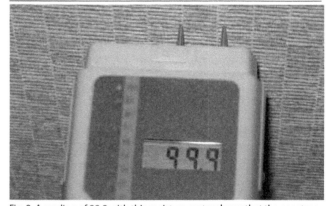

Fig. 9. A reading of 99.9 with this moisture meter shows that the construction material (paper-faced wallboard) behind the vinyl wall covering is moisture saturated.

Fig. 10. A fern grows in masonry in a patio wall, showing that the patio infrastructure is sufficiently saturated with water to support the growth of the plant, as well as fungi and bacteria.

Figs. 11 and 12. Mould growth (arrowed) revealed by inspection on the smooth external surface (Fig. 11) and (arrowed) on lower internal surface of dresser drawer (Fig. 12). Mould spores accumulated at A when drawer was tilted vertically. During inspection, an inventory is made of the extent of mould growth on various kinds of interior surfaces in the room.

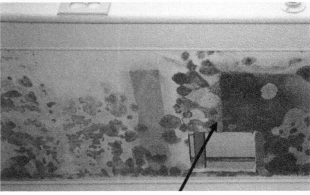

Figs. 13 and 14. Destructive inspection of interior surfaces may be necessary to ascertain the presence or absence of mould colonization, e.g. behind vinyl wall covering (Fig. 13) or in the wall cavity (Fig. 14). Note absence of mould growth on the distal wall surface (arrowed) seen through the opening into the wall cavity.

ceptible to moisture damage, e.g. photographic archives, books and clothes, are stored in chronically damp or wet areas.

9. Assess whether musty odours due to MVOCs emanating from microbial growth are present in damp locations such as in crawlspaces and basements (Horner and Miller 2003).

10. Examine for the presence of plants growing in building components such as masonry or on roofs, as these indicate that substantial moisture is present in building structure (Fig. 10). Overwatering of indoor potted plants can result in water damage to interior finishes such as carpet.

11. Finally, make an inventory of water-damaged materials, distinguishing those that are porous and non-porous. The location and nature of the moisture damage may provide clues as to the source of the water problem For example, bowed ceiling tiles may indicate conditions of chronically elevated RH in interior spaces.

Inspection checklist for visible mould growth

1. Make an inventory of visibly mouldy interior surfaces, noting the extent (surface area in m²) and location of colonized materials. A ruler, a powerful flashlight and a camera are essential inspection tools (Miller 2001).

2. Record the kinds of interior surfaces that are visibly mouldy (Figs. 11 and 12) and the extent of colonization on each type of surface. Mould growth may occur on the smooth, finished surface of furniture (Fig. 11); on the semi-porous surfaces within drawers (Fig. 12) or on the unfinished undersurface of tables; or on porous contents such as carpet, upholstery and clothes. It is also important to document separately the extent of visible

mould growth on wall and ceiling surfaces. The information obtained in this detailed inspection is essential for development of an appropriate mould remediation protocol.

3. Destructive inspection of interior finishes ranging from peeling back vinyl wall coverings to cutting holes in wallboard, with the objective of looking for hidden mould (Figs. 13 and 14), is often necessary in order to document the location of materials to be removed. Dust suppression and use of personal protective equipment are essential during destructive inspection in occupied areas (Chapter 15 in ACGIH 1999, CCA 2004, AIHA 2008, NYC 2008).

4. Destructive inspection of representative areas of the building envelope may be necessary to document the extent of biodeterioration and mould growth (Fig. 15). As with interior finishes, the extent and type of mould growth on various construction materials, e.g. oriented strand board, plywood, paper-faced wallboard, building paper, wood studs, etc., should be documented during destructive inspection (Fig. 15). Documentation of the presence of various kinds of fungi (Figs. 16 and 17) can be forensically important in estimating the longevity of mould growth and the nature of water damage on the construction material (Yang 2007).

5. Determine the extent of corrosion on metal structural components and other materials that may be visible during inspection (Fig. 18). Extensive corrosion of metals, e.g., iron, copper, zinc, etc., can be indicative of repeated moisture incursion (Yang 2007). The soundness of wood framing materials can also provide an indication of the longevity of water damage.

6. An outcome of the inspection is an estimation of the extent (small-, mid-, or large-scale) of visible mould growth in various rooms or zones in the building. An accurate estimate of the location, extent and density of mould growth is the most important consideration in the development of the remediation protocol (Chapter 15 in ACGIH 1999, CCA 2004, AIHA 2008, NYC 2008). It is also important to note that, while the finding of mid- and large-scale mould colonization sets in motion requirements for strong dust suppression/containment methods, the finding of trivial (small-scale) mould and water damage indicates that only simple clean-up actions are necessary.

Figs. 15-17. Destructive inspection of the building envelope revealed extensive biodeterioration of plywood (Fig. 15), building paper and exterior sheathing in a new construction. The presence of rhizomorphs (Fig. 16) and _Doratomyces_ conidiophores (arrowed) growing through a crack in oriented strand board (Fig. 17) indicates chronically wet conditions in the envelope siding.

Fig. 18. Rust (R) is evident on the metal studs in a wall cavity of this four-year-old building. Mould growth (arrowed) is evident on paper-faced wallboard, which has been pulled back to inspect the wall cavity.

Fig. 19. The drain pan in this fan coil unit (FCU), as well as hundreds of other FCUs in the same building, was poorly maintained. Dried biofilm is present in the pan (heating season operation).

The recommendation that regular facilities-maintenance personal can carry out small-scale mould removals was made in the first edition of the NYC Guidelines (1993) and in all subsequent mould remediation guidelines.

Checklist for HVAC inspection

1. Determine whether the HVAC system design relies on large numbers of distributed fan coil units, induction units or unit ventilators. Maintenance of a distributed HVAC system is more difficult than a system with a few large central air-handling units. In a distributed HVAC system, find out if individual units are characterized by water leaks, dampness and musty odours. Water leaks from individual units and associated piping systems may result in hidden water damage and hidden colonization in building components. Fig. 19 shows evidence of poor maintenance in a drain pan in one of the many hundreds of fan coil units in a large building.

2. Inspect airstream surfaces in the HVAC system including air-handling unit return and supply air duct systems for deposits of dirt and dust. Determine the minimum efficiency reporting value (MERV) rating of HVAC filters, and whether or not the filters fit into their holders. Inspection may reveal that substantial dirt and dust have been deposited on airstream surfaces (Fig. 20). Ascertain whether there is access for cleaning, and whether cleaning can be accomplished according to National Air Duct Cleaning Association (NADCA 2006) standards.

3. Determine whether there are problems of water stagnation in and around drain pans and dehumidifying coils. Standing water in drain pans indicates that there is a need for redesign to give complete drainage. Wet or rusty airstream surfaces downstream of coils indicate carry-over of water droplets.

4. Find out whether fibrous materials such as insulation and filters in HVAC systems are wet or chronically damp, and if the fibrous surfaces are heavily impacted by dirt and dust accretions (Fig. 21). Alternatively, note whether airstream surfaces in damp/wet HVAC locations are smooth and easily cleanable. If airstream HVAC surfaces are found to be impacted with mould colonization (Fig. 22), estimate the extent of colonization and follow the third edition of the NYC Guidelines (2008) regarding remediation.

Fig. 20. The cooling coil (CC) section and supply air duct airstream surfaces of this fan coil unit are covered with dirt and dust indicating poor maintenance.

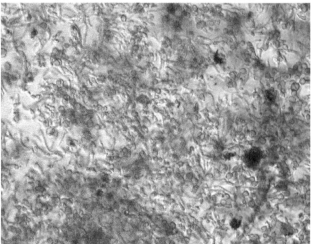

Figs. 21 and 22. Dirt and dust has accumulated on fibrous airstream surfaces (Fig. 21). Under damp/moist conditions, moulds colonize the airstream surface extracting nutrient from the dirt and dust. Spores and hyphae of *Cladosporium* (Fig. 22) dominate the moulds that colonized the dirty/dusty surface when moisture was non-limiting during the cooling (air-conditioning) season.

5. Ascertain whether there are humidifiers in air-handling units or supply ductwork, and whether (a) the operating humidifier wet airstream surfaces; (b) the humidifiers operate by aerosolization of water droplets from open, stagnant water sumps; and (c) humidifiers emit moisture in molecular form or as water droplets. It is also necessary to determine whether HVAC systems in the air-conditioned buildings in humid seasons or climates operate so that unconditioned humid outdoor air enters air-handling units when dehumidifying cooling coils are deactivated. The introduction of humid air into an air-conditioned building can result in massive condensation on cool surfaces and can raise the amount of moisture available for support of mould growth in interior finishes.

6. Determine whether fungal growth occurs on surfaces of diffusers or on surfaces near diffusers. If so, this suggests inadequate dehumidification of room air, localized overcooling of surfaces or infiltration of unconditioned outdoor air into interior zones.

CONCLUSION

This chapter has described general principles for prevention of microbial growth, and for remediation of growth that may have occurred in or on building components. Limiting available moisture is central to preventing moulds colonizing construction and finishing materials. In building systems, including HVAC components and cooling towers, preventive maintenance of a high standard is essential for prevention of microbial amplification. Where prevention has failed in buildings, attention must be focused on establishing preventive maintenance activities, and/or on moisture attenuation. Microbial growth that has occurred, such as mould colonization, must be physically removed using approaches that prevent dispersion of bioaerosols and protect workers performing remediation activities.

REFERENCES

ACGIH (1999) *Bioaerosols: Assessment and Control*, American Conference of Governmental Industrial Hygienists, Cincinnati, OH.

AIHA (2001) *Report of Microbial Growth Task Force*, American Industrial Hygiene Association, Fairfax, VA.

AIHA (2004) *Assessment, Remediation, and Post-Remediation Verification of Mold in Buildings*, American Industrial Hygiene Association, Fairfax, VA.

AIHA (2005) *Field Guide for the Determination of Biological Contaminants in Environmental Samples*, 2nd Ed, L.-L. Hung, J.D. Miller, and H.K. Dillon, (eds.). American Industrial Hygiene Association, Fairfax, VA.

AIHA (2008) *Recognition, Evaluation, and Control of Indoor Mold*, B. Prezant, D.M. Weekes, and J.D. Miller, (eds.). American Industrial Hygiene Association, Fairfax, VA.

Arai, H. (2000) Foxing caused by fungi: twenty-five years of study. *Int. Biodet. Biodeg.*, **46**, 181-188.

Arlian, L.G. (1992) Water balance and humidity requirements of house dust mites. *Exp. Appl. Acarol.*, **16**, 15-35.

ASHRAE (2000) *Minimizing the Risk of Legionellosis Associated with Building Water Systems*, ASHRAE Guideline 12-2000. American Society of Heating, Refrigerating and Air-Conditioning Engineers, Atlanta, GA.

ASHRAE (2005a) Thermal and moisture control in insulated assemblies – fundamentals. *ASHRAE Handbook – Fundamentals*. American Society of Heating, Refrigerating and Air-Conditioning Engineers, Atlanta, GA, chap. 23.

ASHRAE (2005b) Thermal and moisture control in insulated assemblies – applications. *ASHRAE Handbook – Fundamentals*, American Society of Heating, Refrigerating and Air-Conditioning Engineers, Atlanta, GA, chap. 24.

ASHRAE (2007) *Ventilation For Acceptable Indoor Air Quality*, ASHRAE Standard 62.1-2007. American Society of Heating, Refrigerating and Air-Conditioning Engineers, Atlanta, GA.

ASHRAE (2008) Ultraviolet lamp systems. *ASHRAE Handbook – HVAC Systems and Equipment*, American Society of Heating, Refrigerating and Air-Conditioning Engineers, Atlanta, GA, Chapter 16.

Benin, A., Benson, R., and Besser, R. (2002) Trends in Legionnaires' disease 1980-1998: declining mortality and new patterns of diagnosis. *Clin. Infect. Dis.*, **35**, 1039-1046.

Berry, M.A., Bishop, J., Blackburn, C., *et al.* (1994) Suggested guidelines for remediation of damage from sewage backflow into buildings. *J. Environ. Health*, **57**, 9-15.

Blomquist, G. (1994) The Book of Leviticus, Chapter 14. In R.A. Samson, B. Flannigan, M.E. Flannigan, *et al.* (eds.), *Health Implications of Fungi in Indoor Environments*. Elsevier, Amsterdam, p. xiv.

Brundrett, G.W. (1979) *Maintenance of Spray Humidifiers*. The Electricity Council Research Centre, Capenhurst, Chester, UK.

CCA (2004) *Mould Guidelines for the Canadian Construction Industry*, Standard Construction Document CCA 82-2004. Canadian Construction Association, Ottawa.

Conservation Center (1996) *Mold, Managing a Mold Invasion: Guidelines for Disaster Response*, Technical Series No. 1. Conservation Center for Art and Historic Artifacts, Philadelphia, PA.

Crandall, M., and Sieber, W. (1996) The National Institute for Occupational Safety and Health indoor environmental evaluation experience, Part 1: Building environmental evaluations. *Appl. Occup. Environ. Hyg.*, **11**, 533-539.

CRB (2006) *Indoor Mold: A General Guide to Health Effects, Prevention, and Remediation*, CRB 06-001, K. Umbach and P. Davis. California Research Bureau, California State Library, Sacramento, CA.

EPA (2001) *Mold Remediation in Schools and Commercial Buildings*, EPA 402-K-01-001. Environmental Protection Agency, Washington, DC.

Fields, B., and Moore, M. (2006) Control of Legionellae in the environment: a guide to the US Guidelines. *ASHRAE Transactions*, CH-06-12-1, pp. 691-699.

Flannigan, B. (1992) Approaches to assessment of microbial flora in buildings. In *IAQ '92, Environments for People*, American Society of Heating Refrigerating and Air-Conditioning Engineers, Atlanta, GA, pp. 139-145.

Florian, M.-L. (1997) *Heritage Eaters, Insects and Fungi in Heritage Collections*, James and James, London.

Harrison, J., Pickering, A., Finnegan, M., and Austwick, P. (1987) The sick building syndrome: further prevalence studies and investigation of possible causes. In *Indoor Air '87: Proceedings of the Fourth International Conference on Indoor Air Quality and Climate*, Vol. 2. Institut für Wasser-, Boden-, und Lufthygiene, Berlin, pp. 487-491.

Health Canada (1995) *Fungal Contamination in Public Buildings: A Guide in Recognition and Management*, Federal-Provincial Committee on Environmental and Occupational Health, Ottawa.

Health Canada (2004) *Fungal Contamination in Public Buildings: Health Effects and Investigation Method*, H46-2/04-358E. Health Canada, Ottawa.

Heller, R.M., Heller, T., and Sasson, J. (2003) Mold: "tsara'at," Leviticus, and the history of a confusion. *Perspect. Biol. Med.*, **46**, 588-591.

Horner, E., and Miller, J.D. (2003) Microbial volatile organic compounds with emphasis on those arising from filamentous fungal contaminants of buildings. *ASHRAE Trans.*, **109**, 215-231.

Horner, W.E., Worthan, A.W., and Morey, P.R. (2004) Air and dustborne mycoflora in houses free of water damage and fungal growth. *Appl. Environ. Microbiol.*, **70**, 6394-6400.

HSE (1994) *The Control of Legionellosis including Legionnaires' Disease*, Health and Safety Booklet H.S. (G) 70. Health and Safety Executive, Sudbury, UK.

Hunter, C., and Sanders, C. (1991) Mould. In *Annex XIV Condensation and Energy, Sourcebook*, Vol. 1. International Energy Agency, Leuven University, Belgium, pp. 2.1-2.30.

IICRC (2006) *Standard and Reference Guide for Professional Water Damage Restoration*, S500. Institute of Inspection Cleaning and Restoration Certification, Vancouver, WA.

IICRC (2008) *Standard and Reference Guide for Professional Mold Remediation*, S520. Institute of Inspection Cleaning and Restoration Certification, Vancouver, WA.

IOM (1993) *Indoor Allergens – Assessing and Controlling Adverse Health Effects*. Institute of Medicine, National Academy Press, Washington, DC.

IOM (2004) *Damp Indoor Spaces and Health*. Institute of Medicine, National Academies Press, Washington, DC.

ISIAQ (1996) *Control of Moisture Problems Affecting Biological Indoor Air Quality*. International Society of Indoor Air Quality and Climate, Helsinki, Finland.

Lloyd, H., and Singh, J. (1994) Inspection, monitoring and environmental control of timber decay. In J. Singh, (ed.), *Building Mycology*. Spon, London, pp. 159-186.

Lstiburek, J., and Carmody, J. (1994) Moisture control in new residential buildings. In H. Trechsel, (ed.), *Moisture Control in Buildings*. American Society for Testing and Materials, West Conshohocken, PA, pp. 321-347.

Martin, S.B., Dunn, C., Freihaut, J., *et al.* (2008) Ultraviolet germicidal irradiation, current best practices. *ASHRAE J.*, August 2008, 28-36.

Menzies, D., Popa, J., Hanley, J., *et al.* (2003) Effect of ultraviolet germicidal lights installed in office ventilation systems on workers' health and wellbeing: double-blind multiple cross-over trial. *Lancet*, **362**, 1785-1791.

Miller, J.D. (2001) Mycological investigations of indoor environments. In B. Flannigan, R.A. Samson, and J.D. Miller, (eds.), *Microorganisms in Home and Indoor Work Environments*, Taylor & Francis, London, pp. 231-246.

Miller, J.D., Haisley, P.D., and Reinhardt, J.H. (2000) Air sampling results in relation to extent of fungal colonization of building materials in some water-damaged buildings. *Indoor Air*, **10**, 146-151.

Morey, P. (1994) Suggested guidance on prevention of microbial contamination for the next revision of ASHRAE Standard 62. In *IAQ '94: Engineering Indoor Environments*. American Society of Heating, Refrigerating and Air-Conditioning Engineers, Atlanta, GA, pp. 139-148.

Morey, P. (1996) Mold growth in buildings: removal and prevention. In *Indoor Air '96, Proceedings of Seventh International Conference on Indoor Air Quality and Climate, Nagoya*, Vol. 2, pp. 27-36.

Morey, P. (2004) Fungi in buildings. In W. Hansen (ed.) *Infection Control During Construction Manual*, Second Ed. Policies, Procedures, and Strategies for Compliance. Pp. 85-109.

Morey, P.R. (2007a) Microbial sampling strategies in indoor environments. In C. Yang and P. Heinsohn, (eds.), *Sampling and Analysis of Indoor Microorganisms*. John Wiley, New York, pp. 51-74.

Morey, P.R. (2007b) Microbial remediation in non-industrial indoor environments. In C. Yang and P. Heinsohn, (eds.), *Sampling and Analysis of Indoor Microorganisms*, John Wiley, New York, pp. 231-242.

Morey, P.R. (2011) Mold remediation in North American buildings In O.C.G. Adan and R. Samson, (eds.), *Fundamentals of mold growth in indoor environments and strategies for healthy living*. Wageningen Academic Press, The Netherlands. (In press).

Morey, P., and Ansari, S. (1996) Mold remediation protocol with emphasis on earthquake damaged buildings. In *Indoor Air '96, Proceedings of Seventh International Conference on Indoor Air Quality and Climate, Nagoya*, Vol. 3, pp. 399-404.

Morey, P.R., Hull, M.C., and Andrew, M. (2003) El Nino water leaks identify rooms with concealed mould growth and degraded indoor air quality. *Int. Biodet. Biodeg.*, **52**, 197-202.

NADCA (2006) *Assessment, Cleaning, and Restoration of HVAC Systems*, ACR 2006. National Airduct Cleaners Association, Washington, DC.

NYC (1993) *Guidelines on Assessment and Remediation of Stachybotrys atra in Indoor Environments*. New York City Department of Health, New York City Human Resources Administration, and Mount Sinai-Irving J. Selikoff Occupational Health Clinical Center, New York.

NYC (2000) *Guidelines on Assessment and Remediation of Mold in Indoor Environments*, New York Department of Health, New York.

NYC (2008) *Guidelines on Assessment and Remediation of Fungi in Indoor Environments*, New York City Department of Health and Mental Hygiene, New York.

OSHA (2008) *Technical Manual*, Section III, Chapter 7, Occupational Safety and Health Administration, at http://www.osha.gov/dts/osta/otm/otm_iii/otm_iii_7.html (2008H) and etools at http://www.osha.gov/dts/osta/otm/otm/legionnaires/index.html (2008).

Peterson, T.H. (1993) *A Primer on Disaster Preparedness, Management and Response: Paper-Based Materials*. National Archives and Records Administration, Washington, DC.

Rautiala, S., Reponen, T., Hyvärinen, A., *et al.* (1996) Exposure to airborne microbes during the repair of mouldy buildings. *Am. Industr. Hyg. Assoc. J.*, **57**, 279-284.

Rose, C. (1996) Building-related hypersensitivity diseases: sentinel event management and evaluation of building occupants. In R.B. Gammage and B.A. Berven, (eds.), *Indoor Air and Human Health*. CRC Lewis Publishers, Boca Raton, FL, pp. 211-219.

Rose, H., and Hirsch, S. (1979) Filtering hospital air decreases *Aspergillus* spore counts. *Amer. Rev. Resp. Dis.*, **119**, 511-513.

Rose, W.B., and TenWolde, A., (eds.) (1993) *Bugs, Mold & Rot II*. National Institute of Building Sciences, Washington, DC.

Samson, R.A., Flannigan, B., Flannigan, M.E., *et al.*, (eds.) (1994) *Health Implications of Fungi in Indoor Environments*, Elsevier, Amsterdam.

Seppanen, D., and W. Fisk (2002) Association of ventilation system type with SBS symptoms in office workers. *Indoor Air*, **12**, 98-112.

Sieber, W., Stayner, L., Malkin, R., *et al.* (1996) The National Institute for Occupational Safety and Health indoor environmental experience, Part 3: Associations between environmental factors and self-reported health conditions. *Appl. Occup. Environ. Hyg.*, **11**, 1387-1392.

Squier, C., Stout, J., Krsytofiak, S., *et al.* (2005) A proactive approach to prevention of healthcare acquired Legionnaires' disease: The Allegheny County (Pittsburgh) experience. *Am. J. Infect. Control*, **33**, 360-367.

Storey, E., Dangman, K.H., Schenek, P., *et al.* (2004) *Guidance for Clinicians on the Recognition and Management of Health Effects Related to Mold Exposure and Moisture Indoors*, University of Connecticut Health Center, Farmington, CT.

Streifel, A. (1996) Controlling aspergillosis and *Legionella* in hospitals. In R.B. Gammage and B.A. Berven, (eds.), *Indoor Air and Human Health*, CRC Lewis, Boca Raton, FL, pp. 129-139.

Trane Applications Manual (1998) *Managing Building Moisture*, The Trane Company, LaCrosse, WI.

Trechsel, H.R., Ed. (1993) *Moisture Control in Buildings*, ASTM Manual Series, MNL 18, American Society for Testing Materials, West Conshohocken, PA.

Walsh, T., and Dixon, D. (1989) Nosocomial aspergillosis: environmental microbiology, hospital epidemiology, diagnosis and treatment. *Eur. J. Epidemiol.*, **5**, 131-142.

World Health Organization (2007) *Legionella and the Prevention of Legionellosis*. World Health Organization, at http://www.who.int/water_sanitation_health/emerging/legionella/en/index.html.

Yang, C.S. (1996) Fungal colonization of HVAC fiberglass air-duct liner in the U.S.A. In *Indoor Air '96, Proceedings of Seventh International Conference on Indoor Air Quality and Climate, Nagoya*, Vol. 3, pp. 173-177.

Yang, C. (2007) A retrospective and forensic approach to assessment of fungal growth in the indoor environment. In C. Yang and P. Heinsohn, (eds.), *Sampling and Analysis of Indoor Microorganisms*, John Wiley, New York, pp. 215-229.

Yang, C.S., and Ellringer, P.J. (1996) Evaluation of treating and coating HVAC fibrous glass liners for controlling fungal colonization and amplification. In *Indoor Air '96, Proceedings of Seventh International Conference on Indoor Air Quality and Climate, Nagoya*, Vol. 3, pp. 167-172.

Chapter 3. Airborne microorganisms and disease

Chapter 3.1

ALLERGENIC MICROORGANISMS AND HYPERSENSITIVITY

Anne K. Ellis and James H. Day

Division of Allergy and Immunology, Department of Medicine, Queen's University, Kingston, Ontario, Canada.

INTRODUCTION

Allergy may be defined as an untoward response to a specific antigen mediated by an immunological reaction. It can take several forms. The most common clinical allergic reaction to moulds is IgE-mediated, represented by symptoms of rhinitis, conjunctivitis and asthma. Hypersensitivity pneumonitis, another immunological reaction, has both IgG- and cell-mediated features, while other fungal hypersensitivity reactions have even more complex immunological relationships.

The microorganisms capable of producing hypersensitivity responses in humans are mainly fungi, and to a much lesser degree bacteria and algae. Because mites are so frequently encountered in indoor environments, and their allergens are often encountered together with those of fungi, e.g. in house dust, they will be discussed in addition to microbial allergens. As fungi are the main offenders among microorganisms, they will consequently be covered in most detail.

HYPERSENSITIVITY TO FUNGI

The earliest report of an allergic response to moulds occurred in 1726, when a severe asthma attack was reported in a patient who had just visited a wine cellar. In 1873, Blackley later described "bronchial catarrh" and chest tightness following inhalation of *Penicillium* spores (Blackley 1873). Since these early beginnings, research has shown that exposure to fungal products generates a wide range of adverse responses, including hypersensitivity. It has been estimated that approximately 10% of the general population has IgE antibodies to common inhalant moulds (Horner *et al*. 1995) and 15%-50% of the atopic population (depending upon their geographic location) are sensitised to fungi (Institute of Medicine Committee on the Assessment of Asthma and Indoor Air 2000). Most allergenic fungi, including species in the genera *Alternaria*, *Cladosporium*, *Epicoccum* and *Fusarium*, display a seasonal outdoor spore release pattern, but this is less well defined than it is for pollens (Koch *et al*. 2000, de Ana 2006, Gonianakis *et al*. 2006). Indoor fungi are a mixture of those that grow indoors and those that have entered from outside (Burge 1985, Simon-Nobbe *et al*. 2008). Their incidence is influenced by humidity, ventilation, the content of biologically degradable material and the presence of pets, plants and carpets (Dharmage *et al*. 1999). In general, the indoor spore concentration is less than half of that outdoors (unless there is indoor mould growth) and varies from 100 to 1,000 spores m^{-3} (Burge 1985, Dharmage *et al*. 1999, Nevalainen *et al*. 1994). Indoor fungi include species of *Aspergillus* and *Penicillium* (Scott 2001, Górny and Dutkiewicz 2002). In a Danish study of 23 mould-infected buildings (Gravesen *et al*. 1999), the most frequent mould genera encountered were *Penicillium* (68%) and *Aspergillus* (56%), followed by *Chaetomium*, *Ulocladium*, *Stachybotrys* and *Cladosporium* (ranging from 22 to 15%). Various responses occur not only to the spores and mycelia of fungi, but also to the mycotoxins produced as secondary metabolites, the volatile organic compounds (VOC's) emitted by growing fungi, enzymes produced by the organisms and (1→3)-β-D-glucan, a component of the fungal cell wall (Rylander 2005). Rigorous scientific evaluations have confirmed that VOC's play a distinct role in respiratory disease related to moulds (Norbäck *et al*. 1995, Rumchev *et al*. 2007).

EPIDEMIOLOGICAL EVIDENCE FOR MOULD-RELATED RESPIRATORY DISEASE

A number of studies have been conducted on the association of dampness, mould and respiratory health in residential housing. There is now a vast body of literature associating a variety of diagnosable respiratory illnesses (asthma, wheezing and cough), particularly in children, to residence in damp or water-damaged homes (Dales *et al.* 1991, Jaakkola *et al.* 1993, Cuijpers *et al.* 1995, Verhoeff *et al.* 1995, Williamson *et al.* 1997, Andriessen *et al.* 1998, Billings *et al.* 1998, Zacharasiewicz *et al.* 1999, Burr 2001, Kilpeläinen *et al.* 2001, Ren *et al.* 2001, Dharmage *et al.* 2002, Jaakola *et al.* 2002, 2004, Zock *et al.* 2002, Hardin *et al.* 2003, Yazicioglu *et al.* 2004, Hope and Simon 2007). Only a few published studies have failed to confirm this association (Strachan *et al.* 1990, Ceylan *et al.* 2006, Inal *et al.* 2007). In a study of low-income housing residents, self-reported moisture or mildew in homes of asthmatic children was associated with much higher asthma morbidity (Bonner *et al.* 2006). A systematic review of the literature by Denning *et al.* (2006) concluded that there is convincing evidence of a close association between fungal sensitization and asthma severity. A recent meta-analysis of 33 published studies conducted by the Institute of Medicine of the National Academy of Sciences found that building dampness and mould are associated with approximately 30-50% increases in respiratory and asthma-related outcomes (Fisk *et al.* 2007).

Other epidemiological evidence for mould-related respiratory disease comes from evaluations of the effect of outdoor fungal spore concentrations on asthma severity. Sensitization to *Alternaria alternata* in particular has been linked to the presence, persistence and severity of asthma (Bush and Prochnau 2004). A study in Southern California demonstrated that after controlling for weather, total fungal spore concentrations were associated with increased asthma symptom scores, inhaler use and decreases in peak expiratory flow rate (Delfino *et al.* 1997). This study confirmed earlier findings by the same group (Delfino *et al.* 1996). Others have also shown significant correlations between outdoor fungal concentrations and mean monthly rhinitis and asthma symptoms scores, in addition to peak flow measurements (Inal *et al.* 2008). Fungal spore concentrations moving from the lower to the upper quartiles has been linked to emergency room visits for asthma (Atkinson *et al.* 2006), and other workers have found an association between increased outdoor mould spore counts, including Basidiomycetes, Ascomycetes and Deuteromycetes, and asthma hospitalizations (Newson *et al.* 2000, Dales *et al.* 2004).

Newer studies further support a relationship of mould exposure with adverse effects on respiratory health. Cho *et al.* (2006a) visited the homes of 640 infants and determined that visible mould increased the risk of recurrent wheezing nearly two-fold. Another study of patients in a general practice-based asthma cohort found a consistent association between reported moulds and dampness in the living room or the child's bedroom and an increased airway hyperresponsiveness, even following adjustment for gender, presence of inhalant allergy, and use of controller medications (Hagmolen of ten Have *et al.* 2007). A further evaluation of US homes revealed mould and moisture-related problems to be independent predictors of a high allergen burden and an increase in asthma symptoms (Salo *et al.* 2008).

Nevertheless, a position paper published in the *Journal of Allergy and Clinical Immunology* (Bush *et al.* 2006) played down the association between asthma symptoms and exposure to indoor fungi, and stated that there is currently insufficient evidence to support a causal relationship of airborne mould exposure with clinical manifestations of rhinitis. This position was criticized in letters to the editor supported by articles contradicting their conclusions (Goldstein 2006, Kilburn *et al.* 2006, Lieberman *et al.* 2006, Marinkovich 2006, Ponikau and Sherris 2006, Shoemaker *et al.* 2006, Straus and Wilson 2006, Strickland 2006). We believe that, considering the available evidence, there *is* an association between mould exposure and respiratory conditions, including asthma and rhinitis, and quite likely non-respiratory manifestations such as atopic dermatitis and allergic fungal sinusitis, which are detailed later in the Allergic Clinical Syndromes subsection of this chapter.

ALLERGENIC SPECIES OF FUNGI AND THEIR ALLERGENS

A vast number of fungi have been identified in the medical literature as being implicated in allergic disease (Table 1). Most commonly involved are the anamorphic fungi, including species in the genera *Alternaria*, *Aspergillus*, *Cladosporium*, *Epicoccum* and *Penicillium* spp., as well as some members of the Basidiomycota, Ascomycota and Zygomycota.

Fungi produce a number of proteins and glycoproteins that are antigenic (often up to 60 from a single

Table 1. Moulds and other fungi noted in the literature for which IgE-mediated allergy has been demonstrated (adapted from Simon-Nobbe *et al.* 2008).

ASCOMYCOTA	BASIDIOMYCOTA
Pezizomycotina	**Hymenomycetes**
Acremonium (Cephalosporium)	Agaricus (Amanita)
Alternaria	Armillaria
Aspergillus	Boletinellus
Aureobasidium	Boletus
Bipolaris (Drechslera, Helminthosporium)	Calvatia
Chaetomium	Cantharellus
Chrysosporium	Chlorophyllum
Cladosporium	Coprinus
Claviceps	Dacrymyces
Curvularia	Ganoderma
Cylindrocarpon	Geastrum
Daldinia	Hypoloma
Didymella	Inonotus
Embellisia	Lentinus
Epicoccum	Lycoperdon
Epidermophyton	Merulius
Eurotium	Pisolithus
Fusarium	Podaxis
Gliocadium	Polyporus
Leptosphaeria	Pleurotus
Microsphaeria	Psilocybe
Monilia	Schizophyllum
Neurospora	Scleroderma
Nigrospora	Sporotrichum
Nimbya (Macrospora)	Stereum
Paecilomyces	Trichosporon
Penicillium	**Urediniomycetes**
Scopulariopsis	Hemileia
Stemphylium (Pleospora)	Puccinia
Trichoderma	Rhodotorula*
Trichophyton	Sporobolomyces*
Ulocladium	**Ustilaginomycetes**
Xylaria	Malassezia (Pityrosporum)*
Saccharomycotina	Tilletia
Candida	Tilletiopsis
Saccharomyces	Ustilago
Mitosporic Ascomycota	**ZYGOMYCOTA**
Phoma	**Zygomycetes**
Stachybotrys	Absidia
Thermomyces (Humicola)	Mucor
Trichothecium	Rhizopus
Wallemia	

*Yeasts.

species), some of which are also allergenic. Fungal allergens range in molecular weight from roughly 10 to 70 kDa (Larsen 1994, Horner *et al.* 1995). Allergen production varies by isolate, strain, environmental conditions and growth substrates. Even under highly controlled conditions, considerable variability exists in allergen production (Horner *et al.* 1995). Allergens are present in both the fungal mycelium and the spores, and are quickly released from fungal material. However, some allergens require spore germination for release, e.g. *Aspergillus fumigatus* (Rogers 2003). Since the last edition of this text was published, new information regarding the important allergens and epitopes contributing to hypersensitivity responses in humans has emerged. In addition, the nomenclature of allergens has been subject to careful scientific review, clarification, amalgamation and standardisation (IUIS/WHO Allergy Nomenclature Subcommittee 1994, Larsen and Dreborg 2008). We discuss below in detail those species most represented in the literature, either secondary to extensive study or in terms of their relative importance in human disease. The specific allergens and currently accepted nomenclature are summarized in Table 2.

Alternaria alternata

A. alternata is one of the most widely distributed moulds in nature and is one of the commonest fungi associated with asthma (Bush and Prochnau 2004). Additionally, the persistence and severity of asthma have been strongly associated with sensitization and exposure to *A. alternata* (Delfino *et al.* 1997, Halonen *et al.* 1997, Neukirch *et al.* 1999, Black *et al.* 2000, Downs *et al.* 2001a).

Based on IgE binding in sensitised patients Alt a 1, which has both a 16.4 and 15.3 MW band on electrophoresis (nomenclature of Alt a 2 no longer in use), has been identified as the major allergen (De Vouge *et al.* 1996). Alt a 3 (heat shock protein 70), Alt a 4 (disulfide isomerase) and Alt a 5 (ribosomal protein P2) are considered to be minor allergens, although at 33-47% they possess a relatively high degree of IgE binding (Achatz *et al.* 1995, De Vouge *et al.* 1998), whereas Alt a 6 (enolase), Alt a 7 (YCP4 protein), Alt a 8 (mannitol dehydrogenase) and Alt a 10 (aldehyde dehydrogenase) are clearly minor allergens (Breitenbach and Simon-Nobbe 2002). Also described are Alt a 12 (acid ribosomal protein P1) and Alt a 13 (gluathion-*S*-transferase), but data on their allergenicity have not been reported as of January 2009 (Kurup *et al.* 2000, www.allergen.org). Further studies have shown allergens common to *A. alternata*, *Curvularia lunata* and

Table 2: Major allergens of *Alternaria, Aspergillus, Cladosporium, Epicoccum, Malassezia, Penicillium* **and** *Trichophyton* **(adapted from www.allergen. org, Jan 2009).**

Species	Allergen Name	MW (kD)	Biochemical Name	Obsolete Name
Alternaria alternata	Alt a 1	16.4 / 15.3		
	Alt a 3		Heat shock protein 70	
	Alt a 4	57	Disulfide isomerase	
	Alt a 5	11	Ribosomal protein P2	*Alt a 6*
	Alt a 6	45	Enolase	*Alt a 5, 11*
	Alt a 7	22	YCP4 protein	
	Alt a 8		Mannitol dehydrogenase	
	Alt a 10	53	Aldehyde dehydrogenase	
	Alt a 12	11	Acid ribosomal protein P1	
	Alt a 13	53	Glutathione-*S*-transferase	
Aspergillus flavus	Asp fl 13	34	Alkaline serine protease	
A. fumigatus	Asp f 1	18	Mitogillin family	
	Asp f 2	37		
	Asp f 3	19	Peroxysomal protein	
	Asp f 4	30		
	Asp f 5	40	Metalloprotease	
	Asp f 6	26.5	Mn superoxide dismutase	
	Asp f 7	12		
	Asp f 8	11	Ribosomal protein P2	
	Asp f 9	34		
	Asp f 10	34	Aspartate protease	
	Asp f 11	24	Peptidyl-prolyl isomerase	
	Asp f 12	90	Heat shock protein P90	
	Asp f 13	34	Alkaline serine protease	
	Asp f 15	16		*Asp f 13*
	Asp f 16	43		
	Asp f 17	34	Vacuolar serine protease	
	Asp f 18			
	Asp f 22	46	Enolase	
	Asp f 23	44	L3 ribosomal protein	
	Asp f 27	18	Cyclophilin	
	Asp f 28	13	Thioredoxin	
	Asp f 29	13	Thioredoxin	
	Asp f 34	20	PhiA cell wall protein	
A. niger	Asp n 14	105	Beta-xylosidase	
	Asp n 18	34	Vacuolar serine protease	
	Asp n 25	66-100	3-Phytase B	
A. oryzae	Asp o 13	34	Alkaline serine protease	
	Asp o 21	53	TAKA-amylase A	
Cladosporium cladosporioides	Cla c 9	36	Vacuolar serine protease	
C. herbarum	Cla h 2	45		*Ag54*
	Cla h 5	11	Acid ribosomal protein P2	*Cla h 4*
	Cla h 6	46	Enolase	
	Cla h 7	22	YCP4 protein	*Cla h 5*
	Cla h 8		Mannitol dehydrogenase	

	Cla h 9		Vacuolar serine protease	
	Cla h 10	53	Aldehyde dehydrogenase	*Cla h 3*
	Cla h 12	11	Acid ribosomal protein P1	
Epicoccum purpurascens	Epi p 1	30	Serine protease	
Malassezia furfur	Mala f 2	21	Peroxysomal membrane protein	*MF1*
	Mala f 3	20	Peroxysomal membrane protein	*MF2*
	Mala f 4	35	Mitochondrial malate dehydrogenase	
M. sympodialis	Mala s 1			
	Mala s 5			
	Mala s 6		Cyclophilin	
	Mala s 7			
	Mala s 8			
	Mala s 9			
	Mala s 10	86	Heat shock protein 70	
	Mala s 11	23	Manganese superoxide dismutase	
	Mala s 12	67	Glucose-methanol-choline (GMC) oxidoreductase	
	Mala s 13	13	Thioredoxin	
Penicillium brevicompactum	Pen b 13	33	alkaline serine protease	
	Pen b 26	11	acidic ribosomal protein P1	
P. chrysogenum	Pen ch 13	34	alkaline serine protease	
	Pen ch 18	32	vacuolar serine protease	
	Pen ch 20	68	N-acetyl-glucosaminidase	
	Pen ch 31		calreticulin	
	Pen ch 33	16		
P. citrinum	Pen c 3	18	peroxysomal membrane protein	
	Pen c 13	33	alkaline serine protease	
	Pen c 19	70	heat shock protein P70	
	Pen c 22	46	enolase	
	Pen c 24		elongation factor 1 beta	
	Pen c 30	97	catalase	
	Pen c 32	40	pectate lyase	
P. oxalicum	Pen o 18	34	vacuolar serine protease	
Trichophyton rubrum	Tri r 2		Putative secreted alkaline protease Alp1	
	Tri r 4		Serine protease	
T. tonsurans	Tri t 1	30		
	Tri t 4	83	Serine protease	

Epicoccum nigrum (Bisht *et al.* 2002, Gupta *et al.* 2002). Nuclear transport factor 2 and enolase are cross-reactive allergens from *Alternaria* (Weichel *et al.* 2003), while 60S ribosomal protein P2 and MnSOD are from *Penicillium chrysogenum, Fusarium venenatum* and *Aspergillus fumigatus* (Wagner 2001, Flückiger *et al.* 2002a,b, Hoff *et al.* 2003). Glutathione-*S*-transferase has also been demonstrated as a cross-reactive allergen among fungi (Shankar *et al.* 2005, 2006).

Aspergillus

Aspergillus spp. tend to dominate the indoor mould population along with *Penicillium* spp. Approximately 20% of the world's asthmatics suffer from *A. fumigatus*-induced allergic responses (Maurya *et al.* 2005, Gautam *et al.* 2007). Efforts of several groups have resulted in identification of 23 distinct allergens from culture filtrate and/or mycelial extracts of *A. fumigatus* by conventional purification methods, sequencing of cDNA clones and limited immunoproteomics (Nierman *et al.* 2005, Gautam *et al.* 2007).

Asp f 1 is a 20 kDa glycoprotein currently held to be the major allergen. Culture filtrate extracts of this allergen demonstrate more allergenic potential than those extracted from mycelium (Arruda *et al.* 1990, Cruz *et al.* 1995, 1997). Also characterized are Asp f 2 – Asp f 18, although Asp f 4 – Asp f 10 are thought to have much more clinical relevance in allergic bronchopulmonary aspergillosus (ABPA) than in other forms of *Aspergillus*-induced respiratory disease discussed later in the section on Allergic Clinical Syndromes (Kolattukudy *et al.* 1993, Kumar *et al.* 1993, Hemmann *et al.* 1997, Banerjee *et al.* 1998, 2001, Crameri 1998, Kurup *et al.* 2000, 2002, Flückiger *et al.* 2002c). Finally, Table 2 presents a summary of the remaining relevant allergens that have been characterized for this species (Sander *et al.* 1998, Shen *et al.* 1998, Banerjee *et al.* 2001, Lai *et al.* 2002).

Aspergillus-derived enzymes are used as dough improvers in some bakeries. Some of these enzymes are identified as causing IgE-mediated sensitization in bakers with workplace-related symptoms. The first recognized and most frequently reported allergen of these fungal enzymes is α-amylase from *A. oryzae* (Baur *et al.* 1986); the characterized and purified allergen was designated Asp o 2 (Baur *et al.* 1994). In addition, glucoamylase, cellulase, xylanase, and hemicellulase preparations from *A. niger* have been shown to contain allergens (Baur *et al.* 1988, Quirce *et al.* 1992, Sander *et al.* 1998), with multiple proteins in each preparation and high degrees of cross-reactivity. Many of these proteins have now been characterized and their biochemical properties identified (Table 2).

Cladosporium

Members of this genus are among the most common moulds colonizing dying and dead plants and also occur in various soil types. *Cladosporium* spp. are frequently found in uncleaned refrigerators, foodstuffs, on moist window frames, in houses with poor ventilation, houses with straw roofs and those situated in low, damp areas. *C. herbarum* is the most widely studied of the species, and sensitization has been shown to occur in 3–10% of asthmatics, depending on the geographical region of study (Zureik *et al.* 2002, Korhonen *et al.* 2006), and a positive skin prick test to *Cladosporium* (as with *Alternaria*) has been shown to be an independent risk factor of asthma severity (Zureik *et al.* 2002).

Approximately 60 antigens have been revealed in extracts of *C. herbarum* (Gutman and Bush 1993). Two major allergens have been described, viz. Cla h 1 and Cla h 2 (Aukrust and Borch 1979, Swärd-Nordmo *et al.* 1988). Their biological functions are still unknown and their sequences have not yet been determined. Spores contain a greater amount of Cla h 1 than mycelium, and Cla h 2 is found in small amounts in both (Swärd-Nordmo *et al.* 1988). These allergens have been found to vary widely from strain to strain (Day 1996). Minor allergens from *C. herbarum* have been defined as Cla h 5 – Cla h 10 and Cla h 12, and are enzymes and proteins characterized as allergens in *Alternaria* spp. also (Achatz *et al.* 1995; www.allergome. org). Additionally, the "translationally controlled tumor protein" (TCTP) of *C. herbarum* is recognized by IgE antibodies of patients allergic to it (Rid *et al.* 2008). TCTP is a small (19kDa) and versatile protein that has been highly conserved during the evolution of eukaryotes (Venugopal 2005). In one study, about 50% of the patients who recognized this minor allergen also recognized human TCTP (or histamine releasing-factor) in IgE immunoblots (Rid *et al.* 2008).

Epicoccum

Epicoccum nigrum (previously known as *E. purpurascens*), is associated with dead and dying plants and with harvested seeds, and is also found in soil. It has been cited as a cause of severe allergic disorders, including hypersensitivity pneumonitis (Hogan *et al.* 1996) and allergic fungal sinusitis (see later under Allergic Clinical Syndromes) in 5-7% of different populations worldwide (Noble *et al.* 1997, Bisht *et al.* 2000). Epi p 1 is a 33.5 kDa glycoprotein allergen which displays IgE binding to both the carbohydrate and the protein moieties of the structure (Bisht *et al.* 2002), and shows allergenic cross-reactivity with other fungi of clinical relevance (Bisht *et al.* 2004a). This glycoprotein functions as a serine protease, which may enhance the entrance of allergenic proteins by opening of tight junctions of lung epithelium (Bisht *et al.* 2004b).

Penicillium

In this genus, four species have been described as having allergenic potential (*P. brevicompactum, P. chrysogenum, P. citrinum*, and *P. oxalicum*), with a total of 15 distinct allergens having been identified and characterized (www.allergen.org). Several studies on fungal allergens have shown that the alkaline and/or the vacuolar serine proteases are major allergens of eight prevalent species of *Penicillium* and *Aspergillus*, viz. *P. citrinum, P. brevicompactum, P. chrysogenum, P. oxalicum, A. fumigatus, A. niger, A. oryzae* and *A. flavus* (Shen *et al.* 1996, 1997, 1999, 2001, Chou *et al.* 1999, 2002, Lai *et al.* 2004). They have been designated as group 13 (alkaline serine protease) and group 18 (vacuolar serine protease) allergens of both genera (Shen *et al*. 1999, 2001), and are considered to be major allergens. It has also been proposed that the 68 kDa allergen of *P. chrysogenum* is a major allergen (Shen *et al*. 1995), now identified as N-acetyl-glucosaminidase (Pen ch 20).

Dermatophytes

The term dermatophyte is restricted to three anamorphic genera in the family Moniliaceae, *Trichophyton, Epidermophyton* and *Microsporum*, and one teleomorphic genus, *Arthroderma*, in the Ascomycota (Matsumoto and Ajello, 1987). They are highly specialized fungi capable of parasitizing keratinized tissue (hair, skin, nails), but are usually confined to the non-living cornified layer of the epidermis (Weitzmann and Summerbell 1995). These types of infection, or dermatophytoses, are widespread and increasing in prevalence globallly (Woodfolk 2005). Antigens from *Trichophyton* are capable of inducing both an IgE-mediated immediate hypersensitivity response and a delayed-type hypersensitivity (DTH) response. A subset of patients also mount a 'dual' skin test response in which a DTH follows the immediate reaction.

The first *Trichophyton* allergen was isolated from an extract of dried *T. tonsurans* mycelium (Deuell *et al*. 1991), and purification yielded a 20 kDa protein, designated Tri t 1.IgE antibodies to Tri t 1 were measurable in 73% of sera from subjects with asthma, rhinitis or urticaria who were sensitised to *Trichophyton*, thus classing it as a major allergen. Moser and Pollack (1978) reported that DTH reactivity was restricted to a glycopeptide fraction derived from *Trichophyton* and *Microsporum* spp. which contained mannopeptides. Other *Trichophyton* allergens that have been identified and characterized include Tri t 4 (protein IV, a serine proteinase) in *T. tonsurans*, and Tri r 2 and Tri r 4, both serine proteases derived from *T. rubrum* (Slunt *et al*. 1996, Woodfolk *et al*. 1996, 1998, 2000, 2001). Tri t 1 has since been shown to be a glucanase, which not only contributes to the allergenic properties of the species, but also enhances dermatophyte growth by virtue of its biological function in cell expansion, cell-cell fusion during mating, and spore release (Cappellaro *et al*. 1998, Norbeck and Blomberg 1996).

MITE HYPERSENSITIVITY

Two major types of mite are associated with allergy, *viz*. dust mites and storage mites. Dust mite allergy is far more prevalent, owing to ubiquitous exposure, whereas storage mites pose problems only for individuals in certain environments.

That dust mites produce allergic responses in individuals was first suspected in 1928 (Dekker 1928), but not confirmed until more than 30 years later by Voorhorst *et al*. (1964). The most common species is *Dermatophagoides pteronyssinus,* which predominates in most countries. The occurrence of other pyroglyphid mites such as *D. farinae, D. microceras* and *Euroglyphus maynei* varies according to climate (Mehl 1998, Solarz *et al*. 2007).

Exposure to storage (nonpyroglyphid) mites has been increasingly recognized as a cause of asthma and rhinitis (van Hage-Hamsten *et al*. 1988, van Hage-Hamsten and Johansson 1998, Kim and Kim 2002). Several species have been identified, *e.g., Lepidoglyphus destructor, Acarus farris/siro, Tyrophagus* spp., *Glycyphagus domesticus* and *Blomia tjibodas. Blomia tropicalis* predominates in subtropical and tropical areas (van Hage-Hamsten *et al*. 1988). Studies from several countries have shown that IgE-mediated allergy in rural populations is of considerable importance and that storage mites are major allergens. Since these mites are also found in homes, especially in regions with damp housing conditions, certain urban populations are particularly at risk of becoming sensitised to them (Tee 1994). Even though sensitization to storage mites is not restricted to occupational exposure, farmers and their families who are in regular contact with dust in barns containing hay, straw or grain are most commonly affected.

Dust mite allergens

Mite allergens have been divided into groups on the basis of immunological and physicochemical similarities. They have been assigned specific names and seven groups have been delineated by the IUIS/WHO Allergen Nomenclature Subcommittee (1994).

Group 1 allergens

The complete sequences of the group 1 allergens from *D. pteronyssinus* (Der p 1), *D. farinae* (Der f 1), *E. maynei* (Eur m I), *D. microceras* (Der m 1) and *B. tropicalis* (Blo t 1) have been reported (Chua *et al*. 1988, Dilworth *et al*. 1991, Smith *et al*. 1999). Analysis of the available sequence data indicates that the group 1 allergens are very similar (81-84% identity). Homology searches indicate that the group 1 allergens belong to the cysteine group of proteolytic enzymes, i.e. cysteine proteases (Stewart 1995).

Group 2 allergens

The group 2 allergens are 14 kDa non-glycosylated members of the NPC2 family and are also considered as major allergens recognized by the majority of mite-allergic individuals (Thomas *et al.* 2002); they follow the nomenclature of Der f 2, Der p 2, etc. They are neutral-to-basic proteins and are resistant to denaturation by heat, extremes of pH (Lombardero *et al.* 1990) and digestion by proteases (Oshika *et al.* 1994). However, reduction and alkylation of the disulphide bonds significantly reduce the allergenicity of the protein. The complete amino acid sequences of the group 2 allergens from *D. pteronyssinus, D. farinae, E. maynei, B. tropicalis* and the storage mite *L. destructor* have been established either by conventional means or from cDNA clones (Stewart 1995; www.allergen. org).

Analysis of the three sequences indicates that the group 2 allergens from *D. pteronyssinus* and *D. farinae* are around 88% identical, and the overlapping regions of group 2 allergens from the genus *Dermatophagoides* shows approximately 40% identity with the 125-residue sequence of Lep d I, the group 1 allergen of *L. destructor* (Varela *et al.* 1994).

Group 3 allergens

The group 3 allergens are 30 kDa proteins that, on the basis of substrate specificity and amino acid sequence homology, have been shown to be trypsinlike enzymes (Stewart 1995). Most investigators acknowledge the group 3 allergens as being major allergens (Thomas *et al.* 2002). The characterized group 3 allergens include those from *D. pteronyssinus, D. farinae, E. maynei* and *B. tropicalis* (www.allergen.org).

Group 4 allergens

The group 4 allergens are 60 kDa proteins that are intermediate or minor allergens. Physicochemical studies have shown these allergens to be amylases (Lake *et al.* 1991). Thus far, group 4 equivalents have been demonstrated in *D. pteronyssinus, E. maynei* and *B. tropicalis* (Stewart *et al.* 1992, Mills *et al.* 1999, Thomas *et al.* 2002). Studies indicate that the allergenicity of Der p 4 is dependent on the integrity of the disulphide bonds.

Group 5 allergens

The group 5 allergens are ~14 kDa proteins that are recognized by about 50% of mite-allergic individuals (Tovey *et al.* 1989). A cDNA clone for the allergen from *D. pteronyssinus* has been isolated and shown to code for a protein of 148 amino acid residues (Thomas *et al.* 2007). Currently, only *D. pteronyssinus* and *B. tropicalis* appear to express this allergen grouping. The biochemical identity of the group 5 allergens remains unknown. Unlike the pyroglyphid counterparts, Blo t 5 is the major allergen whereas Blo t 1 only has modest allergenicity (Chua *et al.* 2007).

Group 6 allergens

The group 6 allergens are 25 kDa proteins that are recognized by about 40% of mite-allergic individuals (Yasueda *et al.* 1993). They have been shown to be serine proteases with specificities similar to vertebrate and invertebrate chymotrypsins. *D. pteronyssinus, D. farinae* and *B. tropicalis* have been shown to express a group 6 allergen (Yasueda *et al.* 1993; www. allergen.org).

Group 7 allergens

The group 7 allergens are 22 kDa proteins recognized by just over 50% of the mite-allergic population. The cDNA for the group 7 allergen from *D. pteronyssinus* has been cloned and shown to code for a 198-residue mature protein with a predicted molecular mass of 22177 Da (Shen *et al.* 1993). At present, the biochemical identity of this allergen remains unknown, but it has been identified in *D. pteronyssinus* and *D. farinae* (www.allergen.org).

Other Allergens

Less universally expressed are allergens designated group numbers 8 to 23. These include glutathione-S-transferase, tropomyosin, paramyosin, a collagenolytic serine protease, a fatty acid-binding protein, apolipophorin, chitinase, calcium binding protein, gelsolin/villinan and an antimicrobial peptide homologue.

MITE-FUNGUS RELATIONS

The presence and growth of both mites and fungi are enhanced by higher relative humidity (R.H.), and hence these microorganisms can frequently be found together in housing/bedding where there is high humidity (Cho *et al.* 2006b). This can result in increased allergic/asthmatic expression in children exposed to high levels of both viable fungi and house dust mite (Su *et al.* 2005).

It was once thought that fungi enhance rates of mite population by providing a food source. This was disputed by Hay *et al.* (1993), who showed that *D. pteronyssinus* reared over two generations on diets with fungi had reduced survival, prolonged development time and smaller adult size than mites reared on diets without fungi. It is now established that fungi are of no nutritional significance to house dust mites, and the presence of mites does not influence fungal growth (Hay 1991).

In the case of storage mites, however, fungi appear to be a relevant food source. Indeed, *Acarus* and *Tyrophagus* can be reared on a diet consisting exclusively of stored product fungi (Stewart 1995). Others have used mites such as *Tyrophagus* for the purpose of fungal biomass control (Woertz *et al.* 2002). In addition, mite-fungi associations may heighten the risk of occurrence of mycotoxins in food and feed and cause mixed contamination by both fungal and mite allergens, leading to heightened adverse reactivity potential (Hubert *et al.* 2004)

Fungi are an important source of allergens in dust (Bush and Yunginger 1987), and since they are ingested by mites it is possible that fungi could contribute to mite allergenicity if allergenic spores, mycelium or fungal enzymes are present on mites or their faeces. This does not, however, appear to be the case in house dust mites. Hay *et al.* (1992) demonstrated that larval *D. pteronyssinus* which lack fungi have allergen profiles indistinguishable from fungus-bearing adult mites. Whether a similar situation exists for storage mites is unknown.

HYPERSENSITIVITY TO ALGAE, BACTERIA AND BACTERIAL PRODUCTS

As mentioned in Chapter 1.2, algae were first reported as a possible cause of inhalant allergy by McElhenney *et al.* (1962), whose findings were based on the results of skin-prick testing. Since then, there have been reports of positive skin-prick test findings in both atopic children (Tiberg 1987, Tiberg *et al.* 1995) and adults (McGovern *et al.* 1966, Champion 1971, Ng *et al.* 1994). These results are supported by studies utilizing intradermal testing (Mittal *et al.* 1979, Mittal 1981), nasal provocative challenges (Hosen 1968, Lunceford 1968), bronchial provocation challenges (Mittal *et al.* 1979, Henderson *et al.* 1984) and RAST for algae-specific IgE (Tiberg *et al.* 1990, 1995). Negative responses in non-atopic controls supported the reactions as being immunological and not irritative in nature (Mittal *et al.* 1979). The frequency of skin reactivity to algae varies geographically and probably depends on the choice of algal strain, extract concentration, patient specific reactivity and test methodology (McGovern *et al.* 1966). The algal genus most extensively inves-

tigated is the unicellular green alga *Chlorella. Chlorococcum* and the filamentous cyanobacterium, or blue-green alga, *Anabaena* have been studied also (see Chapter 1.2). *Chlorella* has a worldwide distribution, being found in all kinds of environment, often in soils as an aerophyte (Round 1981). At present, however, hypersensitivity to algae remains a minor concern in indoor home and work environments.

Investigations of the role of bacteria in allergic diseases have focused mainly on different respiratory hypersensitivity states. Bacterial species are suspected of playing a significant role in asthma exacerbations and upper respiratory disease. The special role of bacterial endotoxin in hypersensitivity diseases has elicited recent research. Endotoxin, a component of the outer cell wall in Gram-negative bacteria consists of a polysaccharide chain linked with a core polysaccharide to a lipid moiety (lipid A) consisting of fatty acids. Significant epidemiological data exists to support the association between exposure to bacterial endotoxin and severity of asthma (Michel *et al.* 1991, 1996), as demonstrated not only by subjective symptom worsening and use of medication, but also by objective decreases in lung function, i.e. FEV_1 (forced expiratory volume in 1 sec) and VC (vital capacity). While it is apparent that infection with *Chlamydophila* (formerly *Chlamydia*) *pneumoniae* or *Mycoplasma pneumoniae* can precede asthma exacerbation (Sutherland and Martin 2007, Juvonen *et al.* 2008) and worsen severity (Johnston 1997, Laurila *et al.* 1997), the mechanisms driving this association remain unclear (MacDowell and Bacharier 2005, Johnston and Martin 2005). Norn and others have demonstrated that bacteria release histamine from human basophils and mast cells via both IgE-dependent and IgE-independent pathways (Norn *et al.* 1986, 1987). Furthermore, bacteria and endotoxin have been shown to enhance histamine release caused by allergens (Norn 1994). However, while hypersensitivity to bacteria exists, non-allergic mechanisms are thought to predominate, since asthmatic symptoms may occur in non-atopic individuals during infection with bacterial pathogens.

ALLERGIC CLINICAL SYNDROMES

Fungi and fungal products cause a wide range of hypersensitivity responses, some of which have long been recognized, e.g. asthma, and others which have been recognized more recently, e.g. allergic fungal sinusitis. Mite hypersensitivity has also been implicated in a wide range of clinical symptoms, while bacteria and endotoxin are associated with asthma. All are presented in detail below.

Allergic rhinitis

Rhinitis is an expression of nasal mucous membrane inflammation leading to nasal discharge, pruritus, sneezing and congestion. Given their widespread occurrence and their allergenicity, mould spores are an important cause of allergic rhinitis. The particles are much smaller than pollen and are easily inhaled. Allergic rhinitis due to moulds can be seasonal, as found with outdoor species, and/or perennial, as found with indoor and some outdoor species (Hardin *et al.* 2003).

In a large study of 3371 patients, 25.8% of those with a diagnosis of rhinitis were sensitised to moulds, as were 27.4% of those with a combined diagnosis of asthma and allergic rhinitis (Boulet *et al.* 1997). A study of over 200 children referred to a military clinic for evaluation of rhinitis found that 16.3% of children had positive mould skin testing (Calabria and Dice 2007). *Alternaria* (Granel *et al.* 1993, Stark *et al.* 2005), *Aspergillus* (Stark *et al.* 2005), *Penicillium, Cladosporium* (Shah and Sircar 1991), *Fusarium* (Wan Ishlah and Gendeh 2005), *Mucor* and *Aureobasidium* (Collins-Williams *et al.* 1972, Stark *et al.* 2005) have all in varying degrees been implicated in allergic rhinitis. The role of the basidiomycetes is controversial, with the literature presenting conflicting evidence of their involvement (Santilli *et al.* 1985, Lehrer *et al.* 1994), although a recent evaluation of 144 infants enrolled in the Cincinnati Childhood Allergy and Air Pollution Study demonstrated a positive association between basidiospores and rhinitis, yet failed to find a relationship between rhinitis and the more commonly implicated fungi such as *Penicillium* and *Aspergillus* (Osborne *et al.* 2006).

Often, patients with allergic rhinitis symptoms attributed to exposures to substances at work are in fact mould-allergic, and it is the contamination by moulds rather than the suspected agent that leads to symptoms. For example, an investigation of wood furniture workers with rhinitis symptoms showed a higher incidence of sensitivity to the moulds isolated from the factories where they worked than to the wood dust in these workplaces (Wilhelmsson *et al.* 1994).

Prior sensitization to dust mites is a well-documented risk factor in the development of chronic allergic rhinitis (Hagy *et al.* 1976, Skoner 2001, Sanico 2004). In one study of 3371 atopic patients, 54.2% were sensitised to house dust mite antigens (Boulet *et al.* 1997). Of those whose diagnosis was that of allergic rhinitis alone, 52.6% demonstrated dust mite

sensitivity. Among patients with a dual diagnosis of allergic rhinitis and asthma 60.0% were dust mite sensitive.

Bacteria and bacterial products such as endotoxins are rarely cited in the literature as significant players in allergic rhinitis. One study suggested an epidemiological link between the two, as refuse workers who were exposed to significantly more endotoxin than control workers in a water supply plant had an increased prevalence of rhinoconjunctivitis symptoms, itchy eyes and itching nose (Sigsgaard *et al.* 1994). A follow-up evaluation of these workers showed a greater induction of nasal lavage-derived inflammatory cytokines, lymphocytes and albumin after lipopolysaccharide nasal challenge in those garbage workers with occupational symptoms than in those without (Sigsgaard *et al.* 2000). In sawmill workers, the symptom of blocked nose was significantly correlated with respirable endotoxin levels (Mandryk *et al.* 2000). Overall, however, the evidence for the contributory role of bacteria or their products in the pathophysiology of allergic rhinitis is lacking.

Asthma

Asthma is a common respiratory disorder characterized by reversible airflow obstruction, airway inflammation and hyperirritability of the bronchial mucosa with resultant changes in lung volumes and expiratory flow rates. Asthmatic reactions can be immediate, late or dual. Clinically, asthma is manifested by dyspnoea (shortness of breath), wheeze and/or cough.

It has been suggested that as fungal spores are smaller than pollen grains, they can be more readily deposited in smaller bronchi and thus more easily induce asthmatic symptoms.

Not only have multiple mould species been demonstrated to provoke asthma, but their presence has led to respiratory arrest (O'Hollaren *et al.* 1991). A number of studies have documented the link between increases in atmospheric spore counts and hospital admissions for acute asthma exacerbation (Hasnain *et al.* 1985, Packe *et al.* 1985, Delfino *et al.* 1997, Halonen *et al.* 1997, Neukirch *et al.* 1999, Black *et al.* 2000, Downs *et al.* 2001b), and in the number of deaths due to asthma attacks (Targonski *et al.* 1995). Mould spores are also thought to contribute to the phenomenon of "thunderstorm asthma" (Dales *et al.* 2003), the known occurrence of epidemics of asthma exacerbation requiring emergency room visits and/or hospitalization immediately following severe thunderstorms. In particular, *Alternaria* hypersensitivity has shown a strong association with a history of thunderstorm-exacerbated asthma (Pulimood *et al.* 2007).

The mould species implicated in asthma are similar to those associated with allergic rhinitis (Kumar *et al.* 1984, Santilli *et al.* 1985, Gumowksi *et al.* 1987, Granel *et al.* 1993, Lehrer *et al.* 1994), with *Alternaria* having the strongest correlation (Granel *et al.* 1993, Halonen *et al.* 1997, Bush and Prochnau 2004).

The role of dust mites in asthma is considerable, and it has been said that, worldwide, 85% of asthma can be attributed to dust mite exposure (Platts-Mills 1990). Sensitization to house dust mites has been shown to be an independent risk factor for the development of asthma (Platts-Mills and Chapman 1987, Ulrik *et al.* 1996, Wong *et al.* 2002, Cole Johnson *et al.* 2004). The increasing prevalence of asthma has been associated with changes in lifestyle and housing, resulting in increased exposure to indoor allergens, among which dust mite is a primary offender (Custovic and Woodcock 1996). Studies have demonstrated that the severity of asthma in patients sensitised to house dust mites is related to the level of exposure to mite allergens in their beds (Custovic *et al.* 1996). Others have shown that subjects both sensitised and exposed to high levels of dust mite allergen had significantly lower FEV_1 percentage predicted values and more severe airway hyperreactivity (Langley *et al.* 2003).

Although it has been relatively straightforward to demonstrate a dose-response relationship between exposure to indoor allergen and sensitization (Custovic *et al.* 1998a), the relationship between exposure and asthma symptoms in already sensitised individuals is more complex, as in other allergic reactions. Some sensitised patients will react to a very low dose of allergen, while in others the level required to cause symptoms is much higher (Hagy and Settipane 1976, Custovic *et al.* 1996, Custovic and Chapman 1998). Exposure threshold values have been proposed, but the levels of exposure to which individuals are susceptible vary widely, and no absolute value that could generally ensure minimum risk has been identified (Wahn *et al.* 1997, Munir 1998).

Skin test positivity to storage mites is linked to asthmatic symptoms in farmers and others with occupational exposure to storage mites (Kronqvist *et al.* 2001, Parvaneh *et al.* 2002). Also, bronchial provocation studies confirm an important role for storage mites among the allergens responsible for asthmatic reactions (Ingram *et al.* 1979, van Hage-Hamsten *et al.* 1988). Reduction in FEV_1 due to inhalation of storage mite extract is species-specific, and does not occur in asthmatics sensitised to *D. pteronyssinus.*

The contributing role of bacterial endotoxin to asthma is complex. Many cohort studies suggest a protective effect of early bacterial and/or endotoxin exposure from the subsequent development of atopy (Braun-Fahrländer *et al.* 1999, Ernst and Cormier 2000, Gereda *et al.* 2000, Liu and Leung 2000, von Ehrenstein *et al.* 2000, von Mutius *et al.* 2000, Downs *et al.* 2001b), and others have found a protective effect against asthma (Ernst and Cormier 2000, von Ehrenstein *et al.* 2000, Downs *et al.* 2001b), which may only be modest (Lewis 2000, Douwes *et al.* 2002). Once asthma is established, however, exposure to endotoxin produces exacerbations and more severe disease. Several studies have shown that endotoxin in house dust is associated with exacerbation of pre-existing asthma in children and adults (Michel *et al.* 1991, 1996, Rizzo *et al.* 1997, Douwes *et al.* 2000, Park *et al.* 2001).

Subjects exposed to pure endotoxin in inhalation experiments experience acute clinical effects such as fever, influenza-like symptoms, neutrophilic airway inflammation, dry cough, dyspnoea, chest tightness, bronchial obstruction and dose-dependent impairment of lung function (Pernis *et al.* 1961, Michel *et al.* 1992, 1997, Michel 1997, Thorn and Rylander 1998). In addition, naïve subjects challenged with endotoxin-containing cotton dust (Castellan *et al.* 1987) or dust from pig farms (Zhiping *et al.* 1996) developed the same symptoms and lung function changes on exposure to endotoxin, but not to dust alone. Inhalation studies have further shown that subjects with asthma are more sensitive to developing these symptoms (van der Zwan *et al.* 1982, Michel *et al.* 1989, 1992). Endotoxin has also been shown to have non-immune mediated pro-inflammatory effects on airways (Smid *et al.* 1992, Ulmer 1997).

Bacteria have been shown to cause the release of histamine and other mediators from mast cells and basophils in both IgE-dependent and IgE-independent pathways, exacerbating existing allergic respiratory responses (Norn *et al.* 1986, 1987, Norn 1994). When patients hospitalized for acute exacerbation of chronic bronchitis infected with either *Haemophilus influenza* or *Streptococcus pneumoniae* had the bacteria isolated from their expectorate, it was demonstrated that the isolate was able to cause an IgE-mediated histamine release from their own blood leukocytes, indicating sensitization to the bacteria with which they were infected (Norn *et al.* 1994).

Hypersensitivity pneumonitis

Previously referred to as extrinsic allergic alveolitis (EAA), hypersensitivity pneumonitis (HP) is a relatively uncommon syndrome that is caused by the inhalation of a broad spectrum of organic dusts or chemical products that trigger a complex immunological response at the site of deposition within the lungs. The immunological inflammatory response and the attempts by the host to repair the damage lead to a progressive deposition of fibrotic tissue, which may result in permanent dysfunction or disability (Jacobs *et al.* 2005). The clinical expression of HP can vary and includes acute, subacute and chronic presentations (Kurup *et al.* 2006). In an individual patient, the clinical manifestations probably depend on the type of antigen, intensity and duration of exposure, susceptibility of the host, and resulting cellular and humoral immune responses over time.

There are a number of unexplained features of HP, including why it is that so few exposed individuals develop clinical HP, what triggers the acute episode after prolonged periods of sensitization and, additionally, what leads to disease progression (Woda 2008). It is known that following exposure to certain inhaled environmental antigens, most individuals develop precipitating antibodies, but only a few of these individuals will become symptomatic (Kline and Hunninghake 2007). The reason for this is not clear. It is likely that genetic susceptibility is important in determining which individuals are prone to the development of HP. It has been shown, for example, that polymorphisms in the major histocompatibility complex, tumour necrosis factor alpha (TNF-α), and tissue inhibitor of metalloproteinase 3 are associated with the development of, or resistance to, HP (Camarena *et al.* 2001, Hill *et al.* 2004). Despite extensive studies, the exact immunological mechanisms have not been fully elucidated. Antigen exposure is associated with the presence of circulating IgG antibodies in exposed individuals. The acute phase, which is mediated by neutrophils, is followed by a chronic phase mediated by lymphocytes and macrophages, which is thought to represent a delayed-type hypersensitivity response. This may eventually result in the formation of granuloma and in time progress to pulmonary fibrosis (Semenzato *et al.* 2000, Patel *et al.* 2001).

Inciting agents may be derived from fungal, bacterial or animal proteins or reactive chemical sources, and have been recognized in a wide variety of occupations and recreational activities. The most well characterized forms of expression of HP are presented in Table 3. The classic and most widely studied form of HP, known as "Farmer's Lung", is caused by contamination of hay, compost or silage by thermophilic Actinobacteria, with *Saccharopolyspora rectivirgula* being recognized as the major etiological antigen (Fenoglio *et al.* 2007). The moulds *Penicillium brevicompactum* and *P. olivicolor* have also been proposed as probable sources of antigens (Nakagawa-Yoshida *et al.* 1997). Ventilation HP can be caused by thermophilic members of the Actinobacteria, fungi such as *Aspergillus fumigatus* and *Aureobasidium* and possibly Protozoa (Pitcher 1990). HP related to a covered and heated swimming pool environment has been reported, with thermophilic Actinobacteria and a mould, *Neurospora* sp. (*Chrysonilia* sp.), suggested as the causative agents (Moreno-Ancillo *et al.* 1997). A large case series of HP was presented by Hanak *et al.* (2007), in which 34% were due to avian antigens, 21% were considered to be "hot tub lung disease", 9% were attributed to Farmer's Lung, 9% to household mould exposure, and 25% could not be determined. Occupational causes of HP are summarized in Table 4, and reports continue to be published in the medical literature, supporting the ongoing risk posed to those working with organic dusts and compounds susceptible to mould contamination (Galland *et al.* 1991, Halpin *et al.* 1994, Winck *et al.* 2004, Hoy *et al.* 2007, Rydjord *et al.* 2007).

Other causative moulds identified in various case reports include household *Penicillium* spp. (Lee *et al.* 2005), *Paecilomyces*-contaminated wood in a hardwood processing plant (Veillette *et al.* 2006), *Fusarium solani* in a potato and onion sorter (Merget *et al.* 2008) and *Aureobasidium pullulans* in a residence (Temprano *et al.* 2007). Summer-type HP, a unique disease form in Japan, is caused by *Trichosporon* (Yoshida *et al.* 1989, Nishiura *et al.* 1997, Fink and Zacharisen 2003).

Clinical presentation can vary and is affected by the duration and nature of exposure. As mentioned above, three forms have been identified. The acute form includes symptoms of headache, arthralgias, malaise, fever, lethargy, chills, cough and dyspnoea. Typically, 4-6 h after exposure patients experience symptoms, which subside 18-48 h later (Stites *et al.* 1987). Chest radiographs may be normal or show reticulonodular changes in the early stages of disease, and possibly hyperinflation. In the subacute form, symptoms tend to appear over a period of weeks and are characterized by cough and dyspnoea. Tachypnoea, tachycardia, fever and bibasilar respiratory crackles or rales might occur as evidence of acute exacerbation (Stites *et al.* 1987). Chest films may still be normal or show soft reticular or patchy infiltrates with or without small poorly defined nodules. Pulmonary

Table 3. Principal antigens or agents in aerosols from materials associated with hypersensitivity pneumonitis.

Disease/Occupation	Source of Aerosol	Principal Antigen/Agents
Actinomycete-induced		
Farmer's Lung	Mouldy (heated) hay	*Saccharopolyspora rectivirgula, Thermoactinomyces vulgaris*
Bagassosis	Mouldy sugar cane bagasse	*T. vulgaris, T. sacchari, T. thalpophilus*
Mushroom worker's lung	Mouldy mushroom compost	*Excellospora flexuosa, Thermomonospora* spp.
Mould-induced		
Malt-worker's lung	Mouldy malt	*Aspergillus clavatus*
Wood trimmer's disease	Mouldy wood dust	*Aureobasidium pullulans, Rhizopus microsporus, Paecilomyces variotii, Penicillium* spp.
Suberosis (cork worker's lung)	Mouldy cork dust	*Penicillium frequentans (P. glabrum)*
Cheese worker's lung	Cheese mould	*Penicillium casei*
Onion and potato sorters	Mouldy onions/potatoes	*Penicillium* spp., *Fusarium solani*
Sawmill workers	Mouldy wood dust	*Trichoderma koningii, Rhizopus microsporus*
Summer-type pneumonitis	House dust	*Trichosporon* spp.
Ventilation pneumonitis	Contaminated ventilation systems	Various fungi, Actinomycetes and amoebae
Protein-induced		
Bird-breeder's lung	Avian species	Avian dusts
Pituitary snuff-taker's lung	Bovine/porcine pituitary powder	Pituitary proteins
Animal handler's lung	Laboratory animals	Urinary proteins

function tests usually reveal a restrictive defect, decreased compliance, and at the end stage, diminished diffusion capacity. The chronic/insidious form is typified by steadily progressive symptoms with potential intermittent acute periods. Physical findings are compatible with those observed in the subacute form. Chest x-rays indicate interstitial fibrosis, mainly of the peripheral lung fields; honeycombing may occur at the periphery. Prolonged avoidance of the offending antigen is required for effective management of all three forms of the disease.

HP is discussed further in relation to episodes occurring in indoor environments in Chapter 3.2.

Allergic bronchopulmonary aspergillosis (ABPA)

ABPA is a pulmonary disease that is caused almost exclusively by *Aspergillus fumigatus* which has colonized the lower respiratory tract, although *A. niger* is also occasionally implicated (Virnig and Bush 2007). Patients are usually atopics with asthma, and those individuals with cystic fibrosis are particularly at risk (Vaughan 1993, Virnig and Bush 2007). Inhaled conidia are trapped in the respiratory tract, where they are able to persist and germinate. As mycelia develop, the fungi continuously release exoproteases, mycotoxins and other fungal products that further compromise clearance, breach the epithelium and activate immune responses (Moss 2005). Chemotactic cytokines, e.g. IL-8, RANTES and eotaxin, in particular have been implicated in murine models (Schuh *et al.* 2003). Chemokine-mediated recruitment of CD4+Th2 lymphocytes specific for *A. fumigatus* is a crucial feature of ABPA (Moss 2005). Susceptibility also appears to involve immunogenetic factors including atopy and defined major histocompatibility complex-restricted allelic expression on antigen-presenting cells that are permissive for a Th2-predominant immune response. Certain *A. fumigatus* allergens appear more relevant to ABPA than in other *A. fumigatus*-induced respiratory allergy, especially Asp f 4 (Kurup *et al.* 2000) and Asp f 6 (Maurya *et al.* 2005). This may be due in part to differential expression of these allergens during germination (Schwienbacher *et al.* 2005). It has become clear that genetic factors play a role in the pathophysiology of ABPA, as certain HLA haplotypes, especially HLA-DR2/DR5 and possibly DR4/DR7, predispose individuals to ABPA, while HLA-DQ2 seems to have a protective role (Chauhan *et al.* 2000).

In addition, it has recently been shown that serum from patients with ABPA has increased levels of the

Table 4. Minimal essential criteria for the diagnosis of allergic bronchopulmonary aspergillosis (ABPA) in patients with asthma (adapted from Greenberger 2003).

ABPA-Central Bronchiectasis

  Asthma

  Central bronchiectasis

  Immediate cutaneous reactivity to *Aspergillus* species

  Total serum IgE concentration >417 kU l[-1]

  Elevated serum IgE-*Aspergillus fumigatus* or IgG-*A. fumigatus*

ABPA-Seropositive

  Asthma

  Immediate cutaneous reactivity to *Aspergillus* species

  Total serum IgE concentration >417 kU l[-1]

  Elevated serum IgE-*Aspergillus fumigatus* or IgG-*A. fumigatus*

Th2 cytokine TARC (thymus- and activation-regulated chemokine), which may rise and fall with exacerbation of the disease (Hartl *et al*. 2006), and some authors have proposed that this change be included as a potential diagnostic marker for ABPA (Latzin *et al*. 2008).

The classic diagnostic criteria for ABPA include asthma, peripheral blood eosinophilia (>1.0 x 10[9] L[-1]), immediate cutaneous reactivity to *A. fumigatus*, elevated total serum IgE (> 417 kU L[-1]), precipitating antibodies to *A. fumigatus*, proximal bronchiectasis, chest x-ray infiltrates (fixed or transient) and elevated serum IgE and IgG antibodies to *A. fumigatus* (Rosenberg *et al*. 1977). Some asthmatics with ABPA may not meet all these criteria, and thus may be missed, especially if they are diagnosed early or are taking systemic corticosteroids. For this reason, Greenberger (2003) proposed two sets of minimum criteria to diagnose ABPA in asthmatic patients with and without bronchiectasis (Table 4), which can be denoted by ABPA-CB (CB = central bronchiectasis) and ABPA-S (S = serological).

Allergic fungal sinusitis

The combination of nasal polyposis, crust formation and sinus cultures yielding *Aspergillus* was first noted by Safirstein (1976), who observed the clinical similarity that this constellation of findings shared with ABPA. Eventually, this disease came to be known as allergic fungal sinusitis, or AFS (Luong and Marple 2004). It is thought to affect up to 7% of patients with chronic sinusitis (Ence *et al*. 1990), and should be suspected in patients with atopy and chronic (often intractable) sinusitis and nasal polyposis (deShazo *et al*. 1997). Most have pansinusitis, and many have had si-

nus surgeries prior to diagnosis (Goldstein *et al*. 1985, deShazo and Swain 1995). Patients tend to be young and immunocompetent (Spring and Amedee 1995). At surgery, involved sinuses almost invariably contain "allergic mucin" that has the appearance of "peanut butter" (Schubert and Goetz 1998, Gourley *et al*. 1990, Collins *et al*. 2003, Schubert 2004a), which consists of laminated accumulations of intact and degenerating eosinophils, Charcot-Leyden crystals, cellular debris and sparse hyphae rarely visualized with specific fungal stains (Katzenstein *et al*. 1983a, Schubert 2004a, 2004b, 2006). The sinus mucosa contains a mixed cellular infiltrate of eosinophils, plasma cells and lymphocytes. The allergic mucin and polyps obstruct sinus drainage and perpetuate the bacterial sinusitis often associated with AFS (deShazo *et al*. 1997). Elevated levels of fungal-specific IgE help to confirm that AFS is a true allergic, rather than infectious, disease (Manning *et al*. 1993, Mabry and Manning 1995, Schubert 2006).

The dematiaceous fungi *Bipolaris, Curvularia, Exserohilum* and *Alternaria* spp. are most commonly cultured from AFS surgical sinus cultures, followed by *Aspergillus* spp. (Gourley *et al*. 1990, Bent and Kuhn 1994, deShazo and Swain 1995, Schubert and Goetz 1998, Schubert 2004b). *B. spicifera* may predominate in the southwest US and *Curvularia lunata* in southern US (Schubert and Goetz 1998, deShazo and Swain 1995, McCann *et al*. 2002). Patients with AFS often have asthma and usually have allergic rhinitis, eosinophilia and elevated total and fungus-specific IgE concentrations. AFS appears to represent an IgE-mediated hypersensitivity reaction to fungi resembling that occurring in the bronchi in ABPA (Katzenstein *et al*. 1983b, Shah *et al*. 1990, Shah 2008). This may be a locally mediated immunological response, as a recent series of 34 patients with confirmed AFS showed on biopsy that 85% had tissue-bound specific IgE against *Aspergillus*, even though none had positive skin prick tests to this allergen (Chang and Fang 2008).

To date, a full consensus regarding the minimal diagnostic criteria for the diagnosis of AFS has not been reached. The classic and still widely accepted diagnostic criteria for allergic fungal rhinosinusitis (AFRS) were described by Bent and Kuhn (1994), who suggested the following:

(1) type 1 hypersensitivity by history, skin tests, or in-vitro testing;

(2) nasal polyposis;

(3) characteristic computed-tomography scan findings;

(4) eosinophilic mucus without fungal invasion into

sinus tissue; and
(5) positive fungal staining of sinus contents removed at surgery.

In an attempt to clarify some of the observed inconsistencies in the clinical characteristics of some cases, other workers have suggested their own sets of diagnostic criteria (de Shazo and Swain 1995, Ponikau *et al.* 1999). The debate over the value of these added diagnostic criteria has led to increased interest in the disease and helped fuel further investigation (Ryan and Marple 2007). AFS does not become invasive. It must be distinguished from other infectious, neoplastic and inflammatory conditions causing sinusitis (Brandwein 1993). The short-term prognosis for AFS is good, but requires surgical removal of the allergic mucin to resolve. It can also become a chronic disease, with objective CT scan findings recurring prior to subjective clinical symptoms after treatment (Kupferberg *et al.* 1997).

Baker's asthma

Baker's asthma is one of the most common causes of occupational asthma (Brisman 2002), with an annual incidence of 1-10 cases per 1000 bakery workers (Cullinan *et al.* 1994, Brisman *et al.* 2000). Any workers (including confectioners, flour millers and food processors) who are exposed to bakery allergens can develop the disease (Brant 2007). While this occupational illness is most often attributed to cereal flours (wheat, rye and barley) and enzymes, fungal allergens also play a significant role in its etiology. Baker's yeast, *Saccharomyces cerevisiae*, produces a 52 kDa enolase enzyme, which is a potent fungal allergen against which specific IgE has been demonstrated in sera from asthmatic bakers (Baldo and Baker 1988). Alpha-amylase is a starch-cleaving enzyme of fungal origin (*Aspergillus oryzae*) often used as a flour additive and is recognized as an occupational allergen in baker's asthma (Baur *et al.* 1988, Blanco Carmona *et al.* 1991). However, it is important to recognize that cereal α- and β-amylases are more important allergens than fungal α-amylase (Brant 2007), as evidenced by an increased incidence of IgE to cereal amylase, relative to IgE to fungal amylase, in a study of 30 subjects with baker's asthma (Sandiford *et al.* 1994).

In addition to amylase, fungal cellulase has been found to be an occupational allergen by skin testing, histamine-release testing, reverse enzyme-immunoassay for specific IgE antibodies and bronchial provocation (Quirce *et al.* 1992). In two cases of baker's asthma, provocative bronchial and intradermal challenges showed pulmonary hypersensitivity to *Alternaria*

and *Aspergillus*, which were present in the air within the bakeries (Klaustermeyer *et al.* 1977).

Although storage mites may be present as contaminant allergens of the working environment in bakeries, it was considered in a survey of respiratory symptoms and occupational allergens among 226 bakers and pastry makers in Italy that wheat allergens and α-amylase were causative, and sensitization to storage mites was unimportant (De Zotti *et al.* 1994).

Byssinosis

This occupational disease is acquired by cotton workers exposed to aerosols of dust generated in the early stages of processing raw cotton and is thought to be largely due to inhaled endotoxin. Although in the industrialized world there has been a significant decline in the prevalence of cotton dust lung diseases, studies show an increasing incidence in the developing world, notably in India, Pakistan, South Africa, Ethiopia, Sudan, Egypt and China (Parikh 1992). With rapid increases in industrialization, cotton dust-induced lung diseases are poised to become significant globally (Khan and Nanchal 2007).

Although this dust contains many other bioactive impurities, the weight of evidence for the major inciting agent leans heavily towards endotoxin, either alone or in synergy with other agents (Burrell 1994). Several studies have exposed cotton workers (both with and without symptoms) and normal volunteers to a variety of cotton dusts containing different amounts of endotoxin (Castellan *et al.* 1984, 1987, Rylander *et al.* 1985). All these studies demonstrated that the presence/degree of symptoms and decline in FEV_1 were not related to the total dust concentration, but were directly correlated with the endotoxin concentration.

Byssinosis is characterized by symptoms of chest tightness after long-term exposure to cotton dust. It is rare to have this classic form in persons with less than 10 years' exposure. This differentiates it from occupational asthma (Rylander 1990). The classic presentation is a sensation of chest tightness upon return to exposure in the workplace following a holiday or weekend break, and may start either immediately or over the second half of the day (Niven and Pickering 1996), often accompanied by cough, which may be productive, and occasionally shortness of breath. Over the work shift there is often a modest decrease in FEV_1. In most of the affected individuals, these findings disappear or diminish on the second work day. This pattern of symptoms appearing on the first day of the week and resolving by the second work day is termed the cross-shift pattern of symptoms (Khan and Nanchal 2007).

A less common acute form of byssinosis, characterized by an FEV_1 drop exceeding 30%, was demonstrated in volunteers exposed to cotton dust in an enclosed chamber for the very first time (Haglind and Rylander 1984). A study in Finnish cotton spinning mills reported that, as a result of these acute symptoms in the workplace, one in 10 employees had to resign within 2 weeks and one in four within 3 months of taking employment (Koskela *et al.* 1990).

Atopic Dermatitis

Since the previous edition of this book was published, a number of significant and exciting advances have developed regarding *Malassezia* (formerly *Pityrosporum*). Before 1996, this genus of basidiomycetous yeasts comprised only three species, but by 2008 some 13 species had been recognized (Guillot *et al.* 2008). *Malassezia* has been associated with a range of cutaneous and systemic diseases, including pityriasis (tinea) versicolor, seborrhoeic dermatitis, folliculitis, atopic eczema/dermatitis syndromes, catheter-related fungemia, peritonitis and meningitis (Ashbee and Evans 2002, Scheynius *et al.* 2002, Andersson *et al.* 2003). Of relevance to the current discussion is its role in atopic dermatitis (AD).

AD is a multifactorial skin disease characterized by a chronically relapsing course and severe pruritus (Leung 2000, Wollenberg and Bieber 2000). *Malassezia* species in patients with AD serve not only as normal flora but are also associated with exacerbations. Studies have indicated that 40-65% of AD patients have IgE antibodies, positive skin prick tests or positive atopy patch tests to extracts from *Malassezia* spp. (Kieffer *et al.* 1990, Nordvall and Johansson 1990, Nissen *et al.* 1998, Tengvall Linder *et al.* 2000, Johansson *et al.* 2002). Species implicated include *M. globosa* (Takahata *et al.* 2007), *M. restricta* (Sugita *et al.* 2001, Kato *et al.* 2006), *M. sympodialis* (Gupta *et al.* 2001, Faergemann 2002) and *M. furfur* (Nakabayashi *et al.* 2000), with *M. restricta* appearing to be more prominent in children than adults (Takahata *et al.* 2007). Extract from *M. sympodialis* induces a significantly stronger T-cell response in AD patients than in healthy individuals, and this response tends to be associated with production of Th2-like cytokines (Tengvall Linder *et al.* 1996, 1998, Johansson *et al.* 2002). These data suggest that *Malassezia* plays a role in maintaining IgE-mediated skin inflammation in AD.

Treating patients with antifungal agents who are affected mainly in the head and neck regions has led to decreased *Malassezia* colonization and reduced severity of the AD lesions, suggesting that *Malassezia* spp. have an important role in AD (Clemmensen and Hjorth 1983, Nikkels and Piérard 2003). This finding is supported by other antifungal trials that did not specifically examine for *Malassezia* colonization, but nevertheless demonstrated clinical improvement in AD symptoms after receiving treatment (Bäck *et al.* 1995, Broberg and Faergemann 1995, Bäck and Bartosik 2001, Lintu *et al.* 2001, Svejgaard *et al.* 2004).

Nine allergens have so far been cloned from two strains of *Malassezia* (Andersson *et al.* 2003). Some of these allergens were originally cloned from a strain thought to represent *M. furfur*, but this particular strain of *M. furfur* was recently reclassified as *M. sympodialis*. Therefore, in accordance with the allergen nomenclature of the World Health Organization (IUIS/WHO Allergy Nomenclature Subcommittee 1994), the Mala f 1 and Mala f 5 to Mala f 9 allergens were renamed as Mala s 1 and Mala s 5 to Mala s 9. Further details of the characterized allergens of *Malassezia* spp. are outlined in Table 2.

Additionally, evidence is beginning to appear in the literature indicating the role of various other microorganisms and microbial enzymes in the etiology of atopic dermatitis, allergic contact dermatitis and urticaria. A study of 30 children with atopic dermatitis demonstrated sensitization to a mould mixture via patch testing in 50% of those studied (Wananukul *et al.* 1993). A case report described eczema due to airborne *Penicillium* and *Cladosporium* (Kanny *et al.* 1996). Fungal α-amylase has also been put forward as a cause of contact dermatitis in bakers (Morren *et al.* 1993).

Urticaria (hives) can result from exposure to the same fungal α-amylase as implicated in baker's asthma, rhinitis and dermatitis (Kanerva *et al.* 1997). A role for *Candida albicans* as a factor in chronic urticaria, i.e. urticaria lasting longer than 6 weeks, has been suggested (Palma-Carlos and Palma-Carlos 2001, Staubach *et al.* 2008). However, there also exists convincing evidence to the contrary (Ergon *et al.* 2007). Fungal allergy can complicate the clinical picture of tinea pedis, commonly known as "athlete's foot". Tinea pedis can be caused by infection with either *Trichophyton rubrum* or *T. mentagrophytes*. Intense inflammation in *T. mentagrophytes* infections is the result of an immune, contact allergic response to the fungal antigens of that species (Leyden 1994).

The house dust mite *D. pteronyssinus* is suspected of being a factor in the pathogenesis of atopic dermatitis, yet its role remains controversial. Epicutaneous challenge with dust mite allergen in patients with atopic dermatitis may result in an eczematous reaction with allergen-specific CD_4 T-cells, found on biopsy (Reitamo *et al.* 1986, Bruynzeel-Koomen *et al.* 1988). At least one double-blind controlled trial of dust mite allergen avoidance has shown reduction in atopic dermatitis activity in those patients who used impermeable bedcovers, benzyl tannate spray and a high-filtration vacuum cleaner as dust mite reduction/avoidance techniques (Tan *et al.* 1996). In addition, atopy patch testing with dust mite allergen extracts has been shown to be positive in a high percentage of adults with atopic dermatitis (Ingordo *et al.* 2002, Pónyai *et al.* 2008). Thus, dust mite allergy may well be contributory to the symptomatology of eczema.

The prominent role of skin colonization by the Gram-positive bacterium *Staphylococcus aureus* as a contributory factor in atopic dermatitis exacerbation has been confirmed. As demonstrated first by Leyden *et al.* (1974), and confirmed by others (Aly *et al.* 1977, Hanifin and Rogge 1977), *S. aureus* could be isolated from the skin in more than 80% of atopic dermatitis patients, and more recent studies put that proportion at over 90% (Abeck and Mempel 1998). Further studies have shown a link between *S. aureus* colonization and the intensity of skin inflammation (Hauser *et al.* 1985), as the density correlates with severity, exacerbation and IgE levels (Herz *et al.* 1998). In contrast, only about 5-10% of non-atopic individuals carry *S. aureus* on their skin. Also, about 25% of atopic dermatitis patients have circulating IgE antibodies to *S. aureus* cell wall products (Leyden *et al.* 1974), and the staphylococcal enterotoxins A and B, or SEA and SEB (Ide *et al.* 2004). In one investigation, SE sensitization rate increased significantly with increasing eczema severity (Semic-Jusufagic *et al.* 2007). Furthermore, *S. aureus* toxins exacerbate disease activity by both the induction of toxin-specific IgE and the activation of various cell types including Th2 cells, eosinophils and keratinocytes (Baker 2006). The precise pathogenic role of *S. aureus* remains to be fully elucidated, however.

Volatile organic compounds, mycotoxins and β-glucan

Over the last decade, a mounting body of evidence has accumulated that implicates the metabolic products and cell wall components of fungi and bacteria as key players in the allergic response of susceptible individuals. These products are discussed in more detail in Chapters 4.4 and 4.5, but their specific role in allergic disease will be considered here.

Volatile organic compounds (VOCs)
Some VOCs are gaseous metabolic by-products of the metabolism of growing fungi. Not only does dampness in buildings increase the growth potential for moulds and mites, it can also increase the emission of VOCs, due both to degradation of building material and increased microbial activity (Norbäck *et al.* 1993, Nilsson *et al.* 2004). Multiple compounds have been confirmed as being of microbial origin, including hexane, methylene chloride, benzene, 1-octen-3-ol and acetone. VOCs have been demonstrated in experimental settings to induce mucous membrane irritation (Mølhave *et al.* 1986, Fischer and Dott 2003, Hope and Simon 2007) and asthmatic symptoms (Norbäck *et al.* 1995, Becher *et al.* 1996, Rumchev *et al.* 2004, Arif and Shah 2007), and to reduce FEV_1 values significantly (Harving *et al.* 1991). Furthermore, epidemiological evidence supports association between nocturnal breathlessness, wheezing and airway hyperresponsiveness to indoor air VOC exposure (Norbäck *et al.* 1995, Fischer and Dott 2003, Mendell 2007). As Hodgson *et al.* (1994), Nordsrom *et al.* (1994), Anon. (1996) and Sunesson *et al.* (2006) have observed VOCs have also been implicated in the etiology of sick building syndrome (SBS)/building related illness (BRI) (see Chapter 4.4).

Mycotoxins
The spores of some fungal species have a high mycotoxin content, including species producing trichothecenes, ochratoxins and aflatoxins. These have long been known to pose a health hazard to humans and animals in association with mould-contaminated food and feed, and in more recent times, concerns have been raised about exposures to mycotoxins in indoor environment (Jarvis and Miller 2005). These toxins have been shown to disregulate macrophage function *in vitro* and *in vivo* (Sorenson 1990, Jakab *et al.* 1994). In addition, monocyte function has been shown to be impaired (Cusumano *et al.* 1996). *Stachybotrys chartarum* (formerly *S. atra*), a saprophyte that grows on wet cellulose-containing building materials, including drywall, ceiling tiles and cardboard (Andersson *et al.* 1997, Tuomi *et al.* 2000, Boutin-Forzano *et al.* 2004), is often found in low concentrations among the fungi identified in water-damaged buildings (Shelton *et al.* 2002, Kuhn *et al.* 2005) as well as representative US housing (Vesper *et al.* 2007). The

capacity of *Stachybotrys* to produce the most biologically potent members of the large family of mycotoxins, the trichothecenes, has resulted in this fungus being colloquially referred to as "toxic black mould" (Petska *et al.* 2008). During the time period from 1993 to 1996, investigation of cases of acute idiopathic pulmonary haemosiderosis (AIPH) among infants living in Cleveland, Ohio (some of which were fatal) suggested an association between AIPH and exposure to *S. chartarum* (Dearborn *et al.* 1994, 1997, Etzel *et al.* 1998). However, reviews of that investigation by the Centers for Disease Control (CDC) and external consultants identified shortcomings in the methodology and determined that no association between AIPH and exposure to moulds has been established (CDC 2000, 2004).

Case reports have attributed chronic indoor *S. chartarum* exposures to debilitating respiratory symptoms (Johanning 1995, Hodgson *et al.* 1998), as well as non-respiratory effects suggestive of immune dysfunction (Johanning *et al.* 1996). Based on these and other reports, trichothecenes have been implicated as factors in SBS (Croft *et al.* 1986, Smoragiewicz *et al.* 1993) or BRI (Fung *et al.* 1998, Jarvis *et al.* 1998, Dearborn *et al.* 1999, 2002, Hossain *et al.* 2004, Kilburn 2004). The chief arguments against *S. chartarum* causing adverse human effects are based on the predication that the quantities of its conidia or its toxins required for eliciting adverse systemic effects in animal models exceed levels encountered to date in even the most highly contaminated indoor environments (Chapman *et al.* 2003, Hardin *et al.* 2003, Hossain *et al.* 2004, Lai 2006). Local tissue injury, however, has been observed in the immediate deposition site of the inhaled fungal particles containing high concentrations of deleterious toxins (Petska *et al.* 2008). Understanding the details of the pathophysiological changes induced by 'real-life' exposure to *S. chartarum*-derived mycotoxins should be facilitated by referring to Chapter 4.5.

Bacterial endotoxin

This component of the cell walls of Gram-negative bacteria is mentioned in this context for completeness. Full details have been covered in earlier sections of this chapter.

β-Glucan

As discussed in Chapter 4.5, (1→3)-β-D-glucan is a component of the fungal cell wall that possesses potent immunobiological effects (Pretus *et al.* 1991, Thorn 2001). Results from *in vitro* and *in vivo* experi-

ments suggest that glucans can bind to cells by receptors, stimulate bone marrow, activate macrophages and induce production of cytokines such as interleukins IL-1 and IL-2 (Sherwood *et al.* 1987, Williams *et al.* 1991, Adachi *et al.* 1994, Sakurai *et al.* 1994). Pulmonary alveolar macrophage function and the immune system responses to (1→3)-β-D-glucan are only partially understood. Fungal glucan appears to decrease phagocytosis and macrophage numbers (Rylander and Goto 1991, Rylander and Peterson 1993). (1→3)-β-D-glucan induces the synthesis of IL-1, IL-6 and TNF-alpha in human leukocytes (Kubala *et al.* 2003) and importantly in regard to hypersensitivity reactions, (1→3)-β-D-glucan was shown to potentiate IgE-mediated histamine release (Holck *et al.* 2007).

A relation has been found between the indoor level of (1→3)-β-D-glucan and the proportion of persons reporting symptoms indicative of airway inflammation (Rylander 1999, Rylander and Lin 2000). Subjects exposed to the glucan in experimental challenges reported symptoms of thirst and nose irritation, headache, fatigue and cough (Rylander *et al.* 1992, Rylander 1997). Inhalational challenges with (1→3)-β-D-glucan do not induce the robust inflammatory response observed with endotoxin, or indeed any other dramatic effects on cells or secretion of cytokines (Rylander and Lin 2000). These types of study may not represent consequences of long-term airborne exposures (Douwes 2005). Studies examining subjects with longstanding daytime exposures, e.g. schools, day care centres and offices, reported correlations between the extent of nasal and throat irritation, cough and tiredness and levels of airborne (1→3)-β-D-glucan (Rylander *et al.* 1992), and improvements in pulmonary function outcomes after renovation that allowed for a significant decrease in airborne glucan levels (Rylander 1997). The relationship between (1→3)-β-D-glucan and asthma is clearly complex, however, since indoor exposure to high levels of (1→3)-β-D-glucan (concentration >60 μg g^{-1} house dust) was associated with a *decreased* risk for recurrent wheeze among infants born to atopic parents in Cincinnati, USA (Iossifova *et al.* 2007). Supporting this finding is a European evaluation of mattress dust loads of endotoxin, (1→3)-β-D-glucan and extracellular polysaccharides, which similarly demonstrated that higher dust loads of these compounds were associated with a *decreased* risk of sensitization to inhalant allergens (Gehring *et al.* 2007). Additionally, direct intranasal deposition of (1→3)-β-D-glucan induced no change in the percentage of eosinophils and amount of eotaxin in nasal lavage samples either

30 min or 24 h after challenge (Beijer and Rylander 2005). Even more intriguing is a recent report of the use of $(1{\rightarrow}3)$-β-D-glucan derived from the edible mushroom *Lentinus edodes* as a *treatment* for allergic rhinitis (Yamada *et al.* 2007), by inducing a Th1 response with a resultant reduction of Th2-type cytokines (Kirmaz *et al.* 2005).

CLINICAL HISTORY WITH ALLERGENIC MICRO-ORGANISMS

Mould Allergy
In all evaluations of potential allergic disease, the clinical history is extremely important in elucidating the possible causal agents. An exposure history is essential, allowing the clinical symptom profile to match the patterns seen with outdoor and/or indoor mould allergy.

Exposure history
Questions regarding the surrounding vegetation may provide useful information on likely allergenic microorganisms or pathogens. Patients should be asked to describe obvious mouldy odours and/or visible mould growth in the home, work or school environment. Signs of visible water damage in occupied environments and the R.H. in these environments heighten suspicion of mould involvement. The air may feel damp, surfaces may be wet, there may be a history of flooding, or sometimes the mere presence of carpeting in the basement is important. Enquiring about the ventilation of the environment may prove helpful, since not only may a lack of adequate ventilation lead to mould growth, but ventilation sources may be contaminated with fungi, and filters may need to be cleaned or replaced.

Allergy to outdoor airborne moulds
The symptoms of allergy tend to follow a seasonal profile, with certain times of the year (spring and late summer/early autumn) being worse than others. The moulds involved are dominant and ubiquitous. Allergy to these moulds is frequently associated with pollinosis and prolongation of the hay fever season for sufferers. The diagnosis of airborne mould allergy is facilitated by skin prick testing to specific, good quality extracts that are commercially available.

Allergy to moulds in the home
Little or no seasonal fluctuation of symptoms exists in patients with allergies to primarily indoor species.

Table 5. Difficulties encountered with preparation of fungus material for *in vitro* or *in vivo* diagnostic tests.

- Accurate identification of the organism
- Selection and storage of the suitable strain
- Adequate culture methods, metabolism, growth kinetics
- Choice of morphological form: spores or mycelium
- Extraction techniques
- Standardization techniques
- Qualitative variations between batches
- Spontaneous enzymatic degradation
- Choice of substrate and adaptation to specific tests
- Control and elimination of toxins
- Insufficient species studied and characterized

A large number of different fungal species fit into this category. Each habitat has its own characteristic fungal species associated with it, but the commercially available range of extracts remains incomplete and therefore insufficient to categorically exclude the possibility of mould allergy if testing is negative.

Fungal extracts

Part of the challenge in the diagnosis of fungal allergy is the variability of commercially available fungal extracts used in skin-prick testing. The reasons for such wide variation are mycological and immunological. Moulds are highly adaptable, with the result that there are wide variations in cellular and constitutional structures. There is marked pleomorphism in the recognition of these structures by the immune system; not only are many antigenic molecules recognized by allergic patients, but the degree of recognition of each of these antigenic sites differs from one individual to another (Gumowski 1997). Production of a reference antigen is extremely complicated in that, unlike other allergenic agents, moulds are whole living filaments with remarkable powers of adaptation. In addition, a mould genus, such as *Aspergillus* or *Penicillium*, can comprise a range of species, and a number of somewhat different strains may exist within each individual species. Depending on cultural conditions, the available nutrients in the immediate environment and the constitutional and metabolic nature of the fungus, the antigenic expression may change for a single strain. This may be demonstrated by cross-over immunoelectrophoresis (Aukrust 1979). The main problem is that the methods devised by researchers for the adequate production of purified extracts are unusable at a commercial level, since preparation of the starting materials requires an individual and specific protocol for each species (Esch 2004).The challenges faced in extract preparation are listed in Table 5.

Mite allergy

The clinical history in patients with dust mite allergy is typically one of symptoms following a perennial pattern. In temperate climates, seasonal fluctuations in numbers of mites found in house dust are evident, with numbers being low at the beginning of the summer and reaching a peak in late summer before dropping again in late autumn to low winter levels. This fluctuation results from a combination of outdoor temperature and humidity, along with the use of central heating. Despite this, the allergenic material produced by these mites remains in the environment, with the amount fluctuating less than the number of mites present (Platts-Mills *et al.* 1987, Kalra *et al.* 1992). Mite numbers show seasonal trends, but allergic symptoms to mite allergens tend to be perennial in nature, with early morning exacerbations of symptoms resulting from overnight exposure to bedding.

Mite extracts

Extracts used in the testing of allergy to dust and storage mites are free from the component difficulties associated with fungal species. These have proved to be reliable in testing for allergy to dust mite, and are used on occasion in immunotherapy.

Allergy to bacteria and algae

As both of these types of organism are much more uncommon contributors to allergenic hypersensitivity, no unique clinical history is attributed to them, except for obvious specific occupational exposures. As with the dust mite, the extracts used for the diagnosis of algal hypersensitivity appear to be reliable and reproducible.

THERAPEUTIC RECOMMENDATIONS

Allergen Avoidance

Avoidance of allergens is the mainstay of all anti-allergic therapeutic strategies. Given the ubiquitous nature of moulds and mites, this is obviously a challenging undertaking in the case of mould/mite hypersen-

sitivity. However, some personal and environmental modifications can be initiated that may prove beneficial. Firstly, adequate outdoor air ventilation with filtration should be maintained throughout the year. Elimination of dampness and water leaks indoors is essential. The use of dehumidifiers (cleaned regularly) to maintain a R.H. no greater than 50% may assist in inhibiting mould growth, but only if the substrate on which the growth occurs dries out with the dehumidification (see Chapter 2.1). For pre-existing, obviously mouldy areas, cleaning with a hypochlorite bleach and detergent solution will reduce the amount of mycelium that is active, but re-growth may occur if resistant spores remain. It is imperative that persons working with such solutions protect their respiratory system by using a mask in a well-ventilated room. Personal exposure to moulds can be reduced by wearing well-fitted particle masks (1 μm) while handling compost, vacuuming and cleaning fungus-contaminated areas.

Avoidance of dust mite allergens is a widely discussed topic, with many strategies open for employment, so they will be reviewed only briefly here. Allergen avoidance alone should not however be relied on as a sole therapeutic intervention. Dust mite avoidance strategies in the home have previously emphasized very strenuous efforts in cleaning of bedrooms, including removal of carpeting, covering of mattresses and regular washing of bedding at temperatures greater than 70°C (Day and Ellis 2001). Impermeable covers of pillows, mattresses and comforters/duvets have been shown to reduce significantly measurable levels of Der p 1 and Der f 1, the major allergens for *D. pteronyssinus* and *D. farinae*, respectively (Terreehorst *et al.* 2003, Woodcock *et al.* 2003), and seem to produce the best results.

Investigations suggest that even with enthusiastic vacuum cleaning the efficacy is limited to removal of mite allergen on the surface. Mites generally live away from the surface areas and are unaffected by vacuum cleaning. Because mite re-colonization usually occurs within months, replacement or renewal of bedding, mattresses or carpets for mite-allergic patients is only a short-term solution. In a study of allergy avoidance that utilized regular vacuum cleaning of the bedrooms as the major avoidance measure, there was no significant improvement in either the patients' symptoms or the mite antigen levels (Burr *et al.* 1980). Mite-allergic patients who do their own cleaning should wear well-fitting particle masks during cleaning and for 10-15 min afterwards (Day and Ellis 2001).

In undisturbed rooms it is difficult to detect airborne mite allergen even after prolonged sampling. Whole mite bodies and fragments of mite can become airborne during bed-making and vacuuming, but very little allergen is associated with particles that will remain airborne for more than a few minutes. It is therefore unlikely that air filtration can significantly reduce mite allergen exposure. Bowler *et al.* (1985) were unable to show any additional clinical benefit from combining existing dust control measures with electrostatic and high efficiency particulate air (HEPA) filters. HEPA filters may, however, produce clinical benefits for rhinitis and asthma by reducing numbers of small non-allergenic particles with respiratory irritant effects (Evans 1992, Diette *et al.* 2008).

The use of dehumidifiers and/or air conditioning can produce up to 10-fold reductions in mite numbers. As mentioned earlier, limiting R.H. indoors to less than 50% is not only beneficial in reducing mite growth, but may also control the growth of mould species.

Several different acaracidal solutions exist, most containing a combination of products such as benzyl benzoate, pirimiphos methyl, synthetic oxazolidinones and/or other adjuvant substances, e.g. tannic acid (a protein-denaturing substance). All have shown some effect, none achieve complete control of mites, and all would require application every two months to maintain control (Pollart *et al.* 1987, Wickman 1997). One open-label trial demonstrated an improvement in symptoms and rhinitis-specific quality of life scores in patients with dust mite-induced perennial allergic rhinitis who instituted environmental control measures, including monthly application of an acaracide aerosol containing esbiol/benzyl benzoate/piperonyl butoxide/2-phenylphenol (Malet *et al.* 2002). Tannic acid alone has been shown in a randomized controlled trial to be insufficient to produce clinically meaningful reduction in mite allergen levels (Lau *et al.* 2002). In general, however, acaracides are believed to be minimally effective in reducing house dust mite allergen in carpeting (Custovic *et al.* 1998b, Platts-Mills *et al.* 2000), and interventions that include acaracide treatment of carpeting have not affected the outcomes in clinical trials (Ehnert *et al.* 1992, Dietemann *et al.* 1993).

Four Cochrane collaboration systematic reviews have examined controlled trials involving various house dust mite control measures as a treatment strategy for mite-sensitive asthmatics, the most recent review having evaluated 54 studies. All of these meta-analyses concluded that current chemical and physical methods aimed at reducing exposure to house dust mite allergens seem to be ineffective, and cannot be recommended as prophylaxis for mite-sensitive asthmatics (Hammarquist *et al.* 2000, Gøtzsche *et al.* 2001, 2004, Gøtzsche and Johansen 2008). These analyses have been criticized, however, because all avoidance measures were considered equally, e.g. acaracidal treatment *vs* impermeable bedding covers, trials were included with ineffective mite avoidance strategies, and criteria were used that excluded two trials with a striking effect on both exposure and disease activity (Platts-Mills *et al.* 1999, Platts-Mills 2008). The early clinical trials were conducted before the basic elements of mite allergen avoidance were known. As mentioned above, bedding is the most important route of exposure; the use of impermeable mattress and pillow covers together decreased humidity and possibly frequent laundering of bedding can reduce exposure sufficiently to reduce asthma morbidity (Custovic *et al.* 1998b, Tovey and Marks 1999, Platts-Mills *et al.* 2000).

The results of the Cochrane reviews evaluating the same measures in rhinitis patients were more positive, and have suggested that intervention designed to reduce house dust mite exposure in sensitised patients with allergic perennial rhinitis may be of some benefit in reducing rhinitis symptoms. Data were strongest overall for a bedroom-based environmental control program (Sheikh and Hurwitz 2001, Sheikh *et al.* 2007).

Thus, most clinicians would still recommend that impermeable covers on mattresses, pillows and bedding in cases of dust mite allergy.

As bacteria and their products are essentially ubiquitous and unavoidable, no specific strategies other than frequent hand washing and good hygiene will reduce exposure.

Medical therapies

First-line medical therapy for allergic rhinitis includes intranasal corticosteroids and oral antihistamines (H_1 receptor antagonists). Most patients will have already tried at least one antihistamine prior to seeking medical attention because of their widespread over-the-counter availability. The first generation antihistamines, such as diphenhydramine and chlorpheniramine, are usually effective for symptom control, but have a high incidence of side effects, particularly significant drowsiness. The second generation H_1 antagonists, e.g. cetirizine, desloratadine, fexofenadine, levocetirizine and loratadine, are safer and effective non-drowsy alternatives (Day and Ellis 2001, Hu *et al.* 2008). Also of significant benefit are the intranasal glucocorticosteroids, e.g. budesonide, fluticasone propionate/furoate, mometasone furoate and triamcinolone acetonide. These products have low systemic absorption (between 0.1% and 1% depending on the product) and thus provide safe, efficacious therapy in patients with chronic, daily symptoms (Derendorf and Meltzer 2008). In some areas, intranasal antihistamines such as azelastine are available and are also of benefit for persistent symptoms (Kaliner 2007). Intranasal sodium cromoglycate preparations are available, but compared with the steroid preparations have a generally limited efficacy profile.

Patients with asthmatic symptoms in response to allergenic microorganisms will achieve short-term symptomatic relief from the inhaled beta$_2$-agonists, e.g. salbutamol, but regular inhaled corticosteroids are essential to stabilize the underlying inflammatory reaction that produces the airway irritability and narrowing in asthma, e.g. budesonide, fluticasone propionate and mometasone furoate. A long-acting beta-agonist such as formoterol or salmeterol, or a leukotriene-receptor antagonist such as montelukast, may also be added.

Systemic corticosteroids are the mainstay of treatment in HP, following avoidance of the causal agent, as they may prevent fibrotic scarring. Bronchodilators are generally not indicated, since airways are not usually affected. The treatment of ABPA is mainly directed towards the asthmatic component, with systemic and inhaled corticosteroids. The positive effects of corticosteroid therapy in ABPA are thought to result from the dampening of the inflammatory response and an increase in efficiency of killing the causal fungi (Kauffman *et al.* 1995). Antifungal agents such as itraconazole, previously considered controversial (Day 1996), are now generally recommended as adjunctive (and possibly steroid-sparing) therapy (Tillie-Leblond and Tonnel 2005).

Treatment of AFS requires endoscopic removal of polyps and inflammatory material to establish aeration and drainage of the sinuses involved, followed by oral corticosteroids for at least two weeks postoperatively (deShazo *et al.* 1995, Schubert 2007). Patients should receive intranasal corticosteroids on a long-term basis (Corey *et al.* 1995, Schubert 2006). Antifungal agents have not been shown to be useful, but immunotherapy, antihistamines and antileukotrienes may be considered (Schubert 2004b).

Immunotherapy

Even to date, very few controlled studies have examined the efficacy and safety of fungal extract immunotherapy (Martinez-Cañavate Burgos *et al.* 2007). The success of immunotherapy depends on the use of high-quality and adequately standardized allergenic vaccines that can be uniformly produced (Bousquet *et al.* 1998). In recent years, *Alternaria alternata* and *Aspergillus fumigatus* have been the most extensively studied fungal species. Biological standardization has been achieved in these cases (van Ree 2007).

Standardized extracts of *Alternaria* were prepared for diagnosis and immunotherapy (IT) to improve symptoms in 24 patients with allergic rhinitis, with and without asthma (Horst *et al.* 1990). When 39 children with asthma and/or rhinitis were followed prospectively for three years while receiving immunotherapy for *A. alternata* allergy, 80% reported excellent results when they reached the 80,000 PNU maintenance dose, the remaining 20% claiming good results (Cantani *et al.* 1988). Another trial involving 129 subjects demonstrated the benefit of *Alternaria* immunotherapy for patients with asthma and/or rhinitis with appropriate sensitization (Tabar *et al.* 2000). A clinical survey carried out among paediatric patients sensitised to *Alternaria* with confirmed allergic respiratory pathology due to *Alternaria* sensitization were randomized to either *Alternaria* IT or control, and showed a significant reduction in symptom scores and medication use after 1 year of treatment with IT, a benefit not observed in the control group. Open-label trials have suggested symptomatic benefits from sublingual immunotherapy in *A. alternata*-allergic patients, with a tendency for sublingual formulations to show a better safety and tolerability profile than subcutaneous administration (Bernardis *et al.* 1996, Criado Molina *et al.* 2002, Di Rienzo *et al.* 2005).

A beneficial role for *Cladosporium* immunotherapy was demonstrated in a study of 22 adult asthmatics (Malling *et al.* 1986), as well as one involving 30 children with asthma (Dreborg *et al.* 1996).

Mould-specific immunotherapy has also been shown to have a beneficial role in the treatment of AFS, with decreased objective findings in patients as well as diminished need for systemic or topical corticosteroids (Mabry and Mabry 1997, Mabry *et al.* 1998), in addition to a lower rate of relapse compared with patients who did not receive immunotherapy (Bassichis *et al.* 2001). Some authors strongly support the use of specific immunotherapy with appropriate fungal antigens for these patients, as it has been shown to be beneficial when combined with surgery and adjunctive medical management, reducing recurrences and the need for systemic corticosteroids (Mabry and Mabry 2000, Schubert 2004b).

Immunotherapy for dust mite allergic asthmatics has proved effective in reducing asthma symptoms and requirement for medication in a number of studies (Hedlin 1995, Cantani *et al.* 1997, Cools *et al.* 2000, Maestrelli *et al.* 2004, Fernández-Caldas *et al.* 2006, Cevit *et al.* 2007), and others' evaluations have additionally demonstrated reductions in bronchial hyperresponsiveness (either to dust mite allergen provocation or histamine/methacholine challenge) and/or overall lung function (Peroni *et al.* 1995, Olsen *et al.* 1997, Pichler *et al.* 2001, Pifferi *et al.* 2002). A review by Bousquet and Michel (1994) showed that children tended to improve more than adults on mite immunotherapy; the improvement was greater in children who had normal or close to normal lung function. Caution must always be exercised in the implementation of immunotherapy, particularly in asthmatic patients. A prospective study by Oostergaard *et al.* (1986) demonstrated that in asthmatic children receiving immunotherapy for various allergens mould extracts were responsible for the most frequent and serious side effects (*Alternaria,* 3/106 patients; *Cladosporium,* 8/106). Another analysis of children receiving immunotherapy with mould extracts for asthma documented the withdrawal of seven (19%) children because of serious side effects (Kaad and Oostergaard 1982).

MOULD REMEDIATION

If the contribution of mould to the worsening of allergic respiratory disease is as conclusive as that suggested by the literature summarized in this chapter, then the removal of mould contamination from the home or work environment should result in a reduction of such symptomatology, were Koch's postulates to be upheld. Consequently, since the last edition of this text was published, a number of studies have been conducted to evaluate the effects of mould remediation on patient symptomatology.

Bernstein *et al.* (2005) investigated the effects of combined dehumidification and HEPA filtration on airborne mould spores in day-care centres. Fungal analyses demonstrated lower baseline and follow-up mean levels in intervention rooms. Dehumidification with HEPA filtration was effective at controlling indoor dew point in both facilities and at reducing airborne culturable fungal spore levels in one of the two facilities studied. While this appeared promising as a potential intervention technique, Bernstein *et al.* (2005) did not investigate any changes in clinical symptoms experienced by either the children attending or workers staffing these day care centres.

Other workers have, however, gone on to evaluate that next important step. In one trial, 131 houses of asthmatic patients were randomly allocated to an intervention and control group. The intervention consisted of mould removal, fungicide application and the installation of a fan in the loft. In the control group, intervention was delayed for 12 months. At 6 months post-intervention, when compared to the control group, the intervention group showed a net reduction in wheeze, improved breathing and perceived reduction in medication, although there were no changes in objective measures such as peak flow variability or actual medication use (Burr *et al.* 2007). Kercsmar *et al.* (2006) targeted asthmatic children living in homes with documented indoor mould for remediation, which included an action plan, education, individualized problem solving, household repairs, removal of water-damaged building materials and heating/ventilation/air-conditioning alterations. The control group received only home cleaning information. Children in both the intervention and control groups showed improvement in asthma symptoms during the pre-remediation portion of the study, but only the intervention group showed a significant decrease in asthma symptom days relative to pre-remediation, and in the post-remediation period this group had a lower rate of exacerbation compared with control asthmatics (Kercsmar *et al.* 2006).

CONCLUSION

Fungi are the commonest microorganisms responsible for allergic disease. Exposure to various fungal products produces a wide range of clinical hypersensitivity reactions, and these may occur in response to spores, mycelium or components of the cell wall, such as $(1{\rightarrow}3)$-β-D-glucan. Metabolic by-products, including VOC's and mycotoxins, also produce adverse consequences. Dust mites also induce allergic responses, and bacteria and their endotoxins augment hypersensitivity reactions

In spite of important new knowledge, much remains to be learned about the full contribution of indoor microorganisms towards the induction and maintenance of allergic disease. Continuing to improve our understanding of the composition of fungal allergens, the pathophysiology of complex immune responses to mould exposure, and determining the exact contributory role of toxic compounds produced by fungi and bacteria will enhance our ability to accurately diagnose and effectively treat patients that are sensitised to the microorganism(s) involved.

REFERENCES

Abeck, D., and Mempel, M. (1998) *Staphylococcus aureus* colonization in atopic dermatitis and its therapeutic implications. *Br. J. Dermatol.*, **139**, S13-S16.

Achatz, G., Oberkofler, H., Lechenauer, E., *et al.* (1995) Molecular cloning of major and minor allergens of *Alternaria alternata* and *Cladosporium herbarum*. *Mol. Immunol.*, **32**, 213-227.

Adachi, Y., Okazaki, M., Ohno, N., and Yadomae, T. (1994) Enhancement of cytokine production by macrophages stimulated with (1-3)-β-D-glucan, grifolan (GRN), isolated from *Grifola frondosa*. *Biol. Pharm. Bull.*, **17**, 1554-1560.

Allergen Nomenclature (2008) Online at http://www.allergen.org/Allergen.aspx. Last update November 01, 2008.

Allergome (2008) Online at http://www.allergome.com. Accessed November 2008.

Aly, R., Maibach, H., and Shinefield, H. (1977) Microbial flora of atopic dermatitis. *Arch. Dermatol.*, **113**, 780-782.

Anon. (1996) Molds, fungi cause sick building syndrome. *Occup. Health Safety*, **65**, 134.

Andersson, M.A., Nikulin, M., Koljalg, U., *et al.* (1997). Bacteria, molds, and toxins in water-damaged building materials. *Appl. Environ. Microbiol.*, **63**, 387-393.

Andersson, A., Scheynius, A., and Rasool, O. (2003) Detection of Mala f and Mala s allergen sequences within the genus *Malassezia*. *Med. Mycol.*, **41**, 479-485.

Andriessen, J.W., Brunekreef, B., and Roemer, W. (1998) Home dampness and respiratory health status in European children. *Clin. Exp. Allergy*, **28**, 1191-1200.

Arif, A.A., and Shah, S.M. (2007) Association between personal exposure to volatile organic compounds and asthma among US adult population. *Int. Arch. Occup. Environ. Health*, **80**, 711-719.

Arruda, L.K., Platts-Mills, T.A., Fox, J.W., and Chapman, M.D. (1990) *Aspergillus fumigatus* allergen I, a major IgE-binding protein, is a member of the mitogillin family of cytotoxins. *J. Exp. Med.*, 172, 1529-1532.

Ashbee, H.R., and Evans, E.G. (2002) Immunology of diseases associated with *Malassezia* species. *Clin. Microbiol. Rev.*, **15**, 21-57.

Atkinson, R.W., Strachan, D.P., Anderson, H.R., *et al.* (2006) Temporal associations between daily counts of fungal spores and asthma exacerbations. *Occup. Environ. Med.*, **63**, 580-590.

Aukrust, L. (1979) Crossed radioimmunoelectophoretic studies of distinct allergens in two extracts of *Cladosporium herbarum*. *Int. Arch. Allergy Appl. Immunol.*, **58**, 375-390.

Aukrust, L., and Borch, S.B. (1979) Partial purification and characterization of two *Cladosporium herbarum* allergens. *Int. Arch. Allergy Appl. Immunol.*, **60**, 68-79.

Bäck, O., and Bartosik, J., (2001) Systemic ketoconazole for yeast allergic patients with atopic dermatitis. *J. Eur. Acad. Dermatol. Venereol.*, **15**, 34-38.

Bäck, O., Scheynius, A., and Johansson, S.G. (1995) Ketoconazole

in atopic dermatitis: therapeutic response is correlated with decrease in serum IgE. *Arch. Dermatol. Res.*, **287**, 448-451.

Baker, B.S. (2006) The role of microorganisms in atopic dermatitis. *Clin. Exp. Immunol.*, **144**, 1-9.

Baldo, B.A., and Baker, P.S. (1988) Inhalant allergies to fungi: Reactions to baker's yeast (*Saccharomyces cerevisiae*) and identification of baker's yeast enolase as an important allergen. *Int. Arch. Allergy Appl. Immunol.*, **86**, 201-218.

Banerjee, B., Greenberger, P.A., Fink, J.N., and Kurup, V.P. (1998) Immunological characterization of Asp f 2, a major allergen from *Aspergillus fumigatus* associated with allergic bronchopulmonary aspergillosis. *Infect. Immun.*, **66**, 5175-5182.

Banerjee, B., Kurup, V.P., Greenberger, P.A., *et al.* (2001) Cloning and expression of *Aspergillus fumigatus* allergen Asp f 16 mediating both humoral and cell-mediated immunity in allergic bronchopulmonary aspergillosis (ABPA). *Clin. Exp. Allergy*, **31**, 761-770.

Bassichis, B.A., Marple, B.F., Mabry, R.L., *et al.* (2001) Use of immunotherapy in previously treated patients with allergic fungal sinusitis. *Otolaryngol. Head Neck Surg.*, **125**, 487-490.

Baur, X., Fruhmann, G. Haug, B., *et al.* (1986) Role of *Aspergillus amylase* in baker's asthma. *Lancet*, **1**, 43.

Baur, X., Sauer, W., and Weiss, W. (1988) Baking additives as new allergens in baker's asthma. *Respiration*, **54**, 70-72.

Baur, X., Chen, Z., and Sander, I. (1994) Isolation and denomination of an important allergen in baking additives: alpha-amylase from *Aspergillus oryzae* (Asp o II). *Clin. Exp. Allergy*, **24**, 465-470.

Becher, R., Hongslo, J.K., Jantunen, M.J., and Dybing, E. (1996) Environmental chemicals relevant for respiratory hypersensitivity: the indoor environment. *Toxicol. Lett.*, **86**, 155-162.

Beijer, L., and Rylander, R. (2005) (1→3)-β-D-glucan does not induce acute inflammation after nasal deposition. *Mediators Inflamm.*, **1**, 50–52.

Bent, J.P., and Kuhn, F.A. (1994) Diagnosis of allergic fungal sinusitis. *Otolaryngol. Head Neck Surg.*, **111**, 580-588.

Bernardis, P., Agnoletto, M., Puccinelli, P., *et al.* (1996) Injective versus sublingual immunotherapy in *Alternaria tenuis* allergic patients. *J. Invest. Allergol. Clin. Immunol.*, **6**, 55-62.

Bernstein, J.A., Levin, L., Crandall, M.S., *et al.* (2005) A pilot study to investigate the effects of combined dehumidification and HEPA filtration on dew point and airborne mold spore counts in day care centers. *Indoor Air*, **15**, 402-407.

Billings, C.G., Howard, P. (1998) Damp housing and asthma. *Monaldi Arch. Chest Dis.*, **53**, 43-49.

Bisht, V., Sing, B.P., Gaur, S.N., *et al.* (2000) Allergens of *Epicoccum nigrum* grown in different media for quality source material. *Allergy*, **55**, 274-280.

Bisht, V., Singh, B.P., Kumar, R., *et al.* (2002) Culture filtrate antigens and allergens of *Epicoccum nigrum* cultivated in modified semi-synthetic medium. *Med. Microbiol. Immunol.*, **191**, 11-15.

Bisht, V., Arora, N., Singh, B.P., *et al.* (2004a) Purification and characterization of a major cross-reactive allergen from *Epicoccum purpurascens*. *Int. Arch. Allergy Immunol.*, **133**, 217-224.

Bisht, V., Arora, N., Singh, B.P, *et al.* (2004b) Epi p 1, an allergenic glycoprotein of *Epicoccum purpurascens* is a serine protease. *FEMS Immunol. Med. Microbiol.*, **42**, 205-211.

Black, P.N., Udy, A.A., and Brodie, S.M. (2000) Sensitivity to fungal allergens is a risk factor for life-threatening asthma. *Allergy*, **55**, 501-504.

Blackley, D.H. (1873) *Experimental Researches on The Cause and Nature of Catarrhus Aestivus*. Bailliere Tindall and Cox, London.

Blanco Carmona, J.G., Juste Picon, S., and Garces Sotillos, M. (1991) Occupational asthma in bakeries caused by sensitivity to alpha amylase. *Allergy*, **46**, 274-276.

Bonner, S., Matte, T.D., Fagan, J., *et al.* (2006) Self-reported moisture or mildew in the homes of head start children with asthma is associated with greater asthma morbidity. *J. Urban Health*, **83**, 129-137.

Boulet, L.P., Turcotte, H., Laprise, C., *et al.* (1997) Comparative degree and type of sensitization to common indoor and outdoor allergens in subjects with allergic rhinitis and/or asthma. *Clin. Exp. Allergy*, **27**, 52-59.

Bousquet, J., and Michel, F.B. (1994) Specific immuno-therapy in asthma: is it effective? *Allergy Clin. Immunol.*, **94**, 1-11.

Bousquet, J., Lockey, R.F., and Malling, H.G. (1998) WHO Position Paper. Allergen Immunotherapy: therapeutic vaccines for allergic diseases. *Allergy*, **53**, 1-42.

Boutin-Forzano, S., Charpin-Kadouch, C., Chabbi, S., *et al.* (2004) Wall relative humidity: A simple and reliable index for predicting *Stachybotrys chartarum* infestation in dwellings. *Indoor Air*, **14**, 196-199.

Bowler, S.D., Mitchel, C.A., and Miles, J. (1985) House dust control and asthma: a placebo control trial of cleaning air filtration. *Ann. Allergy*, **55**, 498-500.

Brandwein, M. (1993) Histopathology of sinonasal fungal disease. *Otolaryngol. Clin. N. Amer.*, **26**, 949-981.

Brant, A. (2007) Baker's asthma. *Curr. Opin. Allergy Clin. Immunol.*, **7**, 152-155.

Braun-Fahrländer, C.H., Gassner, M., Grize, L., *et al.* (1999) Prevalence of hay fever and allergic sensitization in farmer's children and their peers living in the same rural community. *Clin. Exp. Allergy*, **29**, 28-34.

Breitenbach, M., and Simon-Nobbe, B. (2002) The allergens of *Cladosporium herbarum* and *Alternaria alternata*. *Chem. Immunol.*, **81**, 48-72.

Brisman, J. (2002) Baker's asthma. *Occup. Environ. Med.*, **59**, 498–502.

Brisman, J., Jarvholm, B., Lillienberg, L., *et al.* (2000) Exposure-response relations for self reported asthma and rhinitis in bakers. *Occup. Environ. Med.*, **57**, 335-340.

Broberg, A., and Faergemann, J. (1995) Topical antimycotic treatment of atopic dermatitis in the head/neck area. A double-blind randomized study. *Acta Derm. Venereol.*, **75**, 46-49.

Bruynzeel-Koomen, C.A.F., VanWichen, D.F., Spry, C.J.F., *et al.* (1988) Active participation of eosinophils in patch test reactions to inhalant allergens in patients with atopic dermatitis. *Br. J. Dermatol.*, **118**, 233-247.

Burge, H.A. (1985) Fungus allergens. *Clin. Rev. Allergy*, **3**, 319-329.

Burr, M.L. (2001) Health effects of indoor molds. *Rev. Environ. Health*, **16**, 97-103.

Burr, M.L., Dean, B.V., Merrettt, T.G., *et al.* (1980) Effects of anti-mite measures on children with mite-sensitive asthma. A controlled trial. *Thorax*, **35**, 506-512.

Burr, M.L., Matthews, I.P., Arthur, R.A., *et al.* (2007) Effects on patients with asthma of eradicating visible indoor mould: a randomized controlled trial. *Thorax*, **62**, 767-772.

Burrell, R. (1994) Human responses to bacterial endotoxin. *Circ. Shock*, **43**, 137-153.

Bush, R.K., and Yuninger, J.W. (1987) Standardization of fungal allergens. *Clin. Rev. Allergy*, **5**, 3-21.

Bush, R.K., and Prochnau, J.J. (2004) *Alternaria*-induced asthma. *J. Allergy Clin. Immunol.*, **113**, 227-234.

Bush, R.K., Portnoy, J.M., Saxon, A., *et al.* (2006) The medical effects of mold exposure. *J. Allergy Clin. Immunol.*, **117**, 326-333.

Callabria, C.W., and Dice, J. (2007) Aeroallergen sensitization rates in military children with rhinitis symptoms. *Ann. Allergy Asth-*

ma Immunol., **99**, 161-169.

Camarena, A., Juárez, A., Mejía, M., *et al.* (2001) Major histocompatibility complex and tumor necrosis factor-α polymorphisms in pigeon breeder's disease. *Am. J. Respir. Crit. Care Med.*, **163**, 1528-1533.

Cantani, A., Businco, E., and Maglio, A. (1988) *Alternaria* allergy: a three-year controlled study in children treated with immunotherapy. *Allergol. Immunopathol. (Madrid)*, **16**, 1-4.

Cantani, A., Arcese, G., Lucenti, P., *et al.* (1997) A three-year prospective study of specific immunotherapy to inhalant allergens: evidence of safety and efficacy in 300 children with allergic asthma. *J. Investig. Allergol. Clin. Immunol.*, **7**, 90-97.

Cappellaro, C., Mrsa, V., and Tanner, W. (1998) New potential cell wall glucanases of *Saccharomyces cerevisiae* and their involvement in mating. *J. Bacteriol.*, **180**, 5030-5037.

Castellan, R.M., Olenchock, S.A., Hankinson, J.L., *et al.* (1984) Acute bronchoconstriction induced by cotton dust: dose-related responses to endotoxin and other dust factors. *Ann. Int. Med.*, **101**, 157-163.

Castellan, R.M., Olenchock, S.A., Kinsley, K.B., and Hankinson, J.L. (1987) Inhaled endotoxin and decreased spirometric values. An exposure-response relation for cotton dust. *N. Engl. J. Med.*, **131**, 605-610.

CDC. (2000) Update: Pulmonary hemorrhage/hemosiderosis among infants - Cleveland, Ohio, 1993-1996. *MMWR*, **49**, 180-184.

CDC. (2004) Investigation of acute idiopathic pulmonary hemorrhage among infants - Massachusetts, December 2002-June 2003. *MMWR*, **53**, 817-820.

Cevit, O., Kendirli, S.G., Yilmaz, M., *et al.* (2007) Specific allergen immunotherapy: effect on immunologic markers and clinical parameters in asthmatic children. *J. Investig. Allergol. Clin. Immunol.*, **17**, 286-291.

Ceylan, E., Ozkutuk, A., Ergo, G., *et al.* (2006) Fungi and indoor conditions in asthma patients. *J. Asthma*, **10**, 789-794.

Champion, R.H. (1971) Atopic sensitivity to algae and lichens. *Br. J. Derm.*, **875**, 551-557.

Chang, Y.T., and Fang, S.Y. (2008) Tissue-specific immunoglobulin E in maxillary sinus mucosa of allergic fungal sinusitis. *Rhinology*, **46**, 226-230.

Chapman, J.A., Terr, A.I., Jacobs, R.L., *et al.* (2003) Toxic mold: phantom risk vs science. *Ann. Allergy Asthma Immunol.*, **91**, 222-232.

Chauhan, B., Santiago, L., Hutcheson, P.S., *et al.* (2000) Evidence for the involvement of two different MHC class II regions in susceptibility or protection in allergic bronchopulmonary aspergillosis. *J. Allergy Clin. Immunol.*, **106**, 723-729.

Cho, S.H., Reponen, T., LeMasters, G., *et al.* (2006a) Mold damage in homes and wheezing in infants. *Ann. Allergy Asthma Immunol.*, **97**, 539-545.

Cho, S.H., Reponen, T., Bernstein, D.I., *et al.* (2006b) The effect of home characteristics on dust antigen concentrations and loads in homes. *Sci. Total Environ.*, **371**, 31-43.

Chou, H., Lin, W.L., Tam, M.F., *et al.* (1999) Alkaline serine proteinase is a major allergen of *Aspergillus flavus*, a prevalent airborne *Aspergillus* species in the Taipei area. *Int. Arch. Allergy Immunol.*, **119**, 282-290.

Chou, H., Lai, H.Y., Tam, M.F., *et al.* (2002) cDNA cloning, biological and immunological characterization of the alkaline serine protease major allergen from *Penicillium chrysogenum*. *Int. Arch. Allergy Immunol.*, **127**, 15-26.

Chua, K.Y., Stewart, G.A., Thomas, W.R., *et al.* (1988) Sequence analysis of cDNA coding for a major house dust mite allergen, Der p I. Homology with cysteine proteases. *J. Exp. Med.*, **167**,

175-182.

Chua, K.Y., Cheong, N., Kuo, I.C., *et al.* (2007) The *Blomia tropicalis* allergens. *Protein Pept. Lett.*, **14**, 325-333.

Clemmensen, O.J., and Hjorth, N. (1983) Treatment of dermatitis of the head and neck with ketoconazole in patients with type 1 sensitivity to *Pitryosporum orbiculare*. *Semin. Dermatol.*, **2**, 26-29.

Cole Johnson, C., Ownby, D.R., Havstad, S.L., and Peterson, E.L. (2004) Family history, dust mite exposure in early childhood, and risk for pediatric atopy and asthma. *J. Allergy Clin. Immunol.*, **114**, 105-110.

Collins, M.M., Nair, S.B., and Wormald, P.J. (2003) Prevalence of noninvasive fungal sinusitis in South Australia. *Am. J. Rhinol.*, **17**, 127-132.

Collins-Williams, C., Nizami, R.M., Lamenza, C., and Chiu, A.W. (1972) Nasal provocative testing with molds in the diagnosis of perennial allergic rhinitis. *Ann. Allergy*, **30**, 557-561.

Cools, M., Van Bever, H.P., Weyler, J.J., and Stevens, W.J. (2000) Long-term effects of specific immunotherapy, administered during childhood, in asthmatic patients allergic to either house-dust mite or to both house-dust mite and grass pollen. *Allergy*, **55**, 69-73.

Corey, J.P., Delsupehe, K.G., and Ferguson, B.J. (1995) Allergic fungal sinusitis: allergic, infectious or both? *Otolaryngol. Head Neck Surg.*, **113**, 110-119.

Crameri, R. (1998) Recombinant *Aspergillus fumigatus* allergens: from the nucleotide sequences to clinical applications. *Int. Arch. Allergy Immunol.*, **115**, 99-114.

Criado Molina, A., Guerra Pasadas, F., Daza Muñoz, J.C. *et al.* (2002) Immunotherapy with an oral *Alternaria* extract in childhood asthma. Clinical safety and efficacy and effects on *in vivo* and *in vitro* parameters. *Allergol. Immunopathol. (Madrid)*, **30**, 319-330.

Croft, W.A., Jarvis, B.B., and Yatawara, C.S. (1986) Airborne outbreak of trichothecene toxicosis. *Atmos. Environ.*, **20**, 549-552.

Cruz, A., Saenz de Santamaria, M., Pagan, J., *et al.* (1995) Are fungal spores the main sensitizing components in mold allergy? *Allergy*, **50**, 346. [Abstract].

Cruz, A., Saenz de Santamaria, M., Martinez, J., *et al.* (1997) Fungal allergens from important allergenic fungi imperfecti. *Allergol. Immunopathol.*, **25**, 153-158.

Cuijpers, C.E., Swaen, G.M., Wesseling, G., *et al.* (1995) Adverse effects of the indoor environment on respiratory health in primary school children. *Environ. Res.*, **68**, 11-23.

Cullinan, P., Lowson, D., Nieuwenhuijsen, M.J., *et al.* (1994) Work related symptoms, sensitisation, and estimated exposure in workers not previously exposed to flour. *Occup. Environ. Med.*, **51**, 579-583.

Custovic, A., and Woodcock, A. (1996) Allergen avoidance. *Br. J. Hosp. Med.*, **56**, 409-412.

Custovic, A., Taggart, S.C.O., Francis, H.C., *et al.* (1996) Exposure to house dust mite allergens and the clinical activity of asthma. *J. Allergy Clin. Immunol.*, **98**, 64-72.

Custovic, A., and Chapman, M. (1998) Risk levels for mite allergens. Are they meaningful? *Allergy*, **53**, 71-76.

Custovic, A., Simpson, A., and Woodcock, A. (1998a) Importance of indoor allergens in the induction of allergy and elicitation of allergic disease. *Allergy*, **53**, 115-120.

Custovic, A., Simpson, A., Chapman, M.D., and Woodcock, A. (1998b) Allergen avoidance in the treatment of asthma and atopic disorders. *Thorax*, **53**, 63-72.

Cusumano, V., Rossano, F., Merendino, R.A., *et al.* (1996) Immunobiological activities of mould products: functional impair-

ment of human monocytes exposed to aflatoxin B_1. *Res. Microbiol.*, **147**, 385-391.

Dales, R.E., Burnett, R., and Zwanenburg, H. (1991) Adverse effects in adults exposed to home dampness and molds. *Amer. Rev. Respir. Dis.*, **143**, 505-509.

Dales, R.E., Cakmak, S., Judek, S., *et al.* (2003) The role of fungal spores in thunderstorm asthma. *Chest*, **123**, 745-750.

Dales, R.E., Cakmak S., Judek S., *et al.* (2004) Influence of outdoor aeroallergens on hospitalization for asthma in Canada. *J. Allergy Clin. Immunol.*, **113**, 303-306.

Day, J.H. (1996) Allergic responses to fungi. In D. Howard and J.D. Miller, (eds.), *The Mycota*, Vol. 6, *Human and Animal Relationships*. Springer Verlag, Berlin, pp. 173-192.

Day, J.H., and Ellis, A.K. (2001) Allergenic microorganisms and hypersensitivity. In B. Flannigan, R.A. Samson, and J.D. Miller, (eds.), *Microorganisms in Home and Indoor Work Environments*. Taylor & Francis, London, pp. 103-127.

de Ana, S.G., Torres-Rodríguez, J.M., Ramírez, E.A., *et al.* (2006) Seasonal distribution of *Alternaria, Aspergillus, Cladosporium* and *Penicillium* species isolated in homes of fungal allergic patients. *J. Investig. Allergol. Clin. Immunol.*, **16**, 357-363.

Dearborn, D.G., Infield, M.D., Smith, P., *et al.* (1994) Acute pulmonary hemorrhage/hemosiderosis among infants – Cleveland, January 1993-November 1994. *MMWR*, **43**, 881-883.

Dearborn, D.G., Infield, M.D., Smith, P., *et al.* (1997) Update: pulmonary hemorrhage/hemosiderosis among infants – Cleveland, Ohio, 1993-1996. *MMWR*, **46**, 33-35.

Dearborn, D.G., Yike, I., Sorenson, W.G., *et al.* (1999) Overview of investigations into pulmonary hemorrhage among infants in Cleveland, Ohio. *Environ. Health Perspect.*, **107**, 495-499.

Dearborn, D.G., Smith, P.G., Dahms, B.B., *et al.* (2002) Clinical profile of 30 infants with acute pulmonary hemorrhage in Cleveland. *Pediatrics*, **110**, 627-637.

Dekker, H. (1928) Asthma und Milben. *Münch. Med. Wochensch.*, **75**, 515-516.

Delfino, R.J., Coate, B.D., Zeiger, R.S., *et al.* (1996) Daily asthma severity in relation to personal ozone exposure and outdoor fungal spores. *Amer. J. Respir. Crit. Care Med.*, **154**, 633-641.

Delfino, R.J., Zeiger, R.S., Seltzer, J.M., *et al.* (1997) The effect of outdoor fungal spore concentrations on daily asthma severity. *Environ. Health Perspect.*, **105**, 622-635.

Denning, D.W., O'Driscoll, B.R., Hogaboam, C., *et al.* (2006) The link between fungi and severe asthma: a summary of the evidence. *Eur. Respir. J.*, **27**, 615-626.

Derendorf, H., and Meltzer, E.O. (2008) Molecular and clinical pharmacology of intranasal corticosteroids: clinical and therapeutic implications. *Allergy*, **63**, 1292-1300.

deShazo, R.D., and Swain, R.E. (1995) Diagnostic criteria for allergic fungal sinusitis. *J. Allergy Clin. Immunol.*, **96**, 24-35.

deShazo, R.D., Chapin, K., and Swain, R.E. (1997) Fungal sinusitis. *N. Engl. J. Med.*, **337**, 254-259.

Deuell, B., Arruda, L.K., Hayden, M.L., *et al.* (1991) *Trichophyton tonsurans* allergen I: characterization of a protein that causes immediate but not delayed hypersensitivity. *J. Immunol.*, **147**, 96-101.

De Vouge, M.W., Thaker, A.J., Curran, I.H., *et al.* (1996) Isolation and expression of a cDNA clone encoding an *Alternaria alternata* Alt a 1 subunit. *Int. Arch. Allergy Immunol.*, **111**, 385-395.

De Vouge, M.W., Thaker, A.J., Zhang, L., *et al.* (1998) Molecular cloning of IgE-binding fragments of *Alternaria alternata* allergens. *Int. Arch. Allergy Immunol.*, **116**, 261-268.

De Zotti, R., Larese, F., Bovenzi, M., *et al.* (1994) Allergic airway disease in Italian bakers and pastry makers. *Environ. Med.*, **51**,

548-552.

Dharmage, S., Bailey, M., Raven, J., *et al.* (1999) Prevalence and residential determinants of fungi within homes in Melbourne. Australia. *Clin. Exp. Allergy*, **29**, 1481-1489.

Dharmage, S., Bailey, M., Raven, J., *et al.* (2002) Mouldy houses influence symptoms of asthma among atopic individuals. *Clin. Exp. Allergy*, **32**, 714-720.

Di Rienzo, V.D., Minelli, M., Musarra, A., *et al.* (2005) Post-marketing survey on the safety of sublingual immunotherapy in children below the age of 5 years. *Clin. Exp. Allergy*, **35**, 560-564.

Dietemann, A., Bessot, J.C., Hoyer, C., *et al.* (1993) A double blind placebo controlled trial of solidified benzyl benzoate applied to dwellings of asthmatic patients sensitive to mites: clinical efficacy and effect on mite allergens. *J. Allergy Clin. Immunol.*, **91**, 738-746.

Diette, G.B., McCormack, M.C., Hansel, N.N., *et al.* (2008) Environmental issues in managing asthma. *Respir. Care*, **53**, 602-615.

Dilworth, R.J., Chua, K.Y., and Thomas, W.R. (1991) Sequence analysis of cDNA coding for a major house dust mite allergen, Der f 1. *J. Clin. Exp. Allergy*, **21**, 25-32.

Douwes, J. (2005) (1→3)-β-D-glucans and respiratory health: a review of the scientific evidence. *Indoor Air*, **15**, 160-169.

Douwes, J., Zuidhof, A., Doekes, G., *et al.* (2000) (1→3)-β-D-glucan and endotoxin in house dust and peak flow variability in children. *Am. J. Respir. Crit. Care Med.*, **162**, 1348-1354.

Douwes, J., Pearce, N., and Heederik, D. (2002) Does environmental endotoxin exposure prevent asthma? *Thorax*, **57**, 86-90.

Downs, S.H., Mitakakis, T.Z., Marks, G.B., *et al.* (2001a) Clinical importance of *Alternaria* exposure in children. *Am. J. Respir. Crit. Care Med.*, **164**, 455-459.

Downs, S.H., Marks, G.B., Mitakakis, T.Z., *et al.* (2001b) Having lived on a farm and protection against allergic diseases in Australia. *Clin. Exp. Allergy*, **31**, 570-575.

Dreborg, S., Agrell, B., Foucard, T., *et al.* (1996) A double-blind, multi-center immunotherapy trial in children, using a purified and standardized *Cladosporium herbarum* preparation. *Allergy*, **41**, 131-140.

Ehnert, B., Lau-Schadendorf, S., Weber, A., *et al.* (1992) Reducing domestic exposure to dust mite allergen reduces bronchial hyperreactivity in sensitive children with asthma. *J. Allergy Clin. Immunol.*, **90**, 135-138.

Ence, B.K., Gourley, D.S., Jorgensen, N.L., *et al.* (1990) Allergic fungal sinusitis. *Am. J. Rhinol.*, **4**, 169-178.

Ergon, M.C., Ílknur, T., Yućesoy, M., and Oźkan, S. (2007) *Candida* spp. colonization and serum anticandidal antibody levels in patients with chronic urticaria. *Clin. Exp. Dermatol.*, **32**, 740-743.

Ernst, P., and Cormier, Y. (2000) Relative scarcity of asthma and atopy among rural residents raised on a farm. *Am. J. Respir. Crit. Care Med.*, **161**, 1563-1566.

Esch, R.E. (2004) Manufacturing and standardizing fungal allergen products. *J. Allergy Clin. Immunol.*, **113**, 210-215.

Etzel, R., Montana, E., Sorenson, W.G., *et al.* (1998) Acute pulmonary hemorrhage in infants associated with exposure to *Stachybotrys atra* and other fungi. *Arch. Pediatr. Adolesc. Med.*, **152**, 757-762.

Evans, R. (1992) Environmental control and immunotherapy for allergic disease. *J. Allergy Clin. Immunol.*, **90**, 462-468.

Faergemann, J. (2002) Atopic dermatitis and fungi. *Clin. Microbiol.*, **15**, 545-563.

Fenoglio, C.-M., Reboux, G., Sudre, B., *et al.* (2007) Diagnostic value of serum precipitins to mould antigens in active hypersensitivity pneumonitis. *Eur. Resp. J.*, **29**, 706-712.

Fernández-Caldas, E., Iraola, V., Boquete, M., *et al.* (2006) Mite immunotherapy. *Curr. Allergy Asthma Rep.*, **6**, 413-419.

Fink, J.N., and Zacharisen, M.C. (2003) Hypersensitivity pneumonitis. In N.F. Adkinson Jr., J.W. Yunginger, W.W. Busse, *et al.*, (eds.), *Allergy: Principles and Practice*. Mosby, St. Louis, MO, pp. 1373-1390.

Fischer, G., and Dott, W. (2003) Relevance of airborne fungi and their secondary metabolites for environmental, occupational and indoor hygiene. *Arch. Microbiol.*, **179**, 75-82.

Fisk, W.J., Lei-Gomez, Q., and Mendell, M.J. (2007) Meta-analyses of the associations of respiratory health effects with dampness and mold in homes. *Indoor Air*, **17**, 284-296.

Flückiger, S., Mittl, P.R., Scapozza, L., *et al.* (2002a) Comparison of the crystal structures of the human manganese superoxide dismutase and the homologous *Aspergillus fumigatus* allergen at 2-Å resolution. *J. Immunol.*, **168**, 1267-1272.

Flückiger, S., Scapozza, L., Mayer, C., *et al.* (2002b) Immunological and structural analysis of IgE-mediated cross-reactivity between manganese superoxide dismutases. *Int. Arch. Allergy Immunol.*, **128**, 292-303.

Flückiger, S., Fijten, H., Whitley, P., *et al.* (2002c) Cyclophilins, a new family of cross-reactive allergens. *Eur. J. Immunol.*, **32**, 10-17.

Fung, F., Clark, R., and Williams, S. (1998) *Stachybotrys*, a mycotoxin-producing fungus of increasing toxicologic importance. *J. Toxicol. Clin. Toxicol.*, **36**, 79-86.

Galland, C., Reynoud, C., De Haller, R., *et al.* (1991) Cheese-washer's disease. A current stable form of extrinsic allergic alveolitis in a rural setting. *Rev. Mal. Respir.*, **8**, 381-386. [In French].

Gautam, P., Sundaram, C.S., Madan, T., *et al.* (2007) Identification of novel allergens of *Aspergillus fumigatus* using immunoproteomics approach. *Clin. Exp. Allergy.*, **37**, 1239-1249.

Gehring, U., Heinrich, J., Hoek, G., *et al.* (2007) Bacteria and mould components in house dust and children's allergic sensitization. *Eur. Resp. J.*, **29**, 1144-1153.

Gereda, J.E., Leung, D.Y.M., Thatayatikom, A., *et al.* (2000) Relation between house-dust endotoxin exposure, type 1 T-cell development, and allergen sensitisation in infants at high risk of asthma. *Lancet*, **355**, 1680-1683.

Goldstein, G.B. (2006) Adverse reactions to fungal metabolic products in mold-contaminated areas. *J. Allergy Clin. Immunol.*, **118**, 760-761.

Goldstein, M.F., Atkins, P.C., Cogen, F.C., *et al.* (1985) Allergic *Aspergillus* sinusitis. *J. Allergy Clin. Immunol.*, **76**, 515-524.

Gonianakis, M.I., Neonakis, I.K., Gonianakis, I.M., *et al.* (2006) Mold allergy in the Mediterranean Island of Crete, Greece: a 10-year volumetric, aerobiological study with dermal sensitization correlations. *Allergy Asthma Proc.*, **27**, 354-362.

Gøtzsche, P.C., and Johansen, H.K. (2008) House dust mite control measures for asthma. *Cochrane Database Syst. Rev.*, **2**, CD001187.

Gøtzsche, P.C., Johansen, H.K., Burr, M.L., and Hammarquist, C. (2001) House dust mite control measures for asthma. *Cochrane Database Syst. Rev.*, **3**, CD001187.

Gøtzsche, P.C., Johansen, H.K., Schmidt, L.M., and Burr, M.L. (2004) House dust mite control measures for asthma. *Cochrane Database Syst. Rev.*, 4, CD001187.

Górny, R.L., and Dutkiewicz, J. (2002) Bacterial and fungal aerosols in indoor environment in Central and Eastern European countries. *Ann. Agric. Environ. Med.*, **9**, 17-23.

Gourley, D.S., Whisman, B.A., Jorgensen, N.L., *et al.* (1990) Allergic *Bipolaris* sinusitis: clinical and immunopathologic characteristics. *J. Allergy Clin. Immunol.*, **85**, 583-591.

Granel, C., Tapias, M., Valencia, L., *et al.* (1993) Allergy to *Alternaria*.

I. Clinical aspects. *Allergol. Immunopathol.*, **21**, 15-19.

Gravesen, S., Nielsen, P.A., Iversen, R., and Nielsen, K.F. (1999) Microfungal contamination of damp buildings – examples of risk constructions and risk materials. *Environ. Health Perspect.*, **107**, S505-S508.

Greenberger, P.A. (2003) Allergic bronchopulmonary aspergillosis. In D.A. Stevens, R.B. Moss, and V.P. Kurup, (eds.), *Middleton's Allergy: Principles and Practice*, 6th ed., Mosby, Philadelphia, PA, pp. 1353-1371.

Guillot, J., Hadina, S., and Guého, E. (2008) The genus *Malassezia*: old facts and new concepts. *Parasitologia*, **50**, 77-79.

Gumowski, P.I. (1997) Hypersensitivity to airborne moulds: diagnostics and therapeutic approaches. *Expressions*, **5**, 9-14.

Gumowksi, P., Lech, B., Chaves, I., and Girard, J.-P. (1987) Chronic asthma and rhinitis due to *Candida albicans*, *Epidermophyton*, and *Trichophyton*. *Ann. Allergy*, **59**, 48-51.

Gupta, A.K., Kohli, Y., Summerbell, R.C., and Faergemann, J. (2001) Quantitative culture of *Malassezia* species from different body sites of individuals with or without dermatoses. *Med. Mycol.*, **39**, 243-251.

Gupta, R., Singh, B.P., Sridhara, S., *et al.* (2002) Allergenic cross-reactivity of *Curvularia lunata* with other airborne fungal species. *Allergy*, **57**, 636-640.

Gutman, A.A., and Bush, R.K. (1993) Allergens and other factors important in atopic disease. In R. Patterson, C.R. Zeiss, L.C. Grammar, and P.A. Greenberger, (eds.), *Allergic Disease: Diagnosis and Management*, 4th ed., Lippincott, Philadelphia, PA, pp. 93-158.

Haglind, P., and Rylander, R. (1984) Exposure to cotton dust in an experimental cardroom. *Br. J. Ind. Med.*, **41**, 340-345.

Hagmolen of ten Have, W., van den Berg, N.J., van der Palen, J., *et al.* (2007) Residential exposure to mould and dampness is associated with adverse respiratory health. *Clin. Exp. Allergy*, **37**, 1827-1832.

Hagy, G.W., and Settipane, G.A. (1976) Risk factors for developing asthma and allergic rhinitis. *J. Allergy Clin. Immunol.*, **58**, 330-336.

Halonen, M., Stern, D.A., Wright, A.L., *et al.* (1997) *Alternaria* as a major allergen for asthma in children raised in a desert environment. *Am. J. Respir. Crit. Care Med.*, **155**, 1356-1361.

Halpin, D.M., Graneek, B.J., Lacey, J., *et al.* (1994) Respiratory symptoms, immunologic responses, and aeroallergen concentrations at a saw mill. *Occup. Environ. Med.*, **51**, 165-172.

Hammarquist, C., Burr, M.L., and Gøtzsche, P.C. (2000) House dust mite control measures for asthma. *Cochrane Database Syst. Rev.*, **2**, CD001187.

Hanak, V., Golbin, J.M., and Ryu, J.H. (2007) Causes and presenting features in 85 consecutive patients with hypersensitivity pneumonitis. *Mayo Clin. Proc.*, **87**, 812-816.

Hanifin, J.M., and Rogge J.L. (1977) Staphylococcal infections in patients with atopic dermatitis. *Arch. Dermatol.*, **113**, 1383-1386.

Hardin, B.D., Kelman, B.J., and Saxon, A. (2003) Adverse human health effects associated with molds in the indoor environment. *J. Occup. Environ. Med.*, **45**, 470-478.

Hartl, D., Latzin, P., Zissel, G., *et al.* (2006) Chemokines indicate allergic bronchopulmonary aspergillosis in patients with cystic fibrosis. *Am. J. Resp. Crit. Care Med.*, **173**, 1370-1376.

Harving, H., Dahl, R., and Mølhave, L. (1991) Lung function and bronchial reactivity in asthmatics during exposure to volatile organic compounds. *Am. Rev. Respir. Dis.*, **143**, 751-754.

Hasnain, S.M., Wilson, J.D., and Newhook, F. (1985) Fungal allergy and respiratory disease. *N. Z. Med. J.*, **98**, 342-346.

Hauser, C., Wuethrich, B., Matter, L., *et al.* (1985) *Staphylococcus aureus* skin colonization in atopic dermatitis patients. *Dermatologica*, **170**, 35-39.

Hay, D.B. (1991) Ecology of the house dust mite *Dermatophagoides pteronyssinus* (Trouessart). DPhil Thesis, University of Oxford, UK.

Hay, D.B., Hart, B.J., and Douglas, A.E. (1992) Evidence refuting the contribution of the fungus *Aspergillus penicillioides* to the allergenicity of the house dust mite *Dermatophagoides pteronyssinus*. *Int. Arch. Allergy Immunol.*, **97**, 86-88.

Hay, D.B., Hart, B.J., and Douglas, A.E. (1993) Effects of the fungus *Aspergillus penicillioides* on the house dust mite *Dermatophagoides pteronyssinus*. *Med. Vet. Entomol.*, **7**, 271-274.

Hedlin, G. (1995) The role of immunotherapy in pediatric allergic disease. *Curr. Opinion Pediatr.*, **7**, 676-682.

Hemmann, S., Blaser, K., and Crameri, R. (1997) Allergens of *Aspergillus fumigatus* and *Candida boidinii* share IgE-binding epitopes. *Am. J. Respir. Crit. Care Med.*, **156**, 1956-1962.

Henderson, A.K., Ranger, A.F., Lloyd, J., *et al.* (1984) Pulmonary hypersensitivity in the alginate industry. *Scot. Med. J.*, **29**, 90-95.

Herz, U., Bunikowski, R., and Renz, H. (1998) Role of T cells in atopic dermatitis. *Int. Arch. Allergy Immunol.*, **115**, 179-190.

Hill, M.R., Briggs, L., Montaño, M.M., *et al.* (2004) Promoter variants in tissue inhibitor of metalloproteinase-3 (TIMP-3) protect against susceptibility in pigeon's breeders' disease. *Thorax*, **59**, 586-590.

Hodgson, M., Levin, H., and Wolkoff, P. (1994) Volatile organic compounds and indoor air. *J. Allergy Clin. Immunol.*, **94**, 296-303.

Hodgson, M.J., Morey, P., Leung, W.Y., *et al.* (1998) Building-associated pulmonary disease from exposure to *Stachybotrys chartarum* and *Aspergillus versicolor*. *J. Occup. Environ. Med.*, **40**, 241-249.

Hoff, M., Ballmer-Weber, B.K., Niggemann, B., *et al.* (2003) Molecular cloning and immunological characterisation of potential allergens from the mould *Fusarium culmorum*. *Mol. Immunol.*, **39**, 965-975.

Hogan, M.B., Patterson, R., Poer, R.S., *et al.* (1996) Basement shower hypersensitivity pneumonitis secondary to *Epicoccum nigrum*. *Chest*, **110**, 854-856.

Holck, P., Sletmoen, M., Stokke, B.T., *et al.* (2007) Potentiation of histamine release by microfungal $(1\rightarrow3)$- and $(1\rightarrow6)$-β-D-glucans. *Basic Clin. Pharmacol. Toxicol.*, **101**, 455-458.

Hope, A.P., and Simon, R.A. (2007) Excess dampness and mold growth in homes: an evidence-based review of the aeroirritant effect and its potential causes. *Allergy Asthma Proc.*, **28**, 262-270.

Horner, W.E., Helbling, A., Salvaggio, J.E., and Lehrer, S.H. (1995) Fungal allergens. *Clin. Microbiol. Rev.*, **8**, 161-179.

Horst, M., Hejjaoui, A., Horst, V., *et al.* (1990) Double-blind placebo controlled rush immunotherapy with a standardized extract. *J. Allergy Clin. Immunol.*, **85**, 460-472.

Hosen, H. (1968) Lake algae as a specific allergen in respiratory allergy. *Rev. Allergy*, **22**, 477-482.

Hossain, M.A., Ahmed, M.S., and Ghannoum, M.A. (2004) Attributes of *Stachybotrys chartarum* and its association with human disease. *J. Allergy Clin. Immunol.*, **113**, 200-208.

Hoy, R.F., Pretto, J.J., van Gelderen, D., and McDonald, C.F. (2007) Mushroom worker's lung: organic dust exposure in the spawning shed. *Med. J. Aust.*, **186**, 472-474.

Hu, W., Katelaris, C.H., and Kemp, A.S. (2008) Allergic rhinitis – practical management strategies. *Aust. Fam. Physician*, **37**, 214-220.

Hubert, J., Stejskal, V., Munzbergová, Z., *et al.* (2004) Mites and fungi in heavily infested stores in the Czech Republic. *J. Econ. Entomol.*, **97**, 2144-2153.

Ide, F., Matsubara, T., Kaneko, M., *et al.* (2004) Staphylococcal enterotoxin-specific IgE antibodies in atopic dermatitis. *Pediatr. Int.*, **46**, 337-341.

Inal, A., Karakoc, G.B., Altinatas, D.U., *et al.* (2007) Effect of indoor mold concentrations on daily symptom severity of children with asthma and/or rhinitis monosensitized to molds. *J. Asthma*, **44**, 543-546.

Inal, A., Karakoc, G.B., Altinatas, D.U., *et al.* (2008) Effect of outdoor fungus concentrations on symptom severity of children with asthma and/or rhinitis monosensitized to molds. *Asian Pac. J. Allergy Immunol.*, **26**, 11-17.

Ingordo, V., D'Andria, G., D'Andria, C., and Tortora, A. (2002) Results of atopy patch tests with house dust mites in adults with 'intrinsic' and 'extrinsic' atopic dermatitis. *J. Eur. Acad. Dermatol. Venereol.*, **16**, 450-454.

Ingram, C.G., Jeffrey, I.G., Symington, I.S., and Cuthbert, O.B. (1979) Bronchial provocation studies in farmers allergic to storage mites. *Lancet 2*, 1330-1332.

Institute of Medicine (US) Committee on the Assessment of Asthma and Indoor Air (2000) *Clearing the Air: Asthma and Indoor Air Exposures.* National Academy Press, Washington, DC.

Iossifova, Y.Y., Reponen, T., Bernstein, D.I., *et al.* (2007) House dust $(1\rightarrow3)$-β-D-glucan and wheezing in infants. *Allergy*, **62**, 504-513.

IUIS/WHO Allergy Nomenclature Subcommittee. (1994) Allergen nomenclature. *Bull. World Health Organ.*, **72**, 797-806. [Reprinted with permission in *Clin. Exp. Allergy*, **25**, 27-37 (1995)].

Jaakkola, J.J., Jaakkola, N., and Ruotsalainen, R. (1993) Home dampness and molds as determinants of respiratory symptoms and asthma in pre-school children. *J. Expo. Anal. Environ. Epidemiol.*, **3**, 129-142.

Jaakkola, M.S., Nordman, H., Piipari, R., *et al.* (2002) Indoor dampness and molds and development of adult-onset asthma: a population based incident case-control study. *Environ. Health Perspect.*, **110**, 543-547.

Jaakkola, M.S., and Jaakkol, J.J. (2004) Indoor molds and asthma in adults. *Adv. Appl. Microbiol.*, **55**, 309-339.

Jakab, G.J., Hmieleski, R.R., Zarba, A., *et al.* (1994) Respiratory aflatoxicosis: suppression of pulmonary and systemic defenses in rats and mice. *Toxicol. Appl. Pharmacol.*, **12**, 198-205.

Jacobs, R.L., Andrews, C.P., and Coalson, J.J. (2005) Hypersensitivity pneumonitis: beyond classic occupational disease – changing concepts of diagnosis and management. *Ann. Allergy Asthma. Immunol.*, **95**, 115-128.

Jarvis, B.B., and Miller, J.D. (2005) Mycotoxins as harmful indoor air contaminants. *Appl. Microbiol. Biochem.*, **66**, 267-372.

Jarvis, B.B., Sorenson, W.G., Hintikka, E.L., *et al.* (1998) Study of toxin production by isolates of *Stachybotrys chartarum* and *Memnoniella echinata* isolated during a study of pulmonary hemosidersosis in infants. *Appl. Environ. Microbiol.*, **64**, 3620-3625.

Johanning, E. (1995) Health problems related to fungal exposure: The example of toxigenic *Stachybotrys chartarum atra.* In E. Johanning and C.S.Yang, (eds.), *Fungi and Bacteria in Indoor Air Environments,* Eastern New York Occupational Health Program, Latham, NY, pp. 201-208.

Johanning, E., Biagini, R., Hull, D., *et al.* (1996) Health and immunology study following exposure to toxigenic fungi (*Stachybotrys chartarum*) in a water-damaged office environment. *Int. Arch. Occup. Environ. Health*, **68**, 207-218.

Johansson, C., Eshaghi, H., Linder, M.T., *et al.* (2002) Positive atopy patch test reaction to *Malassezia furfur* in atopic dermatitis correlates with a T helper 2-like peripheral blood mononuclear cells response. *J. Invest. Dermatol.*, **118**, 1044-1051.

Johnston, S.L. (1997) Influence of viral and bacterial respiratory infections on exacerbations and symptom severity in childhood asthma. *Pediatr. Pulmonol.* (Suppl.), **16**, 88-89.

Johnston, S.L., and Martin, R.J. (2005) *Chlamydophila pneumoniae* and *Mycoplasma pneumoniae*: a role in asthma pathogenesis? *Am. J. Respir. Crit. Care Med.*, **172**, 1078-1089.

Juvonen, R., Bloigu, A., Paldanius, M., *et al.* (2008) Acute *Chlamydia pneumoniae* infections in asthmatic and non-asthmatic military conscripts during a non-epidemic period. *Clin. Microbiol. Infect.*, **14**, 207-212.

Kaad, P.H., and Oostergaard, P.A. (1982) The hazard of mould hyposensitization in children with asthma. *Clin. Allergy*, **12**, 317-320.

Kaliner, M.A. (2007) A novel and effective approach to treating rhinitis with nasal antihistamines. *Ann. Allergy Asthma Immunol.*, **99**, 383-390.

Kalra, S., Crank, P., Hepworth, J., *et al.* (1992) Absence of seasonal variation in concentrations of the house dust mite allergen Der p 1 in south Manchester homes. *Thorax*, **47**, 928-931.

Kanerva, L., Vanhanen, M., and Tupasela, O. (1997) Occupational allergic contact urticaria from fungal but not bacterial alpha-amylase. *Contact Dermatitis*, **36**, 306-307.

Kanny, G., Becker. S., deHauteclocque, C., and Moneret-Vautrin, D.A. (1996) Airborne eczema due to mould allergy. *Contact Dermatitis*, **35**, 378.

Kato, H., Sugita, T., Ishibashi, Y., and Nishikawa, A. (2006) Detection and quantification of specific IgE antibodies against eight *Malassezia* species in sera of patients with atopic dermatitis by using an enzyme-linked immunosorbent assay. *Microbiol. Immunol.*, **50**, 851-856.

Katzenstein, A.-L., Sale, S.R., and Greenberger, P.A. (1983a) Allergic *Aspergillus* sinusitis: a newly recognized form of sinusitis. *J. Allergy Clin. Immunol.*, **72**, 89-93.

Katzenstein, A.-L., Sale, S.R., and Greenberger, P.A. (1983b) Pathologic findings in allergic *Aspergillus* sinusitis: a newly recognized form of sinusitis. *Am. J. Surg. Pathol.*, **7**, 439-443.

Kauffman, R.F., Tomee, J.F.C., van der Werff, T.S., *et al.* (1995) Review of fungus-induced asthmatic reactions. *Am. J. Respir. Crit. Care Med.*, **151**, 2109-2116.

Kercsmar, C.M., Dearborn, D.G., Schluchter, M., *et al.* (2006) Reduction in asthma morbidity in children as a result of home remediation aimed at moisture sources. *Environ. Health Perspect.*, **114**, 1574-1580.

Khan, A.J., and Nanchal, R. (2007) Cotton dust lung diseases. *Curr. Opin. Pulm. Med.*, **13**, 137-141.

Kieffer, M., Bergbrant, I.M., Faergemann, J., *et al.* (1990) Immune reactions to *Pityrosporum ovale* in adult patients with atopic and seborrheic dermatitis. *J. Am. Acad. Dermatol.*, **22**, 739-742.

Kilburn, K.H. (2004) Role of molds and mycotoxins in being sick in buildings: neurobehavioural and pulmonary impairment. *Adv. Appl. Microbiol.*, **55**, 339-359.

Kilburn, K.H., Gray, M., and Kramer, S. (2006) Nondisclosure of conflicts of interest is perilous to the advancement of science. *J. Allergy Clin. Immunol.*, **118**, 766-767.

Kilpeläinen, M., Terho, E.O., Helenius, H., and Koskenvuo, M. (2001) Home dampness, current allergic diseases, and respiratory infections among young adults. *Thorax*, **56**, 462-467.

Kim, Y.K., and Kim, Y.Y. (2002) Spider-mite allergy and asthma in fruit growers. *Curr. Opin. Allergy Clin. Immunol.*, **2**, 103-107.

Kirmaz, C., Bayrak, P., Yilmaz, O., and Yuksel, H. (2005) Effects of glucan treatment on the Th1/Th2 balance in patients with allergic rhinitis: a double-blind placebo-controlled study. *Eur. Cytokine Netw.*, **16**, 128-134.

Klaustermeyer, W.B., Bardana, E.J., and Hale, F.C. (1977) Pulmonary hypersensitivity to *Alternaria* and *Aspergillus* in baker's asthma. *Clin. Allergy*, **7**, 227-233.

Kline, J.N., and Hunninghake, G.W. (2007) Hypersensitivity pneumonitis and pulmonary infiltrates with eosinophilia. In D.L. Kasper, E. Braunwald, A.D. Fauci, *et al.*, (eds.), *Harrison's Principles of Internal Medicine*, 16th Ed. McGraw-Hill, New York, pp. 1516-1521.

Koch, A., Heilemann, K.J., Bischof, W., *et al.* (2000) Indoor viable mold spores – a comparison between two cities, Erfurt (eastern Germany) and Hamburg (western Germany). *Allergy*, **55**, 176-180.

Kolattukudy, P.E., Lee, J.D., Rogers, L.M., *et al.* (1993) Evidence for possible involvement of an elastolytic serine protease in aspergillosis. *Infect. Immun.*, **61**, 2357-2368.

Korhonen, K., Mähönen, S., Hyvärinen, A., *et al.* (2006) Skin test reactivity to molds in pre-school children with newly diagnosed asthma. *Pediatr. Int.*, **48**, 577-581.

Koskela, R.S., Klockars, M., and Järvinen, E. (1990) Mortality and disability among cotton mill workers. *Br. J. Ind. Med.*, **47**, 384-391. [Erratum and comment in *Br. J. Ind. Med.* **48**, 143-144 (1991).]

Kronqvist, M., Johansson, E., Pershagen, G., *et al.* (2001) Risk factors associated with asthma and rhinoconjunctivitis among Swedish farmers. *Allergy*, **54**, 1142-1149.

Kubala, L., Ruzickova, J., Nickova, K., *et al.* (2003) The effect of (1→3)-beta-D-glucans, carboxymethylglucan and schizophyllan on human leukocytes *in vitro*. *Carbohydr. Res.*, **338**, 2835-2840.

Kuhn, R.C., Trimble, M.W., Hofer, V., *et al.* (2005) Prevalence and airborne spore levels of *Stachybotrys* spp. in 200 houses with water incursions in Houston, Texas. *Can. J. Microbiol.*, **51**, 25-28.

Kumar, P., Marier, R., and Leech, S.H. (1984) Respiratory allergies related to automobile air conditioners. *N. Eng. J. Med.*, **311**, 1619-1621.

Kumar, A., Reddy, L.V., Sochanik, A., and Kurup, V.P. (1993) Isolation and characterization of a recombinant heat shock protein of *Aspergillus fumigatus*. *J. Allergy Clin. Immunol.*, **91**, 1024-1030.

Kupferberg, S.B., Bent, J.P., and Kuhn, F.A. (1997) Prognosis for allergic fungal sinusitis. *Otolaryngol. Head Neck Surg.*, **117**, 35-41.

Kurup, V.P., Banerjee, B., Hemmann, S., *et al.* (2000) Selected recombinant *Aspergillus fumigatus* allergens bind specifically to IgE in ABPA. *Clin. Exp. Allergy*, **30**, 988-993.

Kurup, V.P., Shen, H.D., and Vijay, H. (2002) Immunobiology of fungal allergens. *Int. Arch. Allergy Immunol.*, **129**, 181-188.

Kurup, V.P., Zacharisen, M.C., and Fink, J.N. (2006) Hypersensitivity pneumonitis. *Ind. J. Chest Dis. Allied Sci.*, **48**, 115-128.

Lai, H.Y., Tam, M.F., Tang, R.B., *et al.* (2002) cDNA cloning and immunological characterization of a newly identified enolase allergen from *Penicillium citrinum* and *Aspergillus fumigatus*. *Int. Arch. Allergy Immunol.*, **127**, 181-190.

Lai, H.Y., Tam, M.F., Chou, H., *et al.* (2004) Molecular and structural analysis of immunoglobulin E-binding epitopes of Pen ch 13, an alkaline serine protease major allergen from *Penicillium chrysogenum*. *Clin. Exp. Allergy*, **34**, 1926-1933.

Lai, K.M. (2006) Hazard identification, dose-response and environmental characteristics of stachybotryotoxins and other health-related products from *Stachybotrys*. *Environ. Technol.*, **27**, 329-335.

Lake, F.R., Ward, L.D., Simpson, R.J., *et al.* (1991) House dust mite-derived amylase: allergenicity and physicochemical characterization. *J. Allergy Clin. Immunol.*, **87**, 1035-1042.

Langley, S.J., Goldthorpe, S., Craven, M., *et al.* (2003) Exposure and sensitization to indoor allergens: Association with lung function, bronchial reactivity, and exhaled nitric oxide measures in asthma. *J. Allergy Clin. Immunol.*, **112**, 362-368.

Larsen, L. (1994) Fungal allergens. In R.A. Samson, B. Flannigan, M.E. Flannigan, *et al.* (eds.), *Health Implications of Fungi in Indoor Environments*, Elsevier, Amsterdam, pp. 215-220.

Larsen, J.N., and Dreborg, S. (2008) Standardization of allergen extracts. *Methods Mol. Med.*, **138**, 133-145.

Latzin, P., Hartl, D., Regamey, N., *et al.* (2008) Comparison of serum markers for allergic bronchopulmonary aspergillosis in cystic fibrosis. *Eur. Respir. J.*, **31**, 36-42.

Lau, S., Wahn, J., Schulz, G., *et al.* (2002) Placebo-controlled study of the mite allergen-reducing effect of tannic acid plus benzyl benzoate on carpets in homes of children with house dust mite sensitization and asthma. *Pediatr. Allergy Immunol.*, **13**, 31-36.

Laurila, A.L., von Hertzen, L., and Saikku, P. (1997) *Chlamydia pneumoniae* and chronic lung disease. *Scand. J. Infect. Dis.* (Suppl.), **104**, 34-36.

Lee, Y.M., Kim, Y.K., Kim, S.O., *et al.* (2005) A case of hypersensitivity pneumonitis caused by *Penicillium* species in a home environment. *J. Korean Med. Sci.*, **20**, 1073-1075.

Lehrer, S.B., Huges, J.M., Altman, L.C., *et al.* (1994) Prevalence of basidiomycete allergy in the USA and Europe and its relationship to allergic respiratory symptoms. *Allergy*, **49**, 460-465.

Leung, D.Y.M. (2000) Atopic dermatitis: new insights and opportunities for therapeutic intervention. *J. Allergy Clin. Immunol.*, **105**, 860-876.

Lewis, S.A. (2000) Animals and allergy. *Clin. Exp. Allergy*, **30**, 153-157.

Leyden, J.L. (1994) Tinea pedis pathophysiology and treatment. *J. Am. Acad. Dermatol.*, **31**, S31-S33.

Leyden, J.E., Marples, R.R., and Kligman, A.M. (1974) *Staphylococcus aureus* in the lesions of atopic dermatitis. *Br. J. Dermatol.*, **90**, 525-530.

Lieberman, A., Rea, W., and Curtis, L. (2006) Adverse health effects of mold exposure. *J. Allergy Clin. Immunol.*, **118**, 763.

Lintu, P., Savolainen, J., Kortkangas-Savolainen, O., and Kalimo, K. (2001) Systemic ketoconazole is an effective treatment of atopic dermatitis with IgE-mediated hypersensitivity to yeasts. *Allergy*, **56**, 512-517.

Liu, A.H., and Leung, Y.M. (2000) Modulating the early allergic response with endotoxin. *Clin. Exp. Allergy*, **30**, 1535-1539.

Lombardero, M., Heymann, P.W., Platts-Mills, T., *et al.* (1990) Conformational stability of B cell epitopes on group I and group II *Dermatophagoides* spp. allergens. *J. Immunol.*, **144**, 1353-1360.

Lunceford, T.M. (1968) Algae as an allergen-provocative nasal inhalation. *J. Kansas Med. Soc.*, **69**, 466-467.

Luong, A., and Marple, B.F. (2004) Allergic fungal rhinosinusitis. *Curr. Allergy Asthma Rep.*, **4**, 465-470.

McCann, W.A., Cromie, M., Chandler, F., *et al.* (2002) Sensitization to recombinant *Aspergillus fumigatus* allergens in allergic fungal sinusitis. *Ann. Allergy Asthma Immunol.*, **89**, 203-208.

McElhenney, T.R., Bold, H.C., Brown, R.M., and McGovern, J.P. (1962) Algae: a cause of inhalant allergy in children. *Ann. Allergy*, **20**, 739-743.

McGovern, J.P., Haywood, T.J., and McElhenney, T.R. (1966) Airborne algae and their allergenicity. II. Clinical and laboratory multiple correlation studies with four genera. *Ann. Allergy*, **24**, 145-149.

Mabry, R.L., and Mabry, C.S. (2000) Allergic fungal sinusitis: the role of immunotherapy. *Otolaryngol. Clin. North Am.*, **33**, 433-440.

Mabry, R.L., and Manning, S. (1995) Radioallergosorbent micro-screen and total immunoglobulin E in allergic fungal sinusitis. *Otolaryngol. Head Neck Surg.*, **113**, 721-723.

Mabry, R.L., and Mabry, C.S. (1997) Immunotherapy for allergic fungal sinusitis: the second year. *Otolaryngol. Head Neck Surg.*, **117**, 367-371.

Mabry, R.L., Marple, B.F., Folker, R.J., and Mabry, C.S. (1998) Immunotherapy for allergic fungal sinusitis: three years' experience. *Otolaryngol. Head Neck Surg.*, **119**, 648-651.

MacDowell, A.L., and Bacharier, L.B. (2005) Infectious triggers of asthma. *Immunol. Allergy Clin. North Am.*, **25**, 45-66.

Maestrelli, P., Zanolla, L., Pozzan, M., *et al.* (2004) Effect of specific immunotherapy added to pharmacologic treatment and allergen avoidance in asthmatic patients allergic to house dust mite. *J. Allergy Clin. Immunol.*, **113**, 643-649.

Malet, A., Cisteró-Bahima, A., Amat, P., *et al.* (2002) Influence in the quality of life of the respiratory patients by environmental control and the acaricide Frontac. *Allergol. Immunopathol. (Madrid)*, **30**, 85-93.

Malling, H.J., Dreborg, S., and Weeke, B. (1986) Diagnosis and immunotherapy of mould allergy. *Allergy*, **41**, 507-519.

Mandryk, J., Alwis, K.J., and Hockings, A.D. (2000) Effects of personal exposures on pulmonary function and work-related symptoms among sawmill workers. *Ann. Occup. Hyg.*, **44**, 281-289.

Manning, S.C., Mabry, R.L., Schaefer, S.D., and Close, L.G. (1993) Evidence of IgE mediated hypersensitivity in allergic fungal sinusitis. *Laryngoscope*, **103**, 717-721.

Marinkovich, V.A. (2006) Position paper on molds is seriously flawed. *J. Allergy Clin. Immunol.*, **118**, 761-762.

Martinez-Cañavate Burgos, A., Valenzuela-Soria, A., and Rojo-Hernadez, A. (2007) Immunotherapy with *Alternaria alternata*: present and future. *Allergol. Immunopathol.*, **35**, 259-263.

Matsumoto, T., and Ajello, L. (1987) Current taxonomic concepts pertaining to the dermatophytes and related fungi. *Int. J. Dermatol.*, **26**, 491-499.

Maurya, V., Gugnani, H.C., Sarma, P.U., *et al.* (2005) Sensitization to *Aspergillus* antigens and occurrence of allergic bronchopulmonary aspergillosis in patients with asthma. *Chest*, **127**, 1252-1259.

Mehl, R. (1998) Occurrence of mites in Norway and the rest of Scandinavia. *Allergy*, **53**, 28-35.

Mendell, M.J. (2007) Indoor residential chemical emissions as risk factors for respiratory for respiratory and allergic effects in children: a review. *Indoor Air*, **17**, 259-277.

Merget, R., Sander, I., Rozynek, P., *et al.* (2008) Occupational hypersensitivity pneumonitis due to molds in an onion and potato sorter. *Am. J. Industr. Med.*, **51**, 117-119.

Michel, O. (1997) Human challenge studies with endotoxins. *Int. J. Occup. Env. Health.*, **3**, S18-S25.

Michel, O., Duchateau, J., and Sergysels, S. (1989) Effect of inhaled endotoxin on bronchial reactivity in asthmatic and normal subjects. *J. Appl. Physiol.*, **66**, 1059-1064.

Michel, O., Ginanni, R., Duchateau, J., *et al.* (1991) Domestic endotoxin exposure and clinical severity of asthma. *Clin. Exp. Allergy*, **21**, 441-448.

Michel, O., Ginanni, R., Le Bon, B., *et al.* (1992) Inflammatory response to acute inhalation of endotoxin in asthmatic patients.

Am. Rev. Respir. Dis., **146**, 352-357.

Michel, O., Kips, J., Duchateau, J., *et al.* (1996) Severity of asthma is related to endotoxin in house dust. *Amer. J. Respir. Crit. Care Med.*, **154**, 1641-1646.

Michel, O., Nagy, A.M., Schroeven, M., *et al.* (1997) Dose-response relationship to inhaled endotoxin in normal subjects. *Am. J. Respir. Crit. Care Med.*, **156**, 157-164.

Mills, K., Hart, B.J., and Lynch, N.R. (1999) Molecular characterization of the group 4 house dust mite allergen from *Dermatophagoides pteronyssinus* and its amylase homologue from *Euroglyphus maynei*. *Int. Arch. Allergy Immunol.*, **120**, 100-107.

Mittal, A. (1981) Algal forms in house dust samples and their role in respiratory allergy – a preliminary report. *J. Assoc. Phys. Ind.*, **29**, 197-200.

Mittal, A., Agarwal, M.K., and Shivpuri, D.N. (1979) Respiratory allergy to algae: clinical aspects. *Ann. Allergy*, **42**, 253-256.

Mølhave, L., Bach, B., and Pedersen, O.F. (1986) Human reactions to low concentrations of volatile organic compounds. *Environment Int.*, **12**, 167-175.

Moser, S.A., and Pollack, J.D. (1978) Isolation of glycopeptides with skin test activity from dermatophytes. *Infect. Immun.*, **19**, 1031-1046.

Moss, R.B. (2005) Pathophysiology and immunology of allergic bronchopulmonary aspergillosis. *Med. Mycol.*, **43**, S203-S206.

Moreno-Ancillo, A., Vicente, J., Gomez, L., *et al.* (1997) Hypersensitivity pneumonitis related to a covered and heated swimming pool environment. *Int. Arch. Allergy Immunol.*, **114**, 205-206.

Morren, M.-A., Janssens, V., Dooms-Goossens, A., *et al.* (1993) Alpha-amylase, a flour additive: an important cause of protein contact dermatitis in bakers. *J. Am. Acad. Dermatol.*, **29**, 723-728.

Munir, A.K. (1998) Risk levels for mite allergen: are they meaningful, where should samples be collected, and how should they be analyzed? *Allergy*, **53**, 84-87.

Nakabayashi, A., Sei, Y., and Guillot, J. (2000) Identification of *Malassezia* species isolated from patients with seborrhoeic dermatitis, atopic dermatitis, pityriasis versicolor and normal subjects. *Med. Mycol.*, **38**, 337-341.

Nakagawa-Yoshida, K., Ando, M., Etches, R.I., and Dosman, J.A. (1997) Fatal cases of farmer's lung in Canadian family: probable new antigens, *Penicillium brevicompactum* and *P. olivicolor*. *Chest*, **111**, 245-248.

Neukirch, C., Henry, C., Leynaert, B., *et al.* (1999) Is sensitization to *Alternaria alternata* a risk factor for severe asthma? A population based study. *J. Allergy Clin. Immunol.*, **103**, 709-711.

Nevalainen, A., Rautiala, S., Hyvarinen, A., *et al.* (1994) Exposure to fungal spores in mouldy houses: effect of remedial work. In S.N. Agashi, (ed.), *Recent Trends in Aerobiology, Allergy and Immunology*, Oxford & IBH, New Delhi, pp. 99-107.

Newson, R., Strachan, D., Corden, J., and Millington, W. (2000) Fungal and other spore counts as predictors of admissions of asthma in the Trent region. *Occup. Environ. Med.*, **57**, 786-792.

Ng, T.P., Tan, W.C., and Lee, Y.K. (1994) Occupational asthma in pharmacist induced by *Chlorella*, a unicellular algae preparation. *Respir. Med.*, **88**, 555-557.

Nierman, W.C., Pain, A., Anderson, M.M., *et al.* (2005) Genomic sequence of the pathogenic and allergenic filamentous fungus *Aspergillus fumigatus*. *Nature*, **22**, 1151-1156.

Nikkels, A.F., and Piérard, G.E. (2003) Framing the future of antifungals in atopic dermatitis. *Dermatology*, **206**, 398-400.

Nilsson, A., Kihlström, E., Lagesson, V., *et al.* (2004) Microorganisms and volatile organic compounds in airborne dust from damp residences. *Indoor Air*, **14**, 74-82.

Nishiura,Y., Nakagaway-Yoshida, K., Suga, M., *et al.* (1997) Assignment and serotyping of *Trichosporon* species: the causative agents of summer-type hypersensitivity pneumonitis. *J. Med. Vet. Mycol.*, **35**, 45-52.

Nissen, D., Petersen, L.J., Esch, R., *et al.* (1998) IgE-sensitization to cellular and culture filtrates of fungal extracts in patients with atopic dermatitis. *Ann. Allergy Asthma Immunol.*, **81**, 247-255.

Niven, R.L., and Pickering, C.A.C. (1996) Byssinosis: a review. *Thorax*, **51**, 632-637.

Noble, J.A., Crow, S.A., Ahearn, D.G., and Kuhn, F.A. (1997) Allergic fungal sinusitis in the southeastern USA: involvement of a new agent *Epicoccum nigrum*. *J. Med. Vet. Mycol.*, **35**, 405-409.

Norbäck, D., Edling, C., Wieslander, G., and Ramadhan, S. (1993) Exposure to volatile organic compounds (VOC) in the general Swedish population and its relation to perceived indoor air quality and sick building syndrome. In J.J.K. Jaakola, R. Ilmarinen, and O. Seppanen, (eds.), *Indoor Air '93, Proceedings of the 6th International Conference on Indoor Air Quality and Climate*, Vol. 1, Helsinki, Finland, pp. 573-578.

Norbäck, D., Bjornsson, E., Janson C., *et al.* (1995) Asthmatic symptoms and volatile organic compounds, formaldehyde, and carbon dioxide in dwellings. *Occup. Environ. Med.*, **52**, 388-395.

Norbeck, J., and Blomberg, A. (1996) Protein expression during exponential growth in 0.7 M NaCl medium of *Saccharomyces cerevisiae*. *FEMS Microbiol. Lett.*, **113**, 1-8.

Nordsrom, K., Norbäck, D., and Akselsson, R. (1994) Effect of humidification on the sick building syndrome and perceived indoor air quality in hospitals: a four month longitudinal study. *Occup. Environ. Med.*, **51**, 683-688.

Nordvall, S.L., and Johansson, S. (1990) IgE antibodies to *Pityrosporum orbiculare* in children with atopic diseases. *Acta Paediatr. Scand.*, **79**, 343-348.

Norn, S. (1994) Micro-organism-induced or enhanced mediator release: a possible mechanism in organic dust related diseases. *Am. J. Industr. Med.*, **25**, 91-95.

Norn, S., Stahl-Skov, P., Jensen, C., *et al.* (1986) Bacteria and their products release histamine and potentiate mediator release: new aspects in airway diseases. *Eur. J. Respir. Dis.*, **69**, 230-234.

Norn, S., Stahl-Skov, P., Jensen, C., *et al.* (1987) Histamine release induced by bacteria. A new mechanism in asthma? *Agents Actions*, **20**, 29-34.

Norn, S., Jensen, L., Kjaergaard, L.L., *et al.* (1994) Bacteria-induced IgE-mediated histamine release: examination of patients with chronic bronchitis (CB) during acute exacerbations. *Agents Actions*. **41**, C22-C23.

O'Hollaren, M.T., Yunginger, S.W., Offord, K.P., *et al.* (1991) Exposure to an aeroallergen as a possible precipitating factor in respiratory arrest in young patients with asthma. *N. Engl. J. Med.*, **324**, 359-363.

Olsen, O.T., Larsen, K.R., Jacobsan, L., and Svendsen, U.G. (1997) A 1-year, placebo-controlled, double-blind house-dust-mite immunotherapy study in asthmatic adults. *Allergy*, **52**, 853-859.

Oostergaard, P.A., Kaad, P.H., and Kristensen, T. (1986) A prospective study on the safety of immunotherapy in children with severe asthma. *Allergy*, **41**, 588-593.

Osbourne, M., Reponen, P., Cho, S.-H., *et al.* (2006) Specific fungal exposures, allergic sensitization, and rhinitis in infants. *Pediatr. Allergy Immunol.*, **17**, 450-457.

Oshika, E., Kuroki, Y., Sakiyama, Y., *et al.* (1994) A study of the binding of immunoglobulin G and immunoglobulin E from children with bronchial asthma to peptides derived from group II antigen of *Dermatophagoides pteronyssinus. Pediatric Res.*, **33**, 209-213.

Packe, G.E., and Ayres, J.G. (1985) Asthma outbreak during a thunderstorm. *Lancet 2*, 199-204.

Palma-Carlos, A.G., and Palma-Carlos, M.L. (2001) Chronic mucocutaneous candidiasis revisited. *Allerg. Immunol. (Paris)*, **33**, 229-232.

Parikh, J.R. (1992) Byssinosis in developing countries. *Br. J. Indust. Med.*, **49**, 217-219.

Park, J.H., Gold, D.R., Spiegelman, D.L., *et al.* (2001) House dust endotoxin and wheeze in the first year of life. *Am. J. Respir. Crit. Care Med.*, **163**, 322-328.

Parvaneh, S., Johansson, E., Elfman, L.H., and van Hage-Hamsten, M. (2002) An ELISA for recombinant *Lepidoglyphus destructor*, Lep d 2, and the monitoring of exposure to dust mite allergens in farming households. *Clin. Exp. Allergy.*, **32**, 80-86.

Patel, A.M., Ryu, J.H., and Reed, C.E. (2001) Hypersensitivity pneumonitis: current concepts and future questions. *J. Allergy Clin. Immunol.*, **108**, 661-670.

Pernis, B., Viglian, E.C., Cavagna, C., *et al.* (1961) The role of bacterial endotoxins in occupational diseases caused by inhaling vegetable dusts. *Br. J. Ind. Med.*, **18**, 120-129.

Peroni, D.G., Piacentini, G.L., Martinati, L.C., *et al.* (1995) Double-blind trial of house-dust mite immunotherapy in asthmatic children resident at high altitude. *Allergy*, **50**, 925-930.

Pestka, J.J. (2008) *Stachybotrys chartarum*, trichothecene mycotoxins, and damp building-related illness: new insights into a public health enigma. *Toxicol. Sci.*, **104**, 4-26.

Pichler, C.E., Helbling, A., and Pichler, W.J. (2001) Three years of specific immunotherapy with house-dust-mite extracts in patients with rhinitis and asthma: significant improvement of allergen-specific parameters and of nonspecific bronchial hyperreactivity. *Allergy*, **56**, 301-306.

Pifferi, M., Baldini, G., Marrazzini, G., *et al.* (2002) Benefits of immunotherapy with a standardized *Dermatophagoides pteronyssinus* extract in asthmatic children: a three-year prospective study. *Allergy*, **57**, 785-790.

Pitcher, W.D. (1990) Hypersensitivity pneumonitis. *Am. J. Med. Sci.*, **300**, 251-266.

Platts-Mills, T.A. (1990) Allergens and asthma. *Allergy Proc.*, **11**, 269-271.

Platts-Mills, T.A. (2008) Allergen avoidance in the treatment of asthma: problems with the meta-analyses. *J. Allergy Clin. Immunol.*, **22**, 694-696.

Platts-Mills, T.A., and Chapman, M.D. (1987) Dust mites: immunology, allergic disease and environmental control. *J. Allergy Clin. Immunol.*, **80**, 755-775. [Published erratum in *J. Allergy Clin. Immunol.*, **82**, 841 (1988).]

Platts-Mills, T.A.E., Hayden, M.L., Chapman, M.D., and Wilkins, S.R. (1987) Seasonal variation in dust mite and grass-pollen allergens in dust from the houses of patients with asthma. *J. Allergy Clin. Immunol.*, **79**, 781-791.

Platts-Mills, T.A.E., Chapman, M.D., and Wheatly, L.M. (1999) Control of house dust mite in managing asthma: conclusions of meta-analysis are wrong. *Br. Med. J.*, **318**, 870-871. [Letter.]

Platts-Mills, T.A.E., Vaughan, J.W., Carter, M.C., and Woodfolk, J.A. (2000) The role of intervention in established allergy: avoidance of indoor allergens in the treatment of chronic allergic disease. *J. Allergy Clin. Immunol.*, **106**, 787-804.

Pollart, S.M., Chapman, M.D., and Platts-Mills, T.A.E. (1987) House dust sensitivity and environmental control. *Prim. Care*, **14**, 591-603.

Ponikau, J.U., and Sherris, D.A. (2006) The role of airborne mold in chronic rhinosinusitis. *J. Allergy Clin. Immunol.*, **118**, 762-763.

Ponikau, J.U., Sherris, D.A., Kern, E.B., *et al.* (1999) The diagnosis and incidence of allergic fungal sinusitis. *Mayo Clin. Proc.* **74**, 877–884.

Pónyai, G., Hidvégi, B., Németh, I., *et al.* (2008) Contact and aeroallergens in adulthood atopic dermatitis. *J. Eur. Acad. Dermatol. Venereol.*, **22**, 1346-1355.

Pretus, H.A., Ensley, H.E., McNamee, R.B., *et al.* (1991) Isolation, physicochemical characterization and preclinical efficacy of soluble scleroglucan. *J. Pharmacol. Exp. Ther.*, **257**, 500-510.

Pulimood, T.B., Corden, J.M., Bryden, C., *et al.* (2007) Epidemic asthma and the role of the fungal mold *Alternaria alternata*. *J. Allergy Clin. Immunol.*, **120**, 610-617.

Quirce, S., Cuevas, M., Diez-Gomez, M., *et al.* (1992) Respiratory allergy to *Aspergillus*-derived enzymes in baker's asthma. *J. Allergy Clin. Immunol.*, **90**, 970-978.

Reitamo, S., Visa, K., Stubbs, S., *et al.* (1986) Eczematous reactions in atopic patients caused by epicutaneous testing with inhalant allergens. *Br. J. Dermatol.*, **114**, 303-309.

Ren, P., Jankun, M., Belanger K., *et al.* (2001) The relation between fungal propagules in indoor air and home characteristics. *Allergy*, **56**, 419-424.

Rid, R., Simon-Nobbe, B., Langdon, J., *et al.* (2008) *Cladosporium herbarum* translationally controlled tumour protein (TCTP) is an IgE-binding antigen and is associated with disease severity. *Mol. Immunol.*, **45**, 406-418.

Rizzo, M.C., Naspitz, C.K., Fernandez-Caldas, E., *et al.* (1997) Endotoxin exposure and symptoms in asthmatic children. *Pediatr. Allergy Immunol.*, **8**, 121-126.

Rogers, C. (2003) Indoor fungal exposure. *Immunol. Allergy Clin. N. Am.*, **23**, 501-518.

Round, F.E. (1981) *The Ecology of Algae*. Cambridge University Press, Cambridge, UK.

Rosenberg, M., Patterson, R., and Roberts, M. (1977) Immunologic responses to therapy in allergic bronchopulmonary aspergillosis: serum IgE value as an indicator and predictor of disease activity. *J. Pediatr.*, **91**, 914-917.

Rumchev, K., Spickett, J., Bulsara, M., *et al.* (2004) Association of domestic exposure to volatile organic compounds with asthma in young children. *Thorax*, **59**, 746-751.

Rumchev, K., Brown, H., and Spickett, J. (2007) Volatile organic compounds: do they present a risk to our health? *Rev. Environ. Health*, **22**, 39-55.

Ryan, M.W., and Marple, B.F. (2007) Allergic fungal rhinosinusitis: diagnosis and management. *Curr. Opin. Otolaryngol. Head Neck Surg.*, **15**, 18-22.

Rydjord, B., Eduard, W., Stensby, B., *et al.* (2007) Antibody response to long-term and high-dose mould-exposed sawmill workers. *Scand. J. Immunol.*, **66**, 711-718.

Rylander, R. (1990) Health effects of cotton dust exposures. *Am. J. Ind. Med.*, **17**, 39-45.

Rylander, R. (1997) Airborne (1→3)-β-D-glucan and airway effects in a day care center before and after renovation. *Arch. Environ. Health*, **52**, 281-285.

Rylander, R. (1999) Indoor air-related effects and airborne (1→3)-β-D-glucan. *Environ. Health Perspect.*, **107**, 501-503.

Rylander, R. (2005) (1→3)-β-D-glucan in the environment. In S.-H. Young and V. Castranova, (eds.), *Toxicology of (1→3)-β-D-glucans*, Taylor & Francis, Boca Raton, FL, pp. 53-64.

Rylander, R., and Goto, H. (1991) *First Glucan Lung Toxicity Workshop, Committee on Organic Dusts*, Report 4/91, International Committee on Occupational Health, Stockholm, pp. 1-20.

Rylander, R., and Peterson, Y. (1993) *Second Glucan Lung Toxicity Workshop, Committee on Organic Dusts*, Report 1/93, International Committee on Occupational Health, Stockholm, pp.

1-53.

Rylander, R., and Lin, R.H. (2000) (1→3)-β-D-glucan-relationship to indoor air-related symptoms, allergy and asthma. *Toxicology*, **152**, 47–52.

Rylander, R., Haglind, P., and Lundholm, M. (1985) Endotoxin in cotton dust and respiratory function decrement among cotton workers in an experimental cardroom. *Am. Rev. Respir. Dis.*, **131**, 209-213.

Rylander, R., Persson, K., Goto, H., *et al.* (1992) Airborne beta-1,3-glucan may be related to symptoms in sick buildings. *Indoor Built Environ.*, **1**, 263-267.

Safirstein, B.H. (1976) Allergic bronchopulmonary aspergillosis with obstruction of the upper respiratory tract. *Chest*, **70**, 788-790.

Sakurai, T., Ohno, N., and Yadomae, T. (1994) Changes in immune mediators in mouse lung produced by administration of soluble (1→3)-β-D-glucan. *Biol. Pharm. Bull.*, **17**, 617-622.

Salo, P.M., Arbes, S.J., Crockett, P.W., *et al.* (2008) Exposure to multiple indoor allergens in US homes and relationship to asthma. *J. Allergy Clin. Immunol.*, **121**, 678-684.

Sander, I., Raulf-Heimsoth, M., Siethoff, C., *et al.* (1998) Allergy to *Aspergillus*-derived enzymes in the baking industry: identification of beta-xylosidase from *Aspergillus niger* as a new allergen (Asp n 14). *J. Allergy Clin. Immunol.*, **102,** 256-264.

Sandiford, C.P., Tee, R.D., and Taylor, A.J. (1994) The role of cereal and fungal amylases in cereal flour hypersensitivity. *Clin. Exp. Allergy*, **24**, 549-551.

Sanico, A.M. (2004) Latest developments in the management of allergic rhinitis. *Clin. Rev. Allergy Immunol.*, **27**, 181-189.

Santilli, J., Rockwell, W.J., and Collins, R.P. (1985) The significance of the spores of the basidiomycetes (mushrooms and their allies) in bronchial asthma and allergic rhinitis. *Ann. Allergy*, **55**, 469-471.

Scheynius, A., Johansson, C., Buentke, E., *et al.* (2002) Atopic eczema/dermatitis syndrome and *Malassezia*. *Int. Arch. Allergy Immunol.*, **127**, 161-169.

Schubert, M.S. (2004a) Allergic fungal sinusitis. *Otolaryngol. Clin. North Am.*, **37**, 310-326.

Schubert, M.S. (2004b) Allergic fungal sinusitis: Pathogenesis and management strategies. *Drugs*, **64**, 363-374.

Schubert, M.S. (2006) Allergic fungal sinusitis. *Clin. Rev. Allergy Immunol.*, **30**, 205-215.

Schubert, M.S. (2007) Allergic fungal sinusitis. *Clin. Allergy Immunol.*, **20**, 263-271.

Schubert, M.S., and Goetz, D.W. (1998) Evaluation and treatment of allergic fungal sinusitis. I. Demographics and diagnosis. *J. Allergy Clin. Immunol.*, **102**, 387-394.

Schuh, J.M., Blease K., Kunkel S.L., *et al.* (2003) Chemokines and cytokines: axis and allies in asthma and allergy. *Cytokine Growth Factor Rev.*, **14**, 503-510.

Schwienbacher, M., Israel, L., Heesemann, J., and Ebel, F. (2005) Asp f6, an *Aspergillus* allergen specifically recognized by IgE from patients with allergic bronchopulmonary aspergillosis, is differentially expressed during germination. *Allergy*, **60**, 1430-1435.

Scott, J.A. (2001) Studies on Indoor Fungi. PhD Thesis, Dept. of Botany, University of Toronto, ON, Canada (http://www.sporometrics.com.libaccess.lib.mcmaster.ca/Thesis/Studies_on_indoor_fungi.pdf.)

Semenzato, G., Adami, F., Maschio, N., and Agostini, C. (2000) Immune mechanisms in interstitial lung diseases. *Allergy*, **55**, 1103–1120.

Semic-Jusufagic, A., Bachert, C., Gevaert, P., *et al.* (2007) *Staphylo-*

coccus aureus* sensitization and allergic disease in early childhood: population-based birth cohort study. *J. Allergy Clin. Immunol.*, **119**, 930-936.

Shah, A. (2008) *Aspergillus*-associated hypersensitivity respiratory disorders. *Ind. J. Chest Dis. Allied Sci.*, **50**, 117-128.

Shah, A., and Sircar, M. (1991) Sensitization to *Aspergillus* antigens in perennial rhinitis. *Asian Pac. J. Allergy Immunol.*, **9**, 137-139.

Shah, A., Khan, Z.U., Sircar, M., *et al.* (1990) Allergic *Aspergillus* sinusitis: an Indian report. *Respir. Med.*, **84**, 249-251.

Shankar, J., Gupta, P.D., Sridhara, S., *et al.* (2005) Immunobiochemical analysis of cross-reactive glutathione-*S*-transferase allergen from different fungal sources. *Immunol. Invest.*, **34**, 37-51.

Shankar, J., Singh B.P., Gaur, S.N., and Arora, N. (2006) Recombinant glutathione-*S*-transferase a major allergen from *Alternaria alternata* for clinical use in allergy patients. *Mol. Immunol.*, **43**, 1927-1932.

Sheikh, A., and Hurwitz, B. (2001) House dust mite avoidance measures for perennial allergic rhinitis. *Cochrane Database Syst. Rev.*, **4**, CD001563.

Sheikh, A., Hurwitz, B., and Shehata, Y. (2007) House dust mite avoidance measures for perennial allergic rhinitis. *Cochrane Database Syst. Rev.*, **1**, CD001563.

Shelton, B.G., Kirkland, K.H., Flanders, W.D., and Morris, G.K. (2002) Profiles of airborne fungi in buildings and outdoor environments in the United States. *Appl. Environ. Microbiol.*, **68**, 1743-1753.

Shen, H.D., Chua, K.Y., Lin, K.L., *et al.* (1993) Molecular cloning of a house dust mite allergen with common antibody binding specificities with multiple components in mite extracts. *Clin. Exp. Allergy*, **23**, 934-940.

Shen, H.D., Liaw, S.F., Lin, W.L., *et al.* (1995) Molecular cloning of cDNA coding for the 68 kDa allergen of *Penicillium notatum* using monoclonal antibodies. *Clin. Exp. Allergy*, **25**, 350-356.

Shen, H.D., Lin, W.L., Tsai, J.J., *et al.* (1996) Allergenic components in three different species of *Penicillium*: crossreactivity among major allergens. *Clin. Exp. Allergy*, **26**, 444-451.

Shen, H.D., Au, L.C., Lin, W.L., *et al.* (1997) Molecular cloning and expression of a *Penicillium citrinum* allergen with sequence homology and antigenic crossreactivity to a hsp 70 human heat shock protein. *Clin. Exp. Allergy*, **27**, 682-690.

Shen, H.D., Lin, W.L., Tam, M.F., *et al.* (1998) Alkaline serine proteinase: a major allergen of *Aspergillus oryzae* and its crossreactivity with *Penicillium citrinum*. *Int. Arch. Allergy Immunol.*, **116**, 29-35.

Shen, H.D., Lin, W.L., Tam, M.F., *et al.* (1999) Characterization of allergens from *Penicillium oxalicum* and *P. notatum* by immunoblotting and N-terminal amino acid sequence analysis. *Clin. Exp. Allergy,* **29**, 642-651.

Shen, H.D., Lin, W.L., Tam, M.F., *et al.* (2001) Identification of vacuolar serine proteinase as a major allergen of *Aspergillus fumigatus* by immunoblotting and N-terminal amino acid sequence analysis. *Clin. Exp. Allergy*, **31**, 295-302.

Sherwood, E.R., Williams, D.L., McNamee, R.B., *et al.* (1987) Enhancement of interleukin 1 and interleukin 2 production by soluble glucan. *Int. J. Immunopharmacol.*, **9**, 261-267.

Shoemaker, R.C., Amman, H., Lipsey, R., and Montz, E. (2006) Rigor, transparency, and disclosure needed in mold position paper. *J. Allergy Clin. Immunol.*, **118**, 764-765.

Sigsgaard, T., Malmros, P., Nersting, L., and Petersen, C. (1994) Respiratory disorders and atopy in Danish refuse workers. *Am. J. Respir. Crit. Care Med.*, **149**, 1407-1412.

Sigsgaard, T., Bonefeld-Jorgensen, E.C., Kjaergaard, S.K., *et al.* (2000) Cytokine release from the nasal mucosa and whole blood after experimental exposures to organic dusts. *Eur. Respir. J.*, **16**, 140-145.

Simon-Nobbe, B., Denk, U., Poll, V., *et al.* (2008) The spectrum of fungal allergy. *Int. Arch. Allergy Immunol.*, **145**, 58-86.

Skoner, D.P. (2001) Allergic rhinitis: definition, epidemiology, pathophysiology, detection, and diagnosis. *J. Allergy Clin. Immunol.*, **108**, S2-S8.

Slunt, J.B., Taketomi, E.A., Woodfolk, J.A., *et al.* (1996) The immune response to *Trichophyton tonsurans*: distinct T cell cytokine profiles to a single protein among subjects with immediate and delayed hypersensitivity. *J. Immunol.*, **157**, 5192-5197.

Smid, T., Heederik, D., Houba, R., and Quanjer, P.H. (1992) Dust and endotoxin-related respiratory effects in the animal feed industry. *Am. Rev. Respir. Dis.*, **146**, 1474-1479.

Smith, W., Mills, K., Hazell, L., *et al.* (1999) Molecular analysis of the group 1 and 2 allergens from the house dust mite, *Euroglyphus maynei*. *Int. Arch. Allergy Immunol.*, **118**, 15-22.

Smoragiewicz, W., Cossette, B., Boutard, A., and Krzystyniak, K. (1993) Trichothecene mycotoxins in the dust of ventilation systems in office buildings. *Int. Arch. Occup. Environ. Health*, **65**, 113-117.

Solarz, K., Senczuk, L., Maniurka, H., *et al.* (2007) Comparisons of the allergenic mite prevalence in dwellings and certain outdoor environments of the Upper Silesia (southwest Poland). *Int. J. Hyg. Environ. Health*, **210**, 715-724.

Sorenson, W.G. (1990) Mycotoxins as potential occupational hazards. *Dev. Industr. Microbiol.*, **31**, 205-211.

Spring, P.M., and Amedee, R.G. (1995) Fungal sinusitis. *J. Louisiana State Med. Soc.*, **147**, 395-398.

Stark, P.C., Celedon, J.C., Chew, G.L., *et al.* (2005) Fungal levels in the home and allergic rhinitis by 5 years of age. *Environ. Health Perspect.*, **113**, 1405-1409.

Staubach, P., Vonend, A., Burow, G., *et al.* (2008) Patients with chronic urticaria exhibit increased rates of sensitisation to *Candida albicans*, but not to common moulds. *Mycoses*, **52**, 334-338.

Stewart, G.A. (1995) Dust mite allergens. *Clin. Rev. Allergy Immunol.*, **13**, 135-150.

Stewart, G.A., Bird, C.H., Krska, K.D., *et al.* (1992) A comparative study of allergenic and potentially allergenic enzymes from *Dermatophagoides pteronyssinus*, *D. farinae* and *Euroglyphus maynei*. *Exp. Appl. Acarol.*, **16**, 165-180.

Stites, D.P., Stobo, J.D., and Wells, J.V. (1987) *Basic and Clinical Immunology*, 6th Ed. Appleton and Lange, Norwalk, CT.

Strachan, D.P., Flannigan, B., McCabe, E.M., and McGarrry, G. (1990) Quantification of airborne moulds in the homes of children with and without asthma. *Thorax*, **45**, 382-387.

Straus, D.C., and Wilson, S.C. (2006) Respirable trichothecene mycotoxins can be demonstrated in the air of *Stachybotrys chartarum*-contaminated buildings. *J. Allergy Clin. Immunol.*, **118**, 760.

Strickland, M.H.V. (2006) How solid is the Academy position paper on mold exposure? *J. Allergy Clin. Immunol.*, **118**, 763-764.

Su, H.J., Wu, P.C., Lei, H.Y., and Wang, J.Y. (2005) Domestic exposure to fungi and total serum IgE levels in asthmatic children. *Mediators Inflamm.*, **2005**, 167-170.

Sugita, T., Suto, H., Unno, T., *et al.* (2001) Molecular analysis of *Malassezia* microflora on the skin of atopic dermatitis patients and healthy subjects. *J. Clin. Microbiol.*, **39**, 3486-3490.

Sunesson, A.L., Rosén, I., Stenberg, B., and Sjöström, M. (2006) Multivariate evaluation of VOCs in buildings where people with non-specific building-related symptoms perceive health problems and in buildings where they do not. *Indoor Air*, **16**,

383-391.

Sutherland, E.R., and Martin, R.J. (2007) Asthma and atypical bacterial infection. *Chest*, **132**, 1962-1966.

Svejgaard, E., Larsen, P.Ø., Deleuran, M., *et al.* (2004) Treatment of head and neck dermatitis comparing itraconazole 200 mg and 400 mg daily for 1 week with placebo. *J. Eur. Acad. Dermatol. Venereol.*, **18**, 445-449.

Swärd-Nordmo, M., Paulsen, B.S., and Wold, J.K. (1988) The glycoprotein allergen Ag54 (Cla h II) from *Cladosporium herbarum*: structural studies of the carbohydrate moiety. *Int. Arch. Allergy Appl. Immunol.*, **85**, 288-294.

Tabar, A.I., Lizaso, M.T., García, B.E., *et al.* (2000) Tolerance of immunotherapy with a standardized extract of *Alternaria tenuis* in patients with rhinitis and bronchial asthma. *J. Investig. Allergol. Clin. Immunol.*, **10**, 327-333.

Takahata, Y., Sugita, T., Kato, H., *et al.* (2007) Cutaneous *Malassezia* flora in atopic dermatitis differs between adults and children. *Br. J. Dermatol.*, **157**, 1178-1182.

Tan, B.B., Weald, D., Strickland, I., and Friedman, P.S. (1996) Double-blind controlled trial of effect of house dust-mite allergen avoidance on atopic dermatitis. *Lancet*, **347**, 15-18.

Targonski, P.V., Persky, V.W., and Ramekrishnan, V. (1995) Effect of environmental molds on risk of death from asthma during the pollen season. *J. Allergy Clin. Immunol.*, **95**, 955-961.

Tee, R.D. (1994) Allergy to storage mites. *Clin. Exp. Allergy.*, **24**, 636-640.

Temprano, J., Becker, B.A., Hutcheson, P.S., *et al.* (2007) Hypersensitivity pneumonitis secondary to residential exposure to *Aureobasidium pullulans* in 2 siblings. *Ann. Allergy Asthma. Immunol.*, **99**, 562-566.

Tengvall Linder, M., Johansson, C., Zargari, A., *et al.* (1996) Detection of *Pityrosporum orbiculare* reactive T cells from skin and blood in atopic dermatitis and characterization of their cytokine profiles. *Clin. Exp. Allergy*, 26, 1286-1297.

Tengvall Linder, M., Johansson, C., Bengtsson, Å., *et al.* (1998) *Pityrosporum orbiculare*-reactive T-cell lines in atopic dermatitis patients and healthy individuals. *Scand. J. Immunol.*, **47**, 152-158.

Tengvall Linder, M., Johansson, C., Scheynius, A., and Wahlgren, C.-F. (2000) Positive atopy patch test reactions to *Pityrosporum orbiculare* in atopic dermatitis patients. *Clin. Exp. Allergy*, **30**, 122-131.

Terreehorst, I., Hak, E., Oosting, A.J., *et al.* (2003) Evaluation of impermeable covers for bedding in patients with allergic rhinitis. *N. Engl. J. Med.*, **349**, 237-246.

Thomas, W.R., Smith, W.A., Hales, B.J., *et al.* (2002) Characterization and immunobiology of house dust mite allergens. *Int. Arch. Allergy Immunol.*, **129**, 1-18.

Thomas, W.R., Heinrich, T.K., Smith, W.A., and Hales, B.J. (2007) Pyroglyphid house dust mite allergens. *Protein Pept. Lett.*, **14**, 943-953.

Thorn, J. (2001) The inflammatory response in humans after inhalation of bacterial endotoxin: a review. *Inflamm. Res.*, **50**, 254-261.

Thorn, J., and Rylander, J. (1998) Inflammatory response after inhalation of bacterial endotoxin assessed by the induced sputum technique. *Thorax*, **53**, 1047-1052.

Tiberg, E. (1987) Microalgae as aeroplankton and allergens. In G. Boehm and R. Leuschner, (eds.), *Advances in Aerobiology, Proceedings of the Third International Conference on Aerobiology, 1986*, Birkhäuser Verlag, Basel, Switzerland, pp. 171-173.

Tiberg, E., Rolfsen. W., Einarsson, R., and Dreborg, S. (1990) Detection of *Chlorella*-specific IgE in mould-sensitized children. *Allergy*, **45**, 481-486.

Tiberg, E., Dreborg, S., and Bjorksten, B. (1995) Allergy to algae (*Chlorella*) among children. *J. Allergy Clin. Immunol.*, **96**, 257-259.

Tillie-Leblond, I., and Tonnel, A.B. (2005) Allergic bronchopulmonary aspergillosis. *Allergy*, **60**, 1004-1013.

Tovey, E.R., Johnson, M.C., Roche, A.L., *et al.* (1989) Cloning and sequencing of a cDNA expressing a recombinant house dust mite protein that binds human IgE and corresponds to an important low molecular weight allergen. *J. Exp. Med.*, **170**, 1457-1462. [Published erratum in *J. Exp. Med.*, **171**, 1387 (1990).]

Tovey, E., and Marks, G. (1999) Methods and effectiveness of environmental control. *J. Allergy Clin. Immunol.*, **103,** 107-191.

Tuomi, T., Reijula, K., Johnsson, T., *et al.* (2000) Mycotoxins in crude building materials from water-damaged buildings. *Appl. Environ. Microbiol.*, **66**, 1899-1904.

Ulmer, A.J. (1997) Biochemistry and cell biology of endotoxins. *Int. J. Occup. Environ. Health*, **3**, S8-S17.

Ulrik, C.S., Backer, V., Hesse, B., and Dirksen, A. (1996) Risk factors for development of asthma in children and adolescents: findings from a longitudinal population study. *Respir. Med.*, **90**, 623-630.

van der Zwan, J.C., Orie, N.M.G., Kaufmann, H.F., *et al.* (1982) Bronchial obstructive reactions after inhalation with endotoxin and precipitinogens of *Haemophilus influenzae* in patients with chronic non-specific lung disease. *Clin. Allergy*, **12**, 547-559.

van Hage-Hamsten, M., Ihre, E., Zettestrom, O., and Johansson, S.G.O. (1988) Bronchial provocation studies in farmers with positive RAST to the storage mite *Lepidoglyphus destructor*. *Allergy*, **43**, 545-551.

van Hage-Hamsten, M., and Johansson, E. (1998) Clinical and immunologic aspects of storage mite allergy. *Allergy*, **53**, S49-S53.

van Ree, R. (2007) Indoor allergens. Relevance of major measurement and standardization. *J. Allergy Clin. Immunol.*, **119**, 270-277.

Varela, J., Ventas, P., Carreria, J., *et al.* (1994) Primary structure of Lep d I, the main *Lepidoglyphus destructor* allergen. *Eur. J. Biochem.*, **225**, 93-98.

Vaughan, L.M. (1993) Allergic bronchopulmonary aspergillosis. *Clin. Pharmacy*, **12**, 24-33.

Veillette, M., Cormier, Y., Israël-Assayaq, E., *et al.* (2006) Hypersensitivity pneumonitis in a hardwood processing plant related to heavy mold exposure. *J. Occup. Environ. Hyg.*, **3**, 301-307.

Venugopal, T. (2005) Evolution and expression of translationally controlled tumour protein (TCTP) of fish. *Comp. Biochem. Physiol. B, Biochem. Mol. Biol.*, **142**, 8-17.

Verhoeff, A.P., van Strien, R.T., van Wignen, J.H., and Brunekreef, B. (1995) Damp housing and childhood respiratory symptoms: the role of sensitization to dust mites and molds. *Am. J. Epidemiol.*, **141**, 103-110.

Vesper, S.J., McKinstry, C., Haugland, R., *et al.* (2007) Development of an environmental relative moldiness index for US homes. *J. Occup. Environ. Med.*, **49**, 829-833.

Virnig, C., and Bush, R.K. (2007) Allergic bronchopulmonary aspergillosis: a US perspective. *Curr. Opin. Pulm. Med.*, **13**, 67-71.

von Ehrenstein, O.S., von Mutius, E., Illi, S., *et al.* (2000) Reduced risk of hay fever and asthma among children of farmers. *Clin. Exp. Allergy*, **30**, 187-193.

von Mutius, E., Braun-Farländer, C., Schierl, R., *et al.* (2000) Exposure to endotoxin or other bacterial components might protect against the development of atopy. *Clin. Exp. Allergy*, **30**,

1230-1234.

Voorhorst, R., Spieksma-Boezeman, M., and Spieksma, F.T.M. (1964) Is a mite (*Dermatophagoides* sp.) the producer of the house-dust allergen? *Allergie Asthma (Leipzig)*, **10**, 329-334.

Wagner, S., Sowka, S., Mayer, C., *et al.* (2001) Identification of a *Hevea brasiliensis* latex manganese superoxide dismutase (Hev b 10) as a cross-reactive allergen. *Int. Arch. Allergy Immunol.*, **125**, 120-127.

Wan Ishlah, L., and Gendeh, B.S. (2005) Skin prick test reactivity to common airborne pollens and molds in allergic rhinitis patients. *Med. J. Malaysia*, **60**, 194-200.

Wahn, U., Lau, S., Bermann, R., *et al.* (1997) Indoor allergen exposure is a risk factor for sensitization during the first three years of life. *J. Allergy Clin. Immunol.*, **99**, 763-769.

Wananukul, S., Huiprasert, P., and Pongprasit, P. (1993) Eczematous skin reaction from patch testing with aeroallergens in atopic children with and without atopic dermatitis. *Pediatr. Dermatol.*, **10**, 209-213.

Weichel, M., Schmid-Grendelmeier, P., Flückiger, S., *et al.* (2003) Nuclear transport factor 2 represents a novel cross-reactive fungal allergen. *Allergy*, **58**, 198-206.

Weitzmann, I., and Summerbell, R.C. (1995) The dermatophytes. *Clin. Microbiol. Rev.*, **8**, 240–259.

Wickman, M. (1997) Prevention and non-pharmacologic treatment of mite allergy. *Allergy*, **52**, 369-373.

Wilhelmsson, B., Jernudd, Y., Ripe, E., and Holmberg, K. (1994) Nasal hypersensitivity in wood furniture workers: an allergological and immunological investigation with special reference to mould and wood. *Allergy*, **49**, 586-595.

Williams, D.L., Pretus, H.A., McNamee, R.B., *et al.* (1991) Development, physicochemical characterization and preclinical efficacy evaluation of a water-soluble glucan sulphate derived from *Saccharaomyces cerevisiae*. *Immunopharmacology*, **22**, 139-155.

Williamson, I.J., Martin, C.J., McGill, G., *et al.* (1997) Damp housing and asthma, a case control study. *Thorax*, **52**, 229-234.

Winck, J.C., Delgado, L., Murta, R., *et al.* (2004) Antigen characterization of major cork molds in suberosis (cork worker's pneumonitis) by immunoblotting. *Allergy*, **59**, 739-745.

Woda, B.A. (2008) Hypersensitivity pneumonitis: an immunopathology review. *Arch. Path. Lab. Med.*, **132**, 204-205.

Woertz, J.R., Kinney, K.A., Kraakman, N.J.R., *et al.* (2002) Mite growth on fungus under various environmental conditions and its potential application to biofilters. *Exp. Appl. Acarol.*, **27**, 265-276.

Wollenberg, A., and Bieber, T. (2000) Atopic dermatitis: from the genes to skin lesions. *Allergy*, **55**, 205-213.

Wong, G.W., Li, S.T., Hui, D.S., *et al.* (2002) Individual allergens as risk factors for asthma and bronchial hyperresponsiveness in Chinese children. *Eur. Respir. J.*, **19**, 288-293.

Woodcock, A., Forster, L., Matthews, E., *et al.* (2003) Control of exposure to mite allergen and allergen-impermeable bed covers for adults with asthma. *N. Engl. J. Med.*, **349**, 225-236.

Woodfolk, J.A. (2005) Allergy and dermatophytes. *Clin. Microbiol. Rev.*, **18**, 30-43.

Woodfolk, J.A., Slunt, J.B., Deuell, B., *et al.* (1996) Definition of a *Trichophyton* protein associated with delayed hypersensitivity in humans: evidence for immediate (IgE and IgG4) and delayed type hypersensitivity to a single protein. *J. Immunol.*, **156**, 1695-1701.

Woodfolk, J.A., Wheatley, L.M., Tyasena, R.V., *et al.* (1998) *Trichophyton* antigens associated with IgE antibodies and delayed type hypersensitivity. Sequence homology to two families of serine proteinases. *J. Biol. Chem.*, **273**, 29489-29496.

Woodfolk, J.A., Sung, S.J., Benjamin, C., *et al.* (2000) Distinct human T cell repertoires mediate immediate and delayed-type hypersensitivity to the *Trichophyton* antigen, Tri r 2. *J. Immunol.*, **165**, 4379-4387.

Woodfolk, J.A., and Platts-Mills, T.A.E. (2001). Diversity of the human allergen-specific T cell repertoire associated with distinct skin test reactions: delayed-type hypersensitivity-associated major epitopes induce Th1- and Th2-dominated responses. *J. Immunol.*, **167**, 5412-5419.

Yamada, J., Hamuro, J., Hatanaka, H., *et al.* (2007) Alleviation of seasonal allergic symptoms with superfine β-1,3-glucan: a randomized study. *J. Allergy Clin. Immunol.*, **119**, 1119-1126.

Yasueda, H., Mita, H., Akiyama, K., *et al.* (1993) Allergens from *Dermatophagoides* mites with chymotryptic activity. *Clin. Exp. Allergy*, **23**, 384-390.

Yazicioglu, M., Asan, A., Anes, U., *et al.* (2004) Indoor airborne fungal spores and home characteristics in asthmatic children from Edirne region of Turkey. *Allergol. Immunopathol. (Madrid)*, **32**, 197-203.

Yoshida, K., Ando, M., Sakata, T., and Araki, S. (1989) Prevention of summer-type hypersensitivity pneumonitis: effect of elimination of *Trichosporon cutaneum* from the patients homes. *Arch. Environ. Health*, **44**, 317-322.

Zacharasiewicz, A., Zidek, T., Haidinger, G., *et al.* (1999) Indoor factors and their association to respiratory symptoms suggestive of asthma in Austrian children aged 6-9 years. *Wien. Klin. Wochenschr.*, **111**, 882-886.

Zhiping, W., Malmber, P., Larsson, B.M., *et al.* (1996) Exposure to bacteria in swine-house dust and acute inflammatory reactions in humans. *Am. J. Respir. Crit. Care Med.*, **154**, 1261-1266.

Zock, J.P., Jarvis, D., Luczynska, C., *et al.* (2002) Housing characteristics, reported mold exposure, and asthma in the European Community Respiratory Health Survey. *J. Allergy Clin. Immunol.*, **110**, 285-292.

Zureik, M., Neukirch, C., Leynaert, B., *et al.* (2002) Sensitization to airborne moulds and severity of asthma: cross-sectional study from European Community respiratory health study. *Br. Med. J.*, **325**, 411-414.

Chapter 3.2

OCCUPATIONAL RESPIRATORY DISEASE: HYPERSENSITIVITY PNEUMO-NITIS AND OTHER FORMS OF INTERSTITIAL LUNG DISEASE

Michael J. Hodgson[1] and Brian Flannigan[2]

[1]Office of Public Health and Environmental Hazards , Veterans Health Administration, Washington, DC, USA;
[2]Scottish Centre for Pollen Studies, Napier University, Edinburgh, UK.

INTRODUCTION

Among the diseases related to the built environment, hypersensitivity pneumonitis (HP) is the one most clearly recognized by health practitioners as attributable to some specific source, despite its rarity. In that, it differs dramatically from asthma, despite evidence that building-related asthma is a far more widespread problem (Sieber *et al*. 1996, Hodgson *et al*. 1998, Fisk *et al*. 2007) and widely acknowledged since the 2004 report from the Institute of the Medicine on Damp Indoor Spaces And Health (IOM 2004). A single case of HP suggests the need for a broader public health approach, as the conditions leading to a single case are likely to put additional members of the exposed cohort at risk (Rutstein *et al*. 1983,1984). In fact, the one well-recognized published report of building-related asthma, among office workers (Hoffmann *et al*. 1993), is associated with additional cases of HP, bronchiolitis obliterans and usual interstitial pneumonitis (UIP). In addition, in at least one outbreak of HP, a substantial proportion of affected individuals developed onset of wheezing 1-3 h after exposure (Hodgson *et al*. 1987). This suggests that outbreaks of one form of building-related allergic disease are not infrequently associated with other forms. In fact, recent publications present evidence indicating that buildings are the single most common source of work-related asthma in the USA (Cooper *et al*. 1997, Jajoski *et al*. 1999 Fisk *et al*. 2007), so pointing to the need for far more intensive scrutiny of building-associated disease. Substantial evidence exists that a similar disease, sarcoidosis, is also related to moisture and fungal exposures indoors. UIP has been associated with several occupations in which bioaerosols exposure is prominent. Each of these will be discussed individually.

Note: The opinions expressed here do not reflect the opinions of the Department of Veterans Affairs or the U.S. Government.

Separately from the recognition of potential causal linkages between disease and environmental exposures, clinicians must address the actual identification of diseases among exposed individuals and and the attribution to a specific exposure. The "sentinel health events" mode of practice in occupational and environmental medicine (Rutstein 1984, Matte *et al*. 1990, Mullan and Murthly 1991) is based on the recognition that diseases may be related to specific risk factors, which may guide specific actions for individual cases and lead to broad-based intervention strategies by identifying both work exposures that are remediable and exposed cohorts that require some screening for further disease. The common failure to seek or recognize additional cases of disease leads to the dramatic under-recognition and reporting of occupational disease throughout the developed countries. A single case of HP should always lead to scrutiny of the group of individuals exposed to the same environment, since identification of the source is of paramount importance. Such scrutiny generally occurs through the use of screening questionnaire surveys, together with physiologic or imaging confirmation of abnormalities. For the diseases in question here, a number of instruments have been used (Fig. 1). They have been used for HP, not just in the built environment, but in other occupational exposures such as metalworking fluids and pigeon breeding.

HYPERSENSITIVITY PNEUMONITIS

History and Epidemiology

Since publication of the report of Fink *et al*. (1971), HP has been recognized as one of the potential consequences of contaminated ventilation systems.

Figure 1. Typical questionnaire (asssembled from Abramson *et al*. 1991, Arnow *et al*. 1978, Finnegan *et al*. 1981).

DATE: ___ ___/___ ___/___ ___   TIME ___ ___:___ ___ AM PM

NAME: _____ _____
                    (last)   (first)
ADDRESS   _____
             _____

II. DEMOGRAPHIC DATA

Age ___ ___   Gender female ___   male ___

Have you ever smoked?   yes___   no___

If yes: How old were you when you first started smoking regularly? ___ ___ (years)

           Do you still smoke?   yes___   no___

           If no: How old were you when you stopped?

           On average how much did you smoke over the whole time you smoked?
           ___ ___ (cigarettes per day)

Total number of years spent in school ___ ___
(for example: high school, college, 1 year of graduate school: 12 + 4 + 1 + 17)

Do you take any medications for chronic health problems?   yes___   no___

     If yes: please list _____, _____

**PLEASE RATE ON AVERAGE HOW SEVERELY YOU HAVE EXPERIENCED EACH OF THE
FOLLOWING SYMPTOMS IN THE COURSE OF THE LAST WEEK BY MAKING A LINE AT
THE POINT THAT BEST DESCRIBES THE LEVEL OF YOUR SYMPTOMS.**

Generalized aching   --

Tiredness at work   --

Feverishness   --

Coughing   --

Chest tightness   --

Wheezing   --

Shortness of breath   --

Sore throat   --
                       none   mild   moderate   severe

In the past 12 months have you had more than TWO episodes of dryness of the eyes?   yes___ no___
          b. if yes, in the winter did you have this   most days   ___1

most weeks	___2
most months	___3
less often	___4

c. if yes: is it now better____   the same__   worse__

In the past 12 months have you had more than TWO episodes of itching or watering of the eyes?

yes__ no__

b. if yes, in the winter did you have this   most days   ___1

most weeks   ___2

most months   ___3

less often   ___4

a.   if yes: is it now   better____   the same__   worse__

In the past 12 months have you had more than TWO episodes of blocked or stuffy nose? yes__ no__

b. if yes, in the winter did you have this   most days   ___1

most weeks   ___2

most months   ___3

less often   ___4

a.   if yes: is it now   better____   the same__   worse__

In the past 12 months have you had more than TWO episodes of runny nose?   Yes__ no__

b. if yes, in the winter did you have this   most days   ___1

most weeks   ___2

most months   ___3

less often   ___4

c. if yes: is it now better____   the same__   worse__

In the past 12 months have you had more than TWO episodes of a dry throat?   yes__ no__

b. if yes, in the winter did you have this   most days   ___1

most weeks   ___2

most months   ___3

less often   ___4

c. if yes: is it now better____   the same__   worse__

In the past 12 months have you had more than TWO episodes of lethargy and/or tiredness? yes__ no__

b. if yes, in the winter did you have this   most days   ___1

most weeks   ___2

most months   ___3

less often   ___4

c. if yes: is it now better____   the same__   worse__

In the past 12 months have you had more than TWO episodes of headache?   yes__ no__

b. if yes, in the winter did you have this   most days   ___1

most weeks   ___2

most months   ___3

less often   ___4

c. if yes: is it now better____   the same__   worse__

In the past 12 months have you had more than TWO episodes of flu-like illness
(including aches in limbs and/or fever)   yes__ no__

b. if yes, in the winter did you have this   most days   ___1

most weeks   ___2

most months   ___3

less often   ___4

c. if yes: is it now better____   the same__   worse__

In the past 12 months have you had more than TWO episodes of difficulty in breathing? yes__ no__

b. if yes, in the winter did you have this   most days   ___1

                                                 most weeks   ___2

                                                 most months   ___3

                                                 less often   ___4

c. if yes: is it now better____   the same__   worse__

In the past 12 months have you had more than TWO episodes of feeling chest tightness? yes__ no__

b. if yes, in the winter did you have this   most days   ___1

                                                   most weeks   ___2

                                                 most months   ___3

                                                 less often   ___4

c. if yes: is it now better____   the same__   worse__

SYMPTOMS AT THE PRESENT TIME

Do you usually have a cough?   yes__ no____

if yes   b. Does this occur?   sometimes__   usually__   rarely__

            c.   Does this occur on weekends or on vacations?   yes____ no____

            d.   Did you have this in the winter?   yes____ no____

            e.   if yes: is it now   better____   the same__   worse__

Do you ever have episodes of shortness of breath unrelated to exertion?   yes____ no____

if yes   b. Does this occur   sometimes__   usually__   rarely__

            c.   Does this occur on weekends or on vacations?   yes____ no____

            d.   Did you have this in the winter?   yes____ no____

            e.   if yes: is it now   better____   the same__   worse__

Do you feel feverish during the day or night?   yes____ no____

if yes,   b.   Does this occur   sometimes__   usually__   rarely__

            c.   Does this occur on weekends or on vacations?   yes____ no____

            d.   Did you have this in the winter?   yes____ no____

            e.   if yes: is it now   better____   the same__   worse__

Do you wheeze during the day or night?   yes____ no____

if yes,   b.   Does this occur   sometimes__   usually__   rarely__

            c.   Does this occur on weekends or on vacations?   yes____ no____

            d.   Did you have this in the winter?   yes____ no____

            e.   if yes: is it now   better____   the same__   worse__

Do you get chills during the day or night?   yes____ no____

if yes,   b.   Does this occur   sometimes__   usually__   rarely__

            c.   Does this occur on weekends or on vacations?   yes____ no____

            d.   Did you have this in the winter?   yes____ no____

            e.   if yes: is it now   better____   the same__   worse__

Do you have a feeling of aching all over?   yes____ no____

if yes,   b. Does this occur   sometimes__   usually__   rarely__

            c.   Does this occur on weekends or on vacations?   yes____ no____

            d.   Did you have this in the winter?   yes____ no____

            f.   if yes: is it now   better____   the same__   worse__

Do you ever become short of breath while walking on the level or up a slight hill?   yes____ no____

Can you keep up with others your own age?   yes____ no____

Do you ever have to stop while walking at your own pace on the level?   yes____ no____

Do you feel you can no longer exercise as hard as you could two years ago?   yes___ no___

Have you had wheezing or whistling in your chest at any time in the last 12 months?
          YES__ NO__
Have you woken up with a feeling of tightness in your chest first thing in the morning at any time
in the last 12 months?
          YES__ NO__
Have you at any time in the last 12 months had an attack of shortness of breath that came on during the day when
you were not doing anything strenuous?
          YES__ NO__
Have you had an attack of shortness of breath that came on after you stopped exercising at any time in the last
12 months?
          YES__ NO__
Have you at any time in the last 12 months been woken at night by an attack of shortness of breath?
          YES__ NO__
Have you at any time in the last 12 months been woken at night by an attack of coughing?
          YES__ NO__
Which of the following statements best describe your breathing?
          ___ I never or only rarely get trouble with my breathing
          ___ I get regular trouble with my breathing but it always gets completely better.
          ___ My breathing is never quite right
When you are in a dusty part of the house or with animals (for instance dogs, cats, or horses) or near pillows
(including pillows, quilts, and eiderdowns) do you ever get a feeling of tightness in your chest?
          YES__ NO__
Have you ever had an attack of asthma?
          YES__ NO__
Have you had an attack of asthma at any time in the last 12 months?
          YES__ NO__
When did you first develop any of the above symptoms regularly? _____ (month)

                                             ___ ___ ___ ___ (year)

The interstitial lung diseases - HP, sarcoidosis and others - occur at a prevalence rate of about 70 per 10^6 persons and an incidence of 30 per 10^6 person years. Less than 3% of these cases represent HP, the disease most clearly associated with moisture and mould; over 80% are considered idiopathic interstitial pneumonitis. Data presented by the National Institute for Occupational Safety and Health (NIOSH) suggest that HP in North America represents primarily a disease of farmers (NIOSH 1996). Pigeon breeding is recognized as the second most frequent cause of disease, at least in the UK. In USA, buildings and ventilation systems are frequently recognized as a source of HP, in parallel with their being commonly contaminated by bioaerosols (Batterman and Burge 1995), although they rarely lead to fatalities.

HP is not a new disease. Ramazzini (1713) recognized that chest symptoms were occupational consequences among workers with organic dust exposures. The clinical syndrome of HP was first described by Campbell (1932). Replacement of square baling of hay with round baling has led to a dramatic de-crease in mouldiness of hay and a dramatic decrease in the incidence of disease. Subsequently, HP has been attributed primarily to large molecular weight antigens associated with bioaerosols in a wide range of environments. Etiological lists often attribute disease to both particular industries, which can range, for example, from floor malting to enzyme detergent manufacture, and to a specific microbial agent. Occupational processes have consistent features from workplace to workplace, so that similarities in materials being processed, the processes themselves and the use of water lead to characteristic combinations of nutrient media and wetness. These similar conditions are then associated with predictably similar microorganisms. As discussed in Chapter 2.1, in buildings where nutrients and moisture conditions (the water activity of the substrate) are similar, predictably similar microorganisms will grow. Residential heating systems, mushroom compost and badly stored (moist) hay or grain may not seem to be similar. However, they can all provide the nutrients, moisture and elevated temperatures which favour the growth of the thermophilic actinobacteria (actinomycetes), *Saccharopolyspora* (*Faenia*) *rectivirgula* and *Thermoactinomyces vulgaris*, responsible for the HP disease known as farmer's lung. Less frequently, small molecular weight antigens are associated with HP. Characteristic of these agents is their reactivity; most are intermediaries or monomers in manufacturing, such as isocyanates and anhydrides.

Mechanisms

The mechanisms underlying HP have been reviewed elsewhere (Salvaggio 1997, Daroowalla and Raghu 1997, Girard *et al*. 2009); they include aspects of both Type III and Type IV reactions. The reaction is steered by Th(1) cells and IgG and, in its chronic form, accompanied by fibrosis. Bronchoalveolar lavage demonstrates lymphocytosis and preponderance of CD8+ cells (Kurup *et al*. 2006). The presence of granulomas clearly documents the importance of cell-mediated phenomena, as do animal models. The immunopathogenesis therefore involves both cellular immunity and antibody responses to inhaled antigens. Still, similar antigens may cause asthma and hypersensitivity pneumonitis. The presence of symptom onset 4-8 h after exposure, complement-fixing antibodies, and Arthus reactions after skin testing support the importance of Type III mechanisms.

Clinical presentation and diagnostic strategies

Patients generally present with one of three forms – acute, subacute or chronic (Hanak *et al*. 2007). Acute episodes occur presenting either as symptoms in a temporal pattern or as acute hospitalization, sometimes with hypoxia (hypoxaemia) requiring mechanical ventilation. With well-established disease, patients generally present with both chest (chest tightness, coughing, dyspnoea, and wheezing) and systemic symptoms (generalized aching, feverishness, and chills). Fatigue is a prominent symptom and in some early disease fatigue is the only presenting symptom, even in individuals who are subsequently shown to have biopsy-confirmed disease. Therefore, prominent fatigue with some nausea should at least trigger the question of moisture and potential exposure to bioaerosols.

The usual diagnostic strategies rely on documentation of disease (restrictive changes on objective lung function testing), documentation of exposure (see later), and linkage.

Documentation of disease

Imaging

Increasing evidence suggests that chest radiography, the standard primary technique for documentation, is quite insensitive (Hodgson *et al*. 1989). In the one population-based investigation, <10% of chest x-rays and <50% of thin-section CT scans were abnormal (Lynch *et al*. 1992). Data from this outbreak suggest that few patients had abnormal lung function tests, the second usual justification for diagnosis (Rose *et al*. 1998). This same phenomenon has been documented in another large outbreak of HP associated with exposure to metal-working fluids (Hodgson *et al*. 2001). Therefore, although most reviews discuss the utility, and need, for radiographic abnormalities (Girard *et al*. 2009) in general, high-resolution thin section CT scanning is the appropriate diagnostic imaging approach to documenting HP (Silva *et al*. 2007).

Physiology

The diagnostic tests most commonly relied upon are those that yield results consistent with restrictive lung disease. In lung function testing, decrements in forced vital capacity, total lung volume, and functional residual capacity are commonly encountered. Decreased carbon monoxide diffusing capacity, either single breath or preferably steady state, are frequently seen. Antibodies to agents in the environment are of interest but remain very difficult to interpret and have been viewed as indicators of exposure rather than of disease.

Biopsy

A relatively invasive technique, lung biopsy, is the only specific diagnostic technique to characterize disease identified through sensitive screening instruments such as questionnaires. Lung biopsy is rarely employed in UK, where the general view of respiratory physicians is summarized by Murphy *et al*. (1995): "Lung biopsy is neither necessary nor justifiable in most cases of allergic alveolitis. However, the disease may occasionally be suspected but no allergen identified. In such cases, a biopsy may be necessary to prove the diagnosis or to exclude other possibilities". Nevertheless, in an outbreak among lifeguards at a recreational pool in USA, a substantial proportion demonstrated disease on biopsy without other evidence of clinical disease, including early airways closure. Similar findings have been encountered in other outbreaks (Hodgson *et al*. 2001), and there is currently disagreement on the need for, and utility of, biopsies across the USA, especially in early cases where disease is most readily completely reversible. In general, the presence of multiple abnormalities on physiological, imaging and laboratory testing in the setting of a likely exposure with reasonably convincing temporal, obviates the need for pulmonary biopsy (Zacharisen *et al.* 1998, Dangmann *et al.* 2004). On the other hand, convincing clinical symptoms without objective evidence, especially in an as yet undocumented etiology or in isolated cases, generally require a tissue diagnosis in USA, particularly because of the legal implications.

Documentation of exposure

The goal of exposure assessment is primarily documenting the presence of a causal agent, not documentation of specific exposure levels. None of the agents to which HP has been attributed have quantitative exposure levels that support the distinction of safe from dangerous environments. Fundamentally, once sensitization has occurred, complete exposure control is generally required. Otherwise, individuals are unable to return to work. In industries clearly recognized to be at risk for HP, the combination of moisture and typical nutrients leads to a characteristic microbial growth. In most forms of HP, either the microbiota or the characteristic agents, such as avian proteins, have been identified and may be sampled in standard ways. Exposure documentation is then useful, primarily in the context of scientific work. On the other hand, outbreaks of HP, and even individual cases (Kreiss and Hodgson 1984), have generally been associated with unwanted moisture in the built environment. As part of the Centers for Disease Control in USA, NIOSH therefore suggested seven steps that should be taken to prevent moisture incursion into buildings (Anon. 1984). Including types other than HP, outbreaks of interstitial lung disease described since 1984 have generally confirmed the paramount importance of moisture. Because many outbreaks actually present with multiple forms of lung disease (asthma, HP, bronchitis, mucosal irritation), it is instructive to examine building factors associated with the various described outbreaks rather than focus on HP alone. Detailed engineering analyses, which systematically reviewed all potential causes of building moisture, have been carried out in only a few outbreaks of HP associated with buildings. Systematic approaches have been discussed elsewhere (Prezant *et al.* 2008). Questionnaire approaches to exposure assessment do exist (Mahooti-Brooks *et al.* 2004, Park *et al.* 2004, 2008) and have been validated through quantitative

sampling approaches. In general, they rely first on documentation of the extent of exposure in simple descriptions of contaminated surfaces in square feet or yards based on an assessment initially described in the New York City Health Department guidelines on mould remediation (10 ft², 30 ft², 100 ft², or 1, 3, 10 m², and contamination of the HVAC system), which were first published more than 10 years ago, but were revised in 2008 (NYC 2008).

Unwanted water may result from failures in construction management, from deterioration in buildings, from inadequate maintenance, from internal water sources, from site planning problems, or from enthalpy problems related to ventilation system sizing. The effect of such moisture problems on microbiological air quality has been discussed in Chapter 2.1. Such problems may be classified according to the system to which they are attributed (mechanical ventilation, building envelope, plumbing, etc.) or to the phase of the building's life (construction, commissioning, operations, maintenance, etc.). As mentioned in Chapter 2.1, common causes of construction moisture include inadequate drying time for wet building materials, inadequate curing times for concrete, and construction defects related to envelope integrity (roof flashing, joint sealing on internal drains, etc.). Inadequate maintenance includes failure to detect and repair breaks in pipework, either within walls or slabs, or in other hidden locations; to detect and remedy contamination in ventilation system filters, sound insulation liners and inadequately draining drain pans; and to deal with carpet contamination. Another recognized problem relates to building siting and drainage, in USA a problem commonly encountered in schools. There, because developers frequently give to the community for the construction of schools the most poorly draining land, as it less desirable for property owners, the schools are often built in areas with high water tables, i.e. swampy areas. Consequently, moisture seeping up through concrete slabs has been identified as a problem in several recent outbreaks of building-related lung disease.

Older investigations generally failed to present detailed descriptions of the HVAC system or of the spectrum of moisture problems in the affected building. In general, they did identify a specific mechanical system or set of conditions that served as a reservoir and amplifier for bioaerosols. Because investigators in the past stopped after identifying a single apparent source, it remains unclear whether other contributory sources would have been revealed by fuller investigation. Careful scrutiny of the built environment in more recent outbreaks has generally identified more than one source, or problem, arising from inadequate moisture control (Welterman *et al.* 1998, Hodgson 1998).

Welterman *et al.* (1998) described a single case of HP associated with a building on the basis of a biopsy and convincing clinical improvement and recurrence associated with removal from and re-exposure to the building. The patient had IgG antibodies to commercial preparations of *Aspergillus* strains that were identified on surface sampling, but not in air samples. Ratios of indoor:outdoor microorganisms in bioaerosols established using an Andersen N-6 sampler were not elevated. In addition, excess symptoms consistent with HP were seen in a comparison with four control buildings, but no clear evidence of additional clinical HP was identified. The building was constructed in an area with a high water table; employed an oversized ventilation system with inadequate enthalpy control; and had water collecting in puddles on the roof, which also had some leaks.

Together with an excess of wide-ranging symptoms consistent with a systemic illness, Hodgson *et al.* (1998) observed obstructive and restrictive lung disease among occupants of a mould-affected building in Florida. This court building had evidence of moisture incursion through the envelope (leaks in flashing on the roof, and in drains within wall cavities); insulation problems, with vinyl wallpaper acting as a moisture barrier on interior surfaces; and ventilation system design problems (inadequate dehumidification), with elevated indoor relative humidity. Growth of *Aspergillus versicolor* and *Stachybotrys chartarum* was widespread under the wallpaper and on surfaces, and spores of these fungi were present in the indoor air. Indoor:outdoor ratios of *Aspergillus* were elevated. "Semi-aggressive" air sampling, i.e. sampling after opening law books and various other routine indoor activities (described in the paper), demonstrated that there were would be substantial increases in indoor bioaerosol concentrations as a result of the natural disturbance created by occupants of the building.

In general, air sampling has not been useful in the diagnosis of immunological lung disease associated with buildings. Walkthroughs have generally allowed the identification of potential reservoirs, either by smell or visual evidence of biological contamination. On the other hand, growth and characterization of the organisms identified in the workplace is often undertaken, but is painstaking, time-consuming and often less than helpful, as is described later.

Fungi can almost always be documented indoors so that exposure to fungi *per se*, without additional information, is not so useful. In some settings, for example where liquid water and its aerosolization are considered a hazard (Nordness 2003), airborne sampling may in fact be quite useful. Similarly, in a recent outbreak of sarcoidosis and asthma (Laney 2009), growth in standing water lines and associated sampling were crucial to documentation of the cause.

Linkage

Disease can be linked to environmental causes in one of three ways, epidemiologically, toxicologically or clinically. Of these, the latter two are most persuasive to physicians, although not necessarily to non-health care practitioners. The classic forms of HP disease, in pigeon breeders and farmers, are generally linked to the cause in a relatively straightforward fashion. Disease attribution related to buildings, where most of us spend over 90% of our time, is far more difficult.

Linkage of acute and subacute forms generally relies on documentation of temporal changes related to exposure. This may require energetic participation on the part of the patient and the employer or the family. The patient should be evaluated, some pertinent outcome measure should be determined, and the patient removed from the source of exposure being re-exposed, with repeated determination of the same outcome. Such strategies may help distinguish home from work-related disease, with major consequences for whichever group of individuals should also be scrutinized (family or work colleagues). This will generally narrow down the site of exposure considerably. In chronic forms, with temporal resolution in several weeks, patients may not present with such clean repeated exposure scenarios and results.

Although some investigators have attempted to use antibodies, these have long been considered markers of exposure rather than effect (Burrell and Rylander 1981). Equally importantly, in some cases, patients with disease related to a specific agent in the workplace have not demonstrated precipitins to agents identified in the air in workplaces. This has occurred for two reasons. Firstly, the immunological characteristics of some agents change as they are cultured, so that any individual strain of organism may produce responses in the laboratory different from those when it is growing in the field. Secondly, some agents are simply not identified because of the time required for sampling for such organisms. Despite frequent assumptions that precipitating antibodies to an agent grown at the implicated site are necessary, the first author has found them less than useful. The linkage to an exposure generally relies on a history of exposure to an agent that has been implicated epidemiologically. This takes a great deal of time and makes the recognition of new causes problematic. Nevertheless, for interstitial lung disease in most of its forms this is the only available technique. For sarcoidosis from beryllium (Newman *et al.* 1997), an alternative relies on the demonstration of the actual immunological response documented through lymphocyte transformation tests. Despite much discussion about the utility of such tests for the diagnosis of cell-mediated immunity in HP, this author is unaware of any cases of HP where such tests have been used or useful.

Clinical Course

When detected early, HP may resolve completely. The more frequently that episodes of acute disease occur, the more likely there is to be some longer-lasting chronic damage, as has been demonstrated in farmer's lung disease. Nevertheless, it is clear that even in the absence of repeated episodes, decrements in carbon monoxide diffusing capacity (Milton *et al.* 1995) and exercise testing (K. Kreiss, personal communication) have been demonstrated.

Interestingly, increasing numbers of "overlap" presentations are recognized. Airways hyper-reactivity, a hallmark of variable airways narrowing and asthma, has been documented in 50%-70% of patients with sarcoidosis and hypersensitivity pneumonitis, respectively. Increasingly, overlaps are seen, both as occurring in the same population (Hoffman *et al.* 1993) and in diagnostic confusion (Allmers 2000). Whether these overlaps represent simple mislabeling of non-specific inflammation associated with interstitial disease as asthma (Rose *et al.* 1998), true occurrence of asthma and interstitial disease in the same population (Hoffman *et al.* 1993, Laney *et al.* 2009), or in fact consequences of protein release and changes in cytokines driving T helper cell, Th1 and Th2, responses (Kurup *et al.* 1996) remains unclear.

SARCOIDOSIS

Over the last 10 years, a number of studies have suggested that sarcoidosis, which resembles hypersensitivity clinically and histologically, similarly represents a disease resulting from fungal exposure and moisture in the indoor environment. Sarcoidosis is meanwhile considered likely to represent an inhalation disorder, with a different immunology from HP (Newman 1997). A report of sarcoidosis appeared related to a wet school (Thorn *et al.* 1996), although the disease in this and a similar case were subsequently called HP after linkage to a building (Forst and Abraham 1993). Two recent outbreaks of sarcoidosis have meanwhile been attributed to moisture in buildings. In a Connecticut school outbreak from the mid-1990s (Hodgson *et al.* 2007) a cluster of teachers with sarcoidosis identified a school with major moisture problems. Symptomatic teachers showed significant decrements in spirometry (both forced vital capacity and forced expiratory volume in the first second), whereas asymptomatic controls did not. A cluster of sarcoidosis in a Vermont office building was attributed to moisture problems in a grossly contaminated water based air-conditioning system (Laney *et al.* 2009). These outbreaks are supported by three case-control studies showing an increased risk of sarcoidosis after indoor and moisture exposures. First, a case-control study using sarcoidosis patients as cases revealed that subjects were 11 times more likely than controls to have exposure to moulds in their workplaces, including those indoors (Ortiz *et al.* 1998). Importantly, exposure to mould in bathrooms and basements in the home was similarly associated with disease. Kucera *et al.* (2003) documented an increased risk of sarcoidosis after exposure to unwanted moisture and fungi in the homes of relatives of sarcoidosis index cases. Finally, in the multi-centre NHLBI-funded ACCESS study, Newman and Rose (2004) demonstrated a twofold risk of sarcoidosis after exposure to damp indoor spaces. These findings have been seen frequently and consistently enough that sarcoidosis should be included in the list of diseases attributed to moisture and fungi. Interestingly, individual cases of sarcoidosis, diagnosed clinically, have often been seen in buildings with active outbreaks of HP.

It remains unclear whether sarcoidosis and hypersensitivity pneumonitis represent differential genetically driven responses to a common set of antigens or whether some other cofactor, such as co-infection with a *Mycobacterium avium* complex, is at issue.

USUAL INTERSTITIAL PNEUMONITIS (UIP)

The first report of the interstitial pulmonary disease associated with fungal exposures in USA appeared in 1975, attributing disease to mycotoxins (Emanuel *et al.* 1975). Since then, controversy has reigned on the underlying etiology of nonspecific interstitial pulmonary disease after fungal exposure. Nevertheless, outbreaks of disease, case-control studies and, most importantly, both animal models and quantitative risk assessment using human cell lines, support a causal association. Outbreaks of the interstitial lung disease, i.e clusters, or even anecdotes, have been associated with fungal exposure. Hodgson *et al.* (1998) reported acute shift changes and asthma in a building contaminated with *Aspergillus versicolor* and *Stachybotrys chartarum*. Later, Hodgson *et al.* (2007) documented forced vital capacity decrements, implying the presence of restrictive disease among teachers in a wet and mouldy school. Individual case reports of nonspecific interstitial pneumonitis, distinct from HP, have been associated with fungal exposure (Lonneux *et al.* 1995). Such clusters have generated interest in documenting associations between mould and moisture and interstitial lung disease.

Several case-control studies have explored that association. Mullen *et al.* (1998) documented a tenfold risk of UIP after exposure to mould and moisture at work and at home. Baumgartner *et al.* (2000) demonstrated an increased risk of UIP in NHLBI-funded multi-centre case-control studies in occupations with predictable bioaerosols exposures, including agriculture, metal working, carpentry/woodworking and painting.

Animal models generally rely on gavage as a delivery route of biological exposures to document pulmonary disease as a consequence. This exposure route is of course overwhelming, does not respect the natural defences of the lung, and can deliver particles of substantially greater size than could be inhaled. Nevertheless, at least three separate groups of investigators have documented similar nonspecific pulmonary disease by this route. Rao *et al.* (2000) showed that spores of *Aspergillus versicolor* were associated with interstitial inflammation and that the inflammatory potential resided in a methanol extractable fraction of eluate. Nikulin *et al.* (1996), Jarvis *et al.* (1998) and Flemming *et al.* (2004) have documented similar disease. However, importantly, mouse immunology differs dramatically from that of humans, so that, in the words of Wenzel and Holgate (2006) "a mouse should know its limitations."

Experimental evidence for the association of mycotoxins and pulmonary disease was first shown by NIOSH investigators in 1987 (Sorenson *et al.* 1987). Those authors assessed cellular and immunological effects in isolated pulmonary cell lines at toxin concentrations likely to occur indoors, using alveolar macrophage survival, thymocyte proliferation, and protein-synthesis inhibition, all outcomes affected by mycotoxins released by *Stachybotrys chartarum* (recently reviewed by Pestka *et al.* 2008). The NIOSH investigators generated airborne dust, collected samples gravimetrically, extracted toxins, and created concentrations of toxins in a fluid bath corresponding to appropriate doses. They showed damage to pulmonary macrophages, implying the possibility of alveolar disease, from exposure to dusts. The ED-50 for protein synthesis is 0.006 micromolar (Sorenson *et al.* 1987, alveolar macrophages). Subsequent work shows that each *Stachybotrys* spore can contain as much as ~1 mM macrocylic trichothecenes (Yike and Dearborn 2004) released quickly (minutes) to the local aqueous environment (Yike *et al.* 2005). This implies effects three orders of magnitude lower than that expected in the immediate environment around spores or spore fragments and, therefore, likely to have very deleterious, local effects in the surrounding lung cells. These local effects appear to be the major source of the lung damage seen in the animal studies, and are the likely initiating factors in the pathophysiology seen in humans.

ORGANIC DUST TOXIC SYNDROME

Some evidence exists that office workers may develop yet another form of interstitial pulmonary response, possibly more frequently than commonly assumed, viz. organic dust toxic syndrome (ODTS), the same disease as humidifier fever (Milton 1996). Beginning with a search for humidifiers as a cause of symptoms noted in office workers (Finnegan *et al.* 1984), excess rates of chest tightness and flu-like illness have been reported. A re-analysis of several older data sets (Apter *et al.* 1997) suggested that a symptom cluster of chest tightness, difficulty in breathing and flu-like illness that was common among office workers without recognized HP may have been associated with exposure to endotoxin (Gyntelberg *et al.* 1994; Teeuw *et al.* 1994) or fungal glucan (Rylander *et al.* 1992).

TYPES OF OCCUPATIONAL HYPERSENSITIVITY PNEUMONITIS

As mentioned earlier in this chapter and Chapter 3.1, and discussed more fully in Murphy *et al.* (1995), the causes of HP are various. The remainder of this chapter is given over to four diseases drawn from the partial list of those caused by microorganisms presented in Table 3 of Chapter 3.1. Three of these diseases are related to each other in being caused by thermophilic actinomycetes (actinobacteria), and the causal organism in the fourth is a mould. The occupational and microbiological aspects of these diseases, and measures which can be taken to reduce or eliminate respiratory exposure to causal organisms in the indoor work environment, are discussed.

Farmer's Lung

Farmer's lung is probably the best-known example of an occupational HP. In a recent literature survey, Wild and Chang (2009) noted that in USA 8-540 cases of farmer's lung per 100,000 farmers have been recorded, and that HP affects 0.4-7% of the farming population. In UK, a prevalence of 420-3000 cases per 100,000 persons at risk has been reported, while epidemiological studies in France and Sweden indicated 2.5-153 cases per 1000 farmers. Fenclová *et al.* (2009) reported that farmer's lung accounted for 70% of the cases of HP in the Czech Republic in the period 1992-2005, when the HP incidence ranged from zero to 0.20 per 100,000 workers. They consider, however, that because of difficulties in diagnosing HP the actual number of cases would have been higher, i.e. HP was under-reported.

HP associated with handling of mouldy hay was first described in the early 1930's (Campbell 1932), but it was not until 30 years later that thermophilic actinomycetes were recognized as a source of "farmer's lung hay" antigens (Pepys *et al.* 1963, Pepys and Jenkins 1965). The disease is caused by inhalation exposure to spores in the dust released not just from "self-heated" or "mouldy" hay but also from similarly deteriorating grain. Hay that is damp when it goes into store, i.e. with a moisture content of 35% or more, provides a substratum on which microbial contaminants acquired during the growing season and at harvesting can grow. The potential health hazards from microorganisms associated with production of composts from various plant materials have been discussed in Chapter 2.2; the biodeterioration, or spoilage, of damp hay is the result of unintentional composting.

The microbiological changes which occur in deteriorating hay (Gregory *et al.* 1963) and moist-stored grain (see Lacey 1989) were established by P.H. Gregory, M.E. Lacey, J. Lacey and other workers at Rothamsted Agricultural Station in UK. Since damp bulk hay or bales provide a well-insulated environ-

ment, the heat generated by microbial metabolism inexorably builds up; the types that are less tolerant of rising temperatures give way to thermotolerant and thermophilic species. Among the thermotolerant fungi which develop, *Aspergillus fumigatus* is prominent, but the upper limit for its growth under ideal circumstances is 50°C or so. However, thermophilic actinomycetes can grow at temperatures above this and become the dominant microorganisms, raising the temperature to around 65°C. *Saccharopolyspora* (*Faenia*) *rectivirgula* and *Thermoactinomyces vulgaris* are usually cited as the principal causative agents of farmer's lung, but *T. vulgaris* has been subdivided into two antigenically different species. One retains that name and the other is *T. thalpophilus* (Lacey, 1989). Two other actinomycete species which have been implicated in the disease are *Saccharomonospora viridis* and *T. candidus* (Kurup 1989).

Huge numbers of spores present in self-heated hay and grain are released into the air during handling. Opening hay bales in an open shed may result in concentrations of nearly 10^8 actinomycetes/eubacteria m^{-3} air; shaking baled hay inside a farm shed may give close to 1.6×10^9 m^{-3} air; and unloading moist-stored barley from an unsealed silo released almost 1.8×10^9 m^{-3} (Lacey 1989). It is in winter when farm workers open bales indoors and fork out rations for stock that they most commonly experience HP symptoms. The incidence of farmer's lung is obviously influenced by climate. For example, in UK the disease is commoner in the wetter, milder west than in the drier east (Grant *et al.* 1972). However, differences in agricultural practice are also important; clearly, the prevalence of farmer's lung among populations engaged in dairy and beef farming is likely to be higher than among those in areas where arable farming predominates.

Typically, farmer's lung is seen as an acute disease that develops a few hours after exposure to the airborne dust from self-heated grain or hay. The initial symptoms – fever, chills, dry cough, joint pains and breathlessness – together resemble influenza, and because they do not occur at the time of the exposure may not be associated in the victim's mind with massive exposure to dust from badly deteriorated hay or grain. Basal crepitant rales, decreased pulmonary diffusing capacity, characteristic changes in radiographs and the development of precipitating IgG antibodies (Pepys and Jenkins 1965) may accompany these symptoms. Repeated exposure to the dust results in loss of weight and increasing breathlessness as the lung is increasingly infiltrated by granulomatous and fibrous tissue and, in extreme cases, is eventually transformed into a multitude of dilated air spaces surrounded by fibrous tissue ("honeycomb lung"). However, it is considered that, depending on the pattern of exposure and natural susceptibility of those exposed to the actinomycete allergens, it is more likely that in a high proportion of exposed subjects the development of the disease is more insidious (Murphy *et al.* 1995). Fatal cases of farmer's lung are rare, but in a case in Saskatchewan, Canada, in which a farmer, his wife and brother died, immunological evidence led the authors to consider that two antigens, *P. brevicompactum and P. olivicolor*, were probably new farmer's lung antigens (Nakagawa-Yoshida *et al.* 1997).

Farmer's lung is, at least theoretically, a wholly avoidable occupational disease. Since 1964, it has been officially recognized as such in UK, where affected workers are eligible for industrial compensation, but in many countries there is no such compensation. The first step to control is to educate and advise the farming community on the hazards and how to avoid them. The prime avoidance measure is the adequate and rapid drying of hay and grain, and where grain for animal feed is stored moist maintenance of a depleted-oxygen atmosphere in the silo, so that microbial growth is largely prevented. Clearly, there are years when it is difficult to dry hay satisfactorily, and in such cases mechanical handling can reduce, and the wearing of effective respirators (HSE 1998a,b) prevent, inhalation exposure to injurious spores. In climatic regions where drying is perpetually a problem, abandoning hay making in favour of silage has been advocated.

As well as stressing the importance of efficient drying of hay, straw and grain before storage, Grant *et al.* (1972) called for more extensive use of silage, better ventilation in farm buildings and introduction of mechanical feeding systems in order to reduce the prevalence of farmer's lung. They assessed the prevalence of the disease in three different areas of Scotland: one being a small-farm dairying area in the wetter west, the second being in the drier east and comprising larger, prosperous arable farms, and the third in the northern islands, with small mixed farms and crofts and rainfall midway between that in the other two areas. The prevalence of the disease was 86 cases per 1,000 farmers in the western and northern areas, but only 23 per 1,000 in the eastern area. The difference could be attributed to fewer farm workers being employed in feeding cattle on the prosperous eastern farms and the use of expensive rapid-drying and mechanical handling systems on these farms, and on the smaller western and northern farms the greater numbers of workers involved in stock feeding without the advantages of rapid drying/handling equipment to deal with crops that would generally have had a higher moisture content at harvest and consequently would suffer microbial deterioration in storage.

Much more recent evidence supporting the strategy of turning from hay to silage for feed comes from Ireland. From 1982 to 1996, when hay was the principal feed for dairy cattle in winter, incidence rates were constant, but from 1997 to 2002 a marked decline was observed with increasing use of silage. A strong positive correlation of farmer's lung with hay production (r = 0.81) and strong negative correlation with silage production (r = -0.82) was recorded. The decline in incidence is considered to be a consequence of changing agricultural practice and increasing awareness of the disease risk (Arya *et al.* 2006).

Mushroom worker's lung

A related HP occurs among workers involved in commercial production of the common cultivated mushroom, *Agaricus bisporus*, was first described in the 1950s (Bringhurst *et al.* 1959). As has been mentioned in Chapter 2.2, it is necessary to produce compost suitable for the growth of the mushroom. In the initial phase of the two-phase process, the usual mixture of moistened cereal straw and horse manure is composted outdoors. It is then heated indoors with humidified air to 60°C in special chambers and composted for a further 10 days or so at this temperature. A thermotolerant/thermophilic microbiota rapidly dominated by actinomycetes develops during this phase, and the temperature of the compost may rise locally to 70°C. The final propagule count for fungi may be around 10^6 g^{-1} compost, and for the actinomycetes and other bacteria as high as 10^{10} g^{-1} compost. The elevated temperature has a pasteurizing effect on the compost, killing mesophilic fungi which might otherwise grow during the *A. bisporus* cultivation stage. The spawn of mushroom mycelium is mechanically mixed with the compost at a rate of approximately 0.5% and incubated at 25°C, the optimum for mycelial growth, for a further two weeks. After this, the compost is dispensed into trays, covered with a layer of peat and chalk and maintained at 16-18°C in a highly humid atmosphere, the first crop being harvested after about 3 weeks and further harvests taken over the next 5 weeks or so (Carlisle and Watkinson 1994).

During the mixing of the mushroom spawn with the compost, huge numbers of thermophilic actinomycete spores may be released into the air, i.e. >10^9 CFU m^{-3} air, but the count for fungi, including the genera *Aspergillus*, *Penicillium* and *Scytalidium*, may be only 10^3 CFU m^{-3} (van den Bogart *et al.* 1993). The predominant thermophilic actinomycetes recorded by van den Bogart *et al.* (1993) were *Excellospora flexuosa*, *Thermomonospora alba*, *T. curvata* and *T. fusca*. In stationary-bed mushroom houses Kleyn *et al.* (1981) found that the total count for spent compost was 16 × 10^8 spores g^{-1}, of which more than 90% were actinomycetes and only about 5% were moulds. The total count for dust emanating from spent compost during dumping was 0.33 × 10^6 m^{-3} air. A worker called upon to dump spent compost could inhale a total of 6.4 × 10^7 microorganisms during the 2.5 h taken to complete the operation (Kleyn *et al.* 1981).

Although Sakula (1967) reported the occurrence of precipitins against *Saccharopolyspora faeni* and *Thermoactinomyces vulgaris* in the sera of mushroom worker's lung patients, when van den Bogart *et al.* (1993) tested sera of 10 Dutch mushroom growers with mushroom worker's lung against spores for antibodies of these microorganisms by a qualitative dot-ELISA all were positive for one or more of the four species mentioned in the previous paragraph. No antibodies were found among these sera in tests against other actinomycete species (*Streptomyces thermovulgaris*, *Thermoactinomyces vulgaris* and *T. sacchari*) or fungi (*Aspergillus fumigatus*, *Penicillium brevicompactum*, *P. chrysogenum*, *Scytalidium thermophilum* and *Trichoderma viride*). While the sera of 19 non-exposed individuals were negative for thermophilic actinomycetes, in 11 of 14 workers routinely involved in spawning the compost in tunnels the sera reacted positively with one actinomycete or more, the titres increasing with length of employment. On inhalation, air rich in spores of *E. flexuosa* and the three species of *Thermomonospora* provoked reaction in workers with mushroom worker's lung. This provocation and the elevated serum titres in the workers led van den Bogart *et al.* (1993) to conclude that these organisms contribute to the occurrence of mushroom worker's lung. In an earlier Dutch study, Cox *et al.* (1991) observed that measures taken to reduce the exposure to these allergenic spores included supplying pretreated compost requiring no preparation by mushroom growing personnel, wetting compost to reduce aerosols, improving air-conditioning at the work sites, and wearing masks.

While the preceding paragraphs have dealt with mushroom worker's lung caused by exposure to actinomycete spores during cultivation of *A. bisporus*, this is not the only type of mushroom cultivated in indoor environments with which respiratory symptoms are associated. It has been pointed out that, with the increasing popularity of different species of mushroom from the Far East, and cultivation of the Shiitaki mushroom (*Lentinula edodes*) in Europe, we should be cognisant of potential health hazards associated with their cultivation (Moore *et al.* 2005). For example, Tarvainen *et al.* (1991) reported that skin and respiratory symptoms developed within 2 months of exposure in a worker engaged in commercial production of Shiitake. Contact urticaria and allergic contact dermatitis, and demonstration of precipitating IgG antibodies to the spores of the mushroom itself and also elevated numbers of inflammatory cells and T lymphocytes in bronchoalveolar lavage indicated HP. On eating the raw mushrooms, another patient developed a widespread rash (exanthema) corresponding to Shiitake-induced toxicodermia previously reported. Kamm et al. (1991) found that clinical diagnosis of HP could be supported by a positive provocation test; four of six Shiitake mushroom workers, as well as all four oyster mushroom (*Pleurotus ostreatus*) workers and 18 of 28 common mushroom (*A. bisporus*) workers, whose medical history indicated possible HP, reacted positively. In Japan a relative of *P. ostreatus*, the cultivated Eringi mushroom, *P. eryngii*, has also been reported as a cause of HP (Saikai *et al.* 2002)

Mushroom worker's lung has also been reported among employees involved with Bunashimeji (*Hypsizigus marmoreus*), a commonly cultivated edible mushroom in Japan (Tanaka *et al.* 2000). A new system of cultivation in which wet wood dust rather than compost is used appears to present a particular risk to respiratory health; the gills of *H. marmoreus* "open" and release spores well before harvesting, so that "the cultivating, harvesting, and packing rooms are filled with the mushroom spores". Over a period of three years, 90% of the workers were sensitized to the spores, but only 3% developed HP (Tanaka *et al.* 2001). Chronic cough was noted in 42 of the 63 workers and, among these 42, six had ODTS, 18 postnasal drip syndrome, 15 cough variant asthma, and three eosinophilic bronchitis (Tanaka *et al.* 2002).

The spores of the Nameko mushroom, *Pholiota nameko*, have also been reported as the cause of HP among Japanese mushroom workers (Nakazawa and Tochigi 1989, Inage *et al.* 1996, Utsugi *et al.* 1999). A 47-year-old woman, who had been engaged in production of the Enoki mushroom (*Flammulina velutipes*) for 22 years, was diagnosed as having mushroom worker's lung caused not by Enoki spores but by the spores of *Penicillium citrinum* associated with the cultivation (Yoshikawa *et al.* 2006). In a follow-up, a further four out of 48 workers were similarly diagnosed (Yoshikawa *et al.* 2007).

Bagassosis

Bagassosis in the cane sugar industry
Bagassosis is another disease in which the presenting symptoms are basically the same as in farmer's lung (Murphy *et al.* 1995). Cases have been reported in USA, UK, Italy, India, Peru and Puerto Rico, and there has been a history of occurrences in the Philippines for over 40 years, the most recent being cited in an anecdotal report by Castaneda (2005). The disease occurs among workers in cane sugar mills or in industries that utilize the fibrous waste product, bagasse, which remains after the sap is extracted from crushed sugar cane (Phoolchund 1991). With around 50% of the fresh weight of bagasse being water and roughly 4% being sugar (Hunter and Perry 1946), bagasse is moist enough and contains sufficient nutrients to support microbial growth. When stacks of the baled by-product, perhaps as much as 1000 tons in a stack (Hearn 1968), are left at the sugar mills to dry outdoors in the hot and humid climate for up to 12 months, microbial activity raises the temperature. As noted in Chapter 2.2, a mixed microbiota of fungi and actinomycetes, but ultimately dominated by thermophilic actinomycetes, develops in the "self-heated" bales (Lacey 1974). If bagasse is later sent for paper or fibreboard manufacture, the badly weathered and mouldered bales to the outside are discarded and the remaining bales are fed manually into an indoor hydraulic ram, which compresses the material into smaller bales for despatch (Hearn 1968).

An investigation of two Indian sugar cane mills by Khan *et al.* (1995) found that the actinomycetes involved were *Thermoactinomyces sacchari, T. vulgaris, T. thalpophilus, Saccharomonospora viridis* and *Saccharopolyspora rectivirgula*. The principal cause of bagassosis is considered to be *T. sacchari*, but *S. rectivirgula* also appears to be involved. Among the 22% of workers with precipitating antibodies against thermophilic actinomycetes, more than one-half had positive precipitin reactions to *T. sacchari* alone, and approaching one-third to *S. rectivirgula*. In symptomatic workers the mean absorbance values for IgG antibody activity against *T. sacchari* and *S. rectivirgula* were significantly higher than in asymptomatic workers and unexposed controls. Khan *et al.* (1995) concluded that *T. sacchari* and *S. rectivirgula* are the major species causing sensi-

tization of bagasse workers in India.

Where most of the fresh bagasse is used immediately as fuel for generating the electricity and steam needed during the seasonal 5-6 months of sugar crushing and extraction, the total airborne bacterial counts are not likely to be excessively high, e.g. in Australia (Dawson *et al.* 1996) and in Japan (Ueda *et al.* 1992). Dawson *et al.* (1996) found that in two Australian mills these counts were in general below the voluntary industry standard of 10^6 m^{-3} air, and even the highest counts were substantially lower than the suggested level of concern for the general population, i.e. 10^8 m^{-3} air. All workers in sugar mills are exposed to spores associated with sugar cane, but it is where they are engaged in re-baling stored bagasse that they have the highest exposure to thermophilic actinomycetes. Hearn (1968) noted in the West Indies that the work of re-baling stored bagasse for onward transport work fell to full-time employees, whereas seasonal workers handled only bales of fresh bagasse and were therefore at less risk of exposure to these agents of bagassosis.

Ueda *et al.* (1992) observed that dust levels in an Okinawan sugar refinery were relatively low, except for a storage room in which 1-3% of the bagasse was processed and bagged for use as fertilizer, feed and a mulch for seedbeds. Those engaged in this work were "exposed to some mouldy bagasse". Although 10% of the refinery workers had abnormal chest radiographs, 6% had positive precipitin tests to bagasse extracts and 20% were positive to at least one of seven antigens (including *Aspergillus fumigatus, S. rectivirgula* and *T. vulgaris*), there were no confirmed cases of bagassosis. In Queensland, when bagasse stored since the end of the previous season was reclaimed for fuel at the beginning of the cane crushing season the total counts for airborne bacteria were >10^6 m^{-3} air (Dawson *et al.* 1996). During the rest of the season the counts were lower than this. In general, high counts were restricted to the boiler station and bagasse transport system. Personal breathing zone spore counts were highest among workers who spent significant periods of time near the transport system. In the cane crushing season, *viable* counts of *T. sacchari* were equivalent to only 0.01-1.0% of the corresponding *total* counts for airborne bacteria. Although *total* counts for *T. sacchari* were not obtained, Dawson *et al.* (1996) concluded from the aerobiological and medical data gathered that workers in Australian sugar mills were not exposed to sufficient spores of this actinomycete to constitute a significant risk of contracting either acute or chronic bagassosis.

As indicated above, using as much of the fresh bagasse as possible for fuel during the crushing season, and thereby avoiding the need to store bagasse, greatly reduces exposure to airborne spores of *T. sacchari* and other thermophilic actinomycetes. Storing fresh bagasse loose rather than in bales improves aeration, and limits the development of these organisms by encouraging some degree of drying and slowing the build-up of heat. Although most fresh bagasse was used as boiler fuel within hours in the mills examined by Dawson *et al.* (1996), the excess was stockpiled on crushed rock and covered by tarpaulins to prevent the saturation by rain responsible for the hazardous state of the outer bales in exposed bagasse stacks (Hearn 1968). Propionic acid may be applied to fresh bagasse to inhibit microbial growth and has been shown to prevent deterioration of the fibre (Wright 1970, Lacey 1974).

Bagassosis in other industries

The use of bagasse in the manufacture of paper and particle board means that the disease is not necessarily confined to regions of the world where sugar cane is grown. For example, it occurred in UK among workers manufacturing board for interior decorating and thermal insulation from bagasse imported from Louisiana (Hunter and Perry 1946). Initially, to keep down dust the imported bales of bagasse were broken under water, but when more tightly packed bales were received this was no longer possible. The bales had to be broken by pick-axe and sections fed dry into a shredder, resulting in a very dusty atmosphere and an outbreak of respiratory illness, largely among those working at this stage of the manufacturing process. After water was routinely directed at the shredder wheel and exhaust ventilation was installed there was only one further case of what was at that time described as bronchiolitis.

In Louisiana itself, Buechner *et al.* (1964) reported that almost immediately after the 1962 opening of a board-manufacturing plant utilizing baled bagasse cases of bagassosis were diagnosed, and by the time that the paper was written it was estimated that there were about 200 such cases. Again, in referring to the first case of bagassosis in Japan, Ueda *et al.* (1992) noted that this occurred within two months of the opening of a particle board factory in 1966. Within a year, 10 of the 134 employees were affected, and a further 24 cases were diagnosed later. Although it was noted that there had been no further cases of bagassosis in Japan since the factory closed, Ueda *et al.* (1992)) indicated that there was a need for surveil-lance in lacquerware factories (where bagasse is used as a cheaper substitute for wood), as well as cane sugar factories.

Lehrer *et al.* (1978) considered that apparent absence of bagassosis among workers in a Louisiana paper mill, where there had been a considerable number affected in the past, was due to increased management awareness of the nature of the problem and the need for its control. Measures were introduced to try to retard microbial growth and reduce the amount of airborne organic dust. These included storing raw bagasse loose or unpacked and continuously sprinkling it with water to reduce airborne dust as it was moved for shredding and hammer milling. The importance of respiratory protection was also recognized and workers were provided with "face masks".

Bagassosis has also been recorded as occurring in three workers exposed to bagasse handled in the manufacture of "mud" for use in oil drilling operations (Jenkins *et al.* 1971).

Malt worker's lung

Although the higher than normal incidence of respiratory problems in the malting industry had been recognized much earlier, it was only in 1968 that the first case of the HP known as malt worker's lung was described (Riddle *et al.* 1968). The malt worker affected exhibited symptoms similar to those in farmer's lung, and it was later shown that about 5% of workers in Scottish maltings had symptoms, although not usually severe (Grant *et al.* 1976). The causative agent was identified as the mould *Aspergillus clavatus*. At 3.0-4.5 × 2.5-4.5 μm, the elliptical conidiospores or conidia of this species are somewhat larger than actinomycete spores (<2 μm), but nevertheless reach and provoke reaction at the acinar (alveolar) level. They are even more readily detached from the conidiophores that bear them than in other species of *Aspergillus*, and have walls that are particularly rich in allergens (Blyth 1978). Cases of malt worker's lung are rare and aside from the study by Grant *et al.* (1976) there is little reference to the prevalence of the disease. However, the report by Fenclová *et al.* (2009), mentioned above in connection with farmer's lung, records that for the period 1992-2005 malt worker's lung was diagnosed in seven out of a total of 72 HP cases, while there were 50 cases of farmer's lung.

The minimum water activity (a_w) for growth of this mould is approximately 0.88, much higher than, for example, *A. versicolor* and more xerophilic species of *Aspergillus* and *Eurotium* (Chapter 2.1). Its optimum a_w

is around 0.98 (Flannigan and Pearce 1994). It is a species, which therefore requires a substrate with a very high level of hydration if it is to colonize the substrate and sporulate on it. The malting of cereals provides a substrate, which is both nutritious and has the degree of hydration needed by *A. clavatus*.

In traditional floor malting (Flannigan and Pearce 1994), high-quality dried barley steeped until it is fully hydrated (M.C. around 45%) is spread over a malting floor as a layer 10-15 cm deep, allowing germination of the grain. At intervals during seven days or longer, maltsmen traverse the floor, turning the germinating grain by shovel or mechanical turner to ensure uniform aeration, prevent matting of developing rootlets and maintain the temperature throughout the bed of grain close to ambient (usually 13-16°C in temperate countries). The conditions during malting are favourable for fungal growth, and both fungi that contaminate the grains as it grows in the field (so called field fungi, or phylloplane fungi) and those that contaminate it during harvesting, post-harvest operations and storage ("storage fungi") develop and even sporulate (Flannigan 2003). The resulting green (fresh) malt is then kilned at temperatures up to 80°C to reduce its M.C. to 3-4% and the rootlets are then screened off mechanically for animal feed.

As *A. clavatus* is rarely found on dry grain elsewhere, individual consignments of substandard grain and the presence of feral pigeons in grain stores have been suggested as sources of contamination in maltings. Whatever the source, *A. clavatus* is extremely difficult to eliminate from maltings. Conventional cleaning procedures and control measures such as adding hypochlorite to steep do not appear to be effective. Consequently, any dry grain arriving or stored at malting premises is likely to become contaminated by airborne *A. clavatus* spores in the air of the premises. The spores on the grain contaminated in this way are not killed at the concentration of hypochlorite that may be added to the steep as a surface disinfectant (Flannigan *et al.* 1984). Although under the almost ideal conditions for growth provided during malting *Alternaria*, *Aureobasidium*, *Fusarium*, *Geotrichum*, *Rhizopus* and other fungi which comprise the normal mycobiota of cereals grow and even sporulate, However, if *A. clavatus* is present it becomes predominant. In extreme cases, large rafts of mycelium bearing blue-green *A. clavatus* spores may appear on the germinating barley (Shlosberg *et al.* 1991). When these areas are disturbed, the visible "smoke" of released spores may be dense enough to impede visibility (Riddle *et al.* 1968). In another case where a worker developed

HP, $>2 \times 10^5$ *A. clavatus* spores m^{-3} air were recorded (Nolard *et al.* 1988). The acts of stripping green malt from the malting floors, loading it into the kiln and then unloading it from the kiln when dry all expose maltsmen to high concentrations of airborne spores. Although mostly inactivated during kilning, the spores are still allergenic, and workers cleaning the residual rootlets from storage bins have been found to have serious respiratory symptoms. Not only that, but the spores of *A. clavatus* are so readily and widely dispersed that they may be isolated from the sputum of office workers and other employees who seldom enter the malthouse (Channell *et al.* 1969).

In recent years, cost-driven mechanization of malting has seen floor malting largely replaced by mechanically turned box (Saladin) malting or enclosed systems such as germination vessels, rotating drums or continuous malting plant. These partially or fully mechanized systems speed up the process, and therefore limit the time available for fungal growth and sporulation, so reducing exposure of workers to this respiratory hazard. Witness to the salutary effects of this increased mechanization is that 1.1% of workers in maltings with enclosed systems showed symptoms of malt worker's lung, compared with 6.8% in floor maltings (Grant *et al.* 1976). However, floor malting has not been eliminated and *A. clavatus* is still a problem in both the Southern (Rabie and Lübben 1984) and Northern Hemisphere (Gilmour *et al.* 1989, Shlosberg *et al.* 1991). In southern Africa, *A. clavatus* is one of the principal moulds encountered during the malting of sorghum in both outdoor commercial floor maltings and in enclosed industrial Saladin maltings (Rabie and Lübben 1984). The abundance of *A. clavatus* in sorghum malt produced outdoors in South Africa is not unexpected as the ambient temperature may reach 28°C, close to the optimum for growth of this species. The profuse growth of *A. clavatus* in a floor malting in Israel appeared to be the result of a heat wave (Shlosberg *et al.* 1991). The first report of malt worker's lung (Riddle *et al.* 1968) noted that the problem arose after the temperature in the green malt was allowed to rise well above the normal 16°C (in order to speed up the malting process) and drew attention to the importance of controlling temperature to limit growth of *A. clavatus*.

In UK, a Health and Safety Executive Guidance Note (HSE 1993) on grain dust in maltings set a maximum exposure limit for inhalable grain dust at 10 mg m^{-3} air for an 8-hour time-weighted average, and recommended engineering and operational control measures for reducing exposure to grain dust. Although

HSE has stated that respiratory protective equipment (RPE) is necessary for plant maintenance and cleaning operations for which other control measures are either not practicable or adequate, it stresses that RPE should normally only be employed as a last resort (HSE 1993). The note does not address the specific problem of *A. clavatus* in maltings, but stresses the importance of health surveillance of workers and the obligation to notify HSE of any cases of occupational asthma or HP. Although it is unlikely that *A. clavatus* can be eliminated entirely from the malting industry, a combination of scrupulous cleanliness, enclosed malting systems, effective temperature control and use of only high-quality, undamaged grain will reduce possible exposure to *A. clavatus* and other airborne moulds.

The importance of controlling this mould is added to by the fact that it can synthesize a range of mycotoxins (Flannigan and Pearce 1994). Among these, patulin and cytochalasin E have been shown to be produced during malting (López-Diaz and Flannigan 1997). The species can also synthesize tremorgens (Flannigan and Pearce 1994), but whether these are actually produced during malting and account for neurological symptoms seen in some victims of malt worker's lung is not known. Since the 1960s it has been known that cereals sprouted hydroponically to provide green fodder (e.g. "barley grass") can cause neuropathological symptoms and death in sheep and cattle if it has become heavily contaminated with *A. clavatus* in the process. An outline history of such outbreaks of mycotoxicosis, and of similar outbreaks caused by contaminated malting by-products (rootlets/sprouts) used in feedstuffs, can be found in Flannigan and Pearce (1994). Since the 1990s there have been other such episodes, caused by feeding of, for example, mixed fodder containing malting by-products (Sabater-Vilar *et al.* 2004) and, especially in arid regions of Australia, hydroponically produced barley and wheat (Anon. 2004, 2009, El-Hage and Lancaster 2004, McKenzie *et al.* 2004). Although there are no reports of respiratory symptoms among workers engaged in indoor hydroponic production of sprouted cereals for stockfeed, it is to be presumed that, similarly to malt workers, these workers will have been exposed to an inhalation hazard if the germinating grain is contaminated with *A. clavatus*, with the health risk being greatest where operating temperatures have risen.

Other microorganisms causing hypersensitivity pneumonitis

The role of thermophilic actinomycetes in HP has been discussed by Kurup (1989) and Murphy *et al.* (1995), but the latter have also reviewed a range of occupation-related types of HP where the causal organisms are fungi, and Flannigan *et al.* (1991) briefly noted cases of HP caused by fungi in domestic environments. These serve to confirm that HP problems arise whatever the environment, work or home, when that environment provides a blend of substrate, moisture and temperature that matches the needs of particular allergenic microorganisms. As indicated earlier, this is perfectly illustrated by thermophilic actinomycetes, which will grow in badly maintained heating, ventilation and air-conditioning systems (Kurup 1989) as well as in hay, grain, mushroom compost and bagasse.

Murphy *et al.* (1995) have reviewed a range of other well-known types of HP caused by exposure to non-microbial and also non-biological agents, as well as by microorganisms or their products, and in another review Kurup *et al.* (2006) has presented a fuller list of antigens associated with HP. However, as new technologies are used, as construction materials and their use changes and as health practitioners remain alert, HP is being recognized in new environments. For example, the disease has been identified in household-waste recycling and attributed to *Aspergillus fumigatus* (Allmers *et al.* 2000); it has been linked to contamination of tatami mats and attributed to *Trichosporon* and *Cryptococcus* species (Ando *et al.* 1991); and it has been associated with avian proteins tracked indoors (Saltoun *et al.* 2000). Disease follows the trails of exposure, which in the case of biological agents are generally moisture and organic matter, including dirt and dust. Of this physicians and patients should be aware.

REFERENCES

Abramson, M.J., Hensley, M.J., Saunders, N.A., and Wlodarczyk, J.H. (1991) Evaluation of a new asthma questionnaire. *J. Asthma*, 28, 129-139.

Allmers, H., Huber, H., and Baur, X. (2000) Two year follow-up of a garbage collector with allergic bronchopulmonary aspergillosis (ABPA). *Am. J. Ind. Med.*, **37**, 438-442.

Ando, M., Arima, K., Yoneda, R., and Tamura M. (1991) Japanese summer-type hypersensitivity pneumonitis. Geographic distribution, home environment, and clinical characteristics of 621 cases. *Am. Rev. Respir. Dis.*, 144, 765-769.

Anon. (1984) Outbreaks of respiratory illness among employees in large office buildings. *MMWR*, **33**, 506-513.

Anon. (2004) Mycotoxicosis. *New South Wales Animal Health Surveillance*, **2004/1**, 3.

Anon. (2009) Mycotoxin-induced staggers and death in yearling steers. *New South Wales Animal Health Surveillance*, **14/1**, 19.

Apter, A., Hodgson, M., Lueng, W.-Y., and Pichnarcik, L. (1997) Nasal symptoms in the "Sick Building Syndrome". *Ann. Allergy Asthma Immunol.*, **78**, 152 (Abstract).

Arnow, P.M., Fink, J.N., Schlueter, D.P., *et al.* (1978). Early detection of hypersensitivity pneumonitis in office workers. *Am. J. Med.*, **64**, 236-242.

Arya, A., Roychoudhury, K., and Bredin, C.P. (2006) Farmer's lung is now in decline. *Ir. Med. J.*, **99**, 203-205.

Batterman, S., and Burge, H.A. (1995) HVAC systems as emission sources affecting indoor air quality: a critical review. *Int. J. HVAC Refrig. Res.*, **1**, 61-81.

Baumgartner, K.B., Samet, J.M., Coultas, D.B., *et al.* (2000) Occupational and environmental risk factors for idiopathic pulmonary fibrosis: a multicenter case-control study. *Am. J. Epidemiol.*, **152**, 307-315.

Blyth, W. (1978) The occurrence and nature of alveolitis-inducing substances in *Aspergillus clavatus*. *Clin. Exp. Immunol.*, **32**, 272-282.

Bringhurst, L.S., Byrne, R.N., and Gershon-Cohen, J. (1959) Respiratory disease of mushroom workers. *J. Am. Med. Assoc.*, **171**, 101-104.

Buechner, H.A., Aucoin, E. Vignes, A.J., and Weill, H. (1964) The resurgence of bagassosis in Louisiana. *J. Occup. Environ. Med.*, **6**, 437-442

Burrell R, and Rylander R. (1981) A critical review of the role of precipitins in hypersensitivity pneumonitis. *Eur. J. Respir. Dis.*, **62**, 332-343.

Campbell, J.M. (1932) Acute symptoms following work with hay. *Br. Med. J.*, **2**, 1143-1144.

Carlisle, M.J., and Watkinson, S.C. (1994) *The Fungi*. Academic Press, London.

Castaneda, D. (2005) The hazards of toiling for the Cojuangcos. *Butalat*, **4**, 51.

Channell, S., Blyth, W., Lloyd, M., *et al.* (1969) Allergic alveolitis in maltworkers. *Q. J. Med.*, **38**, 351-376.

Cooper, K., Demby, S., and Hodgson, M. (1997) Moisture and lung disease: population-attributable risk calculations. In J. Woods, D. Grimsrud and N. Boschi, (eds.), *Proceedings of Healthy Buildings/IAQ'97*, Vol. 1. American Society of Heating, Refrigeration, and Air-Conditioning Engineers, Bethesda, MD, pp. 213-218.

Cox, A.L., van den Bogart, H.G., Folgering, H.T., and van Griensven, L.J. (1991) Mushroom growers' lung; clinical diagnosis and treatment. *Ned. Tijdschr. Geneeskd.*, **135**, 1040-1044 [In Dutch].

Dangman, K.H., Storey, E., Schenck, P., and Hodgson, M.J. (2004) The hypersensitivity pneumonitis diagnostic index: use of non-invasive testing to diagnose hypersensitivity pneumonitis in metalworkers. *Am. J. Ind. Med.*, **45**, 455-467.

Daroowalla, F., and Raghu, G. (1997) Hypersensitivity pneumonitis. *Comprehensive Therapy*, **23**, 244-248.

Dawson, M.W., Scott, J.G., and Cox, L.M. (1996) The medical and epidemiological effects on workers of the levels of *Thermoactinomyces* spp. spores present in Australian raw sugar mills. *Am. Ind. Hyg. Assoc. J.*, **57**, 1002-1012.

El-Hage, C.M., and Lancaster, M.J. (2004) Mycotoxic nervous disease in cattle fed sprouted barley contaminated with *Aspergillus clavatus*. *Austr. Vet. J.*, 82, 639-641.

Emanuel, D.A., Wenzel, F.J., and Lawton, B.R. (1975) Pulmonary mycotoxicosis. *Chest*, **67**, 293-297.

Fenclová, Z., Pelclová, D., Urban, P., *et al.* (2009) Occupational hypersensitivity pneumonitis reported to the Czech National Registry of Occupational Diseases in the period 1992-2005. *Industr. Health*, **47**, 443-448.

Fink, J.N., Thiede, W.H., Banaszak, E.F., and Barboriak, J.J. (1971) Interstitial pneumonitis due to hypersensitivity to an organism contaminating a heating system. *Ann. Intern. Med.*, **74**, 80-83.

Finnegan, M., Pickering, C.A.C., and Burge, P.S. (1984) The sick-building syndrome: prevalence studies. *Br. Med. J.*, **289**, 1573-1575.

Fisk, W.J., Lei-Gomez, Q., Mendell, M.J. (2007) Meta-analyses of the associations of respiratory health effects with dampness and mold in homes. *Indoor Air*, **17**, 284-96.

Flannigan, B. (2003). The microbiota of barley and malt. In F.G. Priest and I. Campbell, *Brewing Microbiology*, 3rd ed. Kluwer/Plenum, New York, pp. 113-180.

Flannigan, B., Day, S.W., Douglas, P.E., and McFarlane, G.B. (1984) Growth of mycotoxin-producing fungi associated with malting of barley. In H. Kurata and Y. Ueno, (eds.), *Toxigenic Fungi – Their Toxins and Health Hazard*, Kodansha/Elsevier, Tokyo, pp. 52-60.

Flannigan, B., McCabe, E.M. and McGarry, F. (1991). Allergenic and toxigenic micro-organisms in houses. *J. Appl. Bact.*, **70**, 61S-73S.

Flannigan, B., and Pearce, A.R. (1994) *Aspergillus* spoilage: spoilage of cereals and cereal products by the hazardous species *A. clavatus*. In K.A. Powell, J. Peberdy, and E. Renwick, (eds.), *Biology of Aspergillus*, Plenum, New York, pp. 115-127.

Flemming, J., Hudson, B., and Rand, T.G. (2004) Comparison of inflammatory and cytotoxic lung responses in mice after intratracheal exposure to spores of two different *Stachybotrys chartarum* strains. *Toxicol. Sci.*, **78**, 267-275.

Forst, L.S., and Abraham, J. (1993) Hypersensitivity pneumonitis presenting as sarcoidosis. *Br. J. Ind. Med.*, **50**, 497-500.

Gilmour, J.S., Inglis, D.M., Robb, J., and Maclean, M. (1989) A fodder mycotoxicosis of ruminants caused by contamination of a distillery by-product with *Aspergillus clavatus*. *Vet. Record*, **124**, 133-135.

Girard, M., Lacasse, Y., and Cormier, Y. (2009) Hypersensitivity pneumonitis. *Allergy*, **64**, 322-334.

Grant, I.W.B., Blyth, W., Wardrop, V.E., *et al.* (1972). Prevalence of farmer's lung in Scotland: a pilot survey. *Br. Med. J.*, **1**, 530-534.

Grant, I.W.B., Blackadder, E.S., Greenberg, M., and Blyth, W. (1976) Extrinsic allergic alveolitis in Scottish maltworkers. *Br. Med. J.*, **1**, 490-493.

Gregory, P.H., Lacey, M.E., Festenstein, G.N., and Skinner, F.A. (1963) Microbial and biochemical changes during the moulding of hay. *J. Gen. Microbiol.*, **33**, 147-174.

Gyntelberg, F., Suadicani, P., Wolkoff, P., *et al.* (1994) Dust and the sick building syndrome. *Indoor Air*, **4**, 223-228.

Hanak, V., Golbin, J.M., and Ryu, J.H. (2007) Causes and presenting features in 85 consecutive patients with hypersensitivity pneumonitis. *Mayo Clin. Proc.*, **82**, 812-816.

Hearn, C.E.D. (1968) Bagassosis: an epidemiological, environmental, and clinical survey. *Br. J. Ind. Med.*, **25**, 267-282.

Hodgson, M.J. (1998) Mycotoxins and building-related illness (letter in response to comments on Hodgson *et al.* (1998) by E. Page and D. Trout). *J. Occup. Environ. Med.*, **40**, 761-764.

Hodgson, M.J., Morey, P.R., Simon, J., *et al.* (1987) Acute and chronic hypersensitivity pneumonitis from the same source. *Am. J. Epidemiol.*, **125**, 631-638.

Hodgson, M.J., Parkinson, D.K., and Karpf, M. (1989) Chest x-rays and hypersensitivity pneumonitis: a secular trend in sensitivity. *Am. J. Ind. Med.*, **16**, 45-63.

Hodgson, M.J., Morey, P., Leung, W.-Y., *et al.* (1998) Pulmonary disease and mycotoxin exposure in Florida associated with *Aspergillus versicolor* and *Stachybotrys atra* exposure. *J. Occup. Environ. Med.*, **40**, 241-249.

Hodgson, M.J., Bracker, A., Yang, C.S., *et al.* (2001) An outbreak of hypersensitivity pneumonitis associated with a metal-working plant. *Am. J. Ind. Med.*, **39**, 616-628.

Hodgson, M., Storey, E., Dangman, K.H., *et al.* (2007) Mixed lung disease in school teachers: an outbreak. *Proceedings, IAQ 2007*, American Society of Heating, Refrigerating, and Air-Conditioning Engineers, Atlanta, GA, pp. 1-8.

Hoffmann, R.E., Wood, R.C., and Kreiss, K. (1993) Building-related asthma in Denver office workers. *Am. J. Public Health*, **83**, 89-93.

HSE (Health and Safety Executive) (1993) *Grain Dust in Maltings (Maximum Exposure Limit)*, Guidance Note EH 67. HSE Books, Sudbury, Suffolk, UK.

HSE (Health and Safety Executive) (1998a) *Grain Dust*, Guidance Note EH 66. HSE Books, Sudbury, Suffolk, UK.

HSE (Health and Safety Executive) (1998b) *The Selection, Use and Maintenance of Repiratory Protective Equipment: A Practical Guide*, HSG 53. HSE Books, Sudbury, Suffolk, UK.

Hunter, D., and Perry, K.M.A. (1946) Bronchiolitis resulting from the handling of bagasse. *Br. J. Ind. Med.*, **3**, 64-74.

Inage, M., Takahashi., H., Nakamura, H., *et al.* (1996) Hypersensitivity pneumonitis induced by spores of *Pholiota nameko*. *Intern. Med.*, **35**, 301-304.

Institute of Medicine (IOM) (2004) *Damp Indoor Spaces and Health*. National Academies of Science Press, Washington, DC.

Jajoski, R.A., Harrison, R., Flattery, J., *et al.* (1999) Surveillance of work-related asthma in selected U.S. states – California, Massachusetts, Michigan, and New Jersey, 1993-1995. *MMWR*, **48** (3),1-20.

Jarvis, B.B., Sorenson, W.G., Hintikka, E.L., *et al.* (1998). Study of toxin production by isolates of *Stachybotrys chartarum* and *Memnoniella echinata* isolated during a study of pulmonary hemosiderosis in infants. *Appl. Environ. Microbiol.*, **64**, 3620-3625

Jenkins, D.E., Malik, S.K., Figueroa-Casas, J.C., and Eichhorn, R.D. (1971) Sequential observations on pulmonary functional derangements in bagassosis. *Arch. Intern. Med.*, **128**, 535-540.

Kamm, Y.J., Folgering, H.T., van den Bogart, H.G., and Cox, A. (1991) Provocation tests in extrinsic allergic alveolitis in mushroom workers. *Neth. J. Med.*, **38**, 59-64.

Khan, Z.U., Gangwar, M., Gaur, S.N., and Randhawa, H.S. (1995) Thermophilic actinomycetes in cane sugar mills: an aeromicrobiologic and seroepidemiologic study. *Antonie van Leeuwenhoek*, **67**, 339-344.

Kleyn, J.G., Johnson, W.M., and Wetzler, T.F. (1981) Microbial aerosols and actinomycetes in etiological considerations of mushroom workers' lungs. *Appl. Environ. Microbiol.*, **41**, 1454-1460.

Kreiss, K., and Hodgson, M.J. (1984) Building-associated epidemics. In C.S. Walsh, P.J. Dudney, and E. Copenhaever, (eds.), *Indoor Air Quality*, CRC Press, Boca Raton, FL, pp. 87-106.

Kucera,G.P., Rybicki, B.A., Kirkey, K.L., *et al.* (2003) Occupational risk factors for sarcoidosis in African-American siblings. *Chest*, **123**,1527-1535.

Kurup, V.P. (1989) Hypersensitivity pneumonitis due to sensitization with thermophilic actinomycetes. *Immunol. Allergy Clin. N. Am.*, **9**, 85-306.

Kurup,V.P., Hari, V., Guo, J., *et al.* (1996) *Aspergillus fumigatus* peptides differentially express Th1 and Th2 cytokines. *Peptides*, **17**, 183-190.

Kurup, V.P., Zacharisen, M.C., and Fink, J.N. (2006) Hypersensitivity pneumonitis. *Ind. Chest Dis. Allied Sci.*, **48**, 115-128.

Lacey, J. (1974) Moulding of sugar-cane bagasse and its prevention. *Ann. Appl. Biol.*, **76**, 63-76.

Lacey, J. (1989) Airborne health hazards from agricultural materials. In B. Flannigan, (ed.), *Airborne Deteriogens and Pathogens*, The Biodeterioration Society, Kew, Surrey, UK, pp. 150-162.

Laney, A.S., Cragin, L.A., Blevins, L.Z., *et al.* (2009) Sarcoidosis, asthma, and asthma-like symptoms among occupants of a historically water-damaged office building. *Indoor Air*, 19, 83-90.

Lehrer, S. B., Turer, E., Weill, H., and Salvaggio, J.E. (1978). Elimination of bagassosis in Louisiana paper manufacturing plant workers. *Clin. Allergy*, **8**, 15-20.

Lonneux, M., Nolard, N., Philippart, I., *et al.* (1995) A case of lymphocytic pneumonitis, myositis, and arthritis associated with exposure to *Aspergillus niger*. *J. Allergy Clin. Immunol.*, **95**, 1047-1049.

López-Diaz, T.M., and Flannigan, B. (1997) Production of patulin and cytochalasin E by *Aspergillus clavatus* during malting of barley and wheat. *Food Microbiol.*, **35**, 129-136.

Lynch, D.A., Way, D., Rose, C.S., and King, T.E. (1992) Hypersensitivity pneumonitis: sensitivity of high-resolution CT in a population-based study. *Am. J. Roentgenol.*, **159**, 469-472.

Mahooti-Brooks, N., Storey, E., Yang, C., *et al.* (2004) Characterization of mold and moisture indicators in the home. *J. Occup. Environ. Hyg.*, **1**, 826-839.

Matte, T.D., Rosenman, K.D., Hoffman, R.E., and Stanbury, M. (1990) Surveillance of occupational asthma under the SENSOR model. *Chest*, **98** (Suppl. 5), 173S-178S.

McKenzie, R.A., Kelly, M.A., Shivas, R.G., *et al.* (2004) *Aspergillus clavatus* tremorgenic neurotoxicosis in cattle fed sprouted grains. *Austr. Vet. J.*, **82**, 635-638.

Milton, D. (1996) Endotoxins. In R. Gammage, (ed.), *Indoor Air Quality II* (Oak Ridge Symposium). Lewis/CRC Press, Boca Raton, FL.

Milton, D.K., Amsel, J., Reed, C.E., *et al.* (1995) Cross-sectional follow-up of a flu-like respiratory illness among fiberglass manufacturing employees: endotoxin exposure associated with two distinct sequelae. *Am. J. Ind. Med.*, **28**, 469-488.

Moore, J.E., Convery, R.P., Millar, B.C., *et al.* (2005) Hypersensitivity pneumonitis associated with mushroom worker's lung: an update on the clinical significance of the importation of exotic mushroom varieties. *Int. Arch. Allergy Immunol.*, **136**, 98-102.

Mullan, R.J., and Murthy, L.I. (1991) Occupational sentinel health events: an up-dated list for physician recognition and public health surveillance. *Am. J. Ind. Med.*, **19**, 775-799.

Mullen, J., Hodgson, M., DeGraff, C.A., and Godar, T. (1998) A case-control study of diffuse interstitial pneumonitis. *J. Occup. En-*

viron. Med., **40**, 1-5.

Murphy, D.M.F., Morgan, W.K.C., and Seaton, A. (1995) Hypersensitivity pneumonitis. In W.K.C. Morgan and A. Seaton, (eds.), *Occupational Lung Diseases,* W.B. Saunders, Philadelphia, pp. 525-567.

Nakagawa-Yoshida, K., Ando, M., Etches, R.I., and Dosman, J.A. (1997) Fatal cases of farmer's lung in Canadian family: probable new antigens, *Penicillium brevicompactum* and *P. olivicolor. Chest*, **111**, 245-248.

Nakazawa, T., and Tochigi, T. (1989) Hypersensitivity pneumonitis due to mushroom (*Pholiota nameko*) spores. *Chest*, **95**, 1149-1151.

National Institute for Occupational Safety and Health (NIOSH). (1996) *Work-Related Lung Disease (WORLD) Surveillance Report*, Publication 96-134. US DHHS/PHS, Atlanta, GA.

Newman, L.S., Rose, C.S., and Maier, L.A. (1997) Sarcoidosis. *N. Engl. J. Med.*, **336**, 1224-1234. Published erratum appears in *N. Engl. J. Med.*, **337**, 139 (1997).

Newman, L.S., Rose, C.S., Bresnitz, E.A., *et al.* (2004) A case control etiologic study of sarcoidosis: environmental and occupational risk factors. *Am. J. Resp. Crit. Care Med.*, **170**, 1324-1330.

Nikulin, M., Reijula, K., Jarvis, B.B., and Hintikka, E.L. (1996) Experimental lung mycotoxicosis in mice induced by *Stachybotrys atra. Int. J. Exp. Pathol.*, **77**, 213-218.

Nolard, N., Detandt, M., and Beguin, H. (1988) Ecology of Aspergillus species in the human environment. In H. Vanden Bosche and D.W.R. Mackenzie, (eds.), *Aspergillus and Aspergillosis*. Plenum, New York, pp. 35-41.

Nordness, M.E., Zacharisen, M.C., Schlueter, D.P., and Fink, J.N. (2003) Occupational lung disease related to cytophaga endotoxin exposure in a nylon plant. *J. Occup. Environ. Med.*, **45**, 385-392.

NYC (2008) *Guidelines on Assessment and Remediation of Fungi in Indoor Environments,* New York City Health Department (http://www.nyc.gov/html/doh/html/epi/moldrpt1.shtml; accessed March 15, 2010).

Ortiz, C., Hodgson, M.J., McNally, D., and Storey, E. (1998) A case-control study of sarcoidosis. In E. Johanning, (ed.), *Bioaerosols, Fungi and Mycotoxins - Health Effects, Assessment, Prevention and Control*, Eastern New York Occupational and Environmental Health Center, Albany/Mount Sinai School of Medicine, New York, pp. 476-481.

Park, J.H., Schleiff, P.L., Attfield, M.D., *et al.* (2004) Building-related respiratory symptoms can be predicted with semi-quantitative indices of exposure to dampness and mold. *Indoor Air*, **14**, 425-433.

Park, J.H., Cox-Ganser, J.M., Kreiss, K., *et al.* (2008) Hydrophilic fungi and ergosterol associated with respiratory illness in a water-damaged building. *Environ. Health Perspect.*, **116**, 45-50.

Pepys, J., and Jenkins, P.A. (1965) Precipitin (F.L.H.) test in farmer's lung. *Thorax*, **20**, 21.

Pepys, J., Jenkins, P.A., Festenstein, G.N., *et al.* (1963) Farmer's lung: thermophilic actinomycetes as a source of 'farmer's lung hay antigen'. *Lancet 2*, 607-611.

Pestka, J.J., Yike, I., Dearborn, D.G., *et al.* (2008) *Stachybotrys chartarum*, trichothecene mycotoxins, and damp building-related illness: new insights into a public health enigma. *Toxicol. Sci.*, **104**, 4-26.

Phoolchund, H.N. (1991) Aspects of occupational health in the sugar cane industry. *J. Soc. Occup. Med.*, **41**, 133-136.

Prezant, B., Weekes, D., and Miller, J.D., (eds.) (2008) *Recognition, Evaluation and Control of Indoor Mold*. American Industrial Hygiene Association, Fairfax, VA.

Rabie. C.J., and Lübben, A. (1984) The mycoflora of sorghum malt. *S. Afr. J. Bot.*, **3**, 251-255.

Ramazzini, B. (1713) *De Morbus Artificum Diatriba* (on the Disease of Occupation), transl. W.C. Wright, Hafner, New York.

Rao, C.Y., Brain, J.D., and Burge, H.A. (2000). Reduction of pulmonary toxicity of *Stachybotrys chartarum* spores by methanol extraction of mycotoxins. *Appl. Environ. Microbiol.* **66**, 2817-2821.

Riddle, H.F.V., Channell, S., Blyth, W., *et al.* (1968) Allergic alveolitis in a maltworker. *Thorax*, **23**, 271-280.

Rose, C.S., Martyny, J.W., Newman, L.S., *et al.* (1998) "Lifeguard lung": endemic granulomatous pneumonitis in an indoor swimming pool. *Am. J. Public Health*, **88**, 1795-1800.

Rutstein, D.D. (1984) The principle of the sentinel health event and its application to the occupational diseases. *Arch. Environ. Health*, **39**, 158.

Rutstein, D.D., Mullan, R.J., Frazier, T.M., *et al.* (1983) Sentinel health events (occupational): a basis for physician recognition and public health surveillance. *Am. J. Public Health*, **73**, 1054-1062.

Rylander, R., Persson, K., Goto, H., *et al.* (1992) Airborne beta-1,3-glucan may be related to symptoms in sick buildings. *Indoor Environ.*, **1**, 263-267.

Sabater-Vilar, M., Maas, R.F.M., De Bosschere, H., *et al.* (2004) Patulin produced by an *Aspergillus clavatus* isolated from feed containing malting residues associated with a lethal neurotoxicosis in cattle. *Mycopathologia*, **158**, 419-426.

Saikai, T., Tanaka, H., Fuji, M., *et al.* (2002) Hypersensitivity pneumonitis induced by the spore of *Pleurotus eryngii* (Eringi). *Int. Med.*, **41**, 571-573.

Sakula, A. (1967) Mushroom worker's lung. *Br. Med. J.*, **2**, 708.

Saltoun, C.A., Harris, K.E., Mathisen, T.L., and Patterson, R. (2000) Hypersensitivity pneumonitis resulting from community exposure to Canada goose droppings: when an external environmental antigen becomes an indoor environmental antigen. *Ann. Allergy Asthma Immunol.*, **84**, 84-86.

Salvaggio, J.E. (1997) Extrinsic allergic alveolitis (hypersensitivity pneumonitis): past, present and future. *Clin. Exp. Allergy*, **1** (Suppl. 1), 18-25.

Shlosberg, A., Zadikov, I., Perl, S., *et al.* (1991) *Aspergillus clavatus* as the probable cause of a lethal mass mycotoxicosis in sheep. *Mycopathologia*, **114**, 35-39.

Sieber, W., Stayner, L., Malkin, R., *et al.* (1996) The NIOSH indoor evaluation experience: associations between environmental factors and self-reported health conditions. *Appl. Occup. Environ. Hygiene*, **11**, 1387-1392.

Silva, C.I., Churg, A., and Müller, N.L. (2007) Hypersensitivity pneumonitis: spectrum of high-resolution CT and pathologic findings. *Am. J. Roentgenol.*, 188, 334-344.

Sorenson, W.G., Frazer, D.G., Jarvis, B.B., *et al.* (1987). Trichothecene mycotoxins in aerosolized conidia of *Stachybotrys atra. Appl. Environ. Microbiol.*, **53**, 1370-1375.

Tanake, H., Sugawara, H., Sakai, T., *et al.* (2000) Mushroom worker's lung caused by spores of *Hypsizigus marmoreus* (Bunashimeji). Elevated serum surfactant protein D levels. *Chest*, **118**, 1506-1509.

Tanaka, H., Saikai, T., Sugawara, H., *et al.* (2001) Three-year follow-up study of allergy in workers in a mushroom factory. *Resp. Med.*, **95**, 943-948.

Tanaka, H., Saikai, T., Sugawara, H., *et al.* (2002) Workplace-related chronic cough on a mushroom farm. *Chest,* **122,** 1080-1085.

Tarvainen, K., Salonen, J.-P., Kanerva, L., *et al.* (1991) Allergy and toxicoderma from shiitake mushrooms. *J. Am. Acad. Dermatol.*, **24**, 64-66.

Teeuw, K.B., Vandenbroucke-Grauls, C.M., and Verhoef, J. (1994) Airborne Gram-negative bacteria and endotoxin in SBS: a study in Dutch office buildings. *Arch. Int. Med.*, **154**, 2339-2345.

Thorn, A., Lewne, M., and Belin, L. (1996) Allergic alveolitis in a school environment. *Scand. J. Work Environ. Health*, **22**, 311-314.

Ueda, A., Aoyama, K., Ueda, T., *et al.* (1992) Recent trends in bagassosis in Japan. *Br. J. Ind. Med.*, **49**, 499-506.

Utsugi, M., Dobashi, K., Ksukakoshi, H., *et al.* (1999) A case of hypersensitivity pneumontis induced by inhalation of spores of *Pholiota nameko* and showed dual decrease of PaO_2 in provocative inhalation test. *Allergy Pract.*, **248**, 516-520.

van den Bogart, H.G., van den Ende, G., van Loon, P.C., and van Griensven, L.J. (1993) Mushroom worker's lung: serologic reactions to thermophilic actinomycetes present in the air of compost tunnels. *Mycopathologia*, **122**, 21-28.

Welterman, B., Hodgson, M.J., Storey, E., *et al.* (1998) Hypersensitivity pneumonitis: sentinel health events in clinical practice. *Am. J. Ind. Med.*, **31**, 499-505.

Wenzel, S., and Holgate, T.S. (2006) Pro-/Con Editorial: Rebuttal from Drs. Wenzel and Holgate. *Am. J. Resp. Crit. Care Med.*, **174**, 1177-1178.

Wild, L.G., and Chang, E.E. (2009) *Farmer's Lung*, CDC Commentary Series (online at http://emedicine.medscape.com/article/298811-print).

Wright, C.L. (1970) Bagasse no longer a health risk. *Board Manuf. Pract.*, **13**, 149-154.

Yike, I., and Dearborn, D. (2004) Pulmonary effects of *Stachybotrys chartarum* in animal studies. *Adv. Appl. Microbiol.*, **55**, 241-273.

Yike, I., Rand, T.G., and Dearborn, D.G. (2005). Acute inflammatory responses to *Stachybotrys chartarum* in the lungs of infant rats: time course and possible mechanisms. *Toxicol. Sci.*, **84**, 408-417.

Yoshikawa, S., Tsushima, K., Koizumi, T., *et al.* (2006) Hypersensitivity pneumonitis induced by spores of *Penicillium citrinum* in a worker cultivating Enoki mushroom. *Internal Med.*, **45**, 537-541.

Yoshikawa, S., Tsushima, K., Yasuo, M., *et al.* (2007) Hypersensitivity pneumonitis caused by *Penicillium citrinum*, not enoki spores. *Am. J. Ind. Med.*, 50, 1010-1017.

Zacharisen, M.C., Kadambi, A.R., Schlueter, D.P., *et al.* (1998) The spectrum of respiratory disease associated with exposure to metal working fluids. *J. Occup. Environ. Med.*, 40, 640-647.

Chapter 3.3

RESPIRATORY TRACT INFECTIONS CAUSED BY INDOOR FUNGI

Richard C. Summerbell

Dalla Lana School of Public Health, University of Toronto; Sporometrics Inc., Toronto, Canada.

RANGE OF FUNGI CAUSING MAMMALIAN RESPIRATORY INFECTIONS

Fungi causing respiratory tract infections in humans and other mammals can be broken down into three categories:

- a small number of species that are capable of causing environmentally acquired, but not usually contagious, respiratory illness in the otherwise healthy, immunocompetent host: *Blastomyces dermatitidis, Histoplasma capsulatum, Paracoccidioides brasiliensis, Coccidioides immitis* and *C. posadasii, Cryptococcus neoformans* and *C. gattii*, and, marginally, the *Sporothrix schenckii* complex, and *Penicillium marneffei*;
- a single genus, *Pneumocystis*, capable of causing contagious respiratory illness, usually in the immunocompromised host; and
- an ever enlarging group of fungi, including *Aspergillus fumigatus* and *Pseudallescheria boydii*, which are unspecialized for vertebrate pathogenesis but which can cause environmentally acquired respiratory disease in unusual situations. Fungi in this category cause invasive disease only in the strongly immunocompromised patient, or rarely in the patient who is both debilitated and heavily exposed, or very rarely in the apparently otherwise normal patient.

Members of the first category are ordinarily referred to as "virulent fungal pathogens." All except *C. neoformans* and the closely related *C. gattii* convert from a mould-like filamentous state on environmental substrata, as well as on ordinary growth media, to a particulate state in the infected host and on special media incubated at 37°C *in vitro*. Species undergoing this conversion are called "dimorphic fungi," or more formally "thermally dimorphic fungi". Particulate states formed in infected hosts are diverse, and include budding yeasts (*B. dermatitidis, H. capsulatum, P. brasiliensis, S. schenckii*), spherules reproducing by means of internal cleavage into endoconidia (*C. immitis, C. posadasii*) and fission yeasts, which might also be conceptualized as self-replicating arthroconidia (*P. marneffei*). *C. neoformans* and *C. gattii*, the two virulent fungal respiratory pathogens that are not ordinarily labelled "dimorphic," grow as budding yeasts both in the host and on ordinary laboratory media, and may also commonly occur in this state in nature. These species, however, can be induced in culture to undergo mating between compatible strains, resulting in the production of mycelium and a sexual reproductive form (teleomorph). The teleomorph of *C. neoformans* is *Filobasidiella neoformans*, while that of *C. gattii* is *F. bacillospora*. In some cases, strains may self-fertilize ('self') and produce structures resembling a sexual state even without being mated (Fraser *et al.* 2005). Genotyping results of strains isolated from nature suggest this mating occurs in nature, though it has not been directly observed (Fraser *et al.* 2005). The alteration of growth forms involved in this sexual cycle may be termed 'dimorphism' in general mycological texts, but is never given this label in medical mycology.

Phylogenetically, *C. neoformans* and *C. gattii* are the only members of phylum *Basidiomycota* recognized as virulent pathogens. Of the remaining virulent fungi mentioned above, all have affinities to the order Onygenales of the phylum Ascomycota except *P. marneffei* and *S. schenckii*, which are related to the orders Eurotiales and Ophiostomatales of the same phylum, respectively.

Members of the second category of fungal respiratory pathogens, the *Pneumocystis* species, were traditionally treated as protozoans. In confirmation of some earlier suggestions arising from ultrastructural and biochemical studies, they were decisively determined as fungi by means of DNA sequencing (Edman *et al.* 1988). These unique organisms, which appear to be remotely related phylogenetically to the budding, usually fermenting yeasts of the Saccharomycetales and the fission yeasts of the Schizosaccharomycetales (Berbee and Taylor 1992), and to other phylogenetically basal fungi, reproduce *in vivo* by means of cysts, visually suggestive of yeast asci, which produce eight intracystic bodies. Whether meiosis occurs in these cysts or not is not definitively established: nuclear associated organelles followed through the life cycle present a picture consistent with sexual conjugation and meiosis (Itatani 1996), whereas nuclei fluorescently stained with 4′,6-diamidino-2-phenylindole, which gives quantitative results on amounts of nucleic acid present, indicated an absence of apparent diploid DNA content in nuclei from all life cycle stages (Wyder *et al.* 1998). In recent decades, evidence for a meiotic cycle has been steadily accumulated in ultrastructural studies (Matsumoto and Yoshida, 1984) and studies on expression of genes related to meiosis (Burgess *et al.* 2008, Aliouat-Denis *et al.* 2009).

Fungi in the third category of fungal respiratory pathogens are generally termed "opportunistic pathogens". The adjectival phrase "of the immunocompromised host" can accurately be added onto that basic term in the great majority of cases. A large and indefinitely expanding diversity of filamentous fungi and yeasts has been implicated in opportunistic respiratory pathogenesis (Table 1). The majority of respiratory opportunists are mould fungi with phylogenetic affinities to the phylum Ascomycota. For example, the two groups containing the most virulent of the filamentous fungal respiratory opportunists are the genus *Aspergillus*, a clade (group of organisms considered to have evolved from a common ancestor) of hyphomycetes with affinities to the ascomycete order Eurotiales, and a clade in the order Microascales containing the ascomycete *Pseudallescheria boydii* and the related hyphomycetes *Scedosporium apiospermum* and *S. prolificans*. The great majority of *Aspergillus* infections are attributable to four species: *Aspergillus fumigatus*, *A. flavus*, *A. terreus* and *A. nidulans*.

Also important as opportunistic pathogens is a group of yeasts normally inhabiting the skin or the mucosal epithelia (oral, vaginal, lower intestinal) of humans and animals as commensal mycota. These yeasts have affinities in the ascomycetous class Endomycetes; *Candida albicans* is the best-known species. These yeasts will not be mentioned further in this review, as their connection to respiratory infections is very limited. These fungi may invade the respiratory tract, but they do this mostly in a context of disseminated infection of the severely immunocompromised patient. Acquisition of such infections is not significantly connected with inhalation of inoculum from environmental sources, indoors or elsewhere. Some members of the order Mucorales of the phylum Zygomycota are also agents of opportunistic respiratory disease of the immunocompromised host: the genera most frequently implicated, in temperate parts of the world, are *Rhizopus* and *Absidia*. Finally, a few members of the phylum Basidiomycota rarely cause opportunistic respiratory infection. Yeasts in the anamorphic genus *Trichosporon*, members of a distinct phylogenetic group related to the virulent *Cryptococcus* species in the order *Tremellales*, are perhaps the most frequently seen from significant cases. Occasionally, significant cases may be caused by a few members of the mushroom forming order Agaricales, namely *Schizophyllum commune* and members of the genus *Coprinus* (which may be recorded under the name of a corresponding anamorphic genus, *Hormographiella*).

ECOLOGY AND EPIDEMIOLOGY OF FUNGAL RESPIRATORY PATHOGENS IN RELATION TO HOMES AND INDOOR WORKPLACES

Virulent pathogens

Virulent fungal pathogens are, fortunately, strongly restricted to specific habitats and geographic areas. Each organism is distinctive in these respects and must be considered separately.

The closely related sibling-species pair *Coccidioides immitis* and *C. posadasii*, arguably the most virulent of these fungi, are restricted to desert soils in the "lower Sonoran life zone" (a phytogeographic zone tending to feature tall species of cactus and the creosote bush) in the southwestern USA and contiguous Mexico (Kwon Chung and Bennett 1992), as well as scattered climatologically similar areas in South America. *Coccidioides immitis* is specifically native to central California, while *C. posadasii* occupies the rest of the range outlined above.

Table 1. Fungi other than endomycetous yeasts which have been confirmed as causing respiratory tract (pulmonary, bronchial) infection, sinusitis or disseminated infections potentially invading the lung in humans.

Species	Virulence category	Frequency of valid respiratory cases or sinusitis
ZYGOMYCOTA		
Absidia corymbifera	Opportunistic	Uncommon, but regular
Actinomucor elegans	Opportunistic	Rare
Apophysomyces elegans	Opportunistic	Rare
Conidiobolus incongruus	Opportunistic	Rare
Cunninghamella bertholletiae	Opportunistic	Uncommon, but regular
Delacroixia coronata	Opportunistic	Rare
Rhizomucor pusillus	Opportunistic	Uncommon
Rhizopus microsporus var. *rhizopodiformis*	Opportunistic	Common
Rhizopus oryzae	Opportunistic	Common
Saksenaea vasiformis	Virulent	Rare (subcutaneous primary infection)
ASCOMYCOTA (and anamorphs)		
Acremonium kiliense	Opportunistic	Rare
Acremonium strictum	Opportunistic	Rare
Acrophialophora fusispora	Opportunistic	Uncommon
Alternaria alternata	Opportunistic	Common (especially sinusitis)
Arthrographis kalrae	Opportunistic	Rare
Ascotricha chartarum	Opportunistic	Rare
Aspergillus calidoustus	Opportunistic	Rare
Aspergillus candidus	Opportunistic	Rare
Aspergillus flavus	Opportunistic	Common
Aspergillus fumigatus	Opportunistic	Common
Aspergillus lentulus	Opportunistic	Common
Aspergillus nidulans	Opportunistic	Uncommon
Aspergillus niger	Opportunistic	Uncommon, but regular
Aspergillus niveus	Opportunistic	Rare
Aspergillus ochraceus	Opportunistic	Rare
Aspergillus terreus	Opportunistic	Uncommon, but regular
Aspergillus tetrazonus	Opportunistic	Rare
Aspergillus ustus	Opportunistic	Rare
Aureobasidium pullulans	Opportunistic	Rare
Bipolaris australiensis	Opportunistic	Rare
Bipolaris hawaiiensis	Opportunistic	Common
Bipolaris spicifera	Opportunistic	Uncommon, but regular
Blastomyces dermatitidis	Virulent	Endemic, common[a]
Chaetomium globosum	Opportunistic	Rare
Chrysosporium zonatum	Opportunistic	Rare
Cladophialophora devriesii	Opportunistic	Rare
Cladosporium cladosporioides	Opportunistic	Rare
Coccidioides immitis	Virulent	Endemic, common
Coccidioides posadasii	Virulent	Endemic, common
Curvularia geniculata	Opportunistic	Uncommon, but regular
Curvularia inaequalis	Opportunistic	Rare
Curvularia lunata	Opportunistic	Common
Curvularia pallescens	Opportunistic	Rare
Cylindrocarpon lichenicola	Opportunistic	Rare
Emmonsia parva	Opportunistic	Rare (obstructive, not invasive)
Epicoccum nigrum	Opportunistic	Rare
Eurotium amstelodami	Opportunistic	Rare

Exophiala dermatitidis	Opportunistic	Uncommon
Exophiala spinifera	Opportunistic	Rare
Exserohilum longirostratum	Opportunistic	Rare
Exserohilum mcginnisii	Opportunistic	Rare
Exserohilum rostratum	Opportunistic	Rare
Fonsacaea pedrosoi	Opportunistic	Rare
Fusarium dimerum	Opportunistic	Rare
Fusarium lichenicola	Opportunistic	Rare
Fusarium napiforme	Opportunistic	Rare
Fusarium oxysporum	Opportunistic	Uncommon, but regular
Fusarium proliferatum	Opportunistic	Uncommon, but regular
Fusarium solani	Opportunistic	Uncommon, but regular
Fusarium verticillioides (= F. moniliforme)	Opportunistic	Uncommon, but regular
Gymnascella hyalinospora	Opportunistic	Rare
Histoplasma capsulatum	Virulent	Endemic, common
Lecythophora hoffmanii	Opportunistic	Rare
Metarrhizium anisopliae	Opportunistic	Rare
Microascus cinereus	Opportunistic	Rare
Microascus cirrosus	Opportunistic	Rare
Microascus trigonosporus	Opportunistic	Rare
Myceliophthora thermophila	Opportunistic	Rare
Neosartorya pseudofischeri	Opportunistic	Rare
Neosartorya udagawae	Opportunistic	Rare (or rarely identified)
Nodulisporium sp.	Opportunistic	Rare
Ochroconis gallopavum	Opportunistic	Rare
Ophiostoma piceae	Opportunistic	Rare
Paecilomyces lilacinus	Opportunistic	Rare
Paecilomyces variotii	Opportunistic	Rare
Paracoccidioides brasiliensis	Virulent	Endemic, common
Penicillium chrysogenum	Opportunistic	Rare
Penicillium citrinum	Opportunistic	Rare
Penicillium decumbens	Opportunistic	Rare
Penicillium marneffei	Virulent	Endemic, common
Penicillium purpurogenum	Opportunistic	Rare
Phaeoacremonium parasiticum	Opportunistic	Rare
Phialemonium obovatum	Opportunistic	Rare
Pleurophomopsis lignicola	Opportunistic	Rare
Pneumocystis jirovecii	Opportunistic	Common
Pseudallescheria boydii	Opportunistic	Common
Scedosporium apiospermum	Opportunistic	Common
Scedosporium prolificans	Opportunistic	Uncommon, but regular
Scopulariopsis acremonium	Opportunistic	Rare
Scopulariopsis brevicaulis	Opportunistic	Rare
Scopulariopsis brumptii	Opportunistic	Rare
Scopulariopsis candida	Opportunistic	Rare
Sporothrix schenckii	Virulent	Uncommon, but regular
Trichoderma longibrachiatum	Opportunistic	Rare

BASIDIOMYCOTA (and anamorphs)

Cryptococcus gattii	Virulent	Regional, common
Cryptococcus neoformans	Virulent	Common
Rhodorula minuta	Opportunistic	Rare
Rhodotorula rubra	Opportunistic	Rare

Schizophyllum commune	Opportunistic	Uncommon, but regular
Scopulariopsis acremonium	Opportunistic	Rare
Trichosporon asahii	Opportunistic	Uncommon, but regular
Trichosporon inkin	Opportunistic	Uncommon
Trichosporon mucoides	Opportunistic	Uncommon

[a] "Endemic, common"- common within endemic area.

Infection is usually acquired by inhalation of conidia from disturbed soil. Dust storms, construction activity, automobile dirt racing and so on are common activities that have resulted in outbreaks or prominent cases. Most cases resolve spontaneously after a few days of respiratory ailment resembling a common "chest cold", but a small proportion of persons experience more severe symptoms, ranging from pneumonia to erythema nodosum (an acute skin rash) to disseminated infection in which the organism spreads to cause skin abscesses, meningitis, and infections of other internal sites such as bones, kidneys, liver and epididymis (Kwon Chung and Bennett 1992). Cases of dissemination are often fatal unless the disease is arrested by antifungal therapy. The fungus has complex interactions with host factors related to sex hormones and pigmentation, with the result that certain groups, such as women in the third trimester of pregnancy and people with high levels of dermal melanin (those of African ancestry, for example), are significantly more likely than others to acquire serious infection (Kwon Chung and Bennett 1992). Adult males in general are about four times as likely as females to experience disseminated infection (Fiese 1958).

Because the fungus is strongly associated with outdoor soils, its connection with indoor environments is desultory, basically consisting of the possibility that outdoor spora entering buildings within the endemic region may include some infectious conidia. On rare occasions, a potted cactus directly lifted from the wild in endemic areas, or personal articles packed in an endemic area, may transport the fungus to a dwelling outside the endemic zone (Kwon Chung and Bennett 1992). For example, the author's laboratory dealt with one case (unpublished) in which an immunocompromised Canadian woman, who had never travelled to an endemic *Coccidioides* zone, became severely infected after receiving a visit from her sister who resided in Arizona. In rare instances, workplaces have also been affected after importation of porous or fibrous material, such as cotton, which happens to have been exposed to sedimentation of *Coccidioides* inoculum in the endemic region (Kwon Chung and Bennett 1992). Despite such occurrences, the outstanding risk

factor for persons living outside the endemic region is travel to that region, particularly California's Death and San Joaquin valleys and certain parts of Arizona. As with tuberculosis, once a person has been exposed to *C. immitis* or *C. posadasii*, a small residue of viable inoculum tends to persist indefinitely in a stable, calcified pulmonary granuloma (Rippon 1988), and immunosuppression later in life may permit the organism to re-emerge and progress to a severe infection. Tobacco smokers residing outside the endemic area also occasionally have such nodules misdiagnosed as carcinoma based on x-ray analysis, and may suffer biopsy or even erroneous surgical removal of the affected lung lobe.

Blastomyces dermatitidis and *Paracoccidioides brasiliensis* have very little connection with indoor environments, and therefore will not be discussed in detail here. The pathogenicity and ecology of the former fungus is summarized by Al Doory and DiSalvo (1992). Both fungi are discussed in detail by Rippon (1988) and Kwon-Chung and Bennett (1992). In essence, *B. dermatitidis* is a riparian, nitrophilous fungus growing mainly on damp debris or soil near waterways in restricted, mostly mountainous and boreal endemic zones within eastern North America. The fungus also uncommonly occurs in parts of Africa and Asia. Very rarely, construction activity in cities in endemic regions has been tentatively linked to small disease outbreaks (Kitchen *et al.* 1977). Occasional evidence, such as infections of urban cats that are always kept indoors, attests to the possibility of domestically acquired infections in highly endemic areas (Baumgardner and Paretsky 2001, Blondin *et al.* 2007). Most commonly, however, blastomycosis appears to be acquired outdoors or sometimes in heavily animal-infested shacks or outbuildings (Bakerspigel *et al.* 1986).

Paracoccidioides brasiliensis is restricted to parts of central and eastern South America and may be most closely associated with soils in zones in which coffee can be grown (Silva Vergara *et al.* 1998). Infection is generally acquired outdoors. *P. marneffei* is also essentially linked to outdoor habitats in montane and high plateau areas of southeast Asia (especially northern Thailand, Burma, Laos and Vietnam, as well as Yunnan, Szechuan, Sichuan, Guangxi and Guangdong provinces of China) which are inhabited by the apparently associated bamboo rat, *Rhizomys sinensis* (Kwon Chung and Bennett 1992).

A dimorphic fungus with considerably more indoor significance is *Histoplasma capsulatum*. This species is found in various parts of the Americas, Africa and Asia, but is uncommon or rare in much of its range. It is, however, abundant in defined endemic areas of North and South America, including most of the watershed of the Mississippi river and its tributaries in central USA (Kwon-Chung and Bennett 1992). There are also marginal endemic areas where it is locally common, such as parts of southeastern USA, as well as the parts of the lower St. Lawrence River valley in Canada and northeastern USA. In its areas of greatest abundance, *H. capsulatum* may be common in outdoor soils, especially when these soils have been subject to nitrogen inputs from migratory and resident bird populations. In marginal areas it is probably seldom associated with such soils, but may be strongly associated with large depositions of bird and bat guano in buildings and caves. Similar sites in highly endemic areas also tend to be infested. Outbreaks frequently occur in groups of spelunkers in caves with resident bats, and in renovation workers or householders cleaning up large depositions of bat or bird guano in attics or belfries which have housed long established breeding colonies (Kwon Chung and Bennett 1992). Exposure levels in such cases can be high, leading to relatively severe disease even in immunocompetent persons. Bat dung in particular may be similar to sawdust in texture, since it is mainly composed of parts of insect exoskeletons, and may sift through porous ceiling boards or be deposited in wall cavities at the edge of the attic, sometimes falling down wall interiors until it reaches lower floors of the building. In an Ontario outbreak investigated by the author's laboratory, workers infected by *H. capsulatum* had worked extensively in the attic, but the fungus was only isolated *in vitro* from bat dung deposits that had fallen into the basement.

A small histoplasmosis outbreak in a Mexican hotel was connected with potted plants, which had been fertilized with contaminated compost (Taylor *et al*. 2005). This is a unique case of direct acquisition from a concentrated environmental source situated in apparently clean indoor environment.

Histoplasmosis, like coccidioidomycosis, most commonly manifests as an insignificant illness resembling a chest cold. In many cases, low level exposure may lead to life-long seroconversion, with the sero-

positive individual remembering no corresponding illness. Such seroconversion, widely studied in the 1950s and 1960s using the histoplasmin skin test, is currently normally elucidated using immunodiffusion or complement fixation serological tests (Kwon Chung and Bennett 1992). Persons who have seroconverted usually have a life-long immunity to the disease. As with coccidioidomycosis, however, the body's successful containment of the infection may leave behind one or more calcified pulmonary nodules. This entails the risk factors mentioned previously for coccidioidomycosis: recrudescence of the disease if the person becomes immunodepressed later in life, and misdiagnosis as pulmonary cancer. Heavily exposed or immunocompromised individuals, including some persons with cellular immune deficits apparently not otherwise of great consequence, may contract a pneumonia-like illness. In immunocompetent persons this acute pulmonary histoplasmosis usually resolves spontaneously, but may continue as a chronic infection in some individuals, typically white, male smokers with pre-existing, incipient chronic bronchitis or emphysema (Goodwin *et al*. 1976, Latham *et al*. 1980). Obstruction of pulmonary blood vessels, bronchi, or the oesophagus as a result of swelling and granulomatous infection of chest lymph nodes is another possible complication, as is pericarditis. In a small proportion of cases, especially in immunocompromised or very young or old patients, the pulmonary infection may be succeeded by dissemination of the organism to internal organs such as the liver, spleen or meninges, as well as to mucous membranes or skin. In HIV-positive people with a CD4+ T-cell count lower than 50 cells μl^{-1}, dissemination is common (Jung and Paauw 1998) and severe. Therapy with antifungal drugs is necessary in chronic pneumonia-like and disseminated infection, and may also be used in some cases of acute histoplasmosis.

In a small proportion of cases, infection has been tentatively associated with ensuing visual problems in the otherwise completely recovered patient. This so-called "presumed ocular histoplasmosis syndrome" (POHS) is controversial. A few case reports have credibly shown *H. capsulatum* in the eye. However, these have mostly involved granulomatous endophthalmitis with pronounced intraocular inflammation, have generally been secondary to significant immune impairment and had near uniformly fatal outcomes (reviewed by Gonzales *et al*. 2000, Knox *et al*. 2003). Only one of the reported cases appears to have applied the gold-standard demonstration of *H. capsulatum* in eye tissue (Hoefnagels and Pijpers 1967) by mouse passage, which is the procedure originally devised by Emmons (1961) and remains the definitive method for demonstrating infection. The first case of ocular involvement attributed to *H. capsulatum* (Reed *et al.* 1942, Woods and Wahlen 1959), occurred in a patient with fatally disseminated disease and compatible ante-mortem fundoscopic findings. Although no eye tissue was collected for verification post mortem, this case clearly resembles that seen in more serious disseminated disease and not that of typical, uncomplicated POHS.

Clinical findings identical to those of POHS have been seen in cases from non-endemic regions suggesting at least that *H. capsulatum* is not solely responsible for this syndrome (Suttorp-Schulten *et al.* 1997, Watzke *et al.* 1998, Ongkosuwito *et al.* 1999). The present author was consulted c. 1995 about an unpublished cluster of such diagnoses in the lower Fraser valley of British Columbia, Canada, an area where *Histoplasma* is unknown and where most residents, including any patients tested, were seronegative for this fungus. Since then, Watzke *et al.* (1998) have confirmed that similar cases are found in subjects never exposed to *Histoplasma* in the immediately adjacent Northwest of the U.S.A. The same ocular syndrome has also been shown to occur in other non-endemic areas such as the Netherlands (Suttorp-Schulten *et al.* 1997) and the São Paulo area of Brazil (Amaro *et al.* 2007). Even in endemic zones, the incidence of positive histoplasmin in uveitis cases has been found to be comparable to that of the general population (Walma and Schlaegel 1964). If *Histoplasma* is causal in this severe ocular syndrome, it is certainly not the only cause (Prasad and van Gelder 2005).

To the extent that there is support for the involvement of *H. capsulatum* in POHS, a hypersensitivity response rather than an infectious process is implied, in which a post-infection autoimmune effect provoked by an immunological cross reaction (the sensitized human body attacking a self-antigen which chemically resembles a fungal epitope). Alternatively, POHS may be an epistemically reified coincidence of an unrelated ocular condition with prior *Histoplasma* exposure. Careful histopathological examinations of histoplasmin-sensitized experimental animals following intraocular challenge with *H. capsulatum* have demonstrated immunogenic macular changes (Okudaira and Schwarz 1962). In the original case series, Woods and Wahlen (1959) reported on one patient who developed acute changes in two retinal lesions within 24 h of receiving an intracutaneous histoplasmin challenge. The recognition of predisposing

genetic polymorphisms in the human leukocyte antigen complex (Dabil *et al.* 2003) further supports an immunogenic basis for POHS.

Sporothrix schenckii is a fungus mainly associated with outdoor vegetative debris, particularly dried straw, straw-like materials, such as dried reeds, or dried *Sphagnum* peat moss (Rippon 1988). It is cosmopolitan, with certain neotropical areas such as parts of Mexico, Brazil, Argentina, Guatemala and Uruguay having unusually high exposure and being considered as endemic regions. The great majority of cases are not respiratory, but rather consist of subcutaneous abscesses caused by minor traumatic implantation, e.g. through a scratch, often accompanied by progressive infection and swelling of a series of proximal lymph nodes. Respiratory cases are very uncommon (Rippon noted about 150 recorded cases up to 1988) and a substantial proportion of cases are in chronic alcoholics. Presumably, in at least some of these cases, a significant inoculum load may be aspirated during extended contact with grass litter or soil while the patient is unconscious. Respiratory cases normally consist of a life-threatening cavitary pneumonia; a less common variant is an often spontaneously resolving infection mostly involving pulmonary lymph nodes (Rippon 1988). Because *S. schenckii* is strongly associated with poorly stored commercial peat moss (Dixon *et al.* 1991) and because this material is often used in potted plants, the rare to occasional occurrence of *S. schenckii* indoors is to be expected. The only recorded indoor isolations of *S. schenckii* are from potted plant soil incorporating oak and birch litter, not peat moss (Mariat 1968) and from the floor around an indoor swimming pool (Staib and Grosse 1983). Clearly, although there may be a small but appreciable risk of subcutaneous sporotrichosis in the indoor environment, especially when poorly stored peat moss is used in domestic cultivation, the probability of respiratory sporotrichosis connected with the presence of the fungus indoors is extremely low.

Recently, Marimon *et al.* (2007) have described two phylogenetically divergent subgroups within *S. schenckii ss. lat.* as separate species, *S. brasiliensis* and *S. globosa*. It is not clear whether or not the latter name is a later synonym of earlier names like *Sporothrix tropicale*, so that, *pro tem*, the entire three-clade *S. schenckii* complex is treated as *S. schenckii* here.

The ecology of *Cryptococcus gattii*, appears not to intersect with indoor habitats, except that in vigorous environmental outbreaks, as seen recently in Vancouver Island, Canada, the possibility of rare airborne exposure indoors cannot be excluded (Kidd *et al.* 2007).

In contrast, *C. neoformans* (now sometimes split into a more narrowly defined type variety and *C. neoformans* var. *grubii*) is unique among the virulent fungi in at least potentially being more strongly associated with indoor than outdoor exposure. Like *H. capsulatum*, it is strongly associated with substantial deposits of bird or bat guano; the two species may occur together in the same deposits in histoplasmosis endemic zones. As a cosmopolitan organism, *C. neoformans* is, however, much more geographically widespread than *H. capsulatum*. It may also require lesser amounts of dung deposition to establish itself, and may infest pigeon dung accumulations on window ledges, bridge girders, and other such unexceptional man-made niches in urban areas. Fortunately, it is much less virulent than *H. capsulatum*, and a large exposure is usually, but not always, needed for an immunocompetent individual to become infected. On infrequent occasions, immunocompetent persons with no marked contact with guano may develop localized pulmonary lesions (Krajden *et al*. 1991). However, people who expose themselves to high inoculum levels by disturbing large quantities of guano without using high efficiency particulate arrestance (HEPA) respiratory filter protection, may or may not develop discernible respiratory illness, but typically develop the very severe condition known as cryptococcal meningitis (Kwon Chung and Bennett 1992). Not only is the meningeal infection severe, but the long-used drug of choice for treatment, amphotericin B, has serious side effects in many patients, most notably rigors (severe stiffening muscle spasms involving much of the body) after intravenous administration, and a risk of severe kidney damage, potentially leading to a requirement for kidney transplant.

The worldwide HIV outbreak enormously increased the number of cryptococcosis cases. In USA, prior to the availability of highly active antiretroviral therapy (HAART), mortality due to cryptococcosis increased fourfold as a result of HIV infection (Selik *et al*. 1997). HIV-positive individuals with CD4+ T-helper lymphocyte counts below 50 cells μl^{-1} are at risk of developing cryptococcal pneumonia, meningitis or disseminated infection (Arasteh *et al*. 1996). Unlike persons with intact cellular immune systems, such persons may be at risk of infection after inhaling a single cryptococcal cell or a small number of cells. In temperate and tropical areas, immunocompromised persons risk exposure to *C. neoformans* whenever they are exposed to bird or bat faecal deposits, regardless of the amount of material present. Aviaries, whether privately owned or in zoos, may have a high

level of risk (Staib 1982, 1985, Staib and Heissenhuber 1989). Persons with AIDS are advised to avoid birds (including domestic pets) when possible, or at the very least not to involve themselves in cleaning cages or other areas where dung might accumulate or where residual traces of dung may be present. In tropical areas, wounds on various tree species may also be reservoirs of *C. neoformans* inoculum (Lazera *et al*. 1996), but it is not clear how wide a range of trees may be involved.

Pneumocystis jirovecii

Although organisms resembling *Pneumocystis jirovecii* (previously called *P. carinii* f. sp. *hominis*) have been detected in a wide range of mammalian species (Laakkonen 1998), molecular and attempted transfection studies have shown that strains which are specific to these organisms make up a species complex with individual species specific to particular host species (Stringer 1996). Also, the degree of genetic difference between strains from different animal species suggests a species level separation (Stringer 1996). *P. jirovecii* itself is only known to infect humans. The study of *P. jirovecii* ecology and epidemiology has been greatly hindered by the failure of this organism to grow in artificial culture. In recent decades, however, molecular biological techniques have made studies of both infecting and environmental inoculum possible (Olsson *et al*. 1996, Bartlett *et al*. 1997, Rabodonirina *et al*. 1997). Pneumocystosis is normally only seen in patients with significant cellular immune deficiencies, and in "premature and marasmic infants between the tenth and twenty-fourth week of life" (Seed and Aikawa 1977). Although the disease may spread from the lungs to various parts of the body, initial infection is clearly respiratory (Hughes 1982). The host specificity of *Pneumocystis* types indicates a close host/pathogen adaptation, making growth and reproduction in the environment unlikely. *P. jirovecii* infection thus appears to be contracted from other infected conspecifics, and outbreaks typical of contagious disease have been reported in immunocompromised humans (Hennequin *et al*. 1995, Yazaki *et al*. 2009) and in experimental animals (Soulez *et al*. 1991). A study in which air from *Pneumocystis*-infected AIDS patients was passed through filters, which were then examined by nested PCR for *Pneumocystis* DNA, found a strong tendency of the air from patients' rooms to contain *P. jirovecii* (Olsson *et al*. 1998). DNA typing by sequencing and allele specific PCR showed that the strain detected by filtration in four of five patient rooms was identical to that found in the corresponding patient. Airborne inoculum thus appears to be shed by infected patients. Air filtering, however, also detected airborne *P. jirovecii* inoculum at sites distant from any person with diagnosed pneumocystosis (Bartlett *et al.* 1997, Olsson *et al.* 1998). Immunocompetent hospital staff studied by serology and PCR showed no signs of becoming infected by *P. jirovecii* from infected patients (Lundgren *et al.* 1997). On the other hand, a survey of 77 HIV negative, demonstrably immunocompetent patients with severe pulmonary disease revealed 5 patients infected with *P. jirovecii*. The infections were subclinical, giving rise to no significant symptoms, and the infected persons were tentatively classed as clinically silent carriers (Armbruster *et al.* 1997). In a *P. jirovecii* outbreak in a renal transplant unit, a single molecular strain type was found in 27 case patients (22 of whom could have the time of transmission from a previous patient inferred from their medical records) as well as in environmental contamination (Yazaki *et al.* 2009).

Widespread infection caused by a related *Pneumocystis* species has also been confirmed in immunocompetent rabbits raised in experimental colonies (Cere *et al.* 1997). Infection is probably transient: for example, in artificially immunosuppressed rats infected with *Pneumocystis carinii* and allowed to recover from immunosuppression, inoculum detectable by histology and PCR steadily decreased, and was eliminated completely from over 75% of rats within one year (Vargas *et al.* 1995). Clinical evidence also suggests that patients are exogenously infected, rather than being infected by recrudescence of a primary infection acquired earlier in life (Latouche *et al.* 1994). Overall, it appears that although suitably immunocompromised hosts are ubiquitously at risk of contracting serious *P. jirovecii* infection, immunocompetent persons are either not at risk or run a small risk of contracting a subclinical infection not producing discernible symptoms. A recent review on nosocomial *P. jirovecii* epidemiology is provided by Nevez *et al.* (2008).

Opportunistic fungi

One matter of human and animal biology that the AIDS pandemic has helped to clarify is that the immune system is highly compartmentalized. Resistance to fungal infections appears to depend largely on the cellular component of the immune system (Levitz 1992). The humoral component, mainly consisting of antibodies and other soluble substances in serum, may be less influential than it is in other types of infection. Persons with a deficit of CD4+ T helper cells, as seen in AIDS, become markedly susceptible to pneumocystosis as well as fulminant cases of infection caused by virulent fungi, such as coccidioidomycosis, cryptococcosis, histoplasmosis, sporotrichosis and *Penicillium marneffei*-penicilliosis. In addition, they typically develop aggravated mucosal candidiases, particularly oropharyngeal, vaginal and oesophageal colonization. They do not, however, generally develop systemic candidiasis or infections caused by opportunistic fungi (Levitz 1992). Only if they later, as a result of AIDS complications, suffer a strong attenuation of another cellular immune component, the neutrophil or granulocyte component, do they have a strongly elevated risk of aspergillosis and other opportunistic mycoses. The complications leading to such a neutropenic (neutrophil deficient) condition in persons with AIDS include B-cell lymphoma and treatment with certain antiviral drugs, such as ganciclovir, which have partial immunodepression as a side effect. Similarly, in people not affected by HIV, the condition generally conducive to opportunistic infection is neutropenia, and this condition is usually induced by (1) leukaemia, lymphoma and related neoplasms; (2) immunosuppressive drugs used in the treatment of these or some other cancers; (3) immunodepressive therapies used in the prevention of rejection of organ transplants; or (4) certain congenital immunodeficiencies. Persons on prolonged, heavy corticosteroid therapy may manifest a partial suppression of neutrophil activity, resulting in an increased susceptibility to opportunistic infections. The opportunistic fungi causing severe infection in neutropenic patients are normally harmless to non-neutropenic persons. A few unusual conditions, such as severe burns and poorly controlled diabetes, may increase the risk of non-neutropenic patients to opportunistic fungal disease.

The above synopsis of conditions leading to opportunistic infection is mainly in reference to invasive conditions. The far less severe colonizing types of opportunistic infection have their own particular predisposing conditions. Allergic bronchopulmonary aspergillosis (ABPA), where the bronchial passages may become colonized with *A. fumigatus* or, rarely, other aspergilli (Summerbell 1998), generally affects persons who have had severe chronic asthma for many years (Varkey 1998) or persons with cystic fibrosis (Schwartz 1998). In both conditions, bronchial passages tend to manifest inflammation and to generate a large amount of mucous material. Long-term asthmatics or persons undergoing long-term systemic corticosteroid treatment may be at increased risk for allergic mycotic sinusitis, an allergy-exacerbated fungal colonization of the sinus walls. Persons such as children or mentally handicapped individuals who put objects into their outer ear canals may be at elevated risk for otomycosis, fungal ear canal colonization.

A very large number of normally non-pathogenic fungi may cause infection of the cornea of the eye, but normally only after traumatic implantation, for example, with trauma to the eye from flying plant debris during agricultural processing. It is extremely rare for such events to occur in the indoor environment.

This explicit breakdown of predisposing conditions is given because members of the public are often confused about indoor moulds and opportunistic infection. Individuals whose homes or workplaces have been reported to contain potentially opportunistic moulds may fear systemic or ocular infections to an unrealistic degree. Also, persons with severe allergies or chemical sensitivities may purport that their immune systems are compromised or (as as the author has often been told) "completely destroyed." Various branches of alternative medicine inform essentially immunologically normal people that they are seriously immunocompromised, and a few even fallaciously assert that patients have yeasts or mould fungi growing in their blood. The conditions or pseudo-conditions diagnosed by these practitioners seldom bear any valid relationship to susceptibility to opportunistic mycosis. With the aid of precise information about predisposing conditions, confusion about susceptibility to fungal infections can be minimized.

Reactions of immunocompetent occupants of buildings to indoor mould proliferation almost invariably consist of a mixture of direct chemical irritation effects and allergic responses. These should not be confused with invasive or colonizing opportunistic infection, which are entirely different phenomena.

Many of the opportunistic mould fungi found in indoor environments are typical inhabitants of natural composts, tropical soils and other heated materials. Fungi such as *Aspergillus terreus, A. fumigatus, A. flavus, Rhizopus microsporus* var. *rhizopodiformis, R. oryzae* and *Rhizomucor pusillus* are thermotolerant, normally growing at temperatures above 45°C, as well as at lower temperatures (Domsch *et al.* 1993). Indoor sites particularly associated with these fungi therefore may be those where humid organic material is exposed to heat (Marsh *et al.* 1979), most no-

tably within poorly maintained heating ducts and attached humidifier structures, in soil of potted plants (Summerbell *et al.* 1989), especially those placed in warm locations (Staib *et al.* 1978), and in indoor composts. The thermotolerant aspergilli, however, are not uncommon in outdoor air, all-year round in tropical areas (Christensen and Tuthill 1985) and seasonally in temperate areas (Kwon Chung and Bennett 1992). Although they are generally quite uncommon in indoor proliferations in homes and non-agricultural workplaces, they are nonetheless a constant hazard for hospitals. Occasional instances of indoor proliferation, along with an association with pigeon dung on building surfaces (Mehta 1990), including areas near building air intakes (Burton *et al.* 1972), and a regular low level presence as dormant propagules in dust combine to ensure that whenever renovation occurs and materials are disturbed, a significant influx of airborne propagules is likely to endanger hospitalized neutropenic patients (Arnow *et al.* 1978, Sarubbi *et al.* 1982, Nolard 1994). This risk can be significantly but not entirely controlled by using HEPA filters for air in areas with high risk patients (Rose 1972, Rosen and Sternberg 1976, Rose and Hirsch 1979), along with other dust control techniques such as having incoming staff and visitors don sterile gowns and hair covers, and also shoe covers. In addition, high-risk patients can only be given food materials that have been suitably processed to inactivate fungal propagules. To be avoided in particular are substances such as spices which tend to have significant *Aspergillus* counts (De Bock *et al.* 1989). Likewise, flowers, tea leaves and other organic or potentially dusty environmental inputs are to be avoided (Nolard 1994). Air filters must be exchanged regularly; such filters may become hazardous sites of *Aspergillus* growth if they are allowed to accumulate sufficient organic material from the air (Arnow *et al.* 1991). Rigorous infection control measures which may be employed to minimize *Aspergillus* in bone marrow transplant units during construction include installation of HEPA filters, applying paint amended with copper-8-quinolinolate (oxine copper) to surfaces, installation of non-perforated ceiling tiles, sealing of windows, replacing potentially dusty horizontal vaned window blinds with vinyl roller shades, and regular cleaning of all surfaces (Loo *et al.* 1996). A portable steam disinfection system has recently been shown to effectively kill *Aspergillus niger* applied to test surfaces (Tanner 2009). Showers utilizing unfiltered tap water can also be a significant source of *Aspergillus* inoculum in some areas (Anaissie *et al.* 2002).

Where the source of contamination is outside the

hospital building, both windows and strategically selected doors may be sealed. Where possible, air handling systems are prepared to filter incoming air and to draw incident air from an area away from the source of contamination or its downwind airstream (Streifel *et al.* 1983). The epidemiology of *Aspergillus* infection as affected by hospital renovation has recently been reviewed by Haiduven (2009); key control strategies such as construction of a "protective environment" are outlined in this review.

These precautions are not applicable to non-neutropenic patients, who have a very low risk of infection by thermotolerant aspergilli and zygomycetes. Many individuals, particularly agricultural workers, may be routinely exposed to such inoculum, putting them at risk of allergy, hypersensitivity reactions or unfavourable chemical exposure, but not significantly at risk of invasive disease (Jappinen *et al.* 1987, Yoshida *et al.* 1993). A few idiopathic cases of invasive pulmonary aspergillosis in non-immunocompromised patients with no significant history of exposure (Karam and Griffith 1986) remain unexplained. Invasive mycotic sinusitis caused by *A. flavus* and other fungi may also occasionally occur in immunocompetent individuals (Hussain *et al.* 1995). As mentioned above, other predisposing conditions leave a few categories of individuals vulnerable to colonizing infections. In addition to the conditions mentioned above (ABPA, sinusitis, otomycosis), aspergilloma or *Aspergillus* fungus ball may occur. The possession of a pre-existing pulmonary cavity, e.g. one caused by previous tuberculosis or by a combat wound, may predispose persons to acquiring this condition.

It is unclear whether patients with predisposition to colonizing infections are placed at particular risk by aggregations of *Aspergillus* inoculum beyond the background level. It is possible that there is sufficient thermotolerant *Aspergillus* inoculum present in normal ambient air and dust to effectively expose all such patients. A study by Radin *et al.* (1983), however, did find a correlation between exacerbations of ABPA and increases in total outdoor mould counts. Similarly, mouldy marijuana, bird dung, horticultural mulch and nearby outdoor dump sites have been apparently linked to initiation or exacerbation of ABPA (Krasnick *et al.* 1995, Summerbell 1998). In most situations, however, the rationale for rigorously eliminating amplifiers of thermotolerant aspergilli and other opportunistic moulds in homes and commercial buildings lies primarily in avoiding the allergenic, toxic and biochemical irritant effects of these fungi, and only to a lesser degree in avoiding risk of invasive

or colonizing disease. At the same time, it should be pointed out that there is an increasing trend not to keep all neutropenic patients in hospital for the full duration of their immunodepression (Freifeld and Pizzo 1996), so a need may arise for certain non-hospital buildings to be as scrupulously clean of such fungi as is reasonably possible. In particular, patients may be sent home, and their homes should therefore be cleared of potted plants, indoor composters, bird cages, spice shakers and grinders, dried grains and other dried plant material, and other such notorious *Aspergillus* sources, during the time in which the patient is vulnerable (Summerbell *et al.* 1992, 1994). The patient is advised to avoid gardening and any construction, renovation or agricultural activities, or activities such as pigeon handling, for the duration of neutropenia. It would be ideal if the environment could be tested for the presence of opportunistic fungi before the patient is discharged to it.

Control of the opportunistic aspergilli is aided by knowing their normal growth sites. In general, most of this subgroup of aspergilli regularly grow in the following sites: composts; herbivore and bird dung accumulations; bird nests; tropical and subtropical soils, especially under cultivation; dead or parasitized insects; human and animal corpses, especially those in which bacterial decomposition is delayed by storage or environmental conditions; and stored seeds of all kinds, most notably spices, groundnuts and edible graminaceous grains such as wheat, barley, oats, sorghum, rice and maize (Domsch *et al.* 1993, Millner *et al.* 1994, Summerbell 1998). Most of these species degrade alkanes and aromatic compounds, and so may be found in association with decaying material adulterated with materials such as fuel oil, tar, organic solvents, phenolic compounds and paraffin (Domsch *et al.* 1993, Summerbell 1998).

Individual thermotolerant opportunists have distinct habitats in which they may become common within buildings. *A. fumigatus* is especially associated with potted plants, although *A. niger* is also common in these habitats (Staib *et al.* 1978, Summerbell *et al.* 1989, Thompson *et al.* 1994). Both these fungi are also strongly associated with improperly stored spices (Domsch *et al.* 1993, Nolard 1994). The popularity of home composting in parts of Europe has provided another locus of *A. fumigatus* growth indoors (Staib 1996). *A. flavus* is particularly associated with poorly stored peanuts and other seeds (Domsch *et al.* 1993), but may be found in a variety of other indoor habitats, e.g. bird dung accumulations (Staib 1985) and food pellets for pet rabbits (Staib 1982). In a case of indoor infestation, this species grew abundantly on moisture damaged, tarred wooden floor tiles (Summerbell 1998), a finding consistent with its conjunction of hydrocarbon degrading and cellulolytic abilities. It may also colonize various sites within hospitals (Arnow *et al.* 1991).

Rhizopus oryzae and another zygomycetous opportunistic fungus, *Absidia corymbifera*, have been found in potted plant soil in a haematology unit where a leukaemia patient developed invasive zygomycosis (Staib 1985). They were also abundantly represented in bird dung accumulations in an aviary (Staib 1985). In hospitals, wooden materials such as tongue depressors have repeatedly been a source of problematic *Rhizopus* inoculum (Mitchell *et al.* 1996, Verweij *et al.* 1997). Domsch *et al.* (1993) record isolation of *R. oryzae* from grains, onions, various nuts and stored sweet potatoes.

Apart from the thermotolerant aspergilli and zygomycetes mentioned above, a few other potentially opportunistic moulds are commonly associated with indoor habitats. *Paecilomyces variotii*, a fungus of much lower virulence than the aspergilli and zygomycetes mentioned, is typically associated with sites high in oils, hydrocarbons, or substances such as chemical preservatives inhibitory to other fungi (Domsch *et al.* 1993). It has been detected in heavy growth in certain indoor situations such as commercial activated charcoal products (George *et al.* 1991), decaying urea formaldehyde foam insulation in walls (Bissett 1987) and deteriorated tarred wooden flooring (Summerbell 1998). Domsch *et al.* (1993) note isolation from compost, peanuts, root vegetables, spoiled olives, decaying plant parts, leather, synthetic rubber, mouldy cigars and ink. The causation of respiratory infection by this fungus is possible but very unlikely in the neutropenic patient, and immunocompetent patients are certainly not at risk of such infections. It may, however, colonize sites where there are barrier breaks in the body integument associated with medical implants, such as implanted catheter sites in chronic ambulatory peritoneal dialysis patients (Crompton *et al.* 1992).

Pseudallescheria boydii and the closely related *Scedosporium apiospermum* (formerly considered conspecific), fungi most typically associated with compost, warm soil and dirty water (Rippon 1988, Domsch *et al.* 1993, Kaltseis *et al.* 2009), have been isolated indoors in potted plants and in water from a canister collecting an air conditioner condensation drip (Summerbell *et al.* 1989). These fungi are capable of causing a range of infections similar to *A. fumigatus*

(Kwon-Chung and Bennett 1992). Significantly less virulent, but still a serious hazard to the neutropenic patient, are *Fusarium oxysporum, F. solani, F. verticillioides* (=*F. moniliforme*)*, F. dimerum* ss. str. (Schroers *et al.* 2009) and *F. proliferatum* (Richardson *et al.* 1988). These species are associated with roots and lower stems of a variety of fresh vegetables, including lettuce and other greens, and with stored tubers and corms (Domsch *et al.* 1993). In addition, *F. solani* and *F. dimerum* are common in fresh water and water pipes (Domsch *et al.* 1993, Schroers *et al.* 2009) and hence with sinks and other sites connected to domestic plumbing. *Fusarium verticillioides* is notoriously associated with decaying maize corn kernels (Wicklow 1988, Domsch *et al.* 1993).

Herpotrichellaceous black yeast anamorphs in the genus *Exophiala* are commonly found in association with sinks and drains indoors (Nishimura *et al.* 1987), a habitat contiguous with their natural habitat in cool or cold, moist soils (Domsch *et al.* 1993). Thermotolerant members of this group have also been found to infest steam baths and hot tubs (Matos *et al.* 2002, Sudhadham *et al.* 2008). These fungi may cause ABPA-like colonizing infections in cystic fibrosis (Haase *et al.* 1991) and long-term asthmatic patients (unpublished data). An *E. dermatitidis* infection marked by recurrent haemoptysis occurred in a patient with a long history of bronchiectasis (Barenfanger *et al.* 1989). Aerosol splash connected with use of plumbing facilities is potentially a significant source of exposure. It should be noted that these same fungi are very common as contaminants of respiratory secretions and skin, and are among the most likely fungi to contaminate bronchioalveolar lavage (BAL) sampling equipment in hospitals.

Some additional marginally opportunistic fungi are very common in the environment, but have only rarely been implicated in causing primary pulmonary disease or invasive sinusitis of severely immunocompromised patients. Included in this group are species in the genus *Scopulariopsis* and a few members of *Acremonium* subgenus *Acremonium*. *Scopulariopsis brevicaulis*, a fungus commonly isolated from indoor dust, is one such fungus, as is the only slightly less environmentally common *S. candida* (Kreisel *et al.* 1994). *Acremonium kiliense* and *A. strictum*, which are readily isolated indoors, e.g. growing within amplifiers of *Stachybotrys* on very moist surfaces, have also been reported to cause pulmonary infection in the severely immunocompromised patient (Boltansky *et al.* 1984). The identities of the isolates assigned to the notoriously difficult genus *Acremonium* must be reassessed with modern techniques. *Paecilomyces lilacinus* is another common indoor dust fungus that may cause invasive mycotic sinusitis (Gucalp *et al.* 1996). *Alternaria alternata*, a species that is relatively common in proliferations on water-damaged wallboard, wallpaper (Bemmann 1979, Miller and Holland 1981), ceiling tiles (Ren *et al.* 1998) and other indoor substrates substantially consisting of finely divided cellulose (Simmons *et al.* 1997), is one of the more common agents of mycotic sinusitis (Morrison and Weisdorf 1993, Iwen *et al.* 1997), whether invasive or allergic. *Alternaria infectoria* may also cause these conditions.

Another group of opportunistic fungi that may become established in the indoor environment and which may rarely have a respiratory infection court consists of basidiomycetous yeast species in the genus *Trichosporon*. Molecular taxonomic studies have shown that the group of isolates formerly considered to be a single species, *T. beigelii*, actually comprises several species, only a small number of which are regularly connected to human systemic invasive or colonizing disease (Guého *et al.* 1992). The most important of these are *T. asahii, T. mucoides* and *T. inkin*. Interestingly, *T. mucoides* and *T. asahii* are among the principal fungi colonizing tatami mats within Japanese style houses, and causing summer type hypersensitivity pneumonitis in some immunocompetent individuals who have experienced prolonged inhalation of high quantities of inoculum (Nishiura *et al.* 1997). *Trichosporon* spp. are environmentally widespread, and the existence of some niches of proliferation within non-Japanese style housing can also be predicted.

There are some significant groups of opportunistic fungi which are not known to have significant proliferation loci indoors. Examples would include *Cochliobolus* anamorphs such as *Curvularia* and *Bipolaris*, as well as diverse fungi such as the basidiomycete *Schizophyllum commune* and species in the zygomycete genus *Conidiobolus*. These will not be mentioned further in this review.

A significant factor in the epidemiology of all these opportunistic filamentous fungi is their mode of dispersal. Hyphomycetes in general can be roughly broken down into species which produce dry conidia specialized for airborne dispersal, and species which produce mucoid conidia specialized for dispersal by water movement, splash aerosolization, arthropod vectors and water surface monolayer spread. The fungi in the former group may greatly predominate in hospital air (Summerbell *et al.* 1992). Members of the dry conidial category, including aspergilli, *Paecilo-*

myces, Scopulariopsis and *Rhizopus*, pose the greatest risk to neutropenic patients both in hospital and at home. Many such fungi, including the groups named above, have conidia with diameters in the range of 2-5 μm, allowing them to penetrate the respiratory tract to the gas exchange region where they are deposited. *Aspergillus fumigatus* in particular has conidia only 2.5-3 μm in diameter, and thus has easy access to deep pulmonary tissue upon inhalation. *Aspergillus flavus*, like some other aspergilli, has somewhat larger conidia (up to 4.5 μm in diameter), which may explain why *A. flavus* forms a much greater percentage of sinus aspergilloses than of pulmonary infections, where small conidial size is of greatest import (Chabra *et al.* 1996). It is worth noting, however, that all aspergilli bear their conidia in dry, unbranched chains that fragment to some extent during aerosolization. It is common in this process for a large proportion of the resulting aerosol particles to consist of short conidial chains rather than solitary conidia. Two potentially opportunistic species of *Fusarium, F. verticillioides* and *F. proliferatum*, although producing mucoid conidia under some conditions, may also produce aerial chains of small conidia under certain substrate conditions. The "wet" conidia of opportunistic fungi such as the *Scedosporium* conidial state of *P. boydii, Absidia*, the black yeasts, *F. oxysporum* and *F. solani, Acremonium* and *Trichosporon*, may be rendered airborne by splash or by the eventual desiccation and fragmentation into dust of mucoid conidial aggregations. Splash aerosolization, a phenomenon most typically associated with raindrops in the outdoor environment, is probably mainly associated with sinks, showers and other aquatic washing facilities indoors. Shower heads, of course, may also directly aerosolize fungi, such as *F. solani*, that may be found in tap water (Anaissie *et al.* 2001).

The present author has observed that many (though by no means all) hospital infection control staff treat all problematic opportunistic fungi, e.g. in outbreaks, pseudo-outbreaks or contamination of material, as if they were dry conidial organisms. Thus, the response to pseudo-outbreaks of *Exophiala* in BAL samples or *Acremonium* contamination of virology sample transport medium may be deployment of air samplers in searching for the source of contamination. Although fungi with wet conidia might, for the reasons mentioned above, derive from an airborne source, it is imperative that infection control staff also account for the reasonable possibility that such fungi have derived from an aqueous source. Such aqueous sources typically include not only bulk fluids but also biofilms attached to wet or frequently moistened surfaces, e.g. within equipment. In the case of BAL contamination, the equipment and fluids used to obtain the bronchial washing samples must be thoroughly tested for contamination, as must disinfection equipment (Hoffman *et al.* 1989). In all such cases of contamination by fungi with wet conidia, particular scrutiny should be given to stock solutions of all kinds containing antifungal drugs such as amphotericin B or other antifungal preservatives and inhibitors. *Acremonium, Fusarium, Pseudallescheria* and *Trichosporon* in particular may be insensitive to amphotericin B and most other known antifungal agents. Some dry conidial fungi, such as *P. variotii* and *P. lilacinus*, may also occur in similar situations. *Paecilomyces* species, especially *P. variotii*, are notoriously resistant to a wide array of noxious materials (Domsch *et al.* 1993) and may be expected in materials such as biomedical solutions or cosmetics containing preservatives. The ability of some fungi such as certain *Paecilomyces* spp. to grow on certain types of plastics, particularly vinyls (Roberts and Davidson 1986, Domsch *et al.* 1993) must also be kept in mind. For example, a severe outbreak of *P. lilacinus* skin infection among neutropenic patients in a Swiss hospital was eventually traced to colonization of plastic skin lotion bottles prior to their being filled with the lotion (Orth *et al.* 1996).

SAMPLING METHODOLOGY FOR FUNGAL RESPIRATORY PATHOGENS

Only fungi which are potentially obtained from indoor growth sources (not incident outdoor dust) are treated below.

Virulent fungi

The dimorphic fungi corresponding or related to the teleomorphic genus *Ajellomyces*, viz. *Histoplasma capsulatum, Blastomyces dermatitidis* and *Paracoccidioides brasiliensis*, pose a unique problem in sampling. Only the first of these fungi is likely to proliferate indoors, but since sampling is similar for all these fungi, general comments will be made. In these fungi, either inoculum is normally extremely sparse even in problem environmental materials, or the propagules of these fungi fare very poorly on isolation plates in competition with antibiotic resistant bacteria and other fungi. Whatever the exact reason or reasons, it is certainly the case that these fungi are not successfully detected in standard culture tests relying on

spreading of diluted environmental materials onto fungal growth medium (Ajello and Weeks 1983). They have not been detected in air samples either. The sole isolation technique repeatedly proven successful, as detailed by Ajello and Weeks (1983), is the injection of suspensions of soil or other environmental material into experimental animals (usually mice), and the subsequent harvesting and culture plating of the livers and spleens onto systemic mycology isolation media. A variation of this test in which the injected mice were tested with a dot-blot ELISA for *Histoplasma* antibodies rather than live organisms was recently published by Reyes-Montes *et al.* (2009). An enrichment-based *in vitro* direct isolation method for *B. dermatitidis* from environmental materials was reported by Baumgardner and Parestky (1999). Although a PCR detection technique for *B. dermatitidis* inoculum soil has been developed, based on the use of spiked samples (Burgess *et al.* 2006), to date no follow-up studies on environmental materials containing *B. dermatitidis* have appeared.

For these fungi in the *Ajellomyces* clade, clinical isolation from mammalian tissues requires media supplemented with cycloheximide and enriched with high protein materials such as brain heart infusion (BHI) and blood (Larsh and Goodman 1980). Examples of such media include BHI sheep blood agar; BHI sheep blood agar plus 3% egg albumin (Chaturvedi *et al.* 1990); Sabhi agar; yeast extract agar; and plain BHI agar. These are all supplemented with cycloheximide, chloramphenicol and gentamicin. The incubation temperature is usually 28-32°C.

In cases where *H. capsulatum* is being sought in guano accumulations, it is useful to recall the high microbial competition and turnover which may occur in this substrate. There is no reason to believe *H. capsulatum* will be strongly persistent at its sites of growth. For this reason, nearby dust, e.g. on ledges or timbers, where *H. capsulatum* conidia may have settled out and remained dormant in relatively dry circumstances, away from the risk of immediate lysis by bacteria, should be sampled in addition to the guano itself. It should be borne in mind that no number of negative sampling results is sufficient to demonstrate absolutely that a site is free of *Histoplasma*, since small or intermittent quantities, or quantities missed by chance alone (or through poor technique), may always be present. The majority of requests for such testing derive from persons wishing a precise assay of their health risk; it must be explained that this is not something this test can accurately provide. It is a "no false positives" type test, in which positive results are of value but negative results are of no practical significance, and its main functions are to confirm the local presence of *Histoplasma* in at least a proportion of suspected environmental foci, and to obtain isolates for possible epidemiological genetics or research related testing. In endemic or marginal endemic zones, the most responsible approach to take is to assume that bird and bat guano deposits are likely to contain *Histoplasma*, and to proceed on that basis (Davies *et al*. 1995). In areas where *Histoplasma* has never been vicinally reported, this assumption is unnecessary; but the probability that *C. neoformans* or large amounts of *Aspergillus* may be present in guano must be borne in mind.

As with *B. dermatitidis*, *H. capsulatum* has been investigated for PCR-based detection, and an assay has been published (Reid and Schafer 1999), but to date this has been subjected to little published evaluation.

Cryptococcus neoformans and *C. gattii*, unlike *Histoplasma*, are readily isolated from environmental materials in ordinary dilution plating. Detection is facilitated by using Staib's phenol oxidase agar (*Guizotia abyssinica* birdseed agar or its refined derivatives such as caffeic acid agar) for detecting, through a brown colour reaction, the strong extracellular phenol oxidase activity peculiar to *C. neoformans* among yeasts (Staib 1962, Salkin 1988, Staib and Heissenkuber 1989). Occasional isolates of non-*C. neoformans* cryptococci may produce a similar reaction, but most do not. Also, some *Trichosporon* species may have similar phenol oxidase activity, but are easily distinguished in a microscopic mount. If this excellent medium is not available, the present author has observed that *C. neoformans* in guano readily grows on Littman oxgall (Difco) streptomycin agar (Kane *et al*. 1997) incubated at 37°C. Most non-*C. neoformans* cryptococci do not grow at this temperature, and a few which do grow poorly. Cryptococci are distinguished on Littman oxgall agar from any background *Candida* species because their colonies remain pale, whereas *Candida* spp. take up the methylene blue dye in the medium and become distinctly tinged with blue, or even bright blue.

Interestingly, a rapid method based on immunomagnetic isolation has been designed for separating *C. neoformans* cells from natural materials (Katsu *et al*. 2003). Magnetic beads are coated with anti-*C. neoformans* (or hypothetically anti-*C. gattii*) antibodies, and then retrieved from the material by means of a magnet. A PCR technique for direct detection of *Cryptococcus* in environmental materials has also been tested (Castañeda *et al*. 2004).

Pneumocystis jirovecii

As mentioned above, *Pneumocystis* cannot be detected by culturing. Environmental detection is achieved by filtering air through cellulose ester filters, extracting DNA, and performing PCR with primers for *P. jirovecii* genes such as the ribosomal internal transcribed spacer (Bartlett *et al*. 1997, Olsson *et al*. 1998), the mitochondrial large subunit ribosomal RNA (Olsson *et al*. 1998) or the thymidylate synthase gene (Olsson *et al*. 1996). Either primers specific to *P. jirovecii*, or more general primers may be used, but in the latter case resulting amplicons must be verified as derived from *P. jirovecii* by rigorous analysis such as sequencing. An improved technique for specifically isolating *Pneumocystis* DNA from environmental sources involves mixing biotinylated pathogen-specific oligonucleotides into the environmental materials, in this case deposits on air filters, and then capturing the resulting annealed DNA with streptavidin-coated magnetic beads (Maher *et al*. 2001).

Opportunistic fungi

In much modern work with indoor moulds, the historic focus on air sampling and airborne spora levels has been strongly diminished in favour of the physical detection of indoor mould amplification sites. These sites are then confirmed and analysed by direct microscopy, e.g. from scrapings of substrate material or from adhesive tape mounts (Samson 1985, Davies *et al*. 1995, Summerbell 1998). Even where reliably culturable moulds are concerned, this practice has the advantage of connecting instances of problematic mould proliferation with specific underlying causes related to building maintenance, and offers more precise analysis and control than air sampling alone does. Although finding indoor growth foci of opportunistic fungi in this way is ideal, and is appropriate when there are no severely immunocompromised persons in the environment, it is inadequate in any situation where neutropenic patients may be involved. The reason is that with neutropenic patients, even small amounts of inoculum, potentially connected not just with on site proliferation but also with normal dust accumulation or even normal penetration of outdoor air, may be life threatening. Also, in contrast to the situation where moulds are evaluated for allergenic or biochemical irritant potential, in which case viability of propagules is not of major significance, in situations where invasive disease is a practical concern, only living propagules constitute a threat. This, then, necessitates a combination of volumetric sampling to monitor the air for culturable propagules. A minimum limit of detection of 1 CFU m^{-3} has been proposed for opportunistic fungal pathogens in critical environments such as health care facilities (Streifel 2004). The additional dilution plating of suspect bulk materials is helpful for identifying foci potentially harbouring inoculum. The small conidial size of the most significant respirable pathogens prevents use of gravitational settle plate techniques for air sampling in connection with opportunistic infection, since these techniques underestimate the relative frequency of small propagules (Sayer *et al.* 1969, Solomon 1975). Settle plates have been demonstrated to miss *Aspergillus fumigatus* in haematology and bone marrow transplant units when it is present at a frequency which is low but sufficient to cause new opportunistic infections (Iwen *et al.* 1994).

The tendency of almost all opportunistic fungi to grow at 37°C can be employed in selective isolation techniques on a variety of growth media. This selective effect can be used to advantage both in air sampling and in bulk sampling. Summerbell *et al.* (1989), for example, using Littman oxgall streptomycin agar incubated at 37°C, obtained significantly higher levels of *A. fumigatus* and *Pseudallescheria boydii* from potted plant soils than were obtained using the same medium incubated at room temperature. Using inhibitory mould agar incubated at 37°C, Streifel *et al.* (1983) successfully monitored *A. fumigatus* concentrations before and after demolition of a hospital building. A selective isolation technique using the commercially unavailable (but still widely stockpiled) fungicide benomyl has facilitated the isolation of *Pseudallescheria, Scedosporium, Petriella* and related fungi (Summerbell 1993, Rainer *et al.* 2008, Kaltseis *et al.* 2009).

Black yeasts and related opportunistic fungi related to the family *Herpotrichiellaceae*, as well as virulent onygenalean fungi, *Sporothrix schenckii, Candida albicans* and several other medically important fungal groups, are noteworthy for their tolerance of cycloheximide. This attribute is routinely used in the design of selective isolation media for these fungi (Dabrowa *et al.* 1964, Summerbell *et al.* 1989). However, see comments above about special difficulties in isolating the *Ajellomyces* clade of the Onygenaceae. This technique is particularly important for *S. schenckii* and herpotrichiellaceous anamorphs, which may grow very poorly at 37°C *in vitro*. Zhang and Andrews (1993) made an improved Sabouraud peptone glucose medium for environmental isolation of *S. schenckii* by adding an

unusually high cycloheximide concentration (800 mg l^{-1}) in addition to 250 mg l^{-1} chloramphenicol, and 20 mg l^{-1} rifampicin. This medium has given very satisfactory results in the present author's laboratory routine for environmental *Sporothrix*.

The fungicide benomyl, mentioned above in connection with *Pseudallescheria*, can still be used where available as a selective agent for isolation of certain opportunistic fungi (Summerbell 1993). It selects for fungi related to the family Microascaceae (e.g. *P. boydii, Scopulariopsis brevicaulis*), the orders Pleosporales (e.g. *Alternaria, Curvularia*), and Endomycetales (e.g. *Candida* spp.), and most members of the phylum Basidiomycota, while inhibiting the great majority of isolates of competing groups such as Eurotiales (e.g. *Aspergillus, Penicillium*), Hypocreales (e.g. *Fusarium, Acremonium, Trichoderma*), Herpotrichiellaceae (e.g. *Exophiala*) and Onygenales (e.g. *Chrysosporium*). Although it has been widely used in phytopathology and wood decay studies for decades, its use in environmental medical mycology is as yet underdeveloped. Regrettably, current difficulty in acquiring it may restrict use to a small number of researchers.

Media with high solute concentrations can be used for the most efficient isolation of osmotolerant and osmophilic fungi. One of the most commonly used media for the efficient isolation of a wide range of fungi including osmotolerant species is dichloran-18% glycerol (DG18) medium (Pitt and Hocking 1985). Among the common respiratory opportunistic pathogens, members of the genera *Aspergillus* and *Fusarium* are the most likely to show improved isolation using such media. No study, however, appears yet to have been made with express reference to isolating a wide range of opportunistic fungi using this technique.

Specialized media have been designed for the selective isolation of *Aspergillus flavus* and the closely related *A. parasiticus* (Bothast and Fennell 1974, Pitt *et al.* 1983) and for the isolation of *Fusarium* spp. (Thrane 1996) from foods. These appear not to have been applied to the study of the same organisms as opportunists, but could well be.

There has been considerable work in the promising area of direct environmental detection of opportunistic filamentous fungi by means of molecular methodology. This work overlaps strongly with the molecular detection of other environmental moulds in indoor situations (see Chapter 4.2), and a review is beyond the scope of this present chapter.

PREVENTION OF SIGNIFICANT EXPOSURE

In many areas, avoidance of significant exposure to indoor sources of virulent fungal pathogens means avoiding aspiration of material from gross accumulations of bird and bat guano. If a building is found to contain such a site, it should be well closed up and temporarily sealed, e.g. with tape around attic trap doors or plastic sheeting over vertical doors. Anyone entering the area is recommended to wear HEPA filter respiratory protection (not a conventional dust mask!) and suitable disposable protective clothing. Precautions normally used for asbestos clean-up and similar procedures should be followed, e.g. disposal of protective suit, showering while still wearing respirator mask until potentially adherent particulates have been washed from hair and skin, and so on. Disposable materials should be placed in suitable heat resistant bags and autoclaved prior to being discarded as autoclaved biohazardous waste. For decontamination of source material, see below. All laboratory work should be carried out in a Class 2 biological cabinet.

Prevention of pneumocystosis in vulnerable patients has been accomplished only by the indefinitely prolonged administration of prophylactic drugs. However, detailed discussion of prophylaxis is beyond the scope of this chapter.

The now widespread recognition that construction or renovation activities in or adjacent to hospitals may pose a profound risk of aspergillosis to neutropenic patients has led to development of stringent infection control procedures. Representative procedures are outlined in the Ecology and Epidemiology section above. Precautions for neutropenic individuals discharged to home environments are also discussed above. Prophylaxis is common in such patients, but as mentioned above, is beyond the scope of this paper. Immunocompetent individuals who are actively disturbing (e.g. renovating) indoor environments known to contain amplifiers (that is, documented *in situ* growth sites) of opportunistic respiratory pathogens, especially *Aspergillus fumigatus, A. flavus, A. terreus, A. nidulans, Neosartorya pseudofischeri, Pseudallescheria boydii, Scedosporium apiospermum, S. prolificans, Rhizopus oryzae, R. microsporus, Absidia corymbifera* or *Rhizomucor pusillus*, would be at an advantage to wear HEPA filter respiratory protection. Specifically, a fit-tested, NIOSH-compliant N100 respirator should be used. This would protect not only against the unlikely possibility of a respiratory infection, but also against the possibility of organic dust toxic syndrome, which may accompany high exposure levels to such fungi. Similarly, if amplifiers of these fungi have not been identified in connection with the building, but unusually high levels suggestive of *in situ* amplification are seen in fungal air samples, the same precautions apply. It should be noted that outdoor levels of *A. fumigatus* may not uncommonly exceed 10 colony forming units (CFU) m^{-3} air in some temperate areas and may seasonally approach 100 CFU m^{-3} (Solomon and Burge 1975, Mullins *et al.* 1976).

As these opportunistic fungi are relatively uncommon in indoor proliferations on structural elements of buildings, except in decaying material associated with a heat source, they need not be assumed to be present in every situation where indoor mould amplification is found. In the vast majority of indoor proliferations, these fungi will be scarce or lacking.

DECONTAMINATION OF AMPLIFIERS AND ACCUMULATIONS

Detailed procedures for disinfecting guano, soil and building structural elements contaminated with *Histoplasma* and *C. neoformans* are given by Ajello and Weeks (1983). Briefly, the procedures involve drenching the contaminated materials with 5% formalin solution on each of three successive days at temperatures over 15°C. Regrettably, despite the carcinogenic nature of formaldehyde, and its propensity to leave potentially noxious residues in many materials (Vink 1986), no properly validated alternative disinfection procedure exists for virulent fungal pathogens in environmental bulk substances. Adequate worker protection must be employed with regard to both pathogen and formaldehyde exposure (see section on prevention, above, and also Ajello and Weeks 1983). In addition, the most up-to-date state/provincial, national or international worker safety standards must always be obtained. Adverse residual formaldehyde effects on building occupants (Wieslander *et al.* 1997) should as far as is possible be minimized.

Opportunistic fungal pathogens growing in buildings can be disinfected in the same fashion as other indoor moulds (see Chapter 2.3), but adequate precautions for prevention of respiratory exposure (see section on Prevention above) should be taken.

REFERENCES

Ajello, L., and Weeks, R.J. (1983) Soil decontamination and other control measures. In A.F. DiSalvo, (ed.), *Occupational Mycoses*, Lea & Febiger, Philadelphia, pp. 229-238.

Aliouat-Denis, C.M., Martinez, A., Aliouat, El M., *et al.* (2009) The *Pneumocystis* life cycle. *Mem. Inst. Oswaldo Cruz*, **104**, 419-426.

Al-Doory, Y., and DiSalvo, A.F. (1992) *Blastomycosis*. Plenum Medical, New York.

Amaro, M.H., Muccioli, C., and Abreu, M.T. (2007) Ocular histoplasmosis-like syndrome: a report from a non-endemic area. *Arg. Bras. Oftalmol.*, **70**, 577-580.

Anaissie, E.J., Kuchar, R.T., Rex, J.H., *et al.* (2001) Fusariosis associated with pathogenic *Fusarium* species colonization of a hospital water system: a new paradigm for the epidemiology of opportunistic mold infections. *Clin. Infect. Dis.*, **33**, 1871-1878.

Anaissie, E.J., Stratton, S.L., Dignani, M.C., *et al.* (2002) Pathogenic *Aspergillus* species recovered from a hospital water system: a 3-year prospective study. *Clin. Infect. Dis.*, **34**, 780-789.

Arasteh, K., Staib, F., Grosse, G., *et al.* (1996) Cryptococcosis in HIV infection of man: an epidemiological and immunological indicator? *Zentralbl. Bakteriol.*, **284**, 153-163.

Armbruster, C., Hassl, A., and Kriwanek, S. (1997) *Pneumocystis carinii* colonization in the absence of immunosuppression. *Scand. J. Infect. Dis.*, **29**, 591-593.

Arnow, P.M., Andersen, R.L., Mainous, P.D., and Smith, E.J. (1978) Pulmonary aspergillosis during hospital renovation. *Am. Rev. Resp. Dis.*, **118**, 49-53.

Arnow, P.M., Sadigh, M., Costas, C., *et al.* (1991) Endemic and epidemic aspergillosis associated with in-hospital replication of *Aspergillus* organisms. *J. Infect. Dis.*, **164**, 998-1002.

Bakerspigel, A., Kane, J., and Schaus, D. (1986) Isolation of *Blastomyces dermatitidis* from an earthen floor in southwestern Ontario, Canada. *J. Clin. Microbiol.*, **24**, 890-891.

Barenfanger, J., Ramirez, F., Tewari, R.P., and Eagleton, L. (1989) Pulmonary phaeohyphomycosis in a patient with hemoptysis. *Chest*, **95**, 1158-1160.

Bartlett, M.S., Vermund, S.H., Jacobs, R., *et al.* (1997) Detection of *Pneumocystis carinii* DNA in air samples: likely environmental risk to susceptible persons. *J. Clin. Microbiol.*, **35**, 2511-2513.

Baumgardner, D.J., and Paretsky, D.P. (1999) The *in vitro* isolation of *Blastomyces dermatitidis* from a woodpile in north central Wisconsin, U.S.A. *Med. Mycol.*, **37**, 163-169.

Baumgardner, D.J., and Paretsky, D.P. (2001) Blastomycosis: more evidence for exposure near one's domicile. *Wisc. Med. J.*, **100**, 43-45.

Bemmann, W. (1979) Studies on the habitat of fungi on glass fibre wallpaper. *Zentralbl. Bakteriol.* (*Naturwiss.*), **134**, 193-197.

Berbee, M., and Taylor, J.W. (1992) Detecting the morphological convergence in true fungi using 18S RNA sequence data. *Biosystems*, **28**, 117-125.

Bissett, J. (1987) Fungi associated with urea-formaldehyde foam insulation in Canada. *Mycopathologia*, **99**, 47-56.

Blondin, N., Baumgardner, D.J., Moore, G.E., and Glickman, L.T. (2007) Blastomycosis in indoor cats: suburban Chicago, Illinois, USA. *Mycopathologia*, **163**, 59-66.

Boltansky, H., Kwon-Chung, K.J., Macher, A.M., and Gallin, J.I. (1984) *Acremonium strictum*-related pulmonary infection in a patient with chronic granulomatous disease. *J. Infect. Dis.*, **149**, 653.

Bothast, R.J., and Fennell, D.I. (1974) A medium for rapid identification and enumeration of *Aspergillus flavus* and related organisms. *Mycologia*, **66**, 365-369.

Burgess, J.W., Kottom, T.J., and Limper, A.H. (2008) *Pneumocystis carinii* exhibits a conserved meiotic control pathway. *Infect. Immun.*, **76**, 417-425.

Burgess, J.W., Schwan, W.R., and Volk, T.J. (2006) PCR-based detection of DNA from the human pathogen *Blastomyces dermatitidis* from natural soil samples. *Med. Mycol.*, **44**, 741-748.

Burton, J.R., Zachery, J.B., Bessin, R., *et al.* (1972) Aspergillosis in four renal transplant recipients: diagnosis and effective treatment with amphotericin B. *Ann. Intern. Med.*, **77**, 383-388.

Castañeda, A., McEwen, J., Hidalgo, M., and Castañeda, E. (2004) Evaluación de varias técnicas de extracción de ADN de *Cryptococcus* spp. a partir de muestras ambientales. *Biomedica*, **24**, 324-331.

Cere, N., Polack, B., Chanteloup, N.K., and Coudert, P. (1997) Natural transmission of *Pneumocystis carinii* in nonimmunosuppressed animals: early contagiousness of experimentally infected rabbits (*Oryctolagus cuniculus*). *J. Clin. Microbiol.*, **35**, 2670-2672.

Chabra, A., Handa, K.K., Chakrabarti, A., *et al.* (1996) Allergic fungal sinusitis: clinicopathological characteristics. *Mycoses*, **39**, 437-441.

Chaturvedi, S., Chaturvedi, H.S., and Khan, Z. (1990) Efficacy of brain heart infusion-egg albumen agar, yeast extract phosphate agar and peptone glucose agar media for isolation of *Blastomyces dermatitidis* from sputum. *Mycopathologia*, **112**, 105-112.

Christensen, M., and Tuthill, D.E. (1985) *Aspergillus*: an overview. In R.A. Samson and J.I. Pitt, (eds.), *Advances in* Penicillium *and* Aspergillus *Systematics*, Plenum Press, New York, pp. 195-209.

Crompton, C.H., Summerbell, R.C., Balfe, J.W., and Silver, M.M. (1992) Peritonitis with *Paecilomyces* complicating peritoneal dialysis. *Pediatr. Inf. Dis. J.*, **10**, 869-871.

Dabil, H., Kaplan, H.J., Duffy, B.F., Phelan, *et al.* (2003) Association of the HLA-DR15/HLA-DQ6 haplotype with development of choroidal neovascular lesions in presumed ocular histoplasmosis syndrome. *Hum. Immunol.*, **64**, 960-964.

Dabrowa, N., Landau, J.W., Newcomer, V.D., and Plunkett, O.A. (1964) A survey of tide-washed coastal areas of southern California for fungi potentially pathogenic to man. *Mycopathologia*, **24**, 137-150.

Davies, R., Summerbell, R.C., Haldane, D., *et al.* (1995) *Fungal Contamination in Public Buildings: A Guide to Recognition and Management*. Environmental Health Directorate, Health Canada, Ottawa. (Full document at http://www.hc-sc.gc.ca/chp/chd/catalogue/bch-pubs/fungal.pdf).

De Bock, R., Gyssens, I., Peetermans, M., and Nolard, N. (1989) *Aspergillus* in pepper. *Lancet 2*, 331-332.

Dixon, D., Salkin, I.F., Duncan, R.A., *et al.* (1991) Isolation and characterization of *Sporothrix schenckii* from clinical and environmental sources associated with the largest U.S. epidemic of sporotrichosis. *J. Clin. Microbiol.*, **29**, 1106-1113.

Domsch, K.H., Gams, W., and Anderson T.-H. (1993) *Compendium of Soil Fungi*. IHW-Verlag, Eching, Germany.

Edman, J.C., Kovacs, J.A., Masur, H., *et al.* (1988) Ribosomal RNA sequence shows *Pneumocystis carinii* to be a member of the fungi. *Nature*, **334**, 519-522.

Emmons, C.W. (1961) Isolation of *Histoplasma capsulatum* from soil in Washington, D.C. *Public Health Rep.*, **76**, 591-596.

Fiese, M.J. (1958) *Coccidioidomycosis*. Charles C. Thomas, Springfield, IL.

Findley, K., Rodriguez-Carres, M., Metin, B., *et al.* (2009) Phylogeny and phenotypic characterization of pathogenic *Cryptococcus* species and closely related saprobic taxa in the *Tremellales*.

Eukaryot. Cell., **8**, 353-361.

Fraser, J.A., Giles, S.S., Wenink, E.C., *et al.* (2005) Same-sex mating and the origin of the Vancouver Island *Cryptococcus gattii* outbreak. *Nature*, **437**,1360-1364.

Freifeld, A.G., and Pizzo, P.A. (1996) The outpatient management of febrile neutropenia in cancer patients. *Oncology (Huntingt.)*, **10**, 599-606.

George, D.L., McLeod, R., and Weinstein, R.A. (1991) Contaminated commercial charcoal as a source of fungi in the respiratory tract. *Infect. Control Hosp. Epidemiol.*, **12**, 732-734.

Gilgado, F., Cano, J., Gené, J., *et al.* (2008) Molecular and phenotypic data supporting distinct species statuses for *Scedosporium apiospermum* and *Pseudallescheria boydii* and the proposed new species *Scedosporium dehoogii. J. Clin. Microbiol.*, **46**, 766-771.

Gonzales, C.A., Scott, I.U., Chaudhry, N.A., *et al.* (2000) Endogenous endophthalmitis caused by *Histoplasma capsulatum* var. *capsulatum* – case report and literature review. *Ophthalmology*, **107**, 625-729.

Goodwin, R.A., Owens, F.T., Snell, J.D., Hubbard, W.W., Buchanan, R.D., Terry, R.T., and Des Prez, R.M. (1976) Chronic pulmonary histoplasmosis. *Medicine (Baltimore)*, **55**, 413-452.

Gucalp, R., Carlisle, P., Gialanella, P., *et al.* (1996) *Paecilomyces* sinusitis in an immunocompromised adult patient: case report and review. *Clin. Infect. Dis.*, **23**, 391-393.

Guého, E., Smith, M.T., de Hoog, G.S., *et al.* (1992) Contributions to a revision of the genus *Trichosporon. Antonie van Leeuwenhoek*, **61**, 289-316.

Haase, G., Skopnick, H., Groten, T., *et al.* (1991) Long-term fungal cultures from sputum of patients with cystic fibrosis. *Mycoses*, **34**, 373-376.

Haiduven, D. (2009) Nosocomial aspergillosis and building construction. *Med. Mycol.*, **47**, (Suppl. 1), S210-S216.

Hennequin, C., Page, B., Roux, P., *et al.* (1995) Outbreak of *Pneumocystis carinii* pneumonia in a renal transplant unit. *Eur. J. Clin. Microbiol. Infect. Dis.*, **14**, 122-126.

Hoefnagels, R.L.J., and Pijpers, P.M. (1967). *Histoplasma capsulatum* in the human eye. *Am. J. Ophthalmol.*, **63**, 715-723.

Hoffman, K.K, Weber, D.J., and Rutala, W.A. (1989) Pseudoepidemic of *Rhodotorula rubra* in patients undergoing fiberoptic bronchoscopy. *Infect. Control Hosp. Epidemiol.*, **10**, 511-514.

Hughes, W.T. (1982) Natural mode of acquisition for *de novo* infection with *Pneumocystis carinii. J. Infect. Dis.*, **145**, 842-848.

Hussain, S., Salahuddin, N., Ahmad, I., *et al.* (1995) Rhinocerebral invasive mycosis: occurrence in immunocompetent individuals. *Eur. J. Radiol.*, **20**, 151-155.

Itatani, C. (1996) Ultrastructural morphology of intermediate forms and forms suggestive of conjugation in the life cycle of *Pneumocystis carinii. J. Parasitol.*, **82**, 163-171.

Iwen, P.C., Davis, J.C., Reed, E.C., *et al.* (1994) Airborne fungal spore monitoring in a protective environment during hospital construction, and correlation with an outbreak of invasive aspergillosis. *Infect. Control Hosp. Epidemiol.*, **15**, 303-306.

Iwen, P.C., Rupp, M.E., and Hinrichs, S.H. (1997) Invasive mold sinusitis: 17 cases in immunocompromised patients and review of the literature. *Clin. Infect. Dis.*, **24**, 1178-1184.

Jappinen, P., Haahtela, T., and Liira, J. (1987) Chip pile workers and mould exposure: a preliminary clinical and hygienic survey. *Allergy*, **42**, 545-548.

Jung, A.C. and Paauw, D.S. (1998) Diagnosing HIV-related disease: using the CD4 count as a guide. *J. Gen. Intern. Med.*, **13**, 131-136.

Kaltseis, J., Rainer, J., and De Hoog, G.S. (2009) Ecology of *Pseudallescheria* and *Scedosporium* species in human-dominated and natural environments and their distribution in clinical samples. *Med. Mycol.*, **47**, 398-405.

Kane, J., Summerbell, R.C., Sigler, L., *et al.* (1997) *Laboratory Handbook of Dermatophytes and Other Filamentous Fungi from Skin, Hair and Nails*. Star Publishing Co., Belmont, CA.

Karam, G.H., and Griffith, F.M. (1986) Invasive pulmonary aspergillosis in non-immunocompromised non-neutropenic hosts. *Rev. Infect. Dis.*, **8**, 357-363.

Katsu, M., Ando, A., Ikeda, R., *et al.* (2003) Immunomagnetic isolation of *Cryptococcus neoformans* by beads coated with anti-*Cryptococcus* serum. *Nippon Ishinkin Gakkai Zasshi.*, **44**, 139-144.

Kidd, S.E., Chow, Y., Mak, S., *et al.* (2007) Characterization of environmental sources of the human and animal pathogen *Cryptococcus gattii* in British Columbia, Canada, and the Pacific Northwest of the United States. *Appl. Environ. Microbiol.*, **73**, 1433-1443.

Kitchen, M.S., Reiber, C.D., and Eastin, G.B. (1977) An urban epidemic of North American blastomycosis. *Am. Rev. Respir. Dis.*, **115**, 1063-1065.

Knox, D.L., O'Brien, T.P., and Green, W.R. (2003) *Histoplasma* granuloma of the conjunctiva. *Ophthalmology*, **110**, 2051-2053.

Krajden, S., Summerbell, R.C., Kane, J., *et al.* (1991) Normally saprobic cryptococci isolated from *Cryptococcus neoformans* infections. *J. Clin. Microbiol.*, **29**, 1883-1887.

Krasnick, J., Patterson, R., and Roberts, M. (1995) Allergic bronchopulmonary aspergillosis presenting with cough variant asthma and identifiable source of *Aspergillus fumigatus. Ann. Allerg. Asthma Immunol.*, **75**, 344-346.

Kreisel, J.D., Adderson, E.E., Gooch, W.M., and Pavia, A.T. (1994) Invasive sinonasal disease due to *Scopulariopsis candida*: case report and review of scopulariopsosis. *Clin. Infect. Dis.*, **19**, 317-319.

Kwon-Chung, K.J., and Bennett, J.E. (1992) *Medical Mycology*. Lea & Febiger, Philadelphia.

Laakkonen, J. (1998) *Pneumocystis carinii* in wildlife. *Int. J. Parasitol.*, **28**, 241-252.

Larsh, H.W., and Goodman, N.L. (1980) Fungi of systemic mycoses. In E.H. Lennette, A. Balows, W.J. Hausler, and J.P. Truant, (eds.), *Manual of Clinical Microbiology*, 3rd Edition, American Society for Microbiology, Washington DC, pp. 577-594

Latham, R.H., Kaiser, A.B., Dupont, W.D., and Dan, B.B. (1980) Chronic pulmonary histoplasmosis following the excavation of a bird roost. *Am. J. Med.*, **68**, 831-834.

Latouche, S., Roux, P., Poirot, J.L., *et al.* (1994) Preliminary results of *Pneumocystis carinii* strain differentiation by using molecular biology. *J. Clin. Microbiol.*, **32**, 3052-3053.

Lazera, M.S., Pires, F.D., Camillo-Coura, L., *et al.* (1996) Natural habitat of *Cryptococcus neoformans* var. *neoformans* in decaying wood forming hollows in living trees. *J. Med. Vet. Mycol.*, **34**, 127-131.

Levitz, S.M. (1992) Overview of host defenses in fungal infections. *Clin. Infect. Dis.*, **14**, (Suppl. 1), S37-S42.

Loo, V.G., Bertrand, C., Dixon, C., *et al.* (1996) Control of construction-associated nosocomial aspergillosis in an antiquated haematology unit. *Infect. Control Hosp. Epidemiol.*, **17**, 360-364.

Lundgren, B., Elvin, K., Rothman, L.P., *et al.* (1997) Transmission of *Pneumocystis carinii* from patients to hospital staff. *Thorax*, **52**, 422-424.

Maher, N., Dillon, H.K., Vermund, S.H., and Unnasch, T.R. (2001) Magnetic bead capture eliminates PCR inhibitors in samples collected from the airborne environment, permitting detection of *Pneumocystis carinii* DNA. *Appl. Environ. Microbiol.*, **67**, 449-452.

Mariat, F. (1968) The epidemiology of sporotrichosis. In G.E.W. Wolstenholme, (ed.), *Systemic Mycoses*, Churchill, London, pp. 144-159.

Marimon, R., Cano, J., Gené, J., *et al.* (2007) *Sporothrix brasiliensis, S. globosa*, and *S. mexicana*, three new *Sporothrix* species of clinical interest. *J. Clin. Microbiol.*, **45**, 3198-3206.

Marsh, P.B., Millner, P.D., and Kla, J.M. (1979) A guide to the recent literature on aspergillosis as caused by *Aspergillus fumigatus*, a fungus frequently found in self-heating organic matter. *Mycopathologia*, **69**, 67-81.

Matos T., de Hoog, G.S., de Boer, A.G., *et al.* (2002) High prevalence of the neurotrope *Exophiala dermatitidis* and related oligotrophic black yeasts in sauna facilities. *Mycoses*, **45**, 373-377.

Matsumoto, Y., and Yoshida, Y. (1984) Sporogony in *Pneumocystis carinii*: synaptonemal complexes and meiotic nuclear divisions observed in precysts. *J. Protozool.*, **31**, 420-428.

Mehta, G. (1990) *Aspergillus* endocarditis after open heart surgery: an epidemiological investigation. *J. Hosp. Infect.*, **15**, 245-253.

Miller, J.D., and Holland, H. (1981) Biodeteriogenic fungi in two Canadian historic houses subjected to different environmental controls. *Int. Biodet. Bull.*, **17**, 39-45.

Millner, P.D., Olenchock, S.A., Epstein, E., *et al.* (1994) Bioaerosols associated with composting facilities. *Compost Sci. Utilization*, **2**, 8-57.

Mitchell, S.J., Gray, J., Morgan, M.E., *et al.* (1996) Nosocomial infection with *Rhizopus microsporus* in preterm infants: association with wooden tongue depressors. *Lancet*, **348**, 441-443.

Morrison, V.A., and Weisdorf, D.J. (1993) *Alternaria*: a sinonasal pathogen of immunocompromised hosts. *Clin. Infect. Dis.*, **16**, 265-270.

Mullins, J., Harvey, R., and Seaton, A. (1976) Sources and incidence of airborne *Aspergillus fumigatus* (Fres.). *Clin. Allerg.*, **6**, 209-217.

Nevez, G., Chabé, M., Rabodonirina, M., *et al.* (2008). Nosocomial *Pneumocystis jirovecii* infections. *Parasite*, **15**, 359-365.

Nishimura, K., Miyaji, M., Taguchi, H., and Tanaka, R. (1987) Fungi in bathwater and sludge of bathroom drainpipes. 1. Frequent isolation of *Exophiala* species. *Mycopathologia*, **97**, 17-23.

Nishiura, Y., Nakagawa-Yoshida, K., Suga, M., *et al.* (1997) Assignment and serotyping of *Trichosporon* species: the causative agents of summer-type hypersensitivity pneumonitis. *J. Med. Vet. Mycol.*, **35**, 45-52.

Nolard, N. (1994) Les liens entre les risques d'aspergillose et la contamination de l'environnement. *Pathol. Biol.*, **42**, 706-710.

Okudaira, M., and Schwarz, J. (1962) Experimental ocular histoplasmosis in rats. A histopathologic study of immunogenic and hypersensitive ophthalmitis produced in rats by *Histoplasma capsulatum* and histoplasmin. *Am. J. Ophthalmol.*, **54**, 427-444.

Olsson, M., Sukura, A., Lindberg, L.A., and Linder, E. (1996) Detection of *Pneumocystis carinii* DNA by filtration of air. *Scand. J. Infect. Dis.*, **28**, 279-282.

Olsson, M., Lidman, C., Latouche, S., *et al.* (1998) Identification of *Pneumocystis carinii* f. sp. *hominis* gene sequences in filtered air in hospital environments. *J. Clin. Microbiol.*, **36**, 1737-1740.

Ongkosuwito, J.V., Kortbeek, L.M., Van der Lelij, A., *et al.* (1999) Aetiological study of the presumed ocular histoplasmosis syndrome in the Netherlands. *Brit. J. Ophthalmol.*, **83**, 535-539.

Orth, B., Frei, R., Itin, P.H., *et al.* (1996) Outbreak of invasive mycoses caused by *Paecilomyces lilacinus* from a contaminated skin

lotion. *Ann. Intern. Med.*, **125**, 799-806.

Pitt, J.I., and Hocking, A.D. (1985) *Fungi and Food Spoilage*. Academic Press, New York.

Pitt, J.I., Hocking, A.D., and Glenn, D.R. (1983) An improved medium for the detection of *Aspergillus flavus* and *A. parasiticus*. *J. Appl. Bacteriol.*, **54**, 109-114.

Prasad, A.G., and Van Gelder, R.N. (2005) Presumed ocular histoplasmosis syndrome. *Curr. Opin. Ophthalmol.*, **16**, 364-368.

Rabodonirina, M., Wilmotte, R., Dannaoui, E., *et al.* (1997) Detection of *Pneumocystis carinii* DNA by PCR amplification in various rat organs in experimental pneumocystosis. *J. Med. Microbiol.*, **46**, 664-668.

Radin, R.C., Greenberger, P.A., Patterson, R., and Ghory, A. (1983) Mould counts and exacerbations of allergic bronchopulmonary aspergillosis. *Clin. Allerg.*, **13**, 271-275.

Rainer, J., Kaltseis, J., de Hoog, S.G., and Summerbell, R.C. (2008). Efficacy of a selective isolation procedure for members of the *Pseudallescheria boydii* complex. *Antonie van Leeuwenhoek*, **93**, 315-322.

Reed, J.D., Scherer, J.H., Herbut, P.A., and Irving, H. (1942) Systemic histoplasmosis diagnosed before death and produced experimentally in guinea pigs. *J. Lab. Clin. Med.*, **27**, 419-434.

Reid, T.M., and Schafer, M.P. (1999) Direct detection of *Histoplasma capsulatum* in soil suspensions by two-stage PCR. *Mol. Cell. Probes*, **13**, 269-273.

Ren, P., Ahearn, D.G., and Crow, S.A. (1998) Mycotoxins of *Alternaria alternata* produced on ceiling tiles. *J. Ind. Microbiol. Biotechnol.*, **20**, 53-54.

Reyes-Montes, M.R., Rodríguez-Arellanes, G., Pérez-Torres, A., *et al.* (2009) Identification of the source of histoplasmosis infection in two captive maras (*Dolichotis patagonum*) from the same colony by using molecular and immunologic assays. *Rev. Argent. Microbiol.*, **41**, 102-104.

Richardson, S.E., Bannatyne, R.M., Summerbell, R.C., *et al.* (1988) Disseminated fusarial infection in the immunocompromised host. *Rev. Infect. Dis.*, **10**, 1171-1181.

Rippon, J.W. (1988) *Medical Mycology. The Pathogenic Fungi and the Pathogenic Actinomycetes*, 3rd Ed, W.B. Saunders, Philadephia.

Roberts, W.T., and Davidson, P.M. (1986) Growth characteristics of selected fungi on polyvinyl chloride film. *Appl. Environ. Microbiol.*, **51**, 673- 676.

Rose, H.D. (1972) Mechanical control of hospital ventilation and *Aspergillus* infections. *Am. Rev. Resp. Dis.*, **105**, 306-307.

Rose, H.D., and Hirsch, S.R. (1979) Filtering hospital air decreases *Aspergillus* spore counts. *Am. Rev. Resp. Dis.*, **119**, 511-513.

Rosen, P.P., and Sternberg, S.S. (1976) Decreased frequency of aspergillosis and mucormycosis. *New Engl. J. Med.*, **295**, 1319-1320.

Salkin, I. (1988) Media and stains for mycology. In B.B. Wentworth, (ed.), *Diagnostic Procedures for Mycotic and Parasitic Infections*, 7th Ed, American Public Health Association, Washington DC, pp. 379-411.

Samson, R. (1985) Occurrence of molds in modern living and working environments. *Eur. J. Epidemiol.*, **1**, 54-61.

Sarubbi, F.A., Kopf, H.B., Wilson, M.B., *et al.* (1982). Increased recovery of *Aspergillus flavus* from respiratory specimens during hospital construction. *Am. Rev. Resp. Dis.*, **125**, 33-38.

Sayer, W., Shean, D.B., and Ghosseiri, J. (1969) Estimation of airborne fungal flora by the Andersen sampler versus the gravity settling culture plate. *J. Allergy*, **44**, 214-227.

Schroers, H.J., O'Donnell, K., Lamprecht, S.C., *et al.* (2009) Taxonomy and phylogeny of the *Fusarium dimerum* species group. *Mycologia*, **101**, 44-70.

Schwartz, H.J. (1998) The association of allergic bronchopulmonary aspergillosis and cystic fibrosis. *Immunol. Allerg. Clinics N. Am.*, **18**, 503-508.

Seed, J.P., and Aikawa, M. (1977) *Pneumocystis*. In J.P. Kreier, (ed.), *Parasitic Protozoa*, Vol. 4, Academic Press, New York, pp. 329-357.

Selik, R.M., Karon, J.M., and Ward, J.W. (1997) Effect of the human immunodeficiency virus epidemic on mortality from opportunistic infections in the United States in 1993. *J. Infect. Dis.*, **176**, 632-636.

Silva-Vergara, M.L., Martinez, R., Chadu, A., *et al.* (1998) Isolation of a *Paracoccidioides brasiliensis* strain from the soil of a coffee plantation in Ibia, State of Minas Gerais, Brazil. *Med. Mycol.*, **36**, 37-42.

Simmons, R.B., Price, D.L., Noble, J.A., *et al.* (1997) Fungal colonization of air filters from hospitals. *Am. Ind. Hyg. Assoc. J.*, **58**, 900-904.

Solomon, W.R. (1975) Assessing fungus prevalence in domestic interiors. *J. Allerg. Clin. Immunol.*, **56**, 235-242.

Solomon, W.R., and Burge, H.P. (1975) *Aspergillus fumigatus* levels in and out of doors in urban air. *J. Allerg. Clin. Immunol.*, **55**, 90-91.

Soulez, B., Palluault, F., Cesbron, J.Y., *et al.* (1991) Introduction of *Pneumocystis carinii* into a colony of SCID mice. *J. Protozool.*, **38**, 123S-125S.

Staib, F. (1962) *Cryptococcus neoformans* und *Guizotia abyssinica* (Syn. *G. oleifera*) Farbreaktion für *Cryptococcus neoformans*. *Z. Hyg.*, **148**, 466-475.

Staib, F. (1982) Mycoses caused by fungal spores in indoor air. *Zbl. Bakt. Hyg.*, *1 Abt., Orig. B*, **176**, 142-154.

Staib, F. (1985) Vorschläge zur Bekämpfung aerogener tiefer Mykosen bei immungeschwächten Personen. *Bundesgesundheitsblatt*, **28**, 132-138.

Staib, F. (1996) Fungi in the home and hospital environment. *Mycoses*, **39**, (Suppl. 1), 26-29.

Staib, F., and Grosse, G. (1983) Isolation of *Sporothrix schenckii* from the floor of an indoor swimming pool. *Zentralbl. Bakt. Hyg.*, *1 Abt., Orig. B*, **177**, 499-506.

Staib, F., and Heissenhuber, M. (1989) *Cryptococcus neoformans* in bird droppings: a hygienic-epidemiological challenge. *AIDS-Forschung*, **12**, 649-655.

Staib, F., Tompak, B., Thiel, D., and Blisse, A. (1978) *Aspergillus fumigatus* and *Aspergillus niger* in two potted ornamental plants, cactus (*Epiphyllum truncatum*) and clivia (*Clivia miniata*). Biological and epidemiological aspects. *Mycopathologia*, **66**, 27-30.

Streifel, A.J. (2004) Design and maintenance of hospital ventilation systems and the prevention of airborne nosocomial infection. In C.G. Mayhall, (ed.), *Hospital Epidemiology and Infection Control*, 3rd Ed, Lippincott Williams and Wilkins, Philadelphia, pp. 1577-1589.

Streifel, A.J., Lauer, J.L., Vesley, D., *et al.* (1983) *Aspergillus fumigatus* and other thermotolerant fungi generated by hospital building demolition. *Appl. Environ. Microbiol.*, **46**, 375-378.

Stringer, J.R. (1996) *Pneumocystis carinii*: what is it, exactly? *Clin. Microbiol. Rev.*, **9**, 489-498.

Sudhadham, M., Prakitsin, S., Sivichai, S., *et al.* (2008) The neurotropic black yeast *Exophiala dermatitidis* has a possible origin in the tropical rain forest. *Stud. Mycol.*, **61**, 145-155.

Summerbell, R.C.(1993) The benomyl test as a fundamental diagnostic method for medical mycology. *J. Clin. Microbiol.*, **31**, 572-577.

Summerbell, R.C. (1998) Taxonomy and ecology of *Aspergillus* species associated with colonizing infections of the respiratory tract. *Immunol. Allerg. Clinics N. Am.*, **18**, 549-573.

Summerbell, R.C., Krajden, S., and Kane, J. (1989) Potted plants in hospitals as reservoirs of pathogenic fungi. *Mycopathologia*, **106**, 13-22.

Summerbell, R.C., Staib, F., Dales, R., *et al.* (1992) Ecology of fungi in human dwellings. *J. Med. Vet. Mycol.*, **30**, (Suppl. 1), 279-285.

Summerbell, R.C., Staib, F., Ahearn, D.G., *et al.* (1994) Household hyphomycetes and other indoor fungi. *J. Med. Vet. Mycol.*, **32**, (Suppl. 1), 277-286.

Suttorp-Schulten, M.S., Bollemeijer, J.G., Bos, P.J., Rothova, A. (1997) Presumed ocular histoplasmosis in The Netherlands – an area without histoplasmosis. *Br. J. Ophthalmol.*, **81**, 7-11.

Tanner, B.D. (2009) Reduction in infection risk through treatment of microbially contaminated surfaces with a novel, portable, saturated steam vapor disinfection system. *Am. J. Infect. Control*, **37**, 20-27.

Taylor, M.L., Ruíz-Palacios, G.M., del Rocío Reyes-Montes, M., *et al.* (2005) Identification of the infectious source of an unusual outbreak of histoplasmosis, in a hotel in Acapulco, state of Guerrero, Mexico. *FEMS Immunol. Med. Microbiol.*, **45**, 435-441.

Thompson, L., Castrillon, M.A., Delgado, M., and Garcia, M. (1994) Isolation of several species of the genus *Aspergillus* from soil of intrahospital ornamental plants. *Rev. Med. Chile*, **122**, 1367-1371.

Thrane, U. (1996) Comparison of three selective media for detecting *Fusarium* species in food: a collaborative study. *Int. J. Food Microbiol.*, **29**, 149-156.

Vargas, S.L., Hughes, W.T., Wakefield, A.E., and Oz, H.S. (1995) Limited persistence in and subsequent elimination of *Pneumocystis carinii* from the lungs after *P. carinii* pneumonia. *J. Infect. Dis.*, **172**, 506-510.

Varkey, B. (1998) Allergic bronchopulmonary aspergillosis. Clinical perspectives. Immunol. *Allerg. Clinics N. Am.*, **18**, 479-501.

Verweij, P.E., Voss, A., Donnelly, J.P., *et al.* (1997) Wooden sticks as the source of a pseudoepidemic of infection with *Rhizopus microsporus* var. *rhizopodiformis* among immunocompromised patients. *J. Clin. Microbiol.*, **35**, 2422-2423.

Vink, P. (1986) Residual formaldehyde in steam-formaldehyde sterilized materials. *Biomaterials*, **7**, 221-224.

Walma, D., and Schlaegel, T.F. (1964) Presumed histoplasmic choroiditis – clinical analysis of 43 cases. *Am. J. Ophthalmol.*, **57**, 107-110.

Watzke, R.C., Klein, M.L., and Wener, M.H. (1998) Histoplasmosis-like choroiditis in a nonendemic area: the northwest United States. *Retina*, **18**, 204-212.

Weislander, G., Norback, D., Bjornsson, E., *et al.* (1997) Asthma and the indoor environment: the significance of emission of formaldehyde and volatile organic compounds from newly painted indoor surfaces. *Int. Arch. Occup. Environ. Health*, **69**, 115-124.

Wicklow, D.T. (1988) Patterns of fungal association within maize kernels harvested in North Carolina. *Plant Dis.*, **72**, 113-115.

Woods, A.C., and Wahlen, H.E. (1959) The probable role of benign histoplasmosis in the etiology of granulomatous uveitis. *Trans. Am. Ophthalmol. Soc.*, **57**, 318-343.

Wyder, M.A., Rasch, E.M., and Kaneshiro, E.S. (1998) Quantitation of absolute *Pneumocystis carinii* nuclear DNA content. Trophic and cystic forms isolated from infected rat lungs are haploid organisms. *J. Eukaryot. Microbiol.*, **45**, 233-239.

Yazaki, H., Goto, N., Uchida, K., *et al.* (2009) Outbreak of *Pneumo-*

cystis jirovecii pneumonia in renal transplant recipients: *P. jirovecii* is contagious to the susceptible host. *Transplantation*, **88**, 380-385.

Yoshida, K., Ueda, A., Yamasaki, H., *et al.* (1993) Hypersensitivity pneumonitis resulting from *Aspergillus fumigatus* in a greenhouse. *Arch. Environ. Health*, **48**, 260-262.

Zhang, X., and Andrews, J.H. (1993) Evidence for growth of *Sporothrix schenckii* on dead but not on living sphagnum moss. *Mycopathologia*, **123**, 87-94.

Chapter 4. Microbiological investigation of indoor environments

Chapter 4.1

MYCOLOGICAL INVESTIGATIONS OF INDOOR ENVIRONMENTS

J. David Miller

Institute of BioChemistry, Carleton University, Ottawa, Ontario, Canada.

INTRODUCTION

In the past few years the firm recognition that saprophytic fungi and bacteria can have serious, even life-threatening, health effects (NAS 2000, 2004, Health Canada 2004) has pointed to the need to investigate residential buildings and the non-industrial workplace for microbiological contamination. However, investigations that have been carried out are of uneven quality, because there are few people with the range of experience necessary to conduct sophisticated investigations, and because of the lack of standardized protocols for sampling, should that be necessary. In 2005, the American Industrial Hygiene Association (AIHA) published the second edition of a manual of consensus methods for microbiological sampling that has addressed this issue in greater depth (Hung *et al.* 2005). This was based in part on principles articulated by the American Conference of Government Industrial Hygienists, ACGIH (Macher *et al.* 1999), a legal cognizant authority in USA and Canada. AIHA recently published a more complete description of the evaluation of mould damage, including some discussion of sampling methods and interpretation (Prezant *et al.* 2008). The present chapter will attempt to put this information into a practical microbiological context and provide information on the limitations of the data obtained from different sampling methods. Sampling for saprophytic bacteria, endotoxin, glucan and peptidyl glucan has been discussed in Hung *et al.* (2005).

The goals of any investigation are to establish the nature and extent of fungal contamination, to minimize the possibility of false negatives and false positives, and to determine the causes and degree of risk borne by the occupant population (Health Canada 2004). At present, assessment of the degree of risk of exposure to saprophytic microorganisms is achieved by determining the area of contamination. This is partly a reductive position: current methods of sampling for saprophytic microbes, including bacterial endotoxin, do not assess exposure in the strict sense. There are, however, positive reasons for this position. Dales *et al.* (1998) showed that children living in more mouldy homes had statistically and clinically significant (but small) changes in lymphocyte cell sub-populations relative to those living in less mouldy homes. This biological difference was associated with visible mould on 1-5% of surfaces in the occupied space. Similarly, reports from the European Community Respiratory Health Survey indicated that countries with a higher prevalence of asthma have a higher percentage of damp homes, and that sensitisation to moulds was a powerful risk factor for severe asthma in adults (Zock *et al.* 2002, Zureik *et al.* 2002).

In Canadian studies, the visible moulds on surfaces (Miller *et al.* 1999) were a mixture of species associated with damp building materials (see Chapter 2.1, and Miller *et al.* 2008). The fungi associated with damp materials in Western Europe (Gravesen *et al.* 1999) mostly overlap with the North American mycobiota, which has been more intensively studied. The principle that area of mould and dampness is associated with occupant risk has been articulated in all versions of the so-called New York guidelines, from the first to the current (Anon. 1993, 2008), and endorsed by the Canadian Federal-Provincial Committee on Occupational Safety and Heath (Health Canada 2004) and the International Society on Indoor Air Quality and Climate (Flannigan and Morey 1996). Recent epidemiology studies support this measure of exposure-outcome (Cho *et al.* 2006, Haverinen-Shaughnessy *et*

Table 1. Analysis of mould damaged buildings in the Pacific Northwest (British Columbia), after Anon. (1996).

Type of fault	Exterior finish		
	Stucco	Vinyl	Wood
Windows: no sealants at frame/cladding joints	9	1	
Windows: no sealants at corner mitre joints	10	1	1
Windows: poor flashing at head or sill	13	3	
Windows: poor paper installation	6	1	
Problems due to windows	27%	19%	16%
Poor deck/walkway/balcony waterproofing: field	12	3	1
Poor deck/walkway/balcony waterproofing:			
Junction with walls	16	1	
Problems due to deck/walkway/balcony	20%	19%	8%
Poor guardrail saddle joints	133	5	1
Poor guardrail cap flashings	8	1	4
Poor parapet cap flashings	7	1	
Problems due to horizontal surface flashings	20%	31%	36%
Poor base/transition/control joint flashings	14	1	
Poor roof/wall joint flashings	3	1	1
Poor concrete slab/wall joints	3	1	1
Poor dryer vents: leaking in wall	6	2	
Poor other details	8	2	1
Material/installation defects: cladding, sheathing	10	2	4
Problems due to other defects	33%	31%	40%

al. 2006, Antova *et al.* 2008, Iossifova *et al.* 2009). To the extent that the area of mould and damage assessments are made very carefully (Haverinen *et al.* 2001, 2003, Foto *et al.* 2005), they are useful for health studies.

Aside from the area of visible mould damage, accumulations of fine dust dominate exposures indoors. This is why symptoms resulting from mould exposure indoors arise from unrepaired or accumulating damage over a period of years. Reduced respiratory symptoms have been shown to correspond with houses having >500 mg dust particles of size <300 μm per square metre, mainly on carpeted surfaces (Krieger *et al.* 2005). The importance of fine settled dust and population health for allergic disease has long been recognized. Important components of fine dust include endotoxin, fungal glucan and various allergens, such as those of fungi, dust mites, insects and furred animals. By the early 1950s, it was understood that sinks of accumulated settled dust were important components of allergen exposure (Maunsell 1952, Swaebly and Christensen 1952). From a health perspective the importance of this was discussed in a review by Colloff *et al.* (1992), who indicated that "it is likely that allergen recovered/unit surface area or airborne al-

lergen measurement will be most acceptable. It is unlikely that allergen expressed on a weight:weight basis gives sufficient information on patient exposure where manipulation of the domestic environment has taken place." For respiratory disease, this prediction has been shown to be the case. In large population studies, when indicators of biological contamination are expressed on a per unit area basis, more robust indications of health have been seen. Douwes *et al.* (2000) found that amounts of both endotoxin and $(1\rightarrow3)$-β-D-glucan m^{-2} living room floor were significantly associated with peak expiratory flow (PEF) variability, but not when expressed per gram of dust sample, particularly in atopic children with asthma symptoms. The amount of any contaminant per square metre is related to the amount of dust sampled (Gehring *et al.* 2001, Lioy *et al.* 2002, Cho *et al.* 2006). A reliable value for weight of settled dust per unit floor area or in soft furniture is predictive of asthma. Elliott *et al.* (2007) found that there were associations of varying strengths between wheezing and asthma and mass loading of dusts on floors, beds and soft furniture. These associations persisted when adjusting for allergen and endotoxin exposures.

Personal exposure to particulates indoors is driven

Table 2. Ascribed sources of water for visible mould growth in 59 single family dwellings in Southern Ontario (after Lawton *et al.* 1998).

Water source	Number of reports	Visible mould (%)
Condensation on envelope	95	36
Condensation on bathroom surfaces, refrigerator pans, sumps	53	34
Wicking from ground	12	16
Rain leaks	5	6
Plumbing accidents	8	5
Condensation on pipes, ducts	13	2

by the personal dust cloud around each person. For example, depending on climate, endotoxin from pets and outdoor sources accumulates indoors (Gehring *et al.* 2001, Dales *et al.* 2006). Rabinovitch *et al.* (2005) showed that personal exposure to endotoxin did not correlate with, and was much higher than, area measurements. This is a generalizable phenomenon for indoor sourced particulate contaminants. Personal exposure indoors is determined by mass loadings of dust and room activity (Ferro *et al.* 2004, Gomes *et al.* 2007, Qian *et al.* 2008). Inspector estimates of cleaning intensity (normal/low/high) have been shown to be related to quantitative dust loadings (Berghout *et al.* 2005; J.D. Miller, L. O'Neil and R. Mazerolle, unpublished data).

INFORMED INSPECTION

The first step in investigating a building for microbial contaminants is an informed inspection, preferably by someone with experience in engineering, architectural and moisture problems in buildings (Miller 1993). For example, in homes in Canada and most of the USA, although human activity can cause moisture problems, the capacity of the house to cope with that moisture is the main determinant of mould growth. Basements and crawl spaces are also frequently damp unless properly designed. In warm or cold climates where the air is conditioned, windows not constructed to reduce condensation will result in water around window frames. Water leaks are a serious problem (Miller 1993). Bathroom and under-sink areas are prime areas for water leakage and mould growth. Undetected water leakage from toilets or bathtub drains frequently results in wood rot, and mould damage becomes evident. The much greater complexity of larger buildings results in a diversity of potential mould damage ranging from the building envelope to interior plumbing and mechanical systems (AIHA 2008a). Architecturally complex buildings

can have more mould and moisture problems than conventional designs (Odom and DuBose, 2000).

Substantial contamination by saprophytic bacteria only occurs in equipment containing stagnant water, such as some humidifiers, drain pans and similar equipment. Fungal contamination, however, can arise from several conditions, including condensation, floods and various types of leakage. Investigation of mould problems requires a thorough knowledge of the design of the building envelope and of the types of failure that result in condensation and water leaks. The design issues have been discussed by several authors (e.g. Trechsel 1994, Lstiburek 2007, 2008, 2009). In areas of North America where subtropical conditions can prevail for some (southern Ontario) or much (New Orleans) of the year, improper installation of the elements that control heat, air and moisture can lead to the growth of moderate xerophilic fungi such as *Aspergillus versicolor* on wallboard or on the inner surface of the wallcovering. These elements include air retarders (unpainted wallboard, spin-bonded polyolifin), vapour barriers (vinyl wallcovering, polyethylene sheeting) and insulation. Since insulation makes one side of the insulated surface cold, steps must be taken to ensure that moisture is kept away from that surface (Odom and DuBose, 2000). Chronic water leaks can lead to growth of, for example, *Chaetomium* and *Stachybotrys* (Morey 1997, Miller *et al.* 2008)

Studies of buildings constructed in the high humidity areas of the USA show that uncontrolled airflow and rainwater barriers are the causes of most mould problems. If the moisture status of buildings is negative to that outside, moist air is pulled into the wall cavity, leading to condensation and mould growth regardless of whether the wall is correctly constructed or not. If the building envelope leaks rainwater, and is negative to the outside, problems will arise very quickly indeed. Perversely, mould growth and the consequent damage and ill-health that can arise from it can be more common in expensive buildings with complex construction requirements than in those where construction is more basic. Common problems are the result of uncontrolled airflow caused by duct leaks and supply/exhaust imbalances bringing in warm wet air to cold surfaces (AIHA 2008b, Chapter 5; Odom and DuBose 2000).

Failures of caulking, joints and other construction details, especially around windows, can lead to the substantive problem of water leaking into wall cavities in buildings. In Canada, a study of mould-damaged buildings in the Pacific Northwest revealed that stucco was particularly likely to suffer from construction detail failures (Table 1; see also Anon. 1996). Poor installation of windows was the top cause of water leaks and mould growth in this set of buildings. Regardless of the type of cladding, the relative prevalence of water/mould damage problems was similar; detailing failures were, for some reason, more common in buildings with stucco cladding (see also Lstiburek 2008).

A study of moisture and mould problems in 59 single-family dwellings in southern Ontario by Lawton *et al.* (1998) showed that visible mould during inspection could not account for objective measurements of mould and other biological contaminants in settled dust. This demonstrated that much of the contamination was hidden in wall cavities. The largest source of mould was explained by condensation on the envelope, closely followed by a variety of sources including mould on the interior walls and ceilings of bathrooms, refrigerator drain pans and sumps. Wicking of water into the basement resulted in 30% of basements showing mould growth (Table 2). A more limited study in eastern Ontario yielded similar results (Miller *et al.* 2007), and data from homes in the Atlantic coastal province of Prince Edward Island (Foto *et al.* 2005) were also similar. In that study, visible mould was common on window sills in each of the rooms (59-71% of windows depending on room). It was also found in bath enclosures (34%), surfaces in bathrooms, principally the underside and back of toilet tanks (64%), and the kitchen (11%, growth on refrigerator pans). Mould growth was seen on 5% of walls, floors and ceilings in all other rooms. Moisture problems in the attic were found in nearly 60% of houses, with some showing evidence of water on the ceiling below. Some 18% of houses had visible mould on below-grade walls and 13% had visible mould on the flooring material in the case of on-grade floors (slab on-grade or basement floors). In contrast to the homes in southern Ontario, the prevalence of mould damage associated with basement leaks was somewhat higher, owing mainly to improper drainage

and waterproofing. Also, in the Prince Edward Island homes the prevalence of moisture damage in the attic was higher, possibly due to moisture entering the attic from the house in the absence of an effective air barrier.

As with buildings in the subtropical areas discussed above, airflow in and out of buildings accounts for an appreciable percentage of water movement in residential housing. Evidence of airflows is indicated by patterns of dirt on the bottom of closed interior doors, and other clues can also shed light on the movement of air (see Lstiburek and Carmody 1994). A final category of damage is unrepaired mould and water damage from the water used to put out fires and from hurricanes, earthquakes, floods or other natural disasters (Morey 1993, Miller *et al.* 2000, AIHA 2008c).

The above-mentioned studies support the inferences drawn in numerous case studies by investigators such as Morey (1997). The physical investigation of moulds in both commercial and domestic buildings requires considerable expertise in the design, construction and operation of these structures. Sampling strategies should not be designed or executed unless a thorough building inspection is made either on a concurrent basis or before the sampling. Sampling is typically carried out in order to document exposure and, importantly, to identify hidden contamination that is not visible without extensive destructive testing. Similarly, after sample results are obtained, the data must be compared with physical inspections, including new inspections indicated by the results. "Are the results plausible?" is a question that must always be asked and answered to assess properly the risk of false-negative and false-positive results for mould contamination (Health Canada 2004). Additionally, documentation of the sources and nature of the contamination allows a failure analysis to be made for the building (or system). This will assist in developing further cost-effective investigation strategies and ultimately any remedial actions that are necessary.

DOCUMENTATION OF MOULD AREA

For both determining the sources and nature of mould contamination and estimating the degree of risk, documentation of the area of contamination is necessary. Within the informed inspection component of the overall investigation, detailed notes of the amounts of mould should be taken on the appropriate perspective of the building plans. The area of mould should be sketched on the plan, making the best effort to document area. Video and photographic evidence can also be collected, but in many cases it is the written documentation and integration of the information that is most useful. Photographic evidence can be used to obtain detailed information when failure analyses are being conducted. Additional detail information on the process can be found in Foto *et al.* (2005) and AIHA (2009c).

Bulk samples (see following) should be taken from the visibly mouldy areas and an equal number of samples should be taken from areas away from the visible contamination. Decisions can be made later whether to process these, but such samples permit an evaluation of the mould just below that visible on the substrate. Usually, coloured conidia determine the colour of visibly mouldy material. It is obvious that the density of conidia on, for example, gypsum wallboard that is not visible to the naked eye in the field is still substantial contamination (see following). Although, depending on the circumstances, typically 0.25-1.0 m of building material is removed beyond the visibly mouldy area, these tests can provide objective data on this point (Miller *et al.* 2000). In some circumstances, the processing of material not visibly mouldy removes the suggestion of bias from the investigation. Formal scientific studies necessarily involve "control" samples and field studies are not relieved of this burden, at least in principle.

Microscopical or agar plating examination should be made of at least some samples to assess whether the colour on the building materials is indeed mould. Often, laboratory examination of building material samples show that mould was present although it was not recorded under field conditions owing to poor light, difficulty in seeing mould through full face respirators, etc. (Miller *et al.* 2000).

Where there is strong probable cause to believe that there is appreciable mould behind wall cavities, estimations of mould area inside can be made by sawing the bottom 0.3 m off one side of interior walls. It is not possible to estimate the true extent of contamination without this measure, which in terms of time and materials is usually no more onerous than using a keyhole saw and a boroscope (Miller *et al.* 2000). Suitable safety procedures for preventing investigator and occupant exposure must be used (AIHA 2008d). The cost of doing such testing is probably less than not doing it at all!

Notes taken should be accurate enough to permit the estimation of the area of contamination (i.e. m²) and allow the proper documentation of where bulk samples were collected. An example of the kind of record that should be taken is found in Fig. 1 (see also Foto *et al.* 2005, AIHA 2008c).

Such destructive testing should be done within simple containments, and appropriate respiratory protection is required (Macher *et al.* 1999, CCA 2004, AIHA 2008d). Polyethylene sheeting and duct tape can be used to make appropriately sized floor-to-ceiling containments (including covering the floor), which can have a simple slit/door made out of an additional piece of polyethylene sheet. A negative air machine or a simple industrial vacuum cleaner with a HEPA filter should be used to make the containment air pressure negative to the occupied space. In cases where these enclosures need to be large, other strategies can be used. As people back out of the containment areas, they can remove their coveralls and shoe covers leaving them inside (AIHA 2008d).

BULK SAMPLES

The two main methods of dealing with the isolation of fungi from bulk materials are destructive samples of building materials or dust samples. Cultural methods, particularly dilution plating methods, are selective and usually fail to discriminate active vs. inactive organisms present in the sample.

Dilution plating of bulk samples

In this method, a measured amount of a powdered material, such as settled dust or ground-up wallboard is suspended in an appropriate diluent and then diluted in 10-fold steps. Aliquots from each dilution are plated on agar media (in at least triplicate) and the liquid evenly spread over the surface of the plates. After incubation of the plates the colonies that have emerged are counted. Mycelium or spores from these colonies are then transferred to agar media appropriate for identification of the species present. The intention is to dilute to extinction, i.e. until there are no or few colonies in the highest dilution plated (Hung *et al.* 2005). Interference competition (see following) between isolates is reduced but not eliminated by this technique. Sometimes commercial laboratories plate single dilutions.

Different species of microorganism have different growth requirements so that the use of any particular medium produces different "recoveries". Spores decline in viability with time; the spores of some species of microorganism remain viable for years, while others are only viable for months. For example, the spores of

Table 3. Model data from plating spores of four taxa on three types of agar: pure cultures.

Species	Spores plated	MEA	CMA	DG18
S. chartarum	1000	1	10	1
P. aurantiogriseum	40	10	20	25
W. sebi	20	1	10	20
T. harzianum	40	10	10	1

Table 4. Model data from plating spores of four taxa on three types of agar: mixture of all species.

Species	Spores plated	MEA	CMA	DG18
S. chartarum	1000	5	5	-
P. aurantiogriseum	40	20	10	20
W. sebi	20	-	-	-
T. harzianum	40	10	10	1

Stachybotrys chartarum, Cladosporium cladosporioides and *Alternaria alternata* have half-lives of months and for those of *Penicillium, Aspergillus,* their sexual states and related taxa the average is ~7 years (Sussman 1968, Miller 1993, Hung *et al.* 2005). In general, the numbers of fungal propagules or colony forming units (CFU), determined by culture, are substantially less (1-50%) than those determined by direct methods such as microscopical examination of air samples collected on sticky surfaces or slides, but this varies with species. This is mainly because of dehydration of spores and the fact that, although dormant, they are respiring and so exhausting endogenous reserves. When there are very large numbers of spores present, it takes many years for all of them to die. At the same time, it means that when the percentage of viable spores is low it is possible to estimate how long it has been at a minimum since there was active growth. The rapid decline in viability of *Cladosporium* spores in settled dust was observed during the seasonal peak in the summer in the northeast USA (Gent *et al.* 2003). Owing to the different summer seasonal pattern, this decline was seen somewhat later in houses in Germany (Heinrich *et al.* 2003). *Stachybotrys* spores that have been dry for years are mostly non-viable, whereas the *Aspergillus/Penicillium* species mostly remain detectable by culture methods (Miller *et al.* 2000, Hung *et al.* 2005). Finally, some species such as *Trichoderma* spp. are very "aggressive" in culture, and producing antifungal agents that affect the growth of other developing colonies on agar plates, i.e. there is what is termed interference competition (see Taylor 1986, Vinale *et al.* 2006).

To illustrate the relevance of this to assessment of contamination, the hypothetical outcome of plating 1,000 spores of *S. chartarum*, 40 of *Penicillium aurantiogriseum*, 20 of *Wallemia sebi* and 40 of *Trichoderma harzianum* on three media: malt extract agar (MEA), corn meal agar (CMA) and DG18 agar (Hung *et al.* 2005) is shown in Table 3. From the 1000 spores plated, 10 colonies of *S. chartarum* (equivalent to 1% of the total spores plated) appear on CMA, the best medium of the three for this species. Recoveries on the

medium for moderate xerophiles, DG18, are very low. In contrast, a high percentage of the spores of the moderate xerophile *P. aurantiogriseum* are culturable on its optimum medium, DG18. For the xerophile *W. sebi*, good recoveries are obtained on DG18. Colonies of the hydrophilic, wood soft rot species *T. harzianum* are difficult to count under ideal circumstances and are poorly recovered on DG18.

Table 4 illustrates the hypothetical response of plating the mixture of species used in the above example. *S. chartarum* recoveries even on the optimal medium used, CMA, are much reduced because of the presence of *T. harzianum*. The xerophile *W. sebi* may not appear at all, and only *P. aurantiogriseum* emerges 'unscathed'.

These examples demonstrate the point that counting CFU from dilution plating of powdered bulk or dust samples over-represents long-lived tolerant *Penicillium* spp. It has long been understood in soil mycology that dilution plating does NOT give a reliable indication of the species active in the ecosystem. In the upper half of Fig. 2, data from dilution plating soil samples from a period of about one year are displayed. The lower half shows measurements of hyphal growth in the soil (after Kjøller and Struwe 1982). Typically, not only are there no correlations between the two measurements, the latter being a direct measure of activity, but 85% of the cultures found by dilution plating are of species not active in the ecosystem (Warcup 1965, Pugh 1980). In house dust samples, Miller *et al.* (1988) and Saraf *et al.* (1997) showed that in settled dust ergosterol (an absolute measure of fungal biomass in the sample) and CFU g^{-1} dust were not well correlated. This means that the actual biomass present was not proportional to the spores detected. Importantly, CFU measurements must be expressed in terms of an amount per gram per unit area (see Colloff *et al.* 1992) OR must involve comparisons of similar data (carpet with carpet, hardwood floor with hardwood floor, etc.).

Other means of assessing least significant differences in CFU data also provide insight into their interpretability. Mould measurements in settled dust (CFU

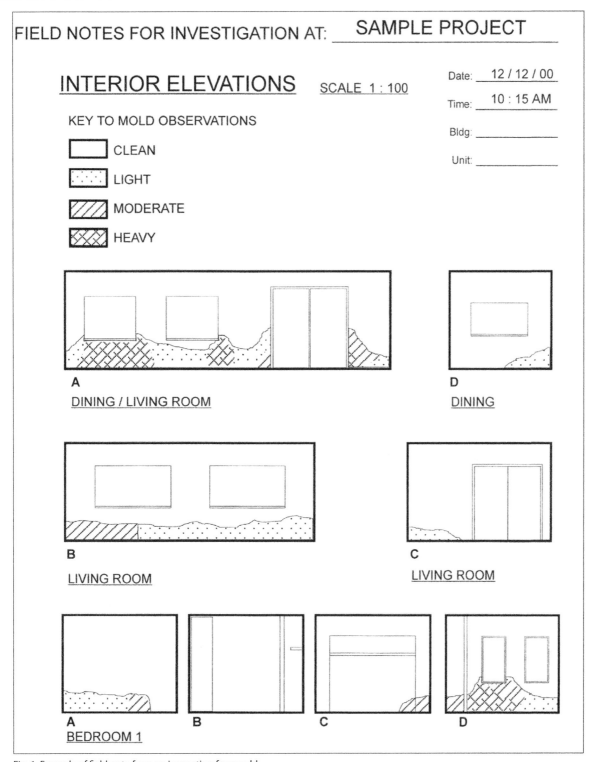

FIELD NOTES FOR INVESTIGATION AT: SAMPLE PROJECT

INTERIOR ELEVATIONS   SCALE 1 : 100

Date: 12 / 12 / 00
Time: 10 : 15 AM
Bldg:
Unit:

KEY TO MOLD OBSERVATIONS

☐ CLEAN
▨ LIGHT
▨ MODERATE
▨ HEAVY

A
DINING / LIVING ROOM

D
DINING

B
LIVING ROOM

C
LIVING ROOM

A   B   C   D
BEDROOM 1

Fig. 1. Example of field note from an inspection for mould.

g^{-1} on a medium for moderate xerophiles, showing a 10-fold increase in the proportion of moderate xerophiles) were weakly correlated with moisture source strength (an estimation of water sources in a house) and related to mould area estimations by the inspectors (Lawton *et al.* 1998). The latter measure, i.e. the ratio of moderate xerophiles to phylloplane species in settled dust was the more sensitive indicator of fungal contamination. In this work the least significant difference of CFU values appeared to be in the order of 100 (Miller *et al.* 1999).

For perspective, *Aspergillus* and *Penicillium* counts for settled dust were higher when occupants reported mould than when they did not. Higher mould counts

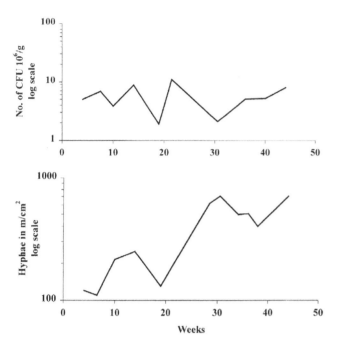

Fig. 2. Comparison of colony forming units (CFU) g⁻¹ soil with *in situ* measurements of hyphal growth (after Kjoller and Struwe, 1982).

in settled dust were associated with occupant reports of mouldy odours. Reported water damage was associated with increased mould counts in settled dust (Dales *et al.* 1997), but again these associations were weak.

Variations in dilution plating, including wiping a sterile cloth or swab over fixed areas of mouldy surfaces and dilution plating the resulting suspension, all suffer from the technical problems described above. CFU counts from dilution plating of bulk samples have no inherent quantitative value. Dilution plating has considerable value in determining the total diversity present in a sample however, and there is some evidence that the results from many replicates produce data that are related to the absolute value (Brewer and Taylor 1980). As observed by Pugh (1980), these methods which do not distinguish those species present as hyphae from those present as dormant spores may be "dangerously liable to waste time and energy and lead to error".

Soil crumb technique

The lower half of Fig. 2 illustrates the value of direct observational methods to reflect activity, but this method does not provide useful evidence of the taxa present. Small pieces of the solid material, e.g. 0.5 g, are plated on different agar media. These are incubated and the colonies that grow out are counted and transferred for identification. When conducting

mycological studies, inoculum is often taken from the leading edge of growing cultures. These are assumed to be active. The notion about the soil crumb technique is the same: the colonies that first emerge from a soil crumb or building material are on balance likely to be most reflective of those active in the ecosystem concerned (Warcup 1965, Pugh 1980).

As suggested in Fig. 1, the direct measurement and plating of bulk samples ("soil crumb method") is the most useful non-biochemical method of measuring fungal activity in samples and this minimizes emphasis on species that are inactive in the system.

Samples of bulk materials that are plated by the soil crumb method should also always be mounted in lactophenol cotton blue and examined under the microscope to determine the presence of organisms that might not be viable (see following). This allows a comparison to be made between viable and non-viable cultures.

Biochemical measurements

As noted above, ergosterol can be used to analyse the fungal burden in dust, and (1→3)-β-D-glucan has also been shown to be a useful measurement. These two types of measurement are discussed more extensively later in the chapter.

A summary of the strengths and weaknesses of the several methods discussed for processing bulk samples is found in Table 5.

Microscopical techniques

There are two basic techniques for the examination of mouldy surfaces by microscopy: sampling by tape, and mounting scrapings of the mould-damaged area. Tape samples are made by pressing the affected surface with good quality adhesive cellulose tape. Upon return to the laboratory, the marked area or approximately 2.5 cm² is cut out of the piece of tape and the adhesive side then rinsed with a few drops of 95% ethanol and drained. The piece of tape is placed, adhesive side down, on one or two drops of lactophenol cotton blue on a microscope slide slide. Another drop of lactophenol is added on top of the piece of tape and a glass coverslip placed on top of that. This preparation must be examined immediately and cannot be satisfactorily preserved. If scrapings are available, these can be mounted on slides and examined. They can be preserved by sealing the coverslips.

As in all microscopical methods, large dark spores are easier to see and, depending on the skill of the mi-

Table 5. Methods of determining fungal contamination of bulk samples.

Method	Characteristics
Dilution of sample to extinction	Over-represents long-lived, aggressive species, tempered by choice of medium, gives best picture of diversity, most taxa found probably not active
Single dilution	As above, reduced diversity, low precision
Soil crumb method	Most reflective of active species, low precision
Ergosterol	Reflects biomass, no qualitative information, good precision
$(1{\to}3)$-β-D-glucan	Reflects biomass, no qualitative information, good precision, biologically active component

croscopist, small hyaline spores are often overlooked. The taxonomic information obtained is limited.

Samples of scrapings from the mould-damaged areas are much more useful than tape samples. Microscopy is more satisfactory in the absence of dissolving cellulose tape. Slides can be kept for other microscopists to examine or for the record. If necessary, residual scrapings can be plated to acquire better taxonomic information and viability data.

AIR SAMPLES

Variability in air samples

Air samplers impacting particles on agar media provide information on "culturable" (hereafter termed "viable") propagules. Air samples impacted on agar media can greatly underestimate the total spore numbers present for the reasons discussed above. In general, the numbers of propagules determined by culture are substantially less (c. 1%) than those determined by direct methods (Flannigan *et al.* 1992, 1996). Variability of airborne propagule concentrations in the air in a building with active mould growth is very large (Miller 1993), whereas the precision of available sampling methods is relatively low (Hung et al. 2005). That is, the ability of current methods to detect environmental variation is limited.

There is considerable variability among airborne propagules (see Fig. 4 in Nevalainen *et al.* 1992). The variability of 1-min samples using an Anderson N6 sampler was found to be more than six times greater than 5-min samples (Stanevich and Petersen 1990). The variation in weekly samples taken in residential housing in Scotland was found to be three orders of magnitude (Hunter *et al.* 1988), and in a room studied for a day the hourly variation was four orders of magnitude (Miller 1993).

The reasons for the variability of spore concentrations include those related to (1) the spores and (2) the building and its equipment. Spores growing on a particular surface such as a wall or HVAC duct will either fall off or be blown off into the air. However, regardless of air flow the ruling relative humidity affects the release of spores from a colony, this varying according to species. Variations in flow/relative humidity can be expected to induce variability in spore concentration (Pasanen *et al.* 1991). As noted in Chapter 1.1, fungal spores have different sedimentation rates once airborne (see also Hunter *et al.* 1988, Flannigan 1993).

In large buildings, a common source of fungi is the HVAC system. Humidifiers, dirty filters, accumulated debris in ducts subject to condensation or leaks can all be sources of fungi. The distribution of fungi into the air is a function of airflow within the system. Accretions of material will be blown out of ducts periodically, rather than continuously. Fungi are also released from contaminated wall cavities, faulty sewer drains, etc. Release from these sources will be affected by the air infiltration rates and pressure differentials, which are in turn affected by weather. Distribution of fungi from carpets or surface contamination is affected by activity in the room (Hung *et al.* 2005).

TYPES OF AIR SAMPLER

There are several kinds of method for taking air samples. These include sticky surface samplers, filters, various devices that impact spores on agar media and liquid impinger type samplers.

Sticky surface samplers

Sticky surface samplers, such as Allergenco™, Burkhart™ and Rotorod™ have been commonly used, but there is little information on their quantitative and qualitative performance (Hung *et al.* 2005). As with other non-culture methods that depend on microscopy, a major limitation is the skill of the microscopist in (a) counting fungal propagules in a field containing other materials and (b) in recognizing small, hyaline spores.

The performance of a common type of sticky surface sampler, the Zefon Air-O-Cell™, has been examined with *P. brevicompactum* spores and with ragweed pollen. The diameter of these pollen grains (16-20 μm) is similar to *Epicoccum* spores. Filters that must be purchased from the manufacturer are connected to a calibrated pump. This device traps the spore in a solid medium suitable for microscopic analysis. The 50% cut off was found to be c. 3 μm (Trakumas *et al.* 1998). With respect to larger spores, the collection efficiency of the Air-O-Cell was found to be lower than the manufacturer states (95%), but still very good (Heffer *et al.* 2005). However, when Foto *et al.* (2005) compared area of visible mould damage with Air-O-Cell data, there was no relationship to either total counts or to the proportion of *Penicillium/Aspergillus* in the sample.

The presence of spore and hyphal fragments in samples indicates sources of indoor mould growth. Buildings generally free of mould growth have fewer fragments of both types. Hyphal fragment numbers do not correlate with total spore counts, reflecting the mainly internal source of such particles (Foto *et al.* 2005).

None of the methods employing sticky surface samplers is sufficiently quantitative to provide exposure measures as strictly defined.

Filter sampling

Fungal spores can be collected on filters and determined in a number of ways, including microscopy after staining or washing the spores off the filters and plating them to make viable counts. The viable count method is limited by the fact that the collection process results in desiccation and rapid reduction in viability (Hung *et al.* 2005). The deposition of fungal spores on open-faced filters has been studied carefully in the field (Eduard *et al.* 1990) and under controlled conditions (Gao *et al.* 1997). Methods have been found to correct problems caused by aggregation of spores on the filters (Eduard and Aalen 1988). These data established that open-faced filter samples are reliable tools for collection of fungal spores. However, the estimations of spore numbers obtained are still dependent on the skill of the microscopist. Epifluoresence microscopy appears to be the best visualization technique (Eduard *et al.* 1990).

Several comparative studies have reported moderate to good correlations between data collected using impaction samplers on agar media and total fungal spores estimated by direct microscopy of micropore filter samples (Palmgren *et al.* 1986a, b, Miller

et al. 1988, Flannigan 1992, Flannigan *et al.* 1996, Tsai *et al.* 1999). Additionally, the distribution of fungal taxa between air samples impacted on agar media appears similar to that obtained on filters (Flannigan *et al.* 1996). Filters have the advantage that they can be run for a long time (days) to attempt to deal with the environmental variability.

The use of filter samplers followed by careful microscopic examination probably gives useful exposure measurements.

Impaction air samplers for culturable propagules

Under realistic conditions, all the proprietary air samplers for culturable fungi are useful for assessing diversity (Laflamme and Miller 1992, Buttner and Stetzenbach 1993, Hung *et al.* 2005). As noted above, the variability of spore concentrations in air is vastly greater than variation in sampler performance. Hung *et al.* (2005) provide detailed methods for the use of such samplers.

Liquid impingers

These samplers have not been used much in building investigations and it is therefore difficult to evaluate their performance. Methods for the use of some liquid impinger samplers are to be found in Hung *et al.* (2005).

COMPARISON OF INDOOR AND OUTDOOR AIR

Unless there are zero viable counts from indoor air samples, it is necessary to determine the fungal species to compare inside vs. outdoor samples. While the CFU in air varies tremendously, the ratio of the taxa present is a more stable property (Miller 1993, Flannigan and Miller 1994). Many of the problems that affect the analyses of dust samples described above apply to the analysis of culturable air samples. The growth of mixtures of fungi on semi-selective agar media (Tables 3 and 4) results in highly variable data, depending on the taxa collected. Duration of sampling has two effects: firstly, the longer the sampling time, the greater the fungal diversity (Miller 1993); secondly, and adversely, the greater the reduction in recovery due to the fall in the water activity (a_w) of the medium caused by dehydration (Saldanha *et al.* 2008). There is another very important problem that applies particularly to the taking of outdoor air samples or samples in air ducts: the movement of air over the sampling plane has profound effects on the recovery of spores from air (Nevalainen *et al.* 1992, Upton *et al.* 1994).

Samples taken in air where there is material air movement, as in outdoor air, cannot be directly compared with other samples. In addition, precipitation also affects the data obtained.

The basis of the current methods for conducting sampling for culturable propagules is a comparison of the indoor value with outdoor air samples (Health Canada 2004, Hung *et al.* 2005). Outdoor air samples must be taken well above grade to avoid problems of collecting windblown soil particles. A dataset on fungal spores in outdoor air that is available was collected in Belgium (Nolard 1997). These data show that the 10-year average airborne concentrations of spores are low in January, February and March. Starting in early spring, there is a shifting array of fungal particles in the air as the season progresses. Order of magnitude changes occur within this qualitative change with average numbers occurring in July of around 20,000 particles m^{-3} air, peaking at twice that value. In the winter, *Aspergillus/Penicillium* (genera associated with soil particles) comprise 10% of the low number of fungi then present in outdoor air. During the rest of the year, the percentage is 0.1% of the much higher numbers of spores and species present (Coates *et al.* 1992, Nolard 1997). Under dry conditions, fungal spores were typically c. 2% of airborne particulate material (PM), and the absolute values were estimated as being similar to the Brussels data (Battarbee *et al.* 1997). After rain, the percentage rose to 27% of particles as a result of a three- to four-fold rise in the numbers of spores, and washout of the larger inorganic particles. These data show that for several days after precipitation, fine PM contains considerable numbers of filamentous fungal particles. The contribution of yeasts to fine PM is also high, particularly after fog or rain. Fuzzi *et al.* (1997) noted that during fog yeast particles increased 100-fold compared with clear air, rising to 5,000 m^{-3} at 10°C.

To accommodate these problems, data from a number of outdoor air samples must be pooled. Percentages of the different taxa in this pool can then be used to make comparisons with indoor samples. Using this approach, there is general agreement that properly conducted air sampling is useful for detecting fungal amplifiers, but not for assessing exposure (Dillon *et al.* 1999). The numerical data obtained from culturable air samples have no value. A study of these issues has been made by Miller *et al.* (2000) in which the data from extensive air sampling in an apartment building were compared with the extent of mould damage revealed by massive destructive inspection to examine wall cavities. Air samples were scored ac-

Table 6. Potential misclassification of spores on sticky surface samplers (adapted from Hung *et al.* 2005).

Alternaria	*Mystrospriella, Ulocladium*
Aspergillus/ Penicillium	*Trichoderma, Acremonium, Paecilomyces, Phialophora, Cladosporium,* basidiospores
Aureobasidium	*Exophiala, Phialophora,* black yeasts
Cladosporium	*Cladophialophora, Exophiala, Fulvia, Gonatobotryum, Mycovellosiella, Periconiella, Phaeoramularia, Septonema,*
Pithomyces	*Ulocladium, Alternaria*
Stachybotrys	*Memnoniella, Gliomastix, Periconia*
Ulocladium	*Alternaria, Monodictys, Pithomyces*

cording to the guidelines in AIHA (Hung *et al* 2005) based on comparison with 50 pooled outdoor air samples taken over the life of the project.

Approximately 25% of the data set is outlined in Tables 6 and 7 (adapted from Miller *et al.* 2000). AIHA "fails" means air samples that were different from pooled outdoor air samples judged by the criteria in Hung *et al.* (2005). Air samples (2 or 3, depending on the size of the apartment), were taken in the morning and afternoon in each room. In this data set, as expected, the numbers of CFU m^{-3} air did not differ significantly between the two data sets using descriptive statistics. However, the proportion of samples that "fail" the above-noted AIHA guidelines is different (p < 0.005). When the data were compared with the area of visible mould revealed by the extensive destructive testing, there was no correlation with CFU m^{-3} values. When the area of fungal damage to proportion of samples was compared, by rank order (Miller *et al.* 2000), to the air sample results (pass/fail) the difference was statistically different (Kruskal-Wallis, p = 0.03, ANOVA, p = 0.10; 264 indoor air and 159 outdoor samples). In this partial data set (and in the full set), the proportion of "fails" was fairly well correlated with mould area (p < 0.11).

Samples estimating viable propagules do not measure exposure.

Biochemical measurement of airborne burden

Because of the need for true exposure measurements, research on the effects of microbes in indoor air has mainly focused on using indicator biochemicals that can be measured from particulate samples. For bacteria, these have included endotoxin and peptidyl glycan for Gram-negative and Gram-positive species, respectively. For fungi, these have included (1→3)-β-D-glucan and ergosterol. Methods for airborne endo-

Table 7. Air sampling data from apartments with large areas of mould in wall cavities (after Miller *et al.* 2000).

Apartment	AIHA "Fails"*	Mean culturable moulds(CFU m⁻³)
1	6-Apr	664
2	4-Mar	142
3	4-Apr	109
4	4-Apr	172
5	4-Apr	203
6	4-Apr	277
7	4-Mar	778
8	6-Apr	146
9	6-May	74
10	6-May	192
11	6-May	201
12	4-Feb	1531
13	6-Mar	134
14	6-Jun	180
15	6-Mar	137
Mean	0.88±0.19	329±389

* AIHA "fails" registered when air samples differed from pooled outdoor air samples according to criteria in Dillon *et al.* (1996). Air samples (2 or 3, depending on the size of the apartment) were taken in the morning and afternoon in each room; number of air samples ,76.

Table 8. Air sampling data from apartments with small areas of mould in wall cavities (after Miller *et al.* 2000).

Apartment	AIHA "Fails"*	Mean culturable moulds (CFU m⁻³)
1	1/4	130
2	0/4	153
3	1/4	332
4	0/4	889
5	0/6	204
6	0/4	194
7	0/4	164
8	0/6	132
9	1/4	128
10	1/4	50
11	0/6	19
12	0/4	214
13	1/4	172
14	0/6	229
15	0/6	197
Mean	0.08±0.12	214±201

*"Fails" registered and samples taken as in the morning and afternoon as in Table 6; number of air samples, 70.

toxin are well evaluated (Hung *et al.* 2005) and widely used in the built environment (Dales *et al.* 2006, Miller *et al.* 2007), but in the case of peptidyl glycan analysis remain immature (Hung *et al.* 2005).

There are large differences between different taxonomic groups of fungi in both glucan type and content. Some basidiomycetes produce (1→3)- and (1→6)-β-D-glucans, the latter of which may also have anti-inflammatory properties (Pacheco-Sanchez *et al.* 2006). Although this is not especially well studied, different fungi produce varying relative amounts of (1→3)-β-D-glucan relative to other glucans (Odabasi *et al.* 2006). Some yeasts make a (1→3)-β-D-glucan 1:1 with a short (1→3)-β-D-glucan (zymosan; Ohno *et al.* 2001). The anamorphic Trichocomaceae (i.e. *Penicillium*, *Aspergillus* and related fungi), often referred to as moulds associated with damp building materials, appear predominantly to contain (1→3)-β-D-glucan (Foto *et al.* 2005, Odabasi *et al.* 2006).

When several species of fungi that grow on building materials, viz. *Eurotium herbariorum* (optimum a_w for growth c. 0.70), *P. aurantiogriseum* (c. 0.82) and *S. chartarum* (c. 0.93; see Chapter 2.1) were cultured on malt extract agar, corn meal agar, Czapek-Dox agar and DG-18 agar (representing a range of a_w and nutrient conditions) and the spores harvested, the glucan levels normalized to measured spore surface areas were not altered in relation to varying nutrients in experiments conducted within normal a_w ranges for the fungi tested (Foto *et al.* 2004). The average molecular weight of the glucan expressed as curdlan equivalents in 10 fungi associated with damp building materials as well as *Cladosporium cladosporioides* was 181 ± 6 kDa calibrated against curdlan with a molecular size of 161 kDa (Foto *et al.* 2005). However, considered when grouped according to taxonomic type, the glucan contents are generally proportional to spore surface area (Foto *et al.* 2005, Iossifova *et al.* 2008). This is also the case with the fungal membrane sterol ergosterol, again when considered within taxonomic groups (Miller and Young 1997).

There are two main ways to measure glucan, but neither of these has been validated in inter-laboratory trials. As with endotoxin, results of different studies from the same laboratory are comparable (Hung *et al.* 2005). The majority of data reported have been based on the so-called factor G of the LAL assay (Foto *et al.* 2004, 2005, Berghout *et al.* 2005, Miller *et al.* 2007, Iossifova *et al.* 2008). The original standard used was curdlan (Rylander and Lin 2000), which was used by others (e.g. Foto *et al.* 2005) although the Glucatel assay from Associates of Cape Cod uses a basidiomycete-derived curdlan (pachyman) as a standard. This glucan is an essentially linear $(1{\rightarrow}3)$-β-D-glucan of degree of polymerisation (DP) ~250, but contains some internal $(1{\rightarrow}6)$-β-linkages and branching, although it can be chemically debranched. If this is done properly, the two standards have similar molecular weights and degrees of polymerization. Curdlan is less soluble in water. There is no difference in response between the two assays when yeast glucan is tested or settled dust samples. A comparison of other fungal-derived glucans indicated the two methods gave identical results. When pachyman is used as a standard instead of curdlan, the assay has a somewhat lower limit of detection (H. Cherid and J.D. Miller, unpublished data). This would be more relevant for clinical rather than environmental use of the assay.

Quite a number of studies have reported using monoclonal antibody-based methods (for a description of the methods see Douwes *et al.* 1996, 1998, Foto *et al.* 2004). However, it appears that many of these data are unreliable (Iossifova *et al.* 2008). There may be other antibody-based methods that prove to be useful (e.g. Sander *et al.* 2008), however, the LAL-based methods are based on well-understood methods and are likely to dominate.

Ergosterol is the primary membrane sterol of filamentous fungi and environmental yeasts and, with rare exceptions, is not accumulated by other organisms. This useful fact has encouraged a great deal of research using ergosterol as a biomass measure (Gessner and Newell 1997). A method was developed for use in house dust (Miller *et al.* 1988) and the approach has since been used by many others (e.g. Saraf *et al.* 1997, Park *et al.* 2008). Ergosterol content of spores is comparable with values from mycelium, and the content of different species that commonly grow on building materials is similar. The analytical performance of the method has been characterized (Miller and Young 1997). Given that correctly exposed open-face filters collect fungal spores in a quantitative manner (see above), ergosterol provides unambiguous exposure measures as long as the spores are of similar size. However, the qualitative features of the exposure can be inferred or known from analysing the fungal spores in dust or other means. A major innovation in the ergosterol method for air samples is microwave-assisted extraction which permits the rapid processing of many samples at once. Fatty acids are also extracted during this process and can be analyzed by GC or CG/MS (Young 1995).

Like ergosterol, glucan can provide information on exposure. In theory, glucan measurements should reflect ergosterol values and *vice versa*. However, yeasts from outdoor air are a major component of settled dust in buildings. Glucan is a crystalline polysaccharide and is stable. However, ergosterol is readily biodegradable and photodegradable. Cells from moulds (also containing ergosterol) are typically coloured, hence the ergosterol in moulds is more stable than that in yeast cells, which are hyaline. It is filamentous fungi that colonize building materials, not yeasts. As a result of these two factors, ergosterol measurements from long-duration air samples have a better correlation with area of visible mould than do glucan measurements (Foto *et al.* 2005). This also explains the contradictory relationships of glucan to health symptoms observed. Many studies report the negative effect of glucan on health (e.g. Rylander and Lin 2000, Douwes 2005). In others, the direction of the effect has been opposite for some outcomes (Schram-Bijkerk *et al.* 2005, 2006, Iossifova *et al.* 2009). It may be that the direction of the effect can change (protective vs. adverse depends on the exposure).

In the late 1970's, Dr David White developed an approach called lipid signature profiles for studying natural bacterial populations by biochemical means. It is based on the fact that for genera and some species within genera, there are characteristic patterns of cell lipids (White 1988). This was suggested as being of potential use in monitoring bacteria in spacecraft and other critical indoor environments (Macnaughton *et al.* 1997). Air samples are extracted and analyzed by CG-MS for signature lipids followed by statistical treatments of the data to convert the analytical results to species profiles. The analysis provides information of the putative identities and biomass of dominant bacteria present in an air sample. Preliminary studies showed that the lipid signature approach for bacteria gives values of "2-3 orders of magnitude" greater than that estimated by cultural techniques (Macnaughton *et al.* 1997). This accords very well with the experience described in the literature that culturable estimates of bacteria in indoor air represent less than 1% of the total (Flannigan *et al.* 1991). Further research indicated that lipid signature profiling was feasible for large samples collected from bulk building materials (Womiloju *et al.* 2005) and high volume samples collected from outdoor air (Womiloju *et al.* 2003).

PCR-BASED METHODS

As PCR-based methods are discussed in some depth in the chapter following this, it is sufficient to say here that, although they have been applied in studies in USA (e.g. Vesper *et al.* 2007) and UK (Vesper *et al.* 2005), they have yet to be widely used. As has been mentioned in connection with outdoor air in Chapter 1.1, what is clear is that the data do not relate to traditional methods (Dillon *et al.* 2007, Pietarinen *et al.* 2008, Pitkäranta *et al.* 2008, Prezant *et al.* 2008, Chapter 12), and they also do not relate to exposure.

FUNGAL ALLERGENS

As noted above, settled dust loading alone predicted exposure and symptoms. This has led to the capacity to sample and rapidly test dust samples for measuring mite and pet animal allergens to aid allergy sufferers in validating the effectiveness of their allergen avoidance measures (Tsay *et al.* 2002, Earle *et al.* 2007). There are also data to indicate that consumers can usefully collect dust samples for this purpose (Arbes *et al.* 2005). These have been extended to measure fungal allergens such as Asp f1 of *Aspergillus fumigatus* (Ryan *et al.* 2001) and antigens such as those from *S. chartarum* (Xu *et al.* 2008, Smith *et al.* 2009), and others such as *P. chrysogenum* (Luo *et al.* 2009) are in development. These in turn can be measured simultaneously (King *et al.* 2007).

Allergen and antigen measurements on a loading basis DO predict exposure.

REFERENCES

AIHA (2008a) Physical inspection of specific building types. In B. Prezant, D. Weekes, and J.D. Miller, (eds.), *Recognition, Evaluation and Control of Indoor Mold*, American Industrial Hygiene Association, Fairfax, VA, pp. 83-96.

AIHA (2008b) Science and buildings. In B. Prezant, D. Weekes, J.D. Miller, (eds.), *Recognition, Evaluation and Control of Indoor Mold*, American Industrial Hygiene Association, Fairfax, VA, pp. 55-60.

AIHA (2008c) Documentation. In B. Prezant, D. Weekes, and J.D. Miller, (eds.), *Recognition, Evaluation and Control of Indoor Mold*. American Industrial Hygiene Association, Fairfax, VA, pp. 105-125.

AIHA (2008d) On-site considerations. In B. Prezant, D. Weekes, and J.D. Miller, (eds.), *Recognition, Evaluation and Control of Indoor Mold*. American Industrial Hygiene Association, Fairfax, VA, pp. 61-81.

Anon. (1993) Guidelines on assessment and remediation of *Stachybotrys atra* in indoor environments. In E. Johanning and C.S. Yang, (eds.), *Fungi and Bacteria in Indoor Air Environments*, Eastern New York Occupational Health Program, Latham, NY, pp. 201-208.

Anon. (1996) Survey of building envelope failures in the coastal climate of British Columbia, Canada. Mortgage and Housing Corporation. 700 Montreal Road, Ottawa. Canada. K1A 0P7.

Anon. (2008) *Guidelines on Assessment and Remediation of Fungi in Indoor Environments*. New York City Department of Health and Mental Hygiene, New York.

Antova, T., Pattenden, S., Brunekreef, B., *et al.* (2008) Exposure to indoor mould and children's respiratory health in the PATY study. *J. Epidemiol. Community Health*, **62**, 708-714.

Arbes, S.J., Sever, M., Vaughn, B., *et al.* (2005) Feasibility of using subject-collected dust samples in epidemiologic and clinical studies of indoor allergens. *Environ. Health Perspect.* 113, 665-669.

Battarbee, J.L., Rose, N.L., and Long, X. (1997) A continuous high resolution record of urban airborne particulates suitable for retrospective microscopical analysis. *Atmos. Environ.*, **31**, 171-181.

Berghout, J., Miller, J.D., Mazerolle, R., *et al.* (2005) Indoor environmental quality in homes of asthmatic children on the Elsipogtog Reserve (NB), Canada. *Int. J. Circumpolar Health*, **64**, 77-85.

Brewer, D., and Taylor, A. (1980) Ovine ill-thrift in Nova Scotia 6, quantitative description of the fungal flora of soils of permanent pasture. *Proc. Nova Scotia Inst. Sci.*, **30**, 101-133.

Buttner, M.P., and Stetzenbach, L. (1993) Monitoring airborne fungal spores in an experimental indoor environment to evaluate sampling methods and the effects of human activity on air sampling. *Appl. Environ. Microbiol.*, **59**, 219-226.

CCA (2004) Mould Guidelines for the Canadian Construction Industry, Standard Construction document CCA 82. Canadian Construction Association, Ottawa, Canada.

Cho, S.H., Reponen, T., LeMasters, G., *et al.* (2006) Mould damage in homes and wheezing in infants. *Ann. Allergy Asthma Immunol.*, **97**, 539-545.

Coates, L.L., Crompton, C.W., Yang, W.H., and Drouin, M.A. (1992) Comparative analysis of the pollen grains and mould spores of Ottawa 1985-1991. *Clin. Invest. Med.*, **S15**, A7.

Colloff, M.J., Ayres, J., Carswell, F., *et al.* (1992) The control of allergens of dust mites and domestic pets: a position paper. *Clin. Exp. Allergy*, **22**, S2, 1-28.

Dales, R.E., Miller, J.D., and McMullen, E. (1997) Indoor air quality and health: validity and determinants of reported home dampness and moulds. *Int. J. Epidemiol.*, **26**, 1-6.

Dales, R.E., Miller, J.D., White, J.M., *et al.* (1998) The influence of residential fungal contamination on peripheral blood lymphocyte populations in children. *Arch. Env. Health*, **53**, 190-195.

Dales, R.E., Miller, J.D., Ruest, K., *et al.* (2006) Airborne endotoxin is associated with respiratory illness in the first two years of life. *Environ. Health Perspect.*, **114**, 610-614.

Dillon, H.K., Miller, J.D., Sorenson, W.G., *et al.* (1999) Assessment of mould exposure in relation to child health. *Environ. Health Perspect.*, **107**, S 3, 473-480.

Dillon, H.K., Boling, D.K., Miller, J.D. (2007) Comparison of detection methods for *Aspergillus fumigatus* in environmental air samples in an occupational environment. *J. Occup. Environ. Hyg.*, **4**, 509-513.

Douwes, J. (2005) Health effects of 1→3-β-D glucans: the epidemiological evidence. In S.-H. Young and V. Castranova, (eds.), *Toxicology of 1→3-Beta-Glucans*, CRC Press, Boca Raton, FL, pp. 35-52.

Douwes.J., Doekes, G., Montijn, R., and Brunekreef, B (1996) Measurement of beta-(1→3)-glucans in the occupational and home environment with an inhibition enzyme immunoassay. *Appl. Environ. Microbiol.*, **62**, 3176-3182.

Douwes, J., van der Sluis, B., Doekes, G., *et al.* (1998) Fungal extracellular polysaccharides in house dust as a marker for exposure to fungi: relations with culturable fungi, reported home dampness and respiratory symptoms. *J. Allergy Clin. Immunol.*, **103**, 494-500.

Douwes, J., Zuidhof, A., Doekes, G., *et al.* (2000) (1→3)-Beta-D-glucan and endotoxin in house dust and peak flow variability in children. *Am. J. Respir. Crit. Care Med.*, **162**, 1348-1354.

Earle, C.D., King ,E.M., Tsay, A., *et al.* (2007) High-throughput fluorescent multiplex array for indoor allergen exposure assessment. *J. Allergy Clin. Immunol.*, **119**, 428-433.

Eduard, W., and Aalen, O. (1988) The effect of aggregation on the counting precision of mould spores on filters. *Ann. Occup. Hyg.*, **32**, 471-479.

Eduard, W., Lacey, J., Karlsson, K., *et al.* (1990) Evaluation of methods for enumerating microorganisms in filter samples from highly contaminated occupational environments. *AIHA J.*, **51**, 427-436.

Elliott, L., Arbes, S.J., Harvey, E.S., *et al.* (2007) Dust weight and asthma prevalence in the National Survey of Lead and Allergens in Housing. *Environ. Health Perspect.*, **115**, 215-220.

Ferro, A.R., Kopperud, R.J., and Hildemann, L.M. (2004) Source strengths for indoor human activities that resuspend particulate matter. *Environ. Sci. Technol.*, **38**, 1759-1764.

Flannigan, B., (1992) Indoor microbiological pollutants - sources, species, characterisation and evaluation. In H. Knöppel, and P. Wolkoff, (eds.), *Chemical, Microbiological, Health and Comfort Aspects of Indoor Air Quality – State of the Art in SBS*. Kluwer, Dordrecht, The Netherlands, pp. 73-98.

Flannigan, B., (1993) Approaches to assessment of the microbial flora of buildings. In *IAQ '92: Environments for People*, American Society of Heating, Refrigerating and Air-conditioning Engineers, Atlanta, GA, pp. 136-146.

Flannigan, B., and Morey, P. (1996) *Control of Moisture Problems Affecting Biological Indoor Air Quality*. International Society of Indoor Air Quality and Climate, Milan, Italy.

Flannigan, B., McCabe, E.M., and McGarry, F. (1991) Allergenic and toxigenic microorganisms in houses. *J. Appl. Bact.*, **79**, 61S-73S.

Flannigan, B., and Miller, J.D. (1994) Health implications of fungi in indoor environments – an overview. In R.A. Samson, B. Flannigan, M.E. Flannigan, *et al.*, (eds.), *Health Implication of Fungi in Indoor Environments*. Elsevier, Amsterdam, pp. 3-28.

Flannigan, B., McCabe, E.M., Jupe, S.V., and Jeffrey, I.G. (1996) Quantification of dust-borne deteriogenic microorganisms in homes. In W. Sand, (ed.), *Biodeterioration and Biodegradation, Papers of the Tenth International Biodeterioration Symposium, Hamburg*, DECHEMA, Frankfurt am Main, Germany, pp. 377-384.

Foto, M., Plett, J., Berghout, J., and Miller, J.D. (2004) Modification of the *Limulus* amebocyte lysate assay for the analysis of glucan in indoor environments. *Anal. Bioanal. Chem.*, **379**, 156-162.

Foto, M., Vrijmoed, L.L.P., Miller, J.D., *et al.* (2005) Comparison of airborne ergosterol, glucan and Air-O-Cell data in relation to physical assessments of mould damage and some other parameters. *Indoor Air*, **15**, 256-266.

Fuzzi, S., Mandrioli, I., and Perfetto, A. (1997) Fog droplets – an atmospheric source of secondary biological aerosol particles. *Atmos. Environ.*, **31**, 287-290.

Gao, P., Dillon, H.K., and Farthing, W.E. (1997) Development and evaluation of an inhalable bioaerosol sampler. *AIHA J.*, **58**, 196-206.

Gehring, U., Douwes, J., Doekes, G., *et al.* (2001) Beta-(1→3)-glucan in house dust of German homes: housing characteristics, occupant behavior, and relations with endotoxins, allergens, and moulds. *Environ. Health Perspect.*, **109**, 139-144.

Gent, J.F., Ren, P., Belanger, K., *et al.* (2003) Levels of household mould associated with respiratory symptoms in the first year of life in a cohort at risk for asthma. Reproducibility of allergen, endotoxin and fungi measurements in the indoor environment. *J. Expo. Anal. Environ. Epidemiol.*, **13**, 152-160.

Gessner, M.O., and Newell, S.Y. (1997) Bulk quantitative methods for the examination of eukaryotic organoosmotrophs in plant litter. In G.R. Knudsen, M.J. McInerney, L.D. Stetzenbach, and M.V. Walter, (eds.), *Manual of Environmental Microbiology*. American Society for Microbiology, Washington, DC, pp. 295-308.

Gomes, C., Freihaut, J., and Bahnfleth, W. (2007) Resuspension of allergen-containing particles under mechanical and aerodynamic disturbances from human walking. *Atmos. Environ.*, **41**, 5257-5270.

Gravesen, S., Nielsen, P.A., Iversen, R., and Nielsen, K.F. (1999) Microfungal contamination of damp buildings – examples of risk constructions and risk materials. *Environ. Health Perspect.*, **107**, 505-508.

Haverinen, U., Husman, T., Vahteristo, M., *et al.* (2001). Comparison of two-level and three-level classifications of moisture-damaged dwellings in relation to health effects. *Indoor Air*, **11**, 192-199.

Haverinen, U., Vahteristo, M., Pekkanen, J., *et al.* (2003) Formulation and validation of an empirical moisture damage index. *Environ. Model. Assess.*, **8**, 303-309.

Haverinen-Shaughnessy, U., Pekkanen, J., Hyvärinen, A., *et al.* (2006) Children's homes – determinants of moisture damage and asthma in Finnish residences. *Indoor Air*, **6**, 248-255.

Health Canada (2004) Fungal Contamination in Public Buildings: Health Effects and Investigation Methods. Health Canada, Ottawa, Ontario.

Heffer, M.J., Ratz, J.D., Miller, J.D., and Day, J.H. (2005) Comparison of the Rotorod to other air samplers for the determination of *Ambrosia artemisiifolia* pollen concentrations conducted in the Environmental Exposure Unit. *Aerobiologia*, **21**, 233-239.

Heinrich, J., Hölscher, B., Douwes, J., *et al.* (2003) Reproducibility of allergen, endotoxin and fungi measurements in the indoor environment. *J. Expo. Anal. Environ. Epidemiol.*, **13**, 152-160.

Hung, L.L., Miller, J.D., Dillon, H.K. (2005) *Field Guide for the Determination of Biological Contaminants in Environmental Samples*, 2nd ed. American Industrial Hygiene Association, Fairfax, VA.

Hunter, C.G., Grant, C., Flannigan, B., and Bravery, A.F. (1988) Mould in buildings: the air spora of domestic dwellings. *Int. Biodet.*, **24**, 84-101.

Iossifova, Y.Y., Reponen, T., Daines, M., Levin, L.A., *et al.* (2008) Comparison of two analytical methods for detecting (1-3)-β-D-glucan in pure fungal cultures and in home dust samples. *The Open Allergy J.*, **1**, 26-34.

Iossifova, Y.Y., Reponen, T., Ryan, P., *et al.* (2009) Mould exposure during infancy as a predictor of potential asthma development. *Ann. Allergy Asthma Immunol.*, **102**, 131-137.

King, E.M., Vailes, L.D., Tsay, A., *et al.* (2007) Simultaneous detection of total and allergen-specific IgE by using purified allergens in a fluorescent multiplex array. *J. Allergy Clin. Immunol.*, **120**, 1126-1131.

Kjøller, A., and Struwe, S. (1982) Microfungi in ecosystems: fungal occurrence and activity in litter and soil. *Oikos*, **39**, 391-422.

Krieger, J.W., Takaro, T.K., Song, L., and Weaver, M. (2005) The Seattle-King County Healthy Homes Project: a randomized, controlled trial of a community health worker intervention to decrease exposure to indoor asthma triggers. *Am. J. Public Health*, **95**, 652-659.

Laflamme, A.M., and Miller, J.D. (1992) Collection of spores of various fungi by a Reuter centrifugal sampler. *Int. Biodet.*, **29**, 101-110.

Lawton, M.D., Dales, R.E., and White, J. (1998) The influence of house characteristics in a Canadian community on microbiological contamination. *Indoor Air*, **8**, 2-11.

Lioy, P.J., Freeman, N.C., and Millette, J.R. (2002) Dust: a metric for use in residential and building exposure assessment and source characterization. *Environ. Health Perspect.*, **110**, 969-983.

Lstiburek, J.W. (2007) The perfect wall. *ASHRAE J.*, **49**, 74-78.

Lstiburek, J.W. (2008) The perfect storm over stucco. *ASHRAE J.*, **50**, 38-42

Lstiburek, J.W. (2009) Building in extreme cold. *ASHRAE J.*, **51**, 56-59.

Lstiburek, J., and Carmody, J. (1994) Moisture control for new residential buildings. In H.R. Trechsel, (ed.), *Moisture Control in Buildings*. American Society for Testing Materials, Philadelphia, pp. 321-347.

Luo, W., Wilson, A.W., and Miller, J.D. (2009) Characterization of a 52 kDa exoantigen of *Penicllium chrysogenum* and monoclonal antibodies suitable for its detection. *Mycopathologia*, **169**, 15-26.

Macher, J., Burge, H.A., Milton, D.K., and Morey, P.R. (1999) *Assessment and Control of Bioaerosols in the Indoor Environment*. American Conference of Industrial Hygienists, Cincinnati, OH.

Macnaughton, S.J., Jenkins, T.L., Alugupalli, S., and White, D.C. (1997) Quantitative sampling of indoor air biomass by signature lipid biomarker analysis: feasibility studies in a model system. *AIHA J.*, **58**, 270-277.

Maunsell, K. (1952) Air-borne fungal spores before and after raising dust; sampling by sedimentation. *Int. Arch. Allergy Appl. Immunol.*, **3**, 93-102.

Miller, J.D. (1993) Fungi and the building engineer. *IAQ '92: Environments for People*, American Society of Heating, Refrigerating and Air-conditioning Engineers, Atlanta, GA, pp. 147-162.

Miller, J.D., and Young, J.C. (1997) The use of ergosterol to measure exposure to fungal propagules. *AIHA J.*, **58**, 39-43.

Miller, J.D., Laflamme, A., Sobol, Y., *et al.* (1988) Fungi and fungal products in some Canadian houses. *Int. Biodet.*, **24**, 103-120.

Miller, J.D., Dales, R.E., and White, J. (1999) Exposure measures for studies of mould and dampness and respiratory health. In E. Johanning, (ed.), *Bioaerosols, Fungi and Mycotoxins: Health Effects, Assessment, Prevention and Control*, Eastern New York Occupational and Environmental Health Center, Albany, NY, pp. 298-305.

Miller, J.D., Haisley, P., and Rhinehardt, J. (2000) Air sampling results in relation to extent of fungal colonization of building materials in some water damaged buildings. *Indoor Air*, **10**, 146-151.

Miller, J.D., Dugandzic, R., Frescura, A.M., and Salares, V. (2007) Indoor and outdoor-derived contaminants in urban and rural homes in Ottawa, Canada. *J. Air Waste Man. Assoc.*, **57**, 297-302.

Miller, J.D., Rand, T.G., McGregor, H., *et al.* (2008) Mould ecology: recovery of fungi from certain mouldy building materials. In B. Prezant, D. Weekes, and J.D. Miller, (eds.), *Recognition, Evaluation and Control of Indoor Mould*, American Industrial Hygiene Association, Fairfax, VA, pp. 43-51.

Morey, P. (1993) Microbiological events after a fire in a high-rise building. *Indoor Air*, **4**, 232-327.

Morey, P. (1997) Fungi and microbial VOC's in indoor air. What do the data mean? How much mould is too much? *Hot and Humid Indoor Environments*, IAQ Publications, Bethesda, MD, pp. 65-77.

NAS (2000) *Clearing the Air: Asthma and Indoor Air Exposures*. Institute of Medicine, National Academy of Sciences, National Academy Press, Washington, DC.

NAS (2004) *Damp Indoor Spaces and Health*. Institute of Medicine, National Academies Press, Washington, DC.

Nevalainen, A., Pastuszka, J., Liebhaber, F., and Willeke, K. (1992) Performance of bioaerosol samplers: collection characteristics and sampler design considerations. *Atmos. Environ.*, **26A**, 531-540.

Nolard, N. (1997) Moulds and respiratory allergies. *Expressions*, **5**, 7-9.

Odabasi, Z., Paetznick, V.L., Rodriguez, J.R., *et al.* (2006) Differences in beta-glucan level in culture supernatants of a variety of fungi. *Med. Mycol.*, **44**, 267-272.

Odom, J.D., and DuBose, G. (2000) *Commissioning Buildings in Hot Humid Climates: Design and Construction Guidelines*. Fairmont Press, Lilburn, GA.

Ohno, N., Miura, T., Miura, N.N., *et al.* (2001) Structure and biological activities of hypochlorite oxidized zymosan. *Carbohydr. Polymers*, **44**, 339-349.

Pacheco-Sanchez, M., Boutin, Y., Angers, P., *et al.* (2006) A bioactive β 1→3, β 1→4-β-D glucan from *Collybia dryophila* and other mushrooms. *Mycologia*, **98**, 180-185.

Palmgren, U., Ström, G., Blomquist, G., and Malmberg, P. (1986a) Collection of airborne micro-organisms on Nucleopore filters, estimation and analysis – CAMNEA method. *J. Appl. Bact.*, **61**, 401-406.

Palmgren, U., Ström, G., Malmberg, P., and Blomquist, G. (1986b) The Nucleopore filter method: a technique for enumeration of viable and nonviable airborne micro-organisms. *Am. J. Ind. Med.*, **10**, 325-327.

Park, J.H., Cox-Ganser, J.M., Kreiss, K., *et al.* (2008) Hydrophilic fungi and ergosterol associated with respiratory illness in a water-damaged building. *Environ. Health Perspect.*, **116**, 45-50.

Pasanen, A.L., Pasanen, P., Jatunen, M.J., and Kallioski, P. (1991) Significance of air humidity and air velocity for fungal spore release into the air. *Atmos. Environ.*, **25A**, 459-462.

Pietarinen, V.M., Rintala, H., Hyvärinen, A., *et al.* (2008) Quantitative PCR analysis of fungi and bacteria in building materials and comparison to culture-based analysis. *J. Environ. Monit.*, **10**, 655-663.

Pitkäranta, M., Meklin, T., Hyvärinen, A., *et al.* (2008) Analysis of fungal flora in indoor dust by ribosomal DNA sequence analysis, quantitative PCR, and culture. *Appl. Environ. Microbiol.*, **74**, 233-244.

Prezant, B., Weekes, D., and Miller, J.D., (eds.) (2008) *Recognition, Evaluation and Control of Indoor Mold*. American Industrial Hygiene Association, Fairfax, VA.

Pugh, G.J.F. (1980) Strategies in fungal ecology. *Trans. Br. Mycol. Soc.*, **75**, 1-14.

Qian, J., Ferro, A.R., and Fowler, K.R. (2008) Estimating the resuspension rate and residence time of indoor particles. *J. Air Waste Manag. Assoc.*, **58**, 502-516.

Rabinovitch, N., Liu, A.H., Zhang, L., *et al.* (2005) Importance of the personal endotoxin cloud in school-age children with asthma. *J. Allergy Clin. Immunol.*, **116**, 1053-1057.

Ryan, T.J., Whitehead, L.W., Connor, T.H., and Burau, K.D. (2001). Survey of the Asp f 1 allergen in office environments. *Appl.*

Occup. Environ. Hygiene, **16**, 679-684.

Rylander, R., and Lin, R.H. (2000) (1→3)-Beta-D-glucan – relationship to indoor air-related symptoms, allergy and asthma. *Toxicology,* **152**, 47-52.

Saldanha, R., Manno, M., Saleh, M., *et al.* (2008) The influence of sampling duration on recovery of viable fungi using the Andersen N6 and RCS Biotest bioaerosol samplers. *Indoor Air,* **18**, 464–472.

Sander, I., Fleische, C., Borowitzki, G., *et al.* (2008) Development of a two-site enzyme immunoassay based on monoclonal antibodies to measure airborne exposure to (1→3)-beta-D-glucan. *J. Immunol. Meth.,* **337**, 55-62.

Saraf, A., Larsson, L., Burge, H., and Milton, D. (1997) Quantification of ergosterol and 3-hydroxy fatty acids in settled house dust: comparison with fungal culture and determination of endotoxin by a *Limulus* amebocyte lysate assay. *Appl. Environ. Microbiol.,* **63**, 2554-2559.

Schram-Bijkerk, D., Doekes, G., Douwes, J., and Boeve, M. (2005) Bacterial and fungal agents in house dust and wheeze in children: the PARSIFAL study. *Clin. Exper. Allergy,* **35**, 1272-1278.

Schram-Bijkerk, D., Doekes, G., Boeve, M., and Douwes, J. (2006) Nonlinear relations between house dust mite allergen levels and mite sensitization in farm and nonfarm children. *Allergy,* **61**, 640-647.

Smith, B., King, E., Belisle, D., *et al.* (2009) Quantitation of *Stachybotrys chartarum* Sch34 antigen using ELISA and fluorescent multiplex array technology. *J. Allergy Clin. Immunol.,* **123**, S172-S172

Stanevich, R., and Petersen, M. (1990) Effect of sampling time on airborne fungal collection. *Indoor Air '90: Proceedings of the 5th International Conference on Indoor Air Quality and Climate,* Vol. 2, CMHC, Ottawa, Canada, pp. 91-95.

Sussman, A.S. (1968) Longevity and survivability of fungi. In G. Ainsworth, and A.S. Sussman, (eds.), *The Fungi,* Vol. 2, Academic Press, New York, pp. 12-20.

Swaebly, M., and Christensen, C.M. (1952) Moulds in house dust, furniture stuffing and the air within houses. *J. Allergy,* **23**, 370-374.

Taylor, A. (1986) Some aspects of the chemistry and biology of the genus *Hypocrea* and its anamorphs, *Trichoderma* and *Gliocladium. Proc. Nova Scotia Inst. Sci.,* **36**, 27-58.

Trakumas, S., Willeke, K., Reponen, T., and Trunov, M. (1998) *Particle Cut-Size Evaluation of the Air-O-Cell Sampler.* Accessible at http://www.zefon.com/analytical/download/cutsize.pdf.

Trechsel, H.R. (1994) *Moisture Control in Buildings.* American Society for Testing Materials, Philadelphia, PA.

Tsai, S.M., Tang, C.S., Moffett, P., and Puccetti, A. (1999) A comparative study of collection efficiency of airborne fungal matter using Andersen single-stage N6 Impactor and the Air-O-Cell cassettes. In G. Raw, C. Aizlewood, and P. Warren, (eds.) *Pro-*

ceedings of Indoor Air '99, Edinburgh, Scotland. The 8th International Conference on Indoor Air Quality and Climate, Vol. 2, Construction Research Communications Ltd, London, pp. 776-781.

Tsay, A., Williams, L., Mitchell, E.B., and Chapman, M.D. (2002) A rapid test for detection of mite allergens in homes. *Clin. Exp. Allergy,* **32**, 1596-1601.

Upton, S.L., Mark, D., Douglass, E.J., *et al.* (1994) A wind tunnel evaluation of the physical sampling efficiencies of three bioaerosol samplers. *J. Aerosol. Sci.,* **25**, 1493-1501.

Vesper, S.J., Wymer, L.J., Meklin, T., *et al.* (2005) Comparison of populations of mould species in homes in the UK and USA using mould-specific quantitative PCR. *Lett. Appl. Microbiol.,* **41**, 367-373.

Vesper, S.J., McKinstry, C., Haugland, R.A., *et al.* (2007) Relative mouldiness index as predictor of childhood respiratory illness. *J. Expo. Sci. Environ. Epidemiol.,* **17**, 88-94.

Vinale, F., Marra, R., and Scala, F. (2006) Major secondary metabolites produced by two commercial *Trichoderma* strains active against different phytopathogens. *Lett. Appl. Microbiol.,* **43**, 143-148.

Warcup, J.H. (1965) Growth and reproduction of soil microorganism in relation to substrate. In R.F. Baker and W.C. Snyder, (eds.), *Ecology of Soil-borne Plant Pathogens,* University of California Press, Berkeley, CA, pp. 52-68.

Womiloju, T.O., Miller, J.D., Mayer, P.M., and Brook, J.R. (2003) Methods to determine the biological composition of particulate matter collected from outdoor air. *Atmos. Environ.,* **37**, 4335-4344.

Womiloju, T.O., Miller, J.D., and Mayer, P. (2005). Phospholipids from some common fungi associated with damp building materials. *Anal. Bioanal. Chem.,* **384**, 972-979.

White, D.C. (1988) Validation of quantitative analysis for microbial biomass, community structure and metabolic activity. *Adv. Limnol.,* **31**, 1-18.

Young, J.C. (1995). Microwave-assisted extraction of the fungal metabolite ergosterol and total fatty acids. *J. Agric. Food Chem.,* **43**, 2904-2910.

Xu, J., Liang, Y., Belisle, D.P., and Miller, J.D. (2008) Characterization of monoclonal antibodies to antiantigenic protein from *Stachybotrys chartarum* and its measurement in house dust. *J. Immunol. Meth.,* **332**, 121-128.

Zock, J.P., Jarvis, D., Luczynska, C., *et al.* (2002) European Community Respiratory Health Survey: housing characteristics, reported mould exposure, and asthma in the European Community Respiratory Health Survey. *J. Allergy Clin. Immunol.,* **110**, 285-292.

Zureik, M., Neukirch, C., Leynaert, B., *et al.* (2002) European Community Respiratory Health Survey. Sensitisation to airborne moulds and severity of asthma: cross sectional study from European Community Respiratory Health Survey. *Br. Med. J.,* **325** (7361), 411-414.

CHAPTER 4.2

Molecular methods for bioaerosol characterization

Richard C. Summerbell[1,2,] Brett J. Green[1,2,] Denis Corr[3] and James A. Scott[1,2]

[1]*Dalla Lana School of Public Health, University of Toronto, Canada;*
[2]*Sporometrics Inc., Toronto, Canada;*
[3]*Corr Associates, Hamilton, Ontario, Canada.*

POLYMERASE CHAIN REACTION

Since the work of Maddox (1857), Cunningham (1873), Miquel (1883) and others in the latter part of the 19th century, the detection and classification of bioaerosols has been dominated by methods based on culture and microscopy. The invention of polymerase chain reaction (PCR) in the 1980s transformed all of biology, including the study of bioaerosols. At first, this technique for amplifying genes was primarily used for single species, sampled as tissue specimens or as living cultures (Sontakke *et al.* 2009). Very rapidly, however, scientists realized that by using conservative gene regions as primer areas, they could amplify an indefinitely large diversity of different species from complex mixtures of organisms without *a priori* knowledge of their DNA sequences (Nocker *et al.* 2007). Thus, a very powerful technique for environmental investigation had been made available. By adjusting the primers used, amplification could be targeted at groups of organisms with any level of specificity: classes, orders, families, genera, species or strains/individuals. In cases where there was a strategic advantage in limiting the diversity elucidated from complex environmental samples, specific primers could be targeted at one, several, or multiple species (Haugland *et al.* 2004). These primers could then be used as a battery to investigate single species or to simultaneously look for two to four species at a time, as is made possible in some quantitative 'real time' PCR techniques.

There were numerous practical problems, as with any newly developed technology. Chemicals in environmental materials often inhibited amplification reactions, and new techniques and kits had to be developed for DNA extraction from various substrates,

e.g. soil (Schneegurt *et al.* 2003). Conservative primers used on environmental materials often elucidated a far greater range of DNA types than investigators were prepared to deal with. Very time-consuming studies involving the cloning of different DNA amplicons from PCR samples often resulted in sequences that were mostly uninterpretable, belonging to little known organisms such as unclassified soil amoeba species. Nonetheless, thanks to the continual deposition of even uninterpretable sequences in GenBank, a baseline was laid down for future comparison, and possibly for eventual taxonomic resolution of unknown types of organism, depending on funding levels for taxonomic work.

Techniques such as denaturing gradient gel electrophoresis (DGGE) and temperature gradient gel electrophoresis (TGGE) were developed that allowed improved resolution of estimates of the numbers of prevalent species in natural materials and their relative abundance (Smit *et al.* 1999, Nakatsu and Marsh 2007). Again, with very delicate and time-consuming techniques, individual sequences could be retrieved from DGGE gels and analysed in order to determine the identity of the organisms. In all environmental PCR-based techniques, the possibility of skewing of results by differential affinity of different organisms for primer sequences was difficult to rule out. This potential for bias was parallel to the potential for sampling bias in culture-based studies of microorganisms, and it was repeatedly observed that the similarity between culture and PCR results in environmental samples was not high (Lynch and Thorn 2006). Nonetheless, it was clear that every sequence elucidated by PCR truly identified a DNA type that was present in the sample, as long as procedural contamination was well-controlled, though the organisms elucidated might be autochthonous or allochthonous in the substrate studied, and might be alive or dead. Indeed, the high ability of PCR to detect dead materials with partially intact DNA diminished its usefulness in some epidemiological studies where only living, disease-causing inoculum was of interest.

Especially initially, there was a strong tendency for such studies based on gene sequences to discover that organisms were misclassified, or that entities thought to be single species or genera were actually multiple taxa, sometimes even very distantly related taxa. Taxonomic work in the last two decades has resolved some of these problems, and moved some remarkably misclassified organisms, such as microsporidia, to the kingdom, class, order, genus, etc., where they belong. In the case of microsporidia, they were misclassified as kingdom Protista but were actually members of kingdom Fungi (Lee *et al.* 2008). All such endeavours have clarified sequence-based environmental analyses and made them more efficient and effective.

In the atmosphere, biogenic particles form a major component of the suspended solid materials that are present (Després *et al.* 2007). These particles include intact biological structures such as fungal spores, pollen grains, microalgae, bacteria, viruses and dormant, desiccated microarthropods and protozoans, as well as partial structures such as broken fragments of fungal spores and hyphae, plant trichomes and fibres, and animal danders. Such particles not only disperse organisms, ranging from ecologically beneficial to epidemic-causing, but also have a significant effect on light dispersion in the atmosphere and in the nucleation of cloud droplets and ice nuclei, essential in the genesis of normal rainfall and other atmospheric precipitation.

A preliminary survey of DNA signatures in fine particulate matter (PM$_{2.5}$, aerodynamic diameter <2.5 μm) in air by Després *et al.* (2007) showed that the majority of sequences that could be obtained were bacterial, particularly α, β, and γ-proteobacteria as well as actinobacteria, followed by fungi (mostly Ascomycota) and plants. One animal sequence was obtained corresponding to a spore-forming protozoan type in the *Alveolata apicomplexa* complex.

Microorganisms, including bacteria, fungi, microalgae and protozoans, often have minute vegetative cells or spores that readily become airborne. For this reason, they are disseminated very readily and, essentially, in many cases may grow anywhere on our planet where conditions for their growth are favourable. Unlike animals and higher plants, they may have very

few limitations in geographic range, other than being limited to growing where conditions are supportive. There are notable exceptions for microorganisms that have a specialized relationship with a particular type of range-limited species, whether it be a plant or an animal. For example, specialized parasites, symbionts and decay organisms may only occur in and around the areas where their partners or hosts live. There are also some known, but poorly understood, range limitations in microorganisms, as seen in disease-causing fungi such as the agents of histoplasmosis and blastomycosis (Rippon 1988).

The state of the art with regard to PCR analysis of organism propagules or parts potentially represented in airborne particulates is reviewed below on a group-by-group basis.

An important extension of PCR analysis has been made possible by the use of high density microarrays (Wilson *et al.* 2002, Brodie *et al.* 2007, De Santis *et al.* 2007). In some microarray techniques, arrays are exposed to DNA obtained from environmental samples using the same sorts of amplification, e.g. 16S rDNA, as are used in the preparation of clone libraries. In others, unamplified genomic DNA obtained directly by extracting the samples may be used (Avarre *et al.* 2007). This latter technique, however, is in the very early stages of development.

Yet another level of development lies in making clone libraries of unamplified genomic DNA from environmental samples and then analyzing large numbers of the clones. The clones can be screened for affinity with taxonomically informative genes such as 16S rDNA or functional genes such as cellulases, or can be randomly sub-sampled for sequencing, or even sequenced *in toto* (Handelsman 2004). This collection of approaches, called 'metagenomics,' can be used to explore microbial diversity or to look for particular groups of enzymes or other genetic features of interest. In addition, when large numbers of sequences are obtained in metagenomic studies, cluster analyses of various kinds can be used to analyse the data (Xu 2006, Li *et al.* 2008). This possibility gives rise to a potential for a very detailed characterization of environments, and the statistical distinction of one environment from another in samples (Tringe *et al.* 2005, Dinsdale *et al.* 2008).

Cutting-edge systems for high-throughput sequencing and bioinformatics processing have further revolutionized metagenomic analyses. For example, Dinsdale *et al.* (2008) compared nearly 15 million sequences from 45 microbiomes (bacterial habitats) and 42 viromes (environmental or medical sites where viruses are present). Currently available technologies like GS FLX Titanium pyrosequencing (Roche Diagnostics) allow the simultaneous sequencing of over 500 Mbp. In a single run the PCR-amplified DNA barcode region can be sequenced to exhaustion in up to 128 community DNA samples with over 10,000 times redundancy. This method is suitable for the metagenomic characterization of low to moderately complex microbial communities. Communities of higher complexity can be sequenced to exhaustion by processing fewer samples simultaneously.

POLLEN, SPORES AND PARTS OF PLANTS

The pollen grains of wind-pollinated (anemophilous) plants are among the most common airborne particulates, and have been intensively studied for decades. The air also includes other plant parts such as pollen grains of insect-pollinated (entomophilous) plants, spores of mosses, ferns and lycopods (clubmosses), and various trichomes (distinctively shaped leaf hairs) and other minute fragments broken away from leaves, stems and decaying wood (Sarna and Govil 1979, Kasprzyk 2004). It is not clear how reliably pollen and trichomes collected from air can be studied using molecular techniques. DNA extraction from such materials often involves distinct processes, since the breakage of heavy and chemically unusual cell walls is usually involved. For example, Zhou *et al.* (2007) picked individual pollen grains from sieved materials and cracked them open between two glass slides prior to doing DNA extraction. Chen *et al.* (2008) developed a protocol for DNA isolation from single pollen grains involving special forceps and alkali/detergent lysis of the walls. Sometimes plant DNA interpreted as deriving from pollen has been detected indirectly; for example, Brodie *et al.* (2007) detected plant chloroplast DNA while DNA microarray probing bacterial-type DNA in urban air (chloroplast DNA is related to cyanobacterial DNA) and stated that this was likely to be from pollen grains. Després *et al.* (2007) detected plant nuclear DNA in urban air samples while using 18S RNA primers intended to amplify animal sequences; these sequences, which included angiosperm, pine, and moss sequences, were only detected in the spring and were interpreted as being from pollen and, in the case of moss, spores. Various studies making bacterial 16S ribosomal DNA clone libraries from collected aerosols have detected plant chloroplast DNA but have not remarked on its probable source (Radosevich *et al.* 2002, Rintala *et al.* 2008). No study has been carried out to determine

the degree of correlation between the chloroplast and nuclear plant sequences obtained and the diversity and number of pollen grains and spores detectable by other techniques.

Where trichomes are concerned, there is no published information relating to their molecular detectability from aerosol sources. Molecular studies on these structures are relatively common, although mostly relating to the genomics of their differentiation. For example, there is a trichome harvesting technique for leaves of the genetic model plant *Arabidopsis thaliana* (Marks *et al.*, 2008), designed to facilitate genomic studies. A typical example of a trichome genomic study involving DNA extraction from these thick-walled structures is that of Liu *et al.* (2006) on proteinase inhibitor IIb expression in trichomes of nightshade, *Solanum americanum*. The more data are collected on DNA extraction techniques for these structures, the more accessible they will become as data sources for molecular investigators of bioaerosols.

The minute spores of non-vascular plants such as bryophytes, ferns and lycopods are capable of long-distance airborne dispersal and thus regularly make up a small but significant component of total airborne particulates (Stoneburner *et al.* 1992, Kasprzyk 2004, Sundberg 2005). About 4% of total pollen belongs to entomophilous vascular plant taxa such as willows, maples, elders (*Sambucus*), herbaceous plantain (*Plantago*), and *Tilia* spp. (basswood, linden, lime). Long range dispersal, however, is inhibited in that pollen grains of these species are often fused together into relatively heavy groups by the sticky material known as pollenkitt (Kasprzyk 2004).

Generally, PCR-based plant identification involves sequencing one or both of the two recognized identification barcode regions, the nuclear ITS region and the chloroplast *trnH-psbA* intergenic spacer (Kress *et al.* 2005).

FUNGI

Fungal spores and fragments are among the most common components of total bioaerosols, and are also among the most intensively studied materials in terms of molecular detection. This level of study can be partly ascribed to the high levels of interest in indoor fungal proliferation and fungal aerosol levels in 'sick' and healthy buildings. Most molecular techniques for studying airborne fungi were developed for indoor materials; however, these can readily be adapted for outdoor materials as well. In many cases, the use of outdoor samples as controls is common in the study of indoor aeromycota.

The identification of indoor fungi by means of diagnostic sequences has been a well-established and commonly used technique since the late 1990s (Haugland *et al.* 2004). Though initially developed for identifying *in vitro* cultures, many approaches can readily be modified to analyse the fungal contents of air samples collected using filter membranes, cyclone samplers or jet-to-plate impactors. Equally, these methods can be used to analyse vacuum-collected dusts. In an early example, Haugland and Heckman (1998) introduced specific primers for the important indoor fungus *Stachybotrys chartarum*. This study was shortly followed by the first of a series of species-specific quantitative PCR (qPCR) studies for indoor fungi, based on use of the TaqMan® fluorogenic probe system combined with the ABI Prism® Model 7700 Sequence Detector (Haugland *et al.* 1999). In this study, *S. chartarum* was again the object of interest; qPCR counts of *S. chartarum* conidia were found to be highly comparable to counts obtained with a haemocytometer. The method was further developed by Roe *et al.* (2001) for direct quantitative analysis of *S. chartarum* in household dust samples.

This TaqMan-based qPCR methodology was extended over subsequent years into a broad ranging methodology encompassing many major indoor air fungal groups (Haugland *et al.*, 2004) and a variety of applications. In conjunction, key studies considered how best to extract DNA for qPCR and related PCR-based analyses of indoor fungi (Williams *et al.* 2001, Haugland *et al.* 2002, Kabir *et al.* 2003). Meklin *et al.* (2004) employed qPCR to evaluate indoor dust from the presence of 82 mould species or species complexes, including members of the genera *Aspergillus, Cladosporium, Penicillium, Trichoderma* and *Ulocladium*, in addition to *Stachybotrys* and the closely related *Memnoniella*. Comparisons between techniques showed that culturing underestimated numbers of key *Aspergillus* species by two to three orders of magnitude. "Mouldy homes" could be distinguished from putatively uncontaminated "reference homes" using mould-specific qPCR (MSQPCR)-based quantification. An online information page about the now widely used technology developed by R.A. Haugland, S.J. Vesper and other members of the US Environmental Protection Agency (US-EPA) group can be found at <http://www.epa.gov/nerlcwww/moldtech.htm>. Currently, 116 primer/probe combinations have been described that target 130 species. Commercial use of these primer sequences for fungal detection requires

a licensing agreement with US-EPA. Similar restrictions do not, however, apply to the use of these sequences for non-commercial, research purposes.

The results of MSQPCR have been used to calculate a ratio of species associated with water damage to those arising from outdoor sources (e.g. phylloplane moulds), unrelated to indoor dampness. This ratio, the Environmental Relative Moldiness Index (ERMI), provides a single value between -10 and +20, that describes the potential of an indoor growth source. Although this approach is aesthetically appealing for its simplicity, apparent objectivity and ease of use, the reductive nature of the approach excludes subtle features of the dataset that may otherwise lead an experienced investigator to a different conclusion. Until more data are available to compare ERMI with the results of building inspections carried out by experienced assessors, the technique must be considered investigational.

More medically oriented environmental studies developed TaqMan qPCR for *Aspergillus fumigatus* conidia in filtered air samples (McDevitt *et al.* 2004, Goebes *et al.* 2007). More recently, the accuracy of qPCR for *A. fumigatus* detection in hospitals has been stringently tested by comparison with green fluorescent protein (GFP)-expressing conidia of this species (McDevitt *et al.* 2005).

In conjunction with other modern techniques such as a quantitative protein translation assay for trichothecene toxicity, qPCR was used in an evaluation of *Stachybotrys* from a house where a case of idiopathic pulmonary haemosiderosis had occurred (Vesper *et al.* 2000). However, haemocytometer counts of the relatively large and conspicuous *Stachybotrys* conidia were relied upon on for quantitation in the data used in the subsequent analysis. A later, much more detailed qPCR analysis of homes where pulmonary haemosiderosis had occurred showed that *S. chartarum* was part of a group of species, also including *A. fumigatus* and several other *Aspergillus* spp., that was significantly elevated in quantity in dust samples in affected homes (Vesper *et al.* 2004). Species abundant in affected homes tended to be haemolytic in *in vitro* testing, whereas the common species associated with reference homes were generally not haemolytic. Another significant application of the qPCR technique was to sensitively monitor *Aspergillus* contamination during hospital renovation (Morrison *et al.* 2004) and related infection control applications.

Relatively recent developments have included detailed studies of the fungal contents of dust from various sources, including studies optimizing qPCR to overcome chemical PCR inhibitors in dust (Keswani *et al.* 2005, Vesper *et al.* 2005). Multi-species qPCR has been applied in comparison of outdoor with indoor air (Meklin *et al.* 2007) and in analysing both fungi and bacteria in building components such as chipboard, paper materials and insulation (Pietarinen *et al.* 2008). These methods have been combined with cloning and sequencing in the analysis of indoor dust to reveal a high prevalence of taxa of the Malasseziales (Ustilaginomycotina), the lipophilic basidiomycetous yeast inhabiting human skin (Pitkäranta *et al.* 2008).

In outdoor air studies, qPCR-based techniques have been used to detect airborne plant pathogens such as *Monilinia fructicola,* the cause of brown rot of stone fruits (Luo *et al.* 2007) and *Fusarium circinatum*, the cause of pitch pine canker (Schweigkofler *et al.* 2004). Spore levels of toxin-producing fungi in airborne grain dust have also been monitored using qPCR (Halstensen *et al.* 2006). In surveys involving large numbers of fungal taxa, cloned libraries have been made from bulk amplification of small subunit ribosomal DNA. The results, however, suggest skewing of the data: a small-subunit library study conducted by Fierer *et al.* (2008) using air samples from Boulder, Colorado, found that fungal sequences from the order Hypocreales (*Fusarium, Trichoderma* and relatives) were overwhelmingly predominant (90% + of sequences) in five libraries derived from aerosol samples. It can readily be seen in conventional air sampling that members of this order never show predominance on this scale; members of the Class Dothideomycetes (*Cladosporium, Alternaria*) are consistently predominant in outdoor air in temperate areas for most of the year (Gregory 1973). Dothideomycetous fungi have, however, spores with resistant, melanized cell walls and may withstand some DNA extraction methodologies even where bead-beating is used to rupture cells, as was done by Fierer *et al.* (2008). In the small number of fungal ribosomal internal transcribed spacer (ITS) sequences obtained by Després *et al.* (2007), *Cladosporium* and Basidiomycota sequences were found, consistent with conventional studies, but no Hypocrealean sequences.

Although qPCR offers major advantages over traditional approaches in the detection and enumeration of indoor fungi, a number of factors influence its performance. Because qPCR directly detects a target gene sequence, strains with variant sequences in either of the primer or probe regions may be missed (false negatives). Conversely, unintended taxa may inadvertently be detected based on their homology to the primer/probe sequences used (false positives).

Validation of the analytical specificity of the primer/probe sequences is greatly hampered by the prevailing lack of knowledge of fungal biodiversity, with an estimated 90% of species yet to be discovered (Hawksworth 2006).

A second limitation of qPCR-based techniques relates to the ability to predict biomass by the number of copies of the target sequence enumerated. In the simplest case in a haploid, uninucleate fungal cell, it is possible to establish a standard curve correlating total biomass to the number of copies of a given gene. The difficulty is that this assumption is not transitive to all cell-types of a given taxon, e.g. monokaryotic vs. dikaryotic hyphae, haploid vs. diploid or polyploid cells, nuclear vs. mitochondrial targets, nor is it directly transferable to other gene targets, which may occur in multiple copies (whose total number is species- or strain-specific) within a single genome copy, e.g. ribosomal subrepeat.

Lastly, qPCR-based methods cannot distinguish viable from non-viable material, or cell-bound from cell-free DNA. However, the condition of cellular material may be relevant to the interpretation of the result, e.g. in the risk assessment of reservoirs of potential agents of nosocomial infection. A recent elegant study by Vesper *et al.* (2008) used propidium monoazide to inactivate DNA from dead cells prior to using qPCR to quantify viable conidia. This approach and similar methods may hold promise in tailoring qPCR-based methods to the detection of cell-bound DNA.

Microarray studies have been outfitted to detect fungal DNA signatures in filtered aerosols from outdoor air. This entails loading the arrays with a wide variety of fungal DNA probe sequences suitable for detecting a reasonable proportion of the known biodiversity. Using such a technique, De Santis *et al.* (2005) found that propagules of the fungal phylum Ascomycota (inclusive of most common mould spores of outdoor air) and Basidiomycota (mainly represented by airborne spores of mushrooms and related fungi) were abundant in outdoor air from southern England. They could only be elucidated efficiently, however, when the filter-collected aerosol sample was subjected to long bead-beating times to release DNA from cells with thick walls. This was not especially inconvenient, however, since similarly long beating times were also required for some bacterial types.

BACTERIA

The organisms that have been most extensively studied in terms of molecular detection in the air are the bacteria. Bacteria are very difficult to identify from their phenotypic characters and, moreover, unculturable bacteria are common; therefore, identification with 16S ribosomal DNA sequences has been the gold standard since the early 1990s (Wilson *et al.* 1990). The application of this identification standard to aerosol samples, e.g. to bacteria on air filters, has developed only in recent years. The development of this area of work has been detailed in a review by Peccia and Hernandez (2006). Techniques used to study bacterial occurrence and diversity in outdoor air have been specifically reviewed by Kuske (2006).

Alvarez *et al.* (1995) developed basic collection and DNA purification techniques for spiked *E. coli* in samples from air, and also developed some dilution procedures for dealing with chemical inhibitors of PCR in the samples. Stark *et al.* (1998) developed a practical application for the technique for monitoring *Mycoplasma hypopneumoniae* levels in commercial swine houses, and refined the techniques for extracting DNA from filter samples. Later, species- or genus-specific techniques were developed for *Legionella* spp. in factory and office air, and *Mycobacterium* spp. above whirlpool facilities in public swimming baths (Pascual *et al.* 2001, Schafer *et al.* 2003). More recent developments on this theme have featured quantitative PCR techniques, to measure, for example, *Staphylococcus* species levels in poultry house air (Oppliger *et al.* 2008), and *Mycobacterium tuberculosis* levels in health care facilities (Chen and Li 2005). *Streptomyces* levels in household dust have also been studied using qPCR (Rintala and Nevalainen 2006). An *et al.* (2006) compared the use of specific primers for *E. coli* with universal primers for total bacterial load in a qPCR study of particulates from laboratory-generated air samples. Rinsoz *et al.* (2008) used three sets of broadly targeted primers to contrast qPCR results with those of epifluorescence microscopy and culture in air samples used to detect total bacterial, staphylococcal and total Gram-negative bacterial loads in poultry houses and wastewater treatment plants.

The sampling and detailed identification of a broad range of bacterial types in air samples began in the current decade, with the publication of studies that featured cloned 16S libraries (Radosevich *et al.* 2002, Maron *et al.* 2005, Després *et al.*, 2007) overlapping in time with studies featuring the use of microarrays (Wilson *et al.* 2002, Brodie *et al.* 2007). Cloned library-based studies were also made of dust as a 'natural' sediment from air (Rintala *et al.* 2008). A very distinctive array of bacteria was obtained by Birenzvige *et al.* (2003) in small-scale cloned libraries from air samples

taken in an urban subway (underground railroad), while another unique perspective was afforded by a cloned 16S library study of bacteria in biogas emitted by anaerobic waste digestors and landfill sites (Moletta *et al.* 2007). The bacterial diversity of the air of swine confinement houses was studied using a 16S cloned library by Nehme *et al.* (2008). An ecologically important extension of the cloned library technique was made by Fierer *et al.* (2008), who compared bacterial biodiversity from samples taken within a single geographic area with that of samples from widely separated geographic areas.

De Santis *et al.* (2007) studied samples of urban aerosol, subsurface soil and subsurface water, and found that high-density microarrays detected more bacterial biodiversity than 16S cloned libraries. Some bacterial phyla such as Nitrospira were only detected on the arrays, signalling that there may have been some bias in the amplifications used to generate the cloned libraries. On the other hand, cloned libraries detected novel bacterial types, something that is not possible with microarrays, where detected species must always match pre-selected oligonucleotide targets linked to the arrays. De Santis *et al.* (2005) found that the bacterial types detected by microarrays also depended on how vigorous cell disruption was in DNA extraction procedures from the aerosol samples. Gentle extractions based on short bead-beating times favoured detection of Mycoplasmatales and Burkholderiales, whereas high bead-beating times favoured Vibrionales, Clostridiales, and Bacillales. This is not surprising, since the former group contains many taxa with notably thin cell walls and the latter group contains endospore-formers and other bacterial types generating thick-walled cells. As mentioned above, fungi detected in the same study tended to be best detected with the long bead-beating times used in vigorous disruption. Various potential biases influencing results of cloned library and microarray studies for complex microbial communities have been reviewed by Avarre *et al.* (2007).

An overview of the diversity of types in air samples was obtained in some studies by means of multispecies DNA profiling techniques such as automated-ribosomal intergenic spacer analysis, A-RISA (Maron *et al.* 2005), single-strand conformational polymorphism (SSCP) profiling (Moletta *et al.* 2007), fluorescent heteroduplex profiling (Merrill *et al.* 2003), DGGE (Nehme *et al.* 2008) and terminal restriction fragment length polymorphism analyses, T-RFLP (Després *et al.* 2007). Such techniques tend to generate one characteristically sized DNA band per species present in a multi-species sample; the analysis thus gives a fingerprint of the entire collection of DNA-extracted, successfully amplified organisms present. Working entirely with the culturable fraction of air samples, Hernlem and Ravva (2007) used flow cytometry to separate viable bacterial cells from other particulates mixed with them in a cyclone-sampled outdoor air sample, and then identified cultures using 16S rDNA sequencing.

VIRUSES

Genetic material, DNA or RNA, makes up a much larger component of viruses than it does in other microorganisms, and the physical and chemical shielding that walls this genetic material off from the outside environment is often easily removed. Viruses thus tend to be very accessible to molecular technologies based on the study of nucleic acids. RNA-based viruses are more difficult to work with than DNA-based viruses because of the chemically labile nature of RNA, and the need to make complementary DNA (cDNA) copies of the genetic material prior to conventional analyses.

The PCR-based study of viruses in air began with the detection of varicella zoster virus (VZV) and cytomegalovirus (CMV, also known as human herpes virus 5, HHV-5) in health care settings (Sawyer *et al.* 1994, McCluskey *et al.* 1996, Suzuki *et al.* 2004). These are both relatively robust DNA viruses in the herpesviruses group. RNA viruses in indoor office air, specifically rhinoviruses causing common cold, were tackled by Myatt *et al.* (2004). These authors used special lysis buffer with carrier RNA to rinse filters, and then converted the captured viral RNA to cDNA with the reverse transcriptase enzyme. A qPCR technique for aerosols of another important RNA virus, influenza virus, was developed by Blachere *et al.* (2007) using the NIOSH miniature cyclone samplers.

Veterinary scientists have taken a leading role in aerosolized virus studies, particular in studies of the recently emerged porcine reproductive and respiratory syndrome virus, PRRSV. Quantitative reverse-transcriptase PCR has been the gold standard in these studies, coupled with impinger air sampling by Hermann *et al.* (2007) and Micro-Tek centrifugal sampling into minimal viral medium by Cho *et al.* (2007).

Several relatively recent crises have also strengthened molecular biological investigation of viruses in aerosols. For example, filters and real-time PCR (rt-PCR) have been coupled with multiplexed immunoassays in an automated pathogen detection system (APDS) designed to monitor air for viruses potentially disseminated by bioterrorists (Hindson *et al.* 2005). The same system was also set up to detect bacteria and toxins. The severe acute respiratory syndrome (SARS) outbreak resulted in PCR-based studies on aerial dissemination of the virus (Xiao *et al.* 2004, Booth *et al.* 2005). Concern about avian influenza led to development of specific rtPCR techniques (Payungporn *et al.* 2004), but there has been very little work done specifically on testing aerosols for these viruses.

Aerosol methods have been developed in studies on viruses in human breath. Techniques have ranged from breathing directly into PCR reaction tubes for detection of HHV-6 (Kelley *et al.* 1994) to use of individual mask samplers for cold viruses (Huynh *et al.* 2008) and of Teflon filters for quantitative PCR of influenza virus coupled with employment of an optical particle counter to measure exhaled particle concentrations (Fabian *et al.* 2008). Some of these techniques aimed at specific types of virus may suggest ways of improving the methods used for studying the overall virus load of aerosol samples.

More general studies have also been conducted, such as that of Sigari *et al.* (2006) based on PCR detection of a number of viruses - especially enteroviruses and reovirus - from aerosols around a sewage treatment facility. The most generalized detection of viruses, however, lies in the metagenomics studies mentioned above, particularly the study of Dinsdale *et al.* (2008). In this study, viral DNA only, not RNA, was isolated from the smallest-size fraction of a serially filtered environmental sample, and included a wide variety of viruses, phages and prophages. This was carried out for 42 distinct sites including habitats like soil, hypersaline water, marine water, fresh water, microbialites (stromatolites, thrombolites), terrestrial animal guts and surfaces, corals and mosquitoes. With suitable filtering, the techniques would be readily adaptable for air as well as for RNA viruses in the airborne particulate material.

MALDI-MS TECHNIQUES

Matrix-assisted laser desorption/ionization (MALDI) is a mass spectrometry (MS) technique that desorbs thermolabile, non-volatile organic compounds including protein, peptide, and glycoprotein biomarkers in addition to oligosaccharides, oligonucleotides and large organic molecules (Claydon *et al.* 1996, Fenselau and Demirev 2001). Unlike other conventional ionization methods that register biomarker ions in a narrow m/z (mass to charge ratio) range, MALDI is a 'softer' methodology in which sample molecules fragment and ionize following bombardment with a laser light. This method allows the analysis of large biomolecules that were not possible to detect using previous MS methods.

Samples are prepared by pre-mixing with a highly absorbing matrix that enables the transfer of laser energy into excitation energy (Domin *et al.* 1999). This process creates gas phase ions from the surface of the mixture; these ions are then pulsed into a vacuum chamber flight tube (Claydon *et al.* 1996). Positively or negatively ionized biomolecules can be generated and these can be analyzed with a time-of-flight (TOF) mass spectrometer equipped with an ion mirror that deflects ions by means of an electric field. Ion mirrors increase the ion flight path and thereby increase the resolving power of the method to measure masses from 2-40 kDa. For a typical MALDI-TOF MS analysis, mass spectra are expressed as a series of peaks that correspond to individual peptides. Each mass spectrum that is acquired using MALDI-TOF MS corresponds to a molecular fragment that has been released from the cell surface during laser desorption (Edwards-Jones 2000) The development of MALDI-TOF MS has provided a unique method for identifying individual microorganisms by rapidly producing representative spectra (Cain *et al.* 1994, Claydon *et al.* 1996, Domin *et al.* 1999, Fenselau and Demirev, 2001) and these can subsequently be used to discriminate between microorganisms within minutes of sample acquisition.

MALDI-TOF MS has been developed as a rapid method to analyze various biomolecules from both Gram-positive and Gram-negative bacterial cell extracts (Claydon *et al.* 1996, Welham *et al.* 1998, Claydon 2000). This technique has also been performed on whole cell extracts and intact cells (Krishnamurthy and Ross 1996, Welham *et al.* 1998, Lynn *et al.* 1999, Lay 2001, Edwards-Jones *et al.* 2000). The detection limits of MALDI-TOF MS for bacterial material are as low as 1×10^3 cells. Bacteria in vegetative growth (without spores) usually contain 20-60 peaks in the mass range 1-25 kDa. Characteristic mass spectrum peaks have been characterized for bacteria belonging to the Enterobacteriaceae (Lynn *et al.* 1999), *Mycobacterium tuberculosis* (Hettick *et al.* 2004) and *Staphylococcus aureus* (Edwards-Jones *et al.* 2000), but mass ranges may extend up to 40kDa (Welham *et al.* 1998). Larger mass ranges may improve the diagnostic discrimination of analyses performed on bacterial isolates. In these studies, Gram-negative bacteria from the Enterobac-

teriaceae (Welham *et al.* 1998, Claydon 2000) produced a range of characteristic biomarkers that could be used for identification. However, the presence of more than one bacterial species in a sample was shown to alter the mass spectral profiles, owing to different growth rates and production of interference-competitive metabolites by the bacteria (Cain *et al.* 1994). The identification of species-specific biomarkers is required to successfully differentiate between bacteria in a sample where mixed populations occur, such as normally unsterile body tissues (e.g. lung or epidermal tissue), contaminated water sources or air samples. Identification using specific biomarkers in heterogeneous samples has recently been achieved in the detection of *Escherichia coli* (Siegrist *et al.* 2007) and *Bacillus* spores (Pribil *et al.* 2005).

MS-based identification of microorganisms has mainly focused on pathogenic bacteria, and comparatively little has been published on MS-based strategies for fungal identification. The applications of physicochemical identification of fungi using MALDI-TOF-MS have recently been realized. The fungal kingdom includes diverse lineages of saprobic heterotrophs that are important biohazards in food-processing, agriculture and health. Traditional methods of identification involve subculturing inoculum onto nutrient media and identifying the resulting culture semi-subjectively according to the determinations of a trained microbiologist. In the clinical setting, this has proved laborious, time-consuming, and, with difficult species, error-prone. Quite often, cultures fail to develop observable phenotypic structures required for subjective classification. The advent of immunodiagnostic and molecular techniques has greatly improved diagnostic intervals for a few important pathogens such as *Cryptococcus neoformans* and *Coccidioides posadasii*. Welham *et al.* (2000) demonstrated the feasibility of identifying a variety of fungi via rapid MALDI-TOF MS-based procedures. Although fungi were shown to produce less complex spectra than bacteria, the results of analysis could still be used for discrimination. Unlike traditional methods, morphologically undistinguished hyphae and other simple structures could be used to make up sample preparations that could still be successfully identified (Hettick *et al.* 2008a,b). This capacity for rapid identification has generated much research attention for the further optimization of MALDI-TOF MS as a diagnostic method for fungal pathogens in the clinical setting (Hettick *et al.* 2008a,b).

For fungi, quality control studies have demonstrated that reproducible mass spectra are obtainable with various *Aspergillus* spp. when mass spectrometric matrices such as α-cyano-4-hydroxycinnamic acid and sinaptic acid are used (Li *et al.* 2000). The spectral profiles of these species are thought to be glycoprotein-based since much of the fungal cell wall consists of polysaccharides. However, smaller concentrations of proteins, lipids and polyphosphates are also present (Welham *et al.* 2000). These structural characteristics produce distinctive spectra in the mass range 2-13 kDa. Such spectral profiles have been used successfully in the food processing and agriculture industries to distinguish aflatoxin-producing and non-producing isolates of *Aspergillus flavus* and *A. parasiticus* (Li *et al.* 2000). Medically important *Aspergillus* spp. implicated in invasive disease have also been shown to have highly reproducible mass spectral fingerprints (Hettick *et al.* 2008a). Similar results have been shown for *Penicillium* spp. contaminating fruit surfaces (Li *et al.* 2000). The specificity of MALDI-TOF MS has stimulated interest in developing the methodology as a diagnostic tool for various fields including the identification of crop contaminants and the diagnosis of invasive fungal disease.

Reproducibility has been demonstrated for a range of sampling and culture conditions for bacteria and fungi. However, maintaining constant culture conditions is pivotal for the production of reproducible spectral profiles (Parisi *et al.* 2008). Variations in growth conditions, nutrient medium and growth time can strongly influence spectral reproducibility (Parisi *et al.* 2008). Standardization of sample preparation methods will enable the future development of algorithms that can be used to discriminate species in 'unknown' samples. This has been demonstrated successfully with the identification of various pathogenic bacteria (Hsieh *et al.* 2007) such as *Mycobacterium tuberculosis* (Hettick *et al.* 2006) and various *E. coli* types (Arnold and Reilly 1998, Bright *et al.* 2002). Although fungal cells at different stages of their lifecycle have been shown to have a similar mass spectra (Hettick *et al.* 2008a,b), a good statistical analysis of spectral stability over the life cycle has yet to be made for any fungus.

Other factors that require standardization include the solvent extraction system and mass spectrometric matrix used. These items may influence the quality of the mass spectral profiles and may significantly affect the quality of the data obtained. To date, there is no universal method recognized for preparing microorganisms for MALDI-TOF MS analysis. Some laboratories prepare samples with extensive washing steps to remove contaminants arising from the growth me-

dium, whereas others use a variety of different matrices and solvents (Domin *et al.* 1999). Matrices such as 2-(4-hydroxyphenylazo)benzoic acid and 2-mercaptobenzothiazole have been shown to yield the best spectral profiles and the best mass range of tested matrices (Domin *et al.* 1999).

Proteins may be released and made available for analysis by physically disrupting fungal or bacterial spores using sonication. This process, in addition to the use of an acid solvent, further lyses the microbial components and releases intracellular proteins that yield additional mass spectra. Intracellular lysates have been successfully used to discriminate between various bacteria (Cain *et al.* 1994) and fungi (Hettick *et al.* 2008a,b). In contrast, the presence of melanin has been shown to interfere with MALDI-TOF MS analysis. It has been hypothesized that the negative ions associated with melanin suppress the ionization that is needed to produce observable mass spectra. Heavily melanized fungi such as *Stachybotrys chartarum*, *Aspergillus niger*, *Alternaria alternata* and *Epicoccum nigrum* produce limited mass spectral profiles that do not enable reliable discrimination.

Although MALDI has emerged as a suitable physicochemical and immunodiagnostic technique for the analysis and identification of environmental microorganisms, the feasibility of this technique for the detection of bioaerosols remains questionable. Species specific biomarkers could be used to discriminate between bioaerosols in air samples; however, few studies have assessed the complex diversity of environmentally ubiquitous bacterial and fungal aerosols.

A very important practical consideration lies in the amounts of material required for successful analysis of airborne particulates. The limit of detection for bacterial and fungal materials in MALDI-TOF MS is 1×10^3 spores. In many environments, target microorganisms that may be specific to those environments (mushroom spores, fungal phytopathogen spores, bacteria) tend to occur in low concentrations relative to ubiquitous airborne taxa. The problems of discriminating between target taxa and common bioaerosols are challenging, requiring the development of complex algorithms to identify biomarker spectral profiles. These environmental surveillance considerations have not been explored to date and represent a significant future workload that will have to be completed before MALDI-TOF MS can be used for the detection of geographically localized bioaerosol sources.

IMMUNOCHEMICAL METHODS

A number of additional diagnostic techniques are available for detecting and quantifying a variety of bioaerosols that are associated with pollution plumes. These alternative techniques range from simple traditional methods, such as direct microscopy, to more advanced methods that utilize emerging quantitative immunodiagnostic technologies. Unlike MALDI-TOF MS and PCR, these methods provide a platform for rapid and reliable quantification of certain bioaerosols originating from the environment.

While direct microscopy has been the traditional method used to identify non-viable airborne particles collected on a filter membrane or adhesive backed tape (Flannigan 1997, Prezant *et al.* 2008), it requires a trained microbiologist to examine deposited particles and reproductive structures and classify and quantify them according to their morphological phenotypic features. This technique has several limitations, such as inter-analyst subjectivity and difficulties associated with the identification of incompletely developed asexual fungal spores and simple amerospores, as well as nondescript fungal particulates like hyphal fragments and submicron particulate material derived from fractured cells (Green *et al.* 2006b). Nonetheless, this direct microscopy is widely used in the fungal exposure assessment and indoor environmental remediation fields.

The detection of viable bacteria and fungi in bioaerosols using culture techniques mentioned in Chapters 1.1 and 4.1 is an even more widely utilized method (Flannigan 1997, Prezant *et al.* 2008). However, it takes several hours or days of exponential growth for development of colonies, from which, if they ultimately produce recognizable reproductive structures such as phialides or spores, a trained microbiologist or technician will be able to identify taxa and incorporate this into quantitation procedures. However, successful growth and the emergence of a colony are dependent on the nutrient medium selected, the incubation conditions and the outcome of inter-colony competition. Successful identification then depends on the professional skills of the technician (Prezant *et al.* 2008).

Viable and non-viable methods of direct identification offer an inexpensive alternative to MALDI-TOF and PCR techniques, are widely utilized in the exposure assessment community and are often regulated by professional organizations such as AIHA and ASTM (Prezant *et al.* 2008). Non-viable methods in particu-

lar would provide a relatively inexpensive alternative for the identification and quantification of bioaerosol sources that have unique morphology and that are localized within regions expelling pollutants.

The advent of immunological detection technologies has, however, enabled the quantification of various microorganisms without the need for subjective visual assessment of phenotypic features (Trout *et al.* 2004, Schmechel *et al.* 2008). These techniques provide quantification of variously sized particles ranging from whole spores but remain validly interpretable to much smaller detection thresholds that include the measurement ranges of typical submicron biological particulates (Douwes *et al.* 1997, Chew *et al.* 2001, Renstrom 2002, Schmechel *et al.* 2003, Green *et al.* 2005a,b,c, Schmechel *et al.* 2005, Green *et al.* 2006a,b,c, Schmechel *et al.* 2008). The development of monoclonal and polyclonal antibodies (mAbs and pAbs) has been instrumental in achieving assay specificity as well as conferring improved sensitivity (Schmechel *et al.* 2003, 2008). Development of mAbs and pAbs has also enabled the formulation of many detection techniques such as enzyme linked immunosorbent assay, ELISA (Trout *et al.* 2004, Schmechel *et al.* 2008), flow cytometry (Rydjord *et al.* 2007), direct and indirect immunostaining (Popp *et al.* 1988, Reijula *et al.* 1991, Takahashi *et al.* 1993, Takahashi and Nilsson 1995, Portnoy *et al.* 1998, Schmechel *et al.* 2003), western blot (Barnes *et al.* 1993, Cruz *et al.* 1997, Bisht *et al.* 2004), and the halogen immunoassay, HIA (Mitakakis *et al.* 2001, Poulos *et al.* 2002, Green *et al.* 2003, Mitakakis *et al.* 2003, Green *et al.* 2006a,c).

ELISAs in combination with mAbs or pAbs have been widely utilized in exposure assessment studies to quantify various microbial pathogens, bioaerosols, and aeroallergens (Schmechel *et al.* 2008). The method is designed for quantifying various biomolecules such as proteins, peptides, antibodies and hormones. In an ELISA, antigens are immobilized on a solid protein-binding surface, and the antigen is then complexed with an antibody conjugated with an enzyme. Detection is accomplished by incubating the enzyme complex with a substrate that produces a quantifiable product. The quantity of antigen in each sample is then determined by reference to a standard curve of controls.

To date, the development of specific mAbs has not kept up with the diverse spectrum of microorganisms in the outdoor environment (Schmechel *et al.* 2003, Trout *et al.* 2004, Schmechel *et al.* 2008). Many developed mAbs also remain uncharacterized and the extent of cross-reactivity with other common bioaerosols is often overlooked. Schmechel and colleagues (Schmechel *et al.* 2003, 2008) recently identified the limitations associated with using uncharacterized antibodies. Prior to this study, an anti-*Alternaria* pAb was used in a multi-centre study to determine the concentration of *A. alternata* in homes throughout the United States (Salo *et al.* 2006). *Alternaria* is an important aeroallergen within USA and personal exposure to its spores has been associated with respiratory morbidity (Licorish *et al.* 1985, O'Hollaren *et al.* 1991, Delfino *et al.* 1997, Downs *et al.* 2001, Andersson *et al.* 2003). This commercially available pAb was not studied for cross-reactions prior to its use. In the Schmechel *et al.* (2008) study, it was shown that the pAb cross-reacted with up to 30 different fungi all belonging to the order Pleosporales, which also contains *Alternaria*. These data demonstrated a number of limitations associated with the detection of specific bioaerosols using antibody-based immunoassays, but also highlighted the care that is required when selecting antibodies for the quantification of bioaerosols. Another factor that requires consideration before utilizing mAb- or pAb-based ELISAs is the differential expression of antigens in different stages of the life cycle of the organism. Relatively recently it was shown that the allergen Alt a 1, which was believed to be a spore-based allergen, was actually expressed in considerably higher concentrations following spore germination (Mitakakis *et al.* 2001, Green *et al.* 2003). Based on these findings, it is critical that all samples following collection are immediately processed to avoid any initiation of germination, as this may result in misleadingly amplified results. These considerations are extremely important when interpreting data derived from microbiologically based ELISAs. However, if the necessary steps are taken, ELISAs can offer a sensitive methodology capable of elucidating submicron-sized, morphologically unidentifiable particles that typically elude direct microscopic characterization by a trained analyst.

Following the development of ELISA techniques, Popp *et al.* (1988) developed an indirect immunostaining method that enabled the identification of morphologically nondescript particulates collected on filter membranes. In this method, fungal surface antigens probed with either human antifungal antibodies or mAbs form immune complexes that are immunostained with a fluorescent conjugate. Positive immunostaining results in the fluorescent illumination of the particle that demonstrates the presence of the specific antigen on the particle (Popp *et al.* 1988). Several variations followed the development of this

methodology and included a press blotting technique (Takahashi *et al.* 1993, Takahashi and Nilsson 1995) and the HIA (Tovey *et al.* 2000). Of these methods, the HIA has provided some of the most exciting developments including paradigm shifts in the exposure assessment field (De Lucca *et al.* 1999a,b, Poulos *et al.* 1999, De Lucca *et al.* 2000, Razmovski *et al.* 2000, De Lucca and Tovey 2001, Poulos *et al.* 2002, Green *et al.* 2005a,b,c, Green *et al.* 2006a,b, Green *et al.* 2009). Briefly, airborne particulates collected by filtration onto a protein binding membrane are laminated with an optically clear, adhesive-backed coverslip to retain the particles on the membrane. Soluble antigens are extracted from the surface of the particle and immobilized on the protein binding membrane in close proximity to the spore (Green *et al.* 2006b). The extracted native antigens are then indirectly immunostained with either pAb or mAbs or human sera, and the immune complexes are immunostained to form a halo of colour around the particle. This immunostaining technique enables the formation of immune complexes where specific antibodies bind with the respective antigens that are immobilized on the surface of the membrane. Immunostained particles can then be quantified and morphologically identified either by direct microscopy or by dual immunostaining with a mAb or pAb (Green *et al.* 2005a,b,c, 2006a, b, 2009). The HIA has enabled the detection and immunolocalization of aeroallergens derived from cat (De Lucca *et al.* 2000), latex (Poulos *et al.* 2002), dust mite (De Lucca *et al.* 1999a), cockroach (De Lucca *et al.* 1999b), pollen (Razmovski *et al.* 2000) and fungi (Green *et al.* 2005c). Unlike other diagnostic methods, HIA allows the identification of morphologically nondescript particulates that are present in the environment of a patient (Green *et al.* 2005c, Green *et al.* 2006b, Green *et al.* 2009). Depending on the basic research needed to generate specifically characterized mAbs, this technique may become very useful for the detection of bioaerosols locally produced in regions emitting pollutant sources.

In addition, other indirect immunostaining methods utilizing field emission scanning electron microscopy (Sercombe *et al.* 2006) and flow cytometry (Rydjord *et al.* 2007) have recently enabled the rapid detection of microbiological submicron particulate material from air samples. Each of these techniques provides sensitive detection and can quantify picogram quantities of antigen in a sample. The improved detection thresholds and specificity may overcome a number of limitations associated with MALDI TOF-MS and PCR especially as large concentrations of whole spores are not required for analysis and quantification. The main limitation is that each type of aerosol specifically identified must have its own specific reagent. These techniques are all most easily used in studies involving one bioaerosol type or a few types.

BIOSENSORS

Various other detection methodologies have been recently developed that utilize either immunodiagnostic or PCR techniques in combination with emerging diagnostic technologies. Airborne biosensors, immunosensors, multiplex immunoassays, remote sensing, and pathogen detection technologies are just a few examples that have significantly improved the rapid detection and identification of bioaerosol sources important to homeland security, infection control and occupational exposure.

Biosensors are analytical devices that combine the recognition of biomolecules with electronics for signal measurement (Carlson *et al.* 2000, Rossi *et al.* 2007). For the first time, these advances provide a real-time remote detection platform for the detection and quantification of dispersed biological agents (Ligler *et al.* 1998). Fibre optic biosensors represent the next generation of biosensor technology (King *et al.* 2000). These biosensors enable fibre optic detection of fluorescent signals following the formation of immune complexes in immunoassays. Such biosensors have been successfully deployed in a remote platform for the detection of pathogens, clinical samples, toxins in food samples, pollutants in ground water and aerosolized biological warfare agents (Ligler *et al.* 1998). Recent developments have also improved the portability of fibre optic biosensors. This has lead to the integration of this technology into a remotely controlled, unmanned aerial vehicle (UAV) designed to allow secure, long distance operation of the detection platform. Such methodologies have enabled the remote sensing and rapid detection of very low concentrations of airborne bacterial bioaerosols using simultaneous immunoassays (Dingus *et al.* 2007, Schmale *et al.* 2007).

Antibody-based identification has been exploited in the development of numerous biosensor platforms because of the high sensitivity, specificity and adaptability to field use. Other biosensors have also been developed that enable the detection of microorganisms important in disease diagnosis, pharmaceutical research, agriculture and homeland security. Some of these methodologies include nanoporous silicon biosensors (Rossi *et al.* 2007), quartz crystal microbal-

ance sensory for the detection of influenza (Owen *et al.* 2007), and multianalyte immunoassays based on surface-enhanced Raman spectrometry (McBride *et al.* 2003). Autonomous molecular platforms, such as some autonomous pathogen detection systems that have been developed utilizing orthogonal and multiplexed PCR, are also capable of continually monitoring the environment for biological agents (Hidson *et al.* 2005, Hofstadler *et al.* 2005). The further development of remotely controlled biosensor platforms will have many future applications in the detection of environmental contaminants, in particular pollutants associated with pollution plumes. Again, however, for these techniques to be successful the development of specific antibodies is critical for the specific detection of each target bioaerosol type.

All of these aforementioned methodologies provide an alternative approach to MALDI-TOF MS or PCR techniques for the identification and quantification of plant, algal, bryophyte, fungal and bacterial bioaerosols. The utilization of antibody-based immunoassays enables the rapid and specific detection of many bacterial and fungal species. These methods provide a cost friendly alternative to the other more expensive biochemical and molecular techniques. However, in the interests of this feasibility study, many of the specific organisms have not been fully characterized immunologically, and antibody-based reagents have not been developed. This is especially the case for many anemophilous plant pollen types, fungal basidiospores and algal propagules. Thus, these alternative techniques may currently only be useful for the detection of a small selection of pathogenic bacteria and fungal spores.

ACKNOWLEDGEMENTS

Funding to support the writing of this chapter was provided by a grant from the Ontario Ministry of Environment Best in Science Program.

REFERENCES

Alvarez, A.J., Buttner M.P., and Stetzenbach, L.D. (1995) PCR for bioaerosol monitoring – sensitivity and environmental interference. *Appl. Environ. Microbiol.*, **61**, 3639-3644.

An, H.R., Mainelis, G., and White, L. (2006) Development and calibration of real-time PCR for quantification of airborne microorganisms in air samples. *Atmos. Environ.*, **40**, 7924-7939.

Andersson, M., Downs, S., Mitakakis, T., *et al.* (2003) Natural exposure to *Alternaria* spores induces allergic rhinitis symptoms in sensitized children. *Pediatr. Allergy Immunol.*, **14**, 100-105.

Arnold, R.J., Karty, J.A., Ellington, A.D., and Reilly, J.P. (1999) Monitoring the growth of a bacteria culture by MALDI-MS of whole cells. *Anal. Chem.*, **71**, 1990-1996.

Avarre, J.C., Lajudie, P., and Béna, G. (2007) Hybridization of genomic DNA to microarrays: a challenge for the analysis of environmental samples. *J. Microbiol. Methods*, **69**, 242-248.

Barnes, C.S., Upadrashta, B., Pacheco, F., and Portnoy, J. (1993) Enhanced sensitivity of immunoblotting with peroxidase-conjugated antibodies using an adsorbed substrate method. *J. .Chromatogr.*, **613**, 281-288.

Birenzvige, A., Eversole, J., Seaver, M., *et al.* (2003) Aerosol characteristics in a subway environment. *Aerosol Sci. Tech.*, **37**, 210-220.

Bisht, V., Arora, N., Singh, B.P., *et al.* (2004) Purification and characterization of a major cross–reactive allergen from *Epicoccum purpurascens*. *Int. Arch. Allergy Immunol.*, **133**, 217-224.

Blachere, F.M., Lindsley, W.G., Slaven, J.E., *et al.* (2007) Bioaerosol sampling for the detection of aerosolized influenza virus. *Influenza Other Resp. Viruses*, **1**, 113-120.

Booth, T.F., Kournikakis, B., Bastien, N., *et al.* (2005) Detection of airborne severe acute respiratory syndrome (SARS) corona-virus and environmental contamination in SARS outbreak units. *J. Infect. Dis.* **191**, 1472-1477.

Bright, J.J., Claydon, M.A., Soufian, M., and Gordon, D.B. (2002) Rapid typing of bacteria using matrix-assisted laser desorption ionisation time-of-flight mass spectrometry and pattern recognition software. *J. Microbiol. Methods*, **48**, 127-138

Brodie, E.L., DeSantis, T.Z., Parker, J.P., *et al.* (2007) Urban aerosols harbor diverse and dynamic bacterial populations. *PNAS*, **104**, 299-304.

Cain, T.C., Lubman, D.M., and Weber, W.J. (1994) Differentiation of bacteria using protein profiles from matrix-assisted laser-desorption ionization time-of-flight mass-spectrometry. *Rapid Commun. Mass Spectrom.*, **8**, 1026-1030.

Carlson, M.A., Bargeron, C.B., Benson, R.C., *et al.* (2000) An automated, handheld biosensor for aflatoxin. *Biosens. Bioelectron.*, **14**, 841-848.

Chen, H.Y., and Chen, Y.C. (2005) Characterization of intact *Penicillium* spores by matrix-assisted laser desorption/ionization mass spectrometry. *Rapid Commun. Mass Spectrom.*, **19**, 3564-3568.

Chen, P.-H., Pan, Y.-B., and Chen, R.-K. (2008) High-throughput procedure for single pollen grain collection and polymerase chain reaction in plants. *J. Integr. Plant Biol.*, **50**, 375-383.

Chew, G.L., Douwes, J., Doekes, G., *et al.* (2001) Fungal extracellular polysaccharides, beta-(1,3)-D-glucans and culturable fungi in repeated sampling of house dust. *Indoor Air*, **11**, 171-178.

Cho, J.G., Deen, J., and Dee, S.A. (2007) Influence of isolate pathogenicity on the aerosol transmission of porcine reproductive and respiratory syndrome virus. *Can. J. Vet. Res.*, **71**, 23-27.

Claydon, M.A. (2000) MALDI-ToF-MS, a new and novel technique for studies of intact cells. *Anaerobe*, **6**, 133-134.

Claydon, M.A., Davey, S.N., Edwards-Jones, V., and Gordon, D.B. (1996) The rapid identification of intact microorganisms using mass spectrometry. *Nature Biotechnol.*, **14**, 1584-1586.

Cruz, A., Saenz de Santamaria, M., Martinez, J., *et al.* (1997) Fungal allergens from important allergenic fungi imperfecti. *Allergol. Immunopathol.*, **25**, 153-158.

Cunningham, D. (1873) *Microscopic Examinations of Air*. Superintendent of Government Printing, Calcutta, India.

Delfino, R.J., Zeiger, R.S., Seltzer, J.M., *et al.* (1997) The effect of outdoor fungal spore concentrations on daily asthma severity. *Environ. Health Persp.*, **105**, 622-635.

De Lucca, S., Sporik, R., O'Meara, T.J., and Tovey, E.R. (1999a) Mite allergen (Der p1) is not only carried on mite feces. *J. Allergy Clin. Immunol.*, **103**, 174-175.

De Lucca, S.D., Taylor, D.J., O'Meara, T.J., *et al.* (1999b) Measurement and characterization of cockroach allergens detected during normal domestic activity. *J. Allergy Clin. Immunol.*, **104**, 672-680.

De Lucca, S.D., O'Meara, T.J., and Tovey, E.R. (2000) Exposure to mite and cat allergens on a range of clothing items at home and the transfer of cat allergen in the workplace. *J. Allergy Clin. Immunol.*, **106**, 874-879.

De Santis, T.Z., Stone, C.E., Murray, S.R., *et al.* (2005) Rapid quantification and taxonomic classification of environmental DNA from both prokaryotic and eukaryotic origins using a microarray. *FEMS Microbiol. Lett.*, **245**, 271-278.

De Santis, T.Z., Brodie, E.L., Moberg, J.P., *et al.* (2007) High-density universal 16S rRNA microarray analysis reveals broader diversity than typical clone library when sampling the environment. *Microbial Biol.*, **53**, 371-383.

Despres, V.R., Nowoisky, J.F., Klose, M., *et al.* (2007) Characterization of primary biogenic aerosol particles in urban, rural, and high-alpine air by DNA sequence andrestriction fragment analysis of ribosomal RNA genes. *Biogeosciences*, **4**, 1127-1141.

Dinsdale, E.A., Edwards, R.A., Hall, D., *et al.* 2008. Functional metagenomic profiling of nine biomes. *Nature*, **452**, 629-632.

Dingus, B.R., Schmale, D.G. , and Reinholtz, C. (2007) Development of an autonomous unmanned aerial vehicle for aerobiological sampling. *Phytopathology*, **97**, S184.

Domin, M.A., Welham, K.J., and Ashton, D.S. (1999) The effect of solvent and matrix combinations on the analysis of bacteria by matrix-assisted laser desorption/ionisation time-of-flight mass spectrometry. *Rapid Commun. Mass Spectrom.*, **13**, 222–226.

Douwes, J., Doekes, G., Montijn, R., *et al.* (1997) An immunoassay for the measurement of beta-(1,3)-D-glucans in the indoor environment. *Mediat. Inflamm.*, **6**, 257-262.

Downs, S.H., Mitakakis, T.Z., Marks, G.B., *et al.* (2001) Clinical importance of *Alternaria* exposure in children. *Am. J. Resp. Crit. Care Med.*, **164**, 455-459.

Edwards-Jones, V., Claydon, M.A., Evason, D.J., *et al.* (2000) Rapid discrimination between methicillin-sensitive and methicillin-resistant *Staphylococcus aureus* by intact cell mass spectrometry. *J. Med. Microbiol.*, **49**, 295-300.

Fabian, P., McDevitt, J.J., DeHaan, W.H., *et al.* (2008). Influenza virus in human exhaled breath: an observational study. *PLoS ONE*, **3**, e2691.

Fenselau, C., and Demirev, P.A. (2001) Characterization of intact microorganisms by MALDI mass spectrometry. *Mass Spectrom. Rev.*, **20**, 157-171.

Fierer, N., Liu, Z., Rodríguez-Hernández, M., *et al.* (2008) Short-term temporal variability in airborne bacterial and fungal populations. *Appl. Environ. Microbiol.*, **74**, 200-207.

Flannigan, B. (1997) Air sampling for fungi in indoor environments. *J. Aerosol Sci.*, **28**, 381-392.

Goebes, M.D., Hildemann, L.M., Kujundzic, E., and Hernandez, M. (2007) Real-time PCR for detection of the *Aspergillus* genus. *J. Environ. Monitor.*, **9**, 599-609.

Green, B.J., Mitakakis, T.Z., and Tovey, E.R. (2003) Allergen detection from 11 fungal species before and after germination. *J. Allergy Clin. Immunol.*, **111**, 285-289.

Green, B.J., Schmechel, D., Sercombe, J.K., and Tovey, E.R. (2005a) Enumeration and detection of aerosolized *Aspergillus fumigatus* and *Penicillium chrysogenum* conidia and hyphae using a novel double immunostaining technique. *J. Immunol. Methods*, **307**, 127-134.

Green, B.J., Schmechel, D. and Tovey, E.R. (2005b) Detection of aerosolized *Alternaria alternata* conidia, hyphae, and fragments by using a novel double-immunostaining technique. *Clin. Diagn. Lab. Immunol.*, **12**, 1114-1116.

Green, B.J., Sercombe, J.K., and Tovey, E.R. (2005c) Fungal fragments and undocumented conidia function as new aeroallergen sources. *J. Allergy Clin. Immunol.*, **115**, 1043-1048.

Green, B.J., Millecchia, L.L., Blachere, F.M., *et al.* (2006a) Dual fluorescent halogen immunoassay for bioaerosols using confocal microscopy. *Anal. Biochem.*, **354**,151-153.

Green, B.J., Tovey, E.R., Sercombe, J.K., *et al.* (2006b) Airborne fungal fragments and allergenicity. *Med. Mycol.*, **44**, S245-S255.

Green, B.J., O'Meara, T., Sercombe, J., and Tovey, E. (2006c) Measurement of personal exposure to outdoor aeromycota in northern New South Wales, Australia. *Ann. Agric. Environ. Med.*, **13**, 225-234.

Green, B.J., Tovey, E.R., Beezhold, D.H. *et al.* (2009) surveillance of fungal allergic sensitization using the fluorescent Halogen Immunoassay. *T. Mycol. Med.*, **19**, 253-261.

Gregory, P.H. (1973) *Microbiology of the Atmosphere*, 2nd Ed. Leonard Hill, Aylesbury, Bucks, UK.

Halstensen, A.S., Nordby, K.C., Eduard, W., and Klemsdal, S.S. (2006) Real-time PCR detection of toxigenic *Fusarium* in airborne and settled grain dust and associations with trichothecene mycotoxins. *J. Environ. Monitor.*, **8**, 1235-1241.

Handelsman, J. (2004) Metagenomics: application of genomics to uncultured microorganisms. *Microbiol. Mol. Biol. Rev.*, **68**, 669-685.

Haugland, R.A., and Heckman, J.L. (1998) Identification of putative sequence specific PCR primers for detection of the toxigenic fungal species *Stachybotrys chartarum*. *Mol. Cell. Probes*, **12**, 387-396.

Haugland, R.A., Vesper, S.J., and Wymer, L.J. (1999) Quantitative measurement of *Stachybotrys chartarum* conidia using real time detection of PCR products with the TaqMan (TM) fluorogenic probe system. *Mol. Cell. Probes*, **13**, 329-340.

Haugland, R.A., Brinkman, N., and Vesper, S.J. (2002) Evaluation of rapid DNA extraction methods for the quantitative detection of fungi using real-time PCR analysis. *J. Microbiol. Methods*, **50**, 319-323.

Haugland, R.A., Varma, M., Wymer, L.J., and Vesper, S.J. (2004) Quantitative PCR analysis of selected *Aspergillus*, *Penicillium* and *Paecilomyces* species. *Syst. Appl. Microbiol.*, **27**, 198-210.

Hawksworth, D.L. (2006) The fungal dimension of biodiversity: magnitude, significance, and conservation. *Mycol. Res.*, **95**, 641-655.

Hernlam, B.J., and Ravva, S.V. (2007) Application of flow cytometry and cell sorting to the bacterial analysis of environmental aerosol samples. *J. Environ. Monitor.*, **9**, 1317-1322.

Hermann, J., Hoff, S., Muñoz-Zanzi, C., *et al.* (2007) Effect of temperature and relative humidity on the stability of infectious porcine reproductive and respiratory syndrome virus in aerosols. *Vet. Res.*, **38**, 81-93.

Hettick, J.M., Green, B.J., Buskirk, A.D., *et al.* (2008a) Discrimination of *Aspergillus* isolates at the species and strain level by matrix-assisted laser desorption/ionization time-of-flight mass spectrometry fingerprinting. *Anal. Biochem.*, **380**, 276-281.

Hettick, J.M., Green, B.J., Buskirk, A.D., *et al.* (2008b) Discrimination of *Penicillium* isolates by matrix-assisted laser desorption/ionization time-of-flight mass spectrometry fingerprinting. *Rapid Commun. Mass Spectrom.*, **22**, 2555-2560.

Hettick, J.M., Kashon, M.L., Simpson, J.P., *et al.* (2004) Proteomic profiling of intact mycobacteria by matrix-assisted laser desorption/ionization time-of-flight mass spectrometry. *Anal. Chem.*, **76**, 5769-5776.

Hettick, J.M., Kashon, M.L., Slaven, J.E., *et al.* (2006) Discrimination of intact mycobacteria at the strain level: a combined MALDI-TOF MS and biostatistical analysis. *Proteomics*, **6**, 6416-6425.

Hindson, B.J., Makarewicz, A.J., Setlur, U.S., *et al.* (2005) APDS: the autonomous pathogen detection system. *Biosens. Bioelectron.*, **20**, 1925-1931.

Hofstadler, S.A., Sampatha, R., Blyna, L.B., *et al.* (2005) TIGER: the universal biosensor. *Int. J. Mass Spectrom.*, **242**, 23-41.

Hsieh, S.Y., Tseng, C.L., Lee, Y.S., *et al.* (2008) Highly efficient classification and identification of human pathogenic bacteria by MALDI–TOF MS. *Mol. Cell. Proteomics*, **7**, 448-456.

Huynh, K.N., Oliver, B.G., Stelzer, S., *et al.* (2008) A new method for sampling and detection of exhaled respiratory virus aerosols. *Clin. Infect. Dis.*, **46**, 93-95.

Kasprzyk, I. (2004) Airborne pollen of entomophilous plants and spores of pteridophytes in Rzeszów and its environs (SE Poland). *Aerobiologia*, **20**, 217-222.

Kelley, P.K., and McClain, K.L. (1994) Respiratory contamination of polymerase chain reactions by human herpesvirus 6. *Am. J. Hematol.*, **47**, 325-327.

Keswani, J., Kashon, M.L., and Chen, B.T. (2005) Evaluation of interference to conventional and real-time PCR for detection and quantification of fungi in dust. *J. Environ. Monitor.*, **7**, 311-318.

King, K.D., Vanniere, J.M., LeBlanc, J.L., *et al.* (2000) Automated fiber optic biosensor for multiplexed immunoassays. *Environ. Sci. Technol.*, **34**, 2845-2850.

Kress, W.J., Wurdack, K.J., Zimmer, E.A., *et al.* (2005) Use of DNA barcodes to identify flowering plants. *PNAS*, **102**, 8369-8374.

Krishnamurthy, T., and Ross, P.L. (1996) Rapid identification of bacteria by direct matrix-assisted laser desorption/ionization mass spectrometric analysis of whole cells. *Rapid Commun. Mass Spectrom.*, **10**, 1992-1996.

Kuske, C.R. (2006) Current and emerging technologies for the study of bacteria in the outdoor air. *Curr. Opin. Biotechnol.*, **17**, 291-296.

Lay, J.O. (2001) MALDI-TOF mass spectrometry of bacteria. *Mass Spectrom. Rev.*, **20**, 172-194.

Lee, S.C., Corradi, N., Byrnes, E.J., *et al.* (2008) Microsporidia evolved from ancestral sexual fungi. *Curr. Biol.*, **18**, 1675-1679.

Li, T.Y., Liu, B.H., and Chen, Y.C. (2000) Characterization of *Aspergillus* spores by matrix-assisted laser desorption/ionization time-of-flight mass spectrometry. *Rapid Commun. Mass Spectrom.*, **14**, 2393-2400.

Li, W., Wooley, J.C., and Godzik, A. (2008) Probing metagenomics by rapid cluster analysis of very large datasets. *PLoS ONE*, **3**, e3375.

Licorish, K., Novey, H.S., Kozak, P., *et al.* (1985) Role of *Alternaria* and *Penicillium* spores in the pathogenesis of asthma. *J. Allergy Clin. Immunol.*, **76**, 819-825.

Ligler, F.S., Anderson, G.P., Davidson, P.T., *et al.* (1998) Remote sensing using an airborne biosensor. *Environ. Sci. Technol.*, **32**, 2461-2466.

Liu, J., Xia, K.-F., Zhu, J.-C., *et al.* (2006) Nightshade proteinase inhibitor IIb gene is constitutively expressed in glandular trichomes. *Plant Cell Physiol.*, **47**, 1274-1284.

Luo, Y., Ma, Z., Reyes, H.C., *et al.* (2007) Quantification of airborne spores of *Monilinia fructicola* in stone fruit orchards of California using real-time PCR. *Eur. J. Plant Pathol.*, **118**, 145-154.

Lynch, M.D., and Thorn, R.G. (2006) Diversity of basidiomycetes in Michigan agricultural soils. *Appl. Environ. Microbiol.*, **72**, 7050-7056.

Lynn, E.C., Chung, M.-C., Tsai, W.-C., and Han, C.-C. (1999) Identification of Enterobacteriaceae bacteria by direct matrix-assisted laser desorption/ionization mass spectrometric analysis of whole cells. *Rapid Commun. Mass Spectrom.*, **13**, 2022-2027.

Maddox, R.L. (1870) On an apparatus for collecting atmospheric particles. *Monthly Microscopy J.*, **3**, 286-290.

Marks, M.D., Betancur, L., Gilding, E., *et al.* (2008) A new method for isolating large quantities of *Arabidopsis* trichomes for transcriptome, cell wall and other types of analyses. *Plant J.*, **56**, 483-492.

Maron, P.-A., Lejon, D.P.H., Carvalho, E., *et al.* (2005) Assessing genetic structure and diversity of airborne bacterial communities by DNA fingerprinting and 16S rDNA clone library. *Atmos. Environ.*, **39**, 3687-3695.

McBride, M.T., Masquelier, D., Hindson, B.J., *et al.* (2003) Autonomous detection of aerosolized *Bacillus anthracis* and *Yersinia pestis*. *Anal. Chem.*, **75**, 5293-5299.

McCluskey, R., Sandin, R., and Greene, J. (1996) Detection of airborne cytomegalovirus in hospital rooms of immuncompromised patients. *J. Virolog. Methods*, **56**, 115-118.

McDevitt, J.J., Lees, P.S.J., Merz, W.G., and Schwab, K.J. (2004) Development of a method to detect and quantify *Aspergillus fumigatus* conidia by quantitative PCR for environmental air samples. *Mycopathologia*, **158**, 325-335.

McDevitt, J.J., Lees, P., Merz, W., and Schwab, K. (2005) Use of green fluorescent protein-expressing *Aspergillus fumigatus* conidia to validate quantitative PCR analysis of air samples collected on filters. *J. Occup. Environ. Hyg.*, **2**, 633-640.

Meklin, T., Haugland, R.A., Reponen, T., *et al.* (2004) Quantitative PCR analysis of house dust can reveal abnormal mold conditions. *J. Environ. Monitor.*, **6**, 615-620.

Merrill, L., Richardson, J., Kuske, C.R., and Dunbar, J. (2003) Fluorescent heteroduplex assay for monitoring *Bacillus anthracis* and close relatives in environmental samples. *Appl. Environ. Microbiol.*, **69**, 3317-3326.

Miquel, P. (1883) *Les Organismes Vivants dans L'Atmosphere*. Gauthier-Villars, Paris.

Mitakakis, T.Z., Barnes, C., and Tovey E.R. (2001) Spore germination increases allergen release from *Alternaria*. *J. Allergy Clin. Immun.*, **107**, 388-390.

Mitakakis, T.Z., O'Meara, T.J., and Tovey, E.R. (2003) The effect of sunlight on allergen release from spores of the fungus *Alternaria*. *Grana*, **42**, 43-46.

Moletta, M., Delgenes, J.P., and Godon, J.J. (2007) Differences in the aerosolization behavior of microorganisms as revealed through their transport by biogas. *Sci. Total Environ.*, **379**, 75-88.

Morrison, C.J., Yang, C., Lin, K.-T., *et al.* (2004) Monitoring *Aspergillus* species by quantitative PCR during construction of a multistorey hospital building. *J. Hosp. Infect.*, **57**, 85-87.

Myatt, T., Johnston, S.L., Zuo, Z.W.M., *et al.* (2004) Detection of airborne rhinovirus and its relation to outdoor air supply in office environments. *Am. J. Resp. Crit. Care Med.*, **169**, 1187-1190.

Nakatsu, C.H., and Marsh, T.L. (2007) Analysis of microbial communities with denaturing gradient gel electrophoresis and terminal restriction fragment length polymorphism. In C.A. Reddy (ed.), *Methods for General and Molecular Microbiology*, 3rd Ed, ASM Press, Washington, DC, pp. 909-923.

Nehme, B., Létourneau, V., Forster, R.J., *et al.* (2008) Culture-independent approach of the bacterial bioaerosol diversity in the standard swine confinement buildings, and assessment of the seasonal effect. *Environ. Microbiol.*, **10**, 665-675.

Nocker, A., Burr, M., and Camper, A.K. (2007) Genotypic microbial community profiling: a critical technical review. *Microbial Ecol.*, **54**, 276-289.

O'Hollaren, M.T., Yunginger, J.W., Offord, K.P., *et al.* (1991) Exposure to an aeroallergen as a possible precipitating factor in respiratory arrest in young patients with asthma. *New Engl. J. Med.*, **324**, 359-363.

Oppliger, A., Charriere, N., Droz, P.O., and Rinsoz, T. (2008) Exposure to bioaerosols in poultry houses at different stages of fattening; use of real-time PCR for airborne bacterial quantification. *Occup. Hyg.*, **52**, 405-412.

Owen, T.W., Al-Kaysi, R.O., Bardeen, C.J., and Cheng, Q. (2007) Microgravimetric immunosensor for direct detection of aerosolized influenza A virus particles. *Sensor. Actuat. B-Chem.*, **126**, 691-699.

Parisi, D., Magliulo, M., Nanni, P., *et al.* (2008) Analysis and classification of bacteria by matrix-assisted laser desorption/ionization time-of-flight mass spectrometry and a chemometric approach. *Anal. Bioanal. Chem.*, **391**, 2127-2134.

Pascual, L., Perez-Luz, S., Moreno, C., *et al.* (2001) Detection of *Legionella pneumophila* in bioaerosols by polymerase chain reaction. *Can. J. Microbiol.*, **47**, 341-347.

Payungporn, S., Phakdeewirot, P., Chutinimitkul, S., *et al.* (2004) Single-step multiplex reverse transcription-polymerase chain reaction (RT-PCR) for influenza A virus subtype H5N1 detection. *Viral Immunol.*, **17**, 588–593.

Peccia, J., and Hernandez, M. (2006) Incorporating polymerase chain reaction–based identification, population characterization, and quantification of microorganisms into aerosol science: a review. *Atmos. Environ.*, **40**, 3941-3961.

Pietarinen, V.M., Rintala, H., Hyvärinen, A., *et al.* (2008) Quantitative PCR analysis of fungi and bacteria in building materials and comparison to culture-based analysis. *J. Environ. Monitor.*, **10**, 655-663.

Pitkäranta, M., Meklin, T., Hyvärinen, A., *et al.* (2008) Analysis of fungal flora in indoor dust by ribosomal DNA sequence analysis, quantitative PCR, and culture. *Appl. Environ. Microbiol.*, **74**, 233-244.

Popp, W., Zwick, H., and Rauscher, H. (1988) Indirect immunofluorecent test on spore sampling preparations: a technique for diagnosis of individual mold allergies. *Stain Technol.*, **63**, 249-253.

Portnoy, J. M., Barnes, C.S., and Kennedy, K. (2004) Sampling for indoor fungi. *J. Allergy Clin. Immunol.*, **113**, 189-198.

Poulos, L.M., O'Meara, T.J., Sporik, R., and Tovey, E.R. (1999) Detection of inhaled Der p 1. *Clin. Exp. Allergy*, **29**, 1232-1238.

Poulos, L.M., O'Meara, T.J., Hamilton, R.G., and Tovey, E.R. (2002) Inhaled latex allergen (Hev b 1). *J. Allergy Clin. Immunol.*, **109**, 701-706.

Prezant, B., Weekes, D.M., and Miller, J.D., (eds.)(2008). *Recognition, Evaluation, and Control of Mold*. AIHA Press, Fairfax, VA.

Pribil, P.A., Patton, E., Black, G., *et al.* (2005) Rapid characterization of *Bacillus* spores targeting species-unique peptides produced with an atmospheric pressure matrix-assisted laser desorption/ionization source. *J. Mass Spectrom.*, **40**, 464-474.

Radosevich, J.L., Wilson, W.J., Shinn, J.H., *et al.* (2002) Development of a high-volume aerosol collection system for the identification of air-borne micro-organisms. *Lett. Appl. Microbiol.* **34**, 162-167.

Razmovski, V., O'Meara, T.J., Taylor, D.J.M., and Tovey, E.R. (2000) A new method for simultaneous immunodetection and morphologic identification of individual sources of pollen allergens. *J. Allergy Clin. Immunol.*, **105**, 725-731.

Reijula, K.E., Kurup, V.P., and Fink, J.N. (1991) Ultrastructural demonstration of specific IgG and IgE antibodies binding to *Aspergillus fumigatus* from patients with aspergillosis. *J. Allergy Clin. Immunol*, **87**, 683-688.

Renstrom, A. (2002) Exposure to airborne allergens: a review of sampling methods. *J. Environ. Monitor.*, **4**, 619-622.

Rintala, H., and Nevalainen, A. (2006) Quantitative measurement of streptomycetes using real-time PCR. *J. Environ. Monitor.*, **8**, 745-749.

Rintala, H., Pitkäranta, M., Toivola, M., *et al.* (2008) Diversity and seasonal dynamics of bacterial community in indoor environment. *BMC Microbiol.*, **8**, 56.

Rinsoz, T., Duquenne, P., Greff-Mirguet, G., and Oppliger, A. (2008) Application of real-time PCR for total airborne bacterial assessment: comparison with epifluorescence microscopy and culture-dependent methods. *Atmos. Environ.*, **42**, 6767-6774.

Rippon, J.W. (1988) *Medical Mycology: The Pathogenic Fungi and the Pathogenic Actinomycetes*, 3rd Ed., Chapter 6. Dermatophytoses. W.B. Saunders, Philadelphia, pp. 105-162.

Roe, J.D., Haugland, R.A., Vesper, S.J., and Wymer, L.J. (2001) Quantification of *Stachybotrys chartarum* conidia in indoor dust using real time, fluorescent probe-based detection of PCR products. *J. Expos. Anal. Environ. Epidemiol.*, **11**, 12-20.

Rossi, A.M., Wang, L., Reipa, V., and Murphy, T.E. (2007) Porous silicon biosensor for detection of viruses. *Biosens. Bioelectron.*, **23**, 741-745.

Rydjord, B., Namork, E., Nygaard, U.C., *et al.* (2007) Quantification and characterisation of IgG binding to mould spores by flow cytometry and scanning electron microscopy. *J. Immunol. Methods*, **323**, 123-131.

Salo, P.M., Arbes, S.J., Sever, M., *et al.* (2006) Exposure to *Alternaria alternata* in US homes is associated with asthma symptoms. *J. Allergy Clin. Immunol.*, **118**, 892-898.

Sarna, N.J., and Govil, C.M. (1979) Survey of airspora of Jaipur City. *J. Indian Bot. Soc.* **58**, 215-225.

Sawyer, M.H., Chamberlin, C.J., Wu, Y.N., *et al.* (1994) Detection of varicella-zoster virus DNA in air samples from hospital rooms. *J. Infect. Dis.*, **169**, 91-94.

Schafer, M.P., Martinez, K.F., and Matthews, E.S. (2003) Rapid detection and determination of the aerodynamic size range of airborne mycobacteria associated with whirlpools. *Appl. Occup. Environ. Hygiene*, **18**, 14-50.

Schmale, D.G. (2007) Use of autonomous unmanned aircraft to validate and improve long distance pathogen transport models. *Phytopathology*, **97**, S147.

Schmechel, D., Górny, R.L., Simpson, J.P., *et al.* (2003) Limitations of monoclonal antibodies for monitoring of fungal aerosols using *Penicillium brevicompactum* as a model fungus. *J. Immunol. Methods*, **283**, 235-245.

Schmechel, D., Green, B.J., Blachere, F.M., *et al.* (2008) Analytical bias of cross-reactive polyclonal antibodies for environmental immunoassays of *Alternaria alternata*. *J. Allergy Clin. Immunol.*, **121**, 763-768.

Schneegurt, M.A., Dore, S.Y., and Kulpa, C.F. (2003) Direct extraction of DNA from soils for studies in microbial ecology. *Curr. Issues Mol. Biol.*, **5**, 1-8.

Schweigkofler, W., O'Donnell, K., and Garbelotto, M. (2004) Detection and quantification of airborne conidia of *Fusarium circinatum*, the causal agent of pine pitch canker, from two California sites by using a real-time PCR approach combined with a simple spore trapping method. *Appl. Environ. Microbiol.*, **70**, 3512-3520.

Sercombe, J.K., Eduard, W., Romeo, T.C., *et al.* (2006) Detection of allergens from *Alternaria alternata* by gold-conjugated anti-human IgE and field emission scanning electron microscopy. *J. Immunol. Methods*, **316**, 167-170.

Siegrist, T.J., Anderson, P.D., Huen, W.H., *et al.* (2007) Discrimination and characterization of environmental strains of *Escherichia coli* by matrix-assisted laser desorption/ionization time-of-flight mass spectrometry (MALDI-TOF-MS). *J. Microbiol.*

Methods, **68**, 554-562.

Sigari, G., Panatto, D., Lai, P., *et al.* (2006) Virological investigation on aerosol from waste depuration plants. *J. Prev. Med. Hyg.*, **47**, 4–7.

Smit, E., Leeflang, P., Glandorf, B., *et al.* (1999) Analysis of fungal diversity in the wheat rhizosphere by sequencing of cloned PCR-amplified genes encoding 18S rRNA and temperature gradient gel electrophoresis. *Appl. Environ. Microbiol.*, **65**, 2614-2621.

Sontakke, S., Cadenas, M.B., Maggi, R.G., *et al.* (2009) Use of broad range 16S rDNA PCR in clinical microbiology. *J. Microbiol. Methods*, **76**, 217-225.

Stark, K.D.C., Nicolet, J., and Frey, J. (1998) Detection of *Mycoplasma hyopneumoniae* by air sampling with a nested PCR assay. *Appl. Environ. Microbiol.*, **64**, 543-548.

Stoneburner, A., Lane, D.M., and Anderson, L.E. (1992) Spore dispersal distances in *Atrichum angustatum* (Polytrichaceae). *Bryologist*, **95**, 324-328.

Sundberg, S. (2005) Larger capsules enhance short-range spore dispersal in Sphagnum, but what happens further away? *Oikos*, **108**, 115-124.

Suzuki, K., Yoshikawa, T., Tomitaka, A., *et al.* (2004) Detection of aerosolized varicella-zoster virus DNA in patients with localized herpes zoster. *J. Infect. Dis.*, **189**, 1009-1012.

Takahashi, Y., and Nilsson, S. (1995) Aeroallergen immunoblotting with human IgE antibody. *Grana*, **34**, 357–360.

Takahashi, Y., Nagoya, T., Watanabe, M., *et al.* (1993) A new method of counting airborne Japanese cedar (*Cryptomeria japonica*) pollen allergens by immunoblotting. *Allergy*, **48**, 94-98.

Tovey, E., Taylor, D., Graham, A., *et al.* (2000) New immunodiagnostic system. *Aerobiologia*, **16**, 113–118.

Tringe, S.G., von Mering, C., Kobayashi, A., et al. (2005) Comparative metagenomics of microbial communities. *Science*, 308, 554-557.

Trout, D.B., Seltzer, J.M., Page, E.H., *et al.* (2004) Clinical use of immunoassays in assessing exposure to fungi and potential health effects related to fungal exposure. *Ann. Allergy Asthma Immunol.*, **92**, 483-492.

Vesper, S.J., Dearborn, D.G., Yike, I., *et al.* (2000) Evaluation of *Stachybotrys chartarum* in the house of an infant with pulmonary hemorrhage: quantitative assessment before, during and after remediation. *J. Urban Health*, **77**, 68-85.

Vesper, S.J., Varma, M., Wymer, L.J., *et al.* (2004) Quantitative PCR analysis of fungi in dust from homes of infants who developed idiopathic pulmonary hemorrhaging. *J. Occup. Environ. Med.*, **46**, 596-601.

Vesper, S.J., Wymer, L.J., Meklin, T., *et al.* (2005) Comparison of populations of mould species in homes in the UK and USA using mould-specific quantitative PCR. *Lett. Appl. Microbiol.*, **41**, 367-373.

Vesper, S., McKinstry, C., Hartmann, C., *et al.* (2008) Quantifying fungal viability in air and water samples using quantitative PCR after treatment with propidium monoazide (PMA). *J. Microbiol. Methods*, **72**,180-184.

Welham, K.J., Domin, M.A., Scannell, D.E., *et al.* (1998) The characterization of micro-organisms by matrix-assisted laser desorption/ionization time-of-flight mass spectrometry. *Rapid Commun. Mass Spectrom.*, **12**, 176-180.

Williams, R.H., Ward, E., and McCartney, H.A. (2001) Methods for integrated air sampling and DNA analysis for detection of airborne fungal spores. *Appl. Environ. Microbiol.*, **67**, 2453-2459.

Wilson, K.H., Blitchington, R.B., and Greene, R.C. (1990) Amplification of bacterial 16S ribosomal DNA with polymerase chain reaction. *J. Clin. Microbiol.*, **28**, 1942-1946.

Wilson, K.H., Wilson, W.J., Radosevich, J.L., *et al.* (2002) High-density microarray of small-subunit ribosomal DNA probes. *Appl. Environ. Microbiol.*, **68**, 2535-2541.

Xiao, W.J., Wang, M.L., Wei, W., *et al.* (2004) Detection of SARS-CoV and RNA on aerosol samples from SARS-patients admitted to hospital. *Zhonghua Liu Xing Bing Xue Za Zhi.*, **25**, 882-885.

Xu, J. (2006) Microbial ecology in the age of genomics and metagenomics: concepts, tools, and recent advances. *Mol. Ecol.*, **15**, 1713-1731.

Zhou, L.-J., Pei, K.-Q., Zhou, B., and Ma, K.-P. (2007) A molecular approach to species identification of Chenopodiaceae pollen grains in surface soil. *Am. J. Bot.*, **3**, 477-481.

Chapter 4.3

LABORATORY ISOLATION AND IDENTIFICATION OF FUNGI

Robert A. Samson and Jos Houbraken

CBS-KNAW Fungal Biodiversity Centre, Utrecht, The Netherlands.

INTRODUCTION

It is estimated that more than 80,000 species of fungi are known, but that our present knowledge includes only a fraction of 10-20% of the species actually occurring in nature. In the indoor environment only a small portion of these fungi occur and it has often been indicated that this mycobiota is very similar to the common taxonomic groups, or taxa, found on food (Pitt and Hocking 2009, Samson *et al.* 2010). In addition to microfungi, a number of Basidiomycetes can be found indoors, growing on wood in buildings. It is important to identify these species, because each has its own characteristics, and these are important for the control of wood rotting and renovation of wooden constructions in damaged buildings.

Fungal taxonomy is rapidly changing through the use of molecular studies. In many common genera the species concept is drastically changed by redefining the species based on sequence data. Particularly in *Fusarium*, *Trichoderma* and *Cladosporium* common species proved to be cryptic, consisting of many taxa defined by sequence data. In *Aspergillus* and *Penicillium* this is also the case, but in these genera the polyphasic approach is now used combining phenotypical, biochemical and molecular characters (Samson and Frisvad 2004, Samson and Varga 2007).

As described in Chapter 4.1, estimating fungal exposure in buildings is best assessed by accurately determining mould area and identifying the predominant species. Unfortunately, isolation methods based on those used in bacteriology or medical mycology are often used, rather than those which are more suitable for the environmental mycobiota.

In Section 5, the morphological features of the most common fungal species are illustrated, together with three species of the Actinomycetes that may be encountered indoors and are mentioned in a short segment on bacteria later in this chapter. The micrographs were prepared from cultures grown under optimum conditions or specimens and mounted in lactic acid or Shear's mounting fluid. The media used for cultivation, which differ according to species, are indicated in the legend.

ISOLATION AND EXAMINATION OF FUNGI FROM SAMPLES

Samples

Any material with suspected mould growth can serve as a sample for further examination and isolation. Samples should be transported and stored in dry conditions, since mould growth can continue to develop under moist conditions. Packaging in clean absorbent paper, such as Kleenex™, or newspaper is recommended, but *not* polyethylene bags. Small plastic boxes or other containers are suitable, particularly in the case of very fragile material. Scrapings of suspected moulds from walls or other surfaces can be taken with a knife or scalpel. A suitable method for sampling is to use transparent adhesive tape (Figs.1-4). After it has been pressed against a contaminated surface, the tape should be transferred onto a glass microscope slide for further microscopical examination (see also Chapter 4.1.). Small pieces can also be cut from the tape bearing mould and transferred directly onto an isolation

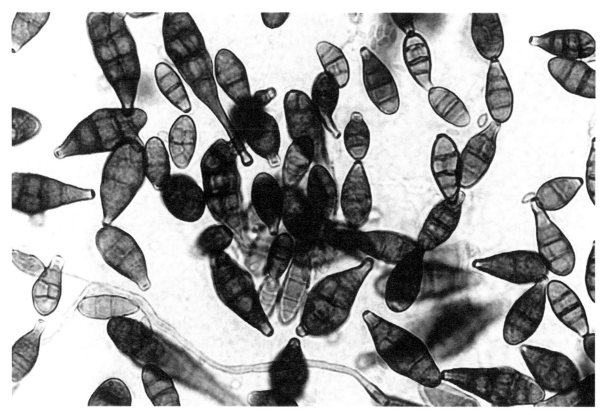

Fig. 1. Transparent adhesive tape preparation showing various spores.

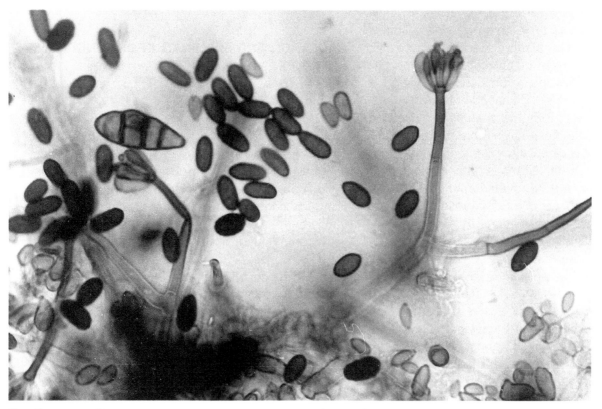

Fig. 2. Transparent adhesive tape preparation showing conidiophores and spores of *Stachybotrys chartarum* and spores of *Cladosporium* and *Alternaria*.

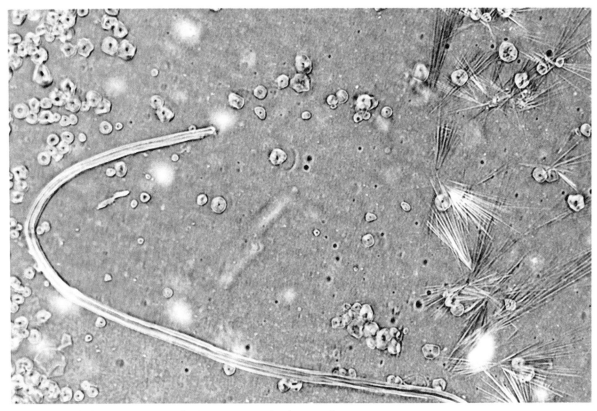

Fig. 3. Crystals of salts in a transparent adhesive tape preparation from a wall.

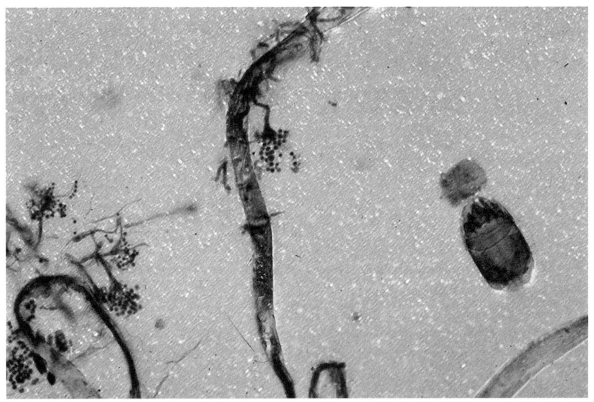

Fig. 4. Fungal structures belonging to *Aspergillus* with textile fibre and a mite.

Table 1. Surface sterilisation of building materials.

Surface disinfection	Particles are surface disinfected by vigorous shaking in freshly prepared sodium hypochlorite solution containing 0.4% chlorine (w/v) for 2 min. If a quantitative estimation is required a minimum of 100 particles should be disinfected. The solution must only be used once.
Rinsing	After pouring off the chlorine, rinse once in sterile distilled or deionised water.
Plating	As quickly as possible, transfer particles with a sterile forceps to previously poured and set plates, at the rate of 5-10 particles per plate.

medium. When sampling in dwellings and other buildings, one should be cautious where *Stachybotrys* and other toxic fungi are present. Protective clothing, gloves and respirators are recommended (see Chapter 2.3).

Isolation in the laboratory

For isolating fungi it is quite suitable to transfer small pieces of the (moulded) material to agar medium and then streak them out over the surface. Direct plating (see also Chapter 4.1) is a technique which is commonly used to isolate fungi growing on food (Samson *et al.* 1992, 2010, Pitt and Hocking, 2009). In some cases, where the responsible mould contamination is hidden or overgrown by secondary invading species, surface sterilisation of the sample (brick, wall material, wood) can be performed. A procedure is described in Table 1.

Media

Many media are used for fungi (Table 2), but for isolation and detection the general purpose media dichloran-18% glycerol agar (DG18 agar) and malt extract agar (MEA) are recommended. For the detection of moulds in wet environments water agar can be used, but it is always preferable to use two media to cover the complete mycobiota, including mesophilic and xerophilic species. Plate count agar and Sabouraud agar are not recommended because several species will not develop on these media. When media which contain rose bengal are used, viz. rose bengal agar or dichloran rose bengal agar (DRBC) (King *et al.* 1979), it should be noted that they only detect mesophilic fungi. Furthermore, exposure to light should be avoided because in light rose bengal becomes increasingly toxic to some fungi, including *Rhizopus* and yeasts.

Many media can be used to isolate yeasts, but Samson *et al.* (1994) recommended malt extract

Table 2. Formulations of some mycological media for isolation of indoor moulds (for more media see Samson *et al.* 2010).

CMA	Cornmeal agar (CBS): Add 60 g freshly ground cornmeal to 1 litre distilled water, heat to boiling and simmer gently for 1 h. Strain through cloth and sterilize at 121°C for 15 min. Fill up to 1 litre with water and add 15 g agar and re-sterilize at 121°C for 15 min. Commercially available.
CYA	Czapek yeast (autolysate) extract agar: $NaNO_3$ 3.0, K_2HPO_4 1.0, KCl 0.5, $MgSO_4.7H_2O$ 0.5, $FeSO_4.7H_2O$ 0.01, Yeast extract 5.0, Sucrose 30.0, Agar 20.0, Distilled water 1000 ml. Final pH 6.0-6.5. NOTE: Addition of trace elements (see above) is recommended. Addition of mineral solution (MS), containing 5g KCL, 5 g $MgSO_4.7H_2O$, 0.1 g $FeSO_4.7H_2O$ per 100ml water is also recommended in some media.
DG18	Dichloran 18% Glycerol agar: Peptone 5.0 g; Glucose 10.0 g; KH_2PO_4 1.0 g $MgSO_4.7H_2$ 00.5g; Dichloran (0.2% in ethanol) 1.0 ml; Glycerol 220.0 g; Chloramphenicol 0.1 g; Agar 15.0 g; Distilled water 1 litre. Add minor ingredients and agar to ca. 800 ml distilled water. Steam to dissolve agar, then make up to 1 litre with distilled water. Add 220 g glycerol and sterilize by autoclaving at 121°C for 15 min. Final a_w is 0.955: final pH 5.6 ± 0.2. Commercially available (Oxoid).
MEA	Malt extract agar: Malt extract 20.0 g; Mycological peptone 5.0 g; Agar 15.0 g; Distilled water 1 litre. Final pH 5.4 ± 0.2. Sterilize by autoclaving at 115°C for 10 min.
OA	Oatmeal agar (CBS): Heat 30 g oat flakes in 1 litre water to boiling and simmer gently for 2 h. Filter through cloth and make up to 1 litre. Add 15 g agar and autoclave at 121°C for 15 min. When using powdered oatmeal, filtering is superfluous. Sterile lupin stems may be placed in slants with oatmeal agar. Also commercially available.
PDA	Potato dextrose agar: Add 200 g scrubbed and diced potatoes to 1 litre water and boil for 1 h. Pass through a fine sieve, add 15 g agar and 20 g dextrose and boil until dissolved. pH 5.6 ± 0.1. Commercially available.
SNA	Synthetischer nährstoffarmer agar: KH_2PO_4 1.0, KNO_3 1.0, $MgSO_4.7H_2O$ 0.5, KCl 0.5, Glucose 0.2, Saccharose 0.2, Agar 20.0, Distilled water 1000 ml. NOTE: Pieces of sterile filter paper may be placed on the agar. Recommended for the cultivation of *Fusarium*, but also for poorly sporulating Deuteromycetes.
WA	Water agar 2%: Agar 20.0 g; Distilled water 1 litre. Steam to dissolve agar and sterilize by autoclaving at 121°C for 15 min.

Table 3. Fungal flora with related water activities (from Samson *et al.* 1994, with slight modification).

Materials with a high water activity (a_w >0.90 - 0.95): *Aspergillus fumigatus, Trichoderma, Exophiala, Stachybotrys, Phialophora, Fusarium, Ulocladium,* yeasts (*Rhodotorula*), Actinomycetes and Gram-negative bacteria (e.g. *Pseudomonas*).

Materials with a moderately high water activity (a_w >0.85 - <0.90): *A. versicolor, A. sydowii, Emericella nidulans.*

Materials with a lower water activity (a_w <0.85): *Aspergillus versicolor, Eurotium, Wallemia,* Penicillia (e.g. *Penicillium chrysogenum, P. brevicompactum*).

agar for bathrooms and cooling water reservoirs in air-conditioning systems, where yeasts usually predominate. Chloramphenicol or oxytetracycline (100 mg L^{-1}) can be added, if necessary, to suppress bacteria. DRBC is preferable when yeasts must be enumerated in the presence of moulds.

Incubation

Plates should be incubated in an upright position at 25°C. For the detection of thermotolerant/thermophilic species, e.g. *Aspergillus fumigatus,* incubation should be at temperatures between 35° and 40°C. Plates can be incubated either in darkness or in diffuse light. When using rose bengal, plates or test strips should always be kept in the dark (see under Media).

Examining the cultures

For identification by microscopic examination it is essential to use appropriate and optimal light microscopic equipment. Good lenses, including x100 objectives will facilitate the investigation of the shape and surface ornamentation of fungal cells and spores. For the examination of colonies and samples with fungal contamination, a dissecting microscope with a total magnification of x100 - x400 and a good light source is recommended. A dissecting microscope is also helpful for making good microscopic slide preparations by enabling sporulating areas of a colony to be selected.

Spore samples from indoors or outdoors, such as those collected on transparent adhesive tape or by Burkard or Rotorod samplers (see Chapters 1.1 and 4.1), can be directly examined by microscope. However, it should be emphasized again that, while species in some genera (e.g. *Alternaria, Epicoccum* and *Curvularia*) can be easily recognized by the size, shape, septation and pigmentation of the spores, species in many other genera have very simple and similar spores (e.g. *Aspergillus* and *Penicillium*). Fig. 1 shows a mixture of spores as collected on a transparent tape preparation from wallpaper. A detailed examination of this preparation shows conidia, of *Alternaria,*

Cladosporium and *Stachybotrys* and ascospores of *Chaetomium*. However, round hyaline spores which belong to either *Penicillium* or *Aspergillus* are also present. More precise identification can be made when structures of a fruiting body or conidiophore can also be observed, as depicted in Fig. 2. Here, typical *Stachybotrys* conidiophores as well as conidia are present. Microscopic examination of preparations made from surfaces often also reveals structures which are not fungal, but may be crystals of salts (Fig. 3), textile fibres (Fig. 4), or mite and insect faeces.

Indicator organisms

All species in the indoor environment have their own individual growth conditions and their presence gives good indications of the condition of the building. Samson *et al. (*1994) suggested that several species can be considered as indicator organisms where moisture and/or health problems are signalled (Table 3). For example, the many species in the genus *Aspergillus* differ from one another in their growth requirements. Consequently, identification only to generic level is not sufficient; identification to species level is essential.

MITES AND OTHER LABORATORY CONTAMINANTS

Various species of mycophagous (fungus-eating) mites can be present on materials and surfaces where mould growth appears. These mites may act as vectors which spread the fungi and therefore it is important to take note of their presence. For experiments in the laboratory and maintenance of fungal cultures, mite infestation should be avoided, because it will hamper the isolation and cultivation of the moulds in pure culture.

When growing pure cultures in the laboratory, besides mites some microorganisms can also cause problems. *Chrysonilia (Monilia) crassa* is a very rapidly growing fungus, invading incubators, Petri dishes and even slant cultures. When this fungal contaminant is found, Petri dishes should be sealed immediately (with Parafilm or similar tape) and autoclaved.

Furthermore, benches and incubators should be cleaned with alcohol and eventually disinfected with formaldehyde.

Fungi can produce volatile compounds and, although little is known about the health implications when individuals are exposed, health complaints are reported. It is recommended that, when fungal strains are being grown in the laboratory, the cultures should be in biosafety cabinets or under hoods with good ventilation.

NOMENCLATURE

The correct name of a taxon reflects the current state of taxonomic knowledge. An important problem which is associated with the identification of moulds is that much confusion exists about the naming of species. A well-known example is *Stachybotrys chartarum* (Ehrenb.) Hughes, which is still, but incorrectly, named as *S. atra* Corda. The species *S. chartarum* was described in 1818 by a German mycologist, Ehrenberg, as *Stilbospora chartarum*. In 1958, Dr S. Hughes examined the type material collected by Ehrenberg and found that it belonged to *Stachybotrys* and therefore proposed the combination *S. chartarum* (Ehrenberg) Hughes. *Stachybotrys atra*, which is identical to *S. chartarum*, but was described later by Corda in 1837, is therefore a synonym. Other incorrect names are *Penicillium frequentans* (correct name *P. glabrum*) and *Epicoccum purpurascens* (correct name *E. nigrum*).

A procedure for stabilising *Aspergillus* and *Penicillium* nomenclature has been developed to protect names in current use. In order to protect these currently used names from being threatened or displaced by names that are no longer in use, Pitt and Samson (1993) and Pitt *et al.* (2000) proposed a list of species names in current use for the family Trichocomaceae (Eurotiales). All published names and synonyms of *Aspergillus* and *Penicillium* are listed in Samson and Pitt (2000).

IDENTIFICATION

Identification of fungi can be based on their physical appearance, i.e. their phenotypic characters. In yeasts, biochemical and physiological profiles are used and this facilitates the use of identification kits such as API and Biolog. Chemotaxonomic approaches such as profiles of secondary metabolites (including

Table 4. Basidiomycetes occurring in buildings.

Dry rot fungus	*Serpula lacrymans*
	Coniophora puteana
	Coniophora marmorata
	Donkioporia expansa
	Phellinus contiguus
	Antrodia xantha
Wet rot fungi	*Antrodia (Fibroporia) valliantii*
	Pleurotus ostreatus
	Sistotrema brinkmannii
	Schizophyllum commune
	Coprinus domesticum

volatiles and mycotoxins) have been proposed for identification (for an overview see Frisvad *et al.* 1998), but have not yet gained any practical significance. Molecular techniques have been tested and are introduced. Depending on the validity of their databases, these may prove to be useful for rapid and objective identification.

IDENTIFICATION OF THE MAJOR FUNGAL GROUPS

Basidiomycetes

The Basidiomycetes form a large group that mostly comprises species with typical fruiting bodies or basidiocarps which are mostly known as toadstools and bracket fungi. In the indoor environment, some species which do not possess macroscopic fruiting bodies, such as *Sistotrema*, may occur in air samples. Typical basidiomycete species in buildings are the wood-rotting fungi, among which *Serpula lacrymans* is the most notorious species, causing dry rot. Airborne spores of *S. lacrymans* can be present in high concentrations and allergenic reactions are reported in the literature (Murphy *et al.* 1995). Table 4 lists the most common species found in North American and European buildings. Most basidiomycetes are difficult to grow on artificial media and they often do not form fruiting bodies which sporulate, but only sterile mycelium.

For identification of wood-rotting species in buildings the publications by Wagenführ and Steiger (1966), Coggins (1980), Bravery *et al.* (1987), Hennebert and Balon (1996) and Schmidt (2006) are recommended.

Zygomycetes

Zygomycetes can be easily recognized because of their fast growth, which covers a Petri dish within 5 days. As mentioned in Chapter 1.2, the mycelium has no or few septa. Asexual reproduction occurs by means of sporangiospores produced endogenously in sporangia, or in merosporangia. Morphological

Table 5. Conidiogenesis of the common indoor moulds

Conidiogenesis	Genera
Phialides with dry conidia	*Aspergillus, Paecilomyces, Penicillium, Stachybotrys (= Memnoniella), Wallemia*
Phialides with slimy conidia	*Acremonium, Clonostachys, Exophiala, Fusarium, Gliomastix, Phialophora, Stachybotrys, Trichoderma, Verticillium*
Phialides in pycnidia	*Phoma*
Annellides	*Scopulariopsis*
Thallic	*Geotrichum*
Blastic arthric	*Aureobasidium, Chrysonilia, Cladosporium, Geomyces, Oidiodendron*
Blastic single	*Botrytis, Epicoccum, Tritirachium*
Poroconidia	*Alternaria, Curvularia, Ulocladium*

features such as an apophysis (a swelling of the sporangiophore just below the sporangium) and stolons which adhere to the substrate by means of rhizoids (root-like structures) are present in some species but not in others and are therefore important in identification. Some species produce chlamydospores or oidia.

Sexual reproduction occurs by fusion of two multinucleate gametangia, resulting in a thick-walled, dark-coloured zygospore, which is, however, rarely found in isolates from indoor environments. Several species, e.g. *Mucor plumbeus, Rhizopus stolonifer* and *Absidia corymbifera,* are frequently encountered in indoor environments, on surfaces and in air samples. In hospitals and buildings in which there are environments at temperatures above 25°C, *Mucor* and *Rhizomucor* species are also sometimes encountered.

For more detailed descriptions and keys to the Zygomycetes the reader is referred to O'Donnell (1979) and von Arx (1981). Keys to the species of *Mucor* are presented by Schipper (1978), while the genus *Rhizopus* has been treated by Scholer and Müller (1971), Schipper (1984) and Liou *et al.* (1990), *Rhizomucor* by Schipper (1978).

Ascomycetes
The Ascomycetes form the largest group in the Fungal Kingdom and most species can be found on plants and other substrates in nature. For the indoor environment most species belong to the Eurotiales with simple and small fruiting bodies. Species of *Chaetomium* are found on material containing cellulose. *Pyronema domesticum* is a species frequently found on fresh plaster in new and renovated buildings. On wood some species with distinct and large fruiting bodies can occur, such as *Peziza domiciliana.*

Asci usually develop in ascomata (ascocarps) and they contain mostly eight ascospores. The morphology of the fruit bodies is important for a systematic division. The fruit bodies may be cup-like apothecia, globose or subglobose cleistothecia, which are non-ostiolate (without a terminal pore), and more or less flask-shaped, mostly ostiolate perithecia. This ascus-producing state, the teleomorph, is often accompanied by one or more asexual reproductive states, the anamorph. The majority of the Deuteromycetes are anamorphs in life cycles of Ascomycetes.

Among ascomycetous fungi found in indoor environments, species of *Chaetomium* are notable on walls and cellulosic materials, as well as being frequently found as storage fungi on, for example, cereals or develop after heat-treatment of a product. *Chaetomium* grows well on oatmeal agar (OA). For details of this genus the reader is referred to von Arx *et al.* (1986). As mentioned in Chapter 2.1, *Eurotium* spp. are the commonest members of the Ascomycetes encountered in indoor environments, on walls and other surfaces and in house dust. For optimal sporulation *Eurotium* should be grown on MEA or Czapek-Dox agar with additional sucrose (20-40%) or NaCl (10-30%). For isolation, DG18 agar can be used. Most of the ascomycetous species treated here mature after 10-14 days. Common indoor ascomycetes are keyed out and described in Samson *et al.* (2010).

Deuteromycetes
Most fungi found in air and in the indoor environment belong to the Deuteromycetes. As mentioned earlier, this group is not a formal Class in the Fungal Kingdom, but is composed of species without a teleomorph (sexual form). However, the species mostly belong *genetically* to the Ascomycetes, and some to the Basidiomycetes. The main criterion used in classifying fungi in the Deuteromycetes is the mode of conidium formation, or **conidiogenesis** (Table 5). For a more

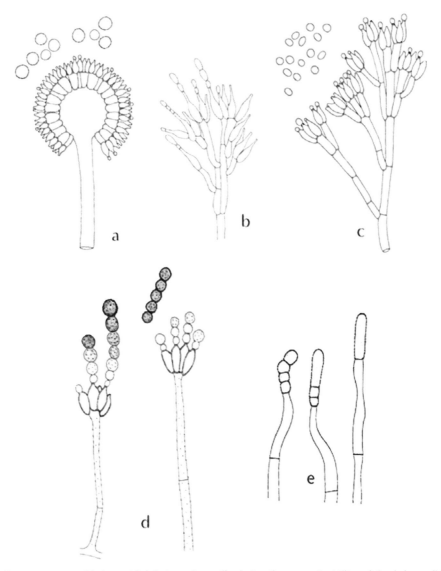

Fig. 5. Examples of phialidic Deuteromycetes with dry conidial chains: a. *Aspergillus*, b. *Paecilomyces*, c. *Penicillium*, d. *Stachybotrys (Memnoniella) echinata*, e. *Wallemia*.

detailed account the reader is referred to Cole and Samson (1979). The specialized non-motile asexual propagules, the **conidia**, are not formed by cleavage like zygomycete sporangiospores. Their shape and colour vary widely and they may arise solitarily, synchronously, in chains or in slimy heads. They are borne on a specialized cell, the **conidiogenous cell.** Such cells can be borne directly in or from a vegetative hypha or on diffentiated supporting structures (stipe and branches). For many genera in the indoor environment the special conidiogenous cell is called a **phialide.** Conidia are mostly produced in high numbers either in dry chains or in slimy heads.

Figs. 5 and 6 illustrate phialidic conidiogenesis in some common genera. The entire system of fertile hyphae is called the **conidiophore**. In Fig. 7 genera with annellides, thallic and various blastic

conidiogenesis are illustrated.

Building materials often provide optimum conditions for production of morphological structures by which these mitosporic fungi are identified, so that individual types can be rapidly recognized by using transparent adhesive tape and examining it by light microscopy. For some difficult genera such as *Aspergillus* and *Penicillium* identification to species level requires isolation and cultivation on special media. Most Deuteromycetes sporulate well on MEA and OA agar. For some genera, e.g. *Fusarium*, SNA agar with a piece of sterile filter paper is very suitable for achieving good sporulation. *Penicillium* and *Aspergillus* require Czapek-yeast extract agar (CYA) for correct identification (see also the description of each genus for culture methods).

Good microscopic mounts are required in order

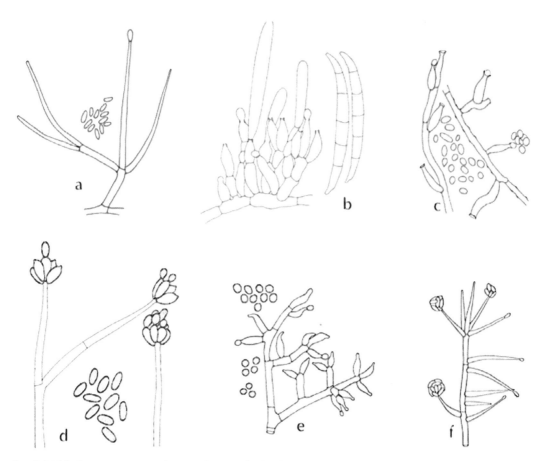

Fig. 6. Examples of phialidic Deuteromycetes with wet or slimy conidial heads: a. *Acremonium*, b. *Fusarium*, c. *Phialophora*, d. *Stachybotrys*, e. *Trichoderma*, f. *Lecanicillium* (formerly *Verticillium*).

to examine conidiogenesis. The mounts are mostly made in lactic acid with aniline blue. For fragile structures such as in *Botrytis, Oidiodendron, Geomyces* and *Cladosporium*, it is often helpful to prepare slides on which a small piece of adhesive tape or a cut-out square of transparent agar is mounted. The production of conidia in chains or slimy heads is best observed in the Petri dish under low power magnification. However, care must be taken to avoid inhalation of fungal products (volatiles) and conidia.

Table 5 lists the common genera of Deuteromycetes likely to be found in indoor environments. For the important genera a short description of characters is given below. The most common genera and species are described in detail by Samson *et al.* (2010). Literature on other deuteromycete genera are described in detail by Carmichael *et al.* (1980), von Arx (1981) and Domsch *et al.* (2007). For special literature on monographic studies see Crous *et al.* (2009).

Aspergillus

Aspergillus spp. are common contaminants on various substrates. In subtropical and tropical regions they occur more commonly than *Penicillium* spp.

Colonies usually grow rapidly, white, yellow, yellow-brown, brown to black or shades of green, mostly consisting of a dense felt of erect conidiophores. Conidiophores are unbranched with a swollen apex (vesicle). Phialides are borne directly on the vesicle (uniseriate) or on metulae (biseriate). Conidia in dry chains form compact columns (columnar) or diverging (radiate), one-celled, smooth or ornamented, hyaline or pigmented. Species may produce Hülle cells (large, thick- and smooth-walled cells) or sclerotia (firm, usually globose, masses of hyphae). Teleomorphs are *Eurotium, Emericella, Neosartorya* and other genera.

Raper and Fennell (1965) subdivided *Aspergillus* into "Groups". This infrageneric classification has no nomenclatural status under the ICBN and Gams *et al.* (1986) replaced the group names by names of subgenera and sections. However, for a more exact identification of important isolates, the recent taxonomic literature (Raper and Fennell 1965, Samson 1992, 1994 a, b, Frisvad and Samson 2000) should be consulted. As already stated above, aspergilli are very common contaminants. Therefore we have not listed every substrate from which each species has been isolated. Data of habitats of many species are given

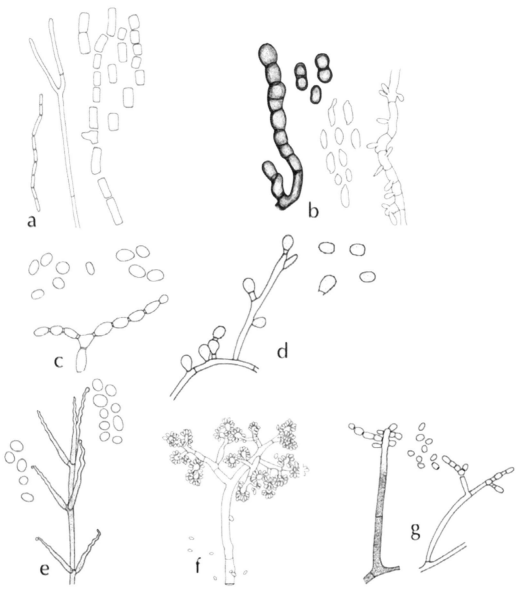

Fig. 7. Examples of thallic and blastic Deuteromycetes: a. *Geotrichum*, b. *Aureobasidium*, c. *Chrysonilia*, d. *Geomyces*, e. *Tritirachium*, f. *Botrytis*, g. *Oidiodendron*.

by Domsch *et al.* (2007).

The taxonomy and classification of *Aspergillus* taxa has been modified in the last five years based on a polyphasic approach. The number of new *Aspergillus* species is growing and the reader should consult Samson and Varga (2007) and recent articles, because the redefinition and delimitation of common indoor species has changed for certain species.

Cladosporium

This genus has a world-wide distribution. Several species are plant pathogens or are saprophytic and more or less host-specific on old or dead plant material.

Colonies are mostly olivaceous-brown to blackish-brown or with a greyish-olive appearance, velvety or floccose becoming powdery due to abundant conidia, and rather slow-growing. Conidiophores are fragile, erect, straight or flexuous, unbranched or branched only in the apical region, with geniculate sympodial elongation in some species. Conidia are in branched chains, one-celled, ellipsoidal, fusiform, ovoid, (sub)globose, often with distinct scars, pale to dark olivaceous-brown, smooth-walled, verrucose or echinulate; (blasto)conidia mostly formed on denticles in groups of 1-3 at the apex of the conidiophore, subapically below a septum or on the tip of previously formed conidia. Owing to the fragile conidiophore structure it is often difficult to obtain good microscope slides. It is helpful to use transparent adhesive tape and lightly touch the colonies.

For more detailed descriptions and keys see

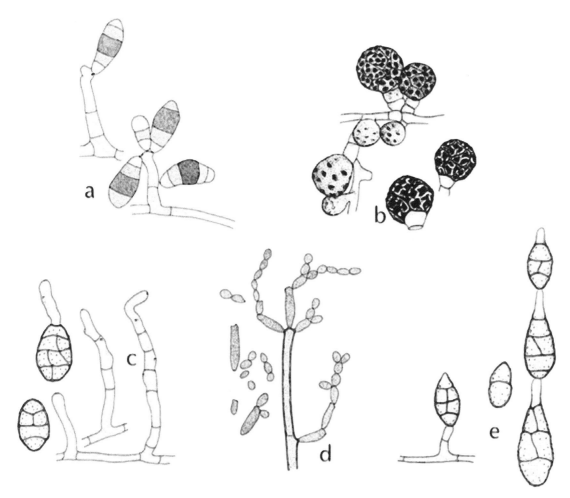

Fig. 8. Examples of blastic Deuteromycetes: a. *Curvularia*, b. *Epicoccum*, c. *Ulocladium*, d. *Cladosporium*, e. *Alternaria*.

De Vries (1952), Ellis (1971), Domsch *et al.* (2007). Molecular studies by Zalar *et al.* (2007), Schubert *et al.* (2007) and Dugan *et al.* (2008) investigated the cryptic species of *C. cladosporioides*, *C. sphaerospermum*, *C. herbarum* and *C. macrocarpum*. More studies are required to determine which species of *Cladosporium* are common in indoor environments.

Fusarium

Most *Fusarium* species occur in soil or on plants and have a world-wide distribution. The genus includes plant pathogens of crops widely grown around the globe, e.g. wheat and maize. A few fusaria can be found in indoor environments and are mostly associated with wet niches, including cooling units of air-conditioning systems. The current taxonomy of *Fusarium* is not clear, and molecular taxonomic data will probably change the classification and number of species radically. There are many monographs which can be used for identification. The following publications are recommended: Gerlach and Nirenberg (1982), Nelson *et al.* (1983), Marasas *et al.* (1984),

Burgess *et al.* (1988), Nirenberg (1989, 1990), Thrane (1989). *Fusarium* systematics is developing rapidly with many new species being discovered by phylogenetic analyses of DNA sequences. A more conservative and broad phenotypic species concept is mostly used for species found in indoor environments. More detailed descriptions and keys can be found in Leslie and Summerell (2006) and Domsch *et al.* (2007).

Penicillium

Many species of *Penicillum* are common contaminants on a wide variety of substrates. Colonies usually grow rapidly, in shades of green, mostly consisting of a dense felt of conidiophores, which arise from the substrate, aerial hyphae or from bundled hyphae. Conidiophores are hyaline, smooth- or rough-walled, single (mononematous) or bundled (synnematous). The penicillus consists of branches and metulae (penultimate branches which bear a whorl of phialides). The branching pattern is either one-stage branched (biverticillate), two-stage branched (terverticillate) or three- (quaterverticillate) to more-stage branched and is important in allocating the isolates to the various subsections. The phialides are mostly flask-shaped or lanceolate (= acerose: more or less narrow basal part tapering to a somewhat pointed apex) and produce long dry chains of conidia.

Because of the widespread occurrence of penicillia and their potential for producing mycotoxins, correct identification is important when studying possible *Penicillium* contamination. Purely morphological characteristics can be used for identification, but the isolates should be grown under standard conditions using malt extract agar and Czapek-yeast extract agar at 25°C for 7 days. For more recent detailed descriptions and keys to penicillia see also Pitt (1979, 2000). Species of *Penicillium* subgenus *Penicillium* are treated in detail by Frisvad and Samson (2004), while the most common *Penicillium* species are described in Samson *et al.* (2010).

For some species, odour and exudate production will help in recognition of the taxa, but it should be stressed that inhalation of conidia and volatiles may cause adverse health effects. However, as in *Aspergillus*, true human pathogenic species are rare in *Penicillium* and limited to *P. marneffei* alone.

A molecular study on the most commonly occurring *Penicillium* species indoors, *P. chrysogenum*, showed a remarkable phylogenetic difference among strains isolated from indoor environment (Scott *et al.* 2004). Another study on occurrence of *P. brevicompactum* indoors showed that this species and the phenotypically similar species *P. bialowiezense* are commonly occurring in indoor environments (Scott *et al.* 2008). It is likely with the introduction of molecular methods, more (cryptic) penicillia will be found as common contaminants of indoor enviroments.

Scopulariopsis

Colonies vary from white, creamish, grey or buff to brown or even blackish, often darkening with age, but never in green like *Penicillium*. The conidiogenous cells are cylindrical or with a slightly swollen base, annellate (rings formed after conidium formation), single or borne on whorls. Conidia are in chains with a broadly truncate base. For a more detailed account of descriptions and key see Morton and Smith (1963), and Domsch *et al.* (2007).

Stachybotrys

Species of *Stachybotrys* have a world-wide distribution and occur on paper, wallpaper, seeds (e.g. wheat, oats), soil, textiles and dead plant material. There are several species of *Stachybotrys* but the most well-

known taxon is *S. chartarum.*

Colonies grow slowly, appearing blackish to blackish-green. *Memnoniella* was considered to be distinct from *Stachybotrys*, by differing in having dry conidia in chains, not in slimy heads like those in *Stachybotrys*. However, phylogenetic studies show that this species clearly belongs in *Stachybotrys* and hence the correct name *Stachybotrys echinata* is now currently used. For detailed descriptions and keys see Ellis (1971) and Domsch *et al.* (2007). Exposure to this fungus on natural substrates or in culture should be avoided.

Yeasts

Yeasts mostly reproduce vegetatively by budding. Exceptions are species of the genus *Schizosaccharomyces*, which reproduces by a fission process, and of the genera *Sporobolomyces*, *Sterigmatomyces* and *Fellomyces*, which form buds on short stalks. The morphology of the unicellular yeast cell indicates the genus, but further identification should be performed using a number of physiological tests, including assimilation of a number of sugars. Commercial identification kits provide computer programs for processing the results obtained with them. These include the API 20C and API YEAST-IDENT™ kits in which growth tests are carried out in small wells on plastic strips. Biolog Inc. markets the Microlog™ system for yeasts which, by using the redox dye tetrazolium violet, detects whether a microorganism can utilize a given carbon source. For a review the reader is referred to Barnett *et al.* (2000a, b). Polyphasic identification systems have been published by Robert and Szoke (2006) and Robert *et al.* (2008). The identification system can be found at http://www.cbs.knaw.nl.

Bacteria

The isolation of bacteria from environmental samples is complicated by the fact that, while many bacteria will grow on a wide range of isolation media, others are extremely fastidious in their nutritional requirements. Where only a general picture of the variety of bacteria and the overall load is required, commercially available general purpose media such as plate count agar, nutrient agar, tryptone soya agar (TSA) or R2A agar can be used, and cycloheximide may be used as an antifungal supplement. Rather than the 37°C used in most bacteriological studies, it is preferable to incubate environmental plates at 25°C, although 15°C may be employed (Austwick *et al.* 1989). For isolation of particular bacteria, media in which their particular nutritional needs are met or selective media which retard or inhibit other bacteria are employed. For example, cetrimide agar or Pseudomonas agar are used for isolation of the pathogen *Pseudomonas aeruginosa*, and buffered charcoal yeast extract agar (Chapter 3.4) is used for *Legionella*.

For general isolation of thermophilic actinomycetes, tryptone soya casein hydrolysate agar (TSC), i.e. half-strength TSA (Oxoid) supplemented with 0.2% casein hydrolysate, incubated at 55°C can be used. TSC can be made selective for *Thermomonospora* and *Saccharomonospora* by addition of 5 µg rifampicin ml⁻¹. For isolation of *Saccharopolyspora rectivirgula*, modified hippurate agar (HAX) can be used, and for *Thermoactinomyces* spp. Czapek yeast casein agar containing tyrosine and novobiocin (CYCTN), again incubated at 55°C (Lacey *et al.* 1992).

Identification of bacteria is highly specialized, and it is not intended to deal with this here. For further information on bacteria, the reader is referred to the *Field Guide* of the American Industrial Hygiene Association (Dillon *et al.* 2005). Illustrated descriptions of selected thermophilic actinomycetes are given in the final section of this book.

REFERENCES

Austwick, P.K.C., Little, S.A., Lawton, L., *et al.* (1989) Microbiology of sick buildings. In B. Flannigan, (ed.), *Airborne Deteriogens and Pathogens*, Biodeterioration Society, Kew, Surrey, UK, pp. 122-128.

Barnett, J.A., Payne, R.W., and Yarrow, D. (2000a). *The Yeasts: Characteristics and Identification*, 2nd Ed. Cambridge University Press, Cambridge.

Barnett, J.A., Payne, R.W., and Yarrow, D. (2000b). *Yeast Identification PC Program*. J.A. Barnett, Norwich, UK.

Bravery, A.F., Berry, R.W., Carey, J.K., *et al.* (1987) *Recognising Wood Rot and Insect Damage in Buildings*. Building Research Establishment, Garston, Watford, UK.

Burgess, L.W., Liddell, C.M., and Summerell, B.A. (1988) *Laboratory Manual for* Fusarium *Research*, 2nd Ed. University of Sydney, Sydney, Australia.

Carmichael, J.W., Kendrick, W.B., Conners, I.L., and Sigler, L. (1980) *Genera of Hyphomycetes*. University of Alberta Press, Edmonton, Alberta, Canada.

Coggins, C.R. (1980) *Decay of Timber in Buildings. Dry Rot, Wet Rot and Other Fungi*. Rentokil Ltd., East Grinstead, UK.

Cole, G.T., and Samson, R.A. (1979) *Patterns of Development in Conidial Fungi*. Pitman, London, UK.

Crous, P.W., Verkley, G.J.M., Groenewald, J.Z., and Samson, R.A., (eds.) (2009) *Fungal Biodiversity*, CBS Laboratory Manual Series 1. CBS-KNAW Fungal Biodiversity Centre, Utrecht, The Netherlands.

De Vries, G.A. (1952) *Contribution to the Knowledge of the Genus* Cladosporium. Dissertation, University of Utrecht, The Netherlands. [Reprinted by J. Cramer, Lehre, Germany, 1967].

Dillon, H.K., Heinsohn, P.A., and Miller, J.D. (2005) *Field Guide for the Determination of Biological Contaminants in Environmental Samples*, 2nd Ed. American Industrial Hygiene Association, Fairfax, VA.

Domsch, K.H., Gams, W., and Anderson, T.-H. (2007). *Compendium of Soil Fungi*, 2nd Ed. IHW-Verlag, Eching, Germany.

Dugan, F.M., Braun, U., Groenewald, J.Z., and Crous, P.W. (2008). Morphological plasticity in *Cladosporium sphaerospermum*. *Persoonia*, **21**, 9-16.

Ellis, M.B. (1971) *Dematiaceous Hyphomycetes*. Commonwealth Mycological Institute, Kew, Surrey, UK.

Frisvad, J.C., Bridge, P.D., and Arora, D.K., (eds.) (1998) *Chemical Fungal Taxonomy*. Marcel Dekker, New York.

Frisvad, J.C., and Samson, R.A. (2004). Polyphasic taxonomy of *Penicillium* subgenus *Penicillium*. A guide to identification of the food and airborne terverticillate Penicillia and their mycotoxins. *Studies in Mycology*, **49**, 1-173.

Gams, W., Christensen, M., Onions, A.H.S., *et al.* (1986). Infrageneric taxa of *Aspergillus*. In R.A. Samson and J.I. Pitt, (eds.), *Advances in* Penicillium *and* Aspergillus *Systematics*, Plenum, New York, pp. 55-62.

Gerlach, W., and Nirenberg, H. (1982) The genus *Fusarium*, a pictorial atlas. *Mitt. Biol. Bundesanst. Land. Forstwissensch., Berlin-Dahlem*, **209**, 1-406.

Hennebert, G.L., and Balon, F. (1996) *Les Mérules des Maisons*, Artel, Namur, France.

King, A.D., Hocking, A.D., and Pitt, J.I. (1979) Dichloran-rose bengal medium for enumeration of molds from foods. *Appl. Environ. Microbiol.*, **37**, 959-964.

Lacey, J., Williamson, P.A.M., and Crook, B. (1992) Microbial emissions from composts made for mushroom production and from domestic waste. In D.V. Jackson, J.M. Merillot and P. L'Hermite, (eds.), *Composting and Compost Quality Assurance Criteria*, EUR14254, Office for Official Publications of the European Community, Luxembourg, pp. 117-130.

Leslie, J. F., Summerell, B.A., Bullock, S. (2006) The *Fusarium* laboratory manual. Blackwell Publishing Ltd., Oxford.

Liou, G.Y., Chen, C.C., Chien, C.Y., *et al.* (1990). *Atlas of the Genus* Rhizopus *and its Allies*. Mycological Monograph No. 3. FIRDI, Hsinchu, Taiwan.

Marasas, W.F.O., and Nelson, P.E. (1984) Mycotoxicology. Introduction to the mycology, plant pathology, chemistry, toxicology and pathology of naturally occurring mycotoxicoses in animals and man. Pennsylvania State University Press, University Park, PA.

Morton, F.J., and Smith, G. (1963) The Genera *Scopulariopsis* Bainier, *Microascus* Zukal, and *Doratomyces* Corda. *Mycol. Papers*, **86**, 1-96.

Murphy, D.M.F., Morgan, W.K.C., and Seaton, A. (1995) Hypersensitivity pneumonitis, In W.K.C Morgan and A. Seaton, (eds.), *Occupational Lung Diseases*, W.B. Saunders Philadelphia, pp. 525-567.

Nelson, P.E., Toussoun, T.A., and Marasas, W.F.O. (1983) Fusarium *Species: An Illustrated Manual for Identification*. Pennsylvania State University Press, University Park, PA.

Nirenberg, H.I. (1989) Identification of Fusaria occurring in Europe on cereals and potatoes. In J. Chelkowski, (ed.), Fusarium: *Mycotoxins, Taxonomy and Pathogenicity*, Elsevier Science, Amsterdam, The Netherlands, pp. 179-193.

Nirenberg, H.I. (1990) Recent Advances in the Taxonomy of *Fusarium. Stud. Mycol.*, **32**, 91-101.

O'Donnell, K.L. (1979) *Zygomycetes in Culture*, Palfrey Contributions in Botany, 2. University of Georgia, Atlanta.

Pitt, J.I. (1979) *The genus* Penicillium *and Its Teleomorphic States* Eupenicillium *and* Talaromyces. Academic Press, London, UK.

Pitt, J.I. (2000) *A Laboratory Guide to Common* Penicillium *Species*, 3rd Ed. CSIRO, Division of Food Processing, North Ryde, Australia.

Pitt, J.I., and Hocking, A.D. (2009) *Fungi and Food Spoilage*, 3rd Ed. Springer, Dordrecht, The Netherlands.

Pitt, J.I., and Samson, R.A. (1993) Species names in current use (NCU) in the Trichocomaceae (Fungi, Eurotiales). *Regnum Vegetabile*, **128**, 13-57.

Pitt, J.I., Samson R.A., and Frisvad, J.C. (2000) List of accepted species and their synonyms in the family *Trichocomaceae*. In R.A. Samson and J.I. Pitt, (eds.) *Integration of Modern Taxonomic Methods for* Penicillium *and* Aspergillus *Classification*, Harwood, Amsterdam, The Netherlands, pp. 9-49.

Raper, K.B., and Fennell, D.I. (1965) *The Genus* Aspergillus. Williams and Wilkins, Baltimore, MD.

Robert, V., and Szoke, S. (2006) BioloMICS Software. Bioaware software SA, Hannut, Belgium.

Robert, V., Groenewald, M., Epping, W., *et al.* (2008) CBS Yeasts Database. Centraalbureau voor Schimmelcultures, Utrecht, The Netherlands.

Samson, R.A. (1992) Current taxonomic schemes of the genus *Aspergillus* and its teleomorphs. In J.W. Bennett and M.A. Klich, (eds.), Aspergillus: *The Biology and Industrial Applications*, Butterworth-Heinemann, Woburn, MA, pp. 355-390.

Samson, R.A. (1994a) Taxonomy – current concepts in *Aspergillus*. In J.E. Smith, (ed.), *Biotechnology Handbooks*: *Aspergillus*, Plenum, London, U.K., pp. 261-276.

Samson, R.A. (1994b) Current systematics of the genus *Aspergillus*. In *The Genus* Aspergillus. *From taxonomy and genetics to industrial applications*. Plenum, New York, pp. 261-276.

Samson, R.A., and Frisvad, J.C. (2004) *Penicillium* subgenus *Penicillium*: new taxonomic schemes, mycotoxins and other extrolites. *Studies in Mycology*, **49**,1-257.

Samson, R.A., and Pitt, J.I. (2000) *Integration of Modern Taxonomic Methods for* Penicillium *and* Aspergillus *Classification*. Harwood, Amsterdam, The Netherlands.

Samson, R.A., and Varga, J. (2007) *Aspergillus* systematics in the genomic era. *Stud. Mycol.*, **59**, 1-206.

Samson, R.A., Hocking, A.D., Pitt. J.I., *et al.* (eds.) (1992) *Modern Methods in Food Mycology*. Elsevier, Amsterdam, The Netherlands.

Samson, R.A., Flannigan, B., Flannigan, M., *et al.* (eds.) (1994) *Health Implications of Fungi in Indoor Environments*. Elsevier, Amsterdam, The Netherlands.

Samson, R.A., Hoekstra, E.S., Frisvad, J.C., *et al.* (eds.) (2000) *Introduction to Food- and Airborne* Fungi, 6th Ed. Centraalbureau voor Schimmelcultures, Baarn, The Netherlands.

Samson, R.A., Houbraken, J., Thrane, U., *et al.* (2010) *Food and Indoor Fungi*. CBS Laboratory Manual Series 2. Centraalbureau voor Schimmelcultures, Utrecht, The Netherlands.

Schipper, M.A.A. (1978) 1. On certain species of *Mucor* with a key to all accepted species. 2. On the genera *Rhizomucor* and *Parasitella*. *Stud. Mycol.*, **17**, 1-70.

Schipper, M.A.A. (1984) Revision of the genus *Rhizopus*. *Stud. Mycol.*, **25**, 1-19.

Schmidt, O. (2006) *Wood and tree fungi. Biology, damage, protection, and use*. Springer, Berlin.

Schöler, H.J., and Muller, E. (1971) *Taxonomy of the pathogenic species of* Rhizopus. 7th Annal Meeting of the British Society of Mycopathology, Edinburgh.

Schubert, K., Groenewald, J.Z., Braun, U., *et al.* (2007) Biodiversity in the *Cladosporium herbarum* complex (*Davidiellaceae, Capnodiales*), with standardisation of methods for *Cladosporium* taxonomy and diagnostics. *Stud. Mycol.*, **58**, 105-156.

Scott, J.A., Untereiner, W.A., Wong, B., *et al.* (2004) Genotypic variation in *Penicillium chysogenum* from indoor environments. *Mycologia*, **96**, 1095-1105.

Scott, J.A., Wong, B., Summerbell, R.C., and Untereiner, W.A. (2008) A survey of *Penicillium brevicompactum* and *P. bialowiezense* from indoor environments, with commentary on the taxonomy of the *P. brevicompactum* group. *Botany*, **86**, 732-741.

Thrane, U. (1989) *Fusarium* species and their specific profiles of secondary metabolites. In J. Chelkowski, (ed.), Fusarium: *Mycotoxins, Taxonomy and Pathogenicity*, Elsevier, Amsterdam, pp. 199-225.

von Arx, J.A. (1981) *The Genera of Fungi Sporulating in Pure Culture*. J. Cramer Verlag, Lehre, Germany.

von Arx, J.A., Guarro, J., and Figueras, M.J. (1986) The ascomycete genus *Chaetomium*. *Nova Hedwigia*, **84**, 1-162.

Wagenführ, R., and Steiger, A. (1966) *Pilze auf Baumholz*. A. Ziemsen Verlag, Wittenberg-Lutherstadt, Germany.

Zalar, P., de Hoog, G.S., Schroers, H.-J., *et al.* (2007) Phylogeny and ecology of the ubiquitous saprobe *Cladosporium sphaerospermum*, with descriptions of seven new species from hypersaline environments. *Stud. Mycol.*, **58**, 157-183.

Chapter 4.4

MICROBIAL VOLATILE ORGANIC COMPOUNDS

Timothy J. Ryan

School of Public Health Sciences and Professions, Ohio University, Athens, OH 45701, USA.

INTRODUCTION

Indoor air has been shown to contain many volatile organic compounds (VOCs), with the vast majority resulting from anthropogenic sources present in such environments. A subset of VOCs is believed to result from the intrusion of outdoor contaminants into the indoor environment. A still smaller group of compounds is considered microbial in origin (MVOCs), which are low molecular weight chemicals thought to be present in the indoor environment largely as the result of the metabolic actions of bacteria and fungi. For almost two decades now there have been suggestions that such volatiles, and the mycotoxins for which they may be indicators (Börjesson *et al.* 1992, Zeringue *et al.* 1993, Jelen *et al.* 1995, Pasanen *et al.* 1996), could be partially to blame for indoor air quality (IAQ) problems caused by fungal growth (Samson 1985, Land *et al.* 1987, Tobin *et al.* 1987, Burge 1989, Sorenson 1989, Flannigan *et al.* 1991, Samson 1992, Miller 1992, Elke *et al.* 1999, Fischer and Dott 2003, Horner and Miller 2003).

The ability of MVOCs to diffuse from enclosed spaces, from behind apparently impermeable vinyl wallpaper and vapour barriers, or off HVAC filters free of visible contamination, has been of interest to investigators seeking to employ MVOCs as indicators of latent mould growth (Elke *et al.* 1999, Ström *et al.* 1994, Ahearn *et al.* 1997). Identification of unique MVOC profiles detected above pure cultures of fungi has been pursued by others with an end toward predicting species-specific mould presence (Larsen and Frisvad 1994, Fiedler *et al.* 2001, Gao and Martin 2002, Schleibinger *et al.* 2005). The innate toxicity of MVOCs has also been of interest to a number of investigators, for their possible role in creating health effects

directly or in conjunction with other indicators such as endotoxin, glucans, and elevated bacterial and mould concentrations. Finally, and because of their significant role in biocorrosion, MVOCs are of interest as chemical methods for the evaluation and prevention of building material destruction (Gutarowska and Piotrowska 2007).

MVOCs are produced during active growth of microorganisms, especially moulds, and are normally found in the indoor environment at very low concentrations. They can be detected in the air of contaminated housing, at commercial composting sites and waste processing facilities, and in many materials with known organic chemical contamination (Kreja and Seidel 2002a; Albrecht *et al.* 2008; Fischer *et al.* 2008).

At present no universally accepted definition or definitive listing of MVOCs exists. Specific compounds emitted and patterns of concentration even under controlled conditions are not always predictable, and generation may be so minimal as to be unable to induce measurable indoor air concentrations (Schleibinger *et al.* 2005). Thus, current research of MVOCs is directly hampered by the lack of a definition of what exactly constitutes a valid MVOC. Nevertheless, an early listing of MVOCs (Wessén and Schoeps 1996) remains unchallenged in the literature and those listings, along with pertinent physical properties of the specific MVOCs, are presented in Tables 1 and 2.

Chemicals held up in the literature as MVOCs are chemically diverse, including alcohols, aldehydes, ketones, esters, ethers, aromatics, amines, terpenes, furans and sulfur-containing compounds. The true diversity of volatiles of microbial origin is probably great, with many volatiles also occurring in indoor air from sources other than microbes. The recent as-

Table 1. Commonly Accepted MVOCs: "Class A" MVOCs[*]

Name	MW	Molecular Formula	Boiling Point °C
2-hexanone	100.16	$C_6H_{12}O$	128
2-heptanone	114.2	$C_7H_{14}O$	151.5
3-methyl-1-butanol	88.15	$C_5H_{12}O$	128.5
3-methylfuran	82.1	C_5H_6O	65.5
1-butanol	74.12	$C_4H_{10}O$	117.2
2-methyl-1-propanol	74.12	$C_4H_{10}O$	108
2-methyl-2-butanol	88.15	$C_5H_{12}O$	102
3-octanone	128.22	$C_8H_{16}O$	167
3-octanol	130.23	$C_8H_{18}O$	174-176
1-octen-3-ol	128.22	$C_8H_{16}O$	174
2-octen-1-ol	128.22	$C_8H_{16}O$	87
2-pentanol	88.15	$C_5H_{12}O$	118-121
dimethyl disulfide	94.19	$C_2H_6S_2$	109.7
1-10-dimethyl-trans-9-decalol (geosmin)	182.31	$C_{12}H_{22}O$	270-271

[*]As defined by Wessen and Schoeps (1996). Data source: NIST 98 spectral library and www.ChemFinder.com.

Table 2. Commonly Accepted MVOCs: "Class B" MVOCs[*]

Name	MW	Molecular Formula	Boiling Point °C
2-pentylfuran	138.1	$C_9H_{14}O$	64-66
2-nonanone	142.24	$C_9H_{18}O$	195,3
4-methyl-3-heptanone	128.22	$C_8H_{16}O$	163-167
fenchone	152.24	$C_{10}H_{16}O$	193-194
endoborneal	154.26	$C_{10}H_{18}O$	212
ethyl isobutyrate	116.16	$C_6H_{12}O_2$	120
karveol	152.2	$C_{10}H_{16}O$	226-227
thujopsene	204.2	$C_{15}H_{24}$	256-257
terpineol	154.3	$C_{10}H_{18}O$	210-218

[*]As defined by Wessen and Schoeps (1996) .Data source: NIST 98 spectral library and www.ChemFinder.com.

sociation of selected MVOCs with specific plasticizers has only further muddied the picture of the possible role MVOCs might play in mould contamination matters. In a study by Kim *et al.* (2007) not only were large variations in MVOC concentrations observed in school classrooms free of any visible signs of dampness or moulds, but some plasticizers and MVOC levels were definitively associated. According to Wessén and Schoeps (1996), an underestimate of the entire microbial impact on a given space when assessed by MVOCs will always occur, since only a few VOCs can pass as originally and uniquely of microbial origin. Re-

ports on pure culture studies of a variety of microbes (Larsen and Frisvad 1995, Wilkins and Larsen 1995, Elke *et al.* 1999, Korpi *et al.* 1999, Wilkins *et al.* 2000, Fiedler *et al.* 2001, Claeson *et al.* 2002, Kreja and Seidel 2002b, Schleibinger *et al.* 2005) reveal anywhere from 25 to 196 MVOCs, belonging to a variety of organic chemical categories, when the organisms are grown on enriched media. The most common MVOCs reported in these studies are shown in Table 3a, listed by type of growth medium and producing fungus.

In addition to their overwhelming diversity, a major hurdle in the use of MVOCs as simple, intuitively understood indicators of contamination was delineated by Schleibinger *et al.* (2005) who found that under controlled pure culture conditions, MVOC production showed a dependency on substrate material, mould genus/species, and strain used. These results echo those of earlier studies as presented in Table 3a, and are notable for the conclusion that only heavy and/or extensive fungal contamination was deemed detectable via MVOCs. This last point is driven home by the results of Claeson *et al.* (2002) who found that MVOC production on the most likely impacted materials, i.e. gypsum board, was very low.

The possible utility of MVOCs with respect to IAQ issues is relatively recent, in that the study of volatiles from biogenic sources has a long history in such diverse interest areas as food and animal feed quality (fermentation processes, taste and aroma), fragrances (plant extracts; perfumes), forestry (emissions and haze), agriculture (insect pheromones, storage spoilage), medicine (biomarkers in blood and urine) and microbial ecology (low-molecular-weight metabolites and intermediates), to name just a few. Occupationally, the inadvertent production of some volatiles has long been recognized viz. hydrogen sulfide or methane with respectively toxic or asphyxiating hazards. Results of early and ongoing IAQ work on fungi provided researchers with a logical group of prospective MVOCs to investigate further as possible indicators of indoor air contamination. For example, the presence of 2-methyl propanol and the "C_8 complex" compounds (so named by Mau *et al.* 1994), e.g. 3-octanol, 3-octanone and 1-octen-3-ol), are readily accepted as MVOCs of fungal origin in the study of feeds contaminated by fungi (Schnürer *et al.* 1999). In both field and pure culture studies, these compounds are now often associated with fungi in the IAQ domain. For a more complete tabulation of the diversity of MVOCs seen in pure culture, see Ammann (1999). To date, the majority of MVOC studies with an IAQ focus have been performed in the laboratory under

Table 3a. MVOCs produced in culture-based studies on enriched media.

Medium	Species	2-alkanones	2-heptanol	2-methyl-1-propanol	2-methyl-1-butanol	3-methyl-1-butanol	3-methyfuran	Dimethyl disulfide	1-butanol	C₈ complex	Limonene
Malt Extract Agar (MEA)	*Aspergillus versicolor*	a	a	c	c	c	c	c		b	
	Penicillium chrysogenum	a	a							b	
	P. commune	a	a	c	c	c	c	c			
	C. cladosporoides			c	c	c	c	c			
	Paecilomyces variotii			c	c	c	c	c			
	Phialophora fastigiata			c	c	c	c	c			
	A. fumigatus			d	d	d				b,d	
	P. brevicompactum			d	d	d				b,d	
Sigma yeast extract sucrose agar (SYES)	*A. versicolor*	a	a	e,f	e,f	e,f				e,f	e,f
	P. chrysogenum	a	a								
	P. commune	a	a								
	Paecilomyces variotti			e,f	e,f	e,f					e,f
	A. fumigatus			e,f	e,f	e,f					e,f
	P. brevicompactum			e,f	e,f	e,f				e,f	e,f
Dichloran 18% glycerol agar (DG-18)	*A. versicolor*			c	c	c	c	c			
	P. chrysogenum										
	P. commune			c	c	c	c	c			
	C. cladosporoides			c	c	c	c	c			
	Paecilomyces variotii			c	c	c	c	c			
	Phialophora fastigiata			c	c	c	c	c			
	A. fumigatus			d	d	d				d	
	P. brevicocompactum			d	d	d			d	d	

Key: a - Wilkins and Larsen (1995); b - Fiedler *et al*. (2001); c - Sunesson *et al*. (1995b); d - Kiviranta *et al*. (1998); e - Fischer *et al*. (1999); f - Fischer *et al*. (2000).

controlled conditions, but a growing number have been completed in the field. It is the field studies that are most germane to the present discussion and so only those findings will be discussed in detail further below.

MVOCS FROM FIELD-BASED STUDIES

A discussion of volatiles of microbial origin under field conditions might logically begin with data collected using the most sensitive detector available, the nose. In fact, the most commonly used method to detect fungal growth in food and feeds is termed "sensory analysis" and is based on the human olfactory system (Schnürer *et al*. 1999). Several MVOCs are readily distinguished by their odour, and for this reason their occurrence is easily associated with damp facilities where mould or moisture are at issue (Dales *et al*.

1991). Perhaps the most characteristic of such odours is that produced by geosmin, with an "earthy" smell often linked to sesquiterpenes or similar compounds (Kiviranta *et al*. 1998). The detection limit of the human olfactory system for geosmin is in the range of 150-200 ng m⁻³. The MVOC 2-methyl-iso-borneol is also perceived as earthy. Other easily recognized odorous MVOCs are 1-octen-3-ol (mushroom-like odour at 10 µg m⁻³) and 2-octen-1-ol (musty odour at 16 µg m⁻³). Chemically related compounds that may contribute to these odour perceptions include 3-octanol and 3-octanone (Ström *et al*. 1994). A listing of MVOCs isolated from pure culture studies on common building materials is shown in Table 3b. As compared with enriched media results (Table 3a) it is clear that the number of MVOCs emanating from building materials is lower and the overlap of types generated (i.e. types consistently detected) is reduced.

In field studies of buildings with visible mould, to-

Table 3b. MVOCs produced on common building materials.

Substrate	Species	2-alkanones	C$_2$-C$_5$ alcohols	2-methyl-1-propanol	2-methyl-1-butanol	3-methyl-1-butanol	2-ethyl-1-hexanol	Dimethyl disulfide	Methyl-alkanones	C$_8$ complex	Geosmin	2-pentylfuran	1-butanol	Terpineol
Wall-paper	*Aspergillus versicolor*		g	g		g				g				
	Penicillium chrysogenum		g	g		g				g				
Card-board	*P. chrysogenum*		h							h				
	Stachybotrys chartarum		h	h										
Gypsum Board/Plasterboard	*Aspergillus versicolor*	i		h,i	i		i			i		i		i
	S. chartarum	j	j	h	j					j			k	
	P. commune	l		l	l	l						l		
	A. fumigatus	i		i	i		i			i		i		i
	P. chrysogenum			h					h	h				
	Streptomyces albidoflavus			m		m			m	m				
	A. sydowi	i		i	i		i			i		i		i
	A. flavus	i		i	i		i			i		i		i
	A. niger	i		i	i		i			i		i		i

Key: g – Larsen and Frisvad (1994); h – Wilkins *et al.* (2000); i – Gao *et al.* (2002); j – Wady *et al.* (2003); k – Gao and Martin (2002); l – Sunesson *et al.* (1996); m – Sunesson *et al.* (1997).

tal concentrations of 18 MVOCs positively correlated with odour perceptions such as those just mentioned when odours were described as belonging to three types (Fischer and Dott 2003). "Fungus-like" odours could not be recognized at <0.035 µg m^{-3}, while a "slight fungal odour" described concentrations of 0.05-1.72 µg m^{-3}, and total concentrations of 0.16-12.3 µg m^{-3} were most often labelled as a "strong" fungus-like odour. These levels are congruent with the finding of Keller *et al.* (1998), who reported the naturally occurring background level for MVOCs as <10 ng m^{-3}. For an additional discussion of MVOCs as they relate to building odour perceptions, see Horner and Miller (2003).

The occurrence of odours as they related to seven selected MVOCs found at composting facilities was studied by Fischer *et al.* (2008). Details are not provided here since that study design did not lend results amenable to statistical analysis. In general, however, MVOC concentration and olfactometry results varied greatly in their association, and coincidence between MVOCs, microbial concentrations and odour was not observed.

Studies of MVOCs at levels below their odour threshold have also been conducted. In their critical examination of 40 homes with and 44 without evident mould damage, Schleibinger *et al.* (2008) provided multiple regression evidence showing no significant association between mould status and the majority (13 of 15) of MVOCs studied. Only 2-methyl-1-butanol and 1-octen-3-ol showed a statistical, albeit weak, association with mould status. The authors concluded that confounders were largely responsible for the difference in MVOC concentrations observed, and therefore the ability of MVOCs to detect mould contamination must be considered with great reservation.

Targeting a list of 14 MVOCs consistently identified in three or more field studies (AIHA 2006), Ryan and Taylor (2006) examined 23 complaint-free Ohio homes for MVOC concentrations. Levels were generally low (5-20 µg m^{-3}) in the houses, and compared with upper floors only 3-octanone was significantly elevated in basements. Owing to their inclusion of basement areas in wet summer months, the authors indicated their results could be taken as worst-case levels in non-problematic homes.

An oft-repeated difficulty encountered in field studies is inherently low MVOC concentrations. Levels of MVOCs found indoors are typically orders of magnitude lower than VOCs in industrial settings, and frequently lower than anthropogenic VOCs in indoor

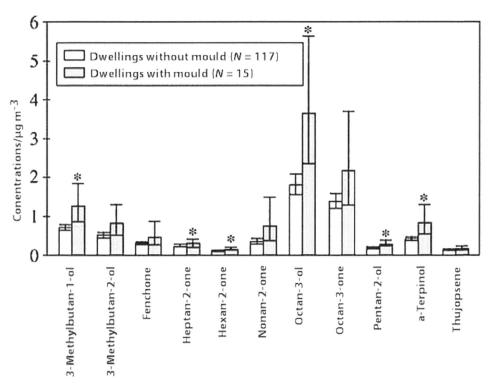

Fig. 1. MVOC concentrations in children's rooms with and without mould formation (geometric means and 95% confidence intervals). Significant differences in t-test (P<0.05) are indicated by asterisk. Reprinted from Elke et al. (1999), with the permission of the Royal Society of Chemistry.

spaces. For example, in waste solvent commingling operations Ryan (2002) found average benzene, toluene, and *m-/p*-xylene levels of 15.3, 33.0 and 36.9 mg m⁻³, respectively. Concentrations of these same three common VOCs averaged only 2.3, 41 and 3.3 μg m⁻³, respectively, in a building free of any IAQ complaints (Ryan *et al.* 2002). Many VOCs in air are the result of the use of gasoline in automobiles and as a result of this use, the microenvironments inside cars in New York City during rush hour were compared and found to contain 9-11 μg m⁻³ benzene, 26-56 μg m⁻³ toluene and 16-23 μg m⁻³ *m-* and *p*-xylenes (Wixtrom and Brown, 1992, Weisel *et al.* 1992). These values can be contrasted with some common MVOCs detected in pure cultures on MEA (Gao *et al.* 2002), where average values ranged from 1.1 (2-pentylfuran) to 19.0 (3-methyl-1-butanol) and 31.8 μg m⁻³ (1-octen-3-ol).

More telling are comparisons with MVOCs determined under field conditions. Based on data from 30 reference buildings generally free of mould or other microbial growth issues, MVOC concentrations of the order of only 2.2-8.8 μg m⁻³ are to be expected. These concentrations are essentially representative of outdoor air, where MVOC values ranged from 1.1-9.5 μg m⁻³, with an average in 27 samples of 4.5 μg m⁻³ (Ström *et al.* 1994). In facilities where microbial contamination is believed to exist, ambient MVOC levels

mimic pure culture results and exhibit a great range of concentrations. Two Swedish homes involved in IAQ complaints were studied by Ström *et al.* (1994) who found total MVOCs in the range of 23.2-26.5 μg m⁻³. Problematic buildings (n=30), i.e. those that suffered from some extent of dampness or water damage, in this same study had average MVOC concentrations of 29.2 μg m⁻³. These levels are over seven times higher than typical outdoor levels, and complaints of odours and eye, nose, and throat irritation were common (although no causal link was established). Mirroring the variability found with viable bioaerosol sampling, data from the contaminated facilities showed much greater variation, with minimum and maximum values of 10.1 and 85.7 μg m⁻³.

Despite such low MVOC concentrations, links between MVOCs and morbidity have been reported. Using passive dosimeters over a one-month sampling period, Elke *et al.* (1999) detected a group of 11 MVOCs, some at quite low levels, in all 132 suspect dwellings studied. For example, geometric means of 0.7 μg m⁻³ for 3-methylbutan-1-ol, 1.8 μg m⁻³ for octan-3-ol and 0.2 μg m⁻³ for pentan-2-ol in rooms without mould contamination were reported. The importance of relative concentrations is demonstrated in this work. Even though levels were low, 6 of the 11 MVOC concentrations differed significantly (p<0.05) between rooms with and without mould (Fig. 1).

These results are difficult to reconcile with those of Schleibinger *et al.* (2008) discussed earlier and Laussman *et al.* (2004) that follow. These works report scientifically valid comparisons that reach essentially the opposite conclusion as to the value of MVOCs. Such divergent results linking (or failing to link) MVOCs to IAQ issues are most likely attributable to variability associated with low concentrations. As there is no other obvious accounting for differences discussed here, the situation poses what might best be characterized as a concentration conundrum.

Despite the positive findings provided by Elke *et al.* (1999) a carefully controlled comparison study by Laussman *et al.* (2004) presented contradictory results, in demonstrating that MVOC results alone were unable to distinguish homes with visible mould growth from control homes clearly free of any contamination. After examining 45 case and 47 control homes, Receiver Operating Characteristic analysis showed conventional culture or spore-based methods readily identified case homes whereas MVOC analysis performed no better than random chance alone in predicting mould issues.

Perhaps because of the concentration conundrum, the prospect of summed MVOC concentrations has been explored. There is limited field data reporting notably higher total VOC (TVOC) concentrations in damp, relatively tightly constructed homes. In his early work on Canadian homes during the winter months, Miller *et al.* (1988) found four houses with TVOC values of nearly 2 mg m^{-3}, with some individual MVOCs more dominant than others. Using the criteria of Seifert and King (1982), only three MVOCs could reasonably be reported by Miller and his colleagues, viz. 3-methyl-1-butanol, 2-hexanone, and 2-heptanone. Some 44% of the 52 homes surveyed contained 3-methyl-1-butanol, while 89% contained 2-hexanone and 89% contained 2-heptanone. The average ratio of relative concentrations of these three was 1:2:4. For these reasons it was suggested that summed totals of common fungal VOCs are unsuitable as indicators of fungal growth. An identical conclusion was reached by Ström *et al.* (1994) who noted the degree of odour problems in the facilities they studied was not directly related to the TVOCs but more to the concentration of individual volatiles (or at least combinations of carefully selected MVOCs). These conclusions were echoed by Korpi *et al.* (1999) who assert that the concept of measuring TVOCs, at least for estimating irritation levels of MVOCs, is of limited value. This is because multiple factors in in-

door air can cause irritation, including dry air, lighting, dust, fibres, ozone, CO_2, tobacco smoke, microorganisms and VOCs (Hempel-Jorgensen *et al.* 1999).

OTHER APPROACHES

There is evidence that all VOCs, and therefore MVOCs, can be adsorbed by a variety of different sized particles from both inorganic and organic sources (Wolkoff and Wilkins 1994). Such particles can adsorb MVOCs present in the environment and later re-emit them by desorption, resulting in elevated concentrations even after the source has been removed. When this happens it is termed the sink effect and has been demonstrated in several studies. Pedersen *et al.* (2002) examined VOCs from dusts and detected MVOCs in samples desorbed under oxidizing conditions (dimethyl disulfide, 2-heptanone and 2-nonanone) as well as under inert conditions (n-butanol). Emissions from the heated (150-200°C) dusts were believed to result from previously adsorbed VOCs, thermal degradation of same, or as the result of oxidation. Dusts heated in helium were noted to emit higher quantities of more compounds, as compared with samples heated in an air stream. Interior surfaces capable of reaching the temperatures of this study include lamps, cooker tops, electric-motor arcs, wood burning fireplaces and furnace or water heater flames.

Insofar as the sink effect is concerned, contradictory results once again exist. It has been demonstrated under laboratory conditions that, while dusts adsorbed five known MVOCs, considerable spontaneous desorption took place within a matter of hours under normal temperatures (Wady and Larsson 2005). One possible explanation of these conflicting results may be linked to different analytical methods, as well as sampling schemes. The authors concluded that MVOC adsorption onto dusts was a reversible process. Given the variability already noted for ambient MVOC studies, sampling dusts would seem to introduce added and unacceptable uncertainty to such applications.

HEALTH EFFECTS

Field Studies

The health hazards of MVOCs *per se* remain of interest to researchers, but also elusive. At present any definitive effects are quite uncertain. Field evidence of a potential link between MVOCs and adverse health effects comes primarily from epidemiological studies examining general indicators of the presence of mould, e.g. "odours", in connection with health outcomes. For example, Engvall *et al.* (2002) found that among 231 Swedish multi-family dwellings, elevated odds ratios (1.2-4.4) for ocular, airway and skin symptoms were statistically associated with signs of dampness and mouldy odour, i.e. only presumptive presence of MVOCs. One in three survey respondents indicated a mouldy odour in their dwelling. Dales *et al.* (1991) conducted a questionnaire-based health effects study and found 32.4% of homes in 30 Canadian communities with mould present, with 14.1% reporting moisture sources such as wet or damp spots in living areas other than basements. Prevalence of respiratory symptoms was consistently higher in moisture or mould affected homes, with significant adjusted odds ratios ranging from 1.32 for bronchitis to 1.89 for cough. Of interest relative to MVOCs, the authors described a dose-response gradient between the number of mould sites and health outcomes, with a maximum odds ratio of 2.55 for cough and the presence of two mould sites. These are only two reports representative of a number of similar studies in Scotland, The Netherlands, Scandinavia, Canada and the U.S. in which dampness and/or mould, determined by questionnaire or building inspection, has been consistently and significantly (p<0.05, minimum) associated with respiratory studies (Dales *et al.* 1991). Although no concentration data were collected, such findings can at least be considered congruent with the assumption that as mould growth increases, so too will MVOC concentrations.

In one of the most all-encompassing epidemiological studies to date, Elke *et al.* (1999) employed passive dosimeters to assess long-term exposures to MVOCs in children's rooms of 132 dwellings. When examined in conjunction with the results from a physician supervised questionnaire, MVOC concentrations were shown to be significantly elevated (p<0.05) in homes reporting a higher prevalence of asthma, hay fever, wheezing and eye irritation. In 15 of the 132 rooms (11.4%) visible mould was reported, and in 19 cases (14.4%) rooms were described as "damp". Confounders such as age, sex, social status, passive smoking and heating type were controlled for in this study. Despite the finding that some MVOC concentrations were significantly higher inside damp or mouldy dwellings (Fig. 1), the association with health effects was not statistically significant.

LABORATORY STUDIES

The ability and potency of three single MVOCs and a mixture of five MVOCs to cause eye and upper respiratory tract irritation in animals was determined by Korpi *et al.* (1999). For all chemicals studied a measurable decrease in respiratory frequency was determined. Individual RD_{50} values were found as follows: 1-octen-3-ol, $RD_{50} = 182$ mg m^{-3}, or 35 ppm; 3-octanol, 1,359 mg m^{-3}, or 256 ppm; and 3-octanone, 17,586 mg m^{-3}, or 3,360 ppm. Thus, the unsaturated 1-octen-3-ol was greater than seven times more potent than the saturated 3-octanol. Histological examination of the lung tissues following the 1-h tests showed no changes even at these levels, i.e. concentrations that are orders of magnitude higher than typically seen in indoor air. The RD_{50} for the five component mixture, containing two of the three aforementioned compounds, was determined to be over 3.5 times lower than could be estimated from the fractional makeup and component-specific RD_{50} values. This led the authors to conclude that some MVOCs may exert a toxic interaction not in an additive fashion, but rather in a synergistic method, insofar as irritation is measured. This synergistic effect was shown to be dominated by 1-octen-3-ol.

From these results, the authors posit that field measurements of single MVOCs as potential causes of occupant sensory irritation, even in the presence of complaints of same, may not be of great value. Furthermore, they note that despite the synergistic effect reported, the recommended indoor air level (RIL) calculated for the mixture tested was 7-17 times greater than the sum of the reported indoor air measurements for the compounds. In addition to work by Pasanen *et al.* (1998) reporting similar outcomes, the utility of measuring total VOCs to estimate sensory irritant levels may be somewhat limited due to the synergism they detected in their study (see the related discussion earlier regarding TVOCs). The authors conclude that typical indoor concentrations of MVOCs are below levels needed for sensory irritation symptoms to be noticed.

At the molecular level there is limited evidence of the genotoxicity of some MVOCs. Luminescent and light absorption *umu* tests confirmed SOS-inducing activity in 15 of 20 compounds studied, two of which were also positive in the Ames test (Nakajima *et al.* 2006). In light of the low MVOC concentrations most often reported, the immediate importance of these findings is unknown.

Some VOCs commonly found in association with MVOCs in the indoor environment include the monoterpenes, emitted from fungi, building materials or furnishings. Mølhave *et al.* (2000), however, ruled out the monoterpene VOCs α-terpineol, 3-carene, n-butanol, limonene and α-pinene as causes of acute eye irritation in indoor air. These results were essentially replicated by Kasanen *et al.* (1999) who reported an RD_{50} for turpentine (mixtures of these and other terpenes) of 1,173 ppm, a level several orders of magnitude higher than usual IAQ concentrations. In the indoor environment persons will typically be exposed to terpenes at only low concentrations (Mølhave *et al.* 2000). Studies of RD_{50} values of the enantiomers of limonene were also confirmed to be quite high relative to IAQ levels of concern. Larsen *et al.* (2000) reported RD_{50} values of 1,000-1,500 ppm in mice, at which levels both species induced only a mild bronchoconstrictive effect. Despite their low potency, there is at least a possibility that the terpenes could have some significance in the residential IAQ discussion, because a broader range of sensitivity is found among domestic indoor occupants as compared with more industrial or occupational populations.

In a unique human volunteer exposure study, Walinder *et al.* (2005) subjected 29 healthy volunteers to a relatively high dose (1,000 µg m⁻³) of a single MVOC, 3-methylfuran. Although the authors conclude that the acute effects observed in the eyes, nose and airways of the subject perhaps resulted from biological activity of the challenge agent, there are several short-comings of their approach. For example, 43% of the test subjects were found to be atopic, and more importantly no positive controls were tested. Given the relatively high exposure dose it cannot be stated with any certainty that the effects elucidated were more than a generalized reaction to a high dose insult.

Although excessive moisture and mould growth are epidemiologically associated with increased symptoms of irritation, allergy, and infection (Fung and Hughson 2003), the importance of MVOCs to such measures remains unclear. Despite these and other associations of health complaints with exposure to fungal volatiles (Hodgson *et al.* 1994), exposure to MVOCs and health effects have not been conclusively linked. This is the case for a variety of reasons, including the possibly episodic nature of MVOC occurrence, widely variable peak exposure times and the extremely low concentrations at which MVOCs are found in indoor environments (as well as the obvious possibility of no actual association). Adverse health

effects of MVOC exposure continue to be researched, but there is no compelling evidence concerning the irritative potency of MVOCs at the concentrations published to date, and the toxicological relevance of these exposures is similarly unclear (Kreja and Seidel, 2002b).

ANALYTICAL METHODS

Thermal desorption of analytes from one or a series of synthetic adsorbents directly onto a capillary GC column, followed by quadrapole mass spectral analysis of the eluting peaks (TD/GC/MS), is the most common analytical method for the detection, identification and quantitation of MVOCs. According to Keller *et al.* (1997) thermodesorption is 100 times more sensitive than solvent extraction/liquid injection of an aliquot of the extract. A variety of instrument manufacturers now offer thermal desorbers and mass spectrometers, so TD/GC/MS is not only commonly commercially available but relatively more affordable. NIOSH has published a method (No. 2549) for the use of such technology in VOC screening (NIOSH 1996), and a review of readily available adsorbents has been published (Sunesson *et al.* 1995a). For these reasons TD/GC/MS has mostly replaced the less qualitative, less sensitive and more cumbersome industrial hygiene analytical technique relying on solvent desorption followed by analysis via a flame ionization detector (FID). That said, Fischer *et al.* (2005) reported comparable results with both methods, concluding that quality assurance issues related to sample handling and calibration must be standardized to ensure acceptable MVOC quantification.

In addition to passive dosimeter and canister collection, adsorption onto a solid phase followed by micro-extraction (SPME) has been employed in several MVOC studies (Nilsson *et al.* 1996, Khaled and Pawliszyn 2000, Koziel and Pawliszyn 2001). Advantages of this approach include the small sampler size, its insertability into small spaces, and ease of use. Limitations are that SPME device exposure is not typically accomplished volumetrically, and so results may be limited to a qualitative report only.

Real-time instruments have long been available for more traditional and well documented industrial or agricultural VOC hazards, e.g. H_2S, methane and combustible gases. Now, with the occurrence of MVOCs in indoor air increasingly better characterized, the emerging development of electronic sensors responsive to MVOCs will almost certainly lead to the eventual availability of hand-held, real-time instruments for their identification and quantitation. Known commonly as "electronic noses", these devices employ a sensor array with different types of nonspecific sensors that, upon exposure to a volatile compound, produce a unique current output pattern. When evaluated by multivariate analysis or processed by an artificial neural network (ANN), the hope is that the collected data can point to a specific fungus as the presumptive cause. However, in their review of the current state of such technology, Kuske *et al.* (2005) point to both low ambient MVOC concentrations as well as interfering compounds as serious limiting factors to the employment of these devices. They go on to note that the great diversity of presumptive MVOCs at this time makes the selection of definite indicators "quasi-impossible".

THE FUTURE OF MVOCS

Compared with other areas of IAQ research, there exist comparatively few field studies of MVOCs. The often-cited work in Swedish houses of Ström *et al.* (1994) is based on only three houses (one of which was a control dwelling). Given the obvious lack of a statistically valid number of observations, the significance of those findings must be regarded with caution. Taken as a whole, the peer-reviewed literature includes relatively few larger field studies (Miller *et al.* 1988, McJilton *et al.* 1990, Bayer and Crow 1993, Schleibing *et al.* 1993, 2005, 2008, Wessén and Schoeps 1996, Morey *et al.* 1997, Elke *et al.* 1999, Fischer *et al.* 2000, Laussman *et al.* 2004, Ryan and Taylor 2006). Field studies which attempted to correlate mould species with predominant MVOCs are also few in number, and those that do exist have yet to be repeated. Therefore, more surveys of MVOCs in relation to the microbial biomass present should be made. A growing database of MVOC concentrations from a variety of environmental conditions, geographic locations, and clean/contaminated facilities will help to elucidate the still elusive utility of MVOCs as indicators of latent mould growth and contamination. It may also be necessary to limit the prospective number of compounds examined, focusing more on those consistently found or found at higher concentrations.

Work with electronic sensors alongside evaluation by ANN approaches in the study of food spoilage has already had a number of successes. For example, ANN results correlated to measured ergosterol and fungal CFUs, and trimethylamine as an indicator of common bunt contamination (Schnürer *et al.* 1999) and volatile sesquiterpenes from fungi in cheeses (Larsen 1997). Owing to the complexity of MVOC patterns possible under field conditions (Larsen and Frisvad 1994, Wilkins *et al.* 2000, Fiedler *et al.* 2001, Gao and Martin 2002), work to establish such ANNs for field-determined MVOCs is necessary (Kenyon *et al.* 1997, Harrington *et al.* 2002). Such studies might begin with experiments like those of Korpi *et al.* (1998) in which mixed cultures of moulds were cultured at various relative humidity levels, but with a less restrictive definition of MVOCs. The advent of more sensitive and qualitative real-time instruments for MVOC determination will open the door to better quantitation of traditional MVOCs. Ripp *et al.* (2003), for example, have reported the development of a bioluminescent bioreporter interfaced with a micro-luminometer to create a sensor capable of detecting the MVOC *p*-cymeme within 3.5 h of its production by *P. roqueforti*.

Finally, almost all researchers of MVOCs list compounds either not detected, or not examined, by other researchers. Conversely, most MVOC works list a variable number of compounds that other authors have also found. This situation mirrors that for VOCs as a whole, where sets of VOCs reported in different studies are inconsistent, and where the criteria on which the authors' selections are based is sometimes unclear (Becher *et al.* 1996). There is no universally accepted listing of MVOCs, perhaps with good reason at this time, but there is a select group of compounds most researchers would accept as arising uniquely from microbial production in indoor environments. The description of Class A and Class B MVOCs (Tables 1 and 2) was a start, but the selection process for these early compilations was based on limited observations under limited environmental conditions. Taking field-identified MVOCs detected by three or more authors as a criterion, the compounds in Table 4 are suggested as a more current compilation of MVOCs (AIHA 2005). With the adoption of a somewhat standard listing of MVOCs, research on the occurrence, prevalence and detectability of such substances can progress in a logical and consistent fashion.

SUMMARY

The ability to detect mould without the need for culture-based methods has long been the promise of microbial volatile organic compounds (MVOCs). Technology is progressing to the point where electronic devices can readily detect such compounds

Table 4. 14 MVOCs isolated in three or more field studies (AIHA 2005).

1-butanol	3-octanone
3-methyl-1-butanol	3-octanol
3-methyl-2-butanol	1-octen-3-ol
3-methyl furan	2-octen-1-ol
2-pentanol	2-nonanone
2-hexanone	borneol
2-heptanone	geosmin

soon after generation or liberation. But MVOCs have as yet not fulfilled their potential as quick, inexpensive analytical indicators of microbial growth. It is becoming clear that simply detecting these compounds in contaminated indoor environments yields limited useful information about the nature of that contamination. There are many reasons for this and chief among them are the low concentrations seen, their variable occurrence, low inherent toxicity, chemical confounding, the arbitrary selection of the MVOCs for inclusion in any given study, and substrate, moisture, and other environmental factor dependence for their appearance. Research on MVOCs should continue as it is possible they will have some use in air quality investigations as technology and understanding advances.

REFERENCES

Ahearn, D.G., Crow, S., Simmons, R. B., *et al.* (1997) Fungal colonization of air filters and insulation in a multi-story office building: production of volatile organics. *Current Microbiol.*, 35, 305-308.

Albrecht, A., Fischer, G., Brunnermann-Stubbe, G., *et al.* (2008) Recommendations for study design and sampling strategies for airborne microorganisms, MVOC and odours in the surrounding of composting facilities. *Int. J. Hyg. Environ. Health*, 211, 121-131.

AIHA (2005) *Field Guide for the Determination of Biological Contaminants in Environmental Samples*, L.-L.Hung, J.D. Miller, and H.K. Dillon, (eds.), American Industrial Hygiene Association, Fairfax, VA.

Ammann, H.M. (1999) Microbial volatile organic compounds. In J. Macher, (ed.), *Bioaerosols Assessment and Control*, American Conference of Governmental Industrial Hygienists, Cincinnati, OH, pp. 26-1 – 26-17.

Bayer, C.W., and Crow, S. (1993) Odorous volatile emissions from fungal contamination. *Proceedings of IAQ '93*, American Society of Heating, Refrigerating, and Air-Conditioning Engineers, Atlanta, GA, pp. 165-170.

Becher, R., Hongslo, J., Jantunen, M., and Dybing, E. (1996) Environmental chemicals relevant for respiratory hypersensitivity: the indoor environment. *Toxicol. Lett.*, 86, 155-162.

Börjesson, T., Stollman, U., and Schnürer, J. (1992) Volatile metabolites produced by six fungal species compared with other

indicators of fungal growth on cereal grains. *Appl. Environ. Microbiol.*, 58, 2599-2605.

Burge, H.A. (1989) Airborne allergenic fungi. Classification, nomenclature, and distribution. *Immunol. Allergy Clin. N. Am.*, 9, 307-319.

Claeson, A.-S., Levin, J.-O., Blomquist, G., and Sunesson, A.-L. (2002) Volatile metabolites from microorganisms grown on humid building materials and synthetic media. *J. Environ. Monit.*, 4, 667-672.

Dales, R.E., Zwanenburg, A., Burnett, R., and Franklin, C.A. (1991) Respiratory health effects of home dampness and molds among Canadian children. *Am. J. Epidemiol.*, 134, 196-203.

Elke, K., Begerow, J., Oppermann, H., *et al.* (1999) Determination of selected microbial volatile organic compounds by diffusion sampling and dual-column capillary GC-FID – a new feasible approach for the detection of an exposure to indoor mould fungi? *J. Environ. Monit.*, 1, 445-452.

Engvall, K., Norrby, C., and Norbäck, D. (2002) Ocular, airway and dermal symptoms related to building dampness and odors in dwellings. *Arch. Environ. Health*, 57, 304-310.

Fiedler, K., Schutz, E., and Geh, S. (2001) Detection of microbial volatile organic compounds (MVOCs) produced by moulds on various materials. *Int. J. Hyg. Environ. Health*, 204, 111-121.

Fischer, G., and Dott, W. (2003) Relevance of airborne fungi and their secondary metabolites for environmental, occupational, and indoor hygiene. *Arch. Microbiol.*, 179, 75-82.

Fischer, G., Schwalbe, R., Möller, M., *et al.* (1999) Species-specific production of microbial volatile organic compounds (MVOC) of airborne fungi from a compost facility. *Chemosphere*, 39, 795-810.

Fischer, G., Müller, M., Schwalbe, R., *et al.* (2000) Exposure to airborne fungi, MVOC and mycotoxins in biowaste-handling facilities. *Int. J. Hyg. Environ. Health*, 203, 97-104.

Fischer, G., Möller, M., Gabrio, T., *et al.* (2005) Comparison of methods for quantification of MVOC in indoor environments. *Bundesgesundheitsbl. Gesundheitsforsch. Gesundheitssch.*, 48, 43-53.

Fischer, G., Albrecht, A., Jäckel, U., and Kämpfer, P. (2008) Analysis of airborne microorganisms, MVOC and odour in the surroundings of composting facilities and implications for future investigations. *Int. J. Hyg. Environ. Health*, 211, 132-142.

Flannigan, B., McCabe, E.M., and McGarry, F. (1991) Allergenic and toxigenic micro-organisms in houses. *J. Appl. Bact.*, 70 (*Symposium Supplement*), 61S-73S.

Fung, F., and Hughson, W.G. (2003) Health effects of indoor fungal bioaerosol exposure. *Appl. Occup. Environ. Hyg.*, 18, 535-544.

Gao, P., and Martin J. (2002) Volatile metabolites produced by three strains of *Stachybotrys chartarum* cultivated on rice and gypsum board. *Appl. Occup. Environ. Hyg.*, 17, 430-436.

Gao, P., Korley, F., Martin, J., and Chen, B.T. (2002) Determination of unique microbial volatile organic compounds produced by five *Aspergillus* species commonly found in problem buildings. *AIHA J.*, 63, 135-140.

Gutarowska, B., and Piotrowska, M. (2007) Methods of mycological analysis in buildings. *Build. Environ.*, 42, 1843-1850.

Harrington, P. de B., Voorhees, K.J., Basile, F., and Hendricker, A.D. (2002) Validation using sensitivity and target factor analyses of neural network models for classifying bacteria from mass spectra. *J. Am. Soc. Mass Spectrom.*, 13, 10-21.

Hempel-Jorgensen, A., Kjaergaard, S., Mølhave, L., and Hudnell, K. (1999) Sensory eye irritation in humans exposed to mixtures of volatile organic compounds. *Arch. Environ. Health*, 54, 416-424.

Hodgson, M., Levin, H., and Wolkoff, P. (1994) Volatile organic compounds and indoor air. *J. Allergy Clin. Immunol.*, **94**, 296-303.

Horner, E., and Miller, J.D. (2003) Microbial volatile organic compounds with emphasis on those arising from filamentous fungal contaminants of buildings. *ASHRAE Transact.*, **109**, 215-231.

Jelen, H.H., Mirocha, C.J., Wasowicz, E., and Kaminski, E. (1995) Production of volatile sesquiterpenes by *Fusarium sambucinum* strains with different abilities to synthesize trichothecenes. *Appl. Environ. Microbiol.*, **61**, 3815-3820.

Kasanen, J.-P., Pasanen, A.-L., Pasanen, P., et al. (1999) Evaluation of sensory irritation of Δ-3-carene and turpentine, and acceptable levels of monoterpenes in occupational and indoor environment. *J. Tox. Environ. Health, Part A*, **56**, 89-114.

Keller, R., Sonnichsen, R., and Ohgke, H. (1997) *Umweltmed. Forsch. Prax.*, **2**, 265. Cited in Elke, K., Begerow, J., Oppermann, H., et al. (1999) Determination of selected microbial volatile organic compounds by diffusion sampling and dual-column capillary GC-FID – a new feasible approach for the detection of an exposure to indoor mould fungi? *J. Environ. Monit.*, **1**, 445-452.

Keller, R., Senkpiel, K., and Ohgke, H. (1998) Geruch als Indicator für Schimmelpilzbelastungen in natürlich belüfteten Innenräumen – Nachweis mit analytischer MVOC-Messung. Gesundheitliche Gefahren durch biogene Luftschadstoffe. *Schriftenreihe des Inst. für Med. Mikrobiologie Hygiene der Med. Universität, Lübeck*, Heft 2. Cited in Fischer, G., and Dott, W. (2003) Relevance of airborne fungi and their secondary metabolites for environmental, occupational and indoor hygiene. *Arch. Microbiol.*, **179**, 75-82.

Kenyon, R.G., Ferguson, E.V., and Ward, A.C. (1997) Application of neural networks to the analysis of pyrolysis mass spectra. *Zentralbl. Bakteriol.*, **285**, 267-277.

Khaled, A., and Pawliszyn, J. (2000) Time-weighted average sampling by volatile and semi-volatile airborne organic compounds by the solid-phase microextraction device. *J. Chromatogr. A*, **892**, 455-467.

Kim, J.L., Elfman, L., Mi, Y., et al. (2007) Indoor molds, bacteria, microbial volatile organic compounds and plasticizers in schools – associations with asthma and respiratory symptoms in pupils. *Indoor Air*, **17**, 153-163.

Kiviranta, H., Tuomainen, A., Reman, M., et al. (1998) Quantitative identification of volatile metabolites from two fungi and three bacteria species cultivated on two media. *Centr. Eur. J. Publ. Health*, **6**, 296-299.

Korpi, A., Pasanen, A.-L., and Pasanen, P. (1998) Volatile compounds originating from mixed microbial cultures on building materials under various humidity conditions. *Appl. Environ. Microbiol.*, **64**, 2914-2919.

Korpi, A., Kasanen, J., Alarie, A., et al. (1999) Sensory irritating potency of some microbial volatile organic compounds (MVOCs) and a mixture of five MVOCs. *Arch. Environ. Health*, **54**, 347-352.

Koziel, J., and Pawliszyn, J. (2001) Air sampling and analysis of volatile organic compounds with solid phase microextraction. *J. Air Waste Manage. Assoc.*, **51**, 173-184.

Kreja, L., and Seidel, H. (2002a) Evaluation of the genotoxic potential of some microbial volatile organic compounds (MVOC) with the comet assay, the micronucleus assay and the HPRT gene mutation assay. *Mut. Res.*, **513**, 143-150.

Kreja, L., and Seidel, H. (2002b) On the cytotoxicity of some microbial volatile organic compounds as studied in the human lung cell line A549. *Chemosphere*, **49**, 105-110.

Kuske, M., Romain, A.-C., and Nicolas, J. (2005) Microbial volatile organic compounds as indicators of fungi. Can an electronic nose detect fungi in indoor environments? *Build. Environ.*, **40**, 824-831.

Land, C.J., Hult, K., Fuchs, K., et al. (1987) Tremorgenic mycotoxins from *Aspergillus fumigatus* as a possible occupational health problem in sawmills. *Appl. Environ. Microbiol.*, **53**, 787-790.

Larsen, T.O. (1997) Identification of cheese-associated fungi using selected ion-monitoring of volatile terpenes. *Lett. Appl. Microbiol.*, **24**, 463-466.

Larsen, T.O., and Frisvad, J.C. (1994) Health Implications of fungi in indoor environments: Production of volatiles and presence of mycotoxins in conidia of common indoor penicillia and aspergilli. In R.A. Samson, B. Flannigan, M.E. Flannigan, et al., (eds.), *Health Implications of Fungi in Indoor Environments*, Elsevier, Amsterdam, pp. 251-279.

Larsen. T.O., and Frisvad, J.C. (1995) Characterization of volatile metabolites from 47 *Penicillium* taxa. *Mycol. Res.*, **99**, 1153-1166.

Larsen, S.T, Hougaard, K.S., Hammer, M., et al. (2000) Effects of R-(+)- and S-(-)-limonene on the respiratory tract in mice. *Human Exp. Toxicol.*, **19**, 456-466.

Laussman, D., Eis, D., and Schleibinger, H. (2004) Comparison of mycological and chemical analytical laboratory methods for detecting mold damage in indoor environments. *Bundesgesundheitsbl. Gesundheitsforsch. Gesundheitssch.*, **47**, 1078-94.

Mau, J., Beelman, R.B., and Ziegler, G.R. (1994) Aroma and flavor components of cultivated mushrooms. In G. Charalambous, (ed.), *Spices, Herbs and Edible Fungi*, Amsterdam, Elsevier, pp. 657-684.

McJilton, C.E., Reynolds, S.J., Streifel, A.J., and Pearson, R.L. (1990) Bacteria and indoor odor problems – three case studies. *Am. Ind. Hyg. Assoc. J.*, **51**, 545-549.

Miller, J.D. (1992) Fungi as contaminants in indoor air. *Atmos. Environ.*, **26A**, 2163-2172.

Miller, J.D., Laflamme, A.M., Sobol, Y., et al. (1988) Fungi and fungal products in some Canadian Houses. *Int. Biodet.*, **24**, 103-120.

Mølhave, L., Kjaergaard, S., Hempel-Jorgensen, A., et al. (2000) The eye irritation and odor potencies of four terpenes which are major constituents of the emissions of VOCs from Nordic soft woods. *Indoor Air*, **10**, 315-318.

Morey, P., Wortham, A., Weber, A., et al. (1997) Microbial VOCs in moisture damaged buildings. *Proceedings of Healthy Buildings/IAQ '97*, Vol. 1, American Society of Heating, Refrigerating, and Air-Conditioning Engineers, Atlanta, GA, pp. 245-250.

Nakajima, D., Ishii, R., Kageyama, S., et al. (2006) Genotoxicity of microbial volatile organic compounds. *J. Health Sci.*, **52**, 148-153.

Nilsson, T., Larsen, T.O., Montanarella, L., and Madsen, J.O. (1996) Application of headspace solid phase microextraction for the analysis of volatile metabolites emitted by *Penicillium* species. *J. Microbiol. Methods*, **25**, 245-255.

NIOSH. (1996) Volatile organic compounds (screening): Method 2549. *Manual of Analytical Methods (NMAM)*, National Institute for Occupational Safety and Health, Cincinnati, OH., pp. 1-8.

Pasanen, A.-L., Lappalainen, S., and Pasanen, P. (1996) Volatile organic metabolites associated with some toxic fungi and their mycotoxins. *Analyst*, **121**, 1949-1953.

Pasanen, A.-L., Korpi, A., and Kasanen, J.-P. (1998) Critical aspects on the significance of microbial volatile metabolites as indoor air pollutants. *Environ. Int.*, **24**, 703-712.

Pedersen, E., Bjørseth, O., and Mathiesen, M. (2002) Emissions from heated indoor dust. *Environ. Int.*, **27**, 579-87.

Ripp, S., Daumer, K.A., McKnight, T., *et al.* (2003) Bioluminescent bioreporter integrated-circuit sensing of microbial volatile organic compounds. *J. Ind. Microbiol. Biotechnol.*, **30**, 636-642.

Ryan, T.J. (2002) Survey of waste comminglers' VOC exposures. *J. Air Waste Manage. Assoc.*, **52**, 1298-1306.

Ryan, T., and Taylor, K. (2006) MVOCs as indicators of prevalent indoor fungi in 23 homes. *American Industrial Hygiene Conference and Exposition, 13-18 May, 2006, Chicago,* Conference Abstracts, p. 28.

Ryan, T.J., Hart, E.H., and Kappler, L.L. (2002) VOC exposures in a mixed-use university art building. *AIHA J.*, **63**, 703-708.

Samson, R.A. (1985) Occurrence of moulds in modern living and working environments. *Eur. J. Epi.*, **1**, 54-61.

Samson, R.A. (1992) Mycotoxins: a mycologist's perspective. *J. Med. Vet. Mycol.*, **30**, 9-18.

Schleibinger, H., Bock, R., and Ruden, H. (1993) Occurrence of microbiologically produced aldehydes and ketones (MVOC) from filter materials of HVAC systems. *Proceedings of IAQ '95*, American Society of Heating, Refrigerating, and Air-Conditioning Engineers, Atlanta, GA, pp. 91-100.

Schleibinger, H., Laussmann, D., Brattig, C., *et al.* (2005) Emission patterns and emission rates of MVOC and the possibility for predicting hidden mold damage? *Indoor Air*, **15** (Suppl. 9), 98-104.

Schleibinger, H., Laussmann, D., Bornehag, C.-G., *et al.* (2008) Microbial volatile organic compounds in the air of moldy and mold-free indoor environments. *Indoor Air*, **18**, 113-124.

Schnürer, J., Olsson, J., and Börjesson, T. (1999) Fungal volatiles as indicators of food and feeds spoilage. *Fungal Gen. Biol.*, **27**, 209-217.

Seifert, R.M., and King, A.D. (1982) Identification of some volatile constituents of *Aspergillus clavatus*. *J. Agric. Chem.*, **39**, 786-790.

Sorenson, W. (1989) Health impact of mycotoxins in the home and workplace – an overview. In C.E. O'Rear and G.C. Llewellyn, (eds.), *Biodeterioration Research*, Plenum Press, New York, pp. 201-215.

Ström, G., West, J., Wessén, B., and Palmgren, U. (1994) Health implications of fungi in indoor environments: quantitative analysis of microbial volatiles in damp Swedish houses. In R.A. Samson, B. Flannigan, M.E. Flannigan, *et al.*, (eds.), *Health Implications of Fungi in Indoor Environments*, Elsevier, Amsterdam, pp. 291-305.

Sunesson, A.-L., Nilsson, C.-A., Blomquist, G., and Andersson, B. (1996) Volatile metabolites produced by two fungal species cultivated on building materials. *Ann. Occup. Hyg.*, **40**, 397-410.

Sunesson, A-L., Nilsson, C.-A., Carlson, R., *et al.* (1997) Production of volatile metabolites from *Streptomyces albidoflavus* cultivated on gypsum board and tryptone glucose extract agar-influence of temperature, oxygen, and carbon dioxide levels. *Ann. Occup. Hyg.*, **41**, 393-413.

Sunesson, A.-L., Nilsson, C.-A., and Andersson, B. (1995a) Evaluation of adsorbents for sampling and quantitative analysis of microbial volatiles using thermal desorption-gas chromatography. *J. Chromatogr. A*, **699**, 203-214.

Sunesson, A.-L., Nilsson, C.-A., Blomquist, G., *et al.* (1995b) Identification of volatile metabolites from five fungal species cultivated on two media. *Appl. Environ. Microbiol.*, **61**, 2911-2918.

Tobin, R.S., Baranowski, E., Gilman, A.P., *et al.* (1987) Significance of fungi in indoor air. *Can. J. Public Health*, **78**, S1-S3.

Wady, L., and Larsson, L. (2005) Determination of microbial volatile organic compounds adsorbed on house dust particles and gypsum board using SPME/GC-MS. *Indoor Air*, **15** (Suppl. 9), 27-32.

Wady, L., Bunte, A., Pehrson, C., and Larsson, L. (2003) Use of gas chromatography-mass spectrometry/solid phase microextraction for the identification of MVOCs from moldy building materials. *J. Microbiol. Methods*, **52**, 325-332.

Walinder, R., Ernstgard, L., Johanson, G., *et al.* (2005) Acute effects of a fungal volatile compound. *Environ. Health Perspectives*, **113**, 1775-1778.

Weisel, C.P., Lawryck, N.J., and Lioy, P.J. (1992) Exposure to emissions from gasoline within automobile cabins. *J. Exp. Anal. Environ. Epidemiol.*, **2**, 79-96.

Wessén, B., and Schoeps, K.-O. (1996) Microbial volatile organic compounds-what substances can be found in sick buildings? *Analyst*, **121**, 1203-1205.

Wilkins, K., and Larsen, K. (1995) Variation of volatile organic compound patterns of mold species from damp building. *Chemosphere*, **31**, 3225-3236.

Wilkins, K., Larsen, K., and Simkus, M. (2000) Volatile metabolites from mold growth on building materials and synthetic media. *Chemosphere*, **41**, 437-446.

Wixtrom, R.N., and Brown, S.L. (1992) Individual and population exposures to gasoline. *J. Exp. Anal. Environ. Epidemiol.*, **2**, 23-78.

Wolkoff, P., and Wilkins, C.K. (1994) Indoor VOCs from household floor dust – comparison to headspace with desorbed VOCs – method for VOC release determination. *Indoor Air*, **4**, 248-54.

Zeringue, H.J., Bhatnager, D., and Cleveland, T.E. (1993) $C_{15}H_{24}$ volatile compounds unique to aflotoxigenic strains of *Aspergillus flavus*. *Appl. Environ. Microbiol.*, **59**, 2264-2270.

Chapter 4.5

ANALYSIS FOR TOXINS AND INFLAMMATORY COMPOUNDS

Thomas G. Rand[1] and J. David Miller[2]

[1]*St Mary's University, Halifax, Nova Scotia, Canada;*
[2]*Institute of BioChemistry, Carleton University, Ottawa, Canada .*

INTRODUCTION

As has been mentioned in earlier chapters, it is well recognized that fungi are common in almost all air in buildings, in settled dusts and in building materials, with few exceptions such as hospital operating rooms and other specialized, well-maintained cleanroom environments. Although hundreds of species from indoor environments have been reported from both temperate and subtropical climates (Samson 1999), only a few dozen are common on mould-damaged materials (Miller *et al.* 2008). Amongst the most widely found moist materials in buildings supporting mould growth are those containing cellulose (see Chapter 2.1), which may be lignified, as in structural and manufactured wood; delignified, e.g. in the surface covering of gypsum wallboard, blown insulation, fibreglass insulation backing, etc.; or in esterified form, such as water-based paint thickeners, wallpaper adhesives, etc. Moulds can also grow on damp carpeting and under-carpet adhesives; some furnishings; books/papers; clothes and other fabrics; and on moist, dirty surfaces such as concrete, fibreglass insulation, and ceramic tiles (Miller *et al.* 2008). The fungi common on mouldy building materials comprise species adapted to the nutrients available and further affected by moisture content. No matter the substrate composition, they are with modest exceptions represented by a narrow range of soil-dwelling species, especially within the genera *Aspergillus, Eurotium* and *Penicillium*.

In a large study in Canada, when there was mould growth indoors air particulates from long duration samples comprised spores (~30% of airborne fungal β-glucan from concurrent Air-O-Cel™ samples), spore and hyphal fragments (~30%) and fragments

smaller than can be recognized by light microscopy (~40%; calculated from Foto *et al.* 2004, 2005). Using a MOUDI air sampler, Salares *et al.* (2009) found 70-80% of airborne β-glucan in 3-h samples. When Reponen *et al.* (2007) measured spores and spore fragments in some homes in Cincinnati and New Orleans using a cyclone sampler, they found spores and fragments and materials in broadly similar proportions (for other data, see Green *et al.* 2006). While exposure to these materials may occur through ingestion, and touching mouldy surfaces, the main route of exposure to mould for people living or working in mouldy indoor environments is inhalation of airborne fungal spores and fragments (Health Canada 2004). Inhaled fungal spores and mycelial fragments contain allergens, β-glucan and low-molecular-weight toxins. Although the largest of these particles neither reach the lungs in high efficiency nor penetrate deeply, the smallest do both. In addition to this, particles are not evenly dispersed throughout the lung as a function of their size, but because of airflow patterns are present in greater amounts at certain spots in the lung (Kleinstreuer *et al.* 2007, 2008, Phalen *et al.* 2008). As Miller *et al.* (2003) observed, overall effects on lung biology at low dose exposures would be consequent on all these components, and some of the effects might be additive. Almost two decades earlier, Miller (1990, 1992) had noted that exposure to low-molecular-weight toxins could be expected to be associated with a number of effects. One of these would be damage to the macrophage system, thus interfering with the particle clearance system of the lung and other processes mediated by lung cells. Several mycotoxins had been shown to interfere with macrophage function (Sorenson 1989, 1999) and the effects were expected to include changes in chemokine expression and to extend beyond the lung (see Miller 1992, Rand *et al.* 2009).

Cells of the respiratory system are therefore among the first cells of the body to be exposed and respond to these airborne environmental contaminants. Consequently, the function of these cells may be altered before cells in other parts of the body are affected. Among others, the National Academy of Sciences (NAS 2004) and Health Canada (2004) have concluded that there is sufficient evidence of an association between exposure to moulds in damp indoor environments and upper respiratory tract symptoms. Exposure symptoms are typically related to allergic and asthmatic reactions, especially upper (proximal) respiratory tract symptoms; asthma in sensitized individuals; rhinosinusitis; and wheeze. These symptoms are largely due to exposure to the mostly unidentified allergens associated with spore and hyphal cell walls that are capable of provoking allergenic and asthmatic responses in humans. Reviews of the epidemiological and experimental evidence for mould-provoked allergic and asthmatic disease have been published elsewhere, and the reader is referred to the chapters in Section 3 for detailed information.

Epidemiological and experimental evidence for mould-related respiratory disease also point to the involvement of non-allergenic mechanisms. These other exposure outcomes, induced by non-allergenic mechanisms, include lower respiratory tract symptoms; cough; headaches, fatigue and memory problems; peripheral appendage numbness and other nervous system effects; and increased susceptibility to lower respiratory tract and other infections. These effects are often dose dependent, in that more visible mould results in more symptoms, and these cannot be attributed to allergy alone (see, for example, Dales *et al.* 1991a,b). It has been speculated that non-allergenic effects are related to exposure to toxin-producing species that grow and sporulate on wet building materials.

The epidemiological signals from moisture-damaged buildings have been linked to exposure to materials from a consortium of toxin-producing species containing the bioactive cell wall polysaccharide $(1{\rightarrow}3)$-β-D glucan, which is inflammatory, and also to spores, mycelial fragments and sequestered in their walls other products containing low-molecular-weight compounds that are either cytotoxic or have some other toxic, including inflammatory properties. These materials contain species- and strain-dependent mixtures of toxins, β-glucan and proteins. Some of these compounds are found in high concentrations

in spores and spore fragments. Amounts of $(1{\rightarrow}3)$-β-D glucan in spores of the species found on building materials range from 1-11 pg spore^{-1} (Foto *et al.* 2004, Iossifova *et al.* 2008a).

Because there are 5-day air measurement data for β-glucan (Foto *et al.* 2005, Miller *et al.* 2007), as well as concurrent PM2.5 data (Miller *et al.* 2007), it is possible to make a reliable assessment of exposure. Brown *et al.* (2005) provided lung PM2.5 dosimetry for resting and moderate nasal breathing and for moderate oral breathing, i.e. all particulate matter ≤2.5 µm in size, which based on Foto *et al.* (2005) and Salares *et al.* (2009) comprise 70-90% of the total airborne β-glucan . On a 24-h basis, lung exposure to β-glucan in materials <2.5 µm for resting and moderate nasal breathing, and moderate oral breathing, would be 10^{-8}-10^{-9} moles. Since the effects of β-glucan exposure persist (Rand *et al.* 2009) a cumulative effect can reasonably be expected.

The reproductive structures of many fungi contain known toxins, usually in high concentrations (Sorenson 1989, 1999). Most is known about the fungi that produce toxins important in agriculture. The conidia of *Aspergillus flavus*, *A. parasiticus*, *Fusarium graminearum* and *F. sporotrichioides* contain very high concentrations of toxins, particularly in the case of the two aflatoxin-producing aspergilli. The spores of the two aspergilli have been reported to contain 100-1100 µg aflatoxin g^{-1}, or ~10^{-4} moles (Wicklow and Shotwell 1983). To put this in perspective, the allowable concentration of aflatoxin in food in Canada is 10 ng g^{-1}, or about 1/10,000 of that concentration. A number of interesting toxins have been found in sclerotia of various aspergilli, again at high concentrations that do not occur either in culture or in affected crops (Gloer *et al.* 1988, Wicklow *et al.* 1988), and some of which are thought to be present in conidia along with kojic acid and other *A. flavus* toxins. Spores of *F. graminearum* contained 30 µg g^{-1} of the regulated trichothecene deoxynivalenol and those of *F. sporotrichioides* 50 µg g^{-1} T-2 toxin (both ~10^{-5} moles). It should be noted that the toxin quantified in these two fusaria is in both cases only one of many of the potent toxins produced by these species (Miller 1992).

With respect to the species associated with the built environment, there are few quantitative data on toxins in spores. *A. versicolor* is known to contain sterigmatocystin and *P. crustosum* contains penetrim A (Fischer *et al.* 2000). Conidia of *A. fumigatus*, which accumulates in settled dust indoors and under some circumstances will grow on building materials, accumulate fumitremorgens A, B and C, verruculogen, tryptacidin and tryptoquivalene (Land *et al.* 1994, Fischer *et al.* 2000). On summing the fumitremorgen B and C and verruculogen values, the total concentrations were similar to the fungi mentioned above, ~1-14 µg g^{-1} or ~10^{-5} moles, depending on strain (Land *et al.* 1994). Conidia of *Stachybotrys chartarum* were found to contain 10-15 µg g^{-1} satratoxin G and trichoverrols A and B, or ~10^{-5} moles (Sorenson *et al.* 1987). *S. chartarum* produces many toxins, but there has been only one report of the spectrum of toxins detected. The spores of one strain produced satratoxins G and H, and compounds of the group of phenylspirodrimanes, viz. stachybotrylactone and stachybotrylactam (Nikulin *et al.* 1997). The concentrations of spirocyclic drimanes present in spores are not known but these compounds are present at levels typically an order of magnitude higher than the macrocyclic trichothecenes. Studies of metabolite production by the less cytotoxic strains have shown that they consistently produce diterpenes called atranones, which are potently inflammatory (Rand *et al.* 2006). Nothing is known about the presence or absence in *S. chartarum* conidia of either important group of toxins.

Based on what is known, toxin concentrations in spore and spore fragments of fungi that grow on damp building materials are in the order of ~10^{-5} moles, and based on the β-glucan exposure data, at least 70% of that is in particles <2.5 µm. As noted, at least for aflatoxin (Jakab *et al.* 1994), atranone and a number of *Penicillium* toxins (Rand *et al.* 2005, 2006), the effects of the toxins on lung biology are persistent. Prolonged exposure would represent an integration of cell response at least for a few days or 7-10 days, if the toxins acted like aflatoxin.

The effects on both immortalized and primary cells for toxins pertinent to agriculture have been very well investigated (Sorenson 1989, 1999), but few studies have been made of pertinent toxins indoors. It is, however, understood that the toxins present in spores diffuse into the surrounding lung cells. Using immunocytochemical methods, Gregory *et al.* (2003, 2004) showed that the macrocyclic trichothecene, satratoxin G, and the protease, haemolysin, were present on the walls of spore and spore fragments of *S. chartarum*, and that these compounds diffused into mouse lung from the site of impaction. Using an immunocytochemical approach, Rand and Miller (2008) located the antigen Sc34 in *S. chartarum* spore walls and also in lung tissues surrounding spores. These few studies are important because they showed that these low-molecular-weight compounds are not only

sequestered at appreciable concentrations in spore walls but also diffuse from the spore walls. Consequently, they are likely to impact on surrounding lung cells and tissues. For example, satratoxin G was localized not only in spore walls but also in lung granulomas, alveolar macrophages and alveolar type II cells associated with spore impaction sites. Haemolysin and Sc34 were detected in granulomatous tissue surrounding impacted spores (Gregory *et al.* 2003, 2004, Rand and Miller 2008).

While epidemiological studies support the association between indoor mould exposures and pulmonary inflammatory disease onset in humans, causality has not been achieved. However, there is compelling experimental data based on animal cell and/or rodent lung studies indicating that exposure to mould spores, spore extracts and purified fungal products results in pro-inflammatory or inflammation response. The responses of these lung disease models to such exposures include hallmarks of inflammation characterized by significant production of various pro-inflammatory cytokines, especially IL-6, TNF-α and MIP-2 among others. These affect leukocyte attraction and activation, and endothelial and alveolar/capillary integrity (Bray and Anderson 1991), and oxygen radicals (NO) in macrophage culture supernatant. In turn, this results in increased concentrations in bronchioalveolar lavage fluid (BALF) of leukocytes (alveolar macrophages and neutrophils), pro-inflammatory cytokines, albumin and other proteins, and lactate dehydrogenase (Health Canada 2004, NAS 2004). As is the case for the allergic effects of mould, the critical problem in resolving the epidemiological association, is that exposure measures are needed. These could be markers of either exposure or effect (Jarvis and Miller 2005).

Despite these results for experimental exposure, there is still little known about the mechanisms of inflammation, and whether the majority of common fungi from moist buildings, and especially their low molecular weight compounds, affect lung function. Lack of scientific understanding of the mechanisms of injury has led to confusing information and opinions in the medical literature. This has prevented development of treatment modalities for patients suffering from mould exposure. Clarification of time and place of injury, and the specific mechanisms of toxicity for various toxins, could lead to the development of agents that intervene in the inflammatory process, can affect immune regulation or lead to means of treating such damage.

In this chapter, we review experimental evidence of the inflammatory potential of (1→3)-β-D-glucan and other low molecular weight compounds from non-infective fungi, isolated from moist buildings, on animal cell and animal models of lung disease. Additionally, we address some of the mechanisms by which spore and spore product exposures may lead to inflammatory pulmonary and other responses in moisture-damaged buildings.

(1-3)-β-D GLUCAN

Although we are considering the inflammatory potential of (1→3)-β-D-glucan, it is worth noting that some higher basidiomycetes produce mixed-linkage glucans that have anti-tumourigenic and anti-inflammatory properties. These bioactive polysaccharides are (1→3)-β-D-glucans with (1→6)-β-linked side branches (Ooi and Liu 1999, Schmid *et al.* 2001) and (13)-,(14)-β-D-glucan (Pacheco-Sanchez *et al.* 2006). The basidiomycetous mould *Wallemia sebi* has been reported as being quite common in some homes in Europe (Rijckaert *et al.* 1981, Lignel *et al.* 2008), Japan (Ara et al. 2004, Sakamoto *et al.* 1989) and North America (Miller and Day 1997, Iossifova *et al.* 2008b). Inhalation of β-glucan from this fungus may then have effects different from those caused by the anamorphic Trichocomaceae (i.e. *Penicillium, Aspergillus* and related hyphomycetes associated with damp building materials), which appear predominantly to contain (1→3)-β-D glucan (Odabasi *et al.* 2006, Foto *et al.* 2005, Iossifova *et al.* 2008a). This β-glucan is considered to have potent inflammatory effects on cells of the respiratory system by targeting the dectin-1 receptor on the surfaces of immuno-sentinel cells (e.g. mast cells, dendritic cells and alveolar macrophages), and plays an important role in the development of non-allergenic respiratory health effects (Rylander and Lin, 2000; Young *et al.* 2001, 2003a,b; Douwes 2005a,b) and in invasive fungal disease (Taylor *et al.* 2006).

There have been two reports in which humans have been challenged with aerosolized curdlan. This is a linear 161 kDa (1→3)-β-D-glucan derived by fermentation from the bacterium *Alcaligenes faecalis* var. *myxogenes* and is in a triple helix configuration closest to the form found in the cell walls of the anamorphic Trichocomaceae. These studies have limitations in terms of size and available description. When a group of 16 non-atopic, asymptomatic subjects were exposed for 4 h to ~200 ng curdlan m^3 air, there was

small but statistically significant reduction in FEV_1 that remained significant for three days after exposure. In the entire population in this study (26 subjects) there was a small, but statistically significant, increase in the severity of nasal and throat irritation (Rylander 1993, 1996).

There are also two human studies involving nasal instillation of curdlan. In a double-blind crossover study, five garbage workers with occupational airway symptoms and five healthy garbage workers were intranasally exposed to 5 mg curdlan in saline for 15 min, with the saline diluent as a control, along with similar treatments involving endotoxin, compost waste dust and *Aspergillus fumigatus* spores. Nasal cavity volume and nasal lavage (NAL) were performed at baseline and 3, 6 and 11 h after exposure. Curdlan induced an increase in albumin and a slight IL-1β increase 6-11 h after exposure. The most pronounced effect on total nasal volume (TNV) was seen after challenge with curdlan, when the authors observed an TNV increase during the whole session, with a significant increase 6 h after the challenge (Sigsgaard *et al.* 2000). In another study, 36 volunteers were exposed to clean air, 332 µg m^3 office dust and 379 µg m^3 dust spiked with ~4 mg β-glucan, with dust spiked with aldehyde as a positive control. Acoustic rhinometry, rhinostereometry, nasal lavage, and lung function tests were applied. After the exposures to dust spiked with the β-glucan, nasal volume had decreased when compared to clean air or office dust (p = 0.036). After 2-3 h, β-glucan-spiked dust produced a 0.7-mm swelling of nasal tissue (p = 0.039). Analysis of nasal lavage pre- and post exposure revealed no change in cytokines, a small but significant change in IL-8, and increased nasal eosinophil cell concentration (p = 0.045; Bonlokke *et al.* 2006). Differences in outcome between Sigsgaard *et al.* (2000) and Bonlokke *et al.* (2006) are probably explained by the earlier work of Rylander (1993, 1996), which indicated that the health effects from β-glucan varied in direction depending on exposure and effect measured. At some exposures, fungal β-glucan clearly has a negative effect in humans (e.g. Rylander 1993, 1996, Bonlokke *et al.* 2006). In addition, on a population level there is evidence of a protective effect in some studies against some disease with a dose response (Schram-Bijkerk *et al.* 2005, 2006, Iossifova *et al.* 2009). This has also been observed with endotoxin. That is, there is an ACGIH threshold limit value (TLV) for endotoxin, but on a population level at some doses and timings of exposure/outcome, the health effects can be positive (Schram-Bijkerk *et al.* 2005,

Table 1. Inflammatory gene expression changes in primary mouse alveolar macrophages exposed to 10^{-8}M/ml ($1{\rightarrow}3$)-β-D-glucan at 2, 4 and 12 h PE. Positive fold values (pale grey) are up-regulated genes; negative value (dark grey) indicates down-regulated gene expression. P values $\leq$ 0.05 are significant.

Gene Symbol	Gene Description	Exposure Time (h)		
		2	**4**	**12**
Cxcl1	Chemokine (C-X-C motif) ligand 1	18.03	16.47	33.94
Cxcl2	Chemokine (C-X-C motif) ligand 2	6.85	5.51	9.0
Cxcl5	Chemokine (C-X-C motif) ligand 5	5.51		14.59
Cxcl10	Chemokine (C-X-C motif) ligand 10		8.42	
Ccl3	Chemokine (C-C motif) ligand 3	1.27	11.54	13.63
Ccl12	Chemokine (C-C motif) ligand 12	5.76		
Ccl20	Chemokine (C-C motif) ligand 20			
Il16	Interleukin 16		2.92	44.14
Il6	Interleukin 6		8.6	
ifi27	Interferon, alpha-inducible protein 27		3.37	205.9
Tnf-α	Tumor necrosis factor alpha	2.27	2.33	43.31
Nos2	Nitric oxide synthase 2, inducible, macrophage			
Blvrb	Biliverdin reductase B (flavin reductase (NADPH))	-2.82	2.14	102.66

2006) or negative (Dales *et al.* 2006).

In vitro and *in vivo* studies have helped to supplement the human exposure experiments by providing a mechanistic basis for some of the outcomes. In a series of experiments, Sorenson *et al.* (1998) compared cell-wall extracts of the yeasts *Pichia fabianii, Candida sake, Trichosporon capitatum, Rhodotorula glutinis* and *Cryptococcus laurentii* with zymosan and curdlan for their ability to stimulate alveolar macrophages (AM) and to activate complement. Wall extracts of all species activated complement. *P. fabianii, C. sake, T. capitatum, R. glutinis, C. laurentii,* and also zymosan and curdlan, stimulated superoxide anion and leukotriene B4 production in a dose-dependent fashion, although *R. glutinis* and *C. laurentii* were much less active. Zymosan, β-glucan, *P. fabianii,* and *R. glutinis* treatment of alveolar macrophages (AM) resulted in increased phagocytosis of labelled sheep RBCs, whereas there was no effect with *C. sake* or *C. laurentii* and *T. capitatum* significantly inhibiting phagocytosis. Kataoka *et al.* (2002) showed that curdlan (100µg ml^{-1}) was a strong activator of iNos, Tnf-α and MIP-2 expression in RAW 264.7 and RAW-R12 cells (mouse leukaemic monocyte macrophage cell line). In a series of experiments, Holck *et al.* (2007) exposed blood leukocytes from healthy volunteers and from patients allergic to house dust mite to ($1{\rightarrow}3$)-β-D-glucans with increasing $1{\rightarrow}6$-branchings: curdlan, laminarin and scleroglucan, and pustulan, a linear ($1{\rightarrow}6$)-β-D-glucan. All the β-glucans investigated led to an enhancement of the IgE-mediated histamine release from cells at con-

centrations in the range of 2-5 × 10^{-5} M, but only after the cells had been previously stimulated by incubation with anti-IgE antibody or allergens, which were unidentified.

Using reverse transcription (RT)-PCR based arrays, Rand *et al.* (2009) evaluated dose (10^{-7} to 10^{-10} moles) and time course (2, 4 and 12 h post-exposure, PE) profiles of inflammation-associated gene transcription in curdlan-exposed primary AM from 10 to 12-week old CFW mice. They showed significant modulation of 13 key genes involved in the inflammatory response. Nine of the 13 inflammation-associated gene transcripts assayed were significantly up-regulated in AM exposed to 10^{-7} and 10^{-8} moles curdlan and on a time-dependent fashion (Table 1). At both concentrations, fewer genes were up-regulated at 2 h PE (7 genes) than at the 12-h time point (9). Five gene mRNA transcripts, CxCl1, Cxcl2, Cxcl5 and Cxcl10 and IL-16, were significantly up-regulated at all time points, while Ifi27 (interferon, alpha-inducible protein) was only significantly up-regulated at 2 h PE at both β-glucan concentrations. Tnf-α and Nos2 mRNA transcripts were significantly expressed in the AM after 4 h PE. Blvrb (a mitochondrial stress gene) was expressed in AM exposed to 10^{-7} M β-glucan and from 4 h to 12 h PE. These data from custom-made mouse inflammatory gene microarrays (SuperArray®) demonstrated that these cells respond rapidly (within 2 h PE) and significantly to β-glucan exposures, and that gene modulation is time dependent. The results also indicated that AM are acutely sensitive to β-glucan exposure, therefore supporting results of previous *in vitro* macrophage studies.

Animal studies also point to the inflammatory potential of (1→3)-β-D-glucan. *In vivo* experiments involving high-dose (1-5 mg kg^{-1} body weight) intratracheal exposure of Sprague Dawley rats to zymosan (= (1→3)-β-D-glucan, (1→6)-β-D-glucan, 1:1) showed that this compound provokes a variety of immunotoxic and immunomodulatory responses (Young and Castranova 2005). These included increased neutrophil and AM abundance and TNF-α, as well as IL-6, IL-10 and IL-12p70 and in a dose-dependent fashion (Young *et al.* 2006). Sjostrand and Rylander (1997) reported that in guinea pigs, inhalation exposure to curdlan caused a small but significant increase in neutrophils in BALF and an increased infiltration of guinea pig airway epithelium by eosinophils. Straszek *et al.* (2007) also reported that in guinea pigs, inhalation exposure to curdlan-spiked dust provoked decreased nasal volume but no changes in BALF cell or IL-8 concentrations. Schuyler *et al.* (1998) reported that male C3H/HeJNCr mice intratracheally exposed to curdlan at concentrations ranging from 0.1 to 2.7 mg kg^{-1} exhibited dose-dependent-like histopathological responses evidenced by infiltration differences in AM and other mononuclear cells in peribronchiolar and adjacent intra-alveolar spaces of treatment animals after exposure for 4 days. However, they did not report changes in any inflammatory mediators. Unfortunately, results of these previous animal studies are contradictory and difficult to compare because of different species, exposure routes and doses (Straszek *et al.* 2007). The need for further work to clarify the role of (1→3)-β-D-glucan on the inflammatory response in rodents was indicated.

Rand *et al.* (2009) exposed groups of 3-week old CFW mice (5 animals per group) intratracheally to 50 µl pure curdlan per animal at doses from 10^{-7} to 10^{-10} moles, or to 50 µl pyrogen-free phosphate-buffered saline, for 4 h and 12 h. At each time point, mice were killed, samples of fresh lung were excised, processed for total RNA extraction using the Trizol reagent and employed for reverse transcription (RT)-PCR based SuperArray® mouse inflammatory gene and receptor array assays (96 well assays). The results demonstrated that with an estimated NOAEL (no observable adverse effect level) of 10^{-11} M concentration in mice this β-glucan was potently immunomodulatory, and showed that immunomodulation was both time and dose dependent (Table 2). Compared with controls, 54 of 83 inflammation-associated gene mRNA transcripts assayed were significantly modulated (p ≤

0.05; ≥1.5-fold or ≤-1.5-fold change) in lungs of curdlan exposed mice (Table 2). Significantly modulated genes exhibited temporal response patterns, with highest numbers of significantly transcribed genes in lungs of animals treated for 4-h. Twenty-one gene transcripts (Ccl1, Ccl2, Ccl3, Ccl4, Ccl9, Ccl11, Ccl12, Ccl17, Ccr9, Ccr10, Crp, Cxcl1, Cxcl9, Cxcr3, Ifng, Il1α, Il20, Itgb2, Tnf-α, Tnfrsf1b and CD40lg) were significantly expressed in animals exposed to 10^{-7} to 10^{-9} moles curdlan while nine genes (Ccl3, Ccl11, Ccl17, Ifng, Il1α, Il-20, TNF-α, Tnfrsf1b and CD40lg) were significantly expressed at all doses. Transcription levels of these nine individual genes in animals exposed to the different curdlan concentrations exhibited dose-response characteristics; the highest expression levels were in lungs of mice treated with 10^{-7} moles β-glucan, and lowest levels in those treated with 10^{-10} moles β-glucan (Table 2). The majority of the nine genes were associated with acute pulmonary inflammation (e.g. Ccl3, Ccl11 (= eotaxin), Ccl17, Il1α, Il20, Tnf-α, Cd40lg), especially in mediating acute inflammatory recruitment response of eosinophils (Ccl11, Ccl3), AM (Il1a, Tnf-α), neutrophils (Ccl17), macrophages (Ccl3, Il1a, Tnf-α) and T-cells (Ccl17, Ifng), and bronchiolar epithelium remodelling (Tnf-α). Up-regulation of Il1α has been implicated in onset of fever (Murphy *et al.* 2008) and biobehavioural modulation in mice (Furuzawa *et al.*, 2002). The latter supports the suggestion that exposure to fungal β-glucan may form a partial basis for the claim that mould exposures result in neurocognitive outcomes in exposed populations (Miller *et al.* 2003), although this should be explored in further animal studies. Significant up-regulation of these gene mRNAs by curdlan treatment helps provide a mechanistic basis for the dose-dependent leukocyte responses in BALF of treated laboratory animals (Fogelmark *et al.* 1997), and in lung pathology as reported by Schuyler *et al.* (1998).

While the array assays revealed that curdlan exposures potentiate pro-inflammatory gene transcription, they did not provide insight into whether the dectin-1 receptor was involved, or in which lung cells it was expressed. Dectin-1 is a type II transmembrane protein receptor with a C-type lectin domain and a cytoplasmic tail with ITAM motif (Adachi *et al.* 2004). The system best understood is the response of a β-glucan sensitive factor (factor G activation) in the blood of horseshoe crabs (*Limulus polyphemus*). The sensitivity of the β-glucan receptor of *Limulus* to curdlan and yeast β-glucan is similar (factor G activation minima = 10^{-11} g ml^{-1} for both; Tanaka *et al.* 1991). Dectin-1 is known to be expressed on a wide range of human

Table 2. Significant inflammatory and inflammatory receptor gene modulation in mouse lungs exposed to 10^{-7} to 10^{-10} M ml^{-1} glucan at 4 and 12 h PE. Significant up- or down-regulated genes have a fold difference ≥ 2 from controls. Pale grey cells are significantly (P ≤ 0.05) up-regulated genes; dark grey cells are significantly down-regulated genes.

Gene Symbol	Gene Description	Glucan Molarity							
		10^{-7}		10^{-8}		10^{-9}		10^{-10}	
		4h	12h	4h	12h	4h	12h	4h	12h
Abcf1	ATP-binding cassette				↓				
Bcl6	B-cell CLL/lymphoma				↑				
Blr1	Burkitt lymphoma receptor				↓				
C3	Complement component								
Casp1	Caspase	↑		↑					↑
Ccl1	Chemokine (C-C motif) ligand			↑					↑
Ccl11	Chemokine (C-C motif) ligand	↑							
Ccl12	Chemokine (C-C motif) ligand								
Ccl17	Chemokine (C-C motif) ligand								
Ccl19	Chemokine (C-C motif) ligand				↑				
Ccl2	Chemokine (C-C motif) ligand	↑							
Ccl20	Chemokine (C-C motif) ligand								
Ccl22	Chemokine (C-C motif) ligand								
Ccl24	Chemokine (C-C motif) ligand								
Ccl25	Chemokine (C-C motif) ligand								
Ccl3	Chemokine (C-C motif) ligand	↑			↓				
Ccl4	Chemokine (C-C motif) ligand								
Ccl5	Chemokine (C-C motif) ligand								↑
Ccl6	Chemokine (C-C motif) ligand								
Ccl7	Chemokine (C-C motif) ligand								
Ccl8	Chemokine (C-C motif) ligand	↑				↑			
Ccl9	Chemokine (C-C motif) ligand	↑							
Ccr1	Chemokine (C-C motif) receptor						↑		
Ccr2	Chemokine (C-C motif) receptor								
Ccr3	Chemokine (C-C motif) receptor								
Ccr4	Chemokine (C-C motif) receptor								
Ccr5	Chemokine (C-C motif) receptor								
Ccr6	Chemokine (C-C motif) receptor	↑							
Ccr7	Chemokine (C-C motif) receptor						↓		
Ccr8	Chemokine (C-C motif) receptor							↓	
Ccr9	Chemokine (C-C motif) receptor	↑			↓				
Crp	C-reactive protein								
Cx3cl1	Chemokine (C-X3-C motif) ligand								
Cxcl1	Chemokine (C-X-C motif) ligand	↑							
Cxcl10	Chemokine (C-X-C motif) ligand				↓				
Cxcl11	Chemokine (C-X-C motif) ligand								
Cxcl12	Chemokine (C-X-C motif) ligand				↓		↓		↓
Cxcl13	Chemokine (C-X-C motif) ligand								
Cxcl15	Chemokine (C-X-C motif) ligand				↓				
Cxcl4	Chemokine (C-X-C motif) ligand	↑							
Cxcl5	Chemokine (C-X-C motif) ligand				↓				
Cxcl9	Chemokine (C-X-C motif) ligand	↑							
Cxcr3	Chemokine (C-X-C motif) receptor								
Ccr10	Chemokine (C-C motif) receptor								
Ifng	Interferon gamma	↑							

Table 2. continued.

Gene Symbol	Gene Description	Glucan Molarity							
		10⁻⁷		10⁻⁸		10⁻⁹		10⁻¹⁰	
		4h	12h	4h	12h	4h	12h	4h	12h
Il11	Interleukin								
Il13	Interleukin								
Il13ra1	Interleukin receptor, alpha								
Il15	Interleukin								
Il16	Interleukin				�				
Il17b	Interleukin								
Il18	Interleukin								
Il1a	Interleukin, alpha	░	░						
Il1b	Interleukin, beta								
Il1f6	Interleukin family (epsilon)								
Il1f8	Interleukin family (eta)								
Il1r1	Interleukin receptor								
Il1r2	Interleukin receptor								
Il20	Interleukin								
Il2rb	Interleukin receptor, beta				■		■		■
Il2rg	Interleukin receptor, gamma				■				
Il3	Interleukin								
Il4	Interleukin								
Il5ra	Interleukin receptor, alpha								
Il6ra	Interleukin receptor, alpha								
Il6st	Interleukin signal transducer				■				
Il8rb	Interleukin receptor, beta								
Itgam	Integrin, alpha M								
Itgb2	Integrin, beta								
Lta	Lymphotoxin alpha								■
Ltb	Lymphotoxin beta	░			■				
Mif	Macrophage migration inhibitory factor								
Scye1	Small inducible cytokine subfamily E								
Spp1	Secreted phosphoprotein								
Tgfb1	Transforming growth factor, beta				■		■		■
Tnf	Tumor necrosis factor	░	░		■				
Tnfrsf1a	Tumour necrosis factor receptor superfamily								
Tnfrsf1b	Tumour necrosis factor receptor superfamily				■				
Cd40lg	CD40 ligand	░			■	░			
Tollip	Toll interacting protein								
Xcr1	Chemokine (C motif) receptor	░						░	
Number of genes significantly regulated		19	13	8	21	6	7	2	9

and animal immunomodulatory cells including macrophages, neutrophils, fibroblasts, dendritic cells and mast cells. It has also been detected in the murine organs (spleen, thymus, lung and intestine) exposed to (1→3)-β-D-glucans (see Rand *et al.* 2009). Extensive reports have shown that zymosan and β-glucan containing fungal organisms bind to the dectin-1 recep-

tor leading to phagocytosis and cytokine production especially by AM (Steele *et al.* 2005), e.g. Tnf-α, IL-1α, IL-1β, Cxcl3, Cxcl2, Ifn-γ, all of which were shown to be transcribed (Rand *et al.* 2009).

Using *in situ* hybridization (ISH) and immunohistochemistry (IHC), Rand *et al.* (2009) showed that low-dose curdlan (10⁻⁹ moles) induces dectin-1 receptor gene transcription and expression *in vivo,* especially in distal lung regions. Moreover, after 4 h and especially 12 h PE, both ISH and IHC showed more intense localization of dectin-1 mRNA transcript and anti-dectin-1 (goat anti-mouse dectin1/CLEC7A polyclonal antibody) expression in non-ciliated, respiratory bronchiolar epithelia (predominately Clara cells) and in some but not all AM and alveolar type II (ATII) cells than in alveolar wall epithelia. These results confirm and supplement the array assays by showing that dectin-1 mRNA transcription is rapid, and by providing more precision on where and in which of the various lung cells mRNA expression is localized, at least in acute exposures. While dectin-1 mRNA expression was expected in AM, localization of dectin-1 mRNA transcript and expression along the respiratory bronchiolar epithelia was interesting, as these lining cells are located at one of the lung hot spots associated with sub-micron particle deposition in the lung (Phalen *et al.* 2008). Dectin-1 mRNA and protein expression by bronchiolar epithelium, AM and ATII by both ISH and IHC indicate that AM are not the only major sources of dectin-1-expressing cells lining the alveoli, and that all these cell types orchestrate and play an important role in the innate immunity of lungs.

Studies of low-molecular-weight fungal compounds

There have been a number of *in vitro* studies involving use of either immortal cell lines and/or primary rodent macrophages that all point to the inflammatory and/or cytotoxic potential of fungal spores and or spore extracts of species found indoors, viz. *Aspergillus versicolor, A. fumigatus, A. niger, A. terreus, A. versicolor, Penicillium spinulosum* and *S. chartarum* (see NAS 2004). Exposure to materials of these species resulted in the production of the cytotoxic marker lactate dehydrogenase (LDH) and/or elevated inflammatory marker changes in MIP-2, KC, Tnf-α, MIP-1 IL-6, and NO and Tnf-α in culture supernatants. The results of these studies suggested that cytotoxic and inflammatory responses *in vitro* were dose dependent and species specific (Shahan *et al.* 1998). However, because spores and spore extracts comprise mixtures of bioactive constituents, often including toxins, it is impossible to resolve which constituent mediates these responses.

There have been only a few studies that provide evidence that pure fungal toxin exposures can provoke inflammatory responses *in vitro.* Exposure to pure macrocyclic trichothecenes, e.g. isosatratoxin-F and satratoxins from *Stachybotrys chartarum* (*sensu lato*) affects a variety of cell functions and biochemical pathways. The effects include depressed macrophage activity (Sorenson *et al.* 1987, Plasencia and Rosenstein 1990), inflammatory response (Routsalainen *et al.* 1998), membrane damage (Peltola et al. 1999), apoptosis (Okumura *et al.* 1999, Yang *et al.* 2000) and cytotoxicity (Nielsen *et al.* 2001). Trichothecenes interact with cell membranes, affecting function prior to lysis (Peltola *et al.* 1999, Riley and Norred 1996). In addition to the macrocyclic trichothecenes, *S. chartarum* (*sensu lato*) isolates can also produce other types of toxin such as atranones. *In vitro* studies employing rodent RAW 264.7 macrophages (Nielsen *et al.* 2001, Huttunen *et al.* 2004) and boar spermatozoa (Peltola *et al.* 2002) have revealed that atranones from *S. chartarum* isolates are only slightly cytotoxic. Nielsen *et al.* (2001) showed that seven of nine atranone-producing *S. chartarum* isolates that they assayed induced inflammatory responses, i.e. Tnf-α and IL-6 induction, in the rat macrophages. The other two isolates, which produced atranones B and D, did not provoke significant *in vitro* pro-inflammatory cytokine production. Differential inflammatory potential among atranone-producing isolates suggests that either toxicological properties of the various atranones isolated from *S. chartarum* vary or that markedly different concentrations in spores produce different responses. For additional details on the effects of the macrocyclic trichothecenes and other toxins from *S. chartarum* on cells, the reader should refer to the reviews by Miller *et al.* (2003), NAS (2004) and Pestka *et al.* (2008).

Miller *et al.* (2009) have recently demonstrated the inflammatory potential of pure fungal metabolites from other species on primary mouse AM also. The toxins used in this study were from common damp building fungi associated with (1) wet conditions, viz. *S. chlorohalonata* (atranone c); (2) intermediate water activities, viz. *Penicillium brevicompactum* (brevianamide), *Aspergillus versicolor* (sterigmatocystin) and *A. insuetus/ A. calidoustus* (= *A. ustus sensu lato;* Slack et al. 2009) (TMC-120A); and (3) low water activities: *Eurotium amstelodami, E. herbariorum* and *E. rubrum* (neoechinulins A and B), and also cladosporin from *E. herbariorum.* Some of these compounds have been detected in fairly high concentrations on building materials, e.g. atranone C, sterigmatocystin, TMC-120A). Echinulin and the related compounds neoechinulins A and B produced by several species of *Eurotium* have also been detected on building materials (Miller *et al.* 2009, Slack et al. 2009). However, except for atranone

Table 3. Significant inflammatory gene modulation in mouse alveolar macrophages exposed to 10^{-8} M ml^{-1} low molecular weight fungal compounds at 2, 4 and 12 h PE. Significant up- or down-regulated genes have a fold difference $\geq$ 2 from controls. Pale grey cells are significantly up-regulated genes (P $\leq$ 0.05) up-regulated genes; dark grey cells are significantly down-regulated genes.

Gene Symbol	Gene Description	Atra C 2	Atra C 4	Atra C 12	Brevi 2	Brevi 4	Brevi 12	Csporin 2	Csporin 4	Csporin 12	Neo A 2	Neo A 4	Neo A 12	Neo B 2	Neo B 4	Neo B 12	Sterig 2	Sterig 4	Sterig 12	TMC 2	TMC 4	TMC 12
Cxcl1	Chemokine (C-X-C motif) ligand																					
Cxcl2	Chemokine (C-X-C motif) ligand																					
Cxcl5	Chemokine (C-X-C motif) ligand																					
Cxcl10	Chemokine (C-X-C motif) ligand																					
Ccl3	Chemokine (C-C motif) ligand																					
Ccl12	Chemokine (C-C motif) ligand																					
Ccl20	Chemokine (C-C motif) ligand																					
Il16	Interleukin-16																					
Il6	Interleukin-6																					
ifi27	interferon, α-inducible protein																					
Tnf	Tumour necrosis factor-α																					
Nos2	Nitric oxide synthase																					
Blvrb	Biliverdin reductase B																					

*Atra = atranone C; Brevi = brevianimide; Csporin = cladosporin; Neo A = neoechinulin A; Neo B = neoechinulin B; Sterig = sterigmatocystin; TMC = TMC 120.

C, which has been reported on building materials contaminated with *S. chartarum* (*sensu lato*), and tested against mouse fibroblasts and an immortalized macrophage cell line (LC50 values, 0.3 to >22 µM; Nielsen *et al.* 2001), the inflammatory potential of these compounds had not been tested on primary rodent lung cells (see Miller *et al.* 2009).

Reverse transcription (RT)-PCR based SuperArray® mouse inflammatory gene transcriptional assays for 13 inflammatory-focused genes (Cxcl1, Cxcl2, Cxcl5, Cxcl10, Ccl3, Ccl12, Ccl20, Il16, Il6, ifi27, Tnf-α, Nos2, Blvrb) demonstrated that low dose (10^{-8} M) concentrations of these purified toxins elicit time-dependent inflammatory transcriptional responses in primary AM's from 10- to 12-week-old male Swiss Webster (CFW) mice. Compared with control treatments, exposures resulted in time dependent modulation of immunogenic gene mRNA transcription (Table 3). Toxin exposures resulted in significant mRNA transcription of all 13 genes assayed. However, not all of these genes were transcribed on exposure to all toxins. Brevianamide exposure resulted in significant transcription of most genes (12/13), whereas sterigmatocystin (9/13) and neoechinulin A (8/13) had the least effect. Patterns of transcription also varied noticeably between toxin treatments and times (Table 3). At 2 h PE there were at least three clusters linking mRNA transcription patterns of atranone C, brevianimide, neoechinulin A and neoechinulin B; control, sterigmatocystin and TMC-120A; and cladosporin. At 12 h PE three clusters emerged linking neoechinulin B to control; brevianimide to cladosporin; and neoechinulin A, sterigmatocystin and TMC.

These results indicated that AM were acutely sensitive to exposure to these toxins from fungi from the built environment, and that exposure to low concentrations results in rapid transcription of a variety of genes involved in pro-inflammatory cytokine expression and lung defence. This helps highlight the position that AM play a central role in innate lung defence, including against exposure to low molecular weight fungal compounds. Moreover, results also demonstrated that patterns of exposure outcome in the AM were remarkably time sensitive and toxin specific. For example, neoechinulin A exposure resulted in cell-stress activated *Nos2* mRNA up-regulation at 2 h PE; *Ccl3* mRNA up-regulation at 4 h PE; and at 12 h PE up-regulation of Cxcl1, Cxcl2, Il-16, ifi27, Tnf-α and Blvrb mRNA. Neoechinulin B, which differs only slightly in structure from neoechinulin A, elicited a very different response in AM, reflecting modest changes in polarity or structure-activity response. At 2 h PE there was up-regulation of the Cc motif family genes *Ccl12* and *Ccl20* mRNA; at 4 h PE Cxcl1, Cxcl2 Ccl3, IL-6, Ifi271, Tnf-α and Blvrb mRNA; and at 12 h PE Ccl12, Tnf-α and Nos2 mRNA. These observations that the neoechinulin toxins exhibit differential pro-inflammatory potential paralleled those of the atranones, which have been shown to have different immunotoxicological properties (Nielsen *et al.* 2001). The data provide further support for previous studies suggesting that low-molecular-weight fungal compounds result in differing inflammatory and/or cytotoxic response patterns. Undoubtedly, there are many factors that can affect these patterns of compound-induced inflammation: amongst others, compounds present and their concentration, and toxicokinetics of inflammatory mediator release, toxin targeting, and different compound clearance rates (Rand *et al.* 2006). Variations in any of these factors would affect the dynamics and outcome of inflammatory responses at the cellular level.

Results of both the β-glucan and toxin exposures have shown that AM produce a variety of cytokines and other inflammatory mediators, the importance of which in innate lung defence is well recognized. However, cytokine immunology involves a cascade of molecular events from the activation of signal transduction pathways to the production of inflammatory mediators, including pro-inflammatory cytokines and chemokines, adhesion molecules, reactive oxygen species and nitric oxide by various other cell types of the lungs (Bray and Anderson 1991). For example, ATII produce surfactant protein-D (SP-D) that modulate AM immune response, and cytokines/chemokines and growth factors that enable the regeneration of alveolar epithelium type I cells (ATIs) in normal and diseased lungs (Vanderbilt *et al.* 2003). Interstitial lung fibroblasts can also produce a variety of cytokines early in the inflammation reaction and may play a role in the perpetuation of the inflammatory response. Moreover, fibroblast proliferation and production of extracellular matrix components contribute to lung fibrosis and therefore lung function perturbation (Jordana *et al.* 2004).

Studies have shown that both of these types of immunomodulatory lung cells respond to fungal toxins. Mason *et al.* (1998) showed that foetal rabbit ATII exposed to isosatratoxin-F in concentrations ranging from 10^{-4} to 10^{-9} M showed a significant reduction in [^{3}H]-choline incorporation into disaturated phospha-

tidylcholine (DSPC). In mice, *S. chartarum* spores (of the trichothecene-chemotype) and iso-satratoxin affected convertase, the enzyme involved (Mason *et al.* 2001). Rand *et al.* (2002) revealed that, compared with controls, exposure to isosatratoxin-F from *S. chartarum* resulted in a number of micro-anatomical changes consistent with oncosis onset in ATII. Recent studies (C. Robbins and T.G. Rand, unpublished data) using custom-made rat RT-PCR gene transcription arrays demonstrated that dose-response (10^{-6} to 10^{-9} M) and time-course (2, 4, 12 and 24 h PE) relationships were associated with surfactant protein and pro-inflammatory transcriptional responses in primary ATII and lung fibroblasts isolated from foetal rats exposed to pure atranones A and C from *S. chartarum* (*sensu lato*), and meleagrin and roquefortine C from *P. chrysogenum*. At the highest concentrations (10^{-6} to 10^{-7} M), toxin exposure resulted in significant Cxcl2, Cxcl5, Tnf-α, iNOS, SP-A, SP-B, and SP-C mRNA up-regulation in ATII and Ccl2, Ccl20, Cxcl2, Cxcl5, iNOS and Tnf-α up-regulation in fibroblasts. Gene mRNA up-regulation was ≥10-fold greater in ATII than in fibroblasts. Compared to carrier controls, atranone A exposures to 10^{-6} to 10^{-8} M resulted in Ccl20, Cxcl2, Cxcl5, IL-6, iNOS, SP-B and SP-C up-regulation and SP-D gene down-regulation in ATII, but only TNF-α up-regulation in fibroblasts. Moreover, inflammatory gene transcription was generally up-regulated in atranone C-treated ATII but down-regulated in atranone A-exposed ATII. In *P. chrysogenum* toxins-exposed ATII, inflammatory genes (especially Cxcl chemokine genes) were transiently up-regulated. This was observed to have a dose- and time-dependence within 2-4 h and de-expressed by 12 h PE, particularly in meleagrin exposed cells. *SP-A, -B* and *-C* genes were down-regulated at 4 h PE. SP-D expression was not significantly affected by the *P. chrysogenum* toxin exposures. These results indicate that ATII and lung fibroblasts are also important immunogenic lung cells, and that they respond on exposure to fungal toxin. They also show that response is both time dependent and toxin specific, which further supports the results of the AM studies. In addition, the response of these three lung cell types to toxin exposure appears to involve the up-regulation of the Ccl and Cxcl motif genes as well as in Tnf-α mRNAs. This suggests that these genes have a central role in the modulation of the inflammatory lung response. These studies showing perturbation of surfactant cell function on exposure to toxins are important because they provide some mechanistic insight into causes of some of the *in vivo* effects of exposure to whole spores. Amongst those *in vivo* effects are increased albumin and total protein concentrations in BALF from treated animals relative to controls. Disturbance of surfactant, which lowers the surface tension at the air/liquid interface, thereby allowing normal breathing and protecting against lung oedema, can lead to lung impairment by increasing capillary leakage, which results in increased total protein and albumin accumulation in BALF among others (Gommers and Lachmann 2007).

In addition to the *in vitro* studies, there are some *in vivo* animal-model studies of a number of pure compounds from fungi common in the built environment. Some of these *in vivo* studies have reported the absence of an inflammatory lung response in animals exposed to pure trichothecenes (Marrs *et al.* 1986, Creasia *et al.* 1987, 1990, Pang *et al.* 1988, Thurman *et al.* 1988, Rand *et al.* 2002). Pang *et al.* (1987) suggested that the absence of an inflammatory response on exposure to T-2 toxin was due to this toxin being polar. This leads to its rapid absorption from lung tissue, whereupon it is metabolized and excreted. However, a feature common to all these studies is that inflammation was based on histological evaluation and not using hallmark markers of inflammation (noted above) that can be readily quantified from the BALF of treated animals. All *in vivo* studies using these markers have provided evidence that all of the compounds tested are potently inflammatory, and apparently immunomodulatory.

Studies by Rand *et al.* (2005, 2006) have shown that *Penicillium* spp. pure toxins (brevianamide C, mycophenolic acid and roquefortine C) and *S. chartarum* toxin (atranones A and C) exposures result in a variety of inflammatory responses in intratracheally instilled 3-week-old Swiss Webster mice. The *Penicillium* toxin study provided evidence that all toxins tested induced significant transient dose- and time-dependent inflammatory responses, which were manifest as differentially elevated and persistent macrophage, neutrophil and also MIP-2 (=Cxcl2), Tnf-α and IL-6 concentrations in the BALF of treated animals. It also demonstrated that brevianamide A induced cytotoxicity, seen from significantly increased LDH concentration in mouse BALF, at doses of 250 nmole brevianimide A per animal and at 6 and 24 h post instillation (PI). Albumin concentrations, measured as a non-specific marker of vascular leakage, were significantly elevated in the BALF of mice treated with ≥ 100 nmole brevianamide A per animal from 6 to 24 h PI and in ≥50 nmole mycophenolic acid-treated animals at 24 h PI. The atranone study examined dose-response

(10, 50, 100, 250 or 1000 nmole atranone per animal) and time-course (3, 6, 24 and 48 h PI) relationships associated with pro-inflammatory and cytotoxic responses. Both atranone A and C induced transient inflammatory responses manifest as elevated MIP-2, Tnf-α, and IL-6 concentrations in the BALF of intratracheally instilled mice. Compared with control values, atranone A induced significant MIP-2 and TNF-α production at 1000 nmoles and IL-6 at ≥100 nmoles per animal. Atranone C induced MIP-2 production at 1000 nmole, Tnf-α at ≥250 nmole and IL-6 at ≥100 nmole per animal. At concentrations ≤100 nM atranone, inflammatory responses manifest in BALF were highest at 6 h PI, while at ≥250 nmole atranone per animal they were maximal at 3 h PI. None of the LDH levels in treated animals was significantly different from those in control animals. These results suggest that the three *Penicillium* toxins and two atranones, from species common on damp materials in residential housing, provoke compound-specific toxic responses with different toxicokinetics. They also support the position that they are not only potent inflammatory mediators, but they also exhibit different toxicological potency.

These experimental findings are also supported by a further study that evaluated inflammation-associated gene transcription, expression and histopathological responses in mouse lungs intratracheally instilled with 50 μl 10^{-7} moles atranone C, brevianamide, cladosporin, mycophenolic acid, neoechinulin A and B, sterigmatocystin and TMC-120A at 4 h and 12 h PE (Miller *et al.*, unpublished data). The dose used was comparable to the lowest effect level for small numbers of these compounds tested in primary cells. Using mouse inflammatory gene and receptor microarrays and (RT)-PCR, all but one of the 83 inflammation-associated genes assayed at 4 h PE were significantly transcribed (at p = ≤0.05, ≥1.5-fold or ≤ minus 1.5-fold change), while at 12 h PE 75 genes were significantly transcribed in the different treatment groups. However, not all of these regulated genes were transcribed in all treated animals. Of the 83 genes assayed, ten (Abcf1, Bir1, Ccl19 Ccl20, Ccr6, Cx3cl1, Cxcl5, Il6st, Scye1, Tnfrsf1a) were significantly transcribed in all treatment groups at least at one time point. Sixteen other genes were significantly transcribed in all but one of the treatment groups (Ccl4, Ccl7, Ccl25, Ccr2, Ccr9, Ccr10, Cxcl1, Cxcl3, Cxcr3, Ilr1, Il2rb, Il5ra, Il10rb, Ltb, Tnf-α, Cd40lg). The majority of those transcribed were genes associated with acute pulmonary inflammation or their receptors, especially in mediating acute inflammatory recruitment response of T-cells, alveolar macrophages, neutrophils and alveolar wall epithelium (Miller *et al.* unpublished data). These results provide a mechanistic basis for the dose response patterns in leukocyte and cytokine production in mouse lungs exposed to the *Penicillium* and *S. chartarum* toxins. Patterns of gene transcription also differed between different treatments. At 4 h PE, there were at least three clusters. One cluster representing all treatments except neoechinulin B comprised Il20, Cxcr3, Ccl19, Cx3cl1, Abcf1, Il10rb1, Mif, and Tnfrsf1a. A second cluster representing all treatments except atranone and neoechinulin B comprised ILr1, Ltb, Ccl25, Cd40lg, Cxcl15, Il6st, Il2rb, Ccr2, Il16, Ccl8 and Il3. The third cluster representing all treatments except neoechinulin B comprised Scye1, Itgb2, Spp1, Il1f6 and tollip. At 12 h PE, signs of clustering were not as apparent as in the 4-h PE treatments. The similarity of the gene response to each toxin, evaluated for significance using cluster analyses, demonstrated that at 12 h PE, three secure patterns of gene response (clusters) were obtained: (1) brevianamide, mycophenolic acid and neoechinulin B, (2) neoechinulin A and sterigmatocystin, and (3) cladosporin, atranone C and TMC 120. A similar pattern was observed 4 h PE, but the statistical significance was marginal. The clustering of brevianamide and neoechinulin B with mycophenolic acid is instructive, because based on an abundance of clinical and experimental data, the latter three are highly immunologically active, whereas the toxicities of the former five have not been explored (Miller *et al.* unpublished data). Similarly, the clustering of cladosporin and atranone C with TMC 120A links these two compounds, previously suggested as being non-toxic, with a compound that is again known to be highly immunologically active in humans. The clustering of the gene expression patterns of neoechinulin A and sterigmatocystin links these compounds to biological effects in the related fungal toxin aflatoxin in humans, and also to sterigmatocystin in *in vivo* and *in vitro* assays. This makes it difficult to rule out the possibility that exposure to very small amounts of fungal products from these species will have an important role in the health effects observed in exposed populations (Miller *et al.* unpublished data).

Another important result of this study is that some, but not all, of the significantly transcribed genes common to the toxin treatments were similar to those expressed in animals exposed to curdlan (described above). Gene transcripts in common were Ccl4, Ccr9, Ccr10, Cxcl1, Cxcr3, Tnf-α, Cd40lg. This suggests that these genes may have a central role in the regulation of the inflammatory response in lungs exposed to all low-molecular-weight fungal compounds, but obviously further study would be required to determine whether this is so. While there were some similarities, differences in the outcome of treatment with these low molecular weight compounds were also apparent. The most obvious difference related to up-regulation of the Tnf-Rsf1a mRNA transcript by toxins and the Tnf-Rsf1β transcript by curdlan. These data not only support for our previous studies suggesting that pulmonary exposure to low-molecular-weight fungal compounds result in different inflammatory response patterns. They also indicate that these effects may be a consequence of the compound action of two or more substances on different cytokine receptors that in turn result in activation of signal transduction pathways and the specific production of inflammatory mediators, including pro-inflammatory cytokines and chemokines.

SUMMARY

It is well understood that exposure to moulds growing in damp buildings can impact on the comfort and health of humans. Respiratory health problems associated with mould exposure can be linked to both allergenic- and non-allergenic respiratory responses (Health Canada 2004; NAS 2004) including non-atopic asthma (see, for example, Cox-Ganser *et al.* 2005). While mechanisms accounting for atopic effects are broadly understood, those associated with non-allergenic effects are less clear. Nevertheless, there is increasing evidence that the non-allergenic effects are linked to exposure to spore and hyphal fragments containing toxins and β-glucan from a consortium of toxin-producing, anamorphic Trichocomaceae (e.g. *Aspergillus*, *Eurotium*, *Penicillium* and related hyphomycetes) that grow on damp building materials. We have reviewed the state of these data up to the present. As with the case of the allergic effects, understanding the "toxic" phenomena requires biomarkers. The intent of this type of research is to identify useful markers of effect (Jarvis and Miller 2005).

REFERENCES

Adachi, Y., Ishii, T., Ikeda, Y., *et al.* (2004) Characterization of β-glucan recognition site on C-Type lectin, dectin 1. *Infect. Immun.*, **72**, 4159-4171.

Ara, K., Aihara, M., Ojima, M., *et al.* (2004) Survey of fungal contamination in ordinary houses in Japan. *Allergol. Int.*, **53**, 369–377.

Bonlokke, J.H., Stridh, G., Sigsgaard, T., *et al.* (2006) Upper-airway inflammation in relation to dust spiked with aldehydes or glucan. *Scand. J. Work Environ. Health*, **32**, 374-382.

Bray, M.A., and Anderson, W.E., (eds.) (1991) *Mediators of Pulmonary Inflammation*. Marcel Dekker, New York.

Brown, J.S., Wilson, W.E., and Grant, L.D. (2005) Dosimetric comparisons of particle deposition and retention in rats and humans. *Inhal. Toxicol.*, **17**, 355-385.

Creasia, D.A., Thurman, J.D., Jones, L.J., *et al.* (1987) Acute inhalation toxicity of T-2 mycotoxin in mice. *Fund. Appl. Tox.*, **8**, 230-235.

Creasia, D.A., Thurman, J.D., Wannemacher, R.W., and Bunner, D.L. (1990) Acute inhalation toxicity of T-2 mycotoxin in the rat and guinea pig. *Fund. Appl. Toxicol.*, **14**, 54-59.

Cox-Ganser, J.M., White, S.K., Jones, R., *et al.* (2005) Respiratory morbidity in office workers in a water-damaged building. *Environ. Health Perspect.*, **113**, 485-490.

Dales, R.E., Zwanenburg, H., Burnett, R., and Franklin, C. (1991a) Respiratory health effects of home dampness and moulds amongst Canadian children. *Am. J. Epidemiol.*, **134**, 196-203.

Dales, R.E., Burnett, R., and Zwanenburg, H. (1991b) Adverse health effects amongst adults exposed to home dampness and moulds. *Am. Rev. Resp. Dis.*, **143**, 505-509.

Dales, R.E., Miller, J.D., Ruest, K., *et al.* (2006) Airborne endotoxin is associated with respiratory illness in the first 2 years of life. *Environ. Health Perspect.*, **114**, 610–614.

Douwes, J. (2005a) Health effects of (1→3)-β-D-glucans: The epidemiological evidence. In S.-H. Young and V. Castranova, (eds.), *Toxicology of (1→3)-beta-D-glucans*. CRC Press, Boca Raton, FL, pp. 35-52.

Douwes, J. (2005b) (1→3)-Beta-D-glucans and respiratory health: a review of the scientific literature. *Indoor Air*, **15**, 160-169.

Fischer, G., Müller, T., Schwalbe, R., *et al.* (2000) Exposure to airborne fungi, MVOC and mycotoxins in biowaste-handling facilities. *Int. J. Hyg. Environ. Health*, **203**, 97-104.

Fogelmark, B., Sjostrand, M., Williams, D., and Rylander, R. (1997) Inhalation toxicity of (1→3)-β-D-glucan: recent advances. *Med. Inflamm.*, **62**, 63-265.

Foto, M., Plett, J., Berghout, J., and Miller, J.D. (2004) Modification of the *Limulus* amebocyte lysate assay for the analysis of glucan in indoor environments. *Anal. Bioanal. Chem.*, **379**, 156-162.

Foto, M., Vrijmoed, L.L.P., Miller, J.D., *et al.* (2005) Comparison of airborne ergosterol, glucan and Air-O-Cell data in relation to physical assessments of mould damage and some other parameters. *Indoor Air*, **15**, 257-266.

Furuzawa,M., Kuwahara, M., Ishii, K., *et al.* (2002) Diurnal variation of heart rate, locomotor activity, and body temperature in interleukin-1α/β doubly deficient mice. *Exp. Animals*, **51**, 49-56.

Gloer, J.B., TePaske, M.R., Sima, J.S., *et al.* (1988) Antiinsectan aflavinine derivatives from the sclerotia of *Aspergillus flavus. J. Organ. Chem.*, **53**, 5457-5460.

Gomers, D., and Lackmann, B. (2007) Role of surfactant in the pathophysiology and therapy of ARDS. *J. Organ Dysfun.*, **4**, 54-65.

Green, B.J., Tovey, E.R., Sercombe, J.K., *et al.* (2006) Airborne fungal fragments and allergenicity. *Med. Mycol.*, **44**, S1, 245-255.

Gregory, L., Rand, T.G., Dearborn, D., *et al.* (2003) Immunocytochemical localization of stachylysin in *Stachybotrys chartarum* spores and spore-impacted mouse and rat lung tissue. *Mycopathologia*, **156**, 109-117.

Gregory, L., Pestka, J.J., Dearborn, D., and Rand, T.G. (2004) Localization of satratoxin-G in *Stachybotrys chartarum* spores and spore-impacted mouse lung tissues using immunocytochemistry. *Tox. Path.*, **32**, 26-34.

Health Canada (2004) *Fungal Contamination in Public Buildings: Health Effects and Investigation Methods*. Health Canada, Ottawa, Ontario.

Holck, P., Sletmoen, M., Stokke, B.T., *et al.* (2007) Potentiation of histamine release by microfungal (1→3)- and (1→6)-β -D-glucans. *Basic Clin. Pharm.Toxicol.*, **101**, 455-458.

Huttunen, K., Pelkonen, J., Nielsen, K.F., *et al.* (2004) Synergistic interaction in simultaneous exposure to *Streptomyces californicus* and *Stachybotrys chartarum*. *Environ. Health Perspect.*, **112**, 659-665.

Iossifova, Y., Reponen, T., Daines, M., *et al.* (2008a) Comparison of two analytical methods for detecting (1→3)-β-D-glucan in pure fungal cultures and in home dust samples. *Open Allergy J.*, **1**, 26-34.

Iossifova,Y., Reponen, T., Sucharew, H., *et al.* (2008b) Use of (1→3)-beta-D-glucan concentrations in dust as a surrogate method for estimating specific fungal exposures. *Indoor Air*, **18**, 225-232.

Iossifova, Y.Y., Reponen, T., Ryan, P.H., *et al.* (2009) Mould exposure during infancy as a predictor of potential asthma development. *Ann. Allergy Asthma Immunol.*, **102**, 131–137.

Jakab, G.J., Hmieleski, R.R., Zarba, A., *et al.* (1994) Respiratory aflatoxicosis: suppression of pulmonary and systemic host defenses in rats and mice. *Toxicol. Appl. Pharmacol.*, **125**, 198-205.

Jarvis, B.B., and Miller, J.D. (2005) Mycotoxins as harmful indoor air contaminants. *Appl. Micro. Biotech.*, **66**, 367-372.

Jordana, M., Sarnstrand, B., Sime, P.J., and Ramis, I. (1994) Immune inflammatory functions of fibroblasts. *Eur. Respir. J.*, **7**, 2212-2222.

Kataoka, K., Muta, T., Yamazaki, S., and Takeshige, K. (2002) Activation of macrophages by linear (1→3)-β-D-glucans. Implications for the recognition of fungi by innate immunity. *J. Biol. Chem.*, **277**, 36825-36831.

Kleinstreuer, C., Zhang, Z., and Kim, C.S. (2007) Combined inertial and gravitational deposition of microparticles in small model airways of the human respiratory system. *J. Aerosol Sci.*, **38**, 1047-1061.

Kleinstreuer, C., Zhang, Z., and Li, Z. (2008) Modeling airflow and particle transport/deposition in pulmonary airways. *Resp. Physiol. Neurobiol.*, **163**, 128-138.

Land, C.J., Rask-Andersen, A., Lundström, H., *et al.* (1994) Tremorgenic mycotoxins and gliotoxin in conidia of *Aspergillus fumigatus*. In R.A. Samson, B. Flannigan, M.E. Flannigan, *et al.*, (eds.), *Health Implications of Fungi in Indoor Environments*. Elsevier, Amsterdam, pp. 307-315.

Létuvé, S., Lajoie-Kadoch, S., Audusseau, S., *et al.* (2006) IL-17E upregulates the expression of proinflammatory cytokines in lung fibroblasts. *J. Allergy Clin. Immunol.*, **117**, 590-596.

Lignell, U., Meklin, T., Rintala, H., *et al.* (2008) Evaluation of quantitative PCR and culture methods for detection of house dust fungi and streptomycetes in relation to moisture damage of the house. *Lett. Appl. Microbiol.*, **47**, 303-307.

Marrs,T.C., Edginton, J.A., Price, P.N., and Upshall, D.C. (1986) Acute toxicity of T-2 mycotoxin to the guinea-pig by inhalation and subcutaneous routes. *Br. J. Exp. Pathol.*, **67,** 259–268.

Mason, C.D., Rand, T.G., Oulton, M., *et al.* (1998) Effects of *Stachybotrys chartarum (atra)* conidia and isolated toxin on lung surfactant production and homeostasis. *Nat. Tox.*, **6**, 27-33.

Mason, C., Rand, T.G., Oulton, G., and MacDonald, J. (2001) The effect of *Stachybotrys chartarum* spores and an isolated trichothecene, isosatratoxin F, on convertase activity in mice. *Tox. Appl. Pharm.*, **172**, 21-28.

Miller, J.D. (1990) Fungi as contaminants of indoor air. In *Proceedings of the 5th International Conference on Indoor Air Quality and Climate, Toronto, Ontario, Vol. 5,* CMHC, Ottawa, Canada, pp. 51-64.

Miller, J.D. (1992) Fungi as contaminants of indoor air. *Atmos. Environ.*, **26A**, 2163-2172.

Miller, J.D., and Day, J.D. (1997) Indoor mould exposure: epidemiology, consequences and immunothapy. *Can. J Allergy Clin. Immunol.*, **2**, 25-32.

Miller, J.D., Rand, T.G., and Jarvis, B.B. (2003) *Stachybotrys chartarum*: cause of human disease or media darling? *Med. Mycol.*, **41**, 271-291.

Miller, J.D., Dugandzic, R., Frescura, A.-M., and Salares, V. (2007) Indoor and outdoor-derived contaminants in urban and rural homes in Ottawa, Canada. *J. Air Waste Man. Assoc.*, **57**, 297-302.

Miller, J.D., Rand, T.G., McGregor, H., *et al.* (2008) Mould ecology: recovery of fungi from certain mouldy building materials. In B. Prezant, D. Weekes, and D.J. Miller, (eds.), *Recognition, Evaluation and Control of Indoor Mould.* American Industrial Hygiene Association, Fairfax, VA, pp. 43-51.

Miller, J.D., Sun, M., Gilyan, A., *et al.* (2009) Inflammation-associated gene transcription and expression in mouse lungs induced by low molecular weight compounds from fungi from the built environment. *Chem. Biol. Interact.*, **183**, 113–124.

Murphy, K., Travers, P., and Walport, M. (2008) *Janeway's Immunobiology*. Garland Science, Taylor & Francis, New York.

NAS (2004) *Damp Indoor Air Spaces and Health*. National Academies Press, Washington, DC.

Nielsen, K.F., Huttunen, K., Hyvärinen, A., *et al.* (2001) Metabolite profiles of *Stachybotrys* isolates from water-damaged buildings and their induction of inflammatory mediators and cytotoxicity in macrophages. *Mycopathologia*, **154**, 201-205.

Nikulin, M., Reijula, K., Jarvis, B., *et al.* (1997) Effects of intranasal exposure to *Stachybotrys atra. Fund. Appl. Toxicol.*, **35**, 182-188.

Odabasi, Z., Paetznick, V.L., Rodriguez, J.R., *et al.* (2006) Differences in beta-glucan levels in culture supernatants of a variety of fungi. *Med. Mycol.*, **44**, 267-272.

Ooi, V.E.C, and Liu, F. (1999) A review of pharmacological activities of mushroom polysaccharides. *Int. J. Med. Mushr.*, **1**, 195-206.

Okumura, H., Yoshino, N., Sugiura, Y., *et al.* (1999) Trichothecenes as potent inducers of apoptosis. In E. Johanning, (ed.), *Bioaerosols, Fungi and Mycotoxins: Health Effects, Assessment, Prevention and Control,* Eastern New York Occupational and Environmental Health Center, Albany, NY, pp. 221-231.

Pacheco-Sanchez, M., Boutin, Y., Angers, P., *et al.* (2006) A bioactive (1→3)-, (1→4)-β-D-glucan from *Collybia dryophila* and other mushrooms. *Mycologia*, **98**, 180-185.

Pang, V.F., Lambert, R.J., Felsburg, P.J., *et al.* (1987). Experimental T-2 toxicosis in swine following inhalation exposure: effects on pulmonary and systemic immunity, and morphologic changes. *Toxicol. Pathol.*, **15,** 308-319.

Peltola, J., Andersson, M., Mikkola, R., *et al.* (1999). Membrane toxic substances in water-damaged construction materials and fungal pure cultures. In E. Johanning, (ed.), *Bioaerosols, Fungi and Mycotoxins: Health Effects, Assessment, Prevention and Control,* Eastern New York Occupational and Environmental Health Center, Albany, NY, pp. 432-443.

Peltola, J., Niessen, L., Nielsen, F., *et al.* (2002) Toxigenic diversity of two different RAPD groups of *Stachybotrys chartarum* isolates analyzed by potential for trichothecene production and for boar sperm cell motility inhibition. *Can. J. Microbiol.*, **48**, 1017–1029.

Pestka, J.J., Yike, I., Dearborn, D.G., *et al.* (2008) *Stachybotrys chartarum*, trichothecene mycotoxins, and damp building–related illness: new insights into a public health enigma. *Tox. Sci.*, **104**, 4-26.

Phalen, R.F., Oldham, M.J., and Wolff, R.K. (2008) The relevance of animal models for aerosol studies. *J. Aerosol Med. Pulm. Drug Deliv.*, **21**, 113-124.

Plasencia, F.J., and Rosenstein, Y. (1990) Effect of *in vivo* administration of T-2 toxin on peritoneal murine macrophages. *Toxicon*, **28**, 559-67.

Rand, T.G., Mahoney, M., White, K., and Oulton, M. (2002) Micro-anatomical changes associated with alveolar type II cells in juvenile mice exposed to *Stachybotrys chartarum* and isolated toxin. *Tox. Sci.* **65**, 239-245.

Rand, T.G., Flemming, J., Giles, S., *et al.* (2005) Inflammatory and cytotoxic responses in mouse lungs exposed to purified toxins from building isolated *Penicillium brevicompactum* Dierckx and *P. chrysogenum* Thom. *Tox. Sci.*, **87**, 213-222.

Rand, T.G., Flemming, J., Miller, J.D., and Womiloju, T.O. (2006) Comparison of inflammatory and cytotoxic responses in mouse lungs exposed to atranone A and C from *Stachybotrys chartarum. Toxicol. Environ. Health*, **69**, 1239-1251.

Rand, T.G., and Miller, J.D. (2008) Immunohistochemical and immunocytochemical detection of SchS34 antigen in *Stachybotrys chartarum* spores and spore impacted mouse lungs. *Mycopathologia*, **165**, 73-80.

Rand, T.G., Sun, M., Gilyan, A., *et al.* (2010) Inflammation-associated gene and dectin-1 receptor transcription and expression in mouse lungs by the (1, 3)-β-D glucan, curdlan. *Arch Toxicol.* 10.1007/s00204-009-0481-4.

Reponen, T., Seo, S.-C., Grimsley, F., *et al.* (2007) Fungal fragments in mouldy houses: a field study in homes in New Orleans and southern Ohio. *Atmos. Environ.*, **41**, 8140-8149.

Rijckaert, G., van Bronswijk, J.E.M., and Linskens, H.F. (1981) House-dust community (fungi, mites) in different climatic regions. *Oecologia*, **48**, 183-185.

Riley, R.T., and Norred, W.P. (1996) Mechanisms of mycotoxicity. In H. Howard and Miller, J.D., (eds.), *The Mycota*, Vol. 6. Springer-Verlag, New York, pp. 193-212.

Ruotsalainen, M., Hirvonen, M.-R., Nevalainen, A., *et al.* (1998) Cytotoxicity, production of reactive oxygen species and cytokines induced by different strains of *Stachybotrys* sp. from mouldy buildings in RAW2647 macrophages. *Env. Tox. Pharm.*, **6**, 193-199.

Rylander, R. (1993) Experimental exposures to 1, 3-beta-D-glucan. *ASHRAE Trans. 1993*, 338-340.

Rylander, R. (1996) Airway responsiveness and chest symptoms after inhalations of endotoxin or (1→3)-beta-D-glucan. *Indoor Built Environ.*, **5**, 106-111.

Rylander, R., and Lin, R.H. (2000) (1→3) beta-D-glucan – relationship to indoor air related symptoms, allergy and asthma. *Toxicology*, **152**, 47-52.

Sakamoto, T., Torii, S., Yamada, M., *et al.* (1989) Allergenic and antigenic activities of the osmophilic fungus *Wallemia sebi* asthmatic patients. *Arerugi*, **38**, 352-359. [In Japanese].

Salares, V.R., Hinde, C.A., and Miller, J.D. (2009) Analysis of settled dust in homes and fungal glucan in air particulate collected during HEPA vacuuming. *Indoor Built Environ.*, **18**, 485-491.

Samson, R.A. (1999) Ecology, detection and identification problems of moulds in indoor environments. In E. Johanning, E., (ed.), *Bioaerosols, Fungi and Mycotoxins: Health Effects, Assessment, Prevention and Control*, Eastern New York Occupational and Environmental Health Center, Albany, NY, pp. 33-37.

Schram-Bijkerk, D., Doekes, G., Douwes, J., *et al.* (2005) Bacterial and fungal agents in house dust and wheeze in children: the PARSIFAL study. *Clin. Exp. Allergy*, **35**, 1272-1278.

Schram-Bijkerk, D., Doekes, G., Boeve, M., *et al.*, PARSIFAL study group. (2006) Nonlinear relations between house dust mite allergen levels and mite sensitization in farm and nonfarm children. *Allergy*, **61**, 640-647.

Schuyler, M., Gott, K., and Cherne, A. (1998) Effect of glucan on murine lungs. *J. Tox. Environ. Health A*, **53**, 493-505.

Shahan, T.A., Sorenson, W.G., Paulauskis, J.D., *et al.* (1998) Concentration and time dependent upregulation and release of the cytokines MIP-2, KC, TNF and MIP1 alpha in rat alveolar macrophages by fungal spores implicated in airway inflammation. *Am. J. Resp. Cell Mol. Biol.*, **18**, 435-440.

Sigsgaard, T., Bonefeld-Jorgensen, E.C., Kjaergaard, S.K., *et al.* (2000) Cytokine release from the nasal mucosa and whole blood after experimental exposures to organic dusts. *Eur. Resp. J.*, **16**, 140-145.

Sjostrand, M., and Rylander, R. (1997). Pulmonary cell infiltration after chronic exposure to $(1\rightarrow3)$-beta-D-glucan and cigarette smoke. *Inflamm. Res.*, **46**, 93-97.

Schmid, F., Stone, B.A., McDougall, B.M., *et al.* (2001) Structure of epiglucan, a highly side-chain/branched $(1,3;1,6)$-β-glucan from the micro fungus *Epicoccum nigrum* Ehrenb. ex Schlecht. *Carbohydr. Res.*, **331**, 163-171.

Slack, G.J., Puniani, E., Frisvad, J.C., *et al.* (2009) Secondary metabolites from *Eurotium* species, *A. calidoustus* and *A. insuetus* common in Canadian homes with a review of their chemistry and biological activities. *Mycol. Res.*, **113**, 480-490.

Sorenson, W. (1989) Health impact of mycotoxins in the home and workplace: an overview. In C.E. O'Rear and G.C. Llewellyn, (eds.), *Bioterioration Research 2*. Plenum, New York, pp. 201-215.

Sorenson, W.G. (1999) Fungal spores: hazardous to health? *Environ. Health Perspect.*, **107**, S3, 469-472

Sorenson, W.G., Frazer, D.G., Jarvis, B., *et al.* (1987) Trichothecene mycotoxins in aerosolized conidia of *Stachybotrys atra*. *Appl. Environ. Microbiol.*, **53**, 1370-1375.

Sorenson, W.G., Shahan, T.A., and Simpson, J. (1998) Cell wall preparations from environmental yeasts: effect on alveolar macrophage function *in vitro*. *Ann. Agric. Environ. Med.*, **5**, 65-71.

Steele, C., Rapaka, R.R., Metz, A., *et al.* (2005) The beta-glucan receptor dectin-1 recognizes specific morphologies of *Aspergillus fumigatus*. *Pathogens*, **1**, 323-334.

Straszek, S.P., Adamcakova-Dodd, A., Metwali, N., *et al.* (2007) Acute effect of glucan-spiked office dust on nasal and pulmonary inflammation in guinea pigs. *J. Tox. Environ. Health A*, **70**, 1923-1928.

Tanaka, S., Aketagawa, J., Takahashi, S., and Shibata, Y. (1991) Activation of *Limulus* coagulation factor G by $(1\rightarrow3)$-beta-D-glucans. *Carbohydr. Res.*, **218**, 167-174.

Taylor, P.R., Tsoni, T.S., Willment, J.A., *et al.* (2006) Dectin-1 is required for glucan recognition and control of fungal infection. *Nat. Immunol.*, **8**, 31-38.

Thurman, J.D., Creasia, D.A., and Trotter. R.W. (1988) Mycotoxicosis caused by aerosolized T-2 toxin administered to female mice. *Am. J. Vet. Res.*, **49**, 1928-1931.

Vanderbilt, J.N., Mager, E.M., Allen, L., *et al.* (2003) CXC chemokines and their receptors are expressed in type II cells and upregulated following lung injury. *Am. J. Resp. Cell Mol. Biol.*, **29**, 661-668.

Wicklow, D.T., and Shotwell, O.L. (1983) Intrafungal distribution of aflatoxins among conidia and sclerotia of *Aspergillus flavus* and *Aspergillus parasiticus*. *Can. J. Microbiol.*, **29**, 1-5.

Wicklow, D.T., Dowd, P.F., Tepaske, M.R., and Gloer, J.B. (1988) Sclerotial metabolites of *Aspergillus flavus* toxic to a detritivorous maize insect (*Carpophilus hemipterus*, Nitidulidae). *Trans. Br. Mycol. Soc.*, **91**, 433-438.

Yang, G.-H., Jarvis, B.B., Chung, Y.-J., and Pestka, J.J. (2000) Apoptosis induction by the satratoxins and other trichothecene mycotoxins: relationship to ERK, p38 MAPK, and SAPK/JNK Activation. *Toxicol. Appl. Pharmacol.*, **164**, 149-160.

Young, S.-H., and Castranova, V. (2005) Animal model of $(1\rightarrow3)$-β-glucan-induced pulmonary inflammation in rats. In S.-H. Young and V. Castranova, (eds.), *Toxicology of $1\rightarrow3$- Beta-Glucans,* CRC Press, Boca Raton, FL, pp. 65-93.

Young, S.-H., Robinson, V.A., Barger, M., *et al.* (2001) Acute inflammation and recovery in rats after intratracheal instillation of a $1\rightarrow3$-beta-glucan (zymosan A). *J. Tox. Environ. Health A*, **64**, 311-325.

Young, S.-H., Robinson, V.A., Barger, M., *et al.* (2003a) Partially opened triple helix is the biologically active conformation of $1\rightarrow3$-beta-glucans that induces pulmonary inflammation in rats. *J. Toxicol. Environ. Health A*, **66**, 551-563.

Young, S.-H., Robinson, V., Barger, M., *et al.* (2003b) Exposure to particulate $1\rightarrow3$-beta-glucans induces greater pulmonary toxicity than soluble $1\rightarrow3$-beta-glucans in rats. *J. Tox. Environ. Health A*, **66**, 25-38.

Young, S.-H., Roberts, J.R., and Antonini, J.M. (2006) Pulmonary exposure to $1\rightarrow3$-beta-glucan alters adaptive immune responses in rats. *Inhal. Toxicol.*, **18**, 865-874.

Chapter 4.6

MICROBIOLOGICAL INVESTIGATIONS OF INDOOR ENVIRONMENTS: INTERPRETING SAMPLING DATA-SELECTED CASE STUDIES

Philip R. Morey

ENVIRON International Corp., Gettysburg, Pennsylvania, USA.

INTRODUCTION

Interpreting microbial sampling data is challenging because there can be a diversity of contaminants in buildings. These include many different culturable and non-culturable fungi and bacteria and their toxins and metabolites, e.g. endotoxins, mycotoxins, microbial volatile organic compounds (MVOCs), and β-(1→3)-glucans. Methods of sampling and analysis for microorganisms and microbial products are reviewed in this volume (Chapters 4.1, 4.3 and 4.5) and elsewhere (ECA 1993, ACGIH 1999, Health Canada 2004, AIHA 2005, 2008, ISIAQ 1996, ASTM 2008, Morey 2007). Past reviews have emphasized specific microorganisms (Health Canada 1995, ASTM 2008), certain kinds of building (ECA 1993, ISIAQ 1996), sampling and analytical methodology (AIHA 2005) and general principles of microbial evaluation (ACGIH 1999, AIHA 2008). This chapter provides examples of practical case studies where the interpretation of sampling data has been a useful component of the building evaluation.

INTERPRETATION OF SAMPLING DATA DURING BUILDING EVALUATION

Sampling must be performed and analytical data must be interpreted in the context of an understanding of how the building is designed, operated and maintained. For example, information on operation of the heating, ventilation and air-conditioning (HVAC) system, the construction of the building and its envelope, and locations and susceptibilities of occupants, must all be understood prior to the development of the sampling plan. Thus, when the sampling objective is to document exposure to *Stachybotrys* or its mycotoxins it is essential to know the location of pre-existing water damage in building systems and the specific location of water-damaged cellulosic materials and occupants who may have been exposed. In a medical centre where the intention in sampling is to estimate risk of potential infection by *Legionella*, an understanding of configuration and operation of potable water services and cooling tower systems as well as the location of the most susceptible patients is essential for the development of the sampling plan.

The limitations of the sampling and analytical method must be understood by the investigator at the beginning of the evaluation. The collection of a single sample, or a few samples, in a building seldom adequately characterizes environmental microbiological conditions (AIHA 2005, ACGIH 1999). In addition, sampling data cannot be used to prove that a building is "safe" or that a microbial agent does not exist. Thus, the inability to detect culturable *Legionella* in a cooling-tower water sample does not mean that these bacteria were absent at any time in the past, or that currently legionellae are absent in unsampled biofilms that may be present on surfaces in water systems.

Investigators must always be aware of limitations in sampling and analytical methodology that affect data interpretation. If the aim of sampling is to determine that *Stachybotrys* is present, then exclusive use of culture-based methods may overlook non-culturable spores detectable only by direct microscopical methods, e.g. cellulose tape and spore-trap sampling (see Chapter 4.1). In evaluations where it is intended to document previous flooding of interior surfaces by sampling for culturable fungi such as *Chaetomium*, *Phoma* or *Trichoderma*, use of isolation media for hy-

Table 1. Air sampling for a broad range of culturable fungi in a building with a past history of chronic roof and window leaks.

Fungi in outdoor air (ranked taxa; % of total)*

Cladosporium cladosporioides (71)

Epicoccum nigrum (14)

Yeasts (5)

Non-sporulating fungi (4)

Penicillium brevicompactum (2)

Cladosporium herbarum (1)

(*Aureobasidium pullulans, P. simplicissimum* ≤0.5%)

Fungi in indoor air (ranked taxa, % of total)*

Cladosporium cladosporioides (56)

Non-sporulating fungi (16)

Epicoccum nigrum (10)

Ulocladium chartarum (8)

Yeasts (3)

Arthrographis sp. (2)

Penicillium brevicompactum (1)

Tritirachium sp. (1)

(*Aureobasidium pullulans, Pithomyces chartarum* ≤0.5%)

* Average total fungi in outdoor air samples, 350 CFU m^{-3} air; average total in indoor air samples, 200 CFU m^{-3}; medium, MEA; N = 6 samples indoors and outdoors.

drophilic fungi, e.g. malt extract agar (MEA), rather than, for example, dichloran 18% glycerol agar (DG-18) as is used for xerophilic species, is essential for appropriate data interpretation.

Sampling in buildings must be preceded by a clear evaluation pathway that outlines how analytical data will be interpreted. For example, when the objective of air sampling is to determine whether exposure conditions in a building are "normal" after clean-up of visible fungal colonization is completed, collection of samples at many locations and at various times indoors, as well as concurrently in the outdoor environment, is the minimum required for data interpretation. In addition, comparison of air sampling data obtained both before and after clean-up and knowledge that moisture problems have been both evaluated and corrected, and visually mouldy materials have been physically removed (Chapter 2.3), add to the strength of possible data interpretation.

INTERPRETATION OF CASE STUDIES

The case studies described here are categorized as "broad", "indicative" or "focused" (ACGIH 1999) according to the methodology used for sample collection and analysis. When the objective of the evaluation is general, such as to determine whether the distribution of phylloplane fungi indoors is typical, sampling and analytical methodology which can potentially identify a wide range of taxa are used. Analysis of collected air or source samples on a medium such as malt extract agar can be expected to provide information on a wide range of fungi that may possibly be present.

Indicator methods are used in sampling where information is sought about a group of organisms or agents that can be used as an indicator of certain environmental conditions. Sampling for *Stachybotrys* and *Chaetomium* can be used as an indicator of soaking wet conditions on cellulosic materials. Sampling for endotoxin can be used as a surrogate indicator for the presence of Gram-negative bacteria.

Focused methods are used when information on specific microorganisms thought to cause disease in a building is sought. Sampling for *Aspergillus fumigatus* and *Legionella pneumophila* in a medical centre following the occurrence of infection among patients is a highly focused building evaluation.

In the case studies that follow, broad (Buildings 1-3), indicative (Buildings 4-10) and focused (Buildings 11-13) approaches to interpretation of sampling data are illustrated.

Building 1

Air sampling was performed in classrooms of a school where the envelope walls (mostly brick, concrete block and hard plaster) had been chronically wet. At the time of sample collection the moisture problem had been corrected. Sampling was undertaken to determine whether airborne fungi in classrooms were similar or not to those present in the outdoor air. The medium used in the culture plate sampler (2% MEA; ECA 1993, AIHA 2005) is supportive of the growth of a wide range of fungi.

Analytical results showed a general similarity between fungi collected indoors and outdoors, with *Cladosporium cladosporioides*, *Epicoccum nigrum*, yeasts and non-sporulating fungi accounting for at least 80% of the taxa found (Table 1). The sampling results suggested that the environment in the classrooms was mycologically "normal". The presence of the hydrophilic species *Ulocladium chartarum* in the classrooms, but not outdoors, might have indicated the presence of water-damaged building materials (Table 1). Concurrent inspection of classrooms and building structural systems did not, however, reveal water damage or visible fungal growth.

Table 2. Surface sampling for a broad range of culturable fungi on louvres of air supply vents in three rooms.

Location and Description	CFU cm^{-2} surface*	Fungal taxa
Room #1 (complainant work area)		
Visibly dirtiest louvre	280	*Cladosporium* spp. (86%), *Penicillium* spp. (9%), *P. corylophilum* (6%)
Visibly cleanest louvre	<5	ND
Room #2 (control area)		
Visibly dirtiest louvre	2×10^4	*Cladosporium* spp. (95%), *Penicillium* spp. (5%),
Visibly cleanest louvre	5	*Penicillium* spp.
Room #3 (control area)		
Visibly dirtiest louvre	1×10^5	*Cladosporium* spp. (98%), *Penicillium* spp. (2%)
Visibly cleanest louvre	1.5×10^3	*Cladosporium* spp. (67%), *P. corylophilum* (33%)

* Sterile 1 × 5 cm template applied to louvre; surface dirt removed by sterile wet cotton swab; dirt eluted in sterile water and plated on DG18 agar medium. Limit of detection 5 CFU cm^{-2}.

In conjunction with the results of the building inspection, the data in Table 1 suggested that exposure to fungi in classrooms was typical of that expected in non-problem schools and outdoor air.

Building 2

Surface sampling of supply air vent (diffuser) louvres was undertaken because several occupants claimed that fungi present in dusts on louvres were the cause of sick building syndrome. While sampling cannot alone prove the presence or absence of building related symptoms, the data from sampling can provide a general indication of the presence of fungal growth in a building. Air vent louvres were surface sampled for culturable fungi at multiple locations in the portion of the building where complaints worked, and in multiple locations in the same building and in other control buildings nearby that had not been visited by the complainants. The objective in sampling was to determine whether the fungi on louvres in locations where complaints worked were suggestive of chronically damp or wet conditions.

Results of surface sampling showed that *Cladosporium* dominated the fungi recovered from both complainant and control locations (data from one complainant and two control locations are shown in Table 2). Sampling data also showed that the concentration of culturable fungi varied by up to three orders of magnitude between clean and dirty (dusty) louvred areas of the same air vent. In addition, concurrent inspection of occupied spaces and the HVAC system of both complaint and control building areas showed an absence of water damage or visible fungal growth on interior surfaces.

Together with the building inspection in this study, the results of sampling in Table 2 suggested

that complaints were not related to either the fungi in surface dust on air supply vents or to the mycological conditions in the indoor environment.

Building 3

Rupture of a pipe resulted in wetting of carpet in several rooms of a residence. Fungi were observed growing on areas of the wet carpet. All carpet which had been wet was removed and replaced with new carpet. Carpet throughout the rest of the house was cleaned with a low efficiency vacuum cleaner. Occupants who had previously vacated the residence after the rupture reported the onset of mucous membrane and upper respiratory tract irritation after returning. The air was sampled for culturable fungi in the residence and also outdoors in order to see if the broad range of taxa indoors indicated a possible hidden mould problem or cleaning that was inadequate. Concurrent visual inspection of the occupied space and HVAC system did not reveal water damage or visible fungal colonization either on interior surfaces or in the HVAC system.

Cladosporium cladosporioides accounted for more than 90% of the fungi recovered in the outdoor air at grade level around the residence (Table 3). Minor amounts of *Penicillium* and *Aspergillus* were also present in the outdoor air. The rank order of taxa present in indoor air was distinct from that outdoors, with *A. versicolor* and *P. corylophilum* recovered as predominant fungi. The isolation of these fungi, and *A. sydowii* and *Wallemia sebi*, from the indoor air suggested hidden dampness in the building infrastructure (Flannigan and Miller 1994, Health Canada 2004, IOM 2004) or the presence of reservoirs of fungal spores not adequately removed during previous clean-up activities. Because a thorough inspection of

Table 3. Air sampling for a broad spectrum of culturable fungi in a residence with a past moisture problem.

Fungi in outdoor air (ranked taxa, %)*

Cladosporium cladosporioides (92)
Penicillium brevicompactum (2)
C. sphaerospermum (1)
P. implicatum (1)
P. sclerotiorum (1)
Yeast (1)
Aspergillus niger (0.5)
(*P. citrinum, Rhodotorula* sp., non-sporulating fungi at ≤0.5%)

Fungi in indoor air (ranked taxa, %)*

A. versicolor (24)
P. corylophilum (21)
C. cladosporioides (14)
P. citrinum (8)
Non-sporulating fungi (8)
Wallemia sebi (7)
A. sydowii (7)
Eurotium herbariorum (2)
P. sclerotiorum (2)
(*P. brevicompactum, P. implicatum, A. niger, A. flavus, Rhodotorula* spp., *C. sphaerospermum* at ≤1%)

*Average total concentration of fungi in outdoor air = 2,000 CFU m^{-3}; mean total concentration of fungi in indoor air = 340 CFU m^{-3}; DG18 agar medium; N=3 indoors and outdoors.

the residence did not reveal a hidden dampness or water problem, it was concluded that the previous clean-up was inadequate. Removal of settled dust from floor and above-floor surfaces and from the HVAC system by means of a high efficiency particulate air (HEPA) vacuum cleaner was recommended for removal of mould spores that had been dispersed during the first, unsuccessful clean-up (see Chapter 2.3).

Building 4

Air was sampled by automatic volumetric spore trap (AVST) in a new building that had suffered water damage. Large areas (>1000 m^2) of visibly mouldy wallpaper, paper-fibre gypsum board, cellulosic fire insulation and floor coverings had previously been removed. At the time of air sampling, water leaks

Table 4. Air sampling for fungal spores in a dusty building with past moisture problems.

Sample Description	Spores m^{-3} air	
	Cladosporium	*Penicillium/ Aspergillus*
Outdoors	120	55
Indoors†	120	11500

Spore trap operating at 0.01 m^3 min^{-1} used to perform air sampling. †Composite sample of settled dusts from floors contained approx. 10^8 CFU g^{-1} (>95% *Penicillium, Aspergillus* and *Eurotium*) on DG18 agar medium.

had been repaired and the building infrastructure had been dried. However, there were considerable amounts of settled dust associated with past clean-up activities on floors and other surfaces. The object of air sampling was to determine whether or not respiratory protection was still needed by clean-up and construction personnel working in the building.

Sampling in the outdoor air showed that the concentration of *Cladosporium* spores was approximately twice that of *Penicillium/Aspergillus* (Table 4). Air sampling indoors under quiescent conditions (dust not being actively stirred up by indoor activities) showed that, while *Cladosporium* levels were similar to those outdoors, *Penicillium/Aspergillus* concentrations exceeded 10^4 spores m^{-3} air. In addition, settled dusts collected from floors and other horizontal surfaces in the building were analyzed for the presence of culturable fungi. The results indicated that *Penicillium/Aspergillus/Eurotium* accounted for >95% of culturable fungi in settled dusts, the total concentration being approximately 10^8 CFU g^{-1} dust (Table 4).

The dominance of *Penicillium* and *Aspergillus* in samples of both air and settled dusts indicated that strong reservoirs of fungi associated with past water damage were still present in the building. Sampling data also showed that it was likely that all persons working in the building would be exposed to *Penicillium/Aspergillus* concentrations approaching those typically found during handling of mouldy agricultural materials. Consequently, use of personal protective equipment such as P-100 respirators by all restoration personnel was required (see Chapter 2.3).

Building 5

Chronic moisture problems and extensive (>30 m^2) visible mould growth occurred in two rooms on the lower floor of a two-storey building. *Aspergillus versicolor* and other non-phylloplane species were identified as dominant fungi on colonized surfaces in rooms. Visibly mouldy materials were removed from the building using full containment remediation techniques (see Chapter 2.3; AIHA 2008). Visual inspection verified that the remediation plan was followed, that visually mouldy materials were removed, and that residual dusts on cleaned surfaces were sufficiently reduced in amount, so that additional HEPA vacuum cleaning was unnecessary.

Air sampling for culturable moulds was carried out before and after remediation to provide an indication of whether or not restoration activities had changed the profile of dominant fungi from non-phylloplane to phylloplane species. For this purpose, phylloplane

Table 5. Air sampling for culturable phylloplane and non-phylloplane fungi as an indication of effectiveness of remediation.

Location of Samples	Culturable Fungi (CFU m⁻³)*			
	Total	AV only	P/Asp/E	C/Alt /Epi[†]
Before clean-up				
Floor 1 (N=10)	650	140	560	45
Floor 2 (N=20)	120	6	60	30
After clean-up				
Floor 1 (N=10)	450	4	75	300
Floor 2 (N=30)	730	2	120	400

*Cellulose medium. AV, *Aspergillus versicolor*; P, *Penicillium*; Asp, *Aspergillus*; E, *Eurotium*; C, *Cladosporium*, Alt, *Alternaria*.; Epi, *Epicoccum*. [†]The dominant taxa in outdoor air both before and after remediation.

fungi were defined as the total of all *Cladosporium*, *Epicoccum* and *Alternaria* counts, and non-phylloplane fungi as the total of all *Penicillium*, *Aspergillus* and *Eurotium* counts.

Air sampling prior to clean-up on the lower floor showed that *A. versicolor* was present at an average concentration of approximately 140 CFU m⁻³ air, or around 22% of the total (Table 5). On this floor the combined *Penicillium/Aspergillus/Eurotium* concentration was about 12 times greater than that of *Cladosporium/Alternaria/Epicoccum*. Non-phylloplane species were also dominant in mycobiota of the air of the upper floor.

When sampling of the lower floor was performed several weeks after the completion of clean-up activities, the *A. versicolor* concentration was only 4 CFU m⁻³ air, or about 1% of the total (Table 5). Phylloplane species now dominated the mycobiota, with *Cladosporium/Alternaria/Epicoccum* accounting for 67% of the fungal count.

The sampling data from this building indicated that restoration activities had successfully changed the profile of airborne fungi from non-phylloplane to phylloplane species. *A. versicolor* was not detected in outdoor air, so that the presence of low concentrations (<1% of the total) of this species indoors after clean-up can be interpreted as indicating that, even with a thorough clean-up, mould spores associated with the past moisture problem will be present in the trivial amount of dust remaining on surfaces after remediation

Building 6

Inspection of a new building revealed that paper-fibre gypsum board was visibly mouldy, and spray-on fire insulation with some cellulosic constituents was discoloured. A dispute arose between parties involved in construction as to the primary cause of the visible damage, i.e. whether it was flooding resulting from leaks or the inability of the building operator to control indoor relative humidity (R.H.) and dampness.

Pieces of discoloured fire insulation and visibly mouldy gypsum board were analyzed by serial dilution for culturable fungi and by direct microscopical observation (cellulose tape lifts; AIHA 2004) to provide insight into the nature of the moisture history. The predominating presence of hydrophilic fungi would have indicated soaking wet conditions (Flannigan and Miller 1994), whereas if damage to the building materials was associated with chronically damp conditions then xerotolerant fungi should have predominated in samples.

Fig. 1 shows fungi directly observable on a representative gypsum board surface. Perithecial hairs and *Chaetomium* spores were predominant on this and other gypsum board surfaces. *Chaetomium* characteristically grows well on wet cellulosic surfaces. *Stachybotrys chartarum* was the dominant fungus found by plating serial dilutions of pieces of fire insulation on MEA (Table 6). Yeasts and *Fusarium*, present as minor contaminants in the samples (Table 6), are characteristically found in waterlogged substrates.

The results in Table 6 and Fig. 1 indicated the mould growth and discoloration were associated with flooding rather than damp operating conditions. This information was used to assign responsibility for damage between the parties involved in building construction and building operation.

Faecal pellets from mycophagous mites or insects can also be seen among the fungal structures in Fig. 1. Such an observation has been used as a biological indicator of long-term colonization of the surface that was sampled (Yang 2007). The faecal pellets in Fig. 1 appear to contain *Aspergillus-Penicillium* spores rather than the larger spores of *Chaetomium*.

Table 6. Presence of culturable *Stachybotrys chartarum* in fire insulation indicating previous flooding conditions.

Sample Description	Fungal taxa present (CFU g⁻¹)*
Composite sample of fire insulation near surface facing return air plenum	6×10^4: 95% *S. chartarum*; 5% yeasts
Composite sample of fire insulation adjacent to ceiling slab	1×10^4: 90% *S. chartarum*; 5% yeasts; 3% *Fusarium*

* MEA; serial dilution.

Building 7

Visible mould colonization was found on wall, ceiling, furniture and other surfaces in guest rooms in a newly renovated hotel. The amount of visible mould in each guest room varied from non-detectable to about 9 m². R.H. in guest rooms was mostly in the 70-90% range, with only a few periods below 70%. Physical inspection did not detect evidence of flooding or water leaks. Guest rooms with more than about 3 m² of visible mould had been taken out of service and were considered non-rentable. Analysis of settled dust from several of the mouldy rooms showed that xerophiles such as *Aspergillus versicolor* and *Wallemia sebi* were prominent culturable moulds.

Air sampling for culturable moulds was carried out in 10 rooms which had been taken out of service because of substantial amounts (2-6 m²) of mould visible on interior surfaces. The objective of air sampling was to estimate exposure that previous guests

in these rooms might have experienced. Accordingly, in each room an air sample was obtained under quiescent conditions, i.e. the sampling person entered the guest room with a portable sampler (see Table 7), stood still holding the sampler and collected the sample. A second sample was then collected when dresser drawers and window drapes were opened, simulating guest check-in activities. At the same time, outdoor control samples were collected on the building roof.

Sampling results showed that *W. sebi* and *Eurotium herbariorum* dominated the air in guest rooms under both quiescent and guest check-in conditions (Table 7). *Cladosporium herbarum* was the dominant taxon in the outdoor air. The results showed that occupant exposure in guest rooms under all conditions was different from that expected in a well-maintained, dry building in which phylloplane species would dominate. Total concentrations of indoor moulds were

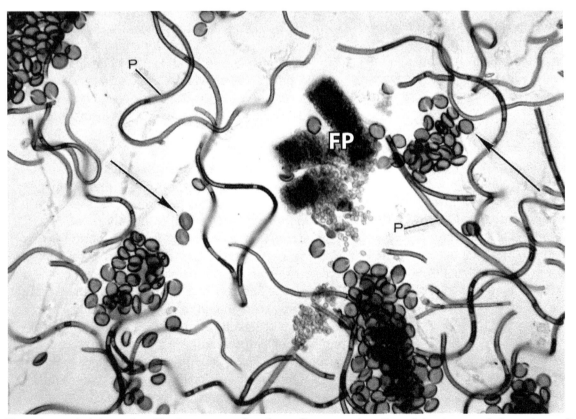

Fig. 1. Direct microscopical observation of fungi present on surface of paper fibre gypsum board in a cellulose tape sample. Spores (arrowed) and hairs (P) from perithecia of the ascomycete, *Chaetomium*, are abundant, indicating past flooding conditions on the paper-fibre surface of the gypsum board. Large faecal pellets (FP) from mycophagous mites or insects are also present.

Table 7. Effect of simulated guest check-in activity on numbers of airborne culturable moulds in visibly mould-affected hotel rooms.

	Location/condition		
	Quiescent	Open drawers and drapes	Outdoor air
Alternaria alternata	-	-	4
Aspergillus niger	3	1	10
A. ochraceus	21	1	-
A. versicolor	22	54	4
Cladosporium herbarum	-	-	382
C. sphaerospermum	20	21	4
Eurotium amstelodami	54	81	1
E. rubrum	2	16	<1
E. herbariorum	505	831	2
Penicillium brevicompactum	36	33	2
Wallemia sebi	3,395	6,142	10

A culture plate impactor (0.18 m^3 min^{-1}) with 219 jets was used for sampling; sample time indoors, 0.25-1.0 min.; sample time outdoors, 1-2 min; N = 8 for outdoor controls; N=10 each for quiescent and open drawers/drapes samples; isolation medium DG18 agar. Data presented as average CFU m^{-3} for each taxon; -, not detected.

around an order of magnitude greater than the outdoor level, with slightly higher concentrations during guest check-in conditions. The practical outcome of this sampling evaluation was closure of the hotel and development of plans to cure the dampness problem and to carry out mould remediation.

Examination of the DG18 culture plates used to collect indoor samples showed that the number of colonies exceeded or equalled 219, the number of jets (holes) in the sieve impactor plate. This suggested that the concentrations of indoor mould taxa listed in Table 7 were underestimates of actual concentrations. The upper limit of detection of the kind of sieve impactor utilized was about 10^4 m^{-3} (AIHA 2005, Morey 2007). A more accurate estimate of the total concentration in the hotel guest rooms would likely have been obtained by use of a liquid impinger (AIHA 2005) with subsequent dilution plating on DG18 agar.

Building 8

Faculty and staff in a school reported health problems (building related symptoms and allergies) possibly related to the workplace environment. Inspection of the occupied space and HVAC system at the school indicated that there was no moisture problem and no mould growth on interior and HVAC airstream surfaces. Additionally, air and dust sampling for moulds in the school indicated an unremarkable distribution of phylloplane species in the indoor environment (results similar to Table 1).

Additional sampling involving the collection of settled dust, and analysis for allergens (mite Der p1 and Der f1, cat Fel d1) and endotoxins was carried out in the school and in residences of 10 symptomatic faculty and staff members. Three hotels which housed building investigators were also evaluated.

Concentrations of Fel d1 in residences were at least two orders of magnitude higher than that found in the school or in the hotels (Table 8), a finding not unexpected as most of the residences contained house cats. Concentrations of mite allergens in both hotels and residences were at least an order of magnitude higher than that found in the school. Endotoxin levels in residential dusts were slightly elevated relative to those found in schools and hotels. While conclusions about health effect cannot be made directly from sampling data alone, the results in Table 8 suggested that building-related symptoms and allergies were most likely associated with residential exposures. A practical recommendation resulting from this study was to use HEPA vacuum cleaners in residences, especially to remove dust containing cat allergen.

Table 8. Allergens and endotoxins in dust in a school, residences and hotels.

Type of building	Allergen level (µg g^{-1})			Endotoxin (EU mg^{-1})
	Der p 1	Der f 1	Fel d 1	
School	0.03	0.86	0.24	14.8
Residences	0.80	12.3	162.0	67.0
Hotels	26.7	22.8	0.9	13.2

Six dust samples from a single school; one sample from each of 10 residences; one sample from each of three hotels. For analytical methods see AIHA (2005).

Table 9. Post-remediation sampling to indicate whether clean-up following a sewage backflow had been adequate or not.

Sample Location	Airborne Endotoxin (EU m^{-3} air)*
Outdoor air (N=2)	0.5
Rooms previously affected by backflow† (N=4)	3.1
Rooms unaffected by back-flow† (N=3)	0.5

†Floor surfaces tested for presence of culturable *Aeromonas, E. coli* O157/H7, *Salmonella, Shigella, Vibrio* and *Streptococcus faecalis.* A 30×30 cm area of floor surface at the junction with a wall was repeatedly swabbed with a sterile gauze pad saturated with water. The gauze pad was transported at 4°C to the laboratory, mixed with additional sterile water to a volume of 50 ml, and aliquots of fluid were plated on various culture media, including MacConkey agar with and without sorbitol; xylose-lysine-desoxycholate agar; thiosulphate-citrate-bile-sucrose agar; and Columbia-nalidixic acid agar. Culturable bacteria characteristic of sewage waters were not detected.*Limulus* assay (Milton *et al.* 1992)

Building 9

A sewage flood occurred in the lower floor of a multi-storey building. Sewage-backflow waters and water-damaged porous finishing materials such as carpet were removed; affected surfaces were disinfected by repeated application of bleach; and construction and finishing materials were dried using dehumidifiers. Restoration efforts notwithstanding, occupants were concerned about possible exposure to harmful organisms, e.g. *Escherichia coli*, that still might be present on surfaces previously contaminated by the sewage-backflow water. Air was sampled for endotoxin both in rooms previously affected by backflow of sewage and in upper-floor rooms which were unaffected by the backflow. Since Gram-negative bacteria such as *E. coli* are predominant in sewage waters and all Gram-negative bacteria contain endotoxin, it was hypothesized that if clean-up was successful levels of endotoxin throughout the entire building would be similar, and that levels of endotoxin indoors would be no higher than those outdoors. Endotoxin sampling was therefore performed as a surrogate indicator for the presence of residual sewage organisms.

Sampling results showed that the average amount of endotoxin in the area of sewage-affected rooms was about 3 endotoxin units (EU) m^{-3} while concentrations in upper-floor rooms and in the outdoor air were about 0.5 EU m^{-3} (Table 9). Relative limit values (RLVs) of 10-100 times the background outdoor level have been suggested for airborne endotoxin (Milton 1999). Adverse health effects such as fatigue, cough and acute airflow obstruction become frequent when measured endotoxin levels approach or exceed RLVs

of 10 to 100 (Milton 1999).

While the average concentration of endotoxin in sewage affected areas was below the suggested RLV, occupants remained unconvinced of the effectiveness of disinfection and clean-up activities. Consequently, additional testing was undertaken for bacteria commonly associated with sewage, e.g. *E. coli* O157/H7, *Streptococcus faecalis* and others listed in the footnote to Table 9. A wet-swab method (Table 9) was used to extract dusts from floors in sewage-affected and sewage-unaffected rooms. Bacteria characteristic of sewage were not detected by culture of any samples (Table 9). Collectively, the data indicated that disinfection and cleaning had probably removed pathogenic bacteria from rooms previously affected by the sewage backflow. IICRC Standard S500 (2006) provides practical recommendations on clean-up of sewage-backflow waters (see also Chapter 2.3).

Building 10

Workers in one area of an industrial building reported a sporadic flu-like illness associated with exposure to water mist. The mist originated from a reservoir containing water that was continuously recirculated with minimum make-up water. Air was sampled for endotoxin in both misty and non-misty indoor areas, as well as outdoors, to indicate whether or not the water reservoir was an amplifier for Gram-negative bacteria and other microorganisms.

Sampling results showed that the endotoxin concentration in the misty area was at least 100 times greater than in non-misty areas or in the outdoor air (Table 10). Thus, the airborne endotoxin level in the misty area exceeded the RLVs suggested by Milton (1999), indicating that there was amplification of Gram-negative bacteria in the water reservoir from which mists originated.

Additional sampling showed that *Pseudomonas*, *Flavobacterium* and other Gram-negative bacteria were the dominant culturable organisms in both the water reservoir and water-mist droplets. A follow-up medical and environmental study suggested that some of the flu-like illness reported by workers was associated with high levels of endotoxin exposure (Milton *et al.* 1995).

Building 11

Occupants of an apartment in a new building reported flooding during rainy periods, when rain water penetrated envelope walls because of inadequate window flashing and cracks in the stucco finish. Destructive inspection of the envelope showed that

Table 10. Air sampling for endotoxin in a misty area of a building where occupants reported a flu-like illness.

Sample Location	Endotoxin (EU m^{-3})*
Outdoor air (N=4)	23
Indoors; misty zone (N=4)	3200
Indoors; dry zone (N=2)	8

Limulus assay (Milton *et al.* 1992.

Table 11. Rank order of mould taxa detected in a two-year-old dust sample determined by different analytical methods.

PCR analysis	
Stachybotrys chartarum	81%
Penicillium roquefortii	3%
Cladosporium cladosporioides	2.5%
Eurotium amstelodami	2.5%
Aspergillus fumigatus	2.5%
Alternaria alternata	2.5%
Dilution Plating on CMA	
Penicillium spp.	70%
Stachybotrys chartarum	20%
Aspergillus niger	10%
Direct Plating on CMA	
Penicillium spp.	50%
Ulocladium botrytis	20%
Stachybotrys chartarum	10%
Chaetomium globosum	10%
Acrodontium spp.	10%

PCR, polymerase chain reaction; CMA, cornmeal agar. Dust sample was sieved and homogenized in the laboratory and aliquots of samplewere processed separately by PCR, direct plating and dilution plating methods.

both the building paper and the exterior sheathing (manufactured wood) were heavily colonized by *Stachybotrys chartarum* and various penicillia and aspergilli. Rainwater entering through the biodeteriorated envelope constantly wetted the carpet in the apartment. After each flooding and subsequent carpet drying, the apartment occupant used a vacuum cleaner to remove dust from carpet and other flooring surfaces.

At the time of destructive inspection of the building envelope, the contents of the vacuum cleaner were collected and archived for future analysis. Two years later, the vacuum cleaner dust was analyzed by culture (direct and dilution plating on cornmeal agar) and polymerase chain reaction (PCR; 24 target list) methodologies (AIHA 2005) to determine the mycobiota of this dust. Both direct and dilution plating showed that unidentified *Penicillium* spp. accounted for at least 50% of the culturable taxa in the vacuum cleaner dust (Table 11), and *Stachybotrys chartarum* accounted for about 10-20%. In contrast, PCR analysis showed that *S. chartarum* accounted for over 80% of the spore equivalents detected (Table 11).

The results in Table 11 show that the method chosen to analyze the collected sample is of great importance to subsequent data interpretation. PCR analysis was the only method that revealed *Stachybotrys* domination in the two-year-old dust sample. This might be anticipated in a sample of this age because of the relatively short half-life (<1 year) of culturable *Stachybotrys* relative to the half-life of *Penicillium* spores (See Table 6.1 in AIHA 2005).

Building 12

Two workers and one office employee in a small industrial facility developed pneumonia and required hospitalization. Laboratory tests performed independently by private physicians showed elevated levels of *Legionella pneumophila* antibodies (serotype not determined) in sera. Sampling focused on detection of legionellae in both the workplace and residential environment was initiated after the management at the facility realized that the three cases had occurred

within the relatively short time period of several months.

An inspection of the facility showed that water sprays were used to suppress dusts in and around buildings. Water for industrial sprays and for potable hot and cold services originated from wells on the property. The water from the wells was initially stored in outdoor tanks above ground and was consequently subject to insolation, the temperature of water in the tanks being c. 30-40°C.

Culture analysis of water samples from spray nozzles and from hot water tanks and piping showed the presence of *L. pneumophila* sero-group 1 at concentrations of 10 to >100 CFU ml^{-1} (Table 12).

This, together with the absence of *Legionella* spp. (detection limit 0.5 CFU ml^{-1}) in potable water systems in the homes of the employees, suggested that pneumonia was likely to be due to legionellae in the workplace environment. The sampling evaluation in this building differed from that in Buildings 1-11 because the pre-existing medical data focused sampling on one culturable bacterial species, *L. pneumophila*. See Chapter 2.3 for emergency *Legionella* disinfection measures.

Building 13

Three bone-marrow transplant patients who had

Table 12. Detection of culturable *Legionella* in water systems in a building where occupants developed Legionnaires' disease.

Sample Locations	*Legionella* (CFU ml⁻¹)	Species and Serotype*
Hot water tank and piping (N=10)	10-1000	*L. pneumophila* serogroup No. 1
Water spray nozzle at end of piping (N=3)	10 -100	*L. pneumophila* serogroup No. 1
Hot water tank and piping in residences of employees (N=6)	ND	—

ND, not detected (LOD = 0.5). Method: culture on buffered charcoal yeast extract agar (AIHA 2005).

Table 13. Air sampling for thermotolerant *Aspergillus* in hospital in which several immunocompromised patients developed aspergillosis.

Location and description of samples	Culturable thermotolerant (37°C) fungi*
Air samples in room that had been occupied by affected patients	One sample with *A. fumigatus* (9 CFU m⁻³) and *A. flavus* (6 CFU m⁻³); two samples without detectable fungi
Settled dust in room	Dust contained *A. fumigatus* (approximately 3,000 CFU g⁻¹)
Adherent/settled dust in air-handling unit and ductwork providing ventilation air to room	All of 5 dust samples collected contained *A. fumigatus*; 2 contained *A. flavus*; concentration of *Aspergillus* spp. in samples varied from 100 to 10,000 CFU g⁻¹

*Isolated on MEA.

sequentially occupied one suite in a medical centre developed aspergillosis. According to the pathology reports, infection was caused by *Aspergillus fumigatus* and/or *A. flavus*. A sampling investigation was undertaken in order to determine whether these *Aspergillus* spp. were present in the patient suite or in the HVAC system serving this area. The evaluation was focused on the recovery of thermotolerant (37°C) *A. fumigatus* and *A. flavus* in air and dust samples, since their presence would have suggested that infections were hospital-acquired.

Three air samples were collected quiescently in the former patient suite (Table 13). Thermotolerant fungi were only detected in one air sample, which contained a total of 15 CFU m⁻³ (*A. fumigatus*, 9 CFU m⁻³; *A. flavus,* 6 CFU m⁻³). Settled dust collected from the floor and from the return register grille in the suite contained *A. fumigatus* at roughly 3,000 CFU g⁻¹ (Table 13).

All five dust samples collected from the air-handling unit and supply ductwork, including the final filter serving the patient suite, contained culturable *A. fumigatus*, but only two yielded *A. flavus* (Table 13). Visible mould growth was observed on the airstream surfaces of the air handling unit serving the patient suite. The collective infection and sampling results showed that patient infections were hospital-acquired, and that the HVAC system was the source of thermotolerant fungi causing infections.

The focused study in Building 13 provides a rare example where environmental organisms (*A. fumigatus* and *A. flavus*) in a building were shown to be the agents causing disease (aspergillosis) in patients.

CONCLUSION

The cases described above illustrate that microbial sampling should be performed in the context of the objective of the overall building investigation. An understanding of building construction and any design, operation, and maintenance failures revealed by an informed inspection is a prerequisite for sampling. The objective of sampling, including an understanding of how data will be interpreted, should be established prior to sample collection. The investigator must select analytical methods appropriate to the study objective and be aware of limitations in sampling methodology. Interpretation of analytical results should be made in the context that sampling is just one component of the informed building evaluation.

REFERENCES

ACGIH (1999) *Bioaerosols: Assessment and Control*. American Conference of Governmental Industrial Hygienists, Cincinnati, OH.

AIHA (2004) *Assessment, Remediation, and Post-Remediation Verification of Mold in Buildings*. American Industrial Hygiene Association, Fairfax, VA.

AIHA (2005) *Field Guide for the Determination of Biological Contaminants in Environmental Samples*, 2nd ed., L.-L. Hung, J.D. Miller, and H.K. Dillon, (eds.), American Industrial Hygiene Association, Fairfax, VA.

AIHA (2008) *Recognition, Evaluation, and Control of Indoor Mold*, B. Prezant, D.M. Weakes, and J.D. Miller, (eds.), American Industrial Hygiene Association, Fairfax, VA.

ASTM (2008) *Standard Guide for the Inspection of Water Systems for Legionella and the Investigation of Possible Outbreaks of Legio-*

nellosis (*Legionnaires' Disease or Pontiac Fever*), 2008 Annual Book of ASTM Standards, 11.07, D5952-08, American Society for Testing and Materials, West Conshohocken, PA.

ECA (1993) *Biological Particles on Indoor Environments*. European Collaborative Action, Report No. 12, Commission of the European Communities, Luxembourg.

Flannigan, B., and Miller, J.D. (1994) Health implications of fungi in indoor environments - an overview. In R.A. Samson, B. Flannigan, M.E. Flannigan, *et al.*, (eds.), *Health Implications of Fungi in Indoor Environments*, Elsevier, Amsterdam, pp. 3-28.

Health Canada (1995) *Fungal Contamination in Public Buildings: A Guide to Recognition and Management*, Federal-Provincial Committee on Environmental and Occupational Health, Ottawa.

Health Canada (2004) *Fungal Contamination in Public Buildings: Health Effects and Investigation Method*, H46-2/04-358E. Ottawa, Canada.

IICRC (2006) *Standard and Reference Guide for Professional Water Damage Restoration*, S500. Institute of Inspection Cleaning and Restoration Certification, Vancouver, WA.

IOM (2004) *Damp Indoor Spaces and Health*. Institute of Medicine, National Academies Press, Washington, DC.

ISIAQ (1996) *Control of Moisture Problems Affecting Biological Indoor Air Quality*. International Society of Indoor Air Quality, Helsinki, Finland.

Milton, D. (1999) Endotoxin and other bacterial cell-wall components. In *Bioaerosols: Assessment and Control*, America Conference of Governmental Industrial Hygienists, Cincinnati, OH, pp. 23.1-23.14.

Milton D., Feldman, H., Neuberg, *et al.* (1992) Environmental endotoxin measurement: The kinetic *Limulus* assay with resistant-parallel-line estimation. *Environ. Res.*, **57**, 212-230.

Milton, D., Amsel, J., Reed, C., *et al.* (1995) Cross-sectional follow-up of a flu-like respiratory illness among fiberglass manufacturing employees: endotoxin exposure associated with two distinct sequelae. *Am. J. Industr. Med.*, **28**, 469-488.

Morey, P.R. (2007) Microbial sampling strategies in indoor environments. In C. Yang and P. Heinsohn, *Sampling and Analysis of Indoor Microorganisms*, John Wiley, New York, pp. 51-74.

Yang, C. (2007) A retrospective and forensic approach to assessment of fungal growth in the indoor environment. In C. Yang and P. Heinsohn, *Sampling and Analysis of Indoor Microorganisms*, John Wiley, New York, pp. 215-229.

Chapter 5. Common and important species of Actinobacteria and fungi in indoor environments

Chapter 5

COMMON AND IMPORTANT SPECIES OF FUNGI AND ACTINOMYCETES IN INDOOR ENVIRONMENTS

Robert A. Samson[1], Jos Houbraken[1], Richard C. Summerbell[2,3], Brian Flannigan[4] and J. David Miller[5]

[1]CBS-KNAW Fungal Biodiversity Centre, Utrecht, The Netherlands;
[2]Dalla Lana School of Public Health, University of Toronto, Canada; [3]Sporometrics Inc., Toronto, Canada;
[4]Scottish Centre for Pollen Studies, Napier University, Edinburgh, UK;
[5]Institute of BioChemistry, Carleton University, Ottawa, Canada.

INTRODUCTION

In the following chapter the most common fungi and actinomycetes are illustrated, together with comments on their occurrence and medical implications.

The micrographs of these microorganisms were made from typical cultures isolated from indoor environments. The microscope slides were prepared in lactic acid, sometimes with the addition of aniline blue. Isolates of the species are in the CBS culture collection (http://www.cbs.knaw.nl).

The health implications of exposure to moulds is discussed below and hazard levels classified and coded. For each of the alphabetically arranged species, the comments on its medical implications include the coded hazard category. Where there have been well established reports of species being allergenic, a note of their allergenicity has been made. However, it should be remembered that all fungi contain proteins and glycoproteins to which humans can be allergic. Therefore, all of the species described (and others not described), whether labelled as allergenic or not, should be regarded as potentially allergenic. Information on relevant mycotoxins produced by toxigenic species occurring indoors is also provided.

A short list of references is given at the end of this introductory section of the chapter, but additional information on, and references to, the species may be found in the preceding chapters.

HEALTH IMPLICATIONS

Definition of normal mould exposure

Moulds may have effects on human health in various different ways: through allergenicity, through direct biochemical effects (irritation, non-allergic immuno-activation or toxicity), or through invasive or colonizing disease. In this section, some perspectives are given about the invasive and colonizing disease hazards posed by moulds. Allergenic and direct biochemical effects are dealt with elsewhere in this volume.

The list of mould species defined as agents of "opportunistic infection" has rapidly expanded in recent years, giving the appearance that humans risk exposure to hundreds of dangerous, disease-causing mould species. This appearance is mostly, although not entirely, illusory. The perspective that is occasionally lost is that opportunistic disease mainly affects limited groups of people with severe immune deficiencies. Of course, such people are important, and their protection is strongly encouraged, but the problems that moulds pose for them are best managed in light of a full knowledge of the ubiquity of most of the organisms involved.

Whereas allergenic and toxic mould hazards are always dependent on the quantity of material present, infectious disease hazards may depend merely on the presence of an infectious organism, regardless of quantity. That is because the more virulent an infecting organism is, the more it has the ability to multiply in the body and increase its population levels after the infected human has been exposed to it, making the actual level of initial exposure relatively unimportant. There is a "grey area", however, in that some infectious disease hazards may also depend on quantity: that is, some organisms that may seldom or never cause disease when contacted in small inoculum quantities may cause disease on heavy exposure to them. This is especially true for partially immunocompromised persons, e.g., persons taking high doses of corticosteroids, or those with HIV infection and reduced CD4+ T-lymphocyte counts. Both the discrete (presence/absence) and the inoculum-dependent effects seen in mould infections are taken into account below.

The above suggestion that many moulds labelled as opportunistic lack the wherewithal to cause disease in most people may be of significant comfort to many persons who, for one reason or another, see lists of moulds isolated or detected at their homes or workplaces. Owners and managers of workplaces, public buildings and rented accommodation, however, should not take similar comfort. Individuals generally know whether they themselves are immunocompromised, since this condition nearly always derives from medical treatment, or from serious genetic or retroviral disease. Building managers, however, must realize that they cannot know whether or not users of their facilities belonged to various categories of immunocompromised patients, and also that there is no effective or legitimate means of excluding such people. People who have no immunocompromising condition today may develop one later and especially with HIV infection, some persons may discover only belatedly that they are immunocompromised. From the building management perspective, therefore, indoor colonization by known, opportunistically invasive moulds are best managed as if immunocompromised persons were present and at risk of exposure; this prudence takes precedence over the knowledge that the great majority of people will suffer only allergic or, at worst, toxic effects from most indoor mould colonization. Precise management in the case of indoor mould colonization may include a request for a building's occupants to disclose any serious medical condition which they have that might predispose them to mould infection. It should always be remembered, however, that persons have no obligation to reveal their confidential medical details in such situations, and in most legal systems they would not be considered to have relieved managers of the burden of responsibility by failing to disclose these details. In any event, they may be insufficiently informed about their condition or its possible relation to mould infection.

After the above practicalities have been taken into account, the actual invasive disease risks posed by moulds can be considered in detail. Some fungi never cause human or animal disease, but many species, if they are able to grow at 37°C, may cause opportunistic infection in humans and animals with compromised immune systems. How alarmed should people be by such fungi? Clearly, since serious fungal infections are quite uncommon in otherwise healthy people, we are not under much threat from the usual moulds found in our clean outdoor air and our tap water. Therefore, it is useful to define a normal mould that people may routinely be exposed to, referred to here as a "table-top mould" (TTM), in terms of what may commonly be found throughout the world in such clean, nonhazardous materials.

The definition of TTM excludes geographically limited "virulent" fungi such as *Coccidioides immitis* that cause mild to severe disease in most typical, immunocompetent people who are exposed to them. These fungi may occasionally be carried to indoor environments in their endemic areas, e.g. deserts in southwest USA in the case of *Coccidioides*, and may also be encountered in "clean" outdoor air in these regions. Clearly, however, any mould that usually causes infections in most people that encounter it (even if these infections are often mild) must be considered hazardous. Likewise, *Cryptococcus neoformans* (not a mould, but a yeast) must be considered to pose a hazard beyond the TTM level. It is common in typical urban outdoor air and poses a negligible hazard to most persons who inhale it daily, but predictably causes severe disease in immunologically normal persons who inhale high inoculum levels, e.g. when breathing heavy dust while cleaning large quantities of pigeon guano from attics. In this case, however, the hazard is dependent on the inoculum level contacted. The fungi defined in this writing as TTMs neither predictably infect immunocompetent people nor pose a known or strongly suspected inoculum-dependent hazard to the normal person. To summarize then, the virulent fungi such as *Coccidioides* are not considered to be everyday, ubiquitous components of outdoor air, while the fungi posing an inoculum-dependent hazard may well be common in typical air, but the inoculum levels that are hazardous are not normally found in outdoor or indoor air. Thus, although both virulent and inoculum-dependent, invasive fungi intersect with normal life to some degree; the hazards they pose cannot be fully synonymized with the hazards usually posed by the common environments and materials of human existence.

TTMs, on the other hand, are very much a part of normal life everywhere. A good illustration of the contrast between the low hazard associated with most TTMs and the seemingly threatening labels they are sometimes given is found in *Aspergillus fumigatus*. Although it is often referred to as the most virulent "opportunistic mould" and is the subject of much medical literature based on cases in immunocompromised patients, it is a common outdoor air fungus all year round in many tropical areas, and becomes common in clean air during the late summer and autumn in most temperate parts of the world. The reason for this is that in the natural environment it colonizes warm piles of decaying vegetation, including fallen leaves and matted grass remains. It is also found in essentially all barns and sheds where plant material is stored on farms, and is present in the soils of most indoor potted plants, where it is often one of the dominant members of the fungal community. There are many other common habitats and materials with relatively high levels of this organism, e.g. composts, small accumulations of bird dung and dried marijuana. Without question, it is therefore a TTM, and everyone on this planet is routinely exposed to it. Although years of constant heavy exposure to it in barns may induce severe allergies in some farmers, this fungus does not usually cause invasive disease even when high inoculum quantities are encountered. Only for certain types of immunocompromised patients does it pose a predictable hazard, and for these patients outdoor air and also typical indoor air are hazardous. Very rare cases of serious disease caused by this organism may be found in seemingly otherwise healthy patients, but such persons may well have an undetected immunodeficiency; and in any case these persons most often appear to have acquired their unusual infections from exposures to completely normal, everyday conditions. Whatever factors occasion infection by such opportunistic fungi, they clearly do not affect the overwhelming majority of other people with comparable exposures. In the category of fungi commonly found in water, *Fusarium oxysporum* and *Fusarium solani* are greatly feared as opportunistic fungi in hospital bone marrow transplant wards, but are normally found in tap water as well as in essentially all fresh plant material, including vegetables. Probably nobody normally goes through a day without ingesting a significant quantity of these fungi.

Even persons with AIDS, unless the AIDS is complicated by certain cancers or use of immunosuppressive antiviral drugs such as ganciclovir, are not sufficiently immunocompromised to be significantly threatened by most nominally opportunistic moulds. This extreme example illustrates the extent to which the distinction between a typical environmental mould exposure and a more hazardous mould exposure is biologically realistic. Humans, like other mammals, are highly adapted to resist infection arising from our normal environmental mould exposures; only a few specialized moulds are able to overcome

our adaptive defences, and even certain major immune system handicaps are insufficient to make us vulnerable to them.

Despite this degree of immune defence, we have some minor, common vulnerabilities. For example, some ubiquitous TTMs may cause painful ear canal colonizations, especially in small children who put objects into their ears, or in farmers exposed to heavy plant dust. The same fungi or others may cause non-invasive nasal sinus infestations in persons with chronic allergic conditions. Again, these hazards are completely attributable to materials found in normal environments: common small objects, typical agricultural materials and normal air. Thus, fungi found in these medical conditions cannot be ascribed an exceptional degree of hazard. Indeed, many are among the most common and widely distributed environmental moulds, guaranteed to be found on every tabletop in the world. For example, one of the major agents of non-invasive fungal sinusitis, *Alternaria alternata*, is the second-most common fungus in almost all surveys of normal outdoor air in temperate latitudes. The most common fungus in such surveys is *Cladosporium herbarum*, which cannot grow at or near body temperature and so never causes infections. We inhale *A. alternata* with each breath we take; exposure to it is thus generally no more hazardous than breathing itself is.

The moulds in this list have been classified with regard to whether they pose any kind of an invasive or colonizing disease hazard. In the former category of hazard, the fungus invades human or other mammalian tissue; in the latter, the fungus grows non-invasively on surfaces within the body cavities such as sinuses, and causes irritation and inflammation. The more hazardous fungi are those that may potentially cause a disease not associated with common, normal TTM exposures. The diseases are categorized according to how they may possibly be acquired, e.g. by inhalation or introduction of fungus under the skin (into the dermis). Secondly, within the hazard levels posed by TTMs, and thus by normal air, water and surfaces, moulds that may cause problems in immunocompromised patients are similarly categorized. It should be noted that so-called "barrier breaks" are regarded as a kind of immunocompromising condition; these are perforations of the body so that fungi can enter into areas that are normally never exposed to such organisms, even in typical cuts and scrapes. Thus, a classification is made for TTMs that can become hazardous if introduced into the body via perforations into the interior of the eye, the peritoneum, the major blood vessels and similar recognized sites of immunologically consequential barrier rupture. Most such perforations occur as part of medical procedures, but other possibilities are deep perforations in accidents, and use of microbiologically unclean injection materials such as street drugs. In the case of the cornea, a deep scratch caused by a piece of sharp plant material or metal may be sufficient to bring about a significant barrier break and introduce a TTM, converting it from innocuous to potentially hazardous.

Finally, some categories are made for TTMs that have never been reported as infecting any kind of human patient or warm-blooded animal, no matter how deeply immunocompromised.

Definition of fungal disease hazards as calibrated to take normal exposure into consideration

Below are listed the TTM-calibrated hazard categories for fungi, from most to least hazardous (A-I). Some of the more serious hazard categories may not be represented by any fungus on the list in this book. *Coccidioides immitis*, a virulent fungus discussed above, would be categorized as ABC. The C is attached to the self-explanatory AB because the severity of inhalation-derived infection is known to be increased in heavily exposed persons, e.g. laboratory workers exposed accidentally to pure cultures. *Penicillium citrinum,* as a contrasting example, is one of the most common indoor/outdoor moulds, and is normally completely harmless. Since, however, it is known from rare eye infections after corneal perforation, plus two opportunistic infections of severely immunocompromised cancer patients and a non-invasive bladder colonization of an otherwise healthy woman (reported in 1951, with no similar cases reported since), it is categorized as EGI. *Aspergillus fumigatus* is categorized EFG(?D). The (?D) is attached because *A. fumigatus* very rarely is reported to cause the subcutaneous infection mycetoma, and it is not clear whether the persons contracting such infections are immunodeficient or merely that their deep skin layer has been inoculated with an unusually heavy quantity of material bearing *A. fumigatus. Pseudallescheria boydii*, a fungus which is often considered similar in medical impact to *A. fumigatus*, but which regularly causes mycetoma and which may cause fatal disease in otherwise healthy persons who have inhaled large quantities of water-borne inoculum in near-drowning experiences, can be categorized as BCG. One might

raise the query "Should the B in that categorization be downgraded to an F because *P. boydii* may commonly be found in typical warm, dirty water, and thus potentially on tabletops or in outdoor air, making its hazard categories by definition fall within the scope of TTM hazards?" The answer to this query is a firm "No", since there is no indication that *P. boydii* mycetoma can be regularly acquired from common materials; in population terms, cases are very rare, whereas *P. boydii* is a relatively common organism (a full discussion of the known virulence gradations in this complex species and the epidemiological probabilities of a healthy person acquiring a serious infection are beyond the scope of this writing). In contrast, neutropenic patients could very readily acquire lethal *A. fumigatus* infection by inhaling typical room dust or being exposed to normal outdoor air. The standard of common, normal life exposures thus excludes perforations inducing *P. boydii* mycetoma, even though the fungus may be commonly contacted, but includes exposure to the microbial conditions that cause opportunistic aspergillosis in neutropenic patients. Compare the above discussion of *C. neoformans*, classed as CDG(?B). As a final example, *Cladosporium herbarum*, an abundant outdoor air species with a maximum temperature for growth of 32°C, is categorized as HI. Like all common contaminant fungi, it has been the subject of occasional erroneous and dubious disease attributions in medical literature, e.g. de Hoog and Guarro (1995); such reports only demonstrate that the fungus can be isolated from medical specimens and misinterpreted, or that other disease-causing fungi may be misidentified and given the name of a nonpathogenic fungus.

Hazard levels above *TTM/normal outdoor air, material surface and tap water level.*

A. Poses an inhalation hazard beyond the hazard conditions typical of normal, clean materials.
B. Poses a deep skin (dermal) inoculation hazard beyond the hazard conditions typical of normal, clean materials.
C. Poses a known or likely inoculum-dependent inhalation hazard beyond the hazard conditions typical of normal, clean materials.
D. Poses a known or likely inoculum-dependent deep skin (dermal) inoculation hazard beyond the hazard conditions typical of normal, clean materials.

Hazard levels within *TTM/normal outdoor air, material surface and tap water level.*

E. Poses an inhalation hazard (e.g., for immunocompromised persons) *within* the hazard conditions typical of normal, clean materials.
F. Poses a deep skin (dermal) inoculation hazard (e.g. for immunocompromised persons) *within* the hazard conditions typical of normal, clean materials.
G. Poses a hazard related to a major barrier break (e.g., corneal perforation, major surgery, peritoneal or venous catheter presence, injection drug use) *within* the hazard conditions typical of normal, clean materials.

Categories describing absence of *known risk.*

H. Poses no known inhalation hazard whatsoever (some highly dubious or demonstrably false reports may be found in medical literature).
I. Poses no known dermal inoculation hazard whatsoever (some highly dubious or demonstrably false reports may be found in medical literature).

Note: Hazard levels indicated by letter(s) are given in the descriptions of the fungi and Actinomycetes starting from page 330.

REFERENCES

Abbott, S.P., Sigler, L., and Currah, R.S. (1998) *Microascus brevicaulis* sp. nov., the teleomorph of *Scopulariopsis brevicaulis*, supports placement of *Scopulariopsis* with the *Microascaceae. Mycologia*, 90, 297-302.

Andersen, B., Nielsen, K.F., and Jarvis, B.B. (2002) Characterization of *Stachybotrys* from water-damaged buildings based on morphology, growth, and metabolite production. *Mycologia*, **94**, 392-403.

Andersen, B., Nielsen, K.F., Thrane, U., *et al.* (2003) Molecular and phenotypic descriptions of *Stachybotrys chlorohalonata* sp. nov. and two chemotypes of *Stachybotrys chartarum* found in water-damaged buildings. *Mycologia*, **95**, 1227-1238.

Autrup, J.L., Schmidt, J., Seremet, T., and Autrup, H. (1993) Exposure to aflatoxin B$_1$ in animal feed production plant workers. *Scand. Environ. Health Perspect.*, **99**, 195-197.

Aveskamp, M., Gruyter, H. de, Woudenberg, J., *et al.* (2009a) Highlights of the *Didymellaceae*: A polyphasic approach to characterise *Phoma* and related pleosporalean genera. *Stud. Mycol.*, **65**, 1-64.

Aveskamp, M.M., Verkley, G.J.M., Gruyter, J. de, *et al.* (2009b) DNA phylogeny reveals polyphyly of *Phoma* section *Peyronellaea* and multiple taxonomic novelties. *Mycologia*, **101**, 363-382.

Balajee, S.A., Baddley, J.W., Peterson, S.W., *et al.* (2009) *Acremomium alabamensis*, a new clinically relevant species in the section *Terrei. Eukaryotic Cell*, **8**, 713-722.

Bensch, K., Groenewald, J.Z., Dijksterhuis, J., *et al.* (2010) Species and ecological diversity within the *Cladosporium cladosporioides* complex (*Davidiellaceae, Capnodiales*). *Stud. Mycol.*, **67**, 1-95.

Day, J.H. (1996) Allergic respiratory responses to fungi. In D.H. Howard and J.D. Miller, (eds.), *The Mycota*, Vol. 6, *Human and Animal Relationships*, Springer-Verlag, Berlin, pp. 173-192.

Domsch, K.H., Gams, W., and Anderson, T.-H. (2007) *Compendium of Soil Fungi*. 2nd Ed. IHW-Verlag, Eching.

Ellis, M.B. (1971) *Dematiaceous Hyphomycetes*. Commonwealth Mycological Institute, Kew, Surrey, UK.

Fischer, G., Muller, T., Ostrowski, R., and Dott, W. (1999) Mycotoxins of *Aspergillus fumigatus* in pure culture and in native bioaerosols from compost facilities. *Chemosphere*, **38**, 1745-1755.

Flannigan, B., and Pearce, A.R. (1994) *Aspergillus* spoilage: spoilage of cereals and cereal products by the hazardous species *A. clavatus*. In K.A. Powell, J. Peberdy and E. Renwick, (eds.), *Biology of Aspergillus*, Plenum, New York, pp. 115-127.

Flannigan, B., McCabe, E.M., and McGarry, F. (1991) Allergenic and toxigenic micro-organisms in houses. *J. Appl. Bact.*, **70**, 61S-73S.

Frankland, A.W. (1983) *Fusarium*: a neglected mould. *Grana*, **22**, 167-170.

Frisvad, J.C. (1987) High performance liquid chromatographic determination of profiles and other secondary metabolites. *J. Chromatog.*, **392**, 333-347.

Frisvad, J.C., and Filtenborg, O. (1989) Terverticillate penicillia: chemotaxonomy and mycotoxin production. *Mycologia*, **81**, 837-861.

Hoffmann, K., Walther, G., and Voigt, K. (2009) *Mycocladus* vs. *Lichtheimia*: a correction (*Lichtheimiaceae* fam. nov., *Mucorales*, *Mucoromycotina*). *Mycological Research News*, **113**, 275-278.

Hoog de, G.S., Guarro, J., Gene, J., and Figueras, M.J. (2000) *Atlas of Clinical Fungi*, 2nd Ed. *Centraalbureau voor Schimmelcultures*,

Utrecht, The Netherlands.

Houbraken, J., Samson, R.A., and Frisvad, J.C. (2006) *Byssochlamys*: significance of heat resistance and mycotoxin production. *Advances in Experimental Medicine and Biology*, **571**, 211–224.

Houbraken, J., Varga, J., Rico-Munoz E., *et al.* (2008) *Byssochlamys spectabilis*, a heterothallic species with *Paecilomyces variotii* as its anamorph. *Appl. Environ. Microbiol.*, **74**, 1613-1619.

Houbraken, J., Frisvad, J.C., and Samson, R.A. (2010) Taxonomy of *Penicillium citrinum* and related species. Fungal Diversity. DOI 10.1007/s13225-010-0047-z

Lacey, J., and Dutkiewicz, J. (1994) Bioaerosols and occupational lung disease. *J. Aerosol Sci.*, **25**, 1371-1404.

Leigh, C., and Taylor, A. (1976) The chemistry of the epipolytbipuperazine 3,6 diones. In Rodricks, J.V. (ed.)., *Mycotoxins and Other Fungal Related Food Problems*, Advances in Chemistry Series 149, ACS, Washington, DC, pp. 228-275.

Luangsa-ard, J.J., Hywel-Jones, N.L., and Samson, R.A. (2004) The polyphyletic nature of *Paecilomyces sensu lato* based on 18S-generated rDNA phylogeny. *Mycologia*, **96**, 773-780.

Lubeck, M., Poulsen, S.Y., Lubeck, P.S., Jensen, D.F., and Thrane, U. (2000) Identification of *Trichoderma* strains from building materials by ITSI ribotyping, UP-PCR fingerprinting and UP-PCR cross hybridization. *FEMS Microbiol. Lett.*, **185**, 129-134.

Miller, J.D. (1999) Mycotoxins. In F.J. Francis, (ed.), *Encyclopedia of Food Science and Technology*, John Wiley, New York, pp 1698-1706.

Morton, F.J., and Smith, G. (1963) The genera *Scopulariopsis* Banier, *Microascus* Zukal and *Doratomyces* Corda. *Mycological Papers*, **86**,1-96.

Nelson, P.E., Dignani, M.C., and Anaisse, E.J. (1994) Taxonomy, biology and clinical aspects of *Fusarium* species. *Clin. Microbiol. Reviews*, **7**, 479-504.

Nelson, P.E., Plattner, R.D., Shackelford, D.D., and Desjardins, A.E. (1991) Production of fumonisins by *Fusarium moniliforme* strains from various substrates and geographic areas. *Appl. Environ. Microbiol.*, **57**, 2410-2412.

Nielsen, K.F., Gravesen, S., Nielson, P.A., Andersen, B., Thrane, U., and Frisvard, J.C. (1999). Production of mycotoxins on artificially and naturally infested building materials. *Mycopathologia*, **145**, 43-56.

O'Donnell, K., Sarver, B.A., Brandt, M., *et al.* (2007) Phylogenetic diversity and microsphere array-based genotyping of human pathogenic fusaria, including isolates from the multistate contact lens-associated U.S. keratitis outbreaks of 2005 and 2006. *J. Clin. Microbiol.*, **45**, 2235-2248.

Pildain, M.B., Frisvad, J.C., Vaamonde, G., *et al.* (2008) Two novel aflatoxin-producing *Aspergillus* species from Argentinean peanuts. *J. Evol.Syst. Microbiol.*, **58**, 725-735.

Sakomoto, T., Urisu, A., and Yamanda, M. (1989) Studies on the osmophilic fungus *Wallemia sebi* as an allergen evaluated by skin prick text and radioallergosorbent test. *Int. Arch. Allergy Appl. Immunol.*, **90**, 368-372.

Samson, R.A. (1974) *Paecilomyces* and some allied hyphomycetes. *Stud. Mycol.*, **6**, 1-119.

Samson, R.A., and Frisvad, J.C. (2004) *Penicillium* subgenus *Penicillium*: new taxonomic schemes, mycotoxins and other extrolites. *Stud. Mycol.*, **49**, 1-174.

Samson, R.A., Hong, S-B., Peterson, S.W., *et al.* (2007a) Polyphasic taxonomy of *Aspergillus* section *Fumigati* and its teleomorph *Neosartorya*. *Stud. Mycol.*, **59**, 147–207.

Samson, R.A., Noonim, P., Meijer, *et al.* (2007b) Diagnostic tools to identify black Aspergilli. *Stud. Mycol.*, **59**, 129-145.

Samson, R.A., Houbraken, J., Varga, J., and Frisvad, J.C. (2009) Po-

lyphasic taxonomy of the heat resistant ascomycete genus *Byssochlamys* and its *Paecilomyces* anamorphs. *Persoonia*, **22**, 14-27.

Samson, R.A., Houbraken, J., Thrane, U., *et al.* (2010). *Food and Indoor Fungi*. CBS Laboratory Manual Series 2. Centraalbureau voor Schimmelcultures, Utrecht, The Netherlands.

Schroers, H.-J. (2001) A monograph of *Bionectria* (Ascomycota, Hypocreales, Bionectriaceae) and its *Clonostachys* anamorphs. *Stud. Mycol.*, **46**, 1-214.

Schubert, K., Groenewald, J.Z., Braun, U., *et al.* (2007) Biodiversity in the *Cladosporium herbarum* complex (*Davidiellaceae, Capnodiales*), with standardisation of methods for *Cladosporium* taxonomy and diagnostics. *Stud. Mycol.*, **58**, 105-156.

Scott, J., Untereiner, W.A., Wong, B., Straus, N.A., and Malloch, D. (2004) Genotypic variation in *Penicillium chysogenum* from indoor environments. *Mycologia*, **96**, 1095-1105.

Scott, J.A., Wong, B., Summerbell, R.C., Untereiner, W.A. (2008) A survey of *Penicillium brevicompactum* and *P. bialowiezense* from indoor environments, with commentary on the taxonomy of the *P. brevicompactum* group. *Botany*, **86**, 732-741.

Summerbell, R.C. (1998) Taxonomy and ecology of *Aspergillus* species associated with colonizing infections of the respiratory tract. *Immunol. Allergy Clin. N. Amer.*, **18**, 449-573.

Taylor, A. (1986) Some aspects of the chemistry and biology of the genus *Hypocrea* and its anamorphs, *Trichoderma* and *Gliocladium*. *Proc. Nova Scotia Inst. Sci.*, **36**, 27-58.

Untereiner, W.A., Angus, A., Réblová, M., and Orr, M-J. (2008) Systematics of the *Phialophora verrucosa* complex: new insights from analyses of β-tubulin, large subunit nuclear rDNA and ITS sequences. *Botany*, **86**, 742-750.

Varga, J., Due, M., Frisvad, J.C., and Samson, R.A. (2007) Taxonomic revision of *Aspergillus* section *Clavati* based on molecular, morphological and physiological data. *Stud. Mycol.*, **59**, 89-106.

Varga, J., Frisvad, J.C., and Samson, R.A. (2007) Polyphasic taxonomy of *Aspergillus* section *Candidi* based on molecular, morphological and physiological data. *Stud. Mycol.*, **59**, 75-88.

Varga J, Houbraken J, van der Lee AL, *et al.*, (2008). *Aspergillus calidoustus* sp. nov., causative agent of human infec¬tions previously assigned to *Aspergillus ustus*. *Eukaryotic Cell* **7**: 630-638.

Visconti, A., and Sibilia, A. (1994) *Alternaria* toxins. In J.D. Miller and H.L. Trenholm, (eds.), *Mycotoxins in Grains: Compounds Other Than Aflatoxin*, Eagan Press, St. Paul, MN, pp. 315-338.

Zalar, P., de Hoog, G.S., Schroers, H.-J., *et al.* (2007) Phylogeny and ecology of the ubiquitous saprobe *Cladosporium sphaerospermum*, with descriptions of seven new species from hypersaline environments. *Stud. Mycol.*, **58**,157-183.

Zalar, P., Gostinčar, C., Hoog, G.S. de, Uršič, V., Sudhadham, M., and Gunde-Cimerman, N. (2008) Redefinition of *Aureobasidium pullulans* and its varieties. *Stud. Mycol.*, **61**, 21–38.

Zhang, N., O'Donnell, K., Sutton, *et al.* (2006) Members of the *Fusarium solani* species complex that cause infections in both humans and plants are common in the environment. *J. Clin. Microbiol.*, **44**, 2186-2190.

COMMON SPECIES OF FUNGI AND ACTINOMYCETES IN INDOOR ENVIRONMENTS

Absidia corymbifera	330	*Geomyces pannorum*	410
Acremonium mumorum	332	*Geotrichum candidum*	412
Acremonium strictum	334	*Lecanicillium (Verticillium) lecanii*	414
Alternaria alternaria	336	*Mucor hiemalis*	416
Aspergillus calidoustus	338	*Mucor plumbeus*	418
Aspergillus candidus	340	*Mucor racemosus*	420
Aspergillus clavatus	342	*Oidiodendron griseum*	422
Aspergillus flavus	344	*Oidiodendron rhodogenum*	424
Aspergillus fumigatus	346	*Paecilomyces lilacinus*	426
Aspergillus niger	348	*Paecilomyces variotii*	428
Aspergillus westerdijkiae	350	*Penicillium brevicompactum*	430
Aspergillus penicillioides	352	*Penicillium chrysogenum*	432
Aspergillus restrictus	254	*Penicillium citrinum*	434
Aspergillus sydowii	356	*Penicillium citreonigrum*	436
Aspergillus terreus	358	*Penicillium commune*	438
Aspergillus versicolor	360	*Penicillium corylophilum*	440
Aureobasidium pullulans	362	*Penicillium crustosum*	442
Botrytis cinerea	364	*Penicillium digitatum*	444
Cadophora (Phialophora) fastigiata	366	*Penicillium expansum*	446
Candida peltata	368	*Penicillium funiculosum*	448
Chaetomium aureum	370	*Penicillium glabrum*	450
Chaetomium globosum	372	*Penicillium olsonii*	452
Chaetomium indicum	374	*Penicillium palitans*	454
Chrysonillia sitophila	376	*Penicillium polonicum*	456
Cladosporium cladosporioides	378	*Penicillium rugulosum*	458
Cladosporium herbarum	380	*Penicillium variabile*	460
Cladosporium sphaerospermum	382	*Phialophora verrucosa*	462
Clonostachys rosea	384	*Phoma glomerata*	464
Coprinus cordisporus	386	*Phoma macrostoma*	466
Cryptococcus laurentii	388	*Pithomyces chartarum*	468
Curvularia lunata	390	*Pyronema domesticum*	470
Emericella nidulans	392	*Rhizopus stolonifer*	472
Epicoccum nigrum	394	*Rhodotorula mucilaginosa*	474
Eurotium amstelodami	396	*Schizophyllum commune*	476
Eurotium chevalieri	398	*Scopulariopsis brevicaulis*	478
Eurotium herbariorum	400	*Scopulariopsis candida*	480
Exophalia dermatitidis	402	*Scopulariopsis fusca*	482
Fusarium culmorum	404	*Serpula lacrymans*	484
Fusarium solani	406	*Sistotrema brinkmanii*	486
Fusarium verticillioides	408	*Sporobolomyces roseus*	488

COMMON SPECIES OF FUNGI AND ACTINOMYCETES IN INDOOR ENVIRONMENTS

Abbreviations used in the species description

MEA = Malt extract agar
Letters indicating the medical significance refer to the hazard levels on page 325.

Descriptions and illustrations of common fungi and actinomycetes

Absidia corymbifera (Cohn) Sacc. & Trotter
= *Absidia ramosa* (Lindt) Lendner; *Mycocladus corymbifer* (Cohn) Vánová; *Lichtheimia corymbifera* (Cohn) Vuillemin

A. Colony on MEA after one week; B. sporangiophore with columella and remnant of sporangium wall after dehiscence, x 920; C. sporangiophores bearing sporangia, x 230; D. sporangiospores, x 920; E. sporangiophore with sporangium (left), sporangiophore with columella (right), x 920.

Occurrence: Worldwide, in soil, decaying plant matter and composts. Found in carpet and mattress dust. Also in hay, cereals, flour; potted plant soil; accrued bird droppings.

Medical significance: EFG. This species can cause (sub)cutaneous infections, and invasive infections in AIDS patients (de Hoog *et al.* 2000).

Notes: The name *Lichtheimia corymbifera* has been proposed for this species (Hoffmann *et al.* 2009); however, for practical uses the name *A. corymbifera* is maintained here. For a more detailed description, see Samson *et al.* (2010).

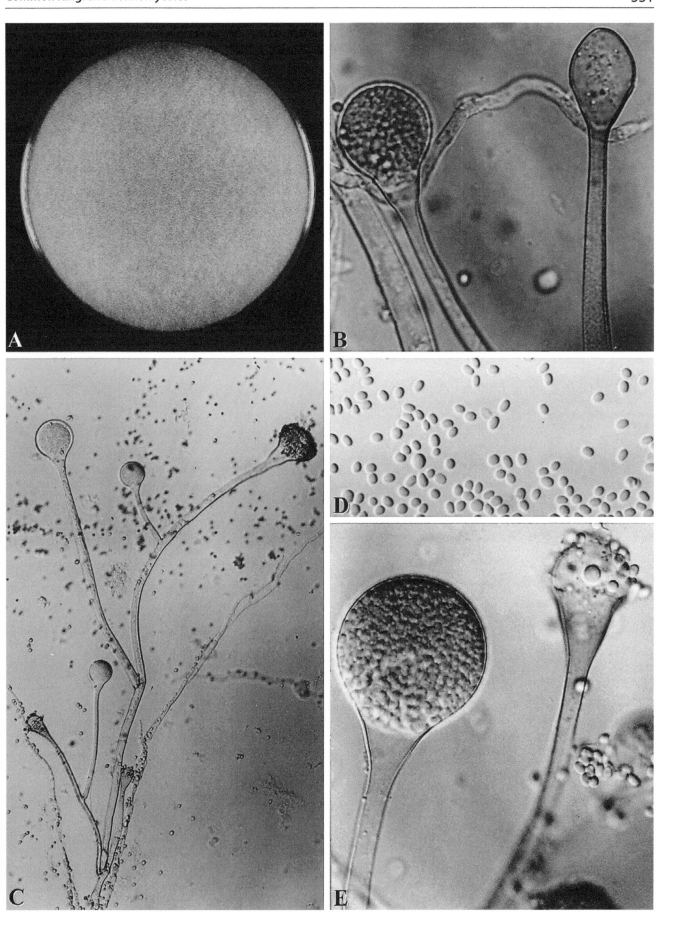

Acremonium murorum (Corda) Gams
= *Gliomastix murorum* (Corda) Hughes

A. Colony on OA after one week, B. conidiophores with conidia in chains, in Petri-dish, x 920; C. conidiophores with conidia, x 920; D. conidia, x 920; E. conidiophores with conidia, x 920.

Occurrence: Isolated from cellar wall, swimming pool floor; cereals.

Medical significance: HI.

Note: For more detailed information, see Domsch *et al.* (2007).

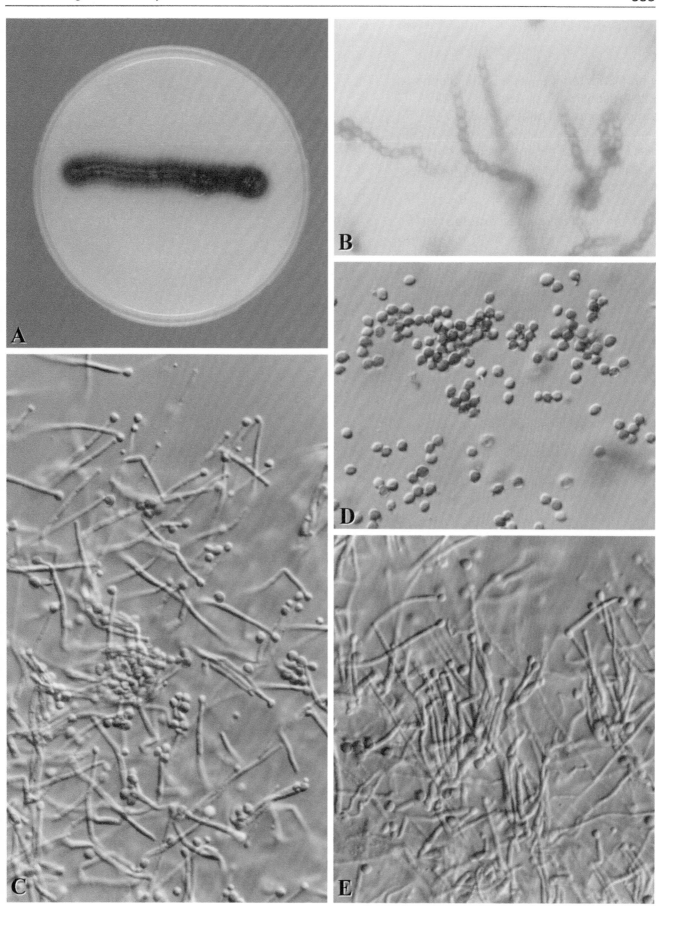

Acremonium strictum W. Gams

A. Colony on MEA after one week; B. conidiophores with conidia in slimy heads, in Petri-dish, x 230; C-E. conidia, conidiophores and phialides, x 920.

Occurrence: Found in carpet and mattress dust, and present on damp or wet (acrylic painted) walls, gypsum board, wallpaper.

Also isolated from humidifier water; exposed wood.

Medical significance: EFG.

Notes: *Acremonium strictum* is a species complex and at least three genotypes are present. A detailed description of this species can be found in Domsch *et al.* (2007).

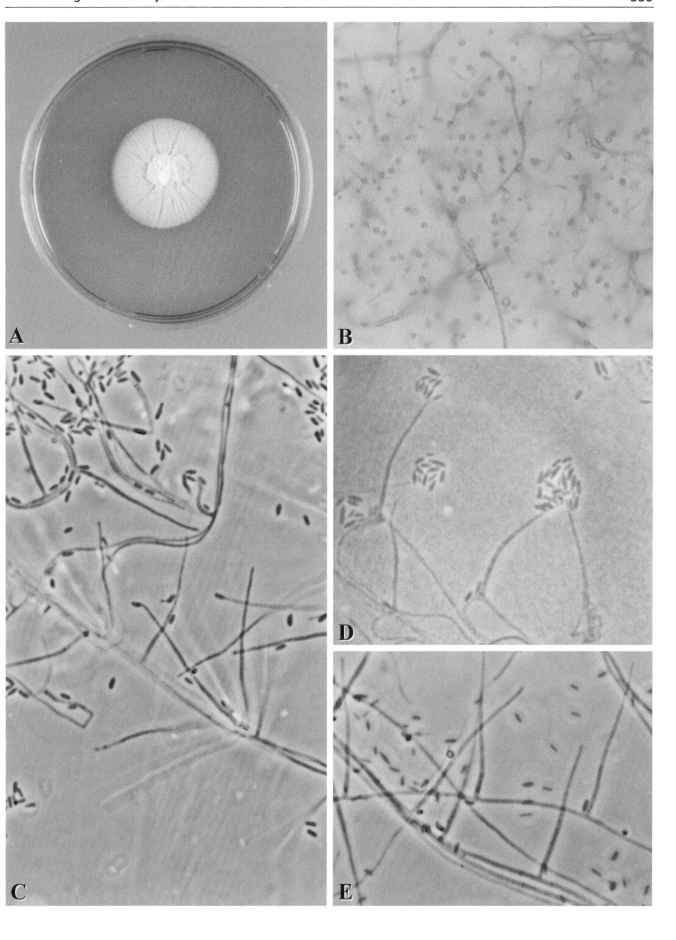

Alternaria Nees

Teleomorph: *Lewia* ME Barr and EG Simmons

A. Colony on MEA after one week; B-E. conidia and conidiophores, x 920.

Occurrence: Worldwide phylloplane species on a broad range of plants, and in soil.

Common in floor, carpet and mattress dust; less common on damp walls, gypsum board and wallpaper. Cause of soft rot and blue stain of wet joinery. Also isolated from UFFI, window putty, (acrylic latex) paint; humidifier water; textiles; hay, cereals, fruit.

Medical significance: EFG. Characterized allergens causing rhinitis, asthma, atopic allergic dermatitis, allergic sinusutis (see Chapter 3.1).

Toxicity: Some strains produce tenuazonic acid, alternariol and other metabolites of low animal toxicity (Visconti 1994, Nielsen *et al.* 1999).

Notes: *Alternaria alternata* is often mentioned as the most common species in indoor environments. Recent data indicate that this species is not commonly occurring and reports of this species are often based on misidentifications (Samson *et al.* 2010).

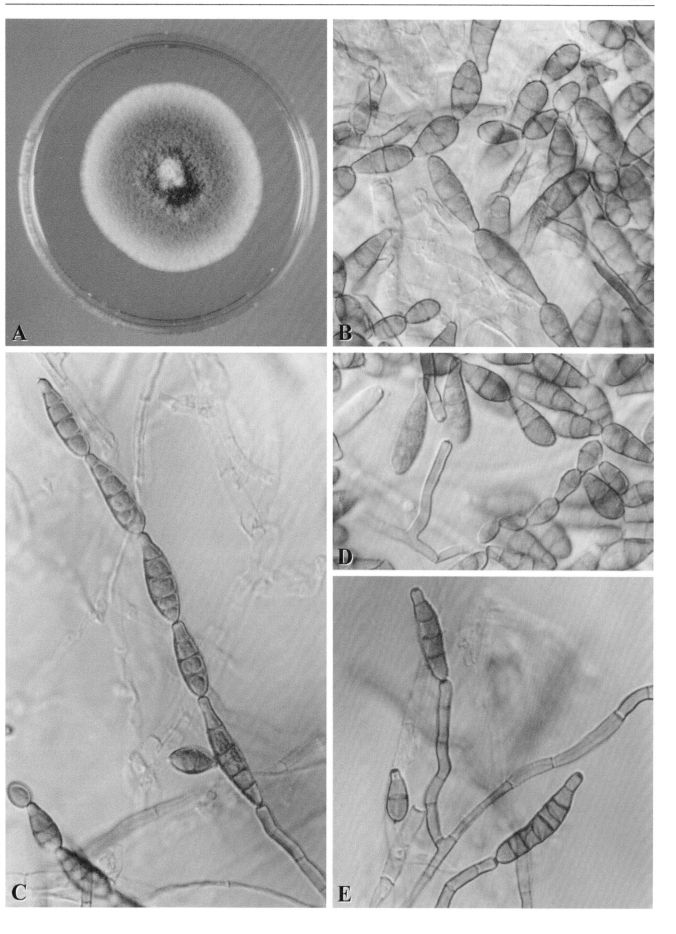

Aspergillus calidoustus Varga, Houbraken & Samson

A. Colonies on CYA after one week; B. conidia, x1600; C. conidial heads, x850; D. conidial head, x1600; E. irregularly shaped Hülle cells, x350

Occurrence: Worldwide in food, soil, and indoor (air) environments.

Frequently isolated from (hospital) air, cereals and groundnuts; uncommonly occurring on building materials, occasionally found on moist substrates, such as wooden construction material, wallpaper and carpets.

Medical significance: EFG. Uncommon cause of aspergillosis and fungal growths, especially in immunocompromised individuals. Antifungal susceptibility tests of *A. calidoustus* isolates show that amphotericin B, echinocandins, and azole derivatives (incl. posaconazole) have limited or no activity (Varga *et al.* 2008).

Notes: *Aspergillus calidoustus* resembles *A. ustus,* but can be distinguished by its ability to grow at or above 37°C; *A. calidoustus* is more commonly occurring in indoor environments than *A. ustus* (Houbraken *et al.* 2007, Varga *et al.* 2008).

Aspergillus candidus Link

A. Colonies on MEA after one week; B. conidial head, x 920; C. atypical reduced conidial head, x 920; D. conidial head, x 920.

Occurrence: Common in soil, dung and crops, predominantly in tropical and subtropical regions, and frequently present in stored products.

Common in floor and carpet and mattress dust. Also isolated from HVAC system air; weaving mill; hay, cereals, flour.

Medical significance: EFG.

Notes: The taxonomy of *A. candidus* has been revised by Varga *et al.* (2007).

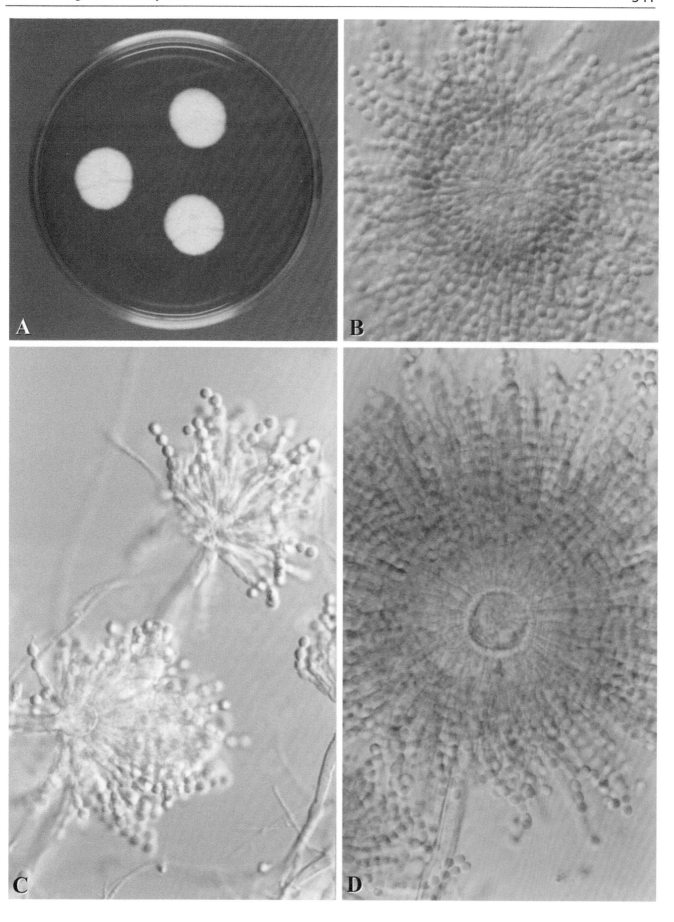

Aspergillus clavatus Desmazières

A. Colonies on MEA after one week; B. conidial heads and tip of conidiophore, x 230; C. conidial head, x 920; D. conidia, x 920; E. conidial head, x 920.

Occurrence: Widely distributed in soil and composts.

Uncommon in non-industrial indoor environments, but has been isolated from mattress dust, window sill, and rarely from indoor air in homes. Associated with bakeries and particularly maltings. Also isolated from cereals, and flour, and also fish meal.

Medical significance: [E?] IG. Allergenic - occupational hypersensitivity pneumonitis known as malt worker's lung (Flannigan and Pearce 1994).

Toxicity: Produces patulin, tryptoquivalines and cytochalasin E; mycotoxicosis in farm stock fed on malting by-products (Flannigan and Pearce 1994).

Notes: The taxonomy of *A. clavatus* has been revised by Varga *et al.* (2007).

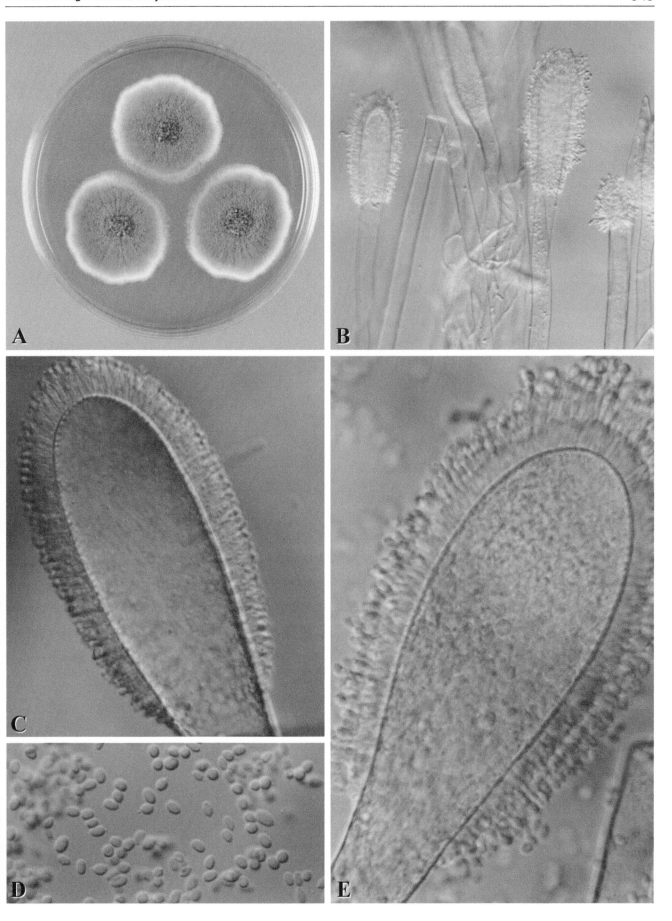

Aspergillus flavus Link

A. Colonies on MEA after one week; B,C. conidial heads, x 920; D. conidia, x 920; E. conidial head, x 920.

Occurrence: Widely distributed in soil, particularly in tropical and subtropical areas, particularly associated with maize, nuts and insects in the field and a vast array of commodities in storage in warm climates (Miller 1999).

Relatively rarely growing on building materials, but may be found on damp walls, wallpaper and glues or plastic cements. Also isolated from floor and carpet dust; tarred wooden flooring, polyurethane foam; particle board frames of filters, humidifiers and HVAC fans; bakeries; shoes, leather, cotton yarn; hay, cereals, flour, bread; bakeries; cosmetics; rabbit food pellets, accrued bird droppings.

Medical significance: EFG. Relatively common cause of aspergillosis (Summerbell 1998).

Toxicity: Strain of *A. flavus* may produce aflatoxin, an IARC Class 1 carcinogen. Occupational inhalation exposure mainly to spores containing aflatoxin in feed mills where grain dusts are tightly managed results in serum aflatoxin adducts associated with increased relative risks of liver cancer (Autrup *et al.* 1993). Besides aflatoxin, also the mycotoxins cyclopiazonic acid and 3-nitropropionic acid may be produced by this species.

Notes: *Aspergillus flavus* is characterized by the formation of yellow-green conidia on MEA. This species belongs to the section *Flavi* and various species, such as *A. parasiticus*, *A. oryzae*, *A. soja*e, *A. archidicola* and *A. minisclerotigenes* are placed in this section. These species are closely related and identification of these species based on phenotypic characters is difficult. More information about the taxonomy of this species can be found in Pildain *et al.* (2008) and Samson *et al.* (2010).

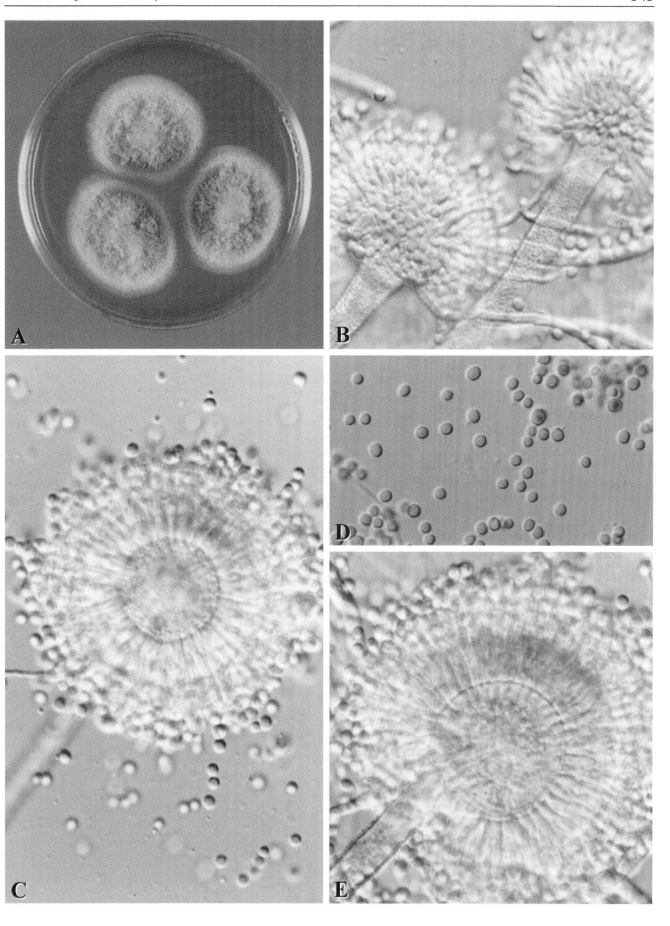

Aspergillus fumigatus Fresenius

A. Colonies on MEA after one week; B. detail of colony showing columnar conidial heads, x 44; C, D. conidial heads, x 920; E. conidia, x 2330.

Occurrence: A common and extremely versatile decay fungus associated with composting vegetation, wood chips, garbage and other materials at temperatures above ambient (Lacey and Dutkiewicz 1994).

Often found in dust infiltrating from outdoor air, particularly when cleaning standards are low, and hence carpet and mattress dust. Can grow on warm, wet building and finishing materials, e.g. wallpaper. Also isolated from polyurethane foam, plastic; particle board; HVAC insulation, fans, filters, humidifier water; air/dust in bakeries, maltings, sawmills; mushroom compost, sawdust, wood chips, bagasse; hay, cereals; potted plant soil.

Medical significance: EFG[D?]. Cause of aspergillosis. Known allergens (Day 1996, Summerbell 1998).

Toxicity: Produces various mycotoxins, including the potent toxin gliotoxin; some have been measured in air (Fischer *et al.* 1999).

Note: This species can be easily recognized by its rapid growth, grey blue colonies and the columnar conidial heads. The taxonomy of this species was revised by Samson *et al.* (2007).

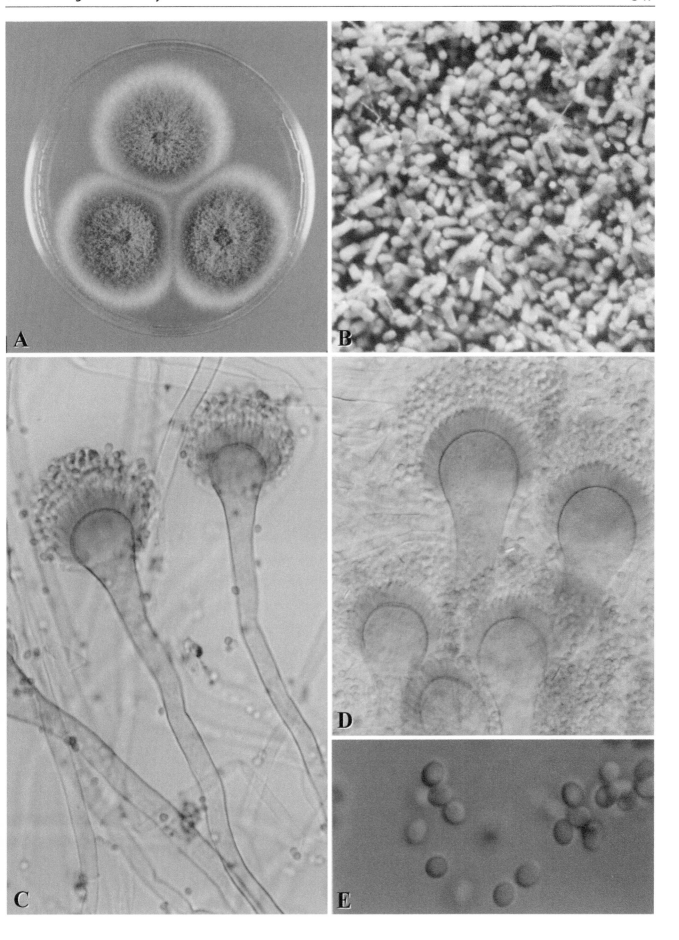

Aspergillus niger van Tieghem

A. Colonies on MEA after one week; B. detail of colony showing radiate conidial heads, x 44; C. conidial head, x 920; D. conidia, x 2330; E. conidial heads and tip of conidiophore, x 230.

Occurrence: Worldwide in soil and plant litter, particularly notable as a contaminant of spices and other sun-dried plant products.

Often found in indoor environments, but not generally associated with contaminated building materials. Found in floor, carpet and mattress dust; acrylic paint, UFFI, polyester polyurethane foam; polyurethane footwear, leather, cosmetics, wood chips; HVAC filters and fans; bakeries, cotton mills, cotton yarn; hay, cereals, cotton seed; potted plant soil.

Medical significance: EFG. An uncommon cause of aspergillosis, but does cause ear, nose and lung infections especially in immunecompromised individuals (Summerbell 1998).

Toxic metabolites: Some strains of *A. niger* are capable of forming ochratoxin A and/or fumonisin B_2 and B_4.

Notes: The taxonomy of this species is subjected to various studies. An overview with diagnostic tools to identify black aspergilli is given by Samson *et al.* (2007b).

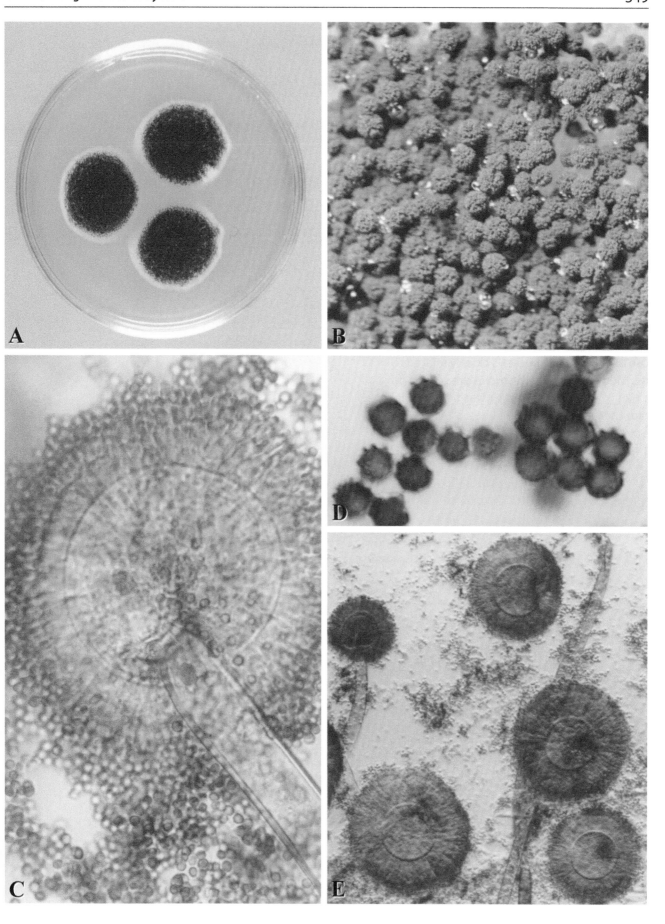

Aspergillus penicillioides Spegazzini

A. Colonies on MEA + 20% sucrose after one week; B. detail of colony showing radiate conidial heads, x 44; C. conidial heads, x 920; D. conidia, x 2330; E. conidial heads, x 920.

Occurrence: A slow growing xerophile that has been isolated from soil, dried foods and air.

Found in floor, carpet and mattress and house dust. Also isolated from glass lens, electronic meter; air from air-conditioning system; dried fish, cereals.

Medical significance: HI.

Note: This species can only be detected when using appropriate low water activity media such as MEA + 20% sucrose or DG18. This species is closely related to *A. restrictus*. The phialides of *A. restrictus* are placed on the upper half of the vesicle, which gives the conidial head a columnar feature; the phialides of *A. penicillioides* are over a larger part of the vesicle.

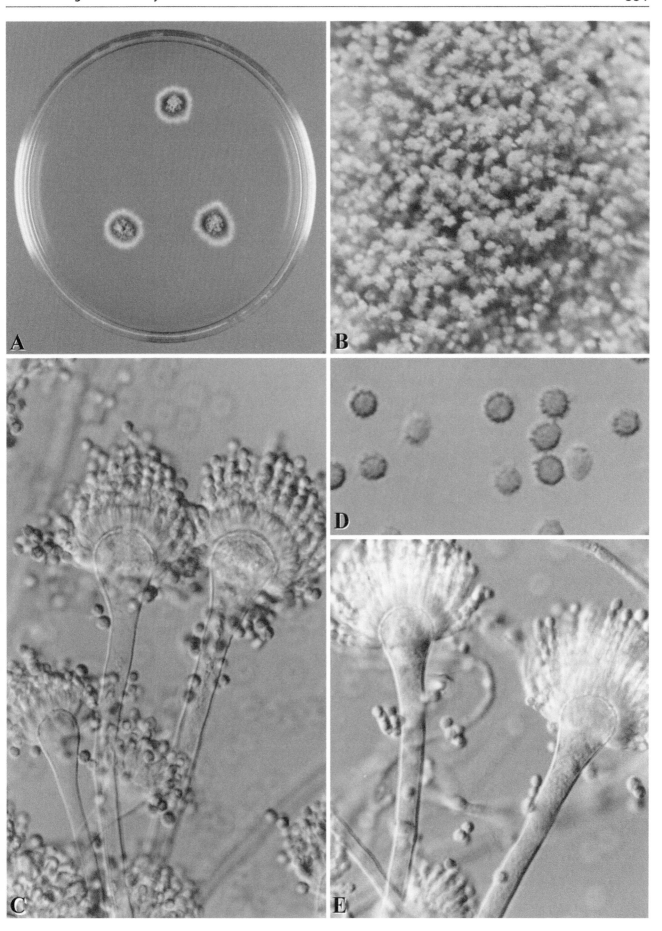

Aspergillus restrictus Smith

A. Colonies on MEA + 20% sucrose after one week; B. detail of colony showing columnar conidial heads, x 44; C. conidial head, x 920; D. conidia, x 2330.

Occurrence: A slow growing xerophile that has been isolated from soil, seeds, fruit juices and air.

Found in floor and carpet, mattress and upholstered-furniture dust. Also isolated from damp walls; leather armchair, cotton fabric, cloth; air from air-conditioning system; cereals.

Medical significance: HI (many medical reports are based on misidentification).

Note: This species can only be detected when using appropriate low water activity media such as MEA + 20% sucrose or DG18. This species is closely related to *A. restrictus*. The phialides of *A. restrictus* are placed on the upper half of the vesicle, which gives the conidial head a columnar feature; the phialides of *A. penicillioides* are over a larger part of the vesicle.

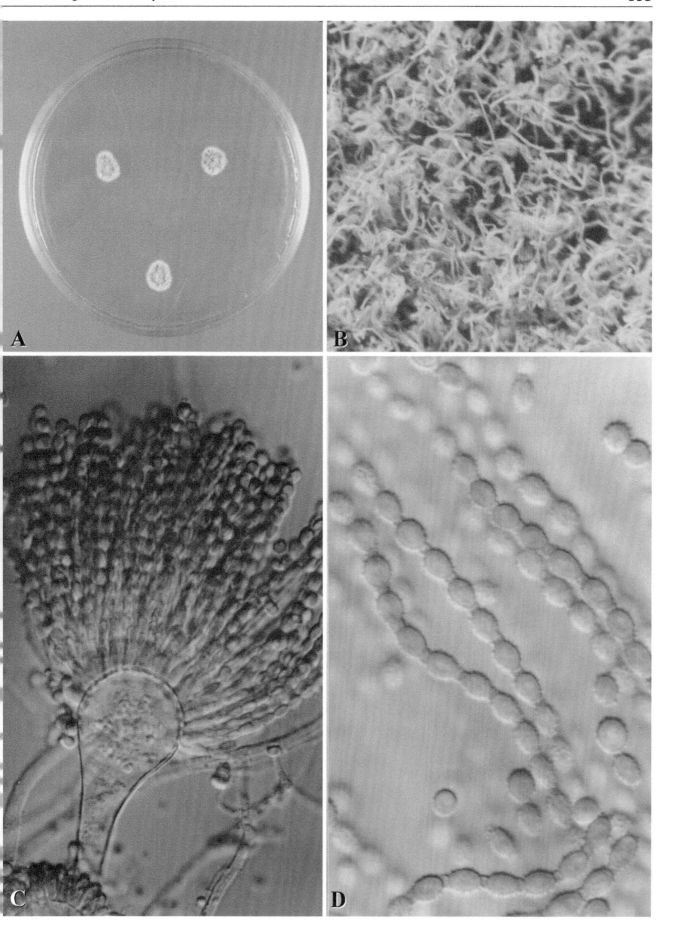

Aspergillus sydowii (Bainier & Sartory) Thom & Church

A. Colonies on MEA after one week; B, C. conidial heads with tip of conidiophore, x 920; D. conidia, x 2330.

Occurrence: Soil fungus with a worldwide distribution.

Very common on mouldy gypsum board and associated wallpaper and paint, and found in floor, carpet and mattress dust. Also isolated from UFFI; cotton yarn, talcum powder; cereals, wheat flour; air in maltings.

Medical significance: HI[G?]. Can cause human mycosis.

Notes: *Aspergillus sydowii* is closely related to *A. versicolor* and can be distinguished by the production of blue green conidia and small reduced conidial heads.

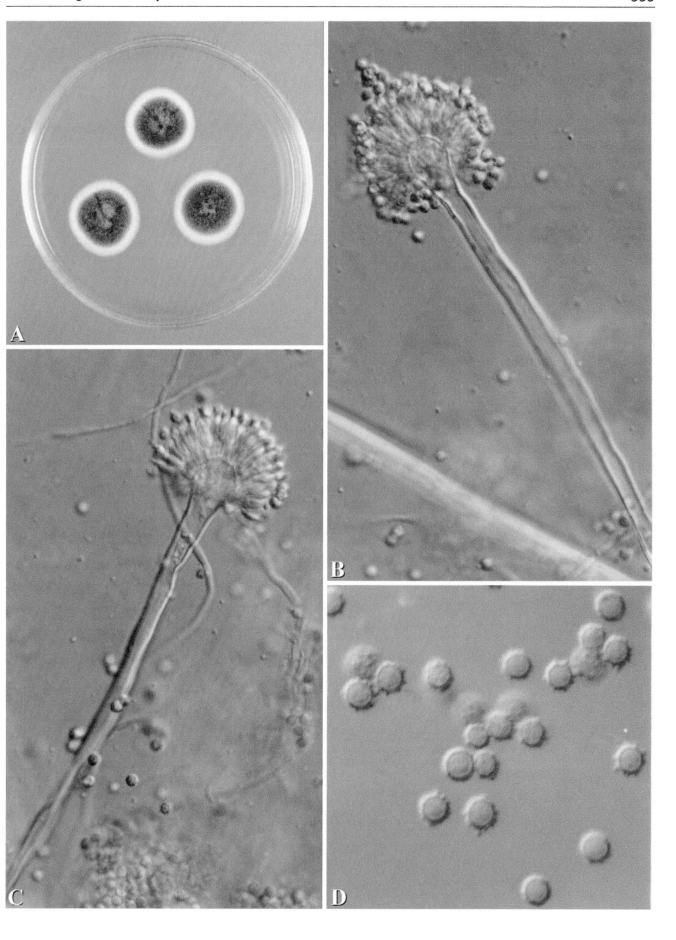

Aspergillus terreus Thom

A. Colonies on MEA after one week; B. detail of colony showing columnar conidial heads, x 44; C.-D. conidial heads with tip of conidiophore, x 920; E. conidia, x 2330.

Occurrence: Worldwide but more abundant in tropical and subtropical areas, this soil-borne species is frequently associated with stored crops.

Uncommonly occurs on building materials, but found in floor, carpet and mattress dust. Also isolated from PVC/paper wallcovering, plastic, decomposing wood; cotton, cotton yarn, cloth; hospital air, bakery; coalmine dust; hay, cereals, flour.

Medical significance: EFG. Uncommon cause of aspergillosis and fungal growths, especially in immunocompromised individuals (Summerbell 1998).

Toxicity: Citreoviridin, occasionally territrem.

Notes: This species is closely related to A. *alabamensis* (Balajee *et al.* 2009).

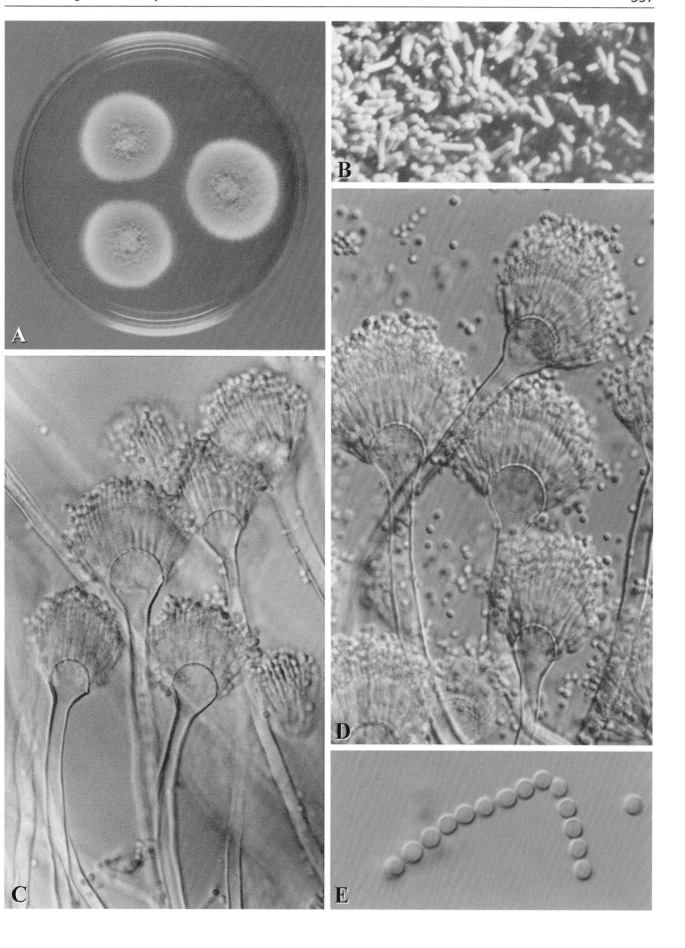

Aspergillus versicolor (Vuill.) Tiraboschi

A. Colonies on MEA after one week, B, C. conidial heads with tip of conidiophore, x 920; D. conidia, x 2330; E. conidial heads, x 920.

Occurrence: Occurs in soils worldwide, and present in many stored food products of plant origin. This species is very common in indoor environments.

Very common on gypsum board and other mouldy building materials, and in floor, carpet, mattress and upholstered-furniture dust; damp walls, painted metal; rock wool/glass fibre insulation, UFFI, polyester polurethane foam; swimming pool floor, tobacco, paraffin wax, talcum powder, cellophane, cotton yarn, cloth; hay, cereals, flour.

Medical significance: [E?]IG (some erroneous medical reports exist).

Toxicity: On building materials produces the potent mycotoxin sterigmatocystin, an IARC Class 2A carcinogen (Nielsen *et al.* 1999).

Notes: *Aspergillus sydowii* is closely related to *A. versicolor*. The former can be distinguished by the production of blue green conidia and small reduced conidial heads.

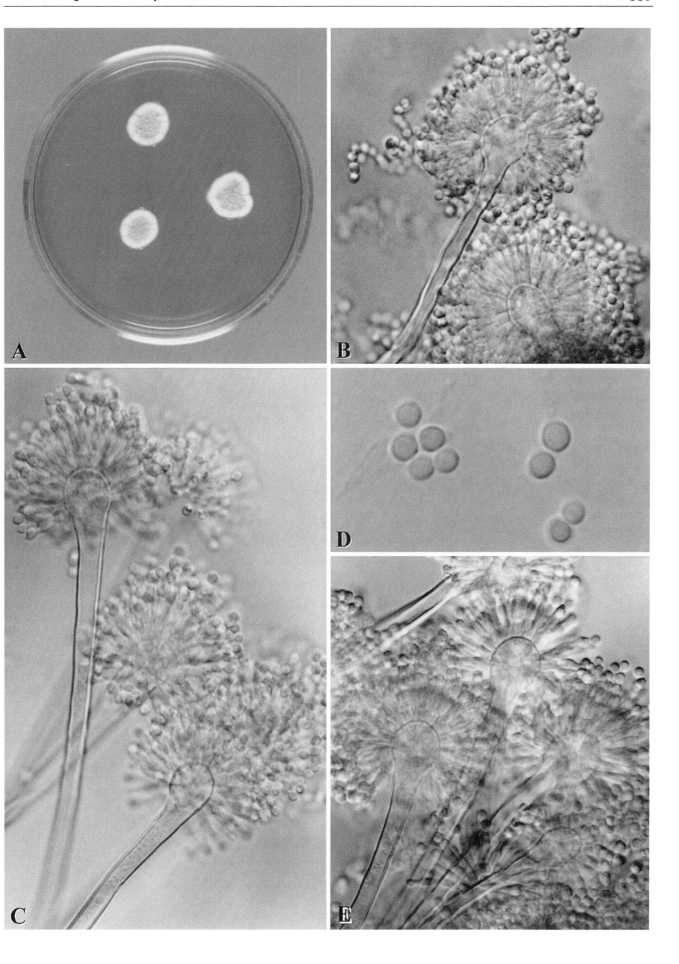

Aspergillus westerdijkiae Frisvad & Samson

A. Colonies on CYA after one week, B. conidia, x 1600; C. conidiophores, x 850; D. conidiophore, x 1600.

Occurrence: Worldwide distribution, commonly associated with stored seeds and other foodstuffs.

Infrequently detected in indoor environments, rather often in air of crawling spaces, isolated from a doorstep of dwelling, salterns, foodstuffs such as coffee beans, rice and wheat. This species also occurs on dead insects and it is speculated that this might be the source of yellow aspergilli in indoor environments.

Medical significance: [E?]IG. This species does not grow or grows very slowly at 37°C.

Toxicity: Ochratoxins, penicillic acid, xanthomegnin, viomellein, vioxanthin are produced by *A. westerdijkiae*.

Note: This species resembles *Aspergillus ochraceus* and other yellow aspergilli. *Aspergillus westerdijkiae* is more commonly occurring in indoor environments than *A. ochraceus*.

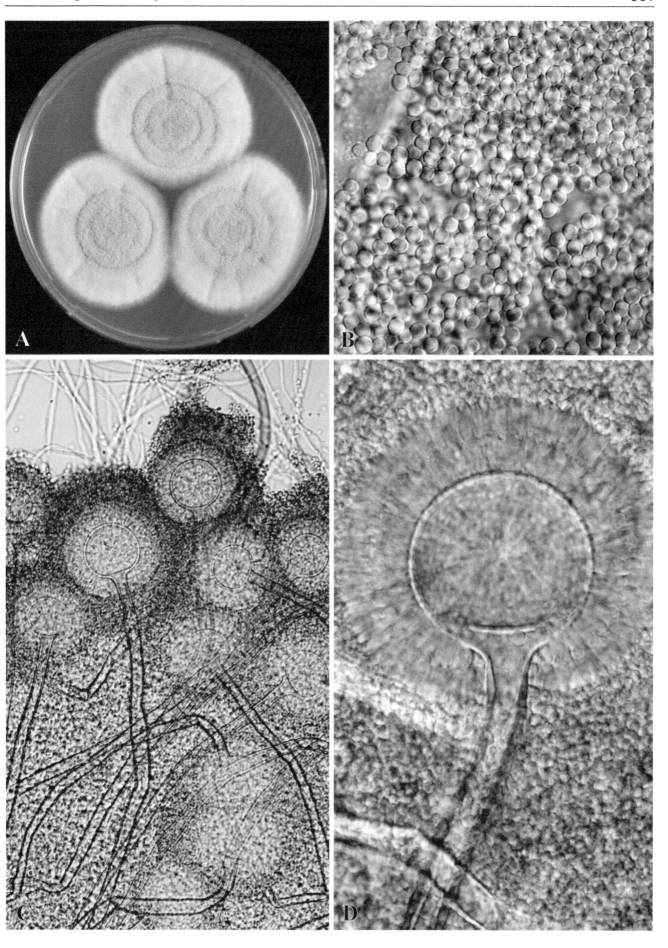

Aureobasidium pullulans (De Bary) Arnaud

A. Colony on MEA after one week; B. conidia, x 920; C. conidiogenous cells producing (blastic) conidia synchronously, x 920; D. arthroconidia, x 920.

Occurrence: Common soil and phylloplane species, also found on both fresh and dried plant foodstuffs.

Grows on very wet wood and window frames, and hence is common in floor, carpet and mattress dust. Also isolated from damp walls, acrylic paint, painted and preservative-treated wood, wood chips; humidifier water; cereals, fruit, vegetables.

Medical significance: EIG. A very rare cause of home-associated hypersensitivity pneumonitis (Flannigan *et al.* 1991), but associated with occupational disease in wood processing (Lacey and Dutkiewicz 1994).

Note: This species does not produce any important toxic metabolites. A multigene phylogeny showed that *Aureobasium pullulans* is species complex and might contain four species (Zalar *et al.* 2008).

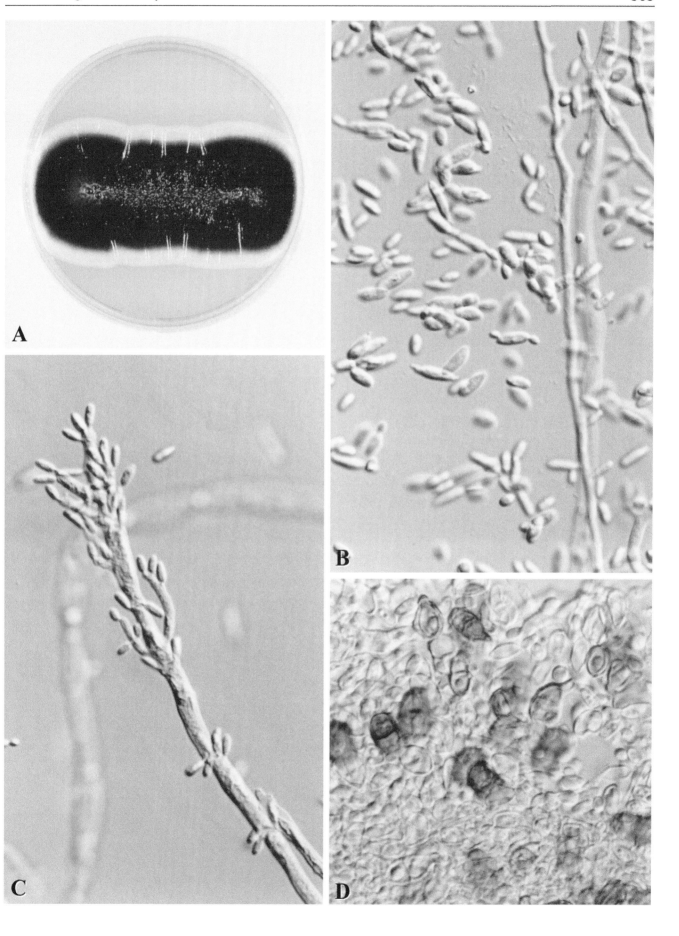

Botrytis cinerea Pers.

Teleomorph: *Botryotinia fuckeliana* (de Bary) Whetzel

A. Colony on MEA after one week, B. conidia, x 920; C, D. conidiophore producing conidia simultaneously from swollen conidiogenous cells, C. x 230, D. x 920.

Occurrence: Soil-borne facultative parasite of a broad range of plants, known as "grey mould".

Found in floor, carpet and mattress dust. Also isolated from mouldy cardboard; hay, cereals, carrots, tomatoes, strawberries.

Medical significance: HI.

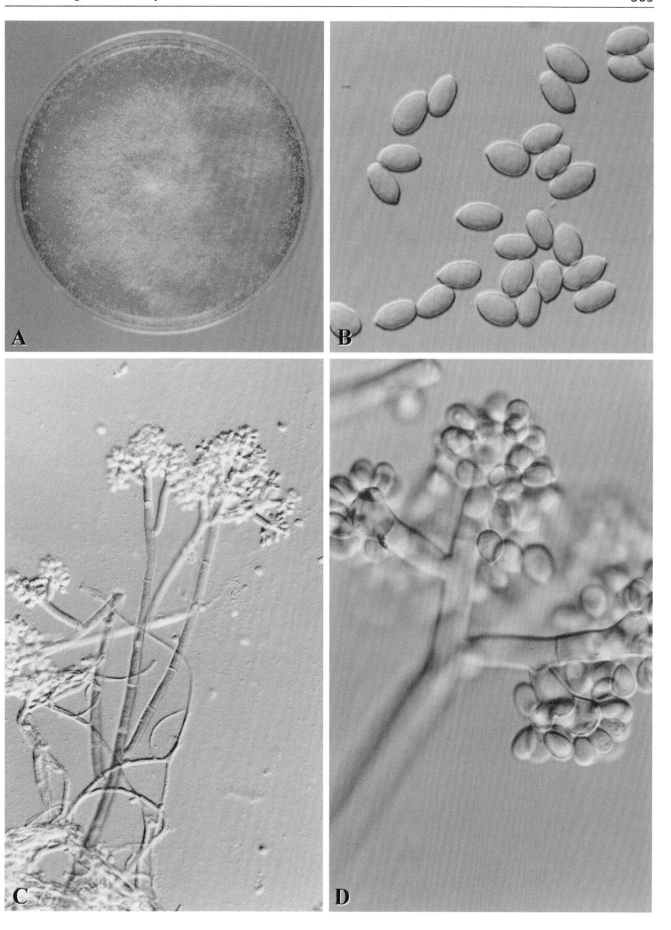

Cadophora fastigiata (Lagerb. & Melin) Conant

A. Colonies on MEA after two weeks; B. conidiophores, x 920; C. phialides with collarettes, x 2330; D. conidiophores, x 2330; E. phialides with collarettes, x 2330; F. conidia, x 2330.

Occurrence: Found in soil, decaying wood and some cereal and grass seed in various regions, including Europe, North America and the Antipodes.

Associated with very wet wood (Ellis 1971), including blue stain wood. Also isolated from humidifier water; paper pulp.

Medical significance: HI.

Note: *Cadophora fastigiata* was previously named *Phialophora fastigiata* (Lagerb. & Melin) Conant. Phylogenetic studies have shown that *Phialophora* is highly polyphyletic genus and it belongs to various genera such as *Cadophora, Lecythophora, Phaeoacremonium* and *Pleurostomophora*.

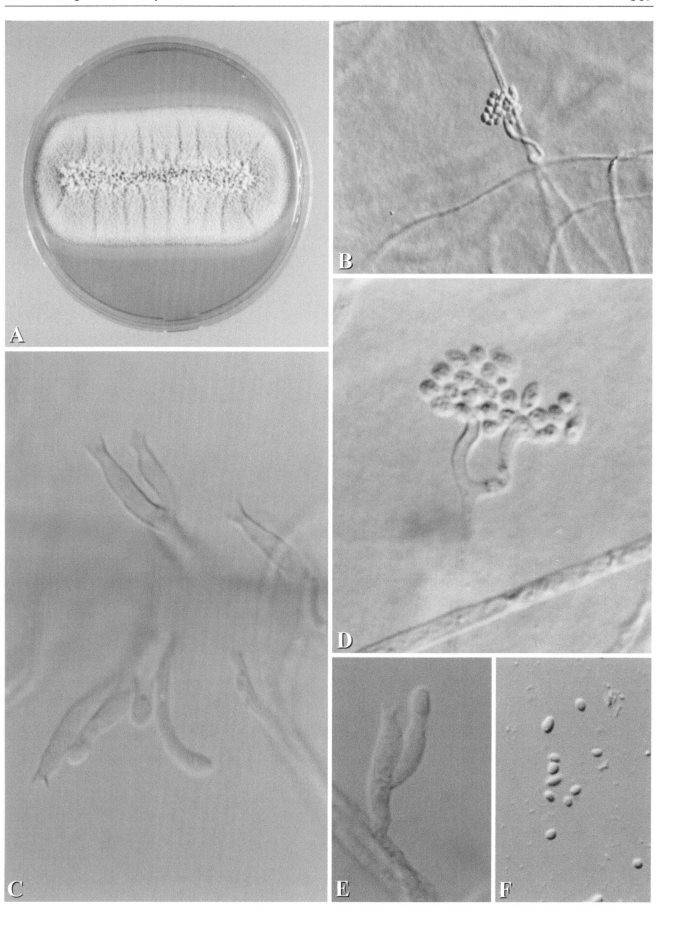

Candida peltata (Yarrow) Meyer & Ahearn

A. Colony on MEA after five days; B. spores, x 920; C, D. spores, x 2330.

Occurrence: Air, milk.

Medical significance: HI.

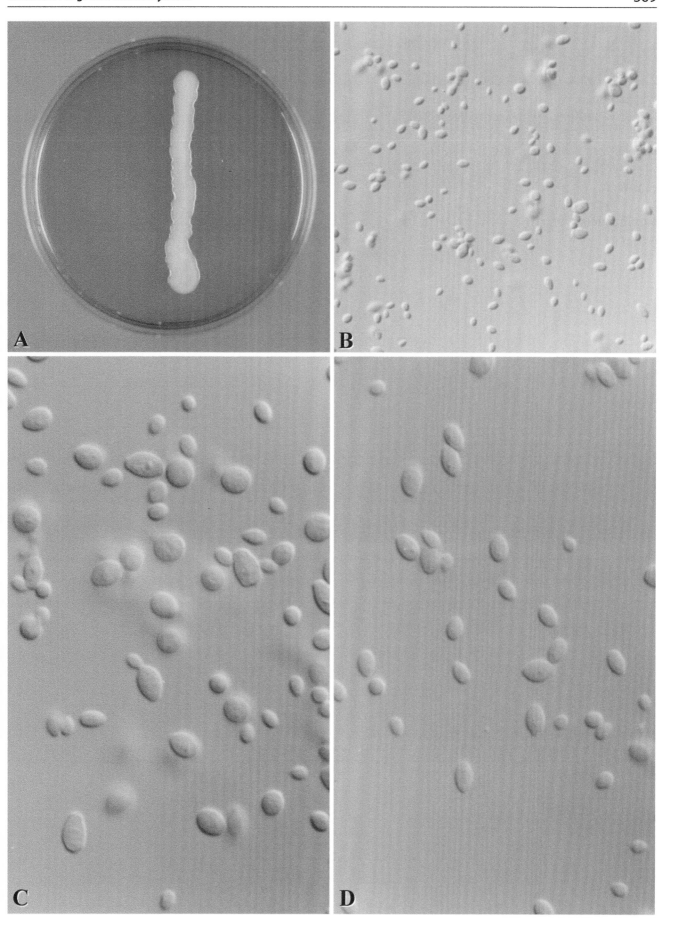

Chaetomium aureum Chivers

A. Colony on MEA after two weeks; B. two perithecia, x 230; C. perithecia with ascospores, x 920; D. ascospores, x 920; E. perithecia with ascospores, x 920.

Occurrence: Soil, cellulosic materials.

Compost, tobacco.

Medical significance: HI.

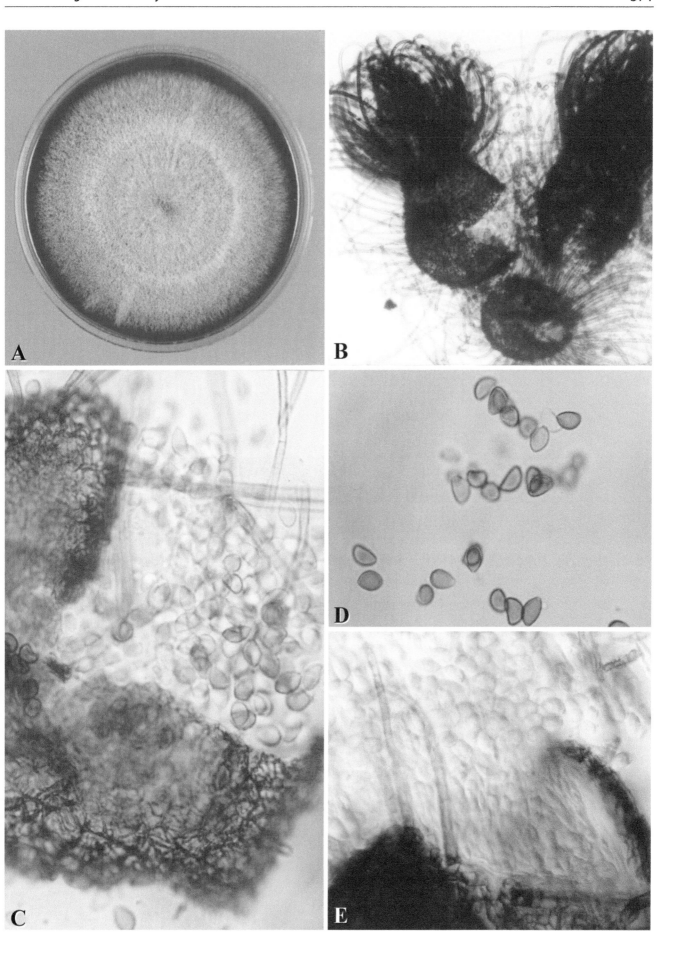

Chaetomium globosum Kunze ex Steud.

A. Colony on MEA after two weeks; B. perithecium, x 230; C, D. ascospores, x 920.

Occurrence: A potent cellulolytic species found on various cellulosic material, e.g. wallpaper.

Very common on mouldy building materials, that have been very wet, including gypsum board, cellulose board and wood. Also isolated from mattress dust; cardboard, paper, polyurethane foam; cotton, cotton yarn, polyurethane-coated fabric; cereals.

Medical significance: EI[G?].

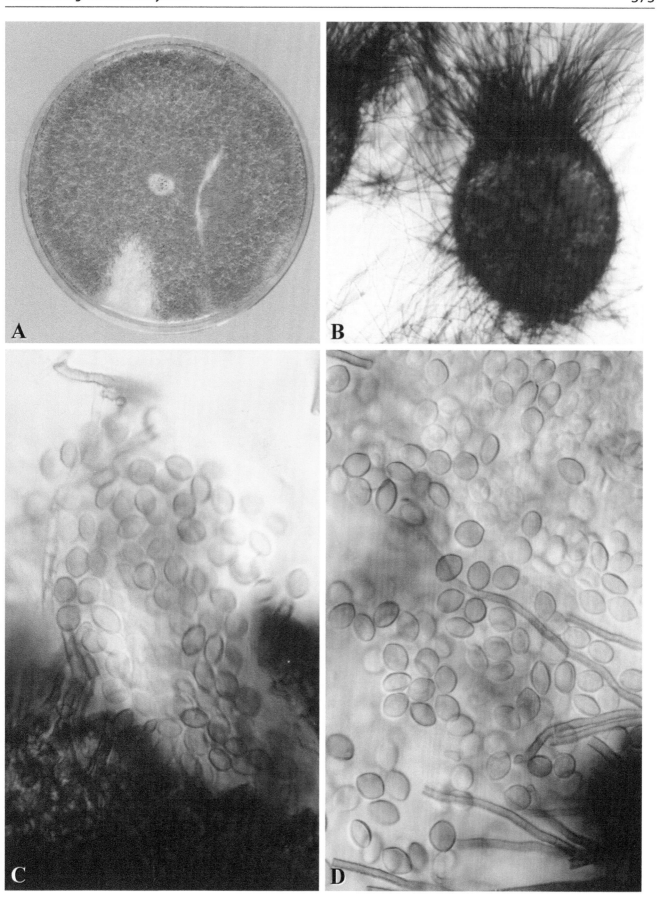

Chaetomium indicum Corda

A. Colony on MEA after two weeks; B. perithecium, x 230; C. ascospores with dichotomously branched hairs, x 920; D. ascospores, x 920; E. dichotomously branched hairs, x 920.

Occurrence: Isolated from mattress dust; hay, cereals.

Medical significance: HI.

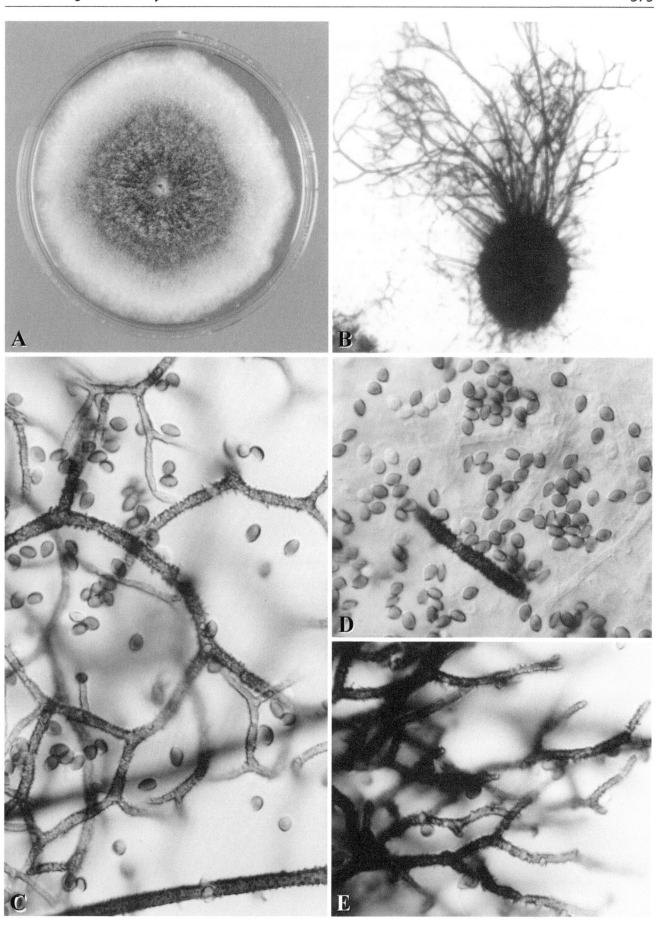

Chrysonilia sitophila (Mont.) v. Arx

    = *Monilia sitophila* (Mont.) Sacc.
    = *Penicillium sitophilum* Mont.

Teleomorph: *Neurospora sitophila* Shear & Dodge.

A. Colony on MEA after five days; B-D. conidiophores, each branch forming chains of conidia, B. x 920, C. x 230, D. x 920.

Occurrence: This "red bread mould" has a wide distribution, and is found on bakery goods and causes a storage rot of soft fruits.

Isolated from carpet and mattress dust; bakeries; burnt wood.

Medical significance: HIG.

Note: *Chrysonilia crassa* (teleomorph *Neurospora crassa*) is related and differs in having smaller conidia (5-8 x 4-6 µm).

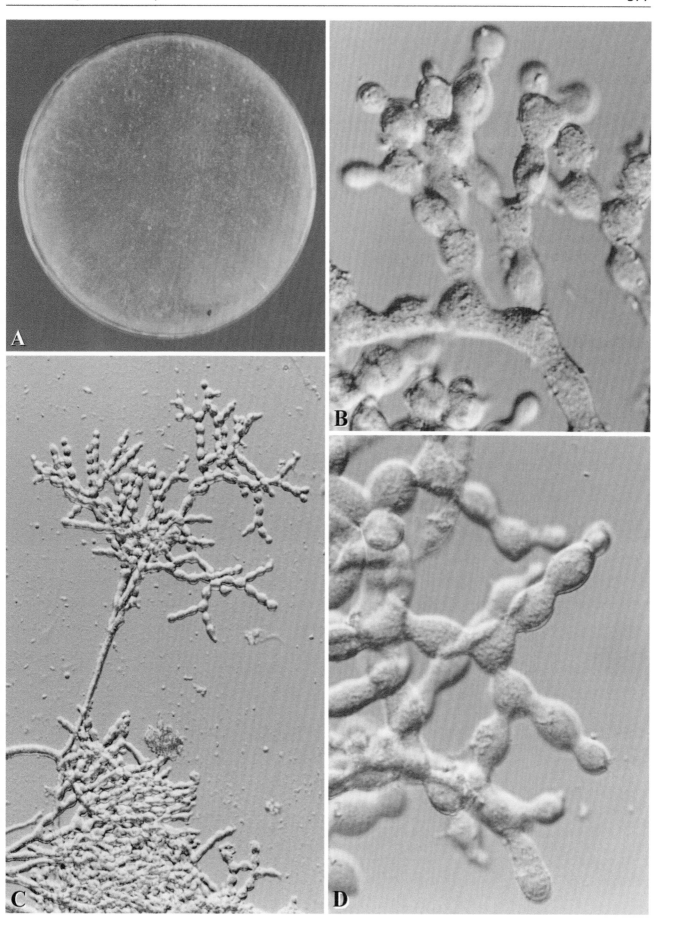

Cladosporium cladosporioides (Fres.) de Vries

A. Colony on MEA after one week; B-D. conidia, x 920.

Occurrence: With a worldwide distribution, this common phylloplane species occurs also in soil and on exposed textiles, and in summer is abundant in air.

Although this species is claimed to *grow* on building materials, it is not always clear whether such strains are truly *C. cladosporioides*, an undescribed species or variants of *C. sphaerospermum*, which does grow on wet building materials. Isolated from floor, carpet and mattress dust; damp PVA emulsion, acrylic painted walls, wallpaper; UFFI; bakery dust; cereals and other foodstuffs.

Medical significance: [E?]IG. Allergenic (Day 1996).

Note: The taxonomy of *C. cladosporioides* is studied by Bensch *et al.* (2010). This study shows that *C. cladosporioides* is a species complex consisting of various species. *Cladosporium pseudocladosporioides*, *C. inversicolor* and *C. delicatulum* are also commonly occurring in indoor and/or outdoor air and construction materials.

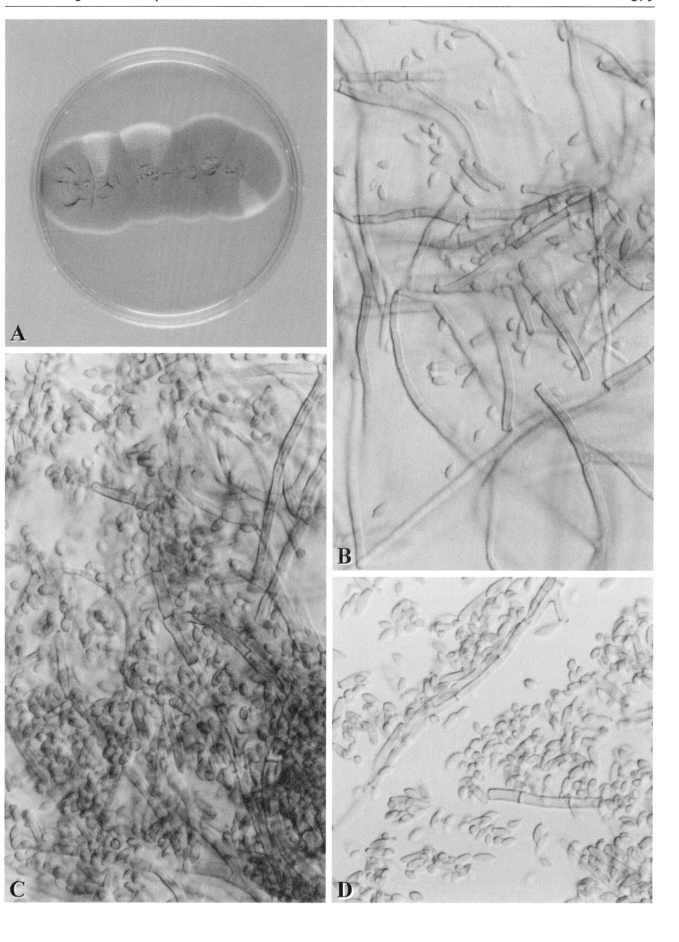

Cladosporium herbarum (Pers.) Link

Teleomorph: *Davidiella tassiana* (de Not.) Crous & U. Braun.

A. Colony on MEA after one week; B-D. conidiophores with conidia, x 920.

Occurrence: Similar to *C. cladosporioides*, and on senescent or dead plant matter.

Less common than *C. cladosporioides* in indoor environments. Isolated from floor, carpet and mattress dust; damp acrylic painted walls, wallpaper; talcum powder; HVAC insulation, filters and fans; hospital air; cereals.

Medical significance: HI. Allergenic (Day 1996).

Note: *C. bruhnei* is closely related to *C. herbarum* and has also been isolated from air and building materials; for detailed descriptions see Schubert *et al.* (2007).

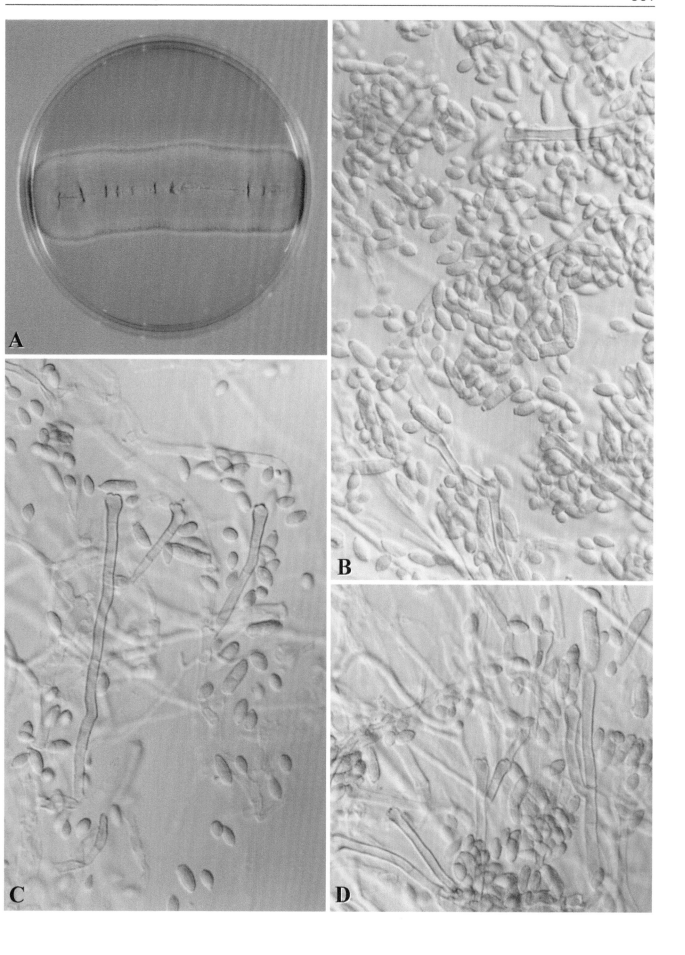

Cladosporium sphaerospermum Penzig

A. Colony on MEA after one week; B-D. conidiophores with conidia, x 920; E. conidiophore with conidia, x 2330.

Occurrence: Similar to *C. cladosporioides*.

Common on wet building elements such as gypsum board, acrylic and oil painted walls, painted wood, and wallpaper. Also isolated from carpet and mattress dust; bakeries; HVAC fans, wet insulation in mechanical cooling units; cloth, pharmaceutical cream; meat (in cold store), bread.

Medical significance: HI. Allergenic (Day 1996).

Notes: *Cladosporium halotolerans* is related to *C. sphaerospermum* and also occurs in air and in bathrooms; see Zalar *et al.* (2007).

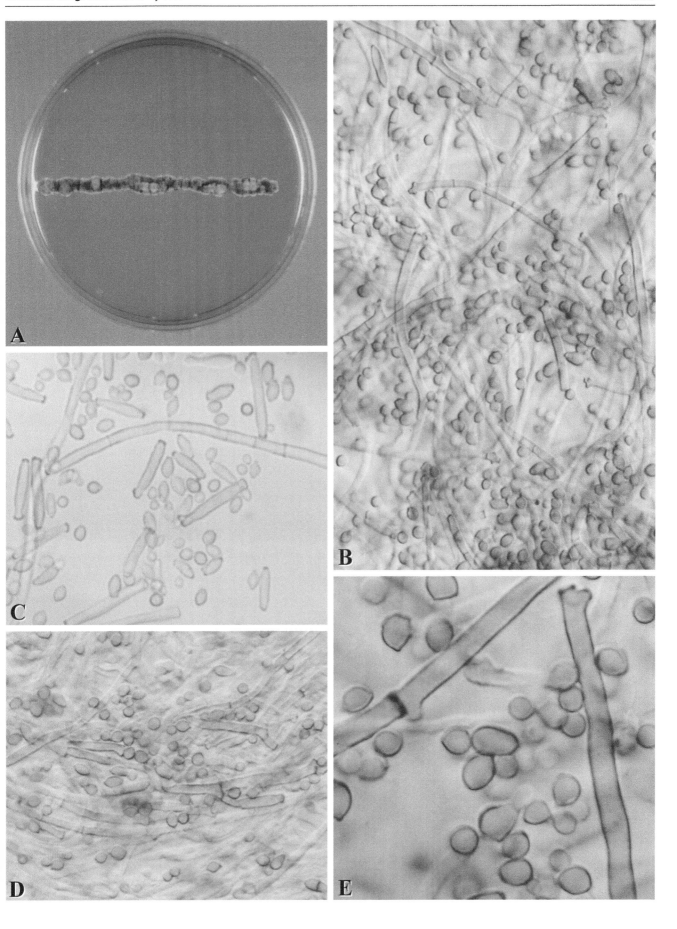

Clonostachys rosea (Link:Fr.) Schroers *et al.*

   = *Gliocladium roseum* (Link) Thom

Teleomorph: *Bionectria ochroleuca* (Schw.) Schroers & Samuels.

A. Colony on MEA after one week; B. conidia, x 920; C.-E. conidiophores with conidia, x 920.

Occurrence: Isolated from house dust; cooling tower slats, gymnasium floor; polyrethanes; greenhouse dusts; cereals.

Medical significance: HI.

Note: For descriptions see Schroers (2001).

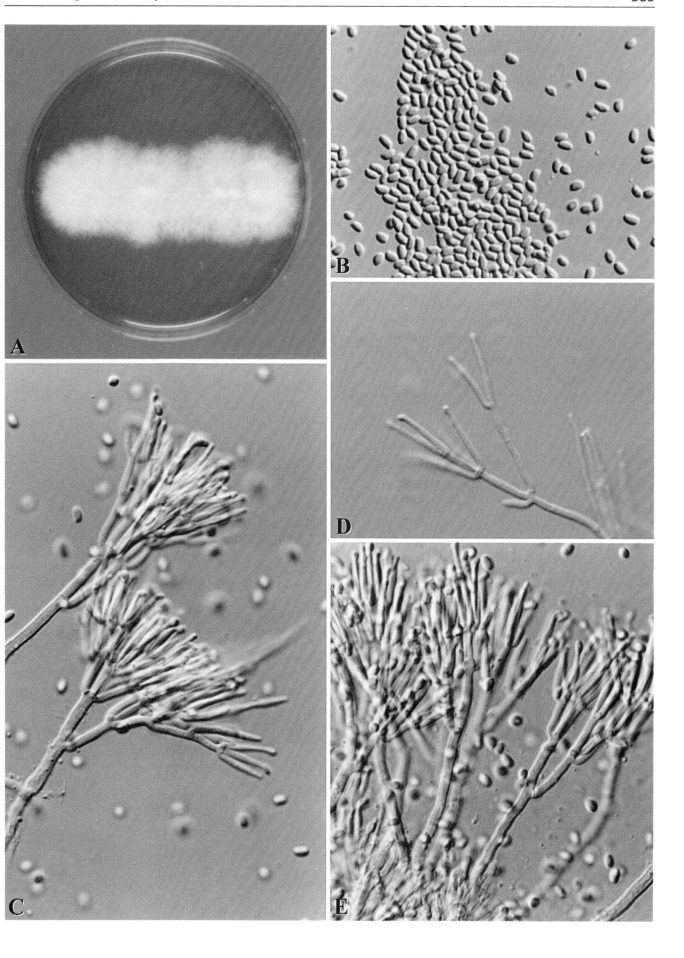

Coprinus cordisporus Gibbs

A.-C. Fruiting bodies of different age; D. basidiospores, x 2330.

Occurrence: Wood, often in crawl spaces.

Medical significance: HI.

Note: Various species of *Coprinus* can occur indoors. The species depicted here is merely an illustration of the morphological features.

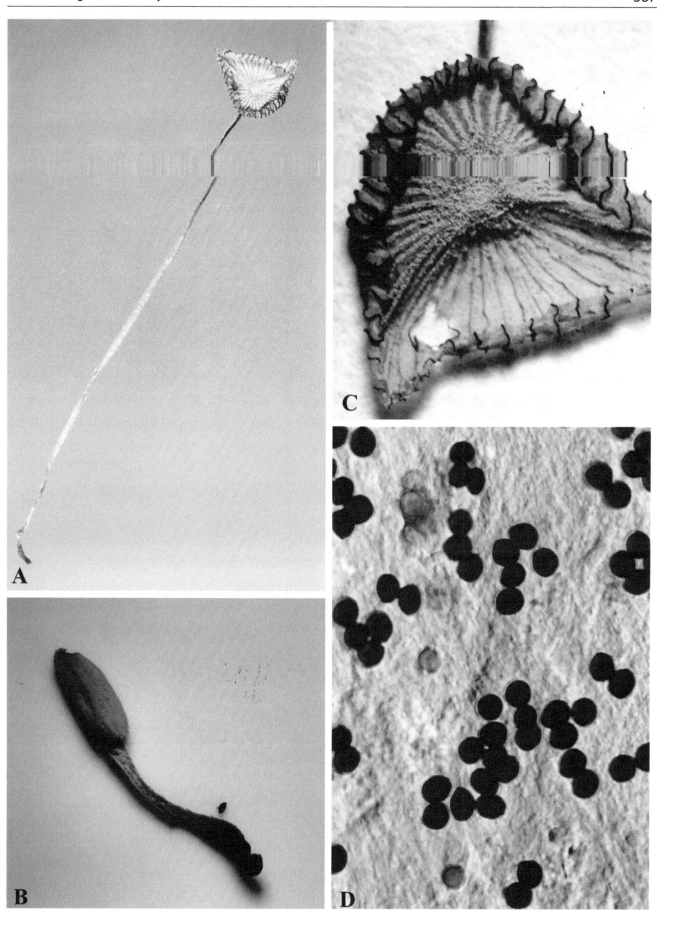

Cryptococcus laurentii (Kufferath) Skinner

A. Colony on MEA after five days; B. spores, x 920; C, D. spores, x 2330.

Occurrence: Indoor air.

Also found in food and beverages, wheat, maize, soil, litter, plants.

Medical significance: [E/][F?][G?]. Many dubious or erroneous medical reports based on incorrect attribution of infection to probable contaminants.

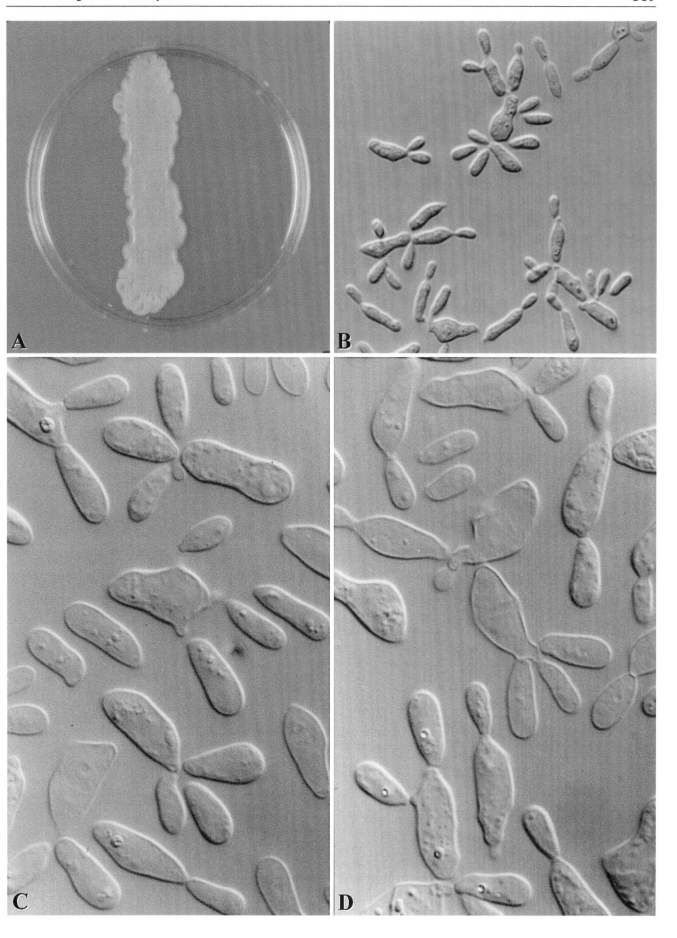

Curvularia lunata (Wakker) Boedijn

A. Colony on MEA after one week; B-D. conidia and conidiophores, B. x 920, C. x 920, D. x 920.

Occurrence: A widely distributed contaminant of seed crops.

Isolated from floor and mattress dust; wallpaper, painted wood; cereals.

Medical significance: EFG.

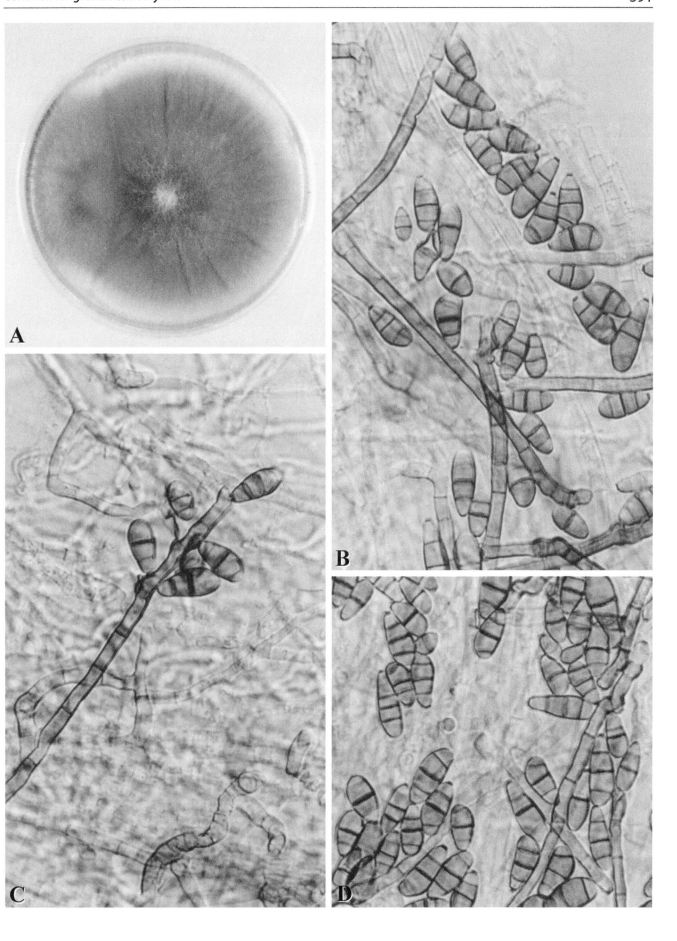

Emericella nidulans (Eidam) Vuill.

Anamorph: *Aspergillus nidulans* (Eidam) Wint.

A. Colonies on OA after one week; B. ascoma surrounded covered by Hülle cells, x 230; C. conidiophores of *Aspergillus nidulans*, x 920; D. conidia of *Aspergillus nidulans*, x 2330; E. Hülle cells, x 920; F. ascospores and conidia, x 2330.

Occurrence: A widespread soil fungus, most common in tropical and subtropical regions, also associated with seeds.

Isolated from carpet and mattress dust; cotton lint, cotton yarn, decomposing wood; hay, cereals. Often in air samples, particularly from compost plants.

Medical significance: EFG.

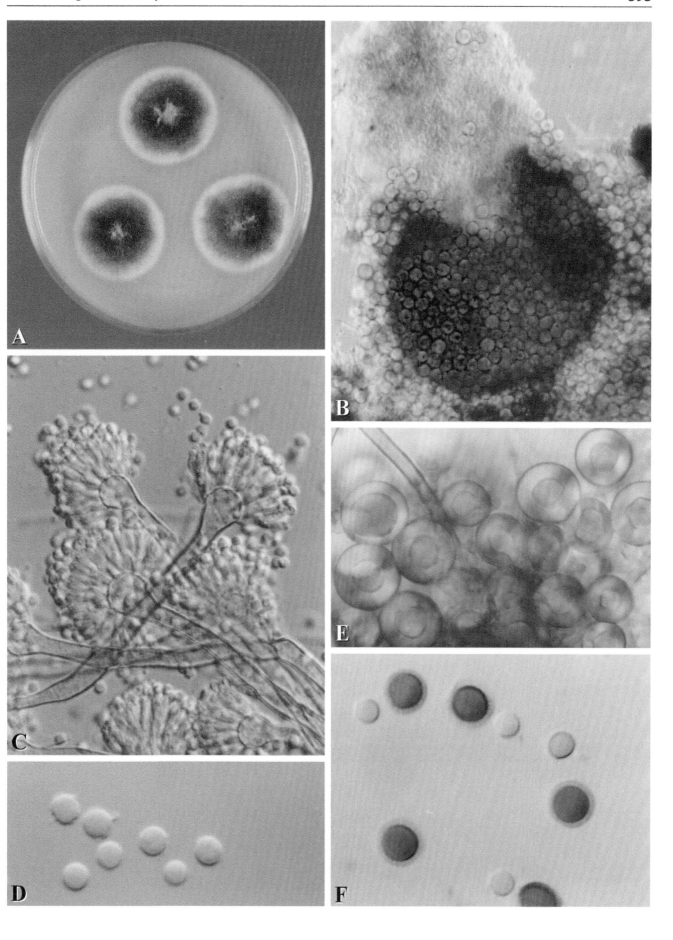

Epicoccum nigrum Link

= *Epicoccum purpurascens* Ehrenb.

A. Colony on MEA after one week; B, C. sporodochia with septate (blasto)conidia, x 920; D. conidiogenous cells with septate (blasto)conidia, x 920.

Occurrence: Universally distributed, this colonizer of senescent and dead plants also occurs in soil and is prominent in freshly harvested seeds.

Isolated from floor, carpet and mattress dust; UFFI; hospital air; exposed acrylic paint, canvas; cereals, beans.

Medical significance: [E?]I.

Notes: *E. nigrum* is phylogenetically embedded in *Phoma* section *Peyronellaea*. *Phoma epicoccina* and *E. nigrum* are claimed to be the same biological species (Aveskamp *et al.* 2009). No pycnidium formation is observed in cultures of *E. nigrum*.

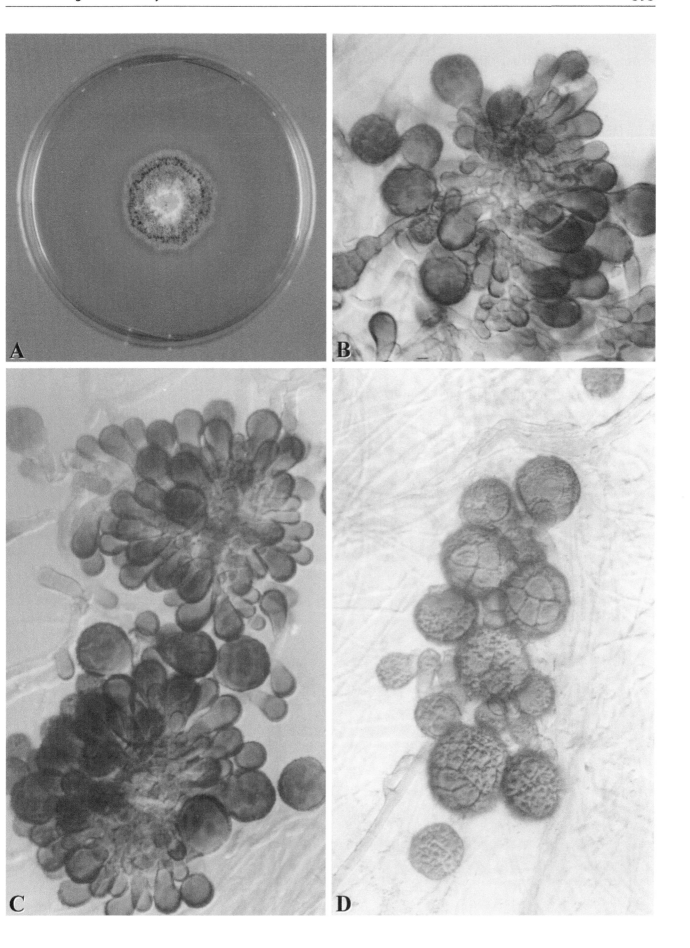

Eurotium amstelodami Mangin

Anamorph: *Aspergillus amstelodami* (Mangin) Thom & Church

A. Colonies on MEA + 20% sucrose after two weeks; B. ascomata, x 40; C. conidia and conidiophore of *Aspergillus amstelodami*, x 920; D. ascospores and conidia, x 2330; E. portion of ascoma with asci, x 920.

Occurrence: In nature, this xerophile is more common in tropical and subtropical soils and is commonly isolated from dried or concentrated food materials.

Found in floor, carpet and mattress dust. Also isolated from hospital air, cloth, shoes; hay, cereals, nuts, fruit juices.

Medical significance: EIG.

Note: *E. amstelodami* is a xerophile and growth is induced on low water activity media, such as DG18 or MEA supplemented with 20% sucrose. A detailed description can be found in Samson *et al.* (2010).

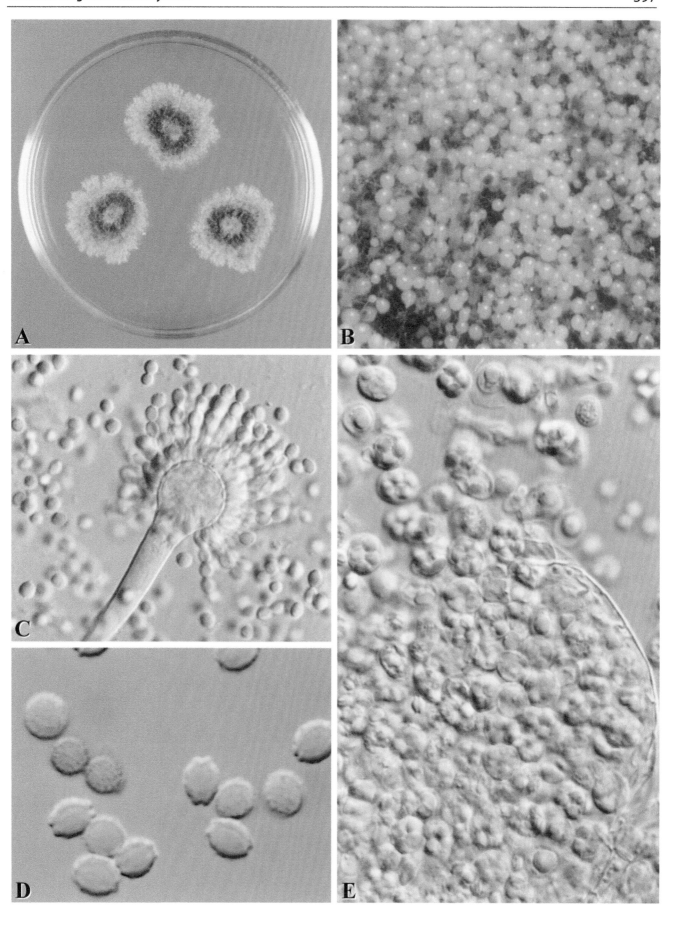

Eurotium chevalieri Mangin

Anamorph: *Aspergillus chevalieri* (Mangin) Thom & Church

A. Colonies on MEA + 20% sucrose after two weeks; B. ascomata, x 40; C. conidiophores of *Aspergillus chevalieri*, x 920; D. ascospores, x 2330; E. ascoma, x 230; F. portion of ascoma with asci and ascospores, x 920.

Occurrence: With a distribution similar to *E. amstelodami*, this ubiquitous soil-borne xerophile is associated with leather, cotton, dried seeds and milled foodstuffs.

Found in floor, carpet and mattress dust. Also isolated from HVAC system air; hay, cereals, flour.

Medical significance: EIG.

Note: *E. chevalieri* is a xerophile and growth is induced on low water activity media, such as DG18 or MEA supplemented with 20% sucrose. A detailed description can be found in Samson *et al.* (2010).

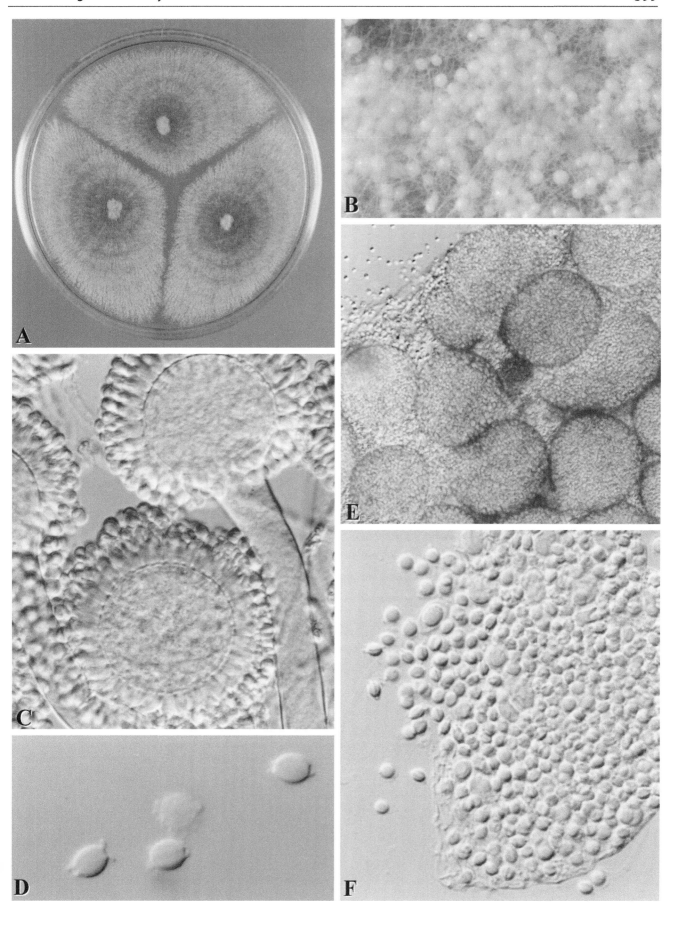

Eurotium herbariorum (Wiggers) Link

Anamorph: *Aspergillus glaucus* Link.

A. Colonies on MEA + 20% sucrose after two weeks; B. ascomata, x 40; C. conidiophore of *Aspergillus glaucus*, x 920; D. conidia of *Aspergillus glaucus*, x 920; E. portion of ascoma with asci, x 920; F. ascospores, x 2330.

Occurrence: This xerophile also has a distribution similar to *E. amstelodami*, and is associated with meat products as well as dried plant products.

Occurs on damp gypsum board and other building materials, and in floor, carpet and mattress dust. Also isolated from HVAC insulation; food store dust, weaving shed wall; hay, cereals, flour, desiccated coconut, nuts.

Medical significance: HI[G?]. Many dubious medical reports based on misidentification or incorrect attribution of disease causation to contaminants.

Note: *Eurotium repens* de Bary is a similar species and is more commonly occurring than *E. herbariorum*. *Eurotium herbariorum* is a xerophile and growth is induced on low water activity media, such as DG18 or MEA supplemented with 20% sucrose. A detailed description can be found in Samson *et al.* (2010).

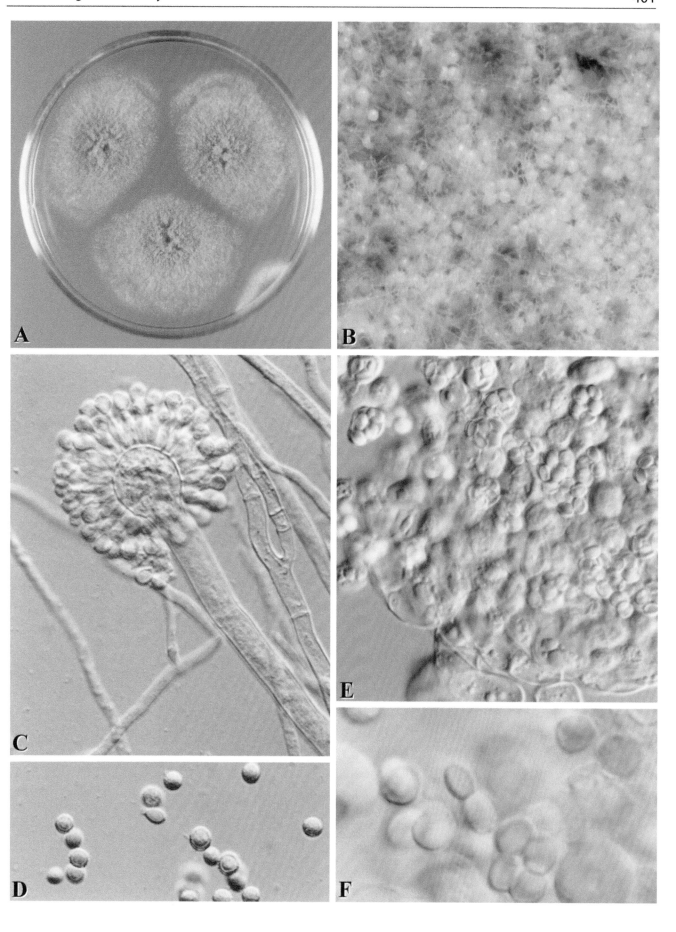

Exophiala dermatitidis (Kano) de Hoog

A. Colony on MEA after one week; B-D. conidiogenous cells with conidia, B. x 920, C, D. x 2330.

Occurrence: An inhabitant of cool moist soils.

Commonly found in sinks, drains and steam baths. Also isolated from bronchioalveolar lavage sampling equipment.

Medical significance: CDG. *Exophiala dermatitidis* is an uncommon etiologic agent of fatal infections of the central nervous system in otherwise healthy, mainly adolescent patients in East Asia.

Note: *Exophiala* is a diverse genus and various genera, such as *Cladophialophora*, *Phialophora*, are linked to this genus.

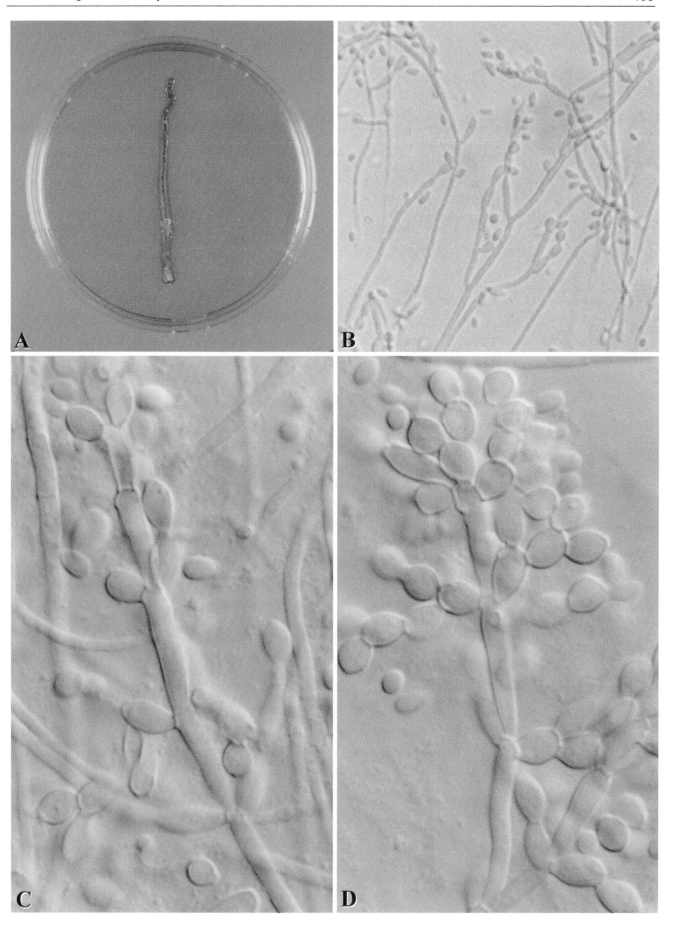

Fusarium culmorum (W.G. Smith) Sacc.

A. Colony on MEA after one week; B-F. conidiophores and macro-conidia, B. in Petri-dish, x 230, C-F. x 920.

Occurrence: With a worldwide distribution, this soil-borne grass and cereal pathogen is more prevalent in temperate regions.

Isolated from floor, carpet and mattress dust; damp wall; polyurethane foam; cereals, apples, potatoes.

Medical significance: HI. Allergenic (Frankland 1983).

Toxicity: Produces zearalenone and deoxynivalenol.

Note: In many areas the related species *F. graminearum* is much commoner. Ascospores of its teleomorph, *Gibberella zeae*, may be common in outdoor air during the growing season where there is a lot of green space and/or wheat or maize fields, but it does not grow on building materials. It also produces deoxynivalenol and zearalenone (Miller 1999).

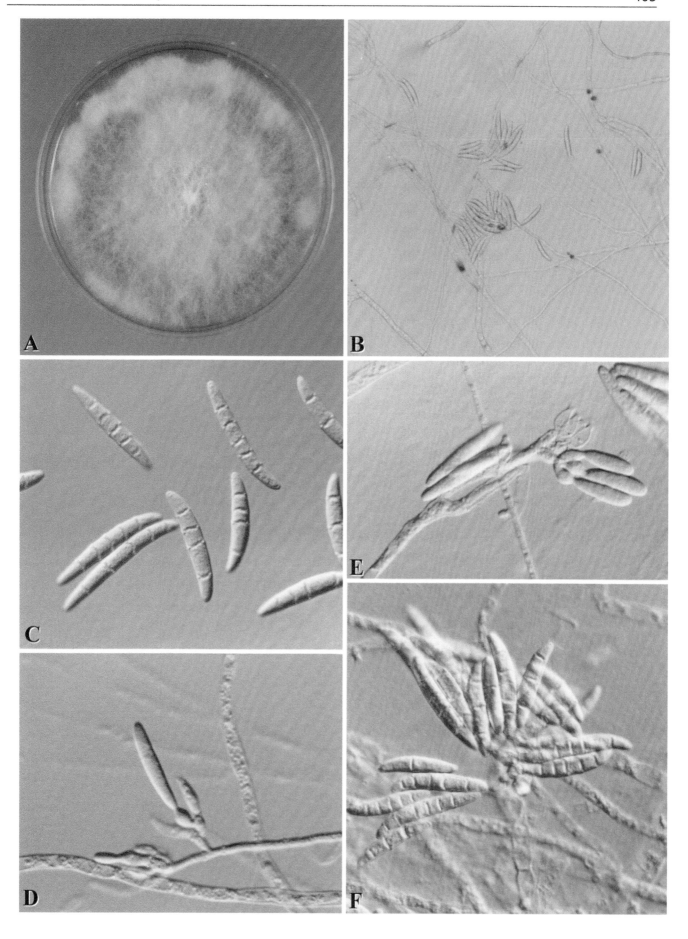

Fusarium solani (Mart.) Sacc.

Teleomorph: *Nectria haematococca* (Berk. & Br.) Samuels & Nirenberg.

A. Colony on MEA after one week; B. conidiophores with conidia in slimy heads, in Petri-dish, x 230; C. elongated conidiophores and phialides forming micro-conidia, x 920; D. micro-conidia, x 920; E. macro-conidia, x 920; F. conidiophores forming macro-conidia, x 920.

Occurrence: A cosmopolitan soil-borne pathogen of a wide range of crop plants. Also common in fresh water.

Isolated from carpet and mattress dust; damp walls, wallpaper; polyester polyurethane foam; cotton of insulating duct liner; water pipes, humidifier water; metal working fluid, coolant; treated wood.

Medical significance: EFG. Keratitis in man; associated with wounds; isolated from eyes and finger nails; see O'Donnell *et al.* (2007), Zhang *et al.* (2006).

Note: More than 40 varieties have been described in *F. solani*, and many phylogenetic species have been proposed. For a detailed description, see Samson *et al.* (2010).

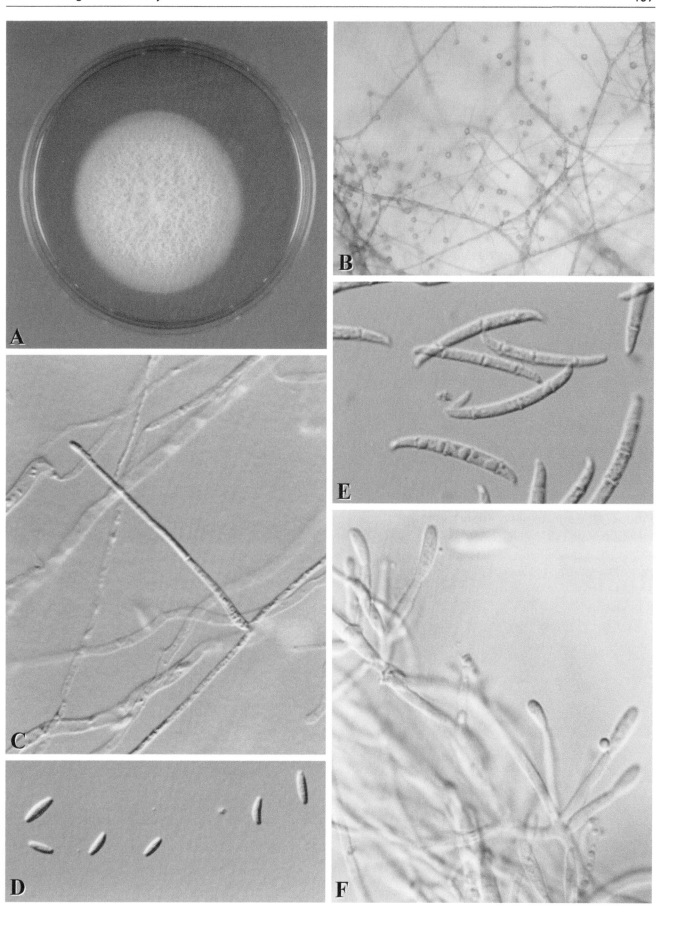

Fusarium verticillioides (Sacc.) Nirenb.

= *Fusarium moniliforme* Sheld. *sensu stricta*

Teleomorph: *Gibberella moniliformis* Winel.

A. Colony on MEA after one week; B. catenulate micro-conidia, in Petri-dish, x 230; C. conidiophores, x 920; D. micro-conidia, x 920; E. macro-conidia, x 920.

Occurrence: Soil-borne pathogen (in North America formerly called one of the mating types of *Fusarium moniliforme* Sheld.) of maize and a wide range of other crops in tropical and subtropical areas, but rare in temperate areas, except under glass.

Can grow in badly maintained humidifier pans and other areas where stagnant water occurs in HVAC systems; such isolates produce fumonisin. Also isolated from mattress dust; damp walls; raw cotton; cereals and other seeds.

Medical significance: EFG. Keratitis in man, and an increasingly observed cause of invasive mycoses in immune compromised individuals (Nelson *et al.* 1994).

Toxicity: Produces the agriculturally important toxin fumonisin (Nelson *et al.* 1991, Miller 1999).

Notes: *F. verticillioides* is a species complex. This species is also referred to as *F. moniliforme*, but this name should not be used.

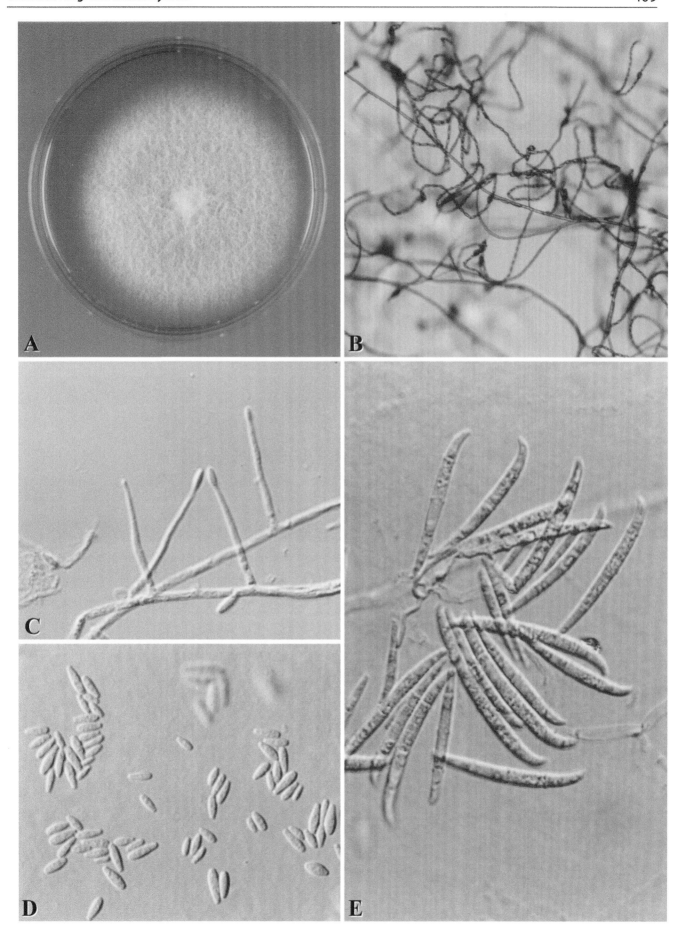

Geomyces pannorum (Link) Sigler & Carmichael

A. Colony on MEA after one week; B. detail of colony, x 40; C. conidiophore and conidia, x 2330; D. conidia, x 2330; E. conidiophores with conidia, x 920.

Occurrence: Isolated from mattress dust; damp walls; soil floor in building, laboratory contaminant, swimming pool floor, gymnasium floor; meat, cod, gelatin, flour.

Often found on paper in archives and libraries.

Medical significance: HI (some erroneous medical reports exist).

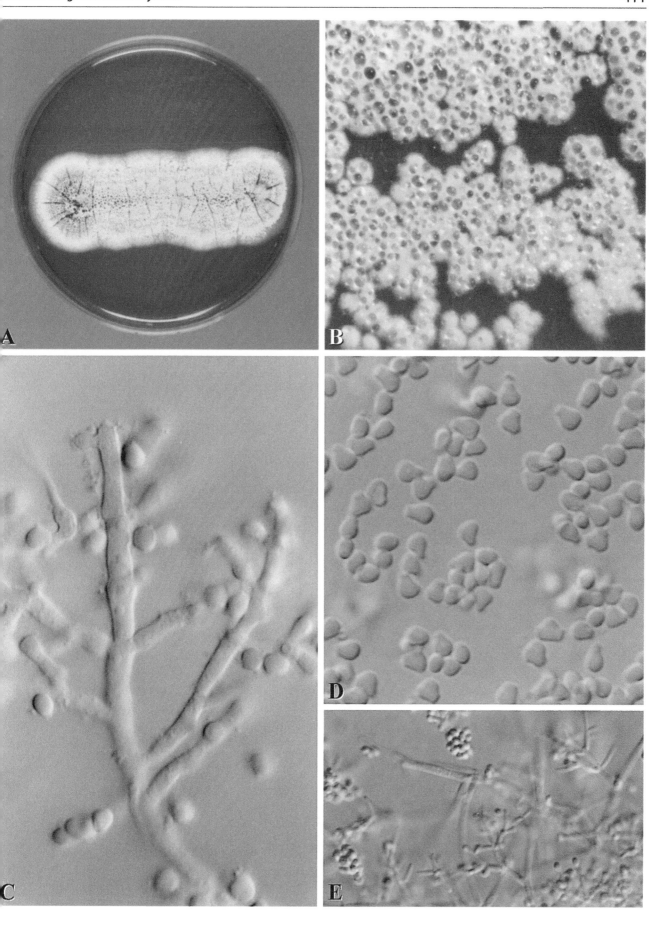

Geotrichum candidum Link

Teleomorph: *Galactomyces geotrichum* (Butler & Petersen) Redhead & Malloch

A. Colony on MEA after one week; B. disarticulation of fertile hyphae into (arthro)conidia, x 920; C, D. dichotomously branched (forked) hyphae, C. x 230, D. x 920; E. (arthro)conidia, x 920.

Occurrence: A cosmopolitan species in soil, water and air, associated with a wide range of seeds and fruits.

Isolated from carpet dust; damp walls; swimming pool floor, gymnasium floor; bakeries.

Medical significance: EIG.

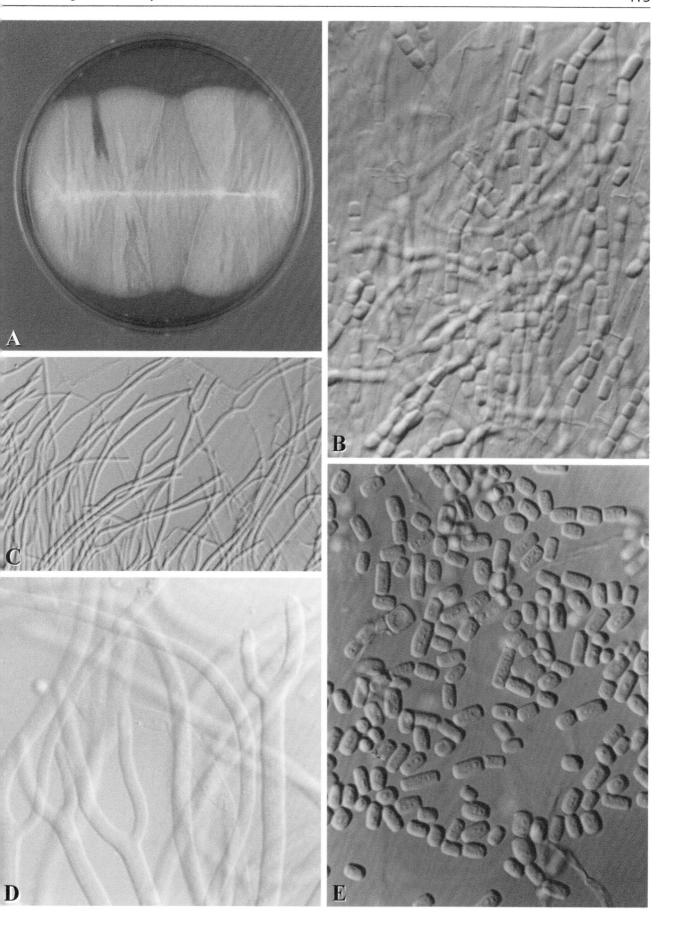

Lecanicillium lecanii (Zimmermann) Zare & W. Gams

= *Verticillium lecanii* Zimmerman

A. Colony on MEA after one week; B, C. verticillate phialides, B. x 2330; C. x 920; D. conidia, x 2330.

Occurrence: Often associated with cereals, hyper-parasitic on other moulds, used in control of insect pests.

Isolated from air in greenhouses, glyphosate solution; contact lens.

Medical significance: HI.

Note: *Verticillium lecanii* is a synonym of this species.

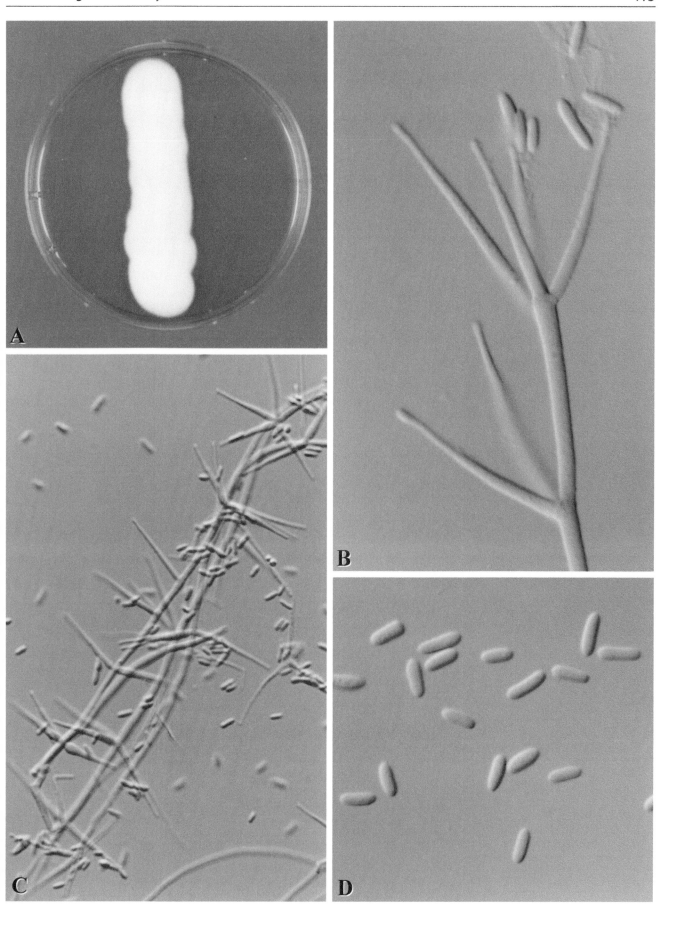

Mucor hiemalis Wehmer

A. Colony on MEA after one week; B. sporangiophore with columella and sporangiospores, x 920; C. sporangiophore with columella, x 920; D. sporangiospores, x 920; E. sporangium containing sporangiospores, x 920.

Occurrence: Cosmopolitan species which is one of the commonest of soil fungi, and associated with a wide range of plant produce, fresh and dried.

Isolated from floor, carpet and mattress dust; bakery dust; hay, cereals.

Medical significance: HI.

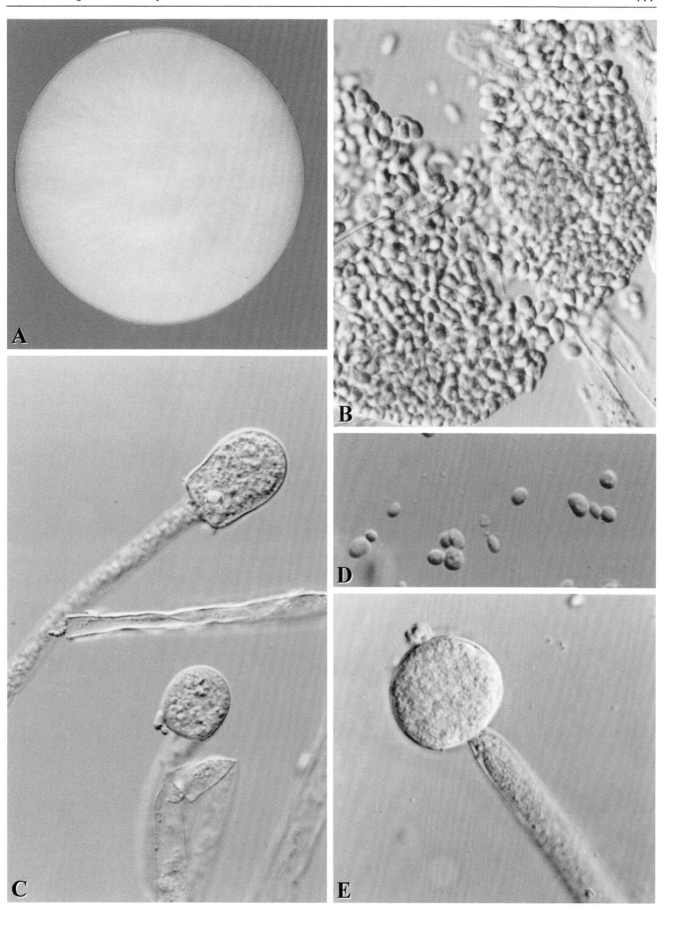

Mucor plumbeus Bon.

A. Colony on MEA after one week; B. sporangiospores, x 920; C. sporangium containing sporangiospores, x 920; D. sporangiophore with columella , x 920.

Occurrence: Soil-borne species with a worldwide distribution, associated with stored plant produce.

Isolated from wood, polyurethane coated fabric; HVAC filter; hospital air; bakeries, cork factories; hay and cereals.

Medical significance: HI.

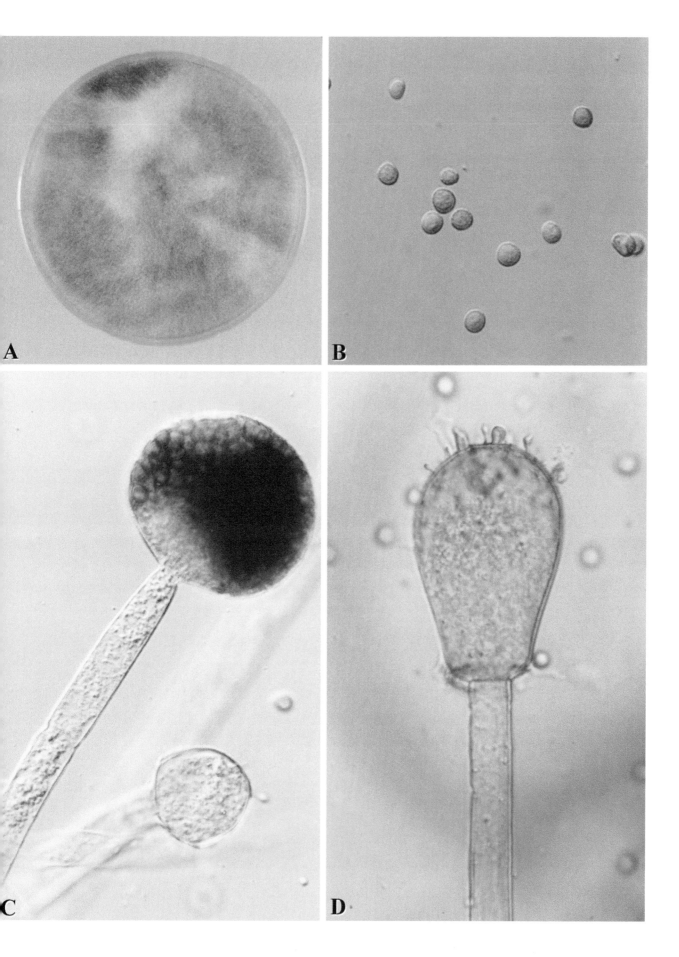

Mucor racemosus Fres.

A. Colony on MEA after one week; B. sporangiophore with columella and part of sporangium wall, x 920; C. sporangiophore with columellae, x 920; D. sporangium, x 920; E. sporangiospores, x 920.

Occurrence: Cosmopolitan soil organism, associated with cereals and other stored foodstuffs.

Isolated from mattress dust; polyurethane-coated fabric; bakeries; wood shavings.

Medical significance: HI.

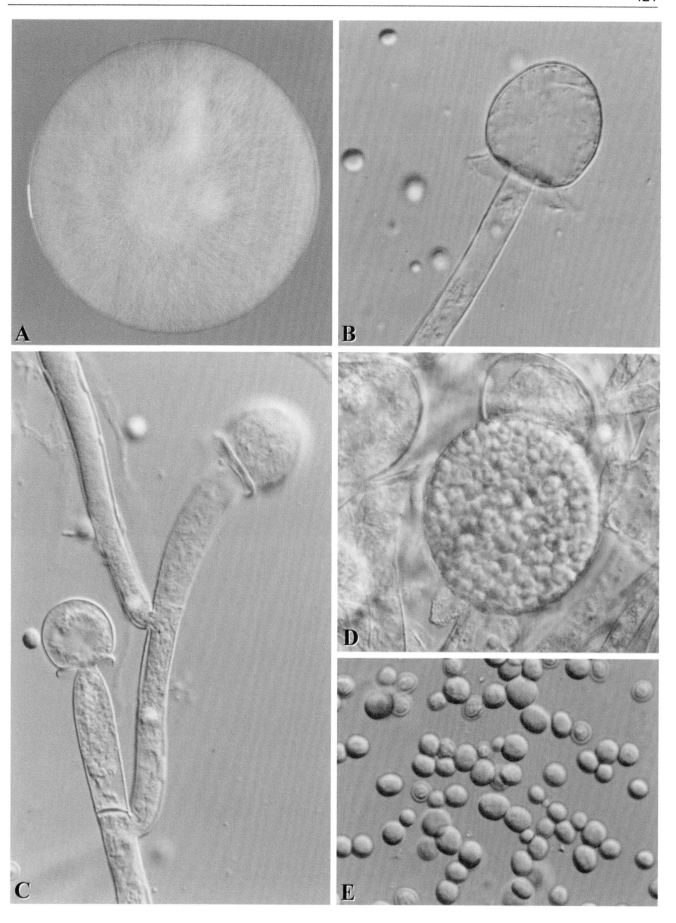

Oidiodendron griseum Robak

A. Colony on MEA after one week; B. conidia, x 2330;
C-E. conidiophores, x 920.

Occurrence: Wood, air and sputum of man.

Isolated from wood pulp, soil.

Medical significance: HI.

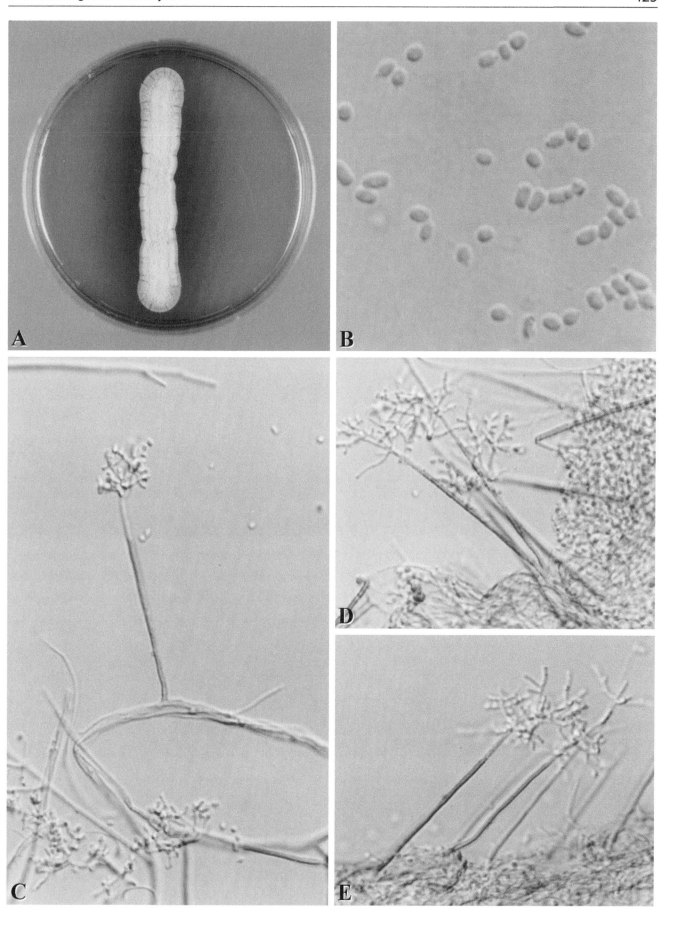

Oidiodendron rhodogenum Robak

A. Colony on MEA after one week; B. conidia, x 2330;
C,D. conidiophores, x 920.

Occurrence: Isolated from air in cork factory; straw, bottle cork; decaying furniture and sputum of man.

Medical significance: HI.

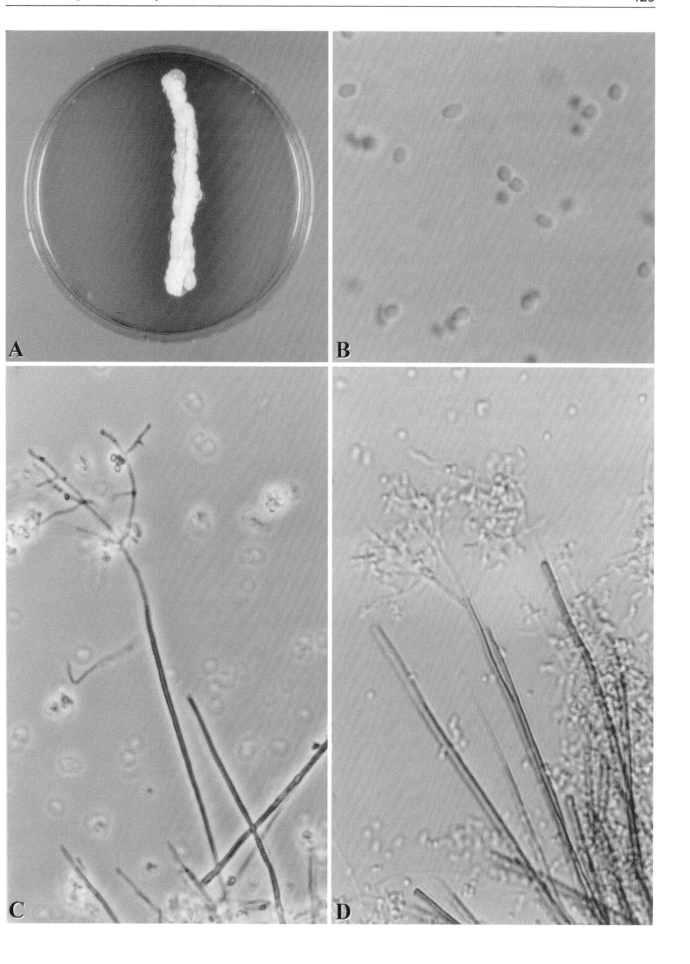

Paecilomyces lilacinus (Thom) Samson

A. Colonies on MEA after one week; B. conidia, x 1000; C. conidiophores, x 500; D. detail of conidiophore, x 1600.

Occurrence: Common in house dust.

Also isolated from humidifier water; swimming pool floor; cotton drill; synthetic rubber, plastic bottles; stored foods.

Medical significance: EFG. Infections in immuno-competent patients are usually associated with foreign bodies or implanted devices, such as intraocular lens implants and peritoneal dialysis devices.

Note: The monographic review of Samson (1974) was based on morphology, culture characteristics, hosts and temperature preferences, and more recently, Luangsa-ard *et al.* (2004) re-examined the genus *Paecilomyces* using molecular data. Consequently, *Paecilomyces* is now restricted to the thermotolerant type species *Paecilomyces variotii* and its allies in the Eurotiales. *Paecilomyces lilacinus* is not related to *P. variotii* and will be accommodated to a new genus.

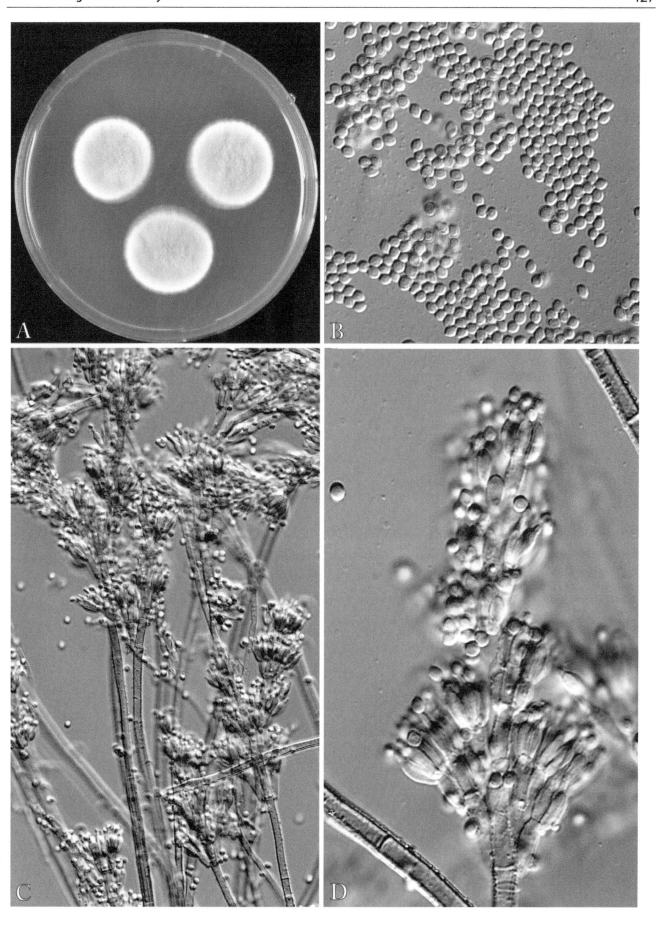

Paecilomyces variotii Bain.

Teleomorph: *Byssochlamys spectabilis* (Udagawa & Suzuki) Houbraken & Samson

A. Colonies on MEA after one week; B. conidia, x 2000; C. conidiophores with conidia x 1000; D. detail of conidiophores, x 2000.

Occurrence: A species associated with composts and other self-heated materials, common in air.

Occurs on damp walls, wet plaster work and in carpet dust. Also isolated from wood, wood chips, tarred wooden flooring, UFFI, synthetic rubber, leather, library paste; HVAC fan; air in bakery; salami, margarine; pectin, hay, grain. Sometimes occurring as a heat-resistant spoilage agent in food.

Medical significance: EFG. Can cause invasive mycoses in compromised individuals.

Toxicity: Viriditoxin is produced by *P. variotii*. Patulin is often reported as a mycotoxin produced by *P. variotii*. However, Houbraken *et al.* (2006, 2008) showed that this compound is produced by a related species, *P. dactylethromorphus*.

Note: The taxonomy of the genus *Paecilomyces* has been revised and nine species are included in this genus. *Paecilomyces variotii* and *P. formosus* are the two most commonly occurring species in air and on building materials (Samson *et al.* 2009, 2010).

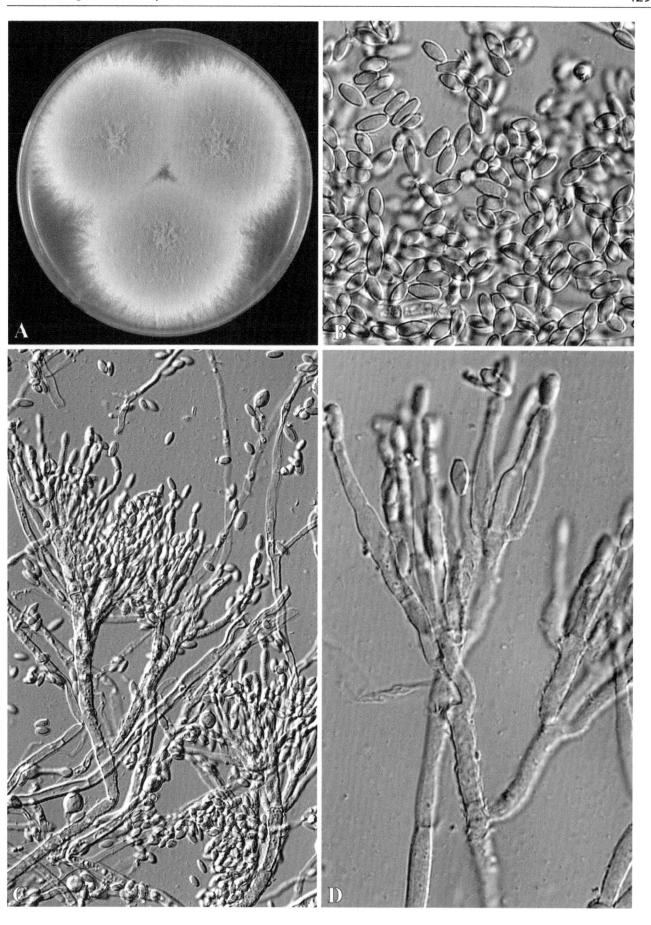

Penicillium brevicompactum Dierckx

A. Colonies on MEA after one week; B. conidia, x 1400;
C. conidiophores, x 1200; D. conidiophores, x 1750.

Occurrence: Widespread in nature, in soil and decaying vegetation.

Common on damp walls and building materials such as gypsum board. Present in floor, carpet, mattress and upholstered furniture dust. Also isolated from UFFI, red lead paint; cotton yarn; hay, cereals; mushroom compost.

Medical significance: HI. No growth at 37°C.

Toxicity: Produces a number of metabolites of moderate toxicity (Nielsen *et al.* 1999); botryodiploidin.

Note: *Penicillium bialowiezense* is closely related to *P. brevicompactum* and both species are commonly occurring in indoor environments (Scott *et al.* 2008).

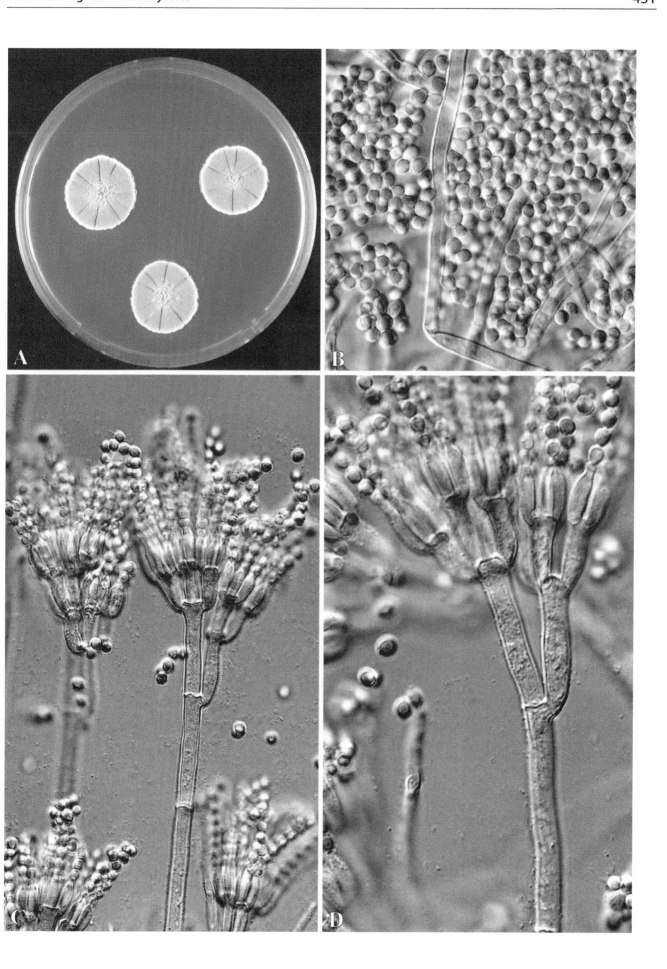

Penicillium chrysogenum Thom

A. Colonies on MEA after one week; B. conidia, x 1400; C. conidiophores, x 1200; D. detail of a conidiophore, x 1750.

Occurrence: A cosmopolitan species found in a wide range of habitats from soils to foodstuffs.

Extremely common on damp building materials, walls and wallpaper and in floor, carpet, mattress and upholstered furniture dust. Also isolated from PVA emulsion paint, oil paint, UFFI; polyester polyurethane foam, polyurethane coated fabric, paper, talcum powder, school ceiling, fruit store air.

Medical significance: EIG.

Toxicity: Produces a number of metabolites of moderate toxicity (Nielsen *et al.* 1999).

Note: *Penicillium chrysogenum* has been subjected to a multigene analysis. This analysis shows that there are four different lineages in this species (Scott *et al.* 2004).

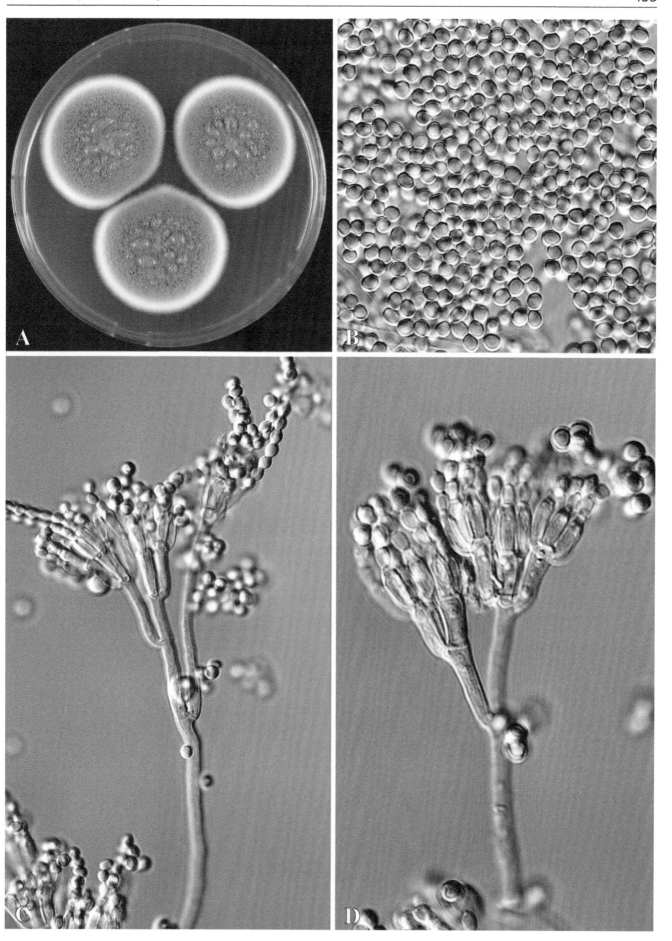

Penicillium citreonigrum Dierckx

A. Colonies on MEA after one week; B. conidia, x 1200;
C, D. conidiophores, x 1200.

Occurrence: Worldwide distribution, in soil and decaying vegetation.

Infrequently detected in indoor environments, occurring on damp building materials such as beaverboard and other wood-derived products; indoor air. Also isolated from soil, stored rice, plant stems and cork.

Medical significance: HI.

Toxicity: This species is claimed to be the causal agent of cardiac beriberi, due to the consumption of "yellow rice" containing citreoviridin.

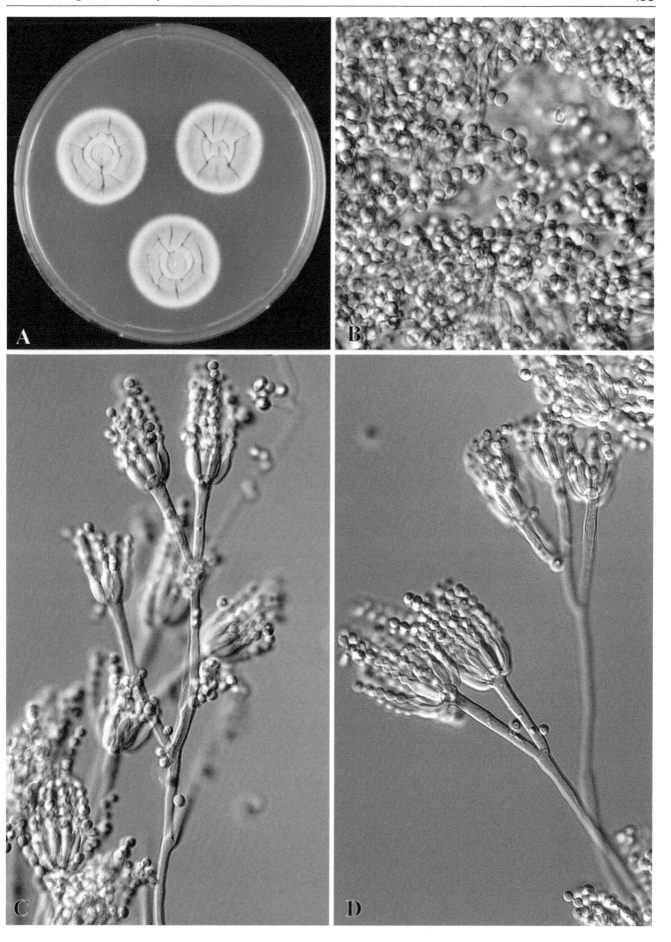

Penicillium citrinum Thom

A. Colonies on MEA after one week; B. conidia, x 1300; C. conidiophores, x 1200; D. conidiophore, x 1650.

Occurrence: Ubiquitous; in soil, decaying vegetation and the air, and an active deteriogen.

Common on damp building materials such as gypsum board, especially in warmer climates, and in settled dust. Also isolated from cotton fabric, textiles; flour.

Medical significance: EIG.

Toxicity: Produces a number of metabolites of moderate toxicity (Frisvad 1987).

Note: *Penicillium steckii* and *P. sizovae* are related to *P. citrinum*, though less common. This species is more predominant in indoor environments in (sub)tropical regions (Houbraken *et al.* 2010).

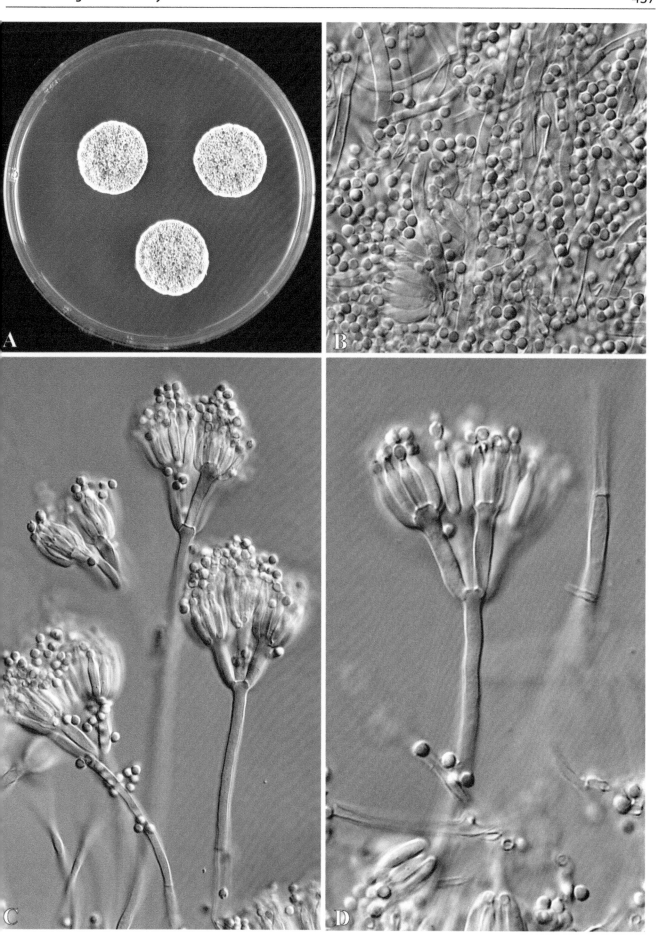

Penicillium commune Thom

A. Colonies on MEA after one week; B. conidia, x 1200; C. conidiophores and conidia, x 950; D. detail of conidiophore, x 2000.

Occurrence: Common on foodstuffs.

Found on damp building materials such as gypsum board, and in house dust. Also isolated from artificial leather cloth.

Medical significance: HI.

Toxicity: Produces a number of metabolites of moderate toxicity (Frisvad and Filtenborg 1989).

Note: For a detailed description see Samson and Frisvad (2004) and Samson *et al.* (2010).

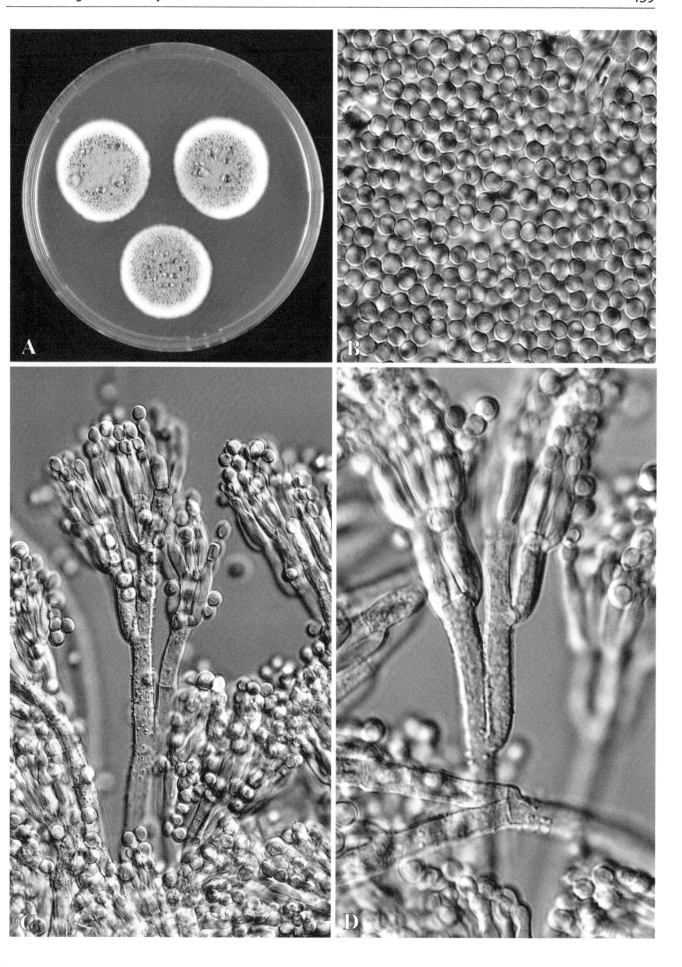

Penicillium corylophilum Dierckx

A. Colonies on MEA after one week; B. conidia, x 1300; C. conidiophores, x 1000; D. conidiophore, x 2000.

Occurrence: In soil and decaying vegetation, and common on cereals and a wide range of other foods.

Found in floor and mattress dust. Also isolated from silk screen, painted wood.

Medical significance: HI.

Note: For a detailed description see Samson *et al.* (2010).

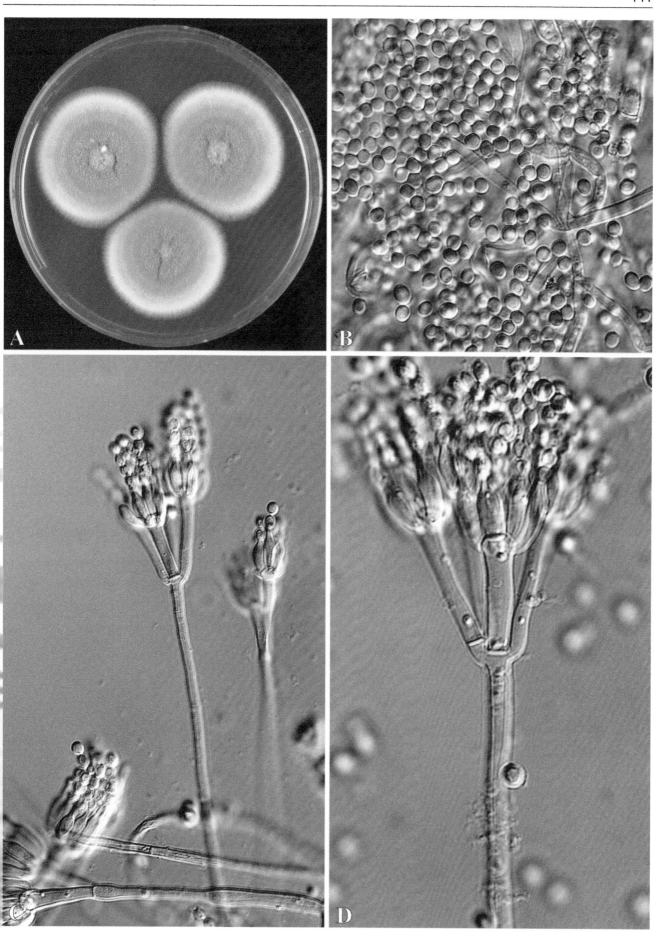

Penicillium crustosum Thom

A. Colonies on MEA after one week; B. conidia, x 1100; C, D. conidiophores, x 1600.

Occurrence: Worldwide distribution, common in soil and foodstuffs.

Frequently detected in indoor environments, isolated from wood and indoor (factory) air. Good growth at low temperatures; isolated as spoilage organism of various foodstuffs such as nuts, cheese, causal agent of rot in apples, pear, plum, cherries and other pomaceous and stone fruits.

Medical significance: HI. No growth at 37°C.

Toxicity: Produces the mycotoxin penitrem A.

Note: For a detailed description see Samson and Frisvad (2004) and Samson *et al.* (2010).

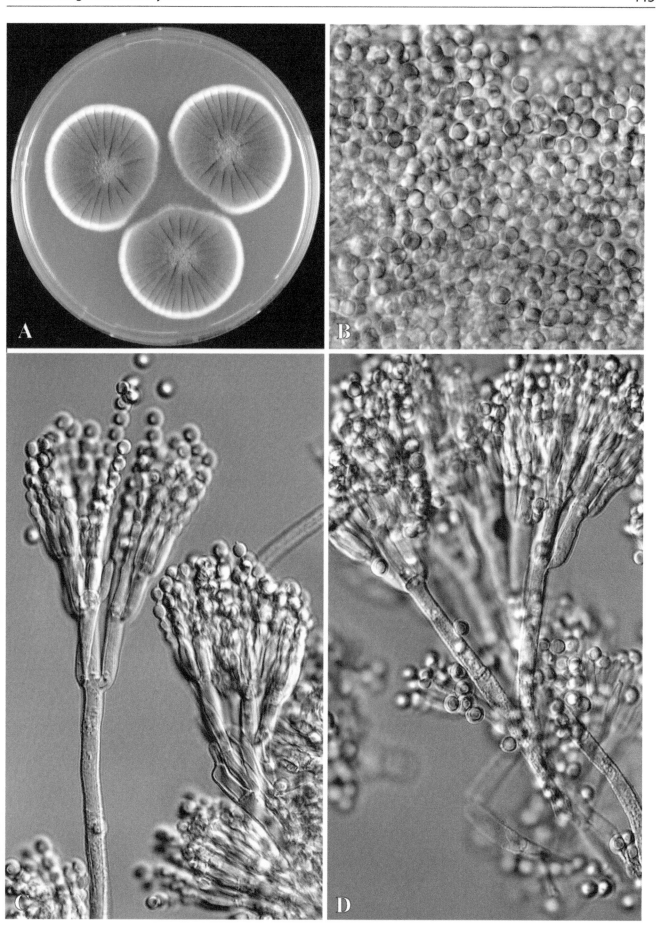

Penicillium digitatum Saccardo

A. Colonies on MEA after one week; B. conidia, x 1100; C, D. conidiophores and conidia, x 2000.

Occurrence: Worldwide distribution, causes destructive rot in citrus fruits.

Occasionally detected in indoor air. The presence of this species indoors often indicates the presence of mouldy citrus fruits. *Penicillium digitatum* is occasionally isolated from (sub)tropical soil or other substrates.

Medical significance: HI. No growth at 37°C.

Note: For a detailed description see Samson and Frisvad (2004) and Samson *et al.* (2010).

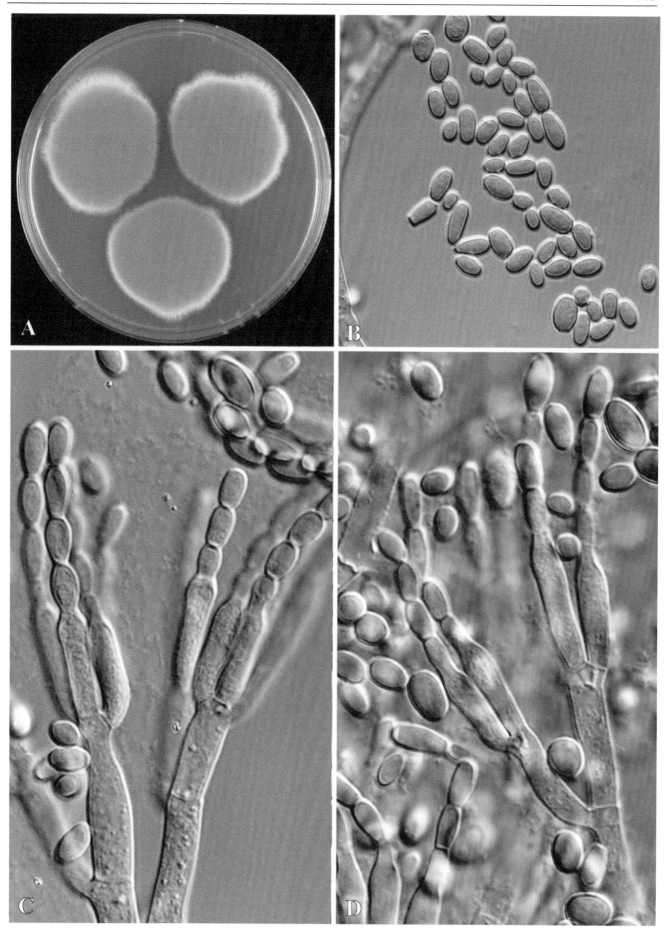

Penicillium expansum Link

A. Colonies on MEA after one week; B. conidia, x 1400; C. conidiophores, x 1000; D. detail of conidiophore, x 2100.

Occurrence: Particularly associated with pomeaceous fruits, such as apple.

House dust; polyurethane foam, red lead paint; HVAC filter; cereals, flour, cake, beer.

Medical significance: HI.

Note: For a detailed description see Samson and Frisvad (2004) and Samson *et al.* (2010).

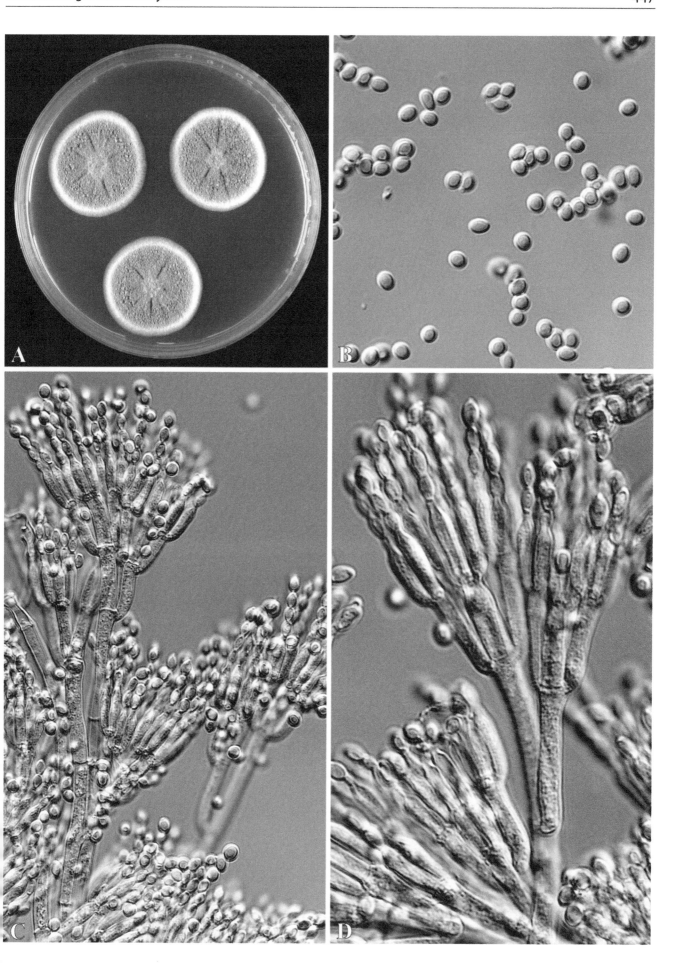

Penicillium funiculosum Thom

A. Colonies on MEA after one week; B. conidia, x 1000; C. conidiophores on a funicle, x 500; D. detail of conidiophores with lanceolate phialides, x 1550.

Occurrence: (Sub)tropics, common in soil and food. Rarely found indoor environments; isolated from tropical fruits, cereals and nuts.

Medical significance: [E?]IG. Good growth at 37°C.

Note: The taxonomy of this species is not elucidated and might be a species complex. For a detailed description, see Samson *et al.* (2010).

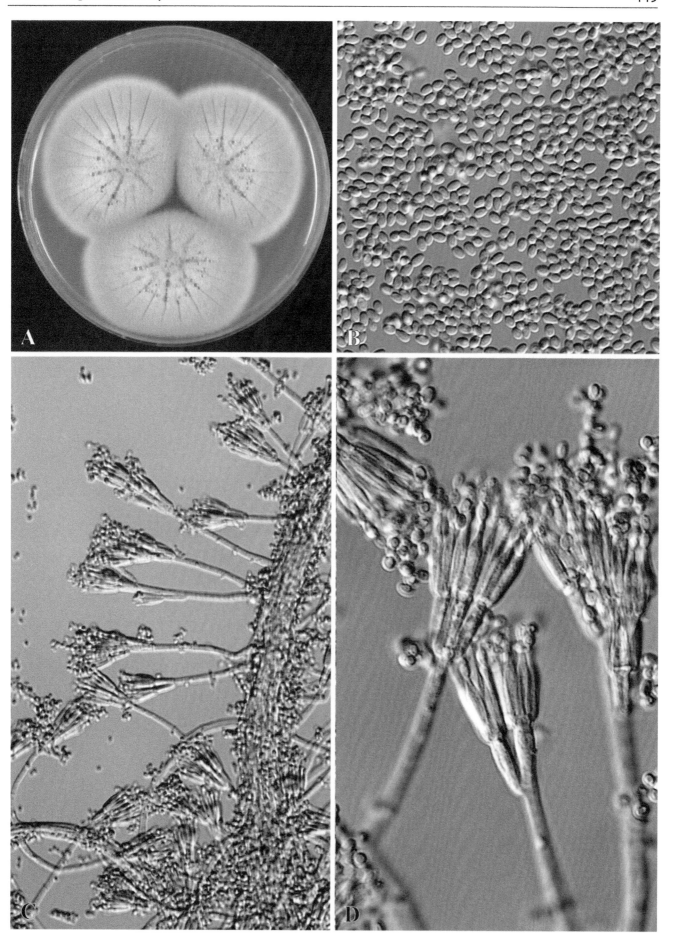

Penicillium glabrum (Wehmer) Westling

= *Penicillium frequentans* Westling

A. Colonies on MEA after one week; B. conidia, x 1100; C. conidiophores, x 1400; D. detail of conidiophores, x 2250.

Occurrence: An important deteriogen widely distributed in soils.

Very common species in floor dust, and also in mattress dust. Present on damp walls, and also isolated from starch paste adhesive, UFFI, wood pulp, cork factories.

Medical significance: HI. Associated with occupational respiratory disease in the cork and wood industries (Lacey and Dutkiewicz 1994).

Note: For a detailed description see Samson *et al.* (2010).

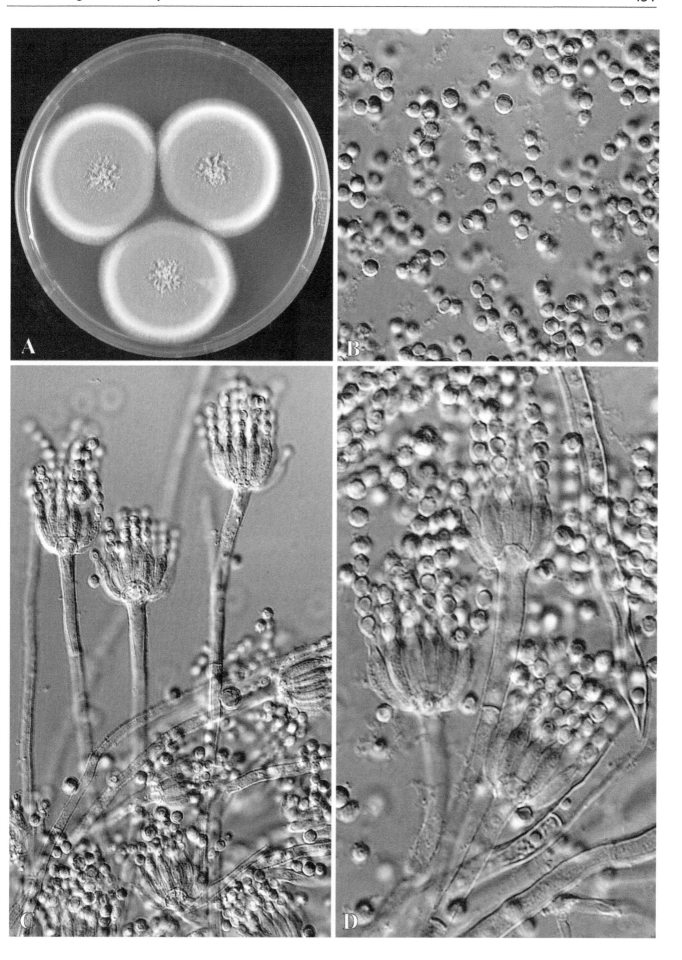

Penicillium olsonii Bainier & Sartory

A. Colonies on MEA after one week; B. conidia, x 1000; C. conidiophore, x 1000; D. conidiophore, x 1250.

Occurrence: A relatively uncommon species, in soil and water.

Found on damp building materials and in carpet dust. Also isolated from straw, bread, jam.

Medical significance: HI.

Note: For a detailed description see Samson and Frisvad (2004) and Samson *et al.* (2010).

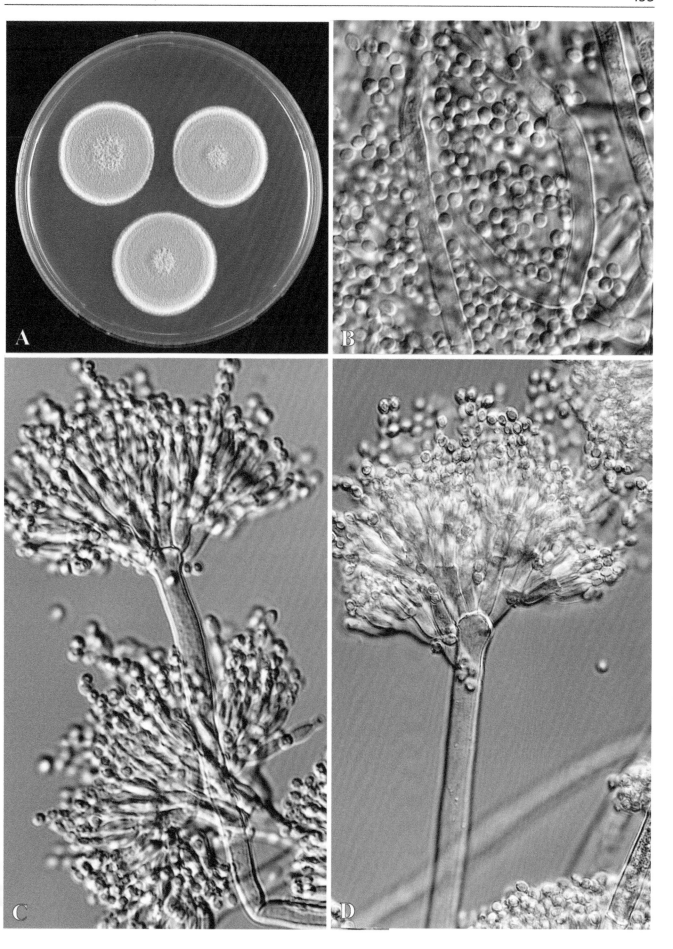

Penicillium palitans Westling

A. Colonies on MEA after one week; B. conidia, x 1375; C. conidiophores, x 1250; D. conidiophore, x 2250.

Occurrence: Common in arctic and temperate regions.

Common on wooden surfaces and in indoor air. It also causes spoilage of cheese and has been isolated from various other foodstuffs such as liver paté, bread and nuts.

Medical significance: HI. No growth at 37°C.

Note: For a detailed description see Samson and Frisvad (2004) and Samson *et al.* (2010).

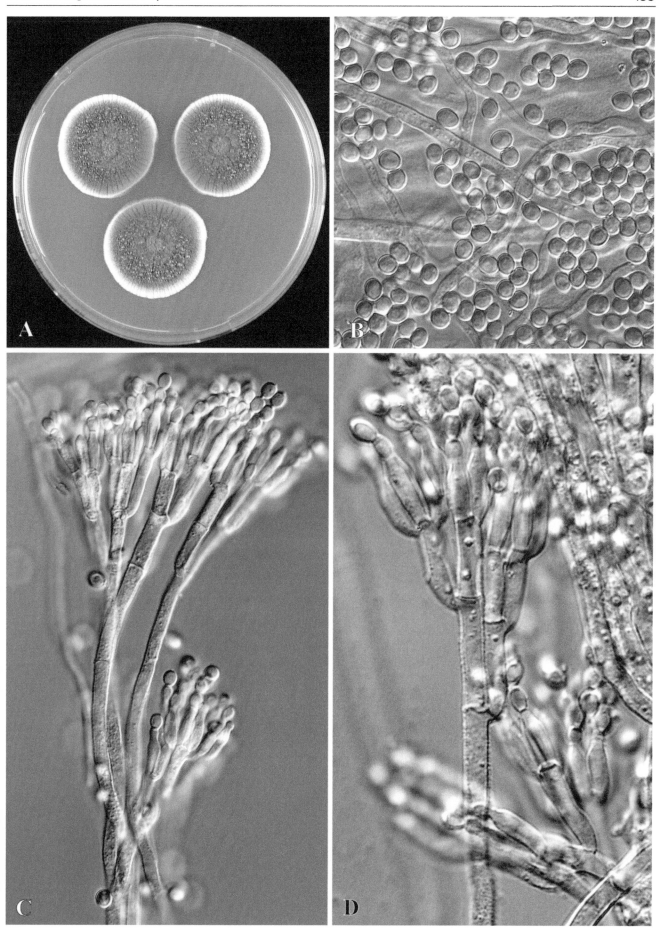

Penicillium polonicum Zaleski

A. Colonies on MEA after one week; B. conidia, x 1250; C. conidiophores and conidia, x 1200; D. conidiophore, x 1750.

Occurrence: Worldwide distribution, common in soil and food.

Occasionally detected in indoor environments, isolated from wood and (factory) air; often found on foodstuffs such as wheat, barley, rye, oats, rice, corn, peanuts, dried meat and onions.

Medical significance: HI. No growth at 37°C.

Toxicity: This species produces penicillic acid, verrucosidin and nephrotoxic glycopeptides. It may play a role in Balkan endemic nephropathy.

Note: For a detailed description see Samson and Frisvad (2004) and Samson *et al.* (2010).

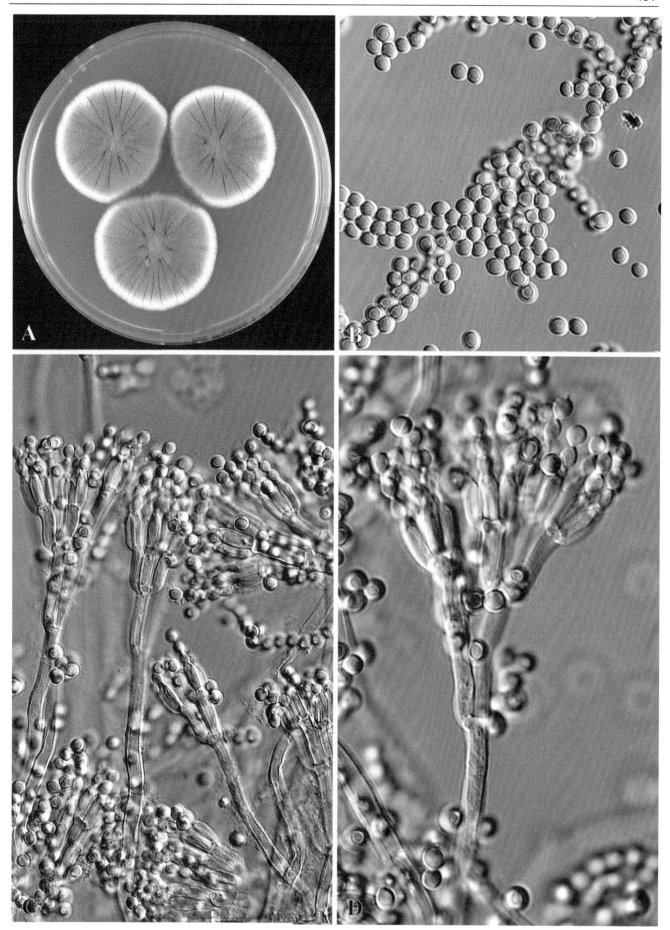

Penicillium rugulosum Thom

A. Colonies on MEA after one week; B. conidia, x 1500;
C. conidiophores, x 1250; D. conidiophore, x 1600.

Occurrence: A slow growing soil-born species thought to be commoner than records of isolations would indicate.

Occurs on damp building materials and walls, and in carpet and mattress dust. Also isolated from outdoor acrylic paint.

Medical significance: HI.

Note: This species is closely related to *P. variabile*. For a detailed description see Samson *et al.* (2010).

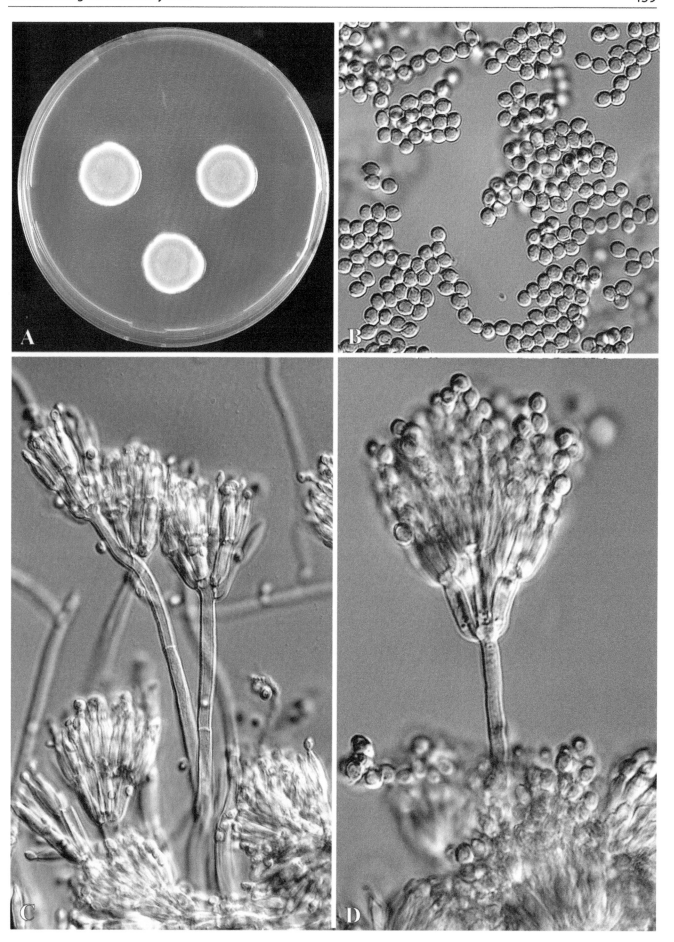

Penicillium variabile Sopp

A. Colonies on MEA after one week; B. conidia, x 2000;
C. conidiophores, x 1150; D. conidiophore, x 1800.

Occurrence: A ubiquitous species frequently isolated from soils and wood.

Found on damp building materials and in house dust. Also isolated from coconut matting; paper pulp.

Medical significance: HI.

Note: This species is closely related to *P. rugulosum*. For a detailed description see Samson *et al.* (2010).

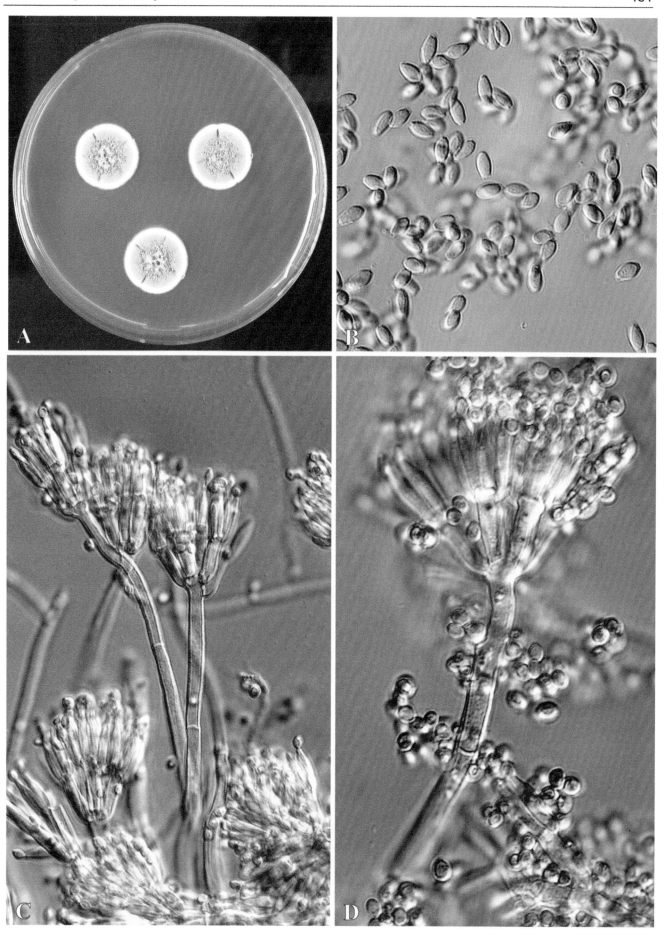

Phialophora verrucosa Medlar

A. Colony on OA after one week; B. conidiophores, x 920; C, D. phialides with collarettes, x 2330; E. conidia, x 2330.

Occurrence: In cooling water of air-conditioning systems, bathrooms, sometimes in air.

Associated with very wet wood (Ellis 1971), e.g. decaying lumber and pulpwood chips.

Medical significance: BEG.

Notes: *Phialophora* is poorly defined and little differentiated. Phylogenetic studies have shown that *Phialophora* is a highly polyphyletic genus, belonging to various genera such as *Cadophora*, *Lecythophora*, *Phaeoacremonium* and *Pleurostomophora*. For a more detailed description on *P. verrucosa*, see de Hoog *et al.* (2000) and Untereiner *et al.* (2008).

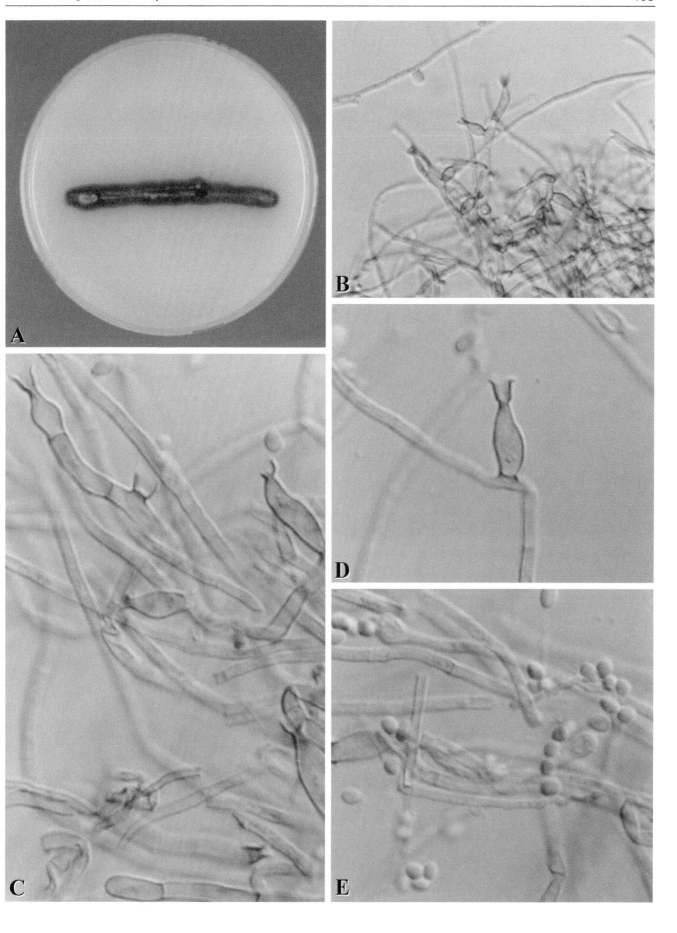

Phoma glomerata (Corda) Wollenweber & Hochapfel

A. Colony on MEA after one week; B. pycnidia with ostioles, in Petri-dish, x 230; C. pycnidia with ostioles, x 920; D. conidia, x 920; E. dictyochlamydospores, x 920.

Occurrence: Cosmopolitan in soil and a broad range of plants and plant materials.

Wood, cement, oil paint; wool.

Medical significance: HI.

Note: For a detailed description, see Aveskamp *et al.* (2009).

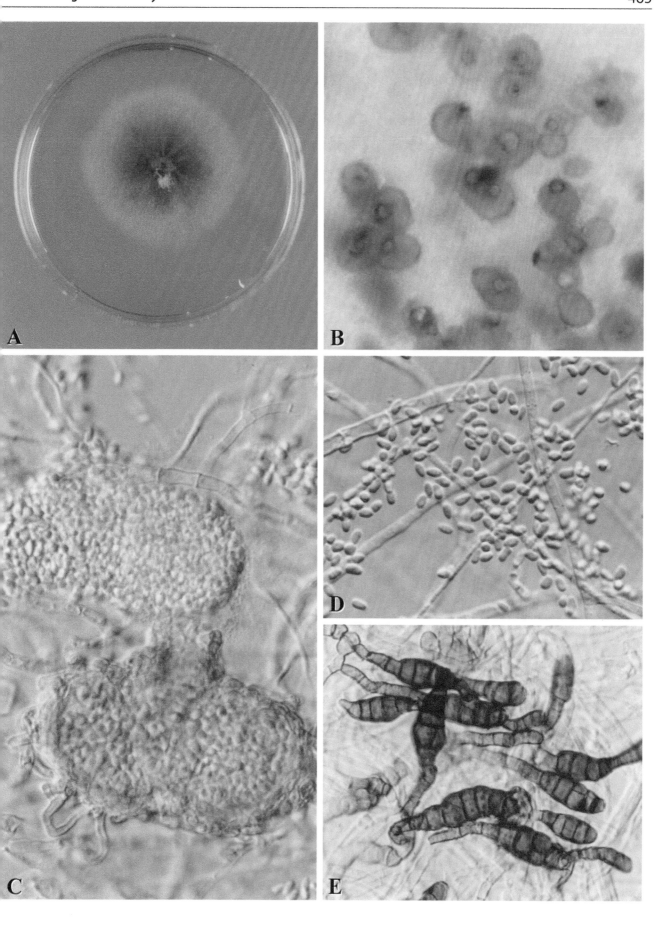

Phoma macrostoma Montagne

A. Colony on MEA after one week; B. pycnidia with ostioles, in Petri-dish, x 230; C. -E. pycnidia, C. x 230, D. x 920, E. x 920; F. conidia, x 920.

Occurrence: Wet environments, bathrooms, kitchen.

Known as a plant pathogen.

Medical significance: HI.

Note: For a detailed description, see Aveskamp *et al.* (2009).

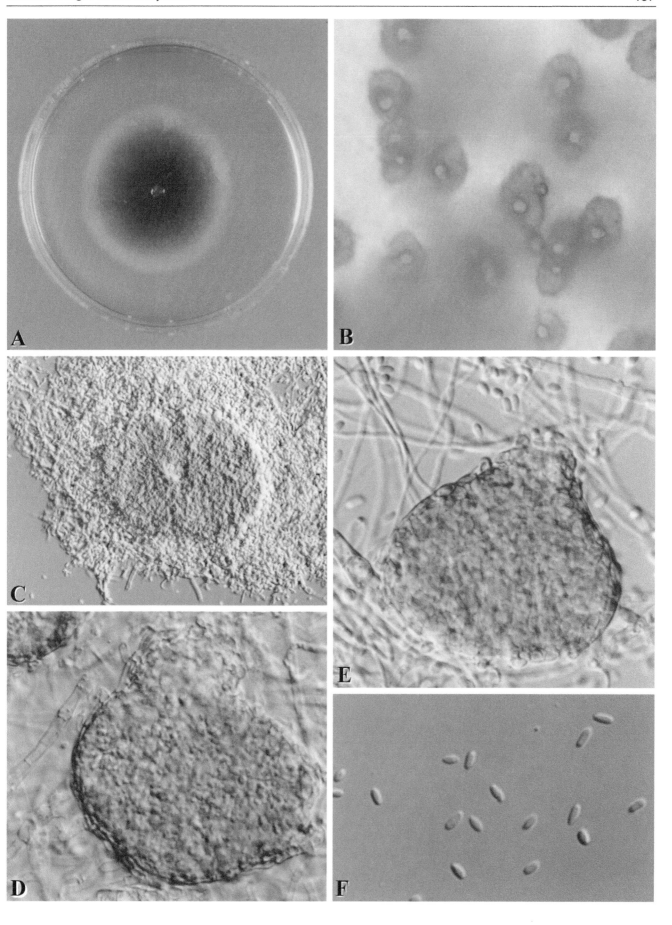

Pithomyces chartarum (Berkeley & Curtis) M.B. Ellis

A. Colony on OA after one week; B-D. conidiophores with conidia, x 920.

Occurrence: A soil-borne toxigenic species associated with forage grasses, particularly in New Zealand.

Found occasionally on paper in indoor environments (Ellis 1971), and also reported on ceiling tiles. May be present in carpet and mattress dust. Also isolated from UFFI; factory air.

Medical significance: HI. This species is involved in facial eczema (pithomycotoxicosis) in cattle.

Toxicity: Produces potent toxins (Leigh and Taylor 1976).

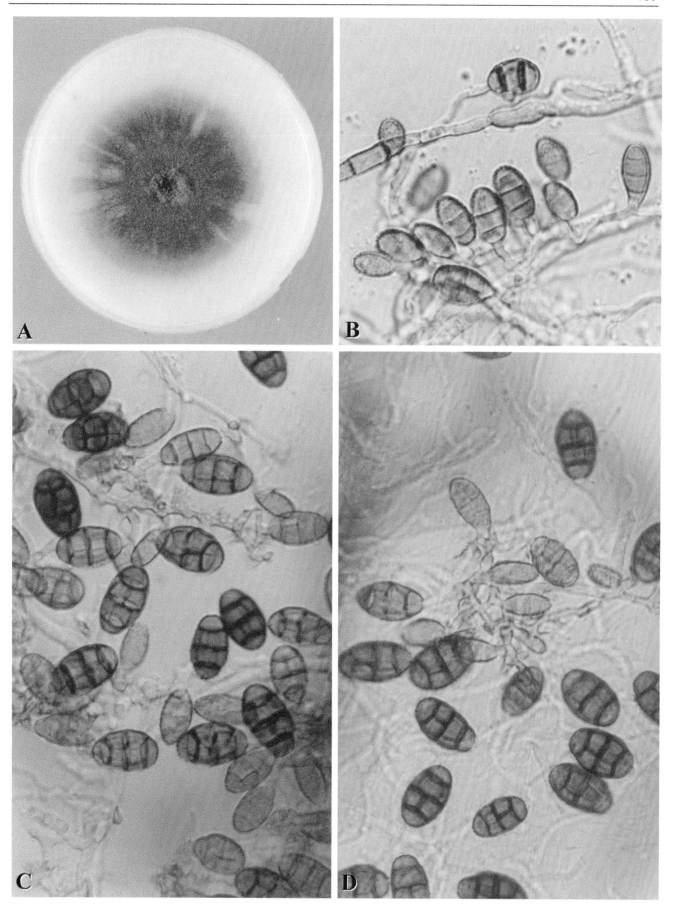

Pyronema domesticum (Sowerby:Fries) Saccardo

A. Colony on MEA after two weeks; B. detail of colony showing apothecia; C. apothecium, x 92; D. detail of apothecium, x 230 ; E. asci with ascospores, x 920.

Occurrence: Wet plasterwork.

Medical significance: HI.

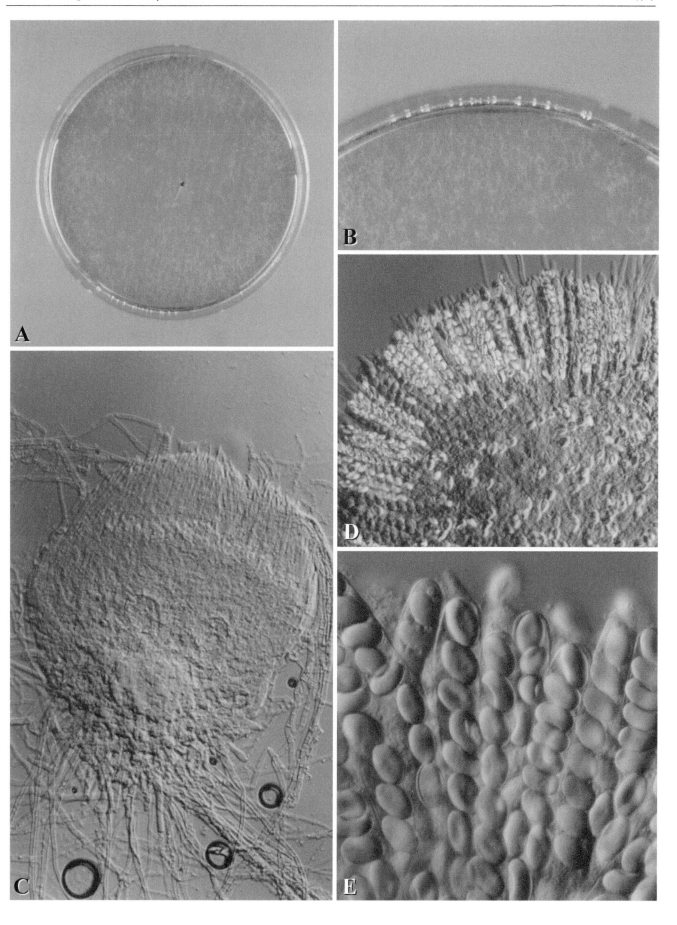

Rhizopus stolonifer (Ehrenb.) Lind.

A. Colony on MEA after one week; B. rhizoid, x 230; C. sporangiophore with sporangium, x 230; D. sporangiospores, x 920; E. sporangiophore with columella, x 230.

Occurrence: A cosmopolitan airborne and soil-borne contaminant of food, exposed surfaces and materials.

Found in carpet and mattress dust. Also isolated from polyurethane-coated fabric; water filter; HVAC filter; bakeries.

Medical significance: HI.

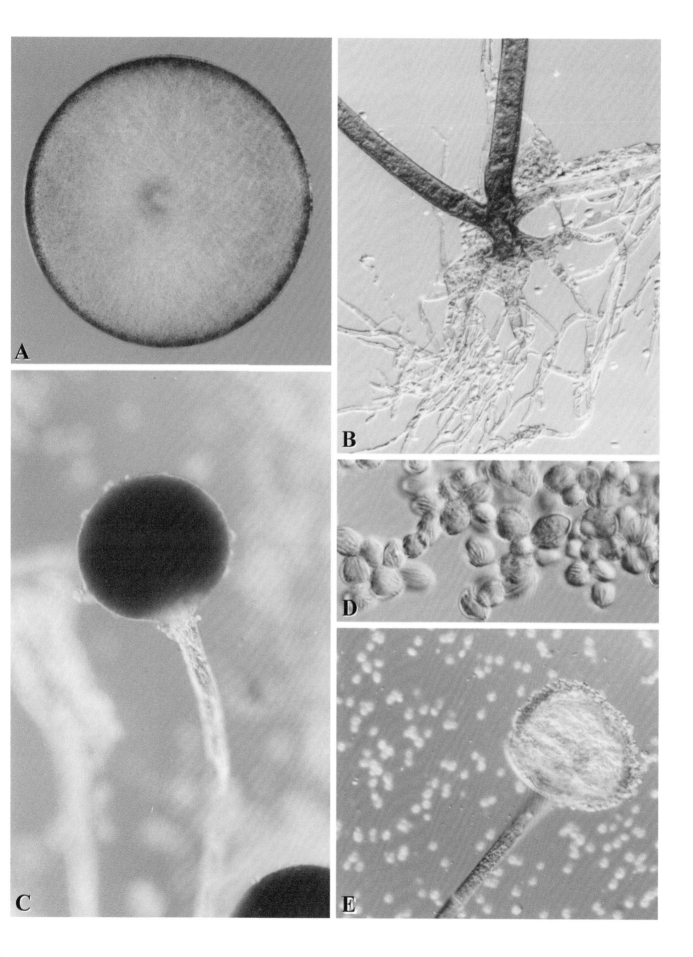

Rhodotorula mucilaginosa (Jörgensen) Harrison

A. Colony on MEA after five days; B. spores, x 920; C, D. sprouting spores, x 2330.

Occurrence: Food, air.

Medical significance: EIG.

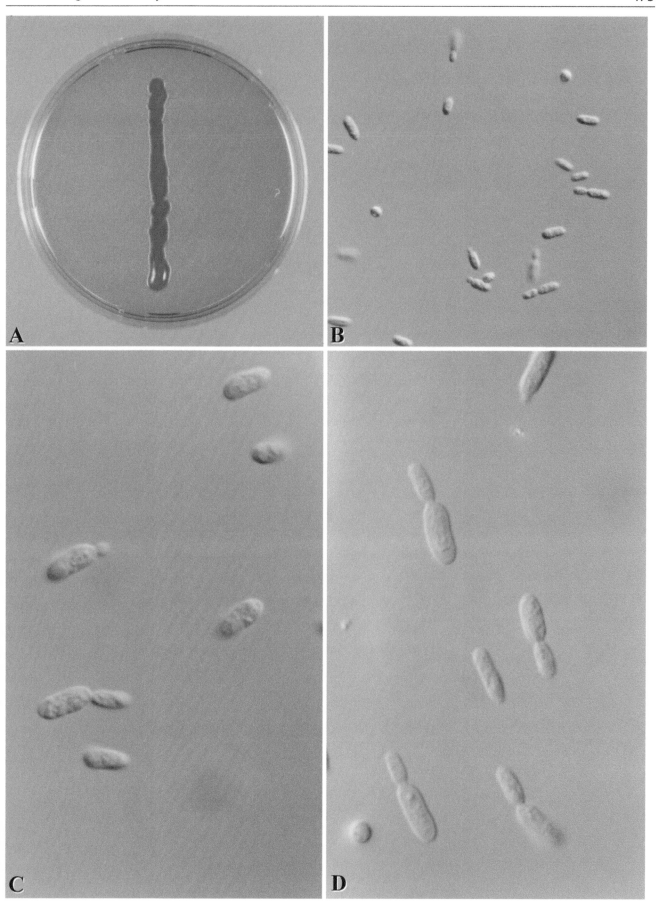

Schizophyllum commune Fries

A. Colony on MEA after three weeks; B. detail of colony showing fruit body; C. section through fruit body showing divided inrolled gills, x 230; D. basidiospores, x 920; E. detail of fruit body, x 920; F. hyphae with clamp , x 2330.

Occurrence: Isolated from (mahogany) wood.

Medical significance: EIG.

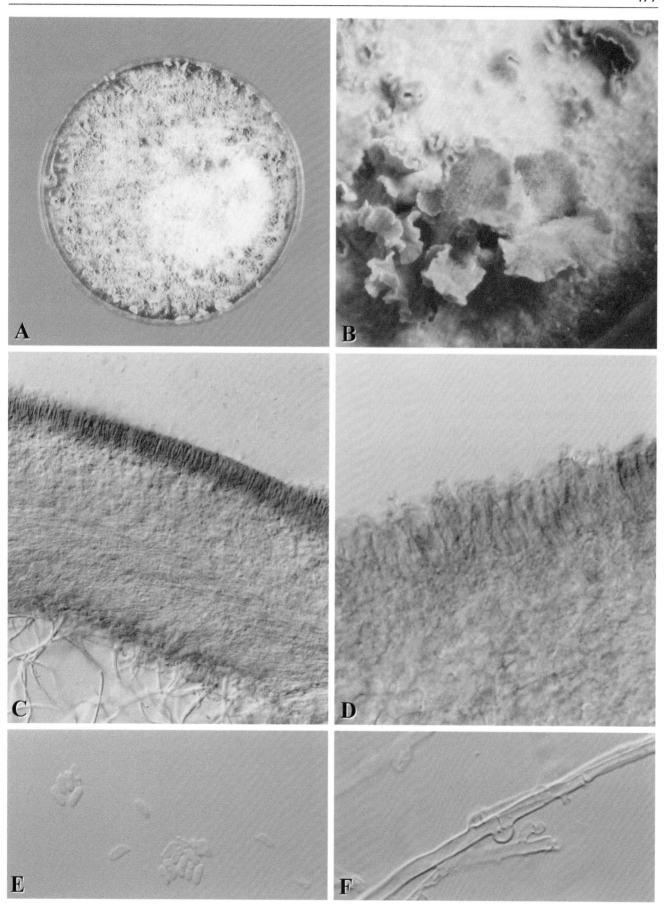

Scopulariopsis brevicaulis (Sacc.) Bain.

Teleomorph: *Microascus brevicaulis* Abbott, Sigler & Currah

A. Colony on MEA after two weeks; B-E. conidiophores with conidia, x 920; F. conidia, with a truncate base, x 2330.

Occurrence: A cosmopolitan species found in soil, and decaying wood, and various other plant and animal products.

Found on damp walls, cellulose board and wallpaper, as well as wood, and common in floor and mattress dust. Also isolated from metalworking fluid; straw, cereals, apples, groundnuts, meat, cheese, butter.

Medical significance: EIG. Skin lesions.

Note: *Scopulariopsis brevicaulis* resembles *S. flava*, though the latter has white colored conidia. The teleomorph of this species was observed after 6-25 weeks of incubation on Oatmeal Salts agar (Abbott *et al.* 1998). For a more detailed description, see Morton and Smith (1963) and Samson *et al.* (2010).

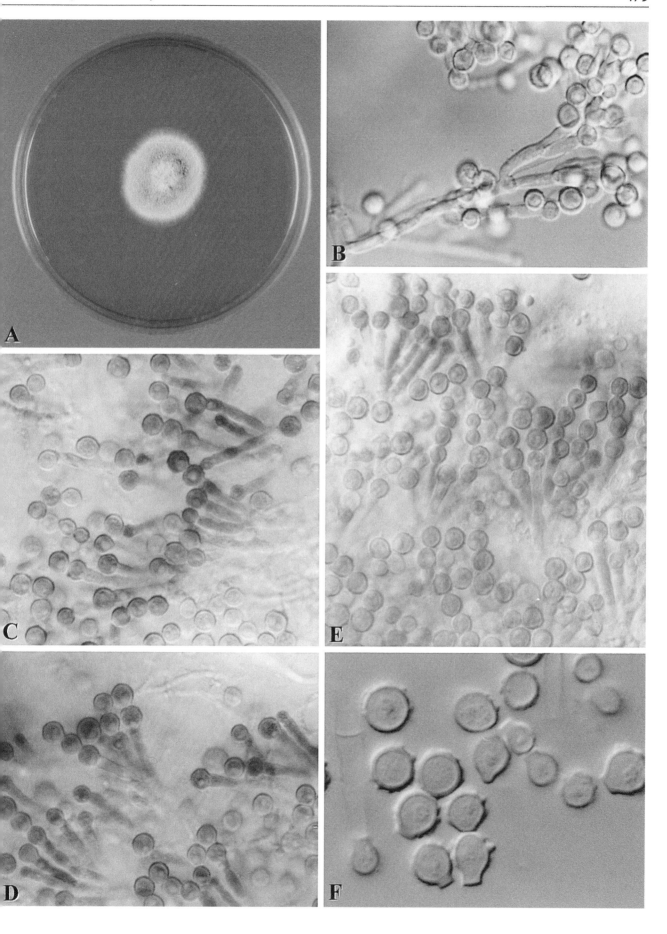

Scopulariopsis candida (Guéguen) Vuill.

A. Colony on MEA after two weeks; B-D. conidiophores with conidia, x 920; E. conidia, x 2330.

Occurrence: A soil-borne fungus also found on plant matter.

Carpet and mattress dust; hospital floor, swimming pool; wooden food packing, shoes; wood pulp; hay, cereals, flour, cheese.

Medical significance: EIG. Skin and nails.

Note: For a more detailed description, see Morton and Smith (1963) and Samson *et al.* (2010).

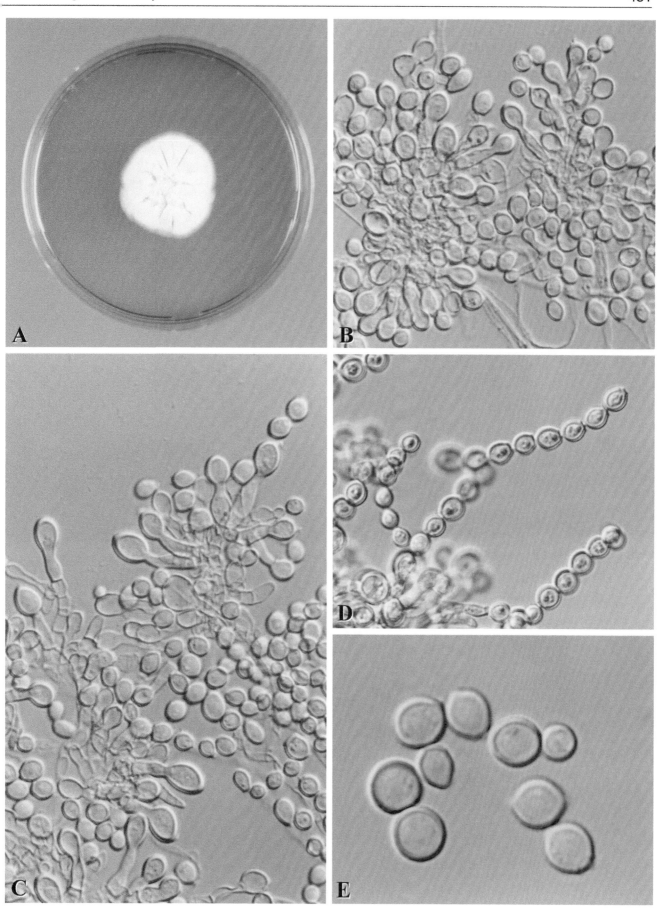

Scopulariopsis fusca Zach

A. Colony on MEA after two weeks; B, C. conidiophores with conidia, x 920; D. annellate conidiogenous cells with conidia, x 2330.

Occurrence: A soil-borne species also associated with decaying plant matter.

Found in mattress dust. Also isolated from straw packing, wooden food packing; cheese, eggs.

Medical significance: [E?]I [G?]. Onychomycosis, dermatomycosis.

Note: For a more detailed description, see Morton and Smith (1963) and Samson *et al.* (2010).

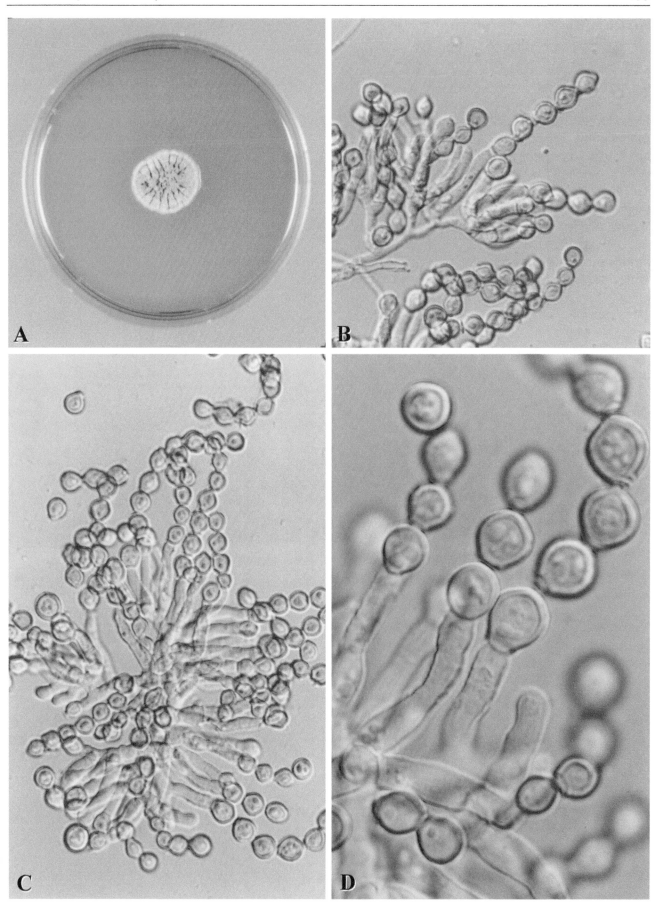

Serpula lacrymans (Wulfen:Fries) Schroeter

Dry rot fungus.

A. Piece of wood with *S. lacrymans*; B. fruiting body; C. detail of fruiting body; D. hyphae with clamp connections, x 2330; E. basidiospores, x 2330.

Occurrence: Most often on softwood in contact with or embedded in wet brickwork where ventilation is poor and natural drying is impeded.

Medical significance: HI. Allergenic – asthma, hypersensitivity pneumonitis.

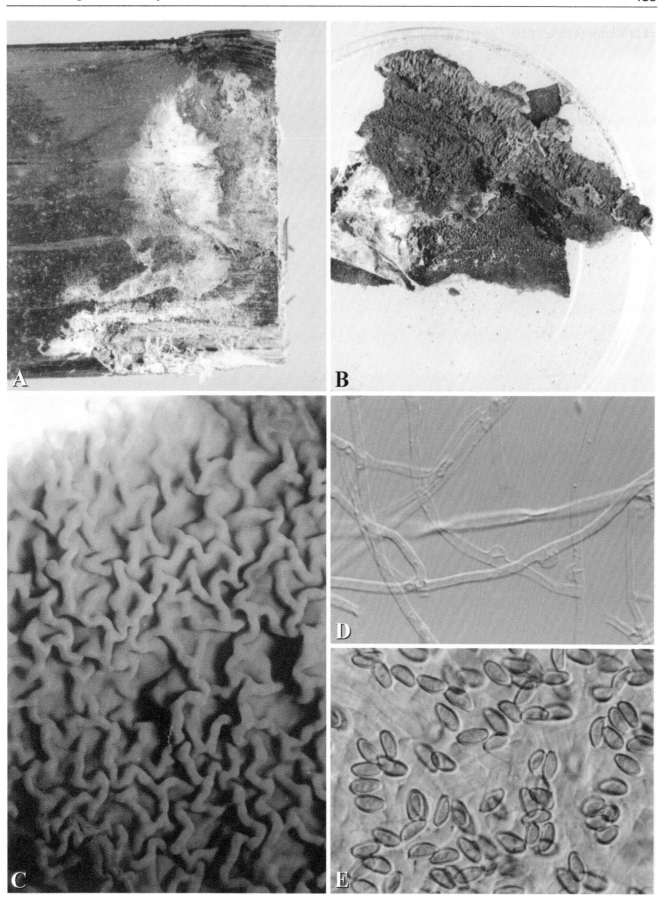

Sistotrema brinkmanii (Bresadola) Eriksson

A. Colony on MEA after three weeks; B. piece of wood with *S. brinkmanii*; C. detail of colony; D. hyphae with clamp, x 2330; E. detail of piece of wood with *S. brinkmanii*.

Occurrence: Commonly isolated from wet, decaying window/door joinery.

Medical significance: HI.

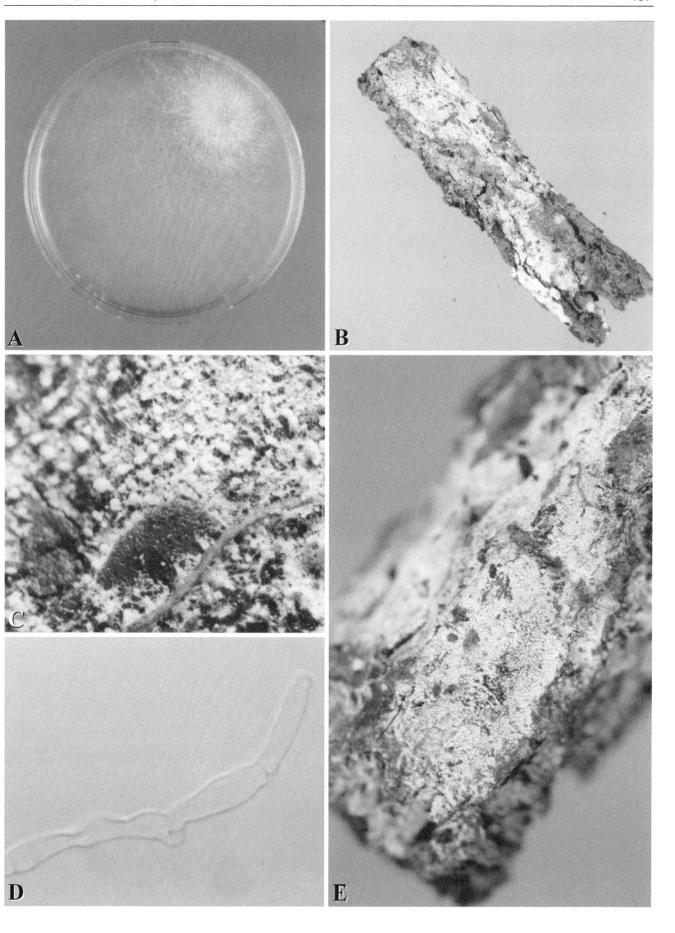

Sporobolomyces roseus Kluyver & Niel

A. Colony on MEA after one week; B. spores, x 920; C. spores, x 2330; D. sterigmata, x 2330; E. ballistoconidia, x 2330.

Occurrence: Air (indoor and outdoor).

Isolated from damp walls; *S. salmonicolor* from air-conditioning humidifier water.

Medical significance: EIG.

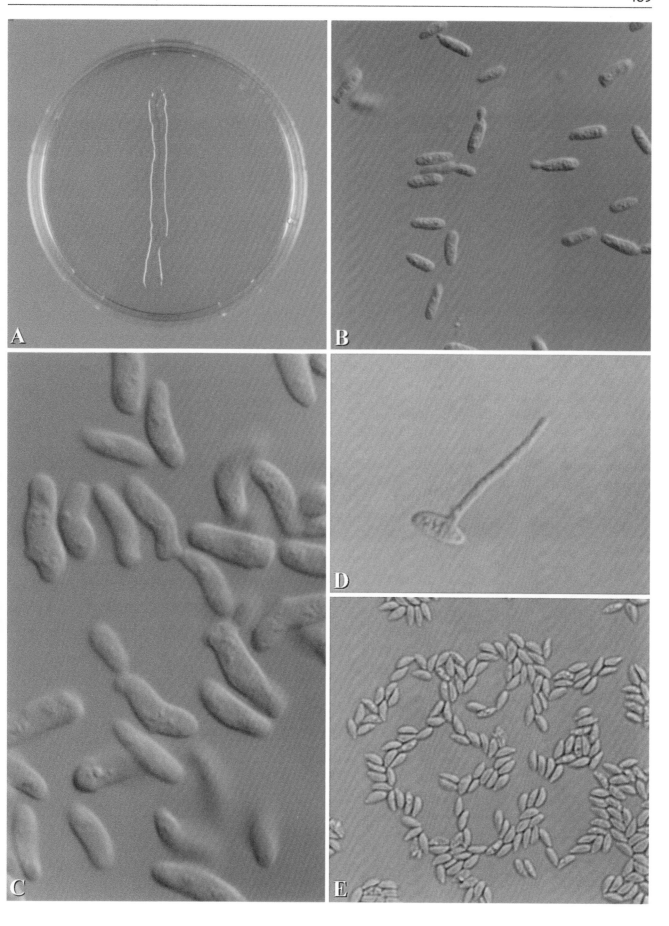

Stachybotrys chartarum (Ehrenb.) Hughes

= *Stachybotrys atra* Corda

A. Colony on MEA after one week; B. conidiophores with conidia in slimy heads, in Petri-dish, x 90; C-D. conidiophores, x 920; E. conidia, x 2330.

Occurrence: A cosmopolitan cellulolytic species isolated from soil and mouldering plant matter, e.g. damp straw.

Commonly on very wet gypsum board/walls and wallpaper, and on cotton fabrics/textiles such as canvas. Also isolated from asbestos building substitute, UFFI; HVAC humidifier water and fan; hospital/factory air.

Medical significance: HI.

Toxicity: *Stachybotrys chartarum* contains two chemotypes (S and A), that share the same morphology, but produce different metabolites. *S. chartarum* (S) produces satratoxins and roridins, while *S. chartarum* (A) produces atranones, dolabellanes and trichodermin (Andersen *et al.* 2002, 2003).

Note: In indoor environments *Stachybotrys chlorohalonata* also occurs; see Samson *et al.* (2010).

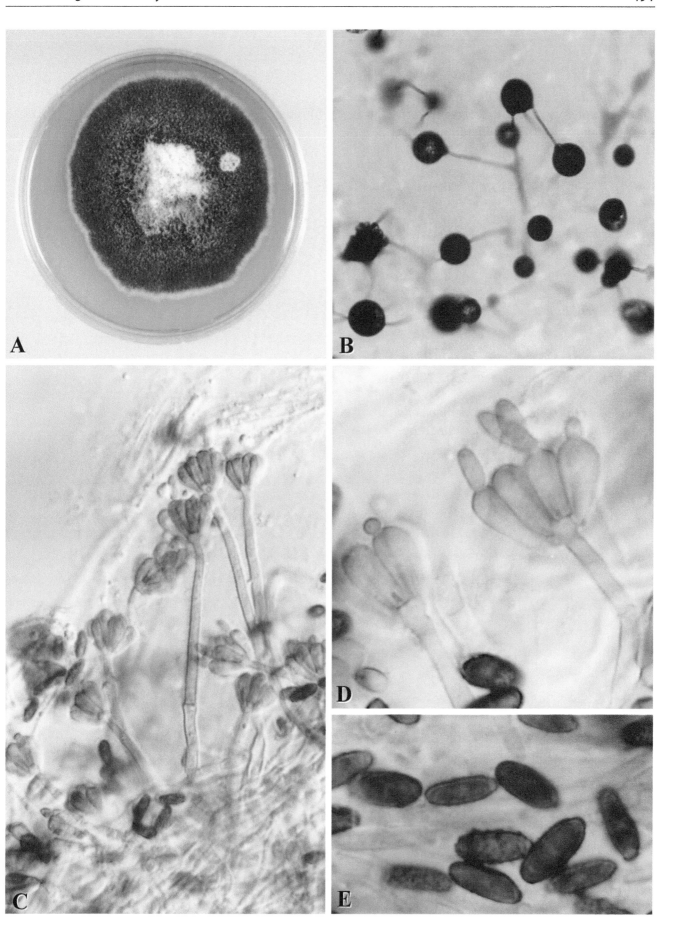

Stachybotrys (Memnoniella) echinata (Rivolta) Smith

A. Colony on MEA after one week; B. conidiophore with conidia in columns, x 230; C. conidiophores with conidia, x 920; D. conidia, x 920; E. conidiophores with conidia, x 920.

Occurrence: A cellulolytic species common on exposed canvas and other cotton fabrics in subtropical and tropical regions.

A common contaminant of very wet gypsum board, particularly in warmer climates.

Medical significance: HI.

Toxicity: Produces griseofulvins, spiriocyclic drimanes and trichodermol (Andersen *et al.* 2002, 2003).

Note: This species is often referred to as *Memnoniella echinata*, but phylogenetic analysis shows that it belongs in *Stachybotrys*.

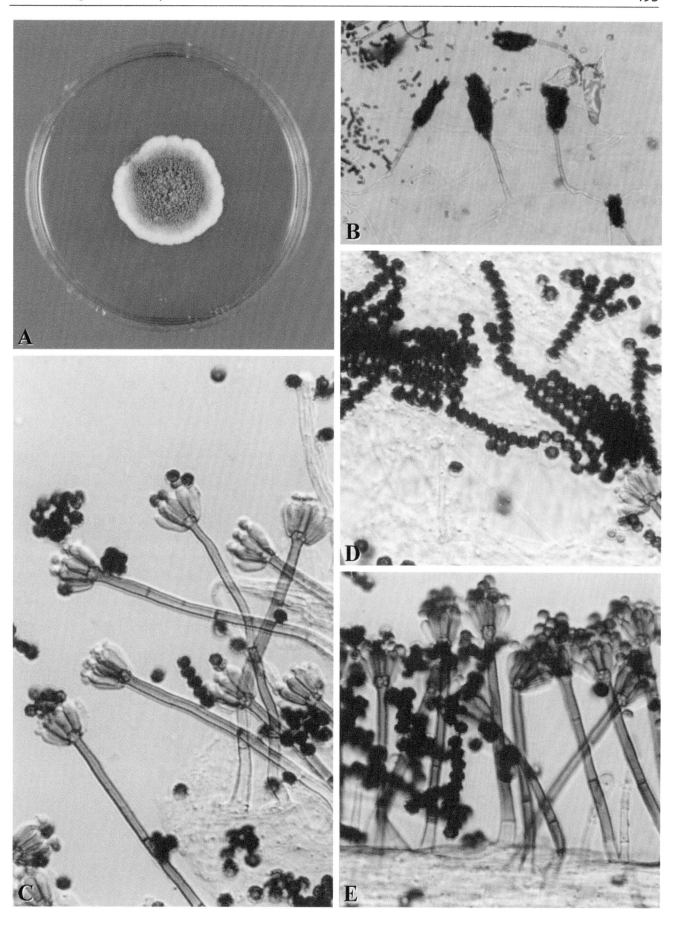

Syncephalastrum racemosum Cohn

Colony on MEA after one week; B. merospores, x 920; C, D. merosporangia with merospores, x 920.

Occurrence: A cosmopolitan species found in soil, dung, cereals and other food products.

Found in mattress dust. Also isolated from cloth, cured tobacco, toilet soap, bottle cork; air-conditioning filter.

Medical significance: EFG.

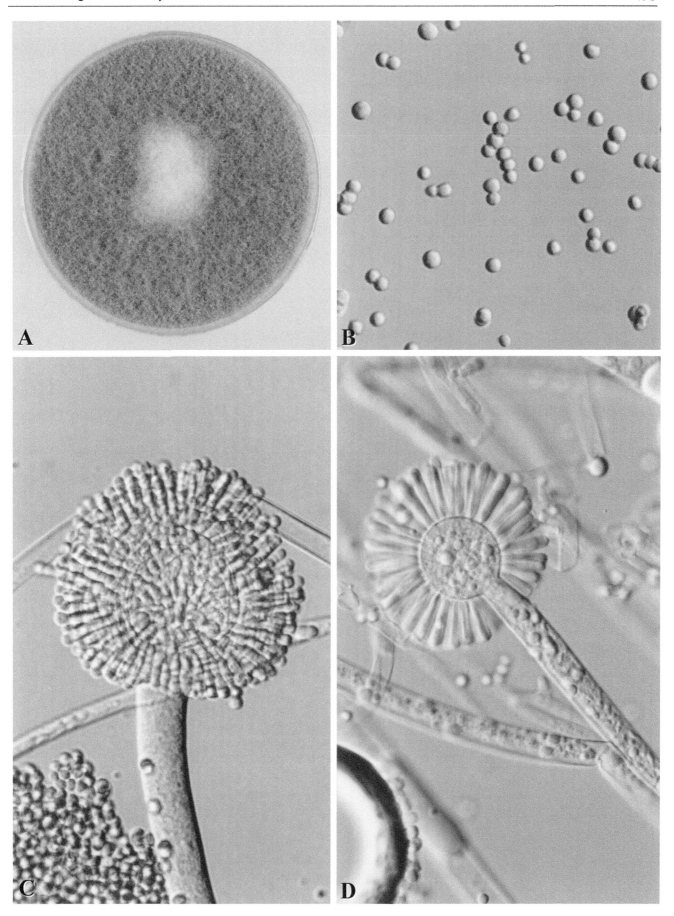

Trichoderma harzianum Rifai

A. Colony on MEA after one week; B. conidia, x 2330; C, D. conidiophores, x 920.

Occurrence: A cellulolytic soil species with a worldwide distribution, also found in cereals and textiles.

Commonly occurs on gypsum board and water-saturated wood (Lubeck *et al.* 2000). Found also in carpet and mattress dust, and isolated from UFFI, polycaprolactone polyurethane, polyester polyurethene foam, polyurethane footwear; air-conditioning filter; bakery, mushroom farms.

Medical significance: HI.

Toxicity: This species produces potent toxins (Taylor 1986).

Note: The genus *Trichoderma* contains 100+ species and identification is mainly based on DNA sequences. Dedicated databases have been developed for identification (TrichOKEY, TrichOBLAST).

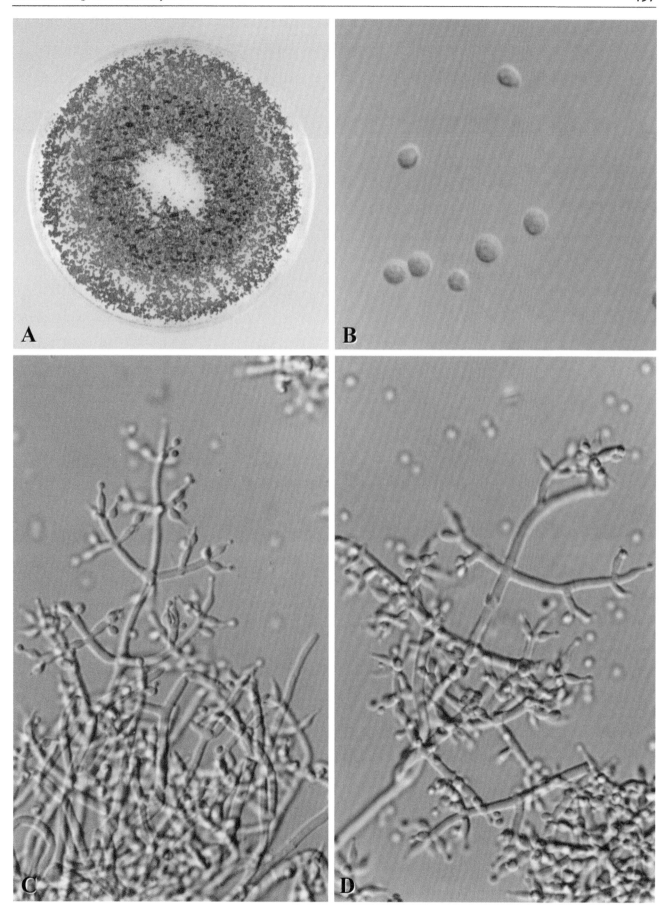

Trichoderma koningii Oudemans

A. Colony on MEA after one week; B, C. conidiophores, x 920; D. conidia, x 2330.

Occurrence: Found in carpet dust.

Also in mushroom spawn/compost, household waste.

Medical significance: HI.

Note: The genus *Trichoderma* contains 100+ species and identification is mainly based on DNA sequences. Dedicated databases have been developed for identification (TrichOKEY, TrichOBLAST).

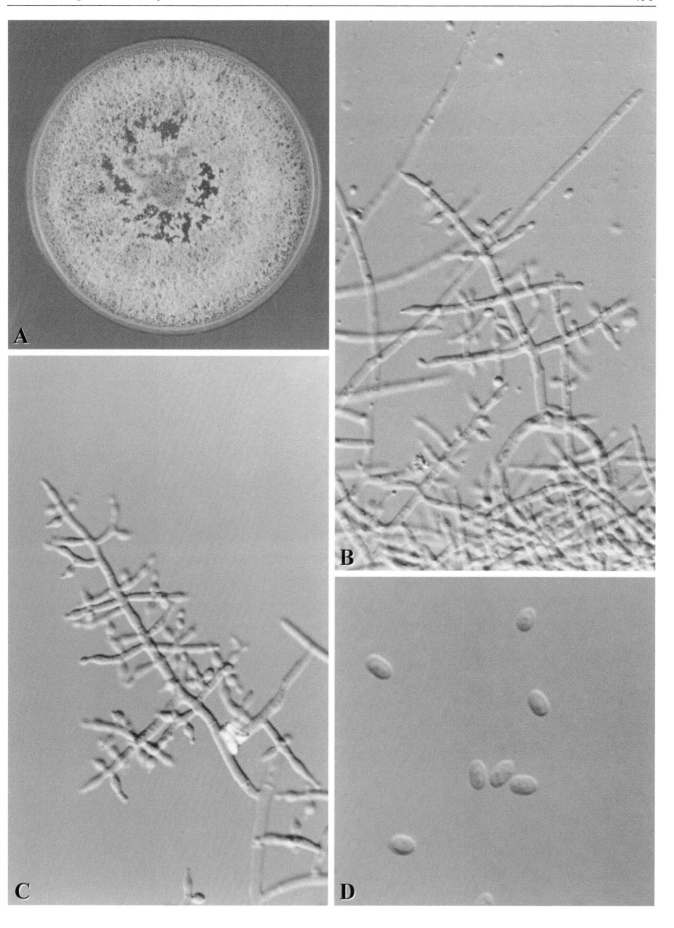

Trichoderma viride Pers.

A. Colony on MEA after one week; B, C. conidiophores, x 920; D. rough-walled conidia, x 2330.

Occurrence: Cosmopolitan soil fungus, this cellulo-lytic species is also found in decaying wood, stored cereals, sweet potatoes and other plant foodstuffs.

Commonly occurs on gypsum board and water-saturated wood (Lubeck *et al.* 2000). Found in floor, carpet and mattress dust. Also isolated from paint, UFFI; domestic water supply, air from HVAC system; wine cellar, grain store dust, mushroom compost, factory air; sawmills; preservative-treated timber.

Medical significance: HI.

Toxicity: Produces potent toxins (Taylor 1986).

Note: The difficult-to-distinguish species *T. citreoviri-de* and *T. longibrachiatum* are also common, but more associated with insulation and roofing material than gypsum wallboard (Lubeck *et al.* 2000). These latter species are associated with invasive mycoses in immune compromised individuals. The genus *Trichoderma* contains 100+ species and identification is mainly based on DNA sequences. Dedicated databases have been developed for identification (TrichOKEY, TrichO-BLAST).

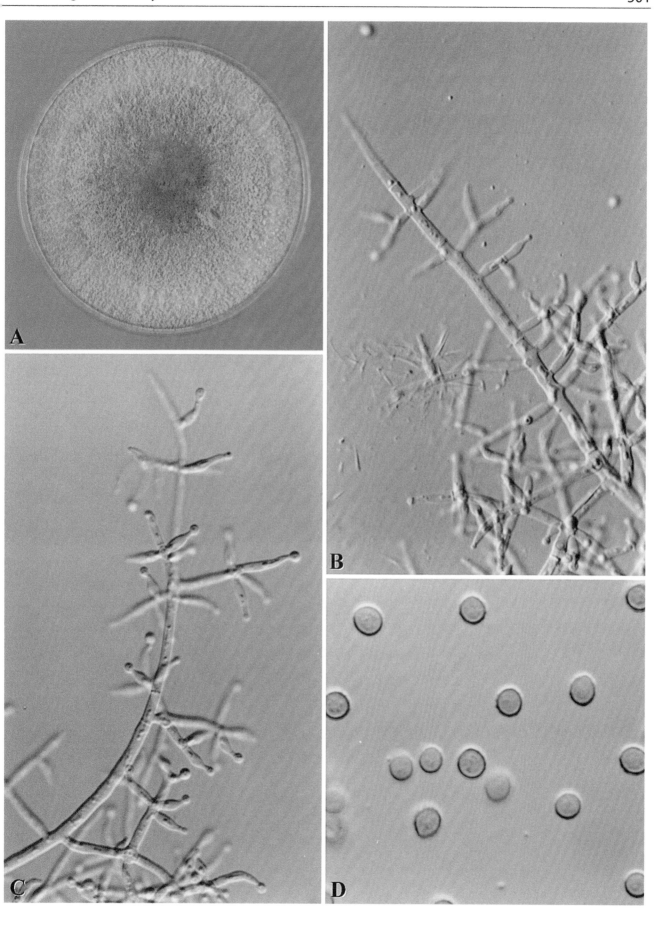

Tritirachium oryzae (Vincens) de Hoog

A. Colony on MEA after one week; B, C. conidiophores
with conidia, x 920; D. conidia, x 2330.

Occurrence: Air and isolated from nylon, soil.

Medical significance: HI.

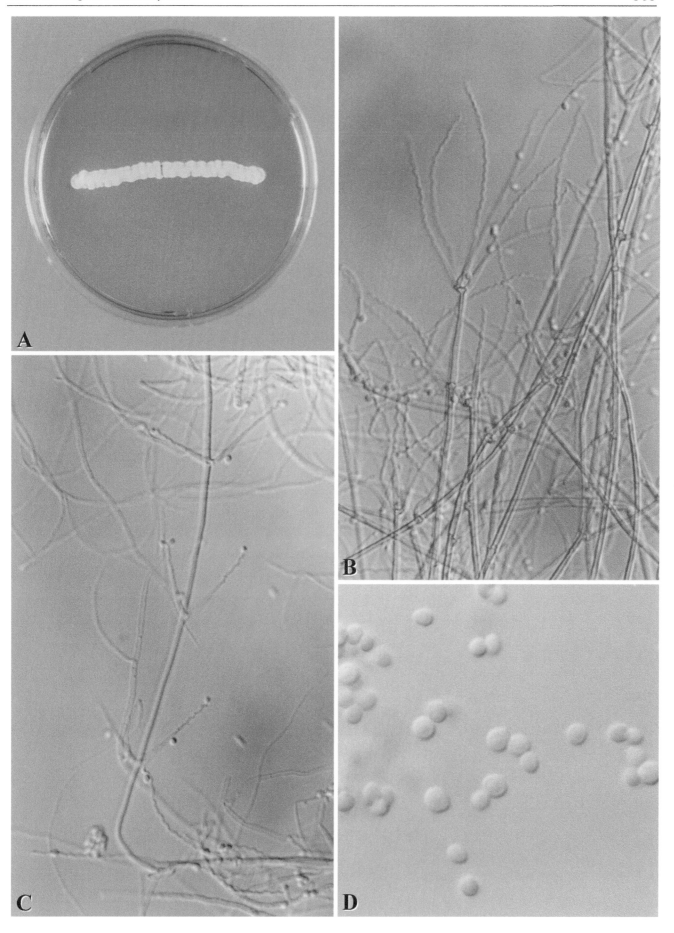

Ulocladium Preuss

A. Colony on MEA after one week; B-D. septate conidia, x 920.

Occurrence: A widespread species found in soil,
dung, plant litter and wood.

Occurs along with *Cladosporium sphaerospermum* on
painted surfaces; on damp walls finished with wallpaper or water-based emulsion paint. Found in floor
and mattress dust. Also isolated from UFFI, manila fibre, plywood.

Medical significance: [E?]F[G?].

Note: *Ulocladium* is morphologically similar to *Alternaria* and *Stemphylium*. *Alternaria* differs in having
ovoid, obpyriform or obclavate conidia that form
branched or unbranched chains. *Stemphylium* has
subglobose to ellipsoidal conidia and annellidic primary conidiophores with a darker, swollen tip. *Ulocladium alternariae* is often found on water-damaged
building materials.

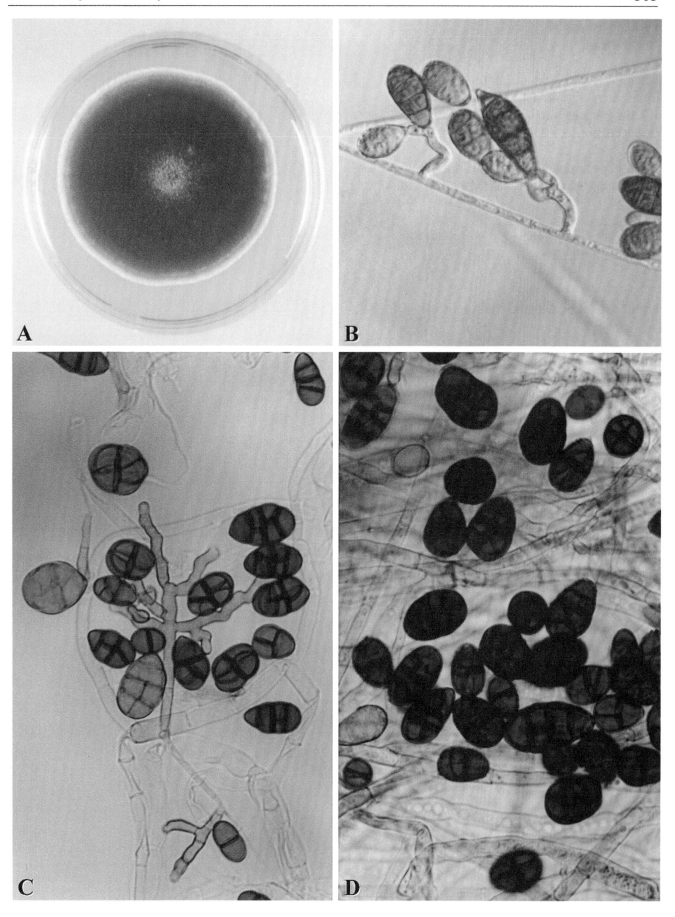

Wallemia sebi (Fries) von Arx

A. Colony on MEA + 20% sucrose after one week; B-D. conidiophores each with an apical fertile, cylindrical hyphae which separates into four conidia, maturing in basipetal succession, B. x 230, C. x 960, D. x 960.

Occurrence: A cosmopolitan xerophile, present in soil and air, and particularly associated with dried foodstuffs.

Found in floor, carpet, mattress and upholstered furniture dust; radiator; hay, cereals, cake, dates, salted fish.

Medical significance: HI. Allergenic (Sakomoto *et al.* 1989).

Note: This xerophilic species is common and can be detected by using appropriate low water actvity media such as DG18 or MEA + 20% sucrose. The genus *Wallemia* consists of three species and two of them (*W. sebi* and *W. muriae*) are commonly occurring in indoor environments.

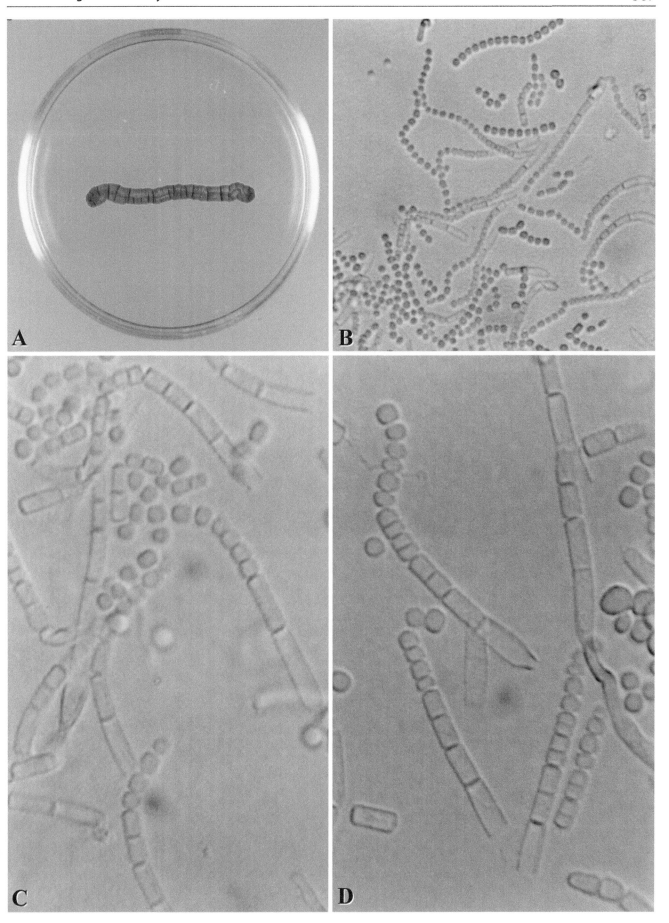

Saccharopolyspora rectivirgula (Krassilnikov and Agre) Korn-Wendisch *et al.*

= *Faenia rectivirgula, Micropolyspora faeni*

A. Colony on OA after one week at 55°C; B. hyphae, x 2330; C, D. spores, C. x 920, D. x 2330.

Occurrence: Cosmopolitan thermophilic species usually associated with composts and self-heating agricultural materials.

Isolated from air-conditioning systems with water spray cooling, furnace dust, humidifiers; hay, cereals; mushroom farms, cotton and sugar cane mills.

Medical significance: Allergens implicated in occupational hypersensitivity pneumonitis known as farmer's lung, but role in bagassosis and mushroom worker's lung in question (see Chapter 3.2).

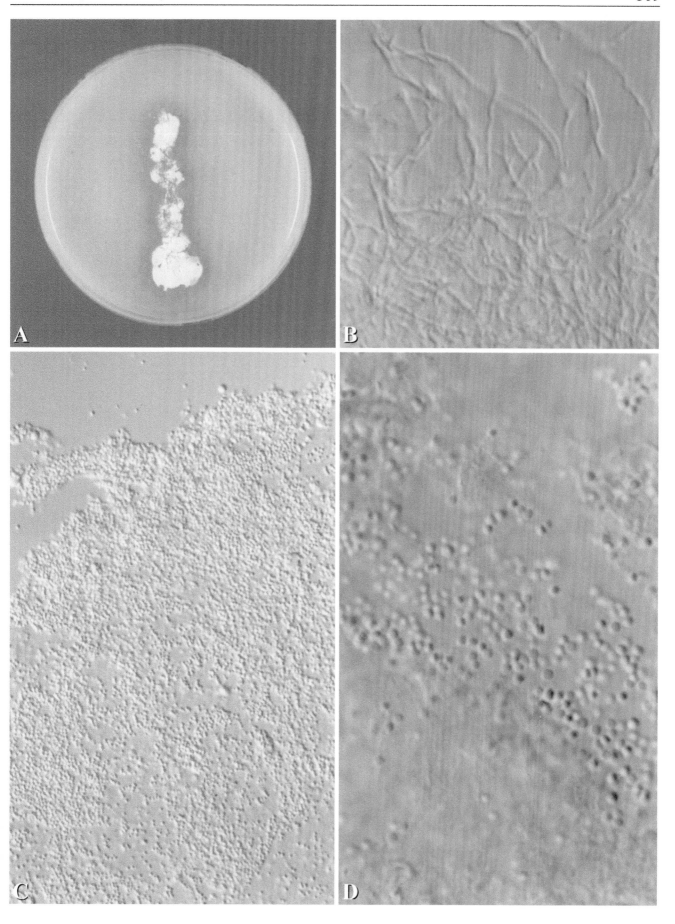

Streptomyces griseus (Krainsky) Waksman and Henrici

A. Colony on OA after one week at 24°C; B. hyphae, x 2330; C-E. spores, x 2330.

Occurrence: A cosmopolitan streptomycin-producing member of a genus associated with, and imparting the characteristic smell (geosmin) to, soil.

Isolated from cork sheeting and other constructional materials; hay, cereals.

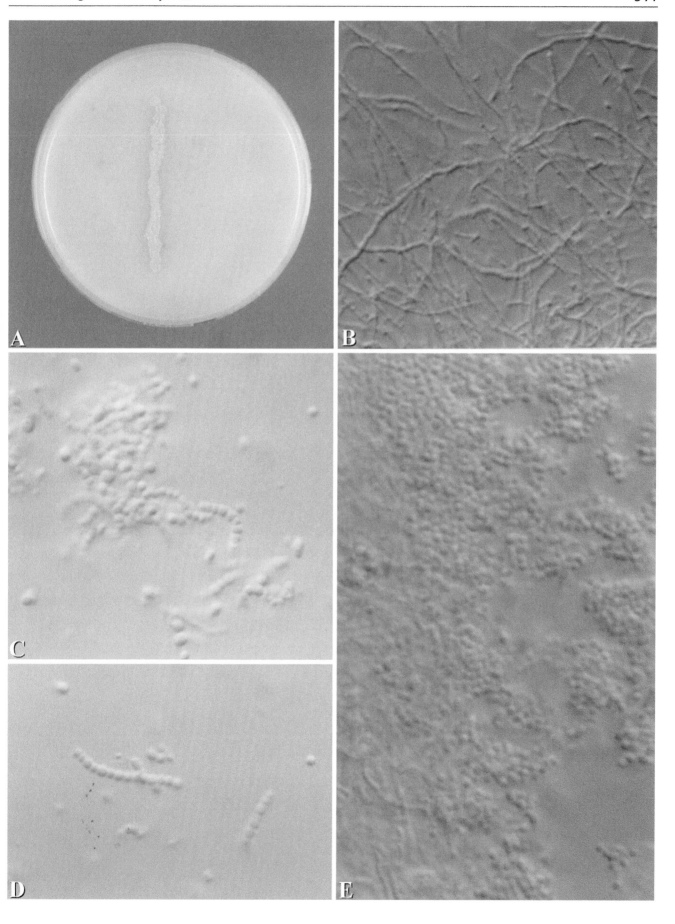

Thermoactinomyces vulgaris Tsilinsky

A. Colony on V8 after one week at 55°C; B. hyphae, x 2330; C, D. spores, x 2330.

Occurrence: Cosmopolitan thermophilic species usually associated with composts and self-heating agricultural materials.

Isolated from humidifiers; wood chips; hay, cereals, bagasse; mushroom farms, cotton and sugar cane mills.

Medical significance: Allergens implicated in occupational hypersensitivity pneumonitis known as farmer's lung, but role in bagassosis and mushroom worker's lung in question (see Chapter 3.2).

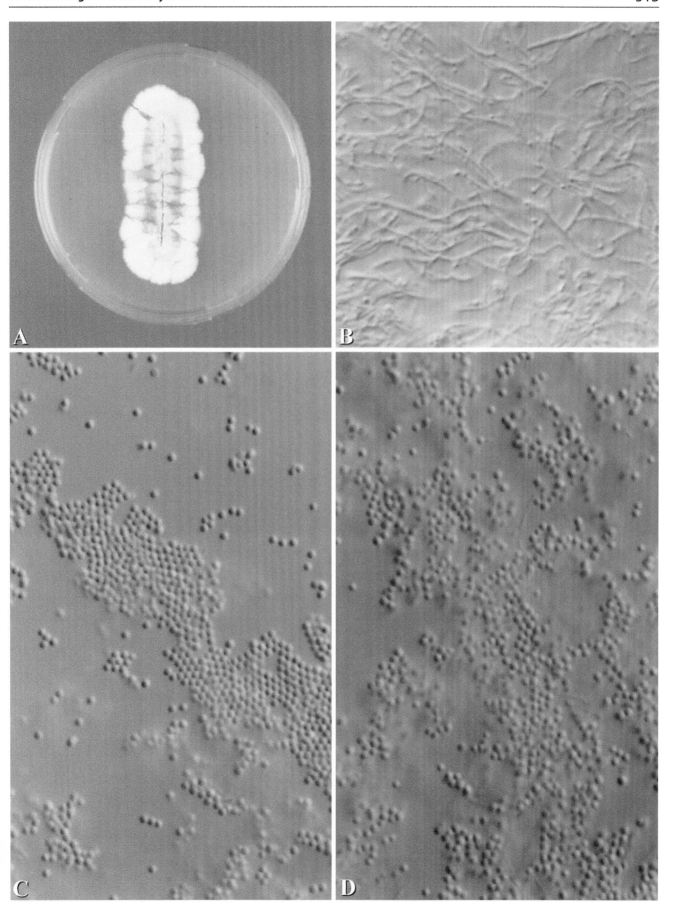

APPENDIX

GLOSSARY OF MYCOLOGICAL TERMS

The following glossary is slightly modified from Samson *et al.*, (2010). For a more detailed dictionary of mycological terms see P.M. Kirk, P.F. Cannon, J.C. David and J.A. Stalpers, (eds.), (2001) *Ainsworth & Bisby's Dictionary of the Fungi*, 9th Ed. CABI Publishing, Wallingford, UK.

---

acerose: needlelike and stiff (like a pine needle)

acervulus: a lens- or cup-shaped fruiting body containing conidiophores and conidia embedded in the host plant

acropetal: describes the succession of conidia arising in chains with the youngest conidium at the top (apex) of the chain

anamorph: imperfect (or asexual state) conidial form of sporulation

annellide: a conidiogenous cell which forms blastoconidia in basipetal succession, each conidium is produced through the same opening of the previously formed conidium and leaves a ringlike band (annellation) at the fertile apex after seceding. The conidiogenous cell elongates during conidiogenesis (progressive)

apophysis: a swelling at the top of the sporangiophore just below the sporangium in Mucorales (e.g. *Absidia*)

apothecium: a cup- or saucer-like ascoma

arthric: (of conidiogenesis), thallic conidiogenesis. Cells are separated from an undifferentiated part of a hypha, and transformed into conidia (see also arthroconidia)

arthroconidia: (= **thalloconidia**) conidia resulting from breaking up of a hypha into separate cells and transformed into conidia, e.g. in *Geotrichum*

ascigerous: producing asci

ascoma (pl. **ascomata**) (= **ascocarp**)**:** a fruiting body in Ascomycetes containing asci and ascospores

ascospores: sexual spores produced in an ascus

ascus: a sac-like organ in which the ascospores are formed after karyogamy and meiosis

aseptate: without a cross-wall

ballistoconidia: see ballistospore

ballistospore: a basidiospore which is forcibly propelled from the sterigma bearing it

basidium: the organ of the Basidiomycetes which bears the basidiospores after karyogamy and meiosis

basipetal: describes the succession of conidia in which the youngest conidium is at the base, e.g. of the chain

basitonous: with branches on main axis confined to the lower part of the axis

biseriate: in two series. In *Aspergillus*: with phialides not formed directly on the vesicle, but on metulae

blastoconidium: conidium in which the wall of the conidiogenous cell has bulged out apically to form the conidial wall. Conidia can be produced solitarily or in acropetal chains, i.e. each new conidium can bud at its tip

budding: a process of vegetative multiplication in which there is a development of a "daughter" cell from a small outgrowth of a "mother" cell

chlamydospore: a thick-walled, thallic, terminal or intercalary asexual resting spore, mostly for longtime survival; usually nondeciduous

cleistothecium: an ascoma enclosing the asci, but lacking a special opening (ostiole)

collarette: a cup-shaped structure at the apex of a phialide

colony: (of mycelial fungi) a group of hyphae (with or without conidia), which arise from one spore or cell. Colony appearance varies, e.g. velvety, floccose (cottony), funiculose (hyphae aggregated into strands), fasciculate (in little groups or bundles), granulose, powdery, or synnematous (compact groups of erect and sometimes fused conidiophores)

columella: a swollen sterile central axis within a sporangium, e.g. in Mucorales

conidiogenesis: process of conidium formation

conidiogenous cell: the fertile cell from or within which conidia are directly produced. Conidiogenous cells may be morphologically identical with or differentiated from vegetative cells (vegetative hyphae)

conidioma: (pl. **conidiomata**); any structure which bears conidia, e.g. separate conidiophores, synnema, acervulus, pycnidium, sporodochium etc.

conidiophore: specialized hypha, simple or branched, on which conidiogenous cells are born

conidium: asexual, vegetative, non-motile propagule, not formed by cleavage (as in sporangiospores). In Deuteromycetes the term "conidium" is recommended and the term "spore" is reserved for zoo-, sporangio-, basidio- and ascospores

conjugation: fusion

diploid: having 2*n* chromosomes

echinulate: with small pointed processes or spines on surface, e.g. of conidia or conidiogenous cells

endoconidium: conidium formed inside a hypha

exudate: droplets excreted by the mycelium. Can be characteristic of a species, e.g. *Penicillium chrysogenum*

fasciculate: hyphae and/or conidiophores in bundles

floccose: cottony

foot cell: basal part of the conidiophore (e.g. in *Aspergillus*) or basal cell of conidium in *Fusarium*

fruiting body (= **fruit-body** or **fructification**)**:** a general term for spore-bearing organs, applicable to micro- and macrofungi, but more often applied to the latter

funiculose: hyphae/conidiophores aggregated into strands

gametangium: a meiosporangium in which generative (sexual) cells (gametes) are formed

geniculate: bent like a knee

hilum: a mark or scar, especially that on a spore at the point of attachment to a conidiophore or sterigma

Hülle cells: terminal or intercalary thick-walled cells surrounding the ascomata (in some *Aspergillus* species)

hyaline: transparent or nearly so

hypha (pl. **hyphae**): (vegetative) filament of a mycelium, without or with crosswalls

intercalary: between two cells, between the apex and base (of cells, spores, etc.)

lateral: at the side

macroconidia: the larger conidia of a fungus, usually multicellular, e.g. in *Fusarium*

microconidia: small conidia, usually one-celled

meiosis: that part of a life cycle in which a diploid nucleus ($2n$) undergoes reduction and becomes haploid (n)

meristem: growing zone

merosporangium: (of Mucorales) a cylindrical outgrowth from a swollen end of the sporangiophore, in which asexual merospores are produced in a linear row

merospores: asexual spores formed in a merosporangium by cytoplasmic cleavage

mesoconidia: in *Fusarium*, conidia intermediate between micro- and macroconidia. Usually one- to several-celled but smaller in size and distinguishable in shape from the macroconidia

metabolite: any substance produced during metabolism

metula (pl. **metulae**): apical branch(es) of conidiophore-bearing phialides, e.g. *Penicillium* and *Aspergillus*

mononematous: with hyphae or conidiophores arising singly from the substrate

monopodial: a way of branching, in which the persistent main axis gives off branches

multinucleate: containing many nuclei

muriform: having longitudinal and transverse septa

mycelium: (vegetative) mass of hyphae; thallus or vegetative body of the fungus

mycotoxins: fungal secondary metabolites which in low concentration are toxic to vertebrates and other animals when introduced via a natural route.

osmophilic: growing under conditions of high osmotic pressure (e.g. on substrate containing 20-40% sucrose or NaCl)

ostiole: any pore by which conidia or spores are freed from a fruiting body, e.g. pycnidium or perithecium

pedicel: a small stalk

penicillate: like a little brush

perithecium: a globose or flask-shaped ostiolate ascoma within which asci are produced

phialide: a conidiogenous cell which produces conidia (phialoconidia) in basipetal succession, without an increase in length of the phialide itself

pionnotes: a spore mass with a fat or grease-like appearance (in *Fusarium*)

polyphialide: conidiogenous cell with more than one

opening from each of which conidia are produced in basipetal succession; the new conidiogenous loci develop by sympodial proliferation of the fertile cell

poroconidia: (= tretic conidia) dark conidiogenous cells with an apical, pigmented wall thickening. The conidium is pierced through, after dehiscence a pore is visible

pseudomycelium: the formation of a filamentous structure consisting of cells which arise exclusively by budding

pseudothecium: an ascoma with asci developing in unwalled cavities (loculi) in a stroma or arising from cushion-like structures in loculi

pycnidium: flask-shaped to globose fruiting body, usually with one apical opening (= ostiole), containing conidiogenous cells

racemose: a way of branching, in which the main axis produces side branches which are shorter than the main axis.

rachis: conidium-bearing extension of a conidiogenous cell (resulting from sympodial development, zigzag geniculate extension)

ramoconidium: (pl. ramoconidia) a fertile apical branch of a conidiophore which produces conidia and later secedes and functions as a conidium itself, e.g. in *Cladosporium*

retrogressive: describes mechanism of conidiogenous cell development in which the fertile cell shortens during basipetal conidium formation, e.g. in *Trichothecium*

rhizoid: hyphal root; a root-like structure, acting as a feeding organ and/or holdfast, e.g. in *Rhizopus*

rostrate: having a beak

sclerotium: a resting body, usually globose, produced by aggregation of hyphae into a firm mass with or without host tissue, normally sterile

septum: a cross-wall

spine: a narrow process with a sharp point

spinulose: delicately spiny

sporangiole: small, usually globose sporangium with a reduced columella and containing one or a few spores

sporangiophore: a specialized hyphal branch, which supports one or more sporangia

sporangiospore: a spore produced within a sporangium

sporangium: asexual reproductive structure, in which spores are produced by cytoplasmic cleavage

spore: a general term for a reproductive structure in fungi, bacteria, and cryptogams. In fungi the term "spore" is used in several combinations, e.g. chlamydospore, ascospore, zoospore or basidiospore

sporodochium: a cushion-like mass of conidiophores; conidia and conidiogenous cells produced above the substrate

sterigma: a conical or elongated process which bears the basidiospore

stipe: stalk

stolon: a "runner" or arching aerial hypha which touches the substratum at intervals, e.g. in *Rhizopus*

striate: marked with lines, grooves or ridges

stroma: a mass or matrix of vegetative hyphae, with or without tissue of the substrate, in or on which fructifications can be produced

substrate: the material on or in which an organism is living

suspensor: a hypha which supports a gamete, gametangium or zygospore

sympodial: describes a mechanism of conidiogenous cell proliferation in which each new growing point appears just behind and to one side of the previous apex, often giving a geniculate cell configuration; of conidiophores characterized by continued growth after the main axis has produced a terminal spore, or by development of a succession of apices each of which originates below and to one side of the previous apex.

synnema: (pl. synnemata) bundles of erect hyphae and conidiogenous cells bearing conidia (in some Hyphomycetes)

teleomorph: perfect (sexual) state of a fungus; ascigerous or basidial form of sporulation

teliospore (teleutospore): one of the spore forms (commonly resting spore) of the Uredinales (rust fungi) or Ustilaginales (smut fungi) from which the basidium is produced

thallic: (of conidiogenesis), a segment of a fertile hypha differentiates by conversion into a conidium (see arthric)

tretic: (see poroconidia)

truncate: ending abruptly, as if cut straight across

uniseriate: in one series. In *Aspergillus* phialides arise directly from the vesicle

velutinous: velvety

verrucose: having small rounded processes (warts)

verruculose: delicately verrucose

verticil: whorl

vesicle: a bladder-like sac; the swollen apex of the conidiophore in *Aspergillus*

zygospore: a sexual spore produced by Zygomycetes, thick-walled, often ornamented and darkly pigmented, formed by fusion of a pair of gametangia

INDEX